Sickness and Health
in America

SICKNESS AND HEALTH IN AMERICA

*Readings in the
History of Medicine and
Public Health*

Third Edition, Revised

Edited by
JUDITH WALZER LEAVITT
and
RONALD L. NUMBERS

The University of Wisconsin Press

The University of Wisconsin Press
114 North Murray Street
Madison, Wisconsin 53715

3 Henrietta Street
London WC2E 8LU, England

2   4   6   8   10   9   7   5   3   1

Printed in the United States of America

Library of Congress Cataloging-in-Publication Data
Sickness and health in America: readings in the history of medicine
and public health / edited by Judith Walzer Leavitt and
Ronald L. Numbers.—3rd ed.
600 pp.   cm.
Includes bibliographical references and index.
ISBN 0-299-15320-7 (cloth: alk. paper).
ISBN 0-299-15324-X (paper: alk. paper).
1. Medicine—United States—History.   2. Public health—
United States—History.   3. Medical care—United States—History.
I. Leavitt, Judith Walzer.   II. Numbers, Ronald L.
R151.S5   1997
362.1′0973—dc20   96-44916

To Our Students
Past, Present, and Future

CONTENTS

Preface to the Third Edition

The history of American medicine and public health is a rapidly expanding field. Because of changing health concerns and because historians are continually publishing new work, revising old conclusions, and charting new territory, we feel that a new version of our 1985 reader, based on a still earlier 1978 anthology, is necessary. For this third edition, we have updated the graphs in our introduction to include data from the 1980s and 1990s. We have, for example, added a graph of the leading causes of death for 1992, which highlights some notable changes in sickness and health in America in the past 20 years, such as the growing mortality from HIV, homicide, and suicide. This edition also recognizes some of the latest scholarship by adding 21 new essays, including entirely

new clusters on Sickness and Health, Early American Medicine, Therapeutics, the Art of Medicine, and Public Health and Personal Hygiene. The explosion of recent books in American medical history is reflected in our Guide to Further Reading at the end of the volume. To make room for new essays, we had to delete a number of excellent articles that appeared in the second edition. While it is rewarding to recognize new work, we are sad to let some of our old favorites go. We hope this edition follows in the tradition of the first two in reflecting the dynamic state of scholarship in the history of American medicine and public health. We are grateful to Sarah K. A. Pfatteicher and Ruth L. Stiling for their assistance in the preparation of this new edition.

Sickness and Health
in America

SICKNESS AND HEALTH IN AMERICA:
AN OVERVIEW

The history of health care in America is much more than the accomplishments of a few prominent physicians. It encompasses all efforts to cure and prevent illness—lay as well as professional, the failures as well as the successes. In recognition of this diversity, we have included in this collection essays ranging from physicians to midwives, from homosexuality to smallpox, from bloodletting to bacteriology. Food and filth often play as important roles as doctors and hospitals. Although, for example, we recognize the skill of surgeons who extended a few lives by dramatically transplanting human hearts, their historical significance pales in comparison with their contemporaries who organized comprehensive health centers in the rural South, markedly reducing infant mortality through unglamorous improvements in diet, sanitation, and preventive medicine.

When we look at the broad picture of sickness and health in America, two trends immediately capture our attention: the conspicuous decline in mortality and the corresponding increase in life expectancy for all ages. Although health records before 1900 are fragmentary and precision is illusory, there is little doubt that average Americans today live more than twice as long as their colonial forebears, whose life expectancy at birth was under 30 years and half of whose children died before their tenth birthdays (see Fig. 1). Unfortunately, not all Americans have benefitted equally from these changes. Women live increasingly longer than men, while whites continue to outlive nonwhites (see Figs. 1 and 2).

The changing disease pattern in America presents several problems for the historian. Health statistics before 1933, when the United States adopted uniform registration procedures for reporting diseases, must be used with caution. They are seldom complete and often inconsistent in classifying diseases. Many of today's clinical distinctions did not exist in the past, and those that did were frequently blurred by practitioners with little diagnostic sophistication. To compound our difficulties, disease patterns varied widely from city to city and state to state, so that what was typical of one region may have been rare in another. Nevertheless, we should not let these problems deter us from attempting qualified generalizations.

Early settlers in America often suffered from malnutrition, which increased their vulnerability to infectious diseases. These maladies, transmitted from one person to another, can be either *endemic*, that is, always present, or *epidemic*, appearing from time to time with great intensity. The gravest threats to life and health were malaria and dysentery in summer and respiratory ailments, like influenza and pneumonia, in winter. Sporadic outbreaks of smallpox, yellow fever, and diphtheria created widespread panic, but over the long run they took far fewer lives than the more familiar scourges.

As living conditions improved during the 18th and early 19th centuries, so apparently did the health of Americans. But with increasing urbanization and industrialization the situation in the cities soon deteriorated. In Boston, New York, Philadelphia, and New Orleans, for example, the death rate per 1,000 rose from 28.1 for the quarter century 1815–1839 to 30.2 for the following 25 years.[1] By mid-century some American cities were scarcely better off than the notorious industrial centers of Europe. Lemuel Shattuck of Massachusetts sadly reported in 1850 that "London, with its imperfect supply of water,—its narrow, crowded streets,— its foul cesspools,—its hopeless pauperism—its crowded grave-yards,—and its other monstrous sanitary evils, is as healthy a city as Boston, and in some respects more so."[2]

3

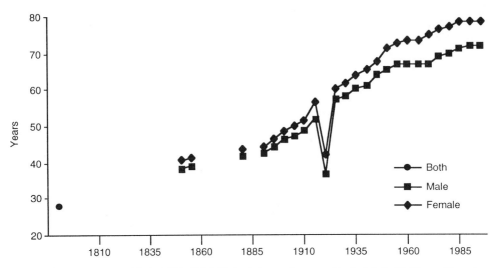

Fig. 1: Life expectancy at birth, 1789–1993, United States. Sources: Frederick L. Hoffman, "American mortality progress during the past half century," in *A Half Century of Public Health*, ed. Mazÿck P. Ravenel (New York: American Public Health Association, 1921), p. 98; U.S. Bureau of the Census, *Historical Statistics of the United States: Colonial Times to 1970* (Washington, D.C.: Government Printing Office, 1975), Part 1, pp. 55–56; U.S. Bureau of the Census, *Statistical Abstract of the United States, 1995* (Washington, D.C.: Government Printing Office, 1995), p. 86. Note: The figure for 1789 is for Massachusetts and New Hampshire only, and the data between 1850 and 1900 are for Massachusetts only. The decrease in life expectancy in 1918–1919 is largely attributable to a severe influenza epidemic.

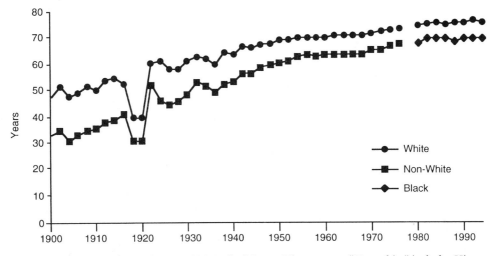

Fig. 2: Life expectancy by race, 1900–1993, United States. The category "Non-white" includes Hispanics, Asians, Indians, and Blacks. From 1975 on, government agencies distinguished between Blacks and other Non-whites; data for the latter are not included here. Sources: U.S. Bureau of the Census, *Historical Statistics of the United States: Colonial Times to 1970* (Washington, D.C.: Government Printing Office, 1975), Part 1, p. 55; U.S. Bureau of the Census, *Statistical Abstract of the United States, 1976* (Washington, D.C.: Government Printing Office, 1976), p. 60; U.S. Bureau of the Census, *Statistical Abstract of the United States, 1995* (Washington, D.C.: Government Printing Office, 1995), p. 86.

The mid-century increases in urban mortality resulted primarily from the numerous infectious diseases that attacked the nation's increasingly dense population centers. Besides intermittent epidemics of yellow fever, cholera, and smallpox, which caused more fear than mortality, there were the ever-present influenza and pneumonia, typhus, typhoid fever, diphtheria, scarlet fever, measles, whooping cough, dysentery, and—above all—tuberculosis. Tuberculosis, sometimes called consumption or phthisis, was the greatest killer of 19th-century Americans. Although present in America since the settling of Jamestown, it did not acquire its deadly reputation until it attacked heavily crowded cities. As early as the 1810s Boston, for example, experienced a tuberculosis mortality rate of 472 per 100,000 inhabitants. By the close of the century this rate had fallen by more than half, and in the mid-1970s tuberculosis killed fewer than two Americans out of every 100,000 (see Fig. 3). Unfortunately, by the 1990s deaths from tuberculosis were increasing among some groups.

The American health picture began to improve by the late 19th century, as evidenced by the declining urban death rate. The cities of Boston, New York, Philadelphia, and New Orleans, which had suffered under a death rate of 30.2 per 1,000 between 1840 and 1864, saw this figure drop to 25.7 between 1865 and 1889 and down to 18.9 between 1890 and 1914.[3] Even more dramatic was the precipitous fall of infant mortality rates. During the first three decades of the 20th century infant mortality decreased by more than 50 percent, and fell at an even greater rate during the next 20 years (see Fig. 4). But despite these improvements, the United States continued to lag behind many other industrialized nations in reducing infant deaths, a symbolic indicator of national health standards.

As the familiar infectious diseases of the 19th century diminished in virulence, a host of chronic, degenerative diseases appeared to take their place (see Figs. 5 and 6). By the mid-1970s tuberculosis, gastritis, and diphtheria no longer ranked among the nation's top ten killers, but heart problems, cancer, and cerebrovascular disorders (e.g., strokes), all typical of an older population, headed the list. By the mid-1990s AIDS, suicide, and homicide had joined the list of leading causes of death (see Fig. 7).

Given the available evidence, it is not easy to find an explanation for the historical decline of infectious diseases and the increase in life expectancy. But the three most likely candidates are medical practice, public-health measures, and improvements in diet, housing, and personal hygiene.

Popularizers of medical history like to glorify the intrepid "doctors on horseback" and the marvelous "magic bullets" of white-coated medical scientists in explaining the miracles of the past hundred years. The historical record, however, suggests a different interpretation. Clearly, the brief improvement of health conditions in the 18th century owed little to the efforts of physicians, who sometimes did their patients more harm than good. Dr. William Douglass, a prominent Boston physician, observed at mid-century that "more die of the practitioner than of the natural course of the disease."[4] If this was true of the 18th century, it was probably even more so in the early 19th century during the heyday of "heroic" medicine, when regular physicians bled, purged, and puked their patients. Until the latter part of the century doctors possessed few specific remedies besides quinine for malaria, digitalis for dropsy, and lime juice for scurvy.

America's experience with three diseases—tuberculosis, diphtheria, and smallpox—further illustrates the limitations of 19th-century medicine. Tuberculosis, the most deadly of the three in Victorian America, declined for almost a century before physicians discovered an effective way to treat or prevent it (see Fig. 3). By the time streptomycin was introduced in 1947, the death rate from tuberculosis had already dropped to 33.5 per 100,000. The use of chemotherapy markedly accelerated the rate of decline, from 44.9 percent during the 1940s to 69.7 percent in the 1950s, but even without this therapy, it seems likely that the death rate would have continued its decline.

For diphtheria, the story is somewhat different. Although the death rate from this disease dropped dramatically after an antitoxin became available in 1894 (see Fig. 8), it is debatable how much of the decline is attributable to this measure. First, the national death rate for diphtheria had been fluctuating downward for almost two decades prior to the introduction of the antitoxin; second, the antitoxin was neither systematically nor consistently used throughout the country. Thus it is probable that other nonmedical factors also contributed to the downfall of diphtheria in the 1890s.

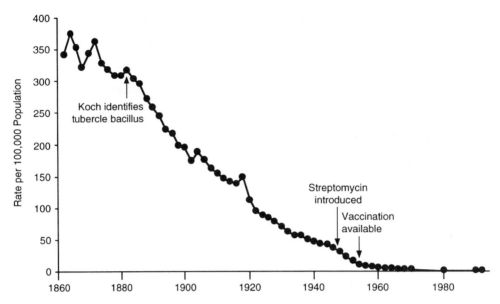

Fig. 3: Death rate for tuberculosis, 1860–1993, United States. Source: U.S. Bureau of the Census, *Historical Statistics of the United States: Colonial Times to 1970* (Washington, D.C.: Government Printing Office, 1975), Part 1, pp. 58, 63; U.S. Bureau of the Census, *Statistical Abstract of the United States, 1995* (Washington, D.C.: Government Printing Office, 1995), p. 92. Note: Data between 1860 and 1900 are for Massachusetts only.

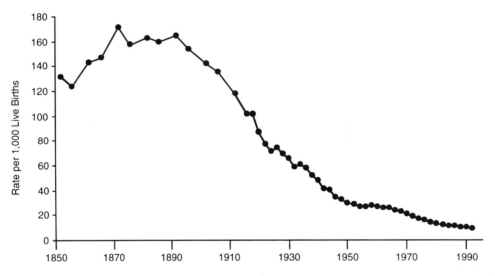

Fig. 4: Infant mortality rate per 1,000 live births, 1851–1993, United States. Sources: U.S. Bureau of the Census, *Historical Statistics of the United States: Colonial Times to 1970* (Washington, D.C.: Government Printing Office, 1975), Part 1, p. 57; U.S. Bureau of the Census, *Statistical Abstract of the United States, 1995* (Washington, D.C.: Government Printing Office, 1995), p. 73. Note: Data between 1851 and 1913 are for Massachusetts only.

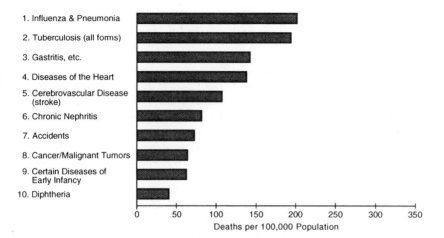

1. Influenza & Pneumonia
2. Tuberculosis (all forms)
3. Gastritis, etc.
4. Diseases of the Heart
5. Cerebrovascular Disease (stroke)
6. Chronic Nephritis
7. Accidents
8. Cancer/Malignant Tumors
9. Certain Diseases of Early Infancy
10. Diphtheria

Deaths per 100,000 Population

Fig. 5: Leading causes of death in the United States, 1900. Source: Monroe Lerner and Odin W. Anderson, *Health Progress in the United States, 1900–1960* (Chicago: Univ. of Chicago Press, 1963), p. 16.

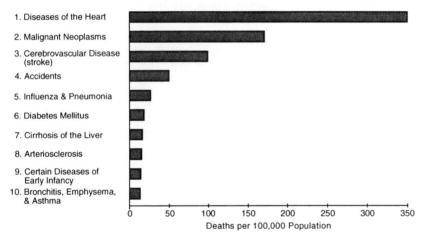

1. Diseases of the Heart
2. Malignant Neoplasms
3. Cerebrovascular Disease (stroke)
4. Accidents
5. Influenza & Pneumonia
6. Diabetes Mellitus
7. Cirrhosis of the Liver
8. Arteriosclerosis
9. Certain Diseases of Early Infancy
10. Bronchitis, Emphysema, & Asthma

Deaths per 100,000 Population

Fig. 6: Leading causes of death in the United States, 1974. Source: U.S. Bureau of the Census, *Statistical Abstract of the United States, 1976* (Washington, D.C.: Government Printing Office, 1976), p. 66.

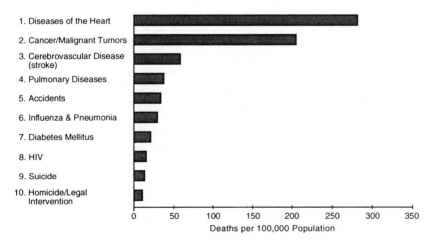

1. Diseases of the Heart
2. Cancer/Malignant Tumors
3. Cerebrovascular Disease (stroke)
4. Pulmonary Diseases
5. Accidents
6. Influenza & Pneumonia
7. Diabetes Mellitus
8. HIV
9. Suicide
10. Homicide/Legal Intervention

Deaths per 100,000 Population

Fig. 7: Leading causes of death in the United States, 1992. Source: U.S. Bureau of the Census, *Statistical Abstract of the United States, 1995* (Washington, D.C.: Government Printing Office, 1995), p. 94.

In the case of smallpox, medicine *did* offer a means of protection as early as the 1720s, when Cotton Mather introduced inoculation (variolation). This method gave treated persons a mild case of smallpox, which provided lifetime immunity. Although inoculation often spread the disease, it also appears to have lessened mortality in the 18th century. At the turn of the 19th century vaccination with cowpox virus, an even safer method, became available. But despite these measures, smallpox continued to plague Americans until the 20th century (see Fig. 9), primarily because many chose to ignore the protection medicine offered them.

If we can generalize from these three examples, it seems that medicine contributed little to the initial decline of infectious diseases in the United States, but sometimes played a crucial role in the 20th century. The discovery in the 1930s and 1940s of powerful new drugs, like the sulphonamides and antibiotics, finally gave physicians effective weapons with which to fight infection.

The public health movement proved more effective than medicine in combatting communicable diseases during the 19th century. It reduced exposure to infectious diseases by cleaning up the physical environment and improving living conditions, and it helped to lower the incidence of waterborne infections like cholera and dysentery by regulating urban water sources and sewerage.

Filth was the premier public health problem in the 19th century. Most physicians thought that dirt caused disease and that cleaning up the cities was the best preventive to high mortality. But as traditional methods of keeping cities clean broke down in the wake of massive population increases, those responsible for public health faced unprecedented problems. Festering piles of garbage littered urban streets, dead animals lay where they fell, privies and cesspools overran into drainless, unpaved streets. Horses defecated indiscriminately. Municipal employees spent countless hours trying to staunch the seemingly endless flow of waste products, and, to varying degrees, they were successful. They removed garbage, emptied privies, drained stagnant pools, improved and extended water and sewer systems, and slowly created an environment in which it was possible to walk down the street without dragging one's skirt in the filth or having to hold a handkerchief over one's nose. Health departments also regulated food and housing, at least

minimally. Undoubtedly these efforts, to the extent they were successful, significantly reduced the spread of infectious diseases that thrived in unhygienic, congested environments. Their precise impact cannot be measured, because conditions at the local level varied so greatly. But it hardly seems coincidental that mortality from communicable diseases declined at the very time American cities were waging vigorous sanitation campaigns.

Improved water supplies and sewerage provided additional protection against infectious diseases. As more and more American cities installed municipal water systems, urbanites no longer had to consume water from contaminated wells and polluted rivers. The construction of sewers, which carried off tons of septic waste products each day, eliminated another source of infection. All of these public-health measures, plus the isolation of the sick during times of epidemics and the regulation of foodstuffs, helped to turn mortality rates downward.

It also seems likely that living conditions—diet, housing, and personal hygiene—contributed to reducing mortality. Recent studies, for example, have demonstrated a high correlation between malnutrition and susceptibility to infectious diseases. Unfortunately, historical data regarding the way people lived are so scarce, we must rely more on inference than hard evidence.

Seventeenth-century colonists frequently suffered from severe food shortages and consequent malnutrition. As one pioneer observed in 1628, "for want of wholesome Diet and convenient Lodgings, many die of Scurvys and other Distempers."[5] These shortages largely disappeared as agricultural production stabilized in the 18th century, but the staple American diet of corn and pork, though ample in quantity, hardly provided a balanced diet.

It was not until the coming of railroads in the 1840s and 1850s and the development of the canning industry after 1860 that products like milk, fruit, and vegetables became readily available year-round, especially to urban dwellers. It is impossible to tell exactly how much the availability of these products increased the consumption of them, but we do know that one early 19th-century family spent 9.7 percent of its food budget on milk, fruit, and vegetables, while a century later another family spent 40.8 percent.[6] This revolution in American eating habits, suggests one historian, "may have contributed as fully to the development of a health-

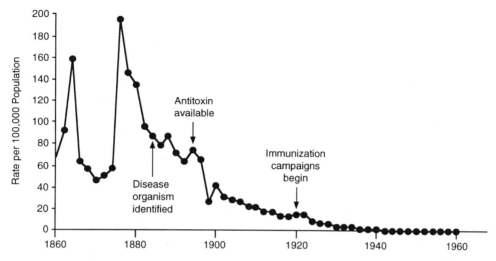

Fig. 8: Death rate for diphtheria, 1860–1960, United States. Source: U.S. Bureau of the Census, *Historical Statistics of the United States: Colonial Times to 1970* (Washington, D.C.: Government Printing Office, 1975), Part 1, pp. 58, 63. Note: Data between 1860 and 1900 are for Massachusetts only.

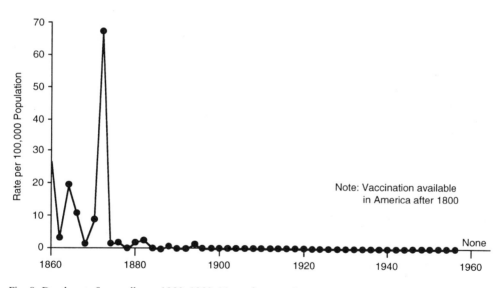

Fig. 9: Death rate for smallpox, 1860–1960, Massachusetts. Source: U.S. Bureau of the Census, *Historical Statistics of the United States: Colonial Times to 1970* (Washington, D.C.: Government Printing Office, 1975), Part 1, p. 63.

ful, vigorous manhood as major measures of sanitary reform and preventive medicine."[7] Certainly it was an important factor.

In the 20th century diet continued to influence the nation's health, though its effect was often negative. As affluent Americans ate more fats and sugar and less fruit, vegetables, and grain products, they fell victim to a form of malnutrition that some authorities predicted might be "as profoundly damaging to the Nation's health as the widespread contagious diseases of the early part of the century." By the mid-1970s six of the ten leading causes of death—heart disease, cancer, cerebrovascular disease, diabetes, arteriosclerosis, and cirrhosis of the liver—could be tied to dietary habits like the over-consumption of saturated fats, cholesterol, sugar, salt, and alcohol.[8] Such evidence clearly demonstrates that to understand sickness and health in America we must study not only medicine, but public health and lifestyle as well.[9]

NOTES

We are indebted to Sarah K. A. Pfatteicher for updating the figures that accompany this essay.

1 Frederick L. Hoffman, "American mortality progress during the last half century," in *A Half Century of Public Health,* ed. Mazÿck P. Ravenel (New York: American Public Health Association, 1921), p. 101.

2 *Report of the Sanitary Commission of Massachusetts* (Boston: Dutton and Wentworth, 1850), p. 281.

3 Hoffman, "American mortality progress," p. 101.

4 Quoted in John Duffy, *Epidemics in Colonial America* (Baton Rouge: Louisiana State Univ. Press, 1971), p. 4.

5 Quoted *ibid.,* p. 11.

6 Chase Going Woodhouse, "The standard of living at the professional level, 1816–17 and 1926–27," *J. Polit. Econ.,* 1929, *37:* 565–567.

7 Richard Osborn Cummings, *The American and His Food: A History of Food Habits in the United States* (Chicago: Univ. of Chicago Press, 1941), p. 52.

8 U.S. Senate Select Committee on Nutrition and Human Needs, *Dietary Goals for the United States* (Washington, D.C.: Government Printing Office, 1977), p. 9.

9 Thomas McKeown's provocative *The Modern Rise of Population* (New York: Academic Press, 1976) inspired this essay.

Sickness and Health

Ideas, like individuals and institutions, have their own histories, and concepts of sickness and health are no exception. What one generation of Americans may have considered as normal, another regarded as pathological. Generally, the more familiar a condition, the less likely it was to be labeled a disease. Malaria, for example, became so common in the Midwest during the mid-19th century, one medical historian, Erwin Ackerknecht, argued that it briefly lost its identity as a disease. Views of childbirth likewise have changed over time. During most of the 18th and 19th centuries, women, who typically bore children at home without the aid of a physician, regarded delivery as a normal—if risky—part of life. But when they began having fewer children, and having them in hospitals attended by physicians, the event took on the characteristics of an illness, customarily reimbursable by health insurance.

Occasionally, once-prominent diseases have been downgraded to the status of mere symptoms. In the late 18th century Benjamin Rush, the foremost physician and medical educator of the early Republic, assured his students that there was "but one disease in the world," characterized by vascular tension. Within a short time, however, diagnostic changes and a new philosophy of medicine had reduced vascular tension to the level of a symptom. Neurasthenia, a common late-19th-century disease associated with physical and mental exhaustion, experienced a similar fate. Once known as "the national disease," it quietly disappeared in the early 20th century, when it no longer served the purposes for which it had been invented a half century earlier. Americans continued to suffer and die from nervous exhaustion, but physicians and patients now attribute this condition to other ailments.

Over the years definitions of health have tended to shift from an emphasis on the absence of sickness to the presence of well-being. In a famous statement written in 1947, the World Health Organization sweepingly characterized health as "a state of complete physical, mental, and social well-being and not merely the absence of disease or infirmity."

The construction and deconstruction of diseases may relate as much to the needs and desires of doctors and patients as to scientific discoveries. During the past century a wide range of "bad habits" have been medicalized as diseases, thereby freeing their sufferers from moral blame. Thus we see drunkenness turning into alcoholism, compulsive stealing into kleptomania, and promiscuity into sexual addiction. In the meantime old diseases such as masturbation and homosexuality have evolved into normal behaviors or lifestyle choices. Both political and medical institutions have participated in this relabeling process. In 1962, for example, the Supreme Court endorsed the view of drug addiction as an illness, not a crime; and in 1973 the American Psychiatric Association voted to stop listing homosexuality as a mental disorder.

Political factors have influenced the definition of such lethal diseases as AIDS. On 1 January 1992 the official number of AIDS cases in the United States jumped significantly, not because large numbers of people had acquired the infection overnight but because the Centers for Disease Control had revised the criteria for diagnosing the ailment. The following two essays, on homosexuality and black lung, further illustrate the roles played by social and biological factors in the creation of diseases.

1

American Physicians' "Discovery" of Homosexuals, 1880–1900: A New Diagnosis in a Changing Society

BERT HANSEN

MEDICAL NOVELTIES

Only in the last third of the 19th century did European and American physicians begin to publish case reports of people who participated in same-sex sexual activity. In 1869 the *Archiv für Psychiatrie und Nervenkrankheiten* in Berlin published the very first medical case report of a homosexual, with six more cases following quickly in the German psychiatric literature.[1] A French report appeared in 1876 and an Italian one in 1878. The first American case report appeared in 1879, the second in 1881, and in the latter year the British journal *Brain* offered an entry, though that report's subject was German and the reporting physician Viennese. (British physicians seem to have been rather reticent on this subject through the 1890s.) The years 1882 and 1883 saw several major American additions to the literature on homosexuals, by which time nearly 20 European cases had been reported.[2] These persons were variously described as exhibiting "sexual inversion," "contrary sexual instinct," or "sexual perversion," though this last term was not common in the narrow sense and usually referred to a wider range of sexual possibilities.

By the century's end, the medical literature on the subject was large and growing fast, with dozens of American contributions and considerably more in Europe. For the most part, these reports treated the problem as a new phenomenon, unprecedented and previously unanalyzed. The case histories read like naturalists' enthusiastic reports of a new biological species. American physicians and their European counterparts not only described their subjects' lives, but also elaborated the novel conception that homosexual behavior was the manifestation of an underlying morbid condition, a view that remained official psychiatric doctrine into the 1970s. A majority of the early reports were of males, but many concerned female homosexuals, and although the social histories of lesbians and gay men have differed in certain periods, this first stage of medicalization treated them simultaneously—and similarly.

In several ways, the new diagnosis of homosexuality—or sexual inversion, as it was commonly labeled until well into the 20th century—resembled other "new" diagnoses of that era, categorizing as medical problems behaviors not previously interpreted as such, for example, masturbatory insanity, neurasthenia, inebriety, anorexia nervosa, sexual psychopathy among women, multiple personality, and kleptomania. "Deviance was increasingly—if by no means universally—being defined as the consequence of disease process and, thus appropriately, the physicians' responsibility."[3] Yet unlike the 19th-century masturbators, kleptomaniacs, and neurasthenics, many of the sexual inverts were securing public space and recognition in a complex process of social and sexual change that was not limited to the medical sphere, however crucial medical diagnosis was to the new group's self-consciousness and the public conception of inversion. The doctors' discovery and framing of this new diagnosis were prompted by changing social phenomena: new kinds of people were in fact gaining public visibility—and consulted physicians who

BERT HANSEN is Associate Professor of History, Baruch College, The City University of New York.

From *Framing Disease: Studies in Cultural History*, edited by Charles E. Rosenberg and Janet Golden (New Brunswick, N.J.: Rutgers University Press, 1992), pp. 134–154.

then discussed cases among themselves and in their journals. The medical formulation of sexual inversion as a disease, in turn, helped shape the behavior and the self-understanding of many inverts. In this interactive process, the physicians' reports of cases came to be normative, as well as descriptive.

The present study aims not only to describe the American medical profession's earliest encounters with homosexuals, but also to set the physicians' case reports and commentaries into the context of broad changes in the social structures of sexuality. It also calls attention to the vivid documents of social history preserved in the literature of clinical medicine; the first generation of medical case reports—in contrast to succeeding materials—naïvely recorded patients' own accounts of their lives, which now give rare glimpses of individuals in the process of creating new social forms.[4]

SOCIAL HISTORY: ACTION AND (MEDICALIZED) REACTION

Earlier historians saw this episode in the history of medicine as an evolution only of attitudes and labels. For them, an unchanging homosexuality was no longer regarded as a sin but simply became a sickness, relabeled, "medicalized," or "morbidified" by Krafft-Ebing, Havelock Ellis, and their less famous European and American colleagues. Such a formulation, however, masks the depth of the change and mistakes its fundamental character. More recent historiography has challenged this picture and reframes the historical process as a metamorphosis of the social and psychological reality of homosexuality itself—due in part to the activities of physicians—from a form of sexual behavior (a pattern of actions) to a condition (a way of being). The condition resides deeply within certain people who eventually came to be known as homosexuals, a group marked off from the rest who (slightly later) came to be known as heterosexuals. The new interpretation is often designated the "historical construction of homosexuality" or the "making of the modern homosexual" and, less formally, the "emergence thesis" or "social constructionism."[5]

An appreciation of this shift will be facilitated by reflecting a little on the modern ways of thinking about types of persons.[6] People's thoughts today commonly proceed, for example, from the observation of a theft to the recognition of a thief, from a

crime to a criminal, and from a homosexual act to a homosexual; but our ancestors did not think that way about those acts. To them theft and sodomy were sins, which anyone might commit. A sinner might subsequently be socially labeled a "thief" or a "sodomite," but this label was only shorthand for "person who perpetrated this sinful act." It was *not* an indication of one's being a fundamentally different kind of person from one's peers. Furthermore, in this traditional viewpoint, a person could not be a thief or a sodomite without having committed the relevant act.

Several features of sexuality's "old regime," which was replaced in America only gradually over the last few decades of the 19th century,[7] are exemplified in a brief newspaper report of 1867 from the *California Police Gazette*, a San Francisco scandal sheet.

> Miss Mary Walker, a fast young lady of Richmond, Va., took it into her head, a few months ago, to don male attire, and engaged as a barman. Somehow, the breeches seem to have put bad notions into her head, for she went making love to the pretty girls who came after the family beer. One of them corresponded with her, and was engaged to be married to her. As making presents costs money, the feminine barman borrowed from the till, was detected, pleaded guilty, and is now waiting for her sentence. Her enamored fiance [*sic*] visited her in jail. They are not to be married at present, as women's rights have not attained to that degree of development.[8]

This account acknowledges (homo)sexual interest only casually and emphasizes instead the anomalous gender role; it does not treat either of the women as a sexually odd sort of person.

When one turns to some of the medical case reports written only 15 years later, one encounters the "modern" way of thinking about persons and their sexuality, in which actions came to be seen as less critical than essence or being. Consider, for example, this excerpt from one of the earliest cases published in America, a long report on "Mr. X" by Dr. G. Alder Blumer, an Assistant Physician at the lunatic asylum in Utica, New York.

> Mr. X. is about twenty-seven years of age, and of high social status. Height, considerably below the average; muscular development, good; hair, dark,

profuse in growth, parted in the middle and brushed well back; eyes, dark brown, large, brilliant and swimming; pupils dilated; long eyelashes; mouth small; teeth, regular and sound; chin, somewhat pointed; general contour of face, oval; expression, womanly. Lower limbs, short in proportion to trunk, but well developed. Sexual apparatus believed to be normal. Gait, precise; strides, quick and short. Voice and intonation, like a woman's; lisps in pronouncing certain words. Capillary circulation, poor; extremities, frequently cold. Head often flushed and hot, and complains of headache.... Probably practiced onanism when younger. Nocturnal emissions of unusual frequency.

... As a child, showed no disposition to mingle with boys of his own age and adopt their pastimes [*sic*]. Very precocious, and developed an early interest in literature, himself writing prose and poetry when quite young. Passionately fond of music, a brilliant pianist and composer of weird-like impromptus. Occupations and tastes essentially womanly. Fond of discussing women's dress, in which he is always *au courant*. His own dress is always precise and natty, showing more especially in pattern, style and arrangement of necktie a taste and deftliness rarely found in men. Conscious of his youthful, unmanlike appearance, he is very sensitive on the subject, and resents imputations of womanliness. Admires manly men, frequently speaks of them, and extols all that is noble in his own sex. Seldom speaks of women, is indifferent to their charms, and expresses a horror of matrimony, the very idea of marital relations being repugnant to him. Admits that he has on several occasions been approached by men of unnatural desire, and declares his unspeakable horror of paederastia. Has in conversation said that these latter individuals are able to recognize each other. Began the study of law, but finding its dry details distasteful, soon abandoned it for literature in which he achieved early success. ... Is good to the sick and poor, and takes pleasure in charitable work.... Punctilious in regard to table etiquette; has a fastidious and capricious appetite, eating in a nibbling, mincing manner like many women.[9]

Like the California newspaper writer quoted a little above, Dr. Blumer attended carefully to X's cross-gender mannerisms and behavior. But Blumer then proceeded to characterize Mr. X as exhibiting "contrary sexual instinct" even while recording that X had never had a sexual experience with another man. Clearly, sexuality was in the process of coming

to stand for something other than a person's sexual behavior. Blumer's 1882 report also illustrates how patient and physician might collaborate in framing (homo)sexuality as a condition, rather than a set of behaviors.

What was the historical process through which previously isolated, individual sexual actions came to be associated with a "type of person" embodying an abnormal "instinct" or "condition"? In my view, this transition can best be described and analyzed as the evolution of four interacting structures: social roles (patterned behaviors and the expectations held by others of these patterns), a self-conscious identity (the personal internalization that one *is* that kind of person, not just acting out a role), social institutions (recognized settings for particular activities), and community (a shared sense of belonging to a group with others who have the same role or identity).[10]

In the decades after 1800 in fast-growing American cities, especially in the Northeast, more and more men were working and living outside of the productive farm family typical of earlier generations. While we have no reason to assume that more of these men manifested homosexual desire than did those in earlier times, their new social circumstances did allow those with such desires more opportunity to find others who shared their tastes. Living outside of the traditional family also allowed such men to spend more time in the pursuit of sex, with the result that specialized meeting places gradually came into existence (certain bars, parks, transportation depots, military installations, etc.). Such institutions fostered a vague sense of community—at first, however, this was only a community of taste, not of identity or style of life. But the growth of such institutions and the more numerous social and sexual encounters they facilitated led to confrontations with the police, with moral reformers, and eventually with doctors. These clashes led to labeling and, over several decades, to some self-consciousness. A tentative sense of identity facilitated further interaction with other members of the community, which then facilitated the formation of a homosexual identity for more individuals. The chain reaction continued, yielding at each step more labeling, further community growth, higher visibility, and the proliferation of institutions.[11]

This general process took different forms and

moved at different rates in different countries, varying also by region, class, level of urbanism, and, of course, sex.[12] It seems that many of these changes occurred about a generation or two later among lesbians in America than among male homosexuals. That the development of homosexual and lesbian identities depended on being able to live and support oneself outside of the family seems partly confirmed by the fact that the first noticeable group of women to do this, the graduates of the new women's colleges, had both a low rate of marriage and a large proportion of lesbian couples.[13]

With the evidence so far available, it seems useful to see the first two thirds of the century as a period characterized by individual explorations of new lifestyles and the next stage as a consolidation of identities and the creation of a new public awareness of sexually deviant social roles. Data for the earlier period are as yet rather limited but nonetheless suggestive. Some records tell of homosexual affairs, meeting places for men, and arrests for sodomy, and hint at a growing awareness that other people shared the same tastes and interests—especially in the eastern seaboard cities and in post–Gold Rush San Francisco. One indication of some level of popular awareness is in the glimpses we get here and there of new stereotypes, for example, of the dry-goods clerk, or "counter-jumper," of the new urban emporia, regarded as an unmanly occupation. Soldiers near military depots and bellboys in grand hotels came to have some notoriety for their sexual accessibility.[14]

Three things made the sexual encounters of this first stage historically different from the couplings of previous centuries. First, they now sometimes occurred in recognized locales. Second, they more frequently involved men on their own, living outside of the productive family units that dominated living space before the rise of wage labor in cities. Third, previously private and personal activities were taking on a social dimension. Men could pursue sexual adventures not as short-term diversions from family responsibilities but as a full-time lifestyle; and by the process of interacting, some of these men came to see themselves not only as having different tastes from the majority, but as being a fundamentally different type of person. In the same era, however, the homosexual experiences of many people did not include even a hint of

the newer consciousness of being a different sort of person.[15] Physicians' case reports, cited below, illustrate the simultaneous presence of both old and new ways of thinking during this transitional period.

One remarkable first-person account can flesh out this rather abstract portrayal of how people in those years articulated a new identity and fashioned a sense of community; it also portrays one of the social institutions that facilitated these developments. Under the auspices of Dr. Alfred W. Herzog, editor of the *Medico-Legal Journal,* a man named Earl Lind, who called himself Jennie June, published his memoir, *Autobiography of an Androgyne,* in the hope that

> every medical man, every lawyer, and every other friend of science who reads this autobiography will thereby be moved to say a kind word for any of the despised and oppressed step-children of Nature—the sexually abnormal by birth—who may happen to be within his field of activity.[16]

The book also solicited reports of similar cases, offering a two-page questionnaire that could be completed by "every friend of science," whether a physician or one of "the homosexualists" himself. (See Fig. 1.) In his sequel, *The Female Impersonators,* edited by Dr. Herzog four years later, Lind reported an experience he had had nearly 30 years before in a New York City bar:

> On one of my earliest visits to Paresis Hall—about January, 1895—I seated myself alone at one of the tables. I had only recently learned that it was the androgyne headquarters—or "fairie" as it was called at the time. Since Nature had consigned me to that class, I was anxious to meet as many examples as possible. . . .
>
> In a few minutes, three short, smooth-faced young men approached and introduced themselves as Roland Reeves, Manon Lescaut, and Prince Pansy—aliases, because few refined androgynes would be so rash as to betray their legal names in the Underworld. Not only from their names, but also from their loud apparel, the timbre of their voices, their frail physique, and their feminesque mannerisms, I discerned they were androgynes. . . .
>
> Roland was chief speaker. The essence of his remarks was something like the following: "Mr.

Werther—or Jennie June, as doubtless you prefer to be addressed—I have seen you at the Hotel Comfort, but you were always engaged. A score of us have formed a little club, the Cercle Hermaphroditos. For we need to unite for defense against the world's bitter persecution of bisexuals. We care to admit only extreme types—such as like to doll themselves up in feminine finery. We sympathize with, but do not care to be intimate with, the mild types, some of whom you see here to-night wearing a disgusting beard! . . .

"We ourselves are in the detested trousers because having only just arrived. We keep our feminine wardrobe in lockers upstairs so that our everyday circles can not suspect us of female-impersonation. For they have such an irrational horror of it!" [17]

This brief extract illuminates how—in one small group at one point in time—community, identity, and a variety of homosexual social roles were being formed and reinforced.

Except for the medical auspices of its eventual publication, this quotation does not, however, make explicit the single most important novelty in the decades after 1870: a new public discourse, primarily medical, about the homosexual as a type of person. Although the popularizing of this medical discourse was slow and uneven, I believe it played a major role in reinforcing gay people's consciousness of being different and also of just how they were different. This process took place directly in the doctors' consulting rooms and indirectly through medical publications and the nonmedical writings they influenced. Even though some of the medical writings were not supposed to be distributed to the public, they were eagerly sought out—and found—by individuals searching for clues to their own nature. [18]

DISCOVERING A NEW DISEASE

In the medical reports about sexual inverts, the clinicians were self-conscious about opening up new territory, and they were proud of being the first generation to study the subject in a scientific manner. [19] The authors of these pioneer case studies agreed as to the history of their collective project. "Casper was the first to call attention to the condition known as sexual perversion," explained one

typical account. "But not until several works had been published by a Hanoverian lawyer, Ulrichs, himself a sufferer from the disease, did the matter become the subject of scientific study on the part of physicians. Westphal was the first to discuss it." [20]

In the 1870s and 1880s, when sexual inversion still seemed a rarity, individual cases were carefully described, numbered, and added to the stock of specimens. In 1883 Dr. J. C. Shaw, a Brooklyn practitioner, and Dr. G. N. Ferris, First Assistant Physician at the Kings County Asylum, announced that theirs was the first case in the United States and the 19th case worldwide. [21] Their report, though shorter than most, is sufficiently typical to be worth examining in full.

Our case . . . came under observation twice in the spring of 1880. The patient has not been seen since; in fact, was so reserved that it has been impossible to find him again.

Case 19.—A German, aged thirty-five years. Height, about five feet four inches. Weight, about one hundred and forty pounds. Black hair. Dark complexion. Well built. Physiognomy of an intelligent expression. Very reticent on some points of his history. Says he is engaged in mercantile business and occupies a good position. For some time past has had an almost uncontrollable desire to embrace men. Fears that some time this horrible morbid desire may overcome him and he will really embrace some of his fellow-clerks. When in the presence of men, he is tormented by constant erections of his penis and a desire to embrace the men. He regards his desire as abnormal, and laments his condition. Remarks that he is ashamed to tell the doctor of his condition, as he must consider him a horrible creature, and look upon him with disgust. Has never given way to his desires, but is afraid that they will overcome him some day and make known his unnatural condition. He has tried to overcome this morbid state by having connection with women, but intercourse with them gave him no pleasure, and he was obliged to force himself to it, and has only tried it on three occasions.

Examination of genital organs shows them to be well developed. At the time of examination penis is in full erection, and this the patient says is the condition of the organ whenever he is near men. He denies nervous disease in his family. Patient is an intelligent man, and perfectly natural in his appearance and manner, except that he is distressed by his abnormal state, and wishes medicine to

QUESTIONNAIRE ON HOMOSEXUALITY

[The governments of all cultured lands take from time to time censuses of the blind, the deaf, and other defective classes. None has ever taken a census of homosexualists, although the latter are fully as numerous as the two definite classes previously named, and their effect on the social body is even more marked. The Medico-Legal Journal, on the basis of the following questionnaire, makes the first essay, in the history of culture, in lining up the defective class in question so that science may have a broader knowledge of them than that afforded by the comparatively few detached biographical and analytical notes at present extant. The reader is therefore requested to fill out the following questionnaire — or have the intelligent homosexualist do so — and mail it to the Medico-Legal Journal, New York. If unable to answer all queries, kindly give as much information as possible. Additional schedules will be furnished on request. Unpublished textual descriptions of cases would be welcome, and will be returned on request. The results of this questionnaire will be published, and the addresses of respondents will be filed for due notification.]

(1) Physical sex of homosexualist...... (2) Age at date....years (or....years at death)
(3) No. of brothers.... Sisters.... (4) Approximate age of father at subject's birth.... Of mother....
(5) Underline applicable physical type: Brunette. Blonde. Red-haired. Not definitely any of these.
(6) Principal occupation as adult...................
(7) Lineage (i.e., from what foreign countries did forebears emigrate) ....
....
(8) Environment in which life principally passed (Indicate by x's):

	Rural or village under 2,500	Municipality 2,500 to 25,000	Municipality 25,000 to 100,000	Municipality 100,000 to 500,000	Municipality over 500,000
Up to 10 years old	.....	.....	.....	.....	.....
11 to 20 years old	.....	.....	.....	.....	.....
21 to 50 years old	.....	.....	.....	.....	.....
After 50 years old	.....	.....	.....	.....	.....

(9) Ever legally married..... Children, how many.....

(10) Plays any musical instrument...
(11) Underline interest in sport: Practically none. Slight. Extensive.
(12) Underline interest in music: Practically none. Slight. Extensive.
(13) Underline interest in other art (designate): Practically none. Slight. Extensive.
(14) Underline interest in religion: Practically none. Slight. Extensive.
(15) Underline applicable schooling: Less than 8 years. High school. Liberal-arts college. Postgraduate. Professional school.
(16) If liberal arts course, favorite subjects in order of preference...................
(17) Number — if any — of foreign languages ever spoken with considerable ability..... Number studied for translation.....
(18) What — if any — mental diseases suffered..... ....
(19) A dipsomaniac..... Other drug addiction.....
(20) What rather serious (excluding the practically universal) bodily diseases suffered...................
.... Particularly underline applicable:
Venereal warts. Syphilis. Gonorrhea. Locality (initial) of last three..........
(21) If dead, cause of death...
(22) What — if any — disease has run in the family of either parent.......................
(23) What other blood relatives have shown sexual abnormality:

Definite relationship   Nature of abnormality
....
....
....

(24) Are sexual organs normal..... Describe any abnormalities ....
....
....

Fig. 1: Questionnaire on homosexuality, from Earl Lind, *Autobiography of an Androgyne* (New York, 1918).

(25) Underline applicable fundamental or original instinct:
Fellatio (active buccal). Passive buccal. Masturbation of other. Mutual onanism. Paedicatio (anal), indicating whether subject is active or passive, or both. Cunnilingus (corresponding to active fellatio in the male). Tribadism (corresponding to masturbation of other in the male).

(26) Approximate age when instinct first manifested itself in the feelings.....years. In actions.....years.

(27) Secondary or acquired methods of coitus...
....

(28) Is subject a psychical hermaphrodite (attracted toward both sexes).........

(29) Does coitus stimulate any erogenous center (i.e., afford a pleasurable titillation of any portion of the body) or is the satisfaction entirely or almost entirely mental ....

(30) State of health day following coitus...

(31) If a male, does coitus induce an emission..... If so, is sensation pleasurable or horrifying............
....

(32) Is there love or adoration for the associate, or is the latter used merely to secure the stimulation of the subject's erogenous center..

(33) Upper and lower age limits of individuals that attract, with indication of subject's own age at the different periods ....

(34) Any quality or apparel (such as plumpness, military uniform) that constitutes a special attraction....
....

(35) If a physical male, is he undersized..... If female, unusually large..... Muscles vigorous or feeble.........

(36) Is there a striking contrast between the real and apparent age.....

(37) Any peculiarity about the hair system, particularly the facial...................................
....

(38) If a physical male, any tendency to wear the hair several inches long...............................

(39) Describe any anatomical approach toward the psychical sex (e.g., milk glands in a male)..............
....

(40) Ever desired to wear apparel of psychical sex..... Ever worn such apparel since childhood.........
....

(41) If a physical male, fondness for loud or fancy apparel..... If female, for plain apparel.............
....

(42) Ever arrested or imprisoned for following instincts of psychical sex..... If so, aggregate number of months or years imprisoned...

(43) Please note any other significant data as to particular homosexualist on separate sheets.

GENERAL QUERIES:

(44) Approximate number of homosexuals — positively known as such — have you encountered in your life-time: Passive inverts.... Active pederasts.... Male psychical hermaphrodites.... Physically female homosexualists....

(45) How many additional individuals have been under your suspicion: Passive inverts.... Active pederasts.... Male psychical hermaphrodites.... Physically female homosexualists....

(46) The author of "Autobiography of an Androgyne" estimates, roughly, *passive inverts* as 1 out of 300 physical males. Please give your estimate: Passive inverts, 1 out of....males. Active pederasts, 1 out of....males. Male psychical hermaphrodites, 1 out of....males. Female psychical hermaphrodites, 1 out of....females. Other physically female homosexualists, 1 out of....females. (In query 46, only adults are to be considered.)

NAME AND ADDRESS OF PHYSICIAN SUBMITTING SCHEDULE. THIS INFORMATION WILL BE CONSIDERED CONFIDENTIAL, AND IS ESSENTIAL.

Name...................
Street and number...................
Post-office...................
Date.........   State...................

overcome it. Would disclose neither his name, residence, or family history.[22]

In 1884 Dr. James G. Kiernan of Chicago brought the total, he claimed, to 27 (with 5 from America) and observed that, with only 4 females and 23 males, men are clearly "predisposed to the affection."[23] Over time as the cases grew more numerous, it was possible for physicians to write about the condition without retelling individual life histories.

The most complete reports of homosexuals and lesbians ran 20 pages or more, but some were as brief as two paragraphs. Extensive case reports included information about age, physique, physiognomy, occupational history, medical history, family history (especially insanity or nervous disorders among relatives), gender attributes, emotional life, and sexual experiences.

Both women and men were described as inverts, although clinicians agreed that men predominated. Chicago neurologist James G. Kiernan suggested that the condition might be noticed more frequently among women if it were not so difficult, in general, to elicit a full sexual history from females.[24] Cases included a full range of occupations and social levels, though clerks and small-business proprietors appear frequently. While a number of the cases were first discovered among asylum patients, most of these exhibited no mental symptoms beyond their odd sexuality, and few of those outside the asylum system had mental problems beyond discontent with their peculiar sexual urges, shame, or fear of exposure and disgrace.

Among those men and women who felt attraction to members of their own sex, a number revealed cross-gender feelings, actions, and even physique, ranging from casual tomboy activities and a somewhat cross-gendered style of attire to the feeling of having a soul mismatched to one's body. Others were "perfectly natural in appearance."[25] While the medical literature of this era also recorded some men having elaborate erotic involvement with female clothing and masquerade, it reported them being sexually oriented toward women rather than other men; they used their feminine finery as a masturbatory stimulus.[26] Sometimes the adoption of an outwardly feminine style served among homosexual men as a means of mutual recognition. No men were recorded as having "passed" consistently in public living as women, though there are numerous cases in which two females lived as husband and wife for years or even decades without detection, sometimes even being legally married.[27] For example, Murray Hall, a prominent Tammany Hall leader for many years, voted regularly as a man and was married twice to other women before her successful masquerade was discovered and revealed by her physician when Hall died of cancer in 1901.[28]

As with gender variation, the actual sexual activity of this first generation of medically observed inverts varied widely. The described sexual relations ranged from the purely genital to feelings and forms of affection far removed from the physical. The unfortunate Mr. X, described above, pursued the affection of his beloved for months without seeking a physical connection, having an "unspeakable horror" of it. An immigrant businessman was tormented by desire "to embrace men," but had "never given way to his desires."[29] One young woman, in the words of Dr. Kiernan,

> feels at times sexually attracted by some of her female friends, with whom she has indulged in mutual masturbation. These feelings come at regular intervals, and are then powerfully excited by the sight of female genitals. . . . She is aware of the fact that while her lascivious dreams and thoughts are excited by females, those of her female friends are excited by males. She regards her feeling as morbid.[30]

In contrast, Lucy Ann Slater, alias "Rev. Joseph Lobdell," and an unnamed cigar dealer were both sexually experienced and demanding. When committed to an asylum, Slater "embraced the female attendant in a lewd manner and came near overpowering her. . . . Her conduct on the ward was characterized by the same lascivious conduct, and she made efforts at various times to have sexual intercourse with her associates."[31] The cigar dealer went to considerable lengths to secure sexual satisfaction on a regular basis, even in a less than ideal relationship.[32]

A number of the inverts described by American physicians acknowledged heterosexual experience, sometimes with husbands or wives, sometimes with prostitutes. While some of these experiences were forced (often at the urging of friends or doctors) and others were reported as unsatisfying, the cases taken together do not actually confirm Krafft-Ebing's characterization, repeated frequently by American doctors,[33] that inverts have a congenital absence of sexual feeling toward the opposite sex.

That the cities might harbor numerous inverts was recognized early by some New York physicians, such as George M. Beard.[34] In 1884 Dr. George F. Shrady, editor of *The Medical Record*, printed without comment the claim of a German homosexual then living in America that an estimate of the rate of sexual inversion at one in 500 was too low, since he was personally acquainted with 12 in his native city of 13,000 and also knew at least 80 in a city of 60,000.[35] When Dr. Blumer's Mr. X rejected the approaches from "men of unnatural desire" out of his distaste for the sexual behavior it implied to him, he noticed that they were "able to recognize each other," and he was aware that these men shared some common nature, even if he thought he was not one of them.

Relatively few lesbians and homosexuals described in the medical literature, however, seem to have been aware of others like themselves. Many felt no possibility of finding like-minded individuals, yet some, especially in larger cities, did locate gathering places and make contacts within tentatively forming communities.[36] It seems likely that they were seeking not only social and sexual relations, but also a confirmation of their odd feelings, for numerous cases refer to persistent efforts toward self-understanding. By voluntarily approaching physicians, many seem to have sought knowledge as much as therapy; case accounts regularly refer to patients' concern for self-justification. Blumer, for example, described a young man's letters and essays which sought to "explain, justify or extenuate his strange feeling."[37] Two decades later, Dr. William Lee Howard of Baltimore observed of homosexuals, "they are well read in literature appertaining to their condition; they search for everything written relating to sexual perversion; and many of them have devoted a life of silent study and struggle to overcome their terrible affliction."[38]

Despite their scientific aspirations, these pioneering clinicians were quick to generalize from a very small number of cases. At first, while the discussion was closely tied to the cases being reported, the diversity of the described experiences slowed efforts to fashion a well-defined syndrome under the rubric of sexual inversion, but after some publications outlined general features, other observers were drawn into elaborating a consistent invert type. Even though George M. Beard had earlier pointed out that most inverts had "no occasion to go to a physician; they enjoy their abnormal life . . . or are too ashamed of it to attempt any treatment,"[39] an increasingly uniform medical image came to stand for the whole group. That the doctors saw sexual inversion as a unitary condition (despite extreme variations), rather than as an accidental collection of disparate phenomena, probably resulted in part from the self-consciousness exhibited by at least some of the inverts themselves.

In searching for the characteristic feature of their sexually inverted patients, the physicians were unselfconsciously formulating the modern notion of a person's "sexuality" as something distinct from that individual's sexual behavior. After the initial group of cases, the reports came to focus less on particular sexual actions than on consistent impulses—today one might say "orientation" or "preference"—and by implication, personality. In moving to a level of characterization that departed from gross sexual behavior, the doctors were not inventing a scheme *ex nihilo;* they were following the lead of their patients, most of whom felt that there was some interior quality that made them "different," whether it appeared in their behavior or not. As the reporting clinicians tried to draw this condition into view for examination—and possibly therapy—they juggled and juxtaposed old and new concepts. For example, when they observed homosexual behavior in persons who seemed to lack this interior condition, they termed it "vice," and condemned it with a vigor that very few of them applied to the behavior of those they considered true inverts.[40] The distinction between being and behavior could even operate in such a way that the sexual partners of the lesbians and homosexuals were in many cases not regarded as inverts by either the doctors or the patients.

PROFESSIONAL REPUTATION AND PATIENT SELF-REFERRAL

Some reports provide considerable evidence of an early pattern of self-referral to appropriate physicians. Looking back from the perspective of 1904, Dr. Howard described it this way:[41]

They have but little faith in the general practitioner; in fact, in our profession, and their past treatment justifies their lack of confidence. Hence it is that when they do find a physician who has taken up a conscientious study of their distressing condition, they open their hearts and minds to him.

And later in the same article, Howard remarked of a patient who was a Princeton student and a music lover:

He was well informed as to the attitude of the family physician in such cases as his, hence had studied up the subject for himself, having quite a library dealing with sexual perversions.

How such people found sympathetic physicians is not indicated in the case records, but since many were aware of the growing medical literature (both European and American) on their condition, it seems probable that some physicians gained a reputation for trust and tolerance. The fact, as noted below, that practitioners of the emerging specialization of outpatient neurology were overrepresented among authors on inversion is another indication that some homosexuals at least were aware of specialty groupings within the profession—and of the neurologists' reputation as healers sympathetic to sufferers from "nervous" ills.

Because, with rare exceptions, the evidence of these men's and women's motivations for self-referral comes only through the physicians' reports, it is impossible to know precisely why they approached physicians for assistance at this particular juncture in time. Yet three general transformations of that era probably helped shape individuals' response to a sense of personal difference. First, science slowly and broadly came to hold a position of social authority by the end of the 19th century in the United States, also enhancing the status of medicine. One consequence for some Americans was that physicians replaced the clergy as authoritative personal consultants in the realm of sex. Second, for contemporaries who observed the behaviors being medicalized in these decades—kleptomania and the judicial defense of "innocent by reason of insanity" were prominently discussed in the press, for example—the shift in status from crime to illness might well have appeared humani-

tarian and progressive. The consequent image of a sympathetic, forward-looking profession probably encouraged individuals to risk sharing their secrets with physicians active in these developments. A third context of self-referral was the contemporary emergence of a public awareness of homosexuals. That novelty showed some people with "odd feelings" that they were not unique in their experience, and while such an awareness would not necessarily lead a troubled person to choose to consult a physician, it promoted the idea that other people might assist in understanding or changing one's feelings. Whether a troubled person turned for help to a physician or to fellow sufferers clearly depended on such factors as the person's occupation, place of residence, social standing, awareness of communities of inverts, and individual psychology, as well as familiarity with the new types of medical practitioners.

The force of these three factors is clear in Earl Lind's *Autobiography of an Androgyne*. When at age 17 Lind first shared concern about his sinful impulses with a minister, he was counseled to see the family doctor. This physician "advised me to enter into courtship with some girl acquaintance, and said that this would render me normal. Like most physicians in 1890, he did not understand the deepseated character of my perversion."[42] Lind later sought specialized help first from Prince A. Morrow, the eminent venereologist, and then from an alienist, Robert S. Newton. Thereafter, Lind read all the medical literature on inversion he could find at the library of the New York Academy of Medicine. For a while he abandoned physicians, who mostly administered anaphrodisiacs. But in time he found a doctor (unnamed) who advised him to follow his sexual desires as a less harmful alternative to the risk of nervous problems from frustrating an exceedingly amorous nature. Through many meetings this physician supported Lind's sense that his peculiar character was natural to him and should be accepted and permitted expression. Eventually Lind published his memoir so that other physicians—and their patients—would suffer less from ignorance of natures like his.

While the pioneer specialists consistently acknowledged social disapproval of the sexual behaviors involved, most tempered moralism with a tone of scientific detachment, exhibiting a tolerance based, at least rhetorically, on a medical materialism powerfully reinforced by contemporary achievements. Reports often included some sort of evaluation, though not always so ferocious as Dr. Monroe's litany: abominable, disgusting, filthy, worse than beastly.[43] Most physician authors, however, distanced themselves from such scornful attitudes—without openly rejecting them—by coupling their disapproval with some justification for the legitimacy of their interest. George F. Shrady's 1884 editorial opened with a typical example:

> Sir Thomas Brown once wrote . . . that the act of procreation was "the foolishest act a wise man commits in all his life. Nor is there anything that will more deject his cooled imagination." The physician learns, however, and finds . . . far down beneath the surface of ordinary social life, currents of human passion and action that would shock and sicken the mind not accustomed to think everything pertaining to living creatures worthy of study. Science has indeed discovered that, amid the lowest forms of bestiality and sensuousness exhibited by debased men, there are phenomena which are truly pathological and which deserve the considerate attention and help of the physician.[44]

Kiernan opened one of his articles in a similar fashion: "The present subject may seem to trench on the prurient, which in medicine does not exist, since 'science like fire, purifies everything.'"[45] In William A. Hammond's monograph on impotence, the section on sexual inversion began with a disclaimer different in tone, but similar in function:

> Several cases of sexual inversion in which the subjects were disposed to form amatory attachments to other men have been under my observation. They are even more distressing and disgusting than cases . . . I have just given; but it is necessary for the elucidation of the subject to bring their details before the practitioner. So long as human nature exists such instances will occur and physicians must be prepared to treat them.[46]

Despite his declaration of disgust and distress, Hammond's reports are quite sympathetic to the two inverts he described; and while such sympathy from Dr. Hammond and his colleagues may not have been unconditional, the humane impulse in many of these pre-1900 cases is prominent. This feature of the new viewpoint may well have arisen from its formation within the clinical context, where categories are not simply abstractions and where "problems" are embodied in persons.

When clinicians asserted that homosexuality should no longer be regarded as criminal and forbidden, but tolerated in private as pathology that might be treated, some of them may at times have been aware that this would expand the medical profession's power.[47] For example, George F. Shrady in *The Medical Record,* which he edited, declared on behalf of the profession:

> we believe it to be demonstrated that conditions once considered criminal are really pathological, and come within the province of the physician. . . . The profession can be trusted to sift the degrading and vicious from what is truly morbid.[48]

But such a motivation seems subordinate to and entirely consistent with their interest in according science and nature a higher status than traditional morality and religion. After the Chicago physician G. Frank Lydston began his 1889 article[49] by quoting Kiernan's statement that "science, like fire, purifies everything," he announced that the subject of sexual perversion was being taken from the "moralist"; it was far better "to attribute the degradation of these poor unfortunates to a physical cause, than to a wilful viciousness" and to think of them as "physically abnormal rather than morally leprous."

Other physicians, nonetheless, saw moral dangers in the profession's accepting the invert as natural, even as a *lusus naturae.* For example, Dr. J. A. De Armand, of Davenport, Iowa, actively opposed the trend toward medicalizing—and thus offering inappropriate sympathy for—sin:

> Sexual perversion is the direct outgrowth of sexual abuse. It is the sensual Alexander seeking new worlds to conquer. It is the legitimate heritage of vicious associations and acquired weakness. The complimentary offering of "mental derangement" which excuses the man who seeks sexual gratification in a manner degrading and inhuman, is seldom more than a cloak within whose folds there is rottenness of the most depraved sort. . . . It surely is unnecessary to complicate medico-legal nomenclature by attributing such conduct to morbid mentality, when it clearly is deviltry. . . . The individual who starts out with jackass proclivities and billy-goat capabilities will soon tire of the normal

sexual act, and the fact that he enters into an era of sexual abandon . . . is no proof that his mentality is at fault. . . . I have no patience with the ready excuse which the medical profession volunteers.[50]

However common De Armand's sentiments may have been among the profession in general, they were not frequently expressed in print.[51] In the published literature on sexual inversion the medical model was becoming dominant by the turn of the century.

If the materialism of a scientific approach within medicine was one of the intellectual supports for reclassifying homosexuality from sin to natural anomaly, the shift was further advanced by a contemporary realignment of power and prominence among specialists in mental disease. An old guard, primarily asylum superintendents who were conservatives on political, religious, moral, and scientific issues, lost ground to neurologists, characterized as a group by private practice, appointments to general hospitals rather than asylums, European training in science, and a tendency toward agnosticism and materialism.[52] The neurologists, including the eminent figures William A. Hammond, Charles H. Hughes, James G. Kiernan, and Edward C. Spitzka, predominated among the mental-disease specialists who published cases of sexual inversion. With their highly visible outpatient practice, they were a natural first recourse for self-referral by troubled men and women.[53]

PERSISTENCE OF OLDER CONCEPTIONS

While some physicians were establishing the invert as a type of person, and others were resisting this reformulation, still others were continuing to deal with sexuality in traditional terms, as simply an aspect of behavior rather than a fundamental aspect of being. For example, when Dr. Randolph Winslow of Philadelphia reported in 1886 on an "epidemic of gonorrhoea contracted from rectal coition" in a boys' reformatory, he described the extent and manner of "buggery" with precision, but without acknowledging any awareness that a burgeoning medical literature was endeavoring to describe the type of person engaged in such practices and to determine whether and when constitu-

tional factors were more significant than depravity in explaining their occurrence.[54]

As late as 1899 George J. Monroe, a Louisville proctologist devoting an entire article to sodomy and pederasty, considered these sexual acts as "habits" ("abominable," "disgusting," "filthy," and "worse than beastly" to be sure) with no attention to what kinds of persons might engage in them and with no psychological etiology. His assumption was that such acts occur either situationally, "where there is enforced abstinence from natural sexual intercourse," such as among "soldiers, sailors, miners, loggers," etc., or in those "satiated with normal intercourse." Without apparent irony, he declared, "There must be something extremely fascinating and satisfactory about this habit; for when once begun it is seldom ever given up."[55]

Noticing these differences in thinking, however, should not lead one to conclude that Winslow, Monroe, and others like them were consciously rejecting the new observations and the style of thinking that defined the homosexual as a particular, if pathological, personality type. Since they neither addressed nor challenged this novel conception, it seems more likely that they were simply unaware of it. In their traditional view, someone need only "tire of the normal sexual act" to engage in homosexual practices.

MEDICAL DISCOVERY, THE DISCOVERED, AND THE WIDER WORLD

Reviewing the ways late 19th-century clinicians discovered homosexuals and then shaped homosexuality into a disease entity brings two points to the fore. First, this initial stage in the morbidification of homosexuality does not fit into a unidirectional model of stigmatization and social control. Patients were often active conspirators with the physicians, not passive victims of the new diagnostic labeling.

Although in the 20th century the homosexuality diagnosis became a central feature in the social oppression of homosexuals, to the benefit of some members of the medical profession offering "cures,"[56] it is wrong to attribute such later baneful developments to the first generation of observers or their self-referring homosexual patients. The actions of those doctors and patients must be appreciated for the ambiguous character they had at the time and not dismissed out of a 20th-century distress over people victimized by medicalization. Carroll Smith-Rosenberg's interpretation of female hysteria is an apt guide here for its emphasis on the collaboration between patients and their physicians, the advantages the diagnosis gave women (at least some of them some of the time), the way both parties' thoughts and actions were shaped by cultural expectations concerning properly gendered norms of behavior, and the way "physicians, especially newly established neurologists with urban practices, were besieged by patients."[57]

Physicians played another and more public role as well. Since few homosexuals before 1900 described their lives publicly from a nonmedical point of view, doctors were, by default, the leaders in accumulating and organizing knowledge of homosexuals for society as a whole. In time, this "knowledge" (even distorted as it was in parts) entered the public domain, where it offered thousands of uncertain people an identity, a way to think of themselves as fundamentally different but neither immoral nor vicious. Charles Rosenberg has described the process this way: "The physician, not the priest or judge, made the most appropriate guardian for the rights of both society and the individual. The sufferer from phobias and anxieties, the victim of sexual incapacity, the man or woman consumed by desire for a socially unacceptable love object could be seen as the product of his or her material condition rather than as an outcast."[58]

As noted by several of the physicians quoted above, some inverts were often quite active in seeking out writings, including medical writings, on their condition. It cannot be proven conclusively that the activities and publications of American physicians and related nonmedical writings significantly affected the self-understanding of perhaps thousands of lesbians and gay men who were not patients, but convincing confirmation comes from numerous individual accounts documenting some of the ways *public discourse* directly and substantially changed the *private experience* of individuals.[59] These accounts, many of which portray both a personal discovery of an identity and an enthusiasm for a new-found sense of community, also indicate the crucial role played by medical models.

From the many personal stories I have collected, two illustrate most graphically the powerful role of public discourse and medical conceptions in the dialectic of gradual change in people's experience of identity, community, social role, and institutions.[60] One of the stories is from the famous autobiography by political activist Emma Goldman and concerns an experience on her lecture tour of 1913:

> Men . . . and women . . . used to come to see me after my lectures on homosexuality, and . . . [they] confided to me their anguish and their isolation. . . . Most of them had reached an adequate understanding of their differentiation only after years of struggle to stifle what they had considered a disease and shameful affliction. One young woman . . . had never met anyone, she told me, who suffered from a similar affliction, nor had she ever read books dealing with the subject. My lecture had set her free; I had given . . . back her self-respect.[61]

The second is by Elsa Gitlow, a lesbian who was born just before the turn of this century. In 1914, at the age of 16, she moved with her family to Montreal. Her recollections in the mid-1970s of her youthful pursuit of other poets includes a remarkable vignette of the process we've been examining.

> When I was about seventeen, I'd been working for a year and a half in tiresome jobs, and I just couldn't see myself going through life that way. So in search of my people—of where, perhaps, I might belong—I had an inspiration one day. Under an assumed name I wrote a letter to the editor's column in the Montreal *Daily Star*, asking, "Is there in Montreal any kind of an organization to which writers, or prospective writers, might belong?" I forget the exact wording, but that was the gist of it. I wrote [that inquiry] under an assumed name; [and] then, under my own name, I sent an answer, saying, "The undersigned knows of no such organization, but one is in the process of being formed and anyone interested should write to . . . ," and I gave my name and address. I received all kinds of letters, and I called a meeting at my parents' house for the formation of the group. One of the men who came was a young man who these days would be called gay. . . . Roswell George Mills—the most extraordinary being I'd ever seen in my life. He was beautiful. About nineteen, exquisitely made up, slightly perfumed, dressed in ordinary men's clothing but a little on the chi-chi side. And he swayed about. . . . We became friends

almost instantly because we were both interested in poetry and the arts.

Roswell apparently recognized immediately my temperament. He said, "Do you know about Sappho?" I don't remember if I'd heard anything about her, but I went to the library, found writings about her and translations of her fragments, and immediately became interested. Through Roswell I started to hear about some literature that would lead me to some knowledge about myself and other people like me. Other than the literary, I think the first books I read were Edward Carpenter's *The Intermediate Sex*, and Kraft-Ebbing [sic], and Lombroso—and all of these were revelatory to me because I could have no doubt, having read them, of where my orientation lay. Though they wrote on the level of morbid psychology, and I couldn't accept the morbidity side of it, it was very interesting to read all this and to find out there had been other people like me in the world.[62]

Elsa Gitlow's finding confirmation of her identity in the new medical discourse on sexually deviant persons may well have been that of thousands of other men and women for several decades after 1880.

Stepping back from the experiences of the individual physicians and the individual homosexuals described here will allow for a more general consideration of how this new disease was framed. While the shaping of the diagnosis stemmed from several factors intrinsic to medicine, the timing was due largely to extrinsic factors, primarily the historical emergence of the social roles and personal identities of homosexual men and women in Europe and North America over the course of the 19th century. Their growing presence might eventually have pro-

voked a spontaneous response from American physicians, but the medical response was unlikely to have gained momentum so rapidly or to have occurred when it did without the particular impetus created by the personal essays of the homosexual jurist Karl Heinrich Ulrichs. Ulrichs's writings engendered much discussion in the medical and psychiatric journals of western Europe and, shortly later, in those of the United States.

As sexual inversion quickly advanced from the subject of almost pure description of odd cases to an articulated disease concept, the form it took was shaped by currents present in American medicine,[63] which was then aspiring to scientific status without relinquishing its humanitarian commitment to patients. American medical thinking of the 1880s and 1890s was dominated by a somatic bias (which would decline after the turn of the century), and this led to linking sexual behavior to defective heredity and its putative physical signs. It was probably the materialism and naturalism of their scientific outlook—so characteristic of the late 19th-century—that helped physicians challenge older views of inverts as sinners or criminals. "A growing secularism paralleled and lent emotional plausibility to this framing in medical terms of matters that had been previously construed as essentially moral. Science not theology should be the arbiter of such questions."[64] But medicine's new outlook and new public status came to frame and legitimate more than a diagnosis and an interpretation of behavior; by helping to give large numbers of people an identity and a name, medicine also helped to shape these people's experience and change their behavior, creating not just a new disease, but a new species of person, "the modern homosexual."

NOTES

1   Westphal, "Die conträre Sexualempfindung: Symptom eines neuropathischen (psychopathischen) Zustandes," *Archiv für Psychiatrie und Nervenkrankheiten,* 1869, *2*: 73–108.
2   References to all these cases can be found in J. C. Shaw and G. N. Ferris, "Perverted sexual instinct (Contrare [sic] Sexualempfindung: Westphal; Inversione dell'instincto sessuale: Arrigo Tamassia; Inversion du sens genital: Charcot et Magnan)," *J. Nerv. & Ment. Dis.,* Apr. 1883, *10*(2) [= n.s. *8*(2)]: 185–204.
3   Charles E. Rosenberg, "Disease and social order in

America: perceptions and expectations," *Milbank Quart.* 1986, *64,* suppl. 1: 34–55, quotation on p. 44.
4   The present article is an expansion of "American physicians' earliest writings about homosexuals, 1880–1900," *Milbank Quart.,* 1989, *67,* suppl. 1: 92–108, which limited itself largely to "what the doctors said" and interpreted their writings only in the context of the profession without attending to American society more broadly.
5   Mary McIntosh, in "The homosexual role," *Social Problems,* 1968, *16*: 182–192, was the first to develop

this point of view, which has now been taken up by many sociologists and historians. A collection of influential essays, edited by Kenneth Plummer, appeared in 1981 under the title *The Making of the Modern Homosexual* (London: Hutchinson); it reprints McIntosh's original article with a new postscript by her (pp. 30–49). Two sociologists have recently addressed these issues in a critical fashion: Stephen O. Murray, *Social Theory, Homosexual Realities* (New York: Gay Academic Union, 1984) and Steven Epstein, "Gay politics, ethnic identity: the limits of social constructionism," *Socialist Review*, 1987, *93/94*: 9–54. The most recent contribution to this school of interpretation is David M. Halperin, *One Hundred Years of Homosexuality and Other Essays on Greek Love* (New York/London: Routledge, 1990), especially the title essay, pp. 15–40, which includes extensive references to the literature.

Some historians of homosexuality ignore or oppose this interpretation, most notably John Boswell; see his "Revolutions, universals and sexual categories," *Salmagundi*, 1982–1983, *58–59*: 89–113, now reprinted in *Hidden from History: Reclaiming the Gay and Lesbian Past*, ed. Martin Bauml Duberman, Martha Vicinus, and George Chauncey, Jr. (New York: New American Library, 1989).

Aspects of the American medical literature on homosexuality have been analyzed in John C. Burnham, "Early references to homosexual communities in American medical writings," *Human Sexuality*, 1973, 7(8): 34–49; Vern L. Bullough, "Homosexuality and the medical model," *J. Homosexuality*, 1974, *1*: 99–110; Lillian Faderman, "The morbidification of love between women by 19th-century sexologists," *J. Homosexuality*, 1978, *4*: 73–90; George Chauncey, Jr., "From sexual inversion to homosexuality: medicine and the changing conceptualization of female deviance," *Salmagundi*, 1982–1983, *58–59*: 114–146; and Vern L. Bullough, "The first clinicians," in *Male and Female Homosexuality: Psychological Approaches*, ed. Louis Diamont (Washington, DC: Hemisphere Publishing Corporation, 1987), pp. 21–30. Much of the medical literature was first reprinted in Jonathan Ned Katz, *Gay American History: Lesbians and Gay Men in the U.S.A.* (New York: Thomas Y. Crowell, 1976).

6   Along with sociologists and historians, some philosophers have taken up the social constructionist position; see, for example, Ian Hacking, "Making up people," in *Reconstructing Individualism: Autonomy, Individuality, and the Self in Western Thought*, ed. Thomas C. Heller, Morton Sosna, and David E. Wellbery (Stanford: Stanford Univ. Press, 1986), pp. 222–236; and Arnold I. Davidson, "Sex and the emergence of sexuality," *Critical Inquiry*, Autumn 1987, *14*: 16–48.

7   Regarding the absence of a concept of homosexual-

as-a-type-of-person in America before the 19th century, one may consider the evidence of sodomy trials in the North American colonies, which furnish important records of homosexual and lesbian behavior under the rubrics of "sodomy," "the crime against nature," "lewdness," etc.

When Edmund Morgan wrote about these cases in the 1940s, he dealt with them using 20th-century sexual categories unselfconsciously; see "The Puritans and sex," *New England Quarterly*, 1942, *15*(4): 591–607. But three more recent studies by Robert F. Oaks—"Perceptions of homosexuality by justices of the peace in colonial Virginia," *Sexualaw Reporter*, Apr./June 1978, *4*(2): 35–36 (reprinted in *J. Homosexuality* 1979–1980, *5*: 35–41); "'Things fearful to name': sodomy and buggery in seventeenth-century New England," *J. Soc. Hist.*, 1978, *12*: 268–281; and "Defining sodomy in seventeenth-century Massachusetts," *J. Homosexuality*, 1980–1981, *6*: 79–83—and Jonathan Ned Katz, *Gay/Lesbian Almanac: A New Documentary* (New York: Harper & Row, 1983) show that the wording of the original documents reveals no conception of the sodomite as a type of person, even when an offender was implicated more than once and was regarded as persisting in the vile behavior.

8   This article was discovered by Allan Bérubé and is quoted with his permission from an unpublished essay, "Lesbians and gay men in early San Francisco: notes toward a social history of lesbians and gay men in America" (1979).

9   G. Alder Blumer, "A case of perverted sexual instinct (*Conträre Sexualempfindung*)," *Am. J. Insanity*, 1882, *39*: 22–35, quotation on pp. 23–25.

10   McIntosh, "The homosexual role"; Lawrence Stone, *The Family, Sex, and Marriage in England, 1500–1800* (New York: Harper & Row, 1977); and Jeffrey Weeks, "'Sins and diseases': some notes on homosexuality in the nineteenth century," *History Workshop*, 1976, *1*: 211–219 were the most important publications in stimulating my own work on the social construction of homosexuality. Conversations with Natalie Zemon Davis, John D'Emilio, Michael Lynch, and Robert Padgug have also played a critical role in helping to shape my interpretation.

11   Evidence on this period's gay history has not yet been collected systematically, and little of it has been published. John D'Emilio, "Capitalism and gay identity," in *Powers of Desire: The Politics of Sexuality*, ed. Ann Snitow, Christine Stansell, and Sharon Thompson (New York: Monthly Review Press, 1983), pp. 100–113, though brief, offers the best published narrative of these developments in American gay history, and the present summary depends on his account. John D'Emilio and Estelle B. Freedman have recently set homosexual experiences into a broader

context in *Intimate Matters: A History of Sexuality in America* (New York: Harper & Row, 1988).

An unpublished paper by Michael Lynch, "The age of adhesiveness: male-male intimacy in New York City, 1830–1880" (presented at American Historical Association Meeting, New York, December 1985), is the richest account of specific developments in New York City at midcentury. George Chauncey has examined the later period in an unpublished Yale dissertation (Ph.D., 1989), forthcoming as *Gay New York: Gender, Urban Culture, and the Making of the Gay Male World, 1890–1940* [New York: Basic Books, 1994].

In addition to my own research, my knowledge of specific details draws on many conversations with friends, who have generously shared information of mutual interest. These include Allan Bérubé, George Chauncey, Lisa Duggan, Joseph Interrante, Jonathan Ned Katz, Gary Kinsman, Michael Lynch, Allan V. Miller, and the late Gregory Sprague. Though not historians, Carole Vance and Ian Hacking have also made critical contributions through our conversations about sex and about "making up people." (On the latter, see Hacking's article cited above.)

12   For aspects of the social history of lesbians in America, see Lillian Faderman, *Surpassing the Love of Men: Romantic Friendships and Love Between Women From the Renaissance to the Present* (New York: William Morrow, 1981); Katz, *Gay American History;* Katz, *Gay/Lesbian Almanac;* Carroll Smith-Rosenberg, *Disorderly Conduct: Visions of Gender in Victorian America* (New York: Alfred A. Knopf, 1985); and *Hidden from History.*

National studies are essential for understanding these developments in a comparative perspective. For Canada, see Gary Kinsman, *The Regulation of Desire: Sexuality in Canada* (Montreal: Black Rose Books, 1987). For England, see R. F. Claus, "Confronting homosexuality: a letter from Francis Wilder," *Signs,* 1977, *2*: 928–933, and Jeffrey Weeks, *Coming Out: Homosexual Politics in Britain, from the Nineteenth Century to the Present* (London: Quartet Books, 1977). For France, see Pierre Hahn, ed., *Nos ancêtres les pervers: La vie des homosexuels sous le Second Empire* (Paris: Olivier Orban, 1979), and Robert A. Nye, "Sex difference and male homosexuality in French medical discourse, 1830–1930," *Bull. Hist. Med.,* 1989, *63*: 32–51. For Germany, see Klaus Pacharzina and Karin Albrecht-Désirat, "Die Last der Ärzte: Homosexualität als klinisches Bild von den Anfängen bis heute," in *Der unterdruckte Sexus,* ed. J. S. Hohmann (Lollar: Achenbach, 1979), pp. 97–112. For Russia, see Laura Engelstein, "Lesbian vignettes: a Russian triptych from the 1890s," *Signs,* 1990, *15*: 813–831, and her forthcoming book on sexuality in late 19th-century

Russia [*The Keys to Happiness: Sex and the Search for Modernity in Fin-de-Siècle Russia* (Ithaca: Cornell Univ. Press, 1992)].

For a transnational perspective, Michel Foucault, *The History of Sexuality. Volume 1: An Introduction,* trans. Robert Hurley (New York: Pantheon, 1979), is stimulating, though my interpretation diverges from his on the exact role doctors' labeling played in the emergence of "the modern homosexual." See also Peter Conrad and Joseph Schneider, *Deviance and Medicalization: From Badness to Sickness* (St. Louis: Mosby, 1980), and David F. Greenberg, *The Construction of Homosexuality* (Chicago: Univ. of Chicago Press, 1988).

13   D'Emilio, "Capitalism and gay identity," p. 106.

14   Dr. Irving C. Rosse mentioned messenger boys, as well as teenagers employed by military personnel; see his "Sexual hypochondriasis and perversion of the genesic instinct," *J. Nerv. & Ment. Dis.,* Nov. 1892, *17*(11): 795–811, pp. 803–804; which was also published in *Va. Med. Mon.,* 1892, *19,* n.s. *17*: 633–649. Dr. William A. Hammond described a bellboy's ready acceptance of his patient's sexual solicitation; see his *Sexual Impotence in the Male* (New York: Bermingham, 1883), pp. 66–67.

Walt Whitman's expansive "Song of Myself" in *Leaves of Grass* was parodied in *Vanity Fair* on March 17, 1860, as follows:

> I am the Counter-jumper, weak and effeminate.
> I love to loaf and lie about dry goods.
>   . . .
> For I am the creature of weak depravities;
> I am the Counter-jumper;
> I sound my feeble yelp over the woofs of the World.

Reprinted in Henry S. Saunders, *Parodies of Walt Whitman* (New York: American Library Service, 1923), p. 18.

15   See, for example, the sexual experiences documented in Martin Bauml Duberman, "'Writhing bedfellows': 1826—two young men from antebellum South Carolina's ruling elite share 'extravagant delight,'" *J. Homosexuality,* 1980–1981, *6*: 85–101, now reprinted in *Hidden From History.*

16   Earl Lind ("Ralph Werther"—"Jennie June"), *Autobiography of an Androgyne,* edited and with an introduction by Alfred W. Herzog (New York: The Medico-Legal Journal, 1918; reprinted New York: Arno Press, 1975), p. 3. Though not published until 1918, this memoir was first written in 1899.

17   Ralph Werther—Jennie June ("Earl Lind"), *The Female Impersonators: A Sequel to the Autobiography of an Androgyne and an Account of Some of the Author's Experiences During His Six Years' Career as Instinctive Female-*

Impersonator in New York's Underworld; Together with the Life Stories of Androgyne Associates and An Outline of His Subsequently Acquired Knowledge of Kindred Phenomena of Human Character and Psychology, edited and with an introduction by Alfred W. Herzog (New York: The Medico-Legal Journal, 1922; reprinted New York: Arno Press, 1975), pp. 150–152; this section is also reprinted in Katz, *Gay American History*, pp. 367–368.

18 In addition to the medical writings, the extensive international publicity in the 1890s surrounding Oscar Wilde's trial in England advanced the process of self-discovery and identity formation for isolated lesbians and gay men, as did the more localized news coverage of raids on homosexual bars and on houses of ill repute that offered "the two prostitutions, feminine and pederastic." When Dr. Irving C. Rosse, in "Sexual hypochondriasis and perversion" (p. 803), reported on the police breakup of such a place, called The Slide, in New York City, he remarked particularly on the publicity this exposure received in the *New York Herald*. For more on The Slide, see Katz, *Gay/Lesbian Almanac*, pp. 219 and 233.

19 The following observations are based on my review of what I believe are all the cases published by physicians in the United States through the end of the 19th century. Bullough, "Homosexuality and the medical model"; Faderman, "The morbidification of love between women"; Chauncey, "From sexual inversion to homosexuality"; and Bullough, "The first clinicians," have analyzed aspects of this medical literature, and much of it was first reprinted in Katz, *Gay American History*. Chauncey's interpretation and mine diverge on the importance of the medical literature in gay people's lives.

20 James G. Kiernan, "Perverted sexual instinct [Report of paper read 7 January 1884]," *Chicago Med. J. & Exam.*, Mar. 1884, 48: 263–265, quotation on p. 263.

The role played by the actions and writings of Karl Heinrich Ulrichs in prompting intense medical study of homosexuality is explained by Vern L. Bullough in "The physician and research into human sexual behavior in nineteenth-century Germany," *Bull. Hist. Med.*, 1989, 63: 247–267, pp. 254–257.

21 Although my own search of the literature revealed that Shaw and Ferris's was in fact the seventh American case report (preceded by the American cases published by Hagenbach, "Dr. H.," Blumer, Hammond, and Wise), their sense of novelty and their enthusiasm for discovery seem justified. These two doctors significantly advanced Americans' familiarity with this new phenomenon by publishing their English versions of 18 prior case reports from medical publications in French, German, and Italian. For the six earlier reports see (1) Allen W. Hagenbach, "Masturbation as a cause of insanity,"

J. Nerv. & Ment. Dis., 1879, 6(n.s. 4): 603–612; (2) "Dr. H," "'Gynomania'—A curious case of masturbation," *Medical Record*, 19 Mar. 1881, 19(12): 336; (3) Blumer, "Case of perverted sexual instinct"; (4 and 5) Hammond, *Sexual Impotence in the Male*, 55ff.; and (6) P. M. Wise, "Case of sexual perversion," *Alienist and Neurologist*, January 1883, 4(1): 87–91.

22 Shaw and Ferris, "Perverted sexual instinct," pp. 202–203.

23 Kiernan, "Perverted sexual instinct," p. 263.

24 *Ibid.*

25 Shaw and Ferris, "Perverted sexual instinct."

26 "Dr. H," "Gynomania."

27 In the 19th century, apparently thousands of women passed as men in public at one time or another. The advantage was significant, in terms of safety, social privilege, and the ability to vote, travel, and work freely, and this clearly motivated many of them. Linda Grant DePauw, "Women in combat: The Revolutionary War experience," *Armed Forces and Society*, 1981, 7: 209–226, reports that during the Civil War as many as 400 Northern women adopted a masculine identity and passed successfully as soldiers: "Some served more than two years before being detected. Of course, those who got away with the masquerade and were never caught are beyond our ability to count" (p. 218). Most of these were probably seeking contact with husbands and boyfriends, not with female lovers, but their ability to pass successfully confirms the feasibility of this option for many women. For a number of detailed, primary source accounts of women passing as men, see Katz, *Gay American History*, pp. 209–279, and San Francisco Lesbian and Gay History Project, "'She even chewed tobacco': a pictorial narrative of passing women in America," in *Hidden From History*, pp. 183–194.

28 Katz, *Gay American History*, pp. 232–238.

29 Shaw and Ferris's case, quoted above in note 22.

30 Kiernan, "Perverted sexual instinct," p. 264.

31 Wise, "Case of sexual perversion," pp. 87–88; also reprinted in Katz, *Gay American History*, pp. 221–225.

32 Hammond, *Sexual Impotence in the Male*, pp. 55–64. These pages on the cigar dealer are of special interest because the man was one of the few American cases to appear in a book instead of a journal. The account by Dr. Hammond includes interesting information on the subject's social life and on the doctor-patient encounter, which often went unrecorded. Additionally, it is striking evidence of the rapid rise of interest in the subject that, although this case of sexual inversion and a second one together occupy only 16 pages in a 274-page book, the book review by William J. Morton singled out for specific mention "those remarkable perversions of the sexual appetite recently studied by Westphal, Charcot and Magnan,

Tamassia, and others." See his "Review of *Sexual Impotence in the Male* by William A. Hammond," *J. Nerv. & Ment. Dis.*, 1883, *10*: 649–651, quotation on p. 650.

33   For example, Shaw and Ferris, "Perverted sexual instinct," p. 203.

34   George M. Beard, *Sexual Neurasthenia (Nervous Exhaustion): Its Hygiene, Causes, Symptoms, and Treatment, with a Chapter on Diet for the Nervous*, ed. A. D. Rockwell (New York: Treat, 1884), p. 102.

35   George F. Shrady, "Editorial: perverted sexual instinct," *Medical Record*, 19 July 1884, *26*: 70–71.

36   We have no evidence that the reticent German immigrant (reported by Shaw and Ferris) regarded himself as a member of a group rather than as a uniquely peculiar individual; he seemed to understand his "abnormal state" as an individual ailment that he simply wished medicine "to overcome." In contrast, when Mr. X rejected approaches from "men of unnatural desire" out of distaste for the implied sexual behavior, he noticed that the men were "able to recognize each other" and shared some common nature, even if he thought he was not one of them. Dr. Hammond gave no evidence that his cigar-dealer patient was aware of a community of like-minded individuals, but this patient's offering the standard formulation of having a woman's soul in a man's body, his ability to find partners, and his use of the name "Lida" with some of his friends all suggest at least a slight involvement in an informal community of inverts.

37   Blumer, "Case of perverted sexual instinct," p. 23.

38   William Lee Howard, "Sexual perversion in America," *Am. J. Dermatology & Genito-Urinary Dis.*, 1904, *8*: 9–14, quotation on p. 11.

39   Beard, *Sexual Neurasthenia*, pp. 101–102.

40   The next section will consider how the physicians distinguished between true inverts and vicious persons: by regarding the former as having a congenital, i.e. inherent if not precisely hereditary, condition and the latter as having an acquired taste.

41   Howard, "Sexual perversion in America," pp. 11 and 13.

42   Lind, *Autobiography*, p. 47.

43   George J. Monroe, "Sodomy—pederasty," *Saint Louis Medical Era*, 1899–1900, *9*: 431–434.

44   Shrady, "Editorial," p. 70.

45   James G. Kiernan, "Sexual perversion and the Whitechapel murders," *Medical Standard*, Nov. 1888, *4*(5): 129–130, and Dec. 1888, *4*(6): 170–172, quotation on p. 129.

46   Hammond, *Sexual Impotence in the Male*, p. 55.

47   See Foucault, *History of Sexuality*.

48   Shrady, "Editorial," p. 71.

49   G. Frank Lydston, "Sexual perversion, satyriasis, and nymphomania," *Med. & Surg. Reptr.*, 7 Sept. 1889, *61*(10): 253–258, and 14 Sept. 1889, *61*(11): 281–285, quotation on pp. 253–254.

50   J. A. Armand, "Sexual perversion in its relation to domestic infelicity," *Am. J. Dermatology & Genito-Urinary Dis.*, 1899, *3*: 24–26, quotation on pp. 24–25.

51   De Armand's views closely resemble the moral stance expressed some years earlier by John Gray and contrast with those of the neurologists, for which see Charles E. Rosenberg, *The Trial of the Assassin Guiteau: Psychiatry and Law in the Gilded Age* (Chicago: Univ. of Chicago Press, 1968), p. 70.

52   On the character and membership of these two groups, see Bonnie Ellen Blustein, "New York neurologists and the specialization of American medicine," *Bull. Hist. Med.*, 1979, *53*: 170–183; Jeanne L. Brand, "Neurology and psychiatry," in *The Education of American Physicians*, ed. R. L. Numbers (Berkeley: Univ. of California Press, 1980), pp. 226–249; F. G. Gosling, *Before Freud: Neurasthenia and the American Medical Community, 1870–1910* (Urbana: Univ. of Illinois Press, 1987); Rosenberg, *Trial;* and Barbara Sicherman, "The uses of a diagnosis: doctors, patients, and neurasthenia," *J. Hist. Med.*, 1977, *32*: 33–54.

53   John Gray, perhaps the most famous representative of the older alienist school, published no articles on inverts, as far as I know. Although three of the authors cited above were employed in asylums (Blumer, Ferris, and Wise), they were all (significantly?) in the post of assistant physician and not superintendent. The conservative professional association of alienists admitted as members only superintendents, not other asylum doctors (Rosenberg, *Trial*, p. 62).

54   Randolph Winslow, "Report of an epidemic of gonorrhoea contracted from rectal coition," *Medical News*, 14 Aug. 1886, *49*: 180–182.

55   Monroe, "Sodomy—pederasty," pp. 432–433.

56   Ronald Bayer, *Homosexuality and American Psychiatry: The Politics of Diagnosis* (New York: Basic Books, 1981).

57   Carroll Smith-Rosenberg, "The hysterical woman: sex roles and role conflict in nineteenth-century America," *Social Research*, 1972, *39*(4): 652–678; reprinted in Smith-Rosenberg, *Disorderly Conduct*, pp. 197–216, quotation on p. 204.

58   Rosenberg, "Disease and social order," p. 45. This passage continues as follows: "By no means all contemporaries accepted such views, of course. But these hypothetical diagnoses may well have been palatable to the stigmatized themselves; given the choice, an individual might well prefer to think of his or her deviant behavior as the product of hereditary endowment or disease process. It might well have offered more comfort than the traditional option of seeing oneself as a reprehensible and culpable actor.

The secular rationalism so prevalent in the late nineteenth century freed many Americans from a measure of personal guilt at the cost of being labelled as sick. Only in the second half of the twentieth century has this come to seem a problematic bargain."

Rosenberg's hypothesis of how these changes "might well have been" received by late 19th-century individuals seems strongly confirmed by many sexual inverts' enthusiasm for the medical literature and their active collaboration with physicians as documented here.

59 See, for example, the unsigned report edited by Charles H. Hughes, "The gentleman degenerate: a homosexualist's self-description and self-applied title: pudic nerve section fails therapeutically," *Alienist and Neurologist,* Feb. 1904, 25(1): 62–70. A section is reprinted in Katz, *Gay American History,* pp. 145–146, where Katz suggests Dr. Hughes himself may be the unnamed doctor reporting this case.

60 See also an early homosexual memoir by Claude Hartland (b. 1871), *The Story of a Life: For the Consider-* ation of the Medical Fraternity (St. Louis, 1901); reprinted with a foreword by C. A. Tripp (San Francisco: Grey Fox Press, 1985). In 1904, when New York neurologist Charles H. Hughes edited and published an unnamed invert's self-description in a medical journal, he mentioned this book and identified Hartland as one of his patients ("The Gentleman Degenerate," p. 68).

61 Quoted in Katz, *Gay American History,* p. 377.

62 *Word Is Out: Stories of Some of Our Lives,* ed. Nancy Adair and Casey Adair (San Francisco: New Glide Publications, 1978), pp. 15–17. Another installment of Elsa Gitlow's recollections, also dealing with Roswell Mills, appeared in *The Body Politic* (Toronto), May 1982. She later published a book-length memoir, *Elsa: I Come With My Songs* (San Francisco: Bootlegger Press, 1986).

63 For a concise review of American medicine's changing features in this era, see Rosenberg, "Disease and social order," especially pp. 44 and 45.

64 *Ibid.,* p. 45.

2

Black Lung: Miners' Militancy and Medical Uncertainty, 1968–1972

DANIEL M. FOX AND JUDITH F. STONE

Miners' respiratory complaints have engaged medical interest and political attention for several hundred years. In his classic study, *The History of Miners' Diseases*, George Rosen concluded that "throughout the nineteenth century the entire body of mining legislation had practically no connection with contemporary medical study of disease among miners."[1] A similar discontinuity between science and social policy characterized the controversy preceding the passage of the Coal Mine Health and Safety Act of 1969 and its amendment in 1972 by the Congress of the United States. These events, however, differed from other episodes in the history of the politics of occupational disease. Such illnesses have, as Rosen declared, a "dual nature."[2] They are both social and medical phenomena. In the United States a decade ago, the medical aspect of the dualism became, for a time, subordinate to the social.

Angry miners persuaded public officials to accept their definition of disease, respiratory impairment resulting from work, and to reject prevailing opinion among pathologists, epidemiologists, and internists. The miners' anger and its translation into effective politics was a result of a series of events which persuaded a significant number of them that they had been betrayed by their employers, their union, and their government. Black Lung, as miners called their disease, did physical

and psychological harm. It was also a symbol of the powerlessness felt by miners as they observed the decline of the coal industry, the corruption of the United Mine Workers of America, the erosion of community life in rural Appalachia as a result of outmigration and war, and the curtailment of the unique health care and pension system that compensated for some of the hardships of mining. Black Lung, unlike other diseases, did not yield precise correlations between clinical and pathological findings. The correlations were instead emotional, social, economic, and political.

This is an essay in contemporary history. Although many of the events we discuss occurred in the past decade, our themes are familiar to historians. The essay is grounded in the history of medicine, of labor unions, of the people of the southern mountains, and of legislative politics in the United States. Without this considerable literature, we would have left the subject to chroniclers and polemicists.

The essay is analytical, although it raises issues about which many people have strong opinions. We intend to disappoint readers who are convinced that there are unambiguous answers to the major questions physicians, epidemiologists, miners, industrialists, and public officials have asked about the relationship between coal dust and disease in this century.

When miners' lung ailments became the focus of regional and national attention in the United States in the late 1960s, medical opinion about them was sharply divided. The most controversial issues were the relationship between coal dust and respiratory distress and the methods of ascertaining the extent of impairment. The mechanism by which coal dust acted on the lungs was unclear. There were con-

DANIEL M. FOX is President, Milbank Memorial Fund, New York, New York.

JUDITH F. STONE is Associate Professor of History, Western Michigan University, Kalamazoo, Michigan.

From the *Bulletin of the History of Medicine*, 1980, *54*: 43–63. Copyright © 1980 by the Johns Hopkins University Press. Reprinted by permission of the Johns Hopkins University Press.

flicting opinions about the effects of the inhalation of coal dust on the development of coal workers' pneumoconiosis, chronic bronchitis, and emphysema. Clinicians disputed the significance of correlations between clinical evidence of respiratory impairment and X-ray findings. Pathologists debated the meaning of considerable blackness in the lungs at autopsy without previous clinical findings of loss of pulmonary function. Epidemiologists could not always find strong independent correlations between respiratory disorders in mining communities and work in the mines.

The consensus among medical investigators in the United States and Europe was that miners had serious respiratory problems which were only in part related to coal dust. Aside from urging the control of exposure to dust in the mines and the provision of care for people in distress, however, medical science offered little guidance for public policy. For instance, Donald Hunter, author of the leading textbook on occupational diseases, concluded that "the reasons for the differences in individual reactions to dust are not accurately known, but it is likely that they depend on anatomical, physiological, and biochemical variations from one person to another."[3] A panel of pathologists, summarizing knowledge about coal workers' pneumoconiosis in 1970, noted that "factors in addition to coal dust may contribute to pulmonary disease in coal workers."[4] In the same year, two German scientists, synthesizing a generation of research in various countries, concluded cautiously that "possible connections between dust exposure, coal miners' pneumoconiosis, and bronchial disease have repeatedly been emphasized."[5] a review of knowledge about the control of environmental lung disease published in the *New England Journal of Medicine* in mid-1970 argued that the "mechanism of action of coal dust on the lungs remains obscure" but that it was "urgently necessary to apply and enforce the safe exposure limits based on information that is already available."[6]

Physicians who were closer to the conditions of miners and mining were impatient with these uncertainties. Lorin Kerr, Director of Occupational Health for the United Mine Workers of America, for example, was aware of a "lack of precise knowledge concerning the lethal effects of coal dust." But Kerr attacked "employer-oriented physicians who avoid facing known facts about the ravages of coal

dust in human lungs because to do otherwise could cost money."[7] As Kerr told the Mine Workers' convention in 1968,

> At work you are covered with dust. It's in your hair, your clothes and your skin. The rims of your eyes are coated with it. It gets between your teeth and you swallow it. You suck so much of it into your lungs that until you die you never stop spitting up coal. . . . Slowly you notice that you are getting short of breath when you walk up a hill. On the job, you stop more often to catch your breath. Finally, just walking across the room at home is an effort. . . . Call it miners' asthma, silicosis, coal workers' pneumoconiosis—they are all dust diseases with the same symptoms.[8]

Donald Rasmussen, a physician at the Appalachian Regional Hospital in Beckley, West Virginia, made the same point in conventional medical language in a paper in the *American Review of Respiratory Disease* the same year:

> Physiologic studies of 192 symptomatic bituminous coal miners from one Southern Appalachian region . . . demonstrated significant pulmonary insufficiency in large numbers of subjects with only minimal roentgenographic evidence of pneumoconiosis.[9]

The physicians who advocated action on behalf of miners became less restrained after a mine explosion at Farmington, West Virginia, in 1968 provoked miners' militancy. I. A. Buff, a Charleston cardiologist, had warned publicly about the dangers of coal dust since the 1950s. In late 1968, Buff, Rasmussen, and Hawey Wells, a physician who was also the son-in-law of a powerful West Virginia Congressman, Harley Staggers, organized a Committee of Physicians for Miners' Health and Safety. Members of the Committee spoke frequently about lung disease over the next several years in mining communities and at Congressional Hearings.[10]

These physician-advocates were less concerned about the limits of medical knowledge than they were about the use of uncertainty to justify political complacency. They knew from both history and personal experience the social and professional pressures that led physicians to minimize the importance of the relationship between coal dust and respiratory ailments. Donald Hunter was bitter about medical complacency in his textbook.[11] Physicians like Kerr and Rasmussen who were close to the UMWA health care system had firsthand experience of the attitude exemplified in a statement by the Cabell County, West Virginia, Medical Society in 1969:

> . . . a consensus of medical authority [considers] simple, uncomplicated, and uninfected coal workers' pneumoconiosis a condition compatible with reasonable health and not associated with significant disabling disease.[12]

Like coal miners themselves, who have traditionally believed that prolonged exposure to coal dust produces wheezing and shortness of breath, the Committee of Physicians for Miners' Health and Safety was more interested in focusing public attention on miners' pain and disability than with precisely what pathology was dose-related.

Appalachian miners were unusually receptive to the warnings of these physicians in the late 1960s. They were worried about a steady loss of jobs and benefits as a result of two decades of decline in the coal industry. Industrial decline, in which the leaders of the United Mine Workers appeared to acquiesce, reduced funds for the health and medical services for miners that had evolved over several generations. The explosion at Farmington, in which 78 miners were killed, led miners throughout the southern coal fields to express their anger publicly. Anger quickly led to political action. Action was most vigorous in West Virginia, where strikes and demonstrations were followed by lobbying of the Legislature and litigation in the courts. These events stimulated and shaped the debates later in 1969 about Federal intervention in mine safety and health.

The events of the late 1960s have an historical background which was very much on the minds of participants. The history of the coal industry, the union, and even the unique system of health care that had evolved in the coal fields has often been described and analyzed orally, in writing, and in music.

Miners are the largest group of Americans for whom prepaid health care has been an entitlement for several generations. Until the late 1940s, physicians under contract to the larger coal companies were paid by a deduction from miners' wages. After the Second World War, the tradition of prepaid medical care was maintained and expanded when the United Mine Workers took responsibility for most health services in the Appalachian coal fields.

In 1948, after a series of bitter strikes, John L. Lewis, president of the UMWA, negotiated a commitment from the mine owners that they would be the sole contributors to a Welfare and Retirement Fund. The contract stipulated that the companies would pay a royalty to the Fund for every ton of coal extracted. Originally the royalty was 5 cents per ton; by 1952 it was raised to 40 cents and remained at that level for the next 20 years. The royalty payments provided sufficient capital for the union to organize clinics on the group-practice model and to construct 10 hospitals. Initially the Fund offered union members a $100/month retirement pension at the age of 65 if they had worked for 20 years in the mines, a disability pension for injuries and sickness, and free medical services for miners and their dependents. A three-member Board, comprised of union, management, and a neutral trustee, administered the Fund. In the late 1940s and early 1950s Lewis' achievement attracted national attention.[13]

In return for the Welfare and Retirement Fund, the United Mine Workers agreed to accept and encourage extensive mechanization of the mining industry and thus a major reduction in the labor force.[14] Lewis was not coerced by the companies to endorse industrial rationalization; on the contrary, he convinced the owners of its efficacy. He believed that "labor is entitled to a participation in the increased productivity due to mechanization . . . It is better to have a half a million men working in the industry at good wages and high standards of living than it is to have a million working in the industry in poverty and degradation." The new "high standards of living" would be provided by the Welfare and Retirement Fund.[15]

The operation of the Fund reflected the UMWA's policy of close collaboration between union and management. The stability and expansion of the income which would eventually be disbursed as pensions and health benefits depended on the uninterrupted flow of ever increasing amounts of coal. Union officials became committed to keeping the mines open so that royalties would be paid.[16] Through the establishment of the Fund, health care in Appalachia became inextricably tied to the health of the mine industry itself. The contract of 1948, which had been intended to free miners' health care from the inadequate system of company doctors and management control, actually tied health services even more closely to fluctuations in the market for coal.

It was soon apparent that management's gains from the exchange of mechanization for fringe benefits clearly outweighed those of the miners. By the early 1960s the benefits of the Fund were no longer secure. Lewis had assumed that a growing economy would easily absorb miners displaced by increasing mechanization. The mechanization of the mines coincided, however, with the recessions of the 1950s and the contraction of the coal market as oil became plentiful. Appalachian miners who chose to remain in the region had difficulty finding employment. The many Appalachians who migrated from the southern coal fields to the industrial cities of the Midwest were often forced to accept poorly paid unskilled jobs.[17] Many of these migrants, unwilling to abandon Appalachia permanently, established a shuttle-migration between their rural homes and the industrial cities which strained family and community life.

The UMWA acquiesced in informal arrangements with mine operators which further reduced wages, health benefits, and even mine safety. "Sweetheart" contracts were signed by many operators of small mines. In these mines, wages were often below union minimum, safety regulations were not enforced, and royalties were not paid to the Fund. To collect a royalty from these "truck" mines might have forced them to shut down, increasing unemployment among miners. Even in mines where the wage and royalty clauses of the contract were enforced, the union often ignored safety standards after 1952. Dust levels are an example. Mechanization increased dust levels and thus the risk of both explosion and lung disease. The only way to force compliance with contractual and governmental safety standards would have been to request inspection teams to close down hazardous mines.[18]

Demand for coal began to increase again in the mid-1960s. Although the acceptance of mechanization by the union had aided the recovery of the industry, unemployment in Appalachia grew. In 1963, 15 percent of the labor force in eastern Kentucky were looking for work.[19] Actual unemploy-

ment was much higher. More jobs were threatened by the introduction of new methods of surface mining which also eroded the land. From 1958 to 1964 no new contract was negotiated, yet wildcat strikes were endemic in the coal fields, as wages, benefits, and working conditions grew worse.[20]

By the late 1950s the Welfare and Retirement Fund began to suffer a chronic liquidity crisis. These financial difficulties were in part a result of changes in the industry during the 1950s and 1960s. With fewer jobs available, an increasing number of union members were drawing their pensions. This situation was exacerbated by the union's acceptance of "sweetheart" contracts, which resulted in fewer owners paying royalties. In addition, the Trustees of the Fund chose unsuccessful strategies for investing its capital.

The union decided to retrench benefits very soon after the Fund was established. As early as 1952 the disability pension was dropped.[21] After 1954, widows, who had been told they would collect their husbands' pensions indefinitely, only received them for one year.

Retrenchment of benefits became more drastic and frequent in the 1960s. In 1960, miners who had been unemployed for more than one year were denied health benefits whether or not they had maintained union membership. The following year the pension was reduced from 100 to 75 dollars a month.[22] The Fund announced new "economies" in 1962 which precipitated violent wildcat strikes in Kentucky. That summer all benefits were suspended to miners whose employers had failed to pay royalties. For a decade the UMWA had accepted the non-compliance of these same mines; miners were now penalized for this practice. In the fall of 1962, the Trustees announced the closing of four of the Appalachian regional hospitals.

The closing of the Appalachian hospitals, though mainly a result of the Fund's financial problems, was also an indirect consequence of the hostility of organized medicine in the region to UMWA health services. Appalachian medical societies, resisting group practice, prepayment, and the recruitment of medical specialists from outside the region, ostracized physicians employed by the Fund. This hostility hastened the turnover of physicians in the miners' clinics and hospitals, adding to the costs of their operation and interfering with continuity of services.[23]

The decline of the UMWA health care system reinforced miners' traditional wariness of physicians. Their wariness was rooted in the pre-1948 system of prepaid-contract practice. Physicians under contract to the mine companies had often minimized miners' complaints or offered behavioral explanations for them in order to reduce absenteeism and limit disability claims. By the 1950s, many mountain physicians had more sophisticated labels for what they had once called malingering. Miners' dust-related symptoms were frequently described as psychosomatic or neurotic illnesses, which were caused by poverty and by dependence on public and union welfare payments. The miners' desire to be compensated for their disabilities was regarded by some physicians as further evidence of their pauperization.[24]

The Farmington disaster precipitated miners' hostility toward the companies, the union, the government, and the medical profession. Whatever the immediate cause of the explosion at Farmington in November, 1968, the root cause was coal dust. Most mine operators and union officials acquiesced in ignoring dust exposure standards set by the contract and by both Federal and State law. Many inspectors were lax in enforcing regulations. Physicians frequently minimized the debilitating effects of dust on the lungs.

Explosions are a traditional cause of miner militancy, public indignation, and action by industry and the government. Many older miners in the 1960s remembered the furor over dust levels after the disaster at Gauley Bridge, West Virginia, three decades earlier. In 1968 and 1969, however, the miners, feeling betrayed by their union and the government, were responsive to leadership from the Black Lung Association, a recently formed group which was led by disabled miners, staffed by young community organizers, and supported by small but intensely committed groups of physicians and lawyers.[25]

Black Lung became the symbol of the miners' hostility in 1968 and 1969. Hostility was expressed through both the traditional techniques of strikes and selective violence and new methods of civil disobedience. Miners and their families had both direct and indirect experience of the civic militancy of the 1960s. Military service in Vietnam and shuttle migration to midwestern cities exposed Appalachians to the new articulateness of blacks involved

in the Civil Rights movement. Television provided national coverage of protests, demonstrations, and civil disturbances. The Appalachian Volunteers, a community organizing and political education program begun in 1964 with Federal government and foundation support, trained and employed a number of disabled and unemployed miners as community organizers.

Forty thousand West Virginia miners shut down the coal industry in February, 1969. Led by the Black Lung Association, and defying union leadership, thousands of them disrupted the State legislative session in Charleston. These West Virginian strikers insisted that the coal industry be forced to maintain its implicit agreement with them. Mechanization was to have been exchanged for pensions and health benefits. If the UMWA could no longer enforce this exchange, the state should at least partially do so through workmen's compensation payments.

The strikers demanded that the West Virginia legislature permit Black Lung to be diagnosed by several methods and not solely by X-ray. The diagnostic methods miners preferred would evaluate functional impairment of breathing capacity rather than evidence of dust accumulation in the lungs. Moreover, any miner who had worked for 10 years or more in the mines and who suffered from respiratory impairment would be presumed to have Black Lung and be eligible for compensation.

After three weeks in which the mines were closed, these demands became state law in March 1969. Legal recognition of Black Lung as a compensable occupational disease was a public condemnation of the eroding medical services of the region, of working conditions in the mines, of corporate priorities, and of the UMWA's inability to represent the interest of its members.

The debate on Black Lung compensation had hardly been completed by the West Virginia legislature when it began in the United States Congress. The rapid transformation of Black Lung into an issue of national politics occurred as a result of several related circumstances. Most directly, Appalachian congressmen, particularly Representatives Kenneth Hechler of West Virginia and Carl D. Perkins of Kentucky, were eager to respond to a vigorous new political force among their constituents. Even before 1969 Hechler had criticized the leadership of the UMWA and advocated more effective regulation of mine health and safety.[26] Perkins, chairman of the House Committee on Education and Labor, could make any domestic issue a high priority for the Congress. Moreover, Black Lung was a dramatic instance of the problems of occupational health and safety which had begun to interest northern liberal politicians who were closely associated with organized labor. Such Senators as Harrison Williams of New Jersey and Jacob Javits of New York viewed the regulation of coal mines as part of a broader effort to reform the industrial environment. Finally, mining, particularly coal mining, had long been regulated by the federal government. The status of mining as a unique occupation, the "mystique" of coal mining, as one congressional aide put it, helped to legitimize the enactment of unprecedented safety regulations and a compensation program.[27]

The Farmington disaster jolted Federal officials, who reexamined mine legislation introduced by the Johnson Administration a year earlier. This bill did not contain benefits for Black Lung victims, but was exclusively concerned with the prevention of dust-related disease by strict enforcement of uniform dust levels in the mines. Because the UMWA did not strongly support this legislation while industrial lobbies opposed it, the bill was never reported onto the floor of either the Senate or the House.[28]

In the late winter of 1969, in the aftermath of Farmington and the miners' protests in West Virginia, broad Administration and Congressional support developed for stronger mine safety legislation. Senators and Representatives most sympathetic to the miners demanded a maximum acceptable level of dust of three milligrams per cubic meter of air. The Administration and its Congressional supporters, more sympathetic to the owners' complaints about the prohibitive cost of meeting stringent standards, argued for a dust level of 4.5 milligrams per cubic meter. The House approved a bill with the higher dust level; the Senate supported the lower standard.[29]

The demand by miners to be compensated for the symptoms of respiratory distress they called Black Lung was more controversial. In response to miners' pressure, Hechler and Perkins amended the House version of the Coal Mine Safety Bill to include a Black Lung benefits program similar to that which had recently been enacted by the West Vir-

ginia Legislature. Senators Jennings Randolph and Robert Byrd of West Virginia, however, amended the Senate Bill in March 1969 to provide compensation only for miners disabled by Coal Workers' Pneumoconiosis, not Black Lung.[30] Although compensation had strong support in both the House and the Senate it was opposed by the Administration. Walter J. Hickel, Secretary of the Interior, testified that compensation for disability "is primarily the responsibility of the operators and the states."[31]

After hearings and debate in both Houses, a Conference Committee completed work on a Coal Mine Health and Safety Act in the fall of 1969. The compromise legislation accepted the lower standards for dust levels proposed by the Senate and the more generous disability payments for Black Lung in the House bill. The Conference agreed on a temporary program of 18 months. President Nixon signed the bill into law, but voiced opposition to the Black Lung program. Fearing a Presidential veto, 1,200 miners in West Virginia had struck.[32]

The Black Lung benefits program of 1969 was a result of both the miners' strength and the shrewd political stance of the mine operators. Appalachian congressmen and Northern liberals found it easier to support Federal compensation for miners' disability because the mine owners focused their attack on safety regulations and dust control. Several members of the Congressional staff believed that the operators favored a Federal compensation program to avoid paying for lung disease through contributions to State workmen's compensation insurance funds.[33] Opposition to the benefits program was left to traditional conservatives: defenders of states' rights and critics of government spending and regulation.[34]

For many Congressional liberals and lobbyists for AFL-CIO unions, the debate about mine health and safety in 1969 was a prelude to more general reform of industrial health and safety. A year later, the same Congressional coalition that sponsored the mine legislation introduced the Occupational Safety and Health Act.[35]

Many miners were, however, dissatisfied with the 1969 law. This dissatisfaction was primarily directed toward Title IV, the Black Lung program. Title IV established a temporary program of compensation for "any coal miner who is . . . unable to be gainfully employed . . . due to complicated

pneumoconiosis which arises out of ... employment in the coal mines."[36] Little attention had been
given to the complexity of the medical issue during
committee hearings; the International Labor Organization's definition of Coal Workers' Pneumoconiosis was accepted without debate. Most Senators
and Representatives and their staffs assumed that
what the medical profession defined as Coal Workers' Pneumoconiosis and identified through X-rays
was the same condition the miners described as
Black Lung. They believed there was a precise
definition of the disease and a simple method of
diagnosis. The miners and the Committee of Physicians had, however, insisted that respiratory
impairment be diagnosed by several methods and
not solely by X-ray. They had also demanded that
any miner who suffered from respiratory impairment and had worked for 10 years or more in the
mines should be presumed to have an occupational
disability.[37]

When extension of the temporary Black Lung
program was debated in 1972, competing pressures
once again focused on the Congress. Most State
governments had not enacted new compensation
and safety programs in response to incentives in
the 1969 Federal legislation.[38] According to the
1969 law, moreover, when the Federal Black Lung
program expired, mine owners or states would be
responsible for benefits to disabled miners. The debate in 1972 focused on the definition of the affliction the miners and their advocates called Black
Lung. In contrast to the Congressional hearing of
1969, physicians from different sectors of the medical community and miners and their advocates debated the etiology and proper diagnosis of miners'
respiratory disabilities.[39]

By 1972, the miners and their supporters had
adopted a new technique, class-action lawsuits based
on community organization, to press for liberalized
diagnostic criteria and extended benefits. In a suit
filed against the Social Security Administration in
1971 by the Appalachian Research and Defense
Fund, the Poor People's Association of Clay County,
Kentucky, claimed that Social Security officials deliberately hampered the filing and processing of
claims. The use of the X-ray as the sole determination of disability was, moreover, discriminatory.[40]

Physicians had become more sharply divided by
1972. Most of them denied that a causal relation
had been demonstrated between underground

mining and respiratory problems. However, physicians who attended miners in union clinics generally disagreed, asserting that Black Lung had a precise etiology and could be diagnosed according to
clinical criteria. Dr. Donald Rasmussen of the Appalachian Laboratory for Occupational Respiratory Diseases, in Beckley, West Virginia, became
the leading advocate for this position, with support
from several prominent pulmonary specialists
outside the region.[41] In his testimony on the 1972
bill to the Senate Committee, Dr. Rasmussen
stated that "the assumption that CWP *per se* is the
only disease related to coal mining is not medically
justified."[42]

The medical uncertainties and political controversy about Black Lung attracted more national attention in 1972 than in 1969. Consumer lobbyists,
notably Ralph Nader, strongly supported extension
of the compensation program.[43] More important,
Black Lung had become an issue in the struggle
for leadership in the miners' union. The Black
Lung movement had, from its origins in 1968, been
closely identified with rank and file insurgency in
the UMWA. The West Virginia Black Lung strike
had contributed to the decision of Joseph Yablonski, a union vice president, to challenge W. A. Boyle,
Lewis' successor, for the presidency of the UMWA in
1969. Yablonski's unsuccessful campaign strengthened the ties between the union dissidents and the
Black Lung movement.[44]

After Yablonski was murdered, early in 1970,
union reform and the Black Lung movement became more closely and formally associated. Miners
for Democracy, the insurgent faction in the UMWA
was, in West Virginia and eastern Kentucky, an extension of the Black Lung Association. The 1972
convention of Miners for Democracy nominated
former BLA president Arnold Miller as its presidential candidate. Miller entered a special UMWA
national election ordered by the United States Department of Labor after an investigation of fraud
in the 1969 election. The Black Lung movement,
which had always been linked to dissatisfaction
with union policies, was reinforced by demands for
greater democracy within the UMWA.[45]

Congress had to determine whether the 1969
federal program should be terminated or extended
in an atmosphere of increasing activism among
miners, continued medical controversy about the
nature of miners' respiratory disease, and the

states' failure to enact compensation legislation for Black Lung. No significant change in political positions in Congress had taken place since the 1969 debates. Moreover, the Administration had not altered its opposition to a Federal compensation program. In 1972 as in 1969, the combined efforts of legislators representing coal-mining constituencies and liberals sympathetic to the demands of labor and committed to federal involvement in the area of occupational health and safety created a majority in favor of a much broader Black Lung benefits act. The Senate unanimously endorsed its version of the legislation as it had in 1969. Critics of the program were again more visible and prominent in the House. There was, however, more opposition to the program in Congress and the Administration than there had been in 1969. Opposition was a result of both a radical departure from traditional concepts of workmen's compensation and occupational disease and the absence of a disaster comparable to Farmington.[46]

The pressure by miners and their medical advocates to revise the definition of coalworkers' lung disease became the most important issue in 1972. In contrast to 1969, the major coal operators' associations vigorously opposed the benefits program. Mine operators feared that the cost of broadly defined compensable illness and disability would ultimately be transferred to them.[47] The Social Security Administration continued to endorse the definition of total disability in the 1969 law— Progressive Massive Fibrosis or complicated Coal Workers' Pneumoconiosis visible on X-ray.[48]

Committees in both Houses of Congress accepted the miners' view of the diagnosis of disability. The House version of the 1972 bill included one of the major demands of the miners. No claim of disability could be rejected only on the basis of a negative X-ray. The Senate Committee went further in reformulating what would be recognized as a compensable disease. The Senate stipulated that anyone who had worked in the mines for 15 years or more and who suffered respiratory impairment would be presumed to have an occupational disease. The Senate bill, moreover, specified that benefits would be extended to anyone disabled by respiratory disease associated with mining. Finally, Senator Vance Hartke of Indiana introduced an amendment which redefined disability as inability to do mine work. Carl Perkins supported a similar

amendment in the House. These amendments reflected the willingness of powerful Senators and Representatives to replace a medically defined condition, Coal Workers' Pneumoconiosis, with concepts closer to what the miners meant by the term Black Lung, and to establish a much broader definition of the relation between the occupational environment and disease.[49]

The opposition saw the Black Lung amendments as a dangerous precedent for other industries and afflictions. The proposed legislation permitted such ailments as emphysema, asthma, and bronchitis to be recognized as occupational diseases. Critics perceived more clearly than did the bill's supporters the implications of such an unorthodox conception of the relation between work and disease. The presumption amendment, coupled with the broadened range of respiratory ailments that could now be regarded as occupationally caused, implied that the occupation of mining was itself a hazard to workers' health. As Representative John Erlenborn of Illinois noted, the experience of mining coal for a particular period of time "would qualify an individual for benefits."[50] The 1972 Congressional debate on the Black Lung Benefits Act focused on a more important issue than whether or not CWP or Black Lung was a disease: Under what circumstances did a physical disability qualify as a compensable occupational disease? Even the strongest advocates of the Black Lung legislation were, however, unwilling to apply the new concept of occupational disease beyond the mining industry.[51]

Congress passed the 1972 act because of the political pressure which miners and their advocates exerted. A congressional staff member recalled that the "decision to extend the benefits . . . was a political decision not a medical one."[52] Another staff member declared that the medical controversy had a "much smaller impact" than the "outrageous" rates of rejection of miners' disability claims by the Social Security Administration.[53] The Director of the Appalachian Research and Defense Fund agreed that "Congress changed the definition of disability because it became politically necessary to do so."[54] Dr. Rasmussen's advocacy legitimized a decision that Congress took in response to pressure from miners whose claims had been denied, to the political potential of a reformed UMWA, and to liberal and labor movement advocates of stricter regulation of occupational safety and health. On May

24, 1972, the Black Lung Benefits Act was signed by a reluctant President. Appalachian miners had, through political action, stimulated social policy to compensate Black Lung as an occupational disease.

In late 1972, the miners and their supporters believed they had achieved a striking victory. The Black Lung Benefits Act imposed on the Social Security Administration, and indirectly on the medical profession, a "new" disease. The miners' conception of Black Lung, grounded in their personal experience, became the legal criterion determining compensation. Moreover, the victory of the Miners for Democracy slate, headed by Arnold Miller, in the special UMWA election appeared to impose rank and file control in the union. Many observers viewed the Black Lung legislation as a first step toward a radical redefinition of occupational disease. Similarly, many believed the victory of the Miners for Democracy foreshadowed renewed union democracy and militancy. However, describing and, more important, explaining what has happened since 1972 is not yet a task for historians.

The history of the Black Lung protest suggests a few broader points for historians of medicine and of politics. Even in a highly specialized society, the definition of disease may be wrested from physicians by consumers and elected officials when medical uncertainty and political turmoil converge. Medical uncertainty can be mistakenly dismissed as a euphemism for opposition to social reform because of physicians' social position in a society stratified by class and income. Moreover, physicians are no better than other professionals at separating their knowledge and their values. Medical uncertainty and physicians' values do not, however, account for the rapidity with which political techniques developed in urban and civil rights protests were transferred to Appalachia. What appeared to be radical demands of a vocal minority became Federal policy because members of Congress coalesced around their interest in reelection and their need to ally with colleagues. Finally, citizens' concerns about their influence on decisions affecting work and health are as important to historians of medicine as they are to our colleagues who study politics.

NOTES

Revised version of a paper read at the 51st annual meeting of the American Association for the History of Medicine, Kansas City, Missouri, May 11, 1978.

1 George Rosen, *The History of Miners' Diseases* (New York: Schuman's, 1943), p. 457.

2 *Ibid.*, p. 309.

3 Donald Hunter, *The Diseases of Occupations* (London: English Universities Press, 1974), p. 929. Hunter summarizes the history of medical research on illnesses related to coal dust, pp. 993ff.

4 Averill A. Liebow, "Pathology of Coal Workers' Pneumoconiosis," *Ind. Med.*, 1970, *39*: 23.

5 G. Reichel and W. T. Ulmer, "The interrelationship of Coalminers' Pneumoconiosis and Bronchitis," *Inhaled Particles and Vapors*, 1970, *2*: 897.

6 Arend Bouhuys and John M. Peters, "Control of environmental lung disease," *New Eng. J. Med.*, 1970, *283*: 574. The classic study is A. G. Heppleston, "Coal Workers' Pneumoconiosis: pathological and etiological considerations," *Archives of Industrial Hygiene and Occupational Medicine*, 1951, *4*: 270–288. Significant work on the pathology of miners' lung disease was reported in the 1960s and early 1970s by W. K. C. Morgan et al., "A comparison of the prevalence of Coal Workers' Pneumoconiosis and respiratory impairment in Pennsylvania bituminous and an-

thracite miners," *Ann. N.Y. Acad. Sci.*, 1972, *200*: 252–259; Richard L. Naeye et al., "Effects of smoking on lung structure of Appalachian coal workers," *Archives of Environmental Health*, 1971, *22*: 190–193; and Philip C. Pratt, "Introductory remarks concerning pathological changes in Coal Workers' Pneumoconiosis," *Ann. N.Y. Acad. Sci.*, 1972, *200*: 342–346. Work in epidemiology was reported in the same years by Philip E. Enterline and William S. Lainhart, "The relationship between coal mining and chronic nonspecific respiratory disease," *Am. J. Public Hlth.*, 1967, *57*: 484–495; Philip E. Enterline, "Epidemiology of Coal Workers' Pneumoconiosis," *Ind. Med.*, 1970, *3*: 20–22; W. S. Lainhart et al., *Pneumoconiosis in Appalachian Bituminous Coal Miners*, U.S. Department of Health, Education and Welfare, Environmental Control Administration (Cincinnati, Ohio, 1969); Ian T. T. Higgins, "Chronic respiratory diseases in mining communities," *Ann. N.Y. Acad. Sci.*, 1972, *200*: 197–220; and George K. Tokuhata et al., "Pneumoconiosis among anthracite coal miners in Pennsylvania," *Am. J. Public Hlth.*, 1970, *60*: 441–451. A more recent survey of the literature is *Mineral Resources and the Environment: Supplementary Report: Coal Workers' Pneumoconiosis—Medical Considerations, Some Social Implications* (Washington, D.C.: National Academy of Sciences, 1976). The authors are grateful to Marvin Kuschner, M.D., of the State University of New York at Stony Brook for guidance in the medical literature.

7   Lorin E. Kerr, "Coal Workers' Pneumoconiosis in an affluent society," *Public Hlth. Rep.*, Oct. 1970, *85*: 850.

8   Lorin E. Kerr, "The facts on Black Lung," in *Black Lung* (Washington, D.C.: United Mine Workers of America, Department of Occupational Health, 1969).

9   D. L. Rasmussen et al., "Pulmonary impairment in southern West Virginia coal miners," *Am. Rev. Resp. Dis.*, 1968, *98*: 665.

10 Robert G. Sherrill, "West Virginia miracle: the Black Lung rebellion," *The Nation*, 1969, *208*: 529–530; Philip Trupp, *Readers' Digest*, June, 1969, *94*: 101–105.

11 Hunter, *Diseases of Occupations*, pp. 992–998.

12 Sherrill, "West Virginia miracle," p. 532.

13 Rosen's study remains the best general history of health care for miners. Also useful is Ludwig Teleky, *History of Factory and Mine Hygiene* (New York: Columbia Univ. Press, 1948). An excellent recent study of health and safety in coal mining in the early 20th century is William Graebner, *Coal-Mining Safety in the Progressive Period: The Political Economy of Reform* (Lexington: Univ. Press of Kentucky, 1976). Less satisfactory is David M. Holford, "Coal miners and Black Lung disease: the origins of awareness," presented at the Missouri Valley History Conference, Omaha, Nebraska, March, 1978. The classic study of prepaid medical care is Pierce Williams et al., *The Purchase of Medical Care through Fixed Periodic Payment* (New York: National Bureau of Economic Research, 1932). Chapters 7 and 8 describe prepaid health services for miners in the Appalachian South. An important study of the national context is Jerome L. Schwartz, "Early history of prepaid medical care plans," *Bull. Hist. Med.*, 1965, *39*: 450–475. A useful study of health problems in the southern mountains before 1948 is M. H. Ross, "The Appalachian coal miner: his way of living, working and relating to others," *Ann. N.Y. Acad. Sci.*, 1972, *200*: 184–196. Conditions in and around the mines, and in particular problems with coal dust, in the 1930s were dramatized in a novel, Hubert Skidmore, *Hawk's Nest* (New York: Doubleday, Doran, 1941). The history of the miners' Welfare and Retirement Fund and its antecedents is described in Leslie A. Falk, "Coal miners' prepaid medical care in the United States—and some British relationships, 1792–1964," *Medical Care*, 1966, *4*: 37–42. The report prepared by Admiral Joel T. Boone and his staff in 1947 remains an important source for health problems and services in Appalachia: *A Medical Survey of the Bituminous-Coal Industry*, Report of the Coal Mines Administration, U.S. Department of the Interior (Washington, D.C.: U.S. Govt. Printing Office, 1947). The most reliable recent secondary source for the history of the UMWA and the Fund is Melvin Dubofsky and Warren Van Tine, *John L. Lewis: A Biography* (New York: Quadrangle, 1977). A narrative of the origins of the Fund and its health care system is in Joseph E. Finley, *The Corrupt Kingdom: The Rise and Fall of the United Mine Workers* (New York: Simon and Schuster, 1972). Other pertinent material on the Fund and the coal companies is in Morton S. Baratz, *The Union and the Coal Industry* (New Haven: Yale Univ. Press, 1955). More analytical studies of the operation and problems of the Fund are: Leslie A. Falk, "Group health plans in coal mining communities," *Journal of Health and Human Behavior*, 1963, *4*: 4–13; Leslie A. Falk, "Comprehensive care in medical programs: the U.M.W.A. Welfare Fund," *Medical Care*, 1968, *6*: 401–411; and Joseph W. Garbarino, *Health Plans and Collective Bargaining* (Berkeley and Los Angeles: Univ. of California Press, 1960).

14 The rapid decline in employment and the simultaneous rise in productivity per miner is exemplified by the following statistics:

1950	416,000	working miners in the U.S.
1959	180,000	working miners in the U.S.
1964	130,000	working miners in the U.S.

1940	4.5	tons of coal mined per day per miner
1961	12.1	tons of coal mined per day per miner
1964	16.84	tons of coal mined per day per miner

Finley, *Corrupt Kingdom*, p. 171. The long-term decline in the labor force in mining from 1923 to 1962 represented a savings of $240 million in wage costs. John Ed Pearce, "The superfluous people of Hazard, Kentucky," *The Reporter*, Jan. 3, 1964, *30*: 34.

15 A. H. Raskin, "John L. Lewis and the mine workers," *Atlantic Monthly*, May, 1963, *211*: 53.

16 The union became reluctant to authorize work stoppages against hazardous conditions, and it discouraged any form of strike activity. The curtailment of strikes was a major concern of the mine owners who sought not only high productivity, but uninterrupted production. As the solvency of the UMWA Fund also depended on these conditions, the union leadership shared this concern. However, neither union nor management was able to eliminate wildcat strikes through which miners demonstrated dissatisfaction with the owners and union officials.

17 In the decade 1950–1960 the net population loss in Appalachia through migration was in excess of one million people. This figure equaled almost one-fifth

of the population of southern Appalachia in 1950. Outmigration was greatest in the mining areas of eastern Kentucky and West Virginia where the population declined 19% during the 1950s. James S. Brown and George A. Hillery, Jr., "The great migration, 1940–1960," in *The Southern Appalachian Region: A Survey,* ed. Thomas R. Ford (Lexington: Univ. Press of Kentucky, 1962), pp. 54, 73.

18 Dubofsky and Van Tine, *John L. Lewis,* pp. 497ff.

19 *U.S. News and World Report,* Apr. 29, 1963, 72.

20 UMWA policy was identified so closely with that of the large corporate mine owners that the leadership failed to maintain miners' benefits on a par with those of steel and auto workers. In the mid-1960s, although the UMWA had pioneered pension and health benefits for members, it was unable or unwilling to demand their expansion. Miners received a pension of $85/month while steel workers received $405/month and members of the United Auto Workers, $500/month. Wages in union mines did remain relatively high. In 1962 the minimum union wage was $118/week. By 1964 it had risen to $120.75 and an increase to $130.75 was planned for the following year. *Business Week,* Nov. 3, 1962, 49; Jan. 4, 1964, 68; and Mar. 28, 1964, 121. However, union practices, at least until the early 1960s, permitted the proliferation of non-union mines where in 1959 real wages were as low as $4/day. Stephanie Gervis, "Gray spring in Hazard," *The Commonweal,* May, 1963, *78:* 220.

21 This pro-rated pension would have covered disability due to respiratory illness.

22 After this initial reduction, the pension was raised in early 1965 to $85/month, in late 1965 to $100, and in 1967 to $115. During the Boyle-Yablonski contest for the union presidency in 1969, Boyle raised the pension to $150/month. Brit Hume, *Death and the Mines: Rebellion and Murder in the UMW* (New York: Grossman, 1971), pp. 33, 34, 62, 176.

23 R. Enroth, "Patterns of response to rural medical practice and rural life in Eastern Kentucky," (Ph.D. dissertation, Univ. of Kentucky, 1967). Also cf. Falk, "Group health plans."

24 Conference on Mental Health in Appalachia, *Mental Health in Appalachia: Problems and Prospects in the Central Highlands,* U.S. Department of Health, Education and Welfare, National Institute of Mental Health (Washington, D.C.: U.S. Govt. Printing Office, 1974); Jerome L. Schwartz, "Rural health problems of isolated Appalachian counties," in R. L. Nolan and J. L. Schwartz, eds., *Rural and Appalachian Health* (Springfield, Ill.: Charles C. Thomas, 1973); Harry M. Caudill, *The Watches of the Night* (Boston: Little, Brown and Co., 1976), *passim.*

25 A history of the Black Lung Association needs to be written. The sources for this history are mainly oral. A number of pamphlets and newsletters exist.

Printed sources for the narrative which follows include numerous articles by Ben A. Franklin in the *New York Times* and by Robert Sherrill in the *Nation* and the lengthy hearings conducted by congressional committees in 1969 and 1972. The importance of the Black Lung Association is assessed in James Branscome, *The Federal Government in Appalachia* (New York: The Field Foundation, 1977); and in J. Davitt McAteer, *Coal Mine Health and Safety: The Case of West Virginia* (New York: Praeger Publishers, 1977). We are grateful to Milton Ogle, founder of the Appalachian Volunteers in 1964 and currently Director of the Appalachian Research and Defense Fund, for many insights into the Black Lung Association, the politics of coal mining, and continuity and change in the Southern Mountains.

26 Both in the Congress and in his West Virginia District, Representative Hechler repeatedly complained of the union's failure to lobby effectively and concertedly for safety and health legislation. Hechler's stand earned him W. A. Boyle's wrath during the UMWA's president's testimony before the Senate subcommittee in 1969. U.S. Congress, Senate, Committee on Labor and Public Welfare, Subcommittee on Labor, *Hearings on S. 355, S. 467, S. 1094, S. 1178, S. 1300, and S. 1907. Bills to improve the health and safety conditions of persons working in the coal mining industry of the U.S. Pt. 1* (Washington, D.C.: U.S. Govt. Printing Office, 1969), p. 456.

27 Eugene Mittleman to Judith F. Stone. Interview, Apr. 11, 1978. Protocols of the interviews are available upon request.

28 *Congressional Quarterly Almanac: 90th Congress, 1969* (Washington, D.C.: Congressional Quarterly, 1970), p. 735.

29 U.S. Congress, Senate, Committee on Labor and Public Welfare, Subcommittee on Labor, *Legislative History of the Federal Coal Mine Health and Safety Act of 1969 (PL 91-173) as Amended through 1974. Including Black Lung Amendments of 1972, Pt. 2* (Washington, D.C.: U.S. Govt. Printing Office, 1975), pp. 2912–2914.

30 Early versions of the bill had indicated that the Senators were aware of the Black Lung issue, but remained uncertain about how to address the plight of its victims. An Administration bill called for Federal encouragement through grants to the states to establish compensation programs covering miners' disability due to occupational disease. A bill submitted by Harrison Williams of New Jersey, who chaired the Senate subcommittee on Labor, included provisions for federal benefits, but only in those states where miners were not already covered by state law. An earlier bill introduced by Senator Randolph made no reference to Black Lung benefits.

31 U.S. Congress, Senate, Committee on Labor and

Public Welfare, Subcommittee on Labor, *Hearings on S. 355, S. 467, S. 1094, S. 1178, S. 1300, and S. 1907. Bills to improve the health and safety conditions of persons working in the coal mining industry of the U.S., Pt. 1* (Washington, D.C.: U.S. Govt. Printing Office, 1969), p. 522.

32 The Senate bill was passed unanimously. The debate in the House was sharper; the Black Lung program evoked the most intense criticism. The House opponents of the program were concerned with the possible implications of establishing a federally administered and financed compensation program even on an interim basis. They viewed Black Lung benefits as a possible first step to a federalized workers' compensation program. Nonetheless the House version passed by a wide margin, 389 to 4.

33 Eugene Mittleman to J. F. Stone. Interview.

34 House debate on amendment submitted by Representative William Scherle of Iowa to delete Black Lung benefits program, October 29, 1969. U.S. Congress, Senate, Subcommittee on Labor, *Legislative History of the Federal Coal Mine Health and Safety Act of 1969*, pp. 2887–2889. These views were reiterated and expanded in the Minority Report of the House Committee on Education and Labor on HR 9212 to amend Title IV of the 1969 Coal Mine Health and Safety Act, Aug. 5, 1971. *Ibid.*, pp. 1746–1757. Also see Representative Erlenborn's statement to Senate, Subcommittee on Labor Hearings to amend Title IV of 1969 Act, January 28, 1972. U.S. Congress, Senate, Committee on Labor and Public Welfare, Subcommittee on Labor, *Hearings on S. 2675, S. 2289, HR 9212 to Amend 1969 Federal Coal Mine Health and Safety Act* (Washington, D.C.: U.S. Govt. Printing Office, 1972), pp. 324–327.

35 From its introduction in 1968, the mining legislation was linked to broader legislation establishing federal regulation of the workplace environment. In his study of the OSHA legislation, Nicholas A. Ashford underscored the relation between the passage of the Federal Coal Mine Health and Safety Act in late 1969 and the enactment of OSHA: *Crisis in the Workplace: Occupational Disease and Injury. A Report to the Ford Foundation* (Cambridge, Mass.: MIT Press, 1976), p. 46. In the area of safety there are clear parallels between the two laws. Although no federal benefits program for any occupational disease similar to the Black Lung program was contemplated during the formulation of OSHA, the entire law was an implicit and explicit condemnation of the states' regulation of occupational health and safety. The Occupational Safety and Health Act did mandate, as proposed by Senator Javits, a federal study and evaluation of the compensatory aspect of state regulatory policies, and the various workers' compensation insurance programs. This report was released in 1972, finding the

states' compensation programs unsatisfactory and outlining 80 specific recommendations for improvement. However, no federal action was taken.

36 U.S. Congress, Senate, Committee on Labor and Public Welfare, Subcommittee on Labor, *Legislative History of the Federal Coal Mine Health and Safety Act of 1969 (PL 91-173) as Amended through 1974. Including Black Lung Amendments of 1972, Pt. 2* (Washington, D.C.: U.S. Govt. Printing Office, 1975), p. 3197.

37 These had been the demands of the BLA in February 1969 and had been incorporated into the West Virginia law. Sherrill, "West Virginia miracle," p. 529. Fred Carter, representing miners and widows of southern West Virginia, reaffirmed these demands in his testimony to the Senate Subcommittee on Labor, December, 1971. U.S. Congress, Senate, Subcommittee on Labor, *Hearings . . . to Amend 1969 Federal Coal Mine Health and Safety Act*, pp. 100–101.

38 None of the mining states which had not provided compensation for Black Lung prior to December, 1969, enacted legislation to do so in 1970–1971. In January, 1972, of 23 coal-mining states only eight had compensation statues which covered Black Lung. U.S. Congress, Senate, Subcommittee on Labor, *Legislative History of the Federal Coal Mine Health and Safety Act of 1969*, p. 1815.

39 Roster of physicians supporting amendments and those critical of amendments to Black Lung benefits program, U.S. Congress, Senate, Subcommittee on Labor, *Legislative History of the Federal Coal Mine Health and Safety Act of 1969*, pp. 1949–1950.

40 Finley, *Corrupt Kingdom*, p. 290. George William Hopkins, "The Miners for Democracy: Insurgency in the United Mine Workers of America, 1970–1972," (Ph.D. dissertation, Univ. of North Carolina, 1976), p. 257.

41 In late 1971, 12 pulmonary specialists met at the Appalachian Regional Hospital in Beckley, West Virginia, to examine miners whose Black Lung claims had been denied. These physicians publicly criticized the exclusive use of the X-ray as a diagnostic method. A few weeks later, under pressure from the Black Lung Association and Dr. Rasmussen, an International Conference on Coal Workers' Pneumoconiosis, sponsored by the New York Academy of Sciences, also called for a revision in the testing for disability in miners claiming Black Lung benefits. Hopkins, "Miners for Democracy," pp. 258–259.

42 U.S. Congress, Senate, Subcommittee on Labor, *Legislative History of the Federal Coal Mine Health and Safety Act of 1969*, p. 1955.

43 Nader had for a number of years prior to 1972 been concerned with the related issues of Black Lung, mine safety, and union reform. As early as 1968, he had written to the Secretary of the Interior criticizing the UMWA for obstructing the improvement of

miners' health and safety conditions. This action may have indirectly influenced the formulation of the 1968 Johnson mine safety bill. Nader was an early vocal supporter of the Black Lung movement. In addition, he became deeply involved in Yablonski's opposition movement within the UMWA. Hume, *Death and the Mines*, p. 161.

44 In the statement announcing his candidacy, Yablonski underscored the significance of the Black Lung movement: "In recent months the shocking ineptitude and passivity of the union's leadership on Black Lung disease—not to mention its ignoring this massive disability of its men for years—became apparent to the nation, not just to those inside the union." Cited in Hume, *Death in the Mines*, p. 173; cf. p. 152. Out of the 25 union districts, Yablonski only won a majority in three during the December 1969 election. One of those three was District 17 in West Virginia, the geographic center of the Black Lung movement. Finley, *Corrupt Kingdom*, p. 270.

45 Hopkins, "Miners for Democracy," pp. 268–271 and chs. 4 and 6, *passim*.

46 In 1972, 79 Representatives voted against the House version of the Black Lung Benefits Act; 311 voted for it. Ninety-nine Republicans and 22 southern Democrats opposed the Conference Report. In a February 1972 letter to Senator Williams, Elliot Richardson, Secretary of Health, Education and Welfare, condemned the proposed amendments. Richardson described the program as an "ill-advised step toward the federalization of Workmen's Compensation at a huge cost to the taxpayers." U.S. Congress, Senate, Committee on Labor and Public Welfare, Subcommittee on Labor, *Hearings on S. 2675, S. 2289, and H.R. 9212, amending certain portions of the 1969 Federal Coal Mine Health and Safety Act* (Washington, D.C.: U.S. Govt. Printing Office, 1972), p. 24.

47 The Bituminous Coal Operators Association, the National Coal Association, the National Independent Coal Operators Association, and the American Mining Congress addressed a joint letter to Senator Randolph voicing their opposition to the proposed amendments. U.S. Congress, Senate, Subcommittee on Labor, *Legislative History of the Federal Coal Mine Health and Safety Act of 1969*, p. 1895. However, one former staff member has suggested that employer opposition to the 1972 amendments was more complex. Some industrial representatives urged the passage of the liberalized Black Lung benefits as long

as the new broad criteria would be dropped when funding reverted to the state or the owners directly. Certain large mine companies actively encouraged their employees to apply for the Black Lung benefits under the Federal program. H. D. Reed to J. F. Stone, Interview, Apr. 6, 1978.

48 U.S. Congress, Senate, Subcommittee on Labor, *Legislative History of the Federal Coal Mine Health and Safety Act of 1969*, p. 32.

49 House and Senate Conference report on 1972 amendments to Title IV of Federal Coal Mine Health and Safety Act of 1969, and Senator Schweiker's statement supporting amendments. U.S. Congress, Senate, Subcommittee on Labor, *Legislative History of the Federal Coal Mine Health and Safety Act of 1969*, pp. 1889, 2014.

50 U.S. Congress, Senate, Subcommittee on Labor, *Legislative History of the Federal Coal Mine Health and Safety Act of 1969*, pp. 1871–1872.

51 When Representative Thompson of Georgia attempted to extend the legislation to cover quarry and textile workers, the House supporters of the bill quickly defeated his proposal. In general, the supporters of the bill insisted that it was to be *sui generis*. U.S. Congress, Senate, Subcommittee on Labor, *Legislative History of the Federal Coal Mine Health and Safety Act of 1969*, p. 1879.

52 Eugene Mittleman to J. F. Stone. Interview.

53 Carl Perkins, Chairman of the House Committee on Education and Labor called his colleagues' attention to the inadequacies of the 1969 law. He documented the striking disparity which existed between the low proportion of approved claims within Appalachia and the much higher percentage outside the region. H. D. Reed to J. F. Stone. The Social Security Administration gave the total number of claims filed between January 1970 and March 1972 as 356,857. (A much greater number, incidentally, than the Congress had anticipated.) Of those, 166,593 had been approved and 169,999 rejected. U.S. Congress, Senate, Subcommittee on Labor, *Legislative History of the Federal Coal Mine Health and Safety Act of 1969*, p. 1948. This was already a large proportion of rejections, but they were by no means evenly distributed within the coal regions. In Pennsylvania, 69% of the claims had been approved, 44% in West Virginia, and only 28% in Kentucky received approval. *Ibid.*, p. 50.

54 Milton Ogle to J. F. Stone, Interview, Apr. 13, 1978.

EARLY AMERICAN MEDICINE

In 1607 the British established their first permanent community in North America, in Jamestown, Virginia. During the next two centuries European immigrants settled up and down the Atlantic seaboard and pushed westward to the Appalachian Mountains. By the time of the American Revolution, in the 1770s, most colonists still lived in rural areas, but thriving towns had appeared from Maine to Georgia. The leading medical center developed in Philadelphia, Pennsylvania, which gave birth to the first general hospital in the British colonies of North America, in 1751, and the first medical school, in 1765.

Sick and injured Americans typically received their medical care at home, often from family members. If they needed additional assistance, they might send for a skilled neighbor, a medically knowledgeable minister, or a midwife. Although best known for delivering babies, early American midwives provided a broad range of medical care, especially for women and children. In view of their diverse activities, Laurel Thatcher Ulrich has suggested that "social healer" might serve as a more accurate title than midwife.

Also scattered throughout the colonies were a number of self-styled doctors, few of whom had set foot inside a medical school. As Eric H. Christianson shows, the typical practitioner in colonial New England learned how to heal while serving for a year or so as an apprentice to an established doctor or by some other, less formal, means. In London and other European cities one might find healers specializing in medicine, surgery, or pharmacy, but colonial American doctors engaged in all three

tasks: practicing medicine, performing surgery, and preparing drugs. Increasingly, toward the end of the 18th century, they delivered babies.

Some historians have argued that this type of general practice resulted from the democratizing effects of the New World environment on Old World distinctions that rigidly separated the duties of physician, surgeon, and apothecary. Other historians have pointed out that even in England such strict divisions of labor seldom held outside London; in the provinces medical practitioners performed all the healing arts. Besides, few members of the upper classes, to which university-educated physicians belonged, immigrated to the colonies. Thus nondomestic medicine in America from the beginning fell largely into the hands of doctors already accustomed to a mixed practice.

We should not be surprised that medical arrangements in the British colonies closely mirrored what the colonists had left in provincial Britain. The same correspondence between Old and New World practices prevailed in New France (present-day Canada) to the north and in New Spain (Mexico) to the south. Medicine changed remarkably little in transit across the Atlantic. New World doctors may have borrowed occasionally from indigenous healers and may have employed a few native remedies, but these local influences scarcely affected the theoretical framework that guided their day-to-day tasks. Even when they prescribed homegrown medicines, the colonial doctors did so for the same reasons that had governed Western medical practice for centuries.

3

Medicine in New England

ERIC H. CHRISTIANSON

In 1734 Dr. John Perkins of Boston visited London to "see what they had in practice new, or better than we had." Despite finding vast environmental differences between the teeming English metropolis and his provincial hometown, he concluded that medical practice was much the same in both places. Other, more recent, observers have argued that the distinctive conditions of New England substantially altered the medical institutions and practices of the Old World.[1] In this essay, based on a survey of existing literature, I have sought to adjudicate between these two views by examining the medical history of colonial New England, particularly that of Massachusetts, against the background of developments in Old England.

ENGLAND

At the time of Queen Elizabeth's death in 1603, England did not possess a profitable and well-organized colonial empire like Spain's.[2] With some Crown control, enterprising individuals and joint-stock companies, rather than the English monarchy itself, colonized and exploited the new lands. The conditions of life in England combined with the prospects of living in new lands to attract thousands of eager settlers for these endeavors. The individualism of the optimist and of the opportunist provided the momentum for English expansion.

Between 1600 and 1700 the population of England and Wales increased from about four million to over six million. By the end of the 18th century

ERIC H. CHRISTIANSON is Associate Professor of History, University of Kentucky, Lexington, Kentucky.

Reprinted with permission from *Medicine in the New World: New Spain, New France, and New England*, edited by Ronald L. Numbers (Knoxville: University of Tennessee Press, 1987), pp. 101–153.

the number of inhabitants had grown to over nine million, with London accounting for about 10 percent of the total.[3] Although the vast majority lived in communities of less than 5,000 inhabitants, a few provincial towns in 1700—Norwich, York, Bristol, Newcastle, and Exeter—boasted populations of between 10,000 and 20,000.[4] From the southwest and southeast came the largest flow of early immigrants to the North American colonies, organized by ambitious businessmen and God-fearing reformers seeking a diversified labor force to support their overseas ventures.[5] Included in the cultural baggage of these immigrants to the New World were their experiences with disease, their medical practices, and their medical organizations.

Diseases

Any attempt to classify the diseases of the 17th and 18th centuries involves hazards for the historian because diagnosis was often imprecise. As one medical historian has noted, it is difficult, if not impossible, even to distinguish between such diseases as diphtheria and scarlet fever "when eighteenth century doctors themselves made no attempt to separate them."[6] Nevertheless, contemporary records and epidemiological patterns reveal much about the medical problems of Stuart and Georgian England. Perhaps the most familiar medical event is the epidemic of bubonic plague that took over 60,000 lives in London in 1665.[7] But the English also suffered from smallpox, syphilis (also called the great pox or French pox), measles, malaria, dysentery (the bloody flux), pneumonia, tuberculosis (consumption), dropsy (congestive heart failure), and stones in the urinary bladder. The last condition was so prevalent in England during the 18th century that one observer speculated that enough bladder stones could be found to "macadamize [pave] one side of Lincoln's Inn Fields."[8] Some diseases, such as smallpox, became endemic and distinguishable from others with somewhat similar symptoms, such as measles; beginning in 1629, London's Bills of Mortality separated the two diseases.[9] By 1700, smallpox had become the most common mortal disease of English childhood, with the possible exception of infantile diarrhea.

Such figures are based on the experience of London, where there were more record-keeping practitioners, governmental agencies, and medical institutions than in the countryside. We do not possess enough evidence at this time to tell if life in the larger provincial towns was more conducive to longevity than life in London. The belief of some contemporaries, however, that living in rural areas favored healthier, if not longer, lives has received qualified endorsement from recent scholars who have proposed that "inhabitants in smaller towns in both America and Europe had greater life expectancies than did those in larger communities."[10]

Medical Practitioners

A variety of healers provided medical care in England: physicians, surgeons, apothecaries, and other ordinary practitioners, as well as ministers, midwives, nurses, grocers, traveling mountebanks, and patients themselves, who bought over-the-counter medicines from chemists or concocted homemade simples.[11] The dominant medical institutions in London were the three corporations that separated members from ordinary practitioners and protected the interests of physicians, apothecaries, and surgeons.

During the first half of the 16th century, King Henry VIII granted charters to the Royal College of Physicians of London (1518) and to the United Company of Barber-Surgeons of London (1540), thus giving them independence and autonomy. Apothecaries won similar concessions from King James I in 1617, when he bestowed corporate privileges on the Guild of Apothecaries of London, separating them from grocers. These institutions created a medical hierarchy with physicians on top. As one early 17th-century observer put it, the physician was the "great commander [who] has subordinates to him, the Cooks for diet, the Surgeons for manual operation, and the Apothecaries for confecting and preparing medicines."[12] By the open-

ing decades of the 17th century, which coincided with the initial phase of English colonization in North America, the essential institutional structure of the London medical community was in place.

This tripartite division of functions, however, could not be maintained strictly by these corporations. Although not necessarily without training or experience, unlicensed practitioners flourished in London, and they were dominant in the provinces.[13] In theory the medical corporations and guilds protected London physicians, surgeons, and apothecaries from would-be competitors in other guilds and from nonmembers in the provinces. According to medical statutes, physicians were to limit their practice to diagnosis, prognosis, and prescription. Apothecaries were to prepare medications prescribed by physicians. The activities of surgeons were limited to extracting teeth, letting blood (phlebotomy), treating skin lesions, and performing such operations as amputations and lithotomies for the removal of stones in the urinary bladder— all of which they could do either on their own or at the direction of a physician. But in 1745, King George II granted surgeons their own London guild, thus depriving barber-surgeons of the right to perform lucrative major operations and to consult with physicians.[14] Although charged with maintaining the boundaries between specialties, the corporations for a number of reasons were unable to do so. Among the destabilizing factors were increased educational opportunities, the ambition of apothecaries and surgeons to usurp the functions of the more prestigious physicians, and the growth of the medical marketplace.

Many barber-surgeons, surgeons, apothecaries, and ordinary practitioners ignored the medical statutes and continued to practice medicine along with surgery or pharmacy. Attempting to guard its regulatory power over this diverse group of healers in London, between 1581 and 1600 the Royal College of Physicians prosecuted at least 236 individuals for practicing medicine without a license.[15] In response to their critics' derisive claim that they were uneducated pretenders (the physician was to be a university medical graduate), apothecaries established in 1627 a seven-year apprenticeship for those hoping to join the ranks. Pointing with pride to their lengthy and supervised period of training, they defended their increasingly physician-like functions. After all, they argued, who knew more than they about the substances, both botanical and chemical, from which prescriptions were prepared and about the actions of these prescriptions on patients. The appearance of standardized pharmacopoeias in English would eventually assist the apothecaries in their quest to expand their role. So, too, did the great London plague in 1665, during which many physicians retreated to the safety of the countryside, leaving ordinary practitioners, including apothecaries, surgeons, and barber-surgeons with most of the responsibility of caring for the sick.[16]

The frequent and bold intrusions of apothecaries into the domain of physicians prompted the College of Physicians to seek legal redress. In the celebrated Rose Case of 1703, pitting an apothecary practicing medicine against the college, the House of Lords granted apothecaries within the jurisdiction of London permission to function as ordinary practitioners of medicine.[17] As apothecaries practiced more medicine and surgery, druggists, or chemists, increasingly took over the preparation and sale of drugs.

During the first half of the 18th century, London's physicians continued to express much concern over the blurring of corporate distinctions. Increased educational opportunities, over which the college exercised little direct control, facilitated the upward mobility of the lower ranks. The medical, surgical, and midwifery personnel at the hospitals and medical schools of Leiden and Paris continued to attract English students well into the 18th century.[18] But the founding of new hospitals in London, such as Guy's, Middlesex, and the Westminster Lying-In, convinced many others to study at home. These hospitals allowed their staff medical personnel to offer instruction and demonstration to private pupils.[19] The availability of this training certainly benefited the surgical specialist and male midwife, both of whom were more likely to reside in larger communities, as well as the apothecary-surgeon. As the dominant type of practitioner in the provinces, apothecary-surgeons practiced medicine, prescribed the drugs that they prepared, and usually limited their surgical intervention to minor operations.[20]

As formal instruction became increasingly important for these medical practitioners, many traveled north to Scotland to obtain an M.D. degree from one of the new medical faculties at Edin-

burgh, Aberdeen, or St. Andrews. These schools did not admit female practitioners, but did accept male dissenters from the Church of England who would not have been admitted to the universities at Cambridge and Oxford. The Scottish schools offered both beginning medical students and veteran apothecary-surgeons, at least the men, a relatively quick way to join the ranks of the physician by obtaining degrees. During its early years the Royal College of Physicians had granted licenses to those who had secured the endorsement of a respected physician and had passed an examination to determine competency, but by 1750 the academic M.D. degree had become virtually the only acceptable qualification for practicing medicine in London.[21] The college responded to these changing educational opportunities for would-be physicians by attempting to prevent degree-holding apothecary-surgeons, who lacked training in the liberal arts, from practicing in London. Presumably more interested in practicing than in being prosecuted, apothecary-surgeons who earned Scottish M.D.'s settled, or relocated, in the villages and towns of the provinces. Some returned to care for their former patients and sought to increase the size and diversity of their clientele in areas where the regulatory powers of the college were weak.

Although the terms "physician" and "doctor of medicine" had always been synonymous, many who called themselves "doctors" did not restrict themselves to diagnosis or prescription. By the second half of the 18th century this situation led some physicians in London, and even in the larger provincial towns, to call for a new name more "descriptive of the physician's circumstances," a replacement for this "antiquated and prostituted title."[22]

Even before the creation of London's medical corporations, a statute enacted in 1511 established a basis for organizing and regulating medical practice throughout provincial England.[23] This law placed the licensing of physicians, surgeons, and other practitioners under the church. In 1643, parliament abolished the ecclesiastical hierarchy, thus suspending episcopal licensing until it was reinstituted in 1660. Licensing under this arrangement proved to be inconsistent, and the level of the bishop's attentiveness in this matter varied considerably throughout the realm. University licenses were not covered by the 1511 statute and remained valid except in London. In effect, access to practice the healing arts was open to all under common law; licensing remained a route to legitimation rather than a means to seek permission to practice as a physician, surgeon, or other practitioner. The right, not the competence to practice, was the issue here. Over time, academic credentials made the distinctions between the titles of physician, surgeon, or apothecary less confusing, but they did not eliminate the controversy over appropriate medical functions.[24]

Studies of the differences, as well as of the similarities, between the organization of medical practice in London and in the provinces have become methodologically refined. Several studies based on fragmentary data have offered approximations of the number of practitioners and their ratios to population in selected locations and periods of time. A conclusion shared by these studies is that the number of practitioners throughout England has been vastly underestimated. The results of some analyses also suggest that differences between the organization and practice of medicine in London and in the provinces have been overestimated. Although admittedly incomplete, these data permit some illumination of the differences and similarities. The research focuses largely on the period 1580–1643, during the latter part of which New England was settled, with much less available for the late 18th century, when the colonies declared their independence.

For the period 1580–1643, recent studies have been concerned with communities in East Anglia, a major supplier of immigrants to New England. Located in the counties of Norfolk and Suffolk, the most thoroughly studied towns are Norwich, King's Lynn, and Ipswich. With a population of about 17,000 in 1600, Norwich was the second largest city in England. At least 73 people have been identified as practitioners in Norwich for 1580–1600.[25] Quite similar in its diversity to other provincial towns, the listing of Norwich practitioners has been reported as 37 surgeons or barber-surgeons, 12 apothecaries, 10 women practitioners, 6 practitioners of physic, 5 academically trained physicians, and 3 miscellaneous practitioners. Comparable to findings for King's Lynn and Ipswich, a Norwich practitioner "might hold a license in medicine or surgery or both from the University of Cambridge, the Bishop of Norwich, the Archbishop of Canterbury

or a variety of agencies in London, including the Bishop of London and the Privy Council."[26] These communities also served as temporary residences for itinerant practitioners who moved about the countryside.[27] Ordinary practitioners, especially surgeons and barber-surgeons, and not physicians, were the largest numerical group in East Anglia.[28] Much the same can be said of London where in 1600 only 50 of its estimated 500 practitioners were physicians.[29] From results obtained from limited data, it would seem that physicians generally constituted less than 20 percent of the medical practitioners in any community. This diversity, if not also the dominance of the ordinary practitioner, has been implicitly suggested by a study of a later period.

Between the accession of King James I in 1603 and the temporary abolition of episcopal licensing in 1643, a date coinciding with the end of the Great Migration to New England, one study reports the presence of 814 men licensed to practice medicine in "the villages and hamlets of provincial England."[30] Of this total, the vast majority (604 or 74.2 percent) had attended college, and 76 percent of this number had received a bachelor's degree. Although nearly one-third (250 or 30.7 percent) possessed an M.D. degree, better than one out of five (22 percent) neither attended college nor received any documentable formal training. While it demonstrates the minority status of physicians, two critics of this study conclude that it does not take into account a much larger group of healers, such as those identified in Norwich. For East Anglia, it is "doubtful whether even freemen, or licentiates, represent a majority of those dispensing medical care.[31]

A crude approximation of the ratio of practitioners to population is possible for the period 1580–1643. In Norwich, Ipswich, and London, the ratios of practitioners to population around 1600 have been estimated at 1:220–250, 1:208, and 1:400, respectively.[32] The evidence remains too sketchy to make the claim that smaller communities generally had more practitioners then did larger ones. Future analyses of practitioners not consistently included in available studies (such as women practitioners, midwives, and nurses) would, however, undoubtedly yield higher ratios of practitioners to population in all communities than have been reported in the literature. Lacking analyses of numerical data for the 18th century, long-term comparisons are not possible at the present time.

Perhaps because of the diversity of practitioners, most of whom were also engaged in other income-producing activities, provincial towns seem to have been less rigid in organizing and regulating medical practice. With membership open to all categories of practitioners, medical societies in the provinces during the 18th century reflected a spirit of cooperation among members that did not exist between London's corporations. In contrast to the Royal College of Physicians of London, preoccupied as it was well into the 18th century with regulating practice, the corporations for apothecaries and surgeons took an active role in the training of their apprentices and the continuing education of their members. London's three royally chartered corporations differed markedly in composition, though not always in all functions, from medical societies created in London and in the provinces.

Between 1757 and 1836 there were many medical societies in London, some of them quite short-lived.[33] A few, such as the Society for the Improvement of Medical and Chirugical Knowledge (1783), catered to the needs of specific medical corporations, yet disregarded their functional distinctions. Three of the societies active during 1795–1815 were rather inclusive in composition within the context of the teaching hospital.

Hospitals like Guy's, Middlesex, and St. Bartholomew's attracted physicians, surgeons, apothecaries, and pupils who enrolled under their supervision to study medicine and its various branches. Advances in knowledge and skill brought them closer together in spite of their separate corporate or noncorporate origins. Records of these societies remain fragmentary. For the Guy's Hospital Physical Society, founded in 1771, the breakdown of each category by percentage of the total membership in 1804 of 659, less six clergymen, was 13 percent M.D.s, 20 percent surgeons, less than 1 percent apothecaries, and 65 percent (431), who "had no professional identification and probably were surgeons, apothecaries, or practicing as both."[34]

Quite similar to contemporary philosophical and natural history societies, mutual interests brought diverse groups together in Guy's Hospital Physical Society. Its members were "desirous of improvements in Medicine, and the other Sciences nearly allied to it, and convinced of the numerous and great Advantages, arising from a free communication of Observations and Opinions."[35] Not all of the members were general practitioners, but they expressed the "general practitioner's interest in, and understanding of, both medical and surgical problems."[36] Provincial societies, open to all established medical practitioners, surgeons, apothecaries, and their apprentices, also stressed the importance of appreciating the complementary nature of the various branches of medicine.

In 1774, Robert Richardson Newell, an apothecary-surgeon practicing in Colchester, East Anglia, organized the first of the provincial societies.[37] Newell's father, also an apothecary-surgeon, practiced in nearly Harwich, and his wife's father was an apothecary in Colchester. Newell apparently served an apprenticeship with his father, a practice more common among apothecary-surgeons in the provinces than among physicians in London. The varied membership of this society suggests an absence of fixed professional boundaries; among the six original members besides Newell, there were two M.D.s (Cambridge and Edinburgh), an apprentice-trained apothecary, an apothecary-surgeon, a surgeon, and a surgeon-midwife. Other provincial societies, such as those in Plymouth (1794) and Leicester (1800), also defied the functional distinctions of London's guilds, a practice that certainly reflected the realities of medical practice in rural areas. These societies assisted in the training of apprentices, fostered professional improvement, and helped solve disputes among members. Their appearance outside of London suggests that even in the provinces the need to improve skills and expand knowledge within a formal context was increasing.

Medical Theory and Practice

At the most general level the practice of 17th-century physicians differed little from that of physicians in ancient Greece and Rome.[38] In both eras physicians possessed three techniques by which to determine the nature of an illness: listening to the patient's own description of symptoms, observing the patient's appearance and behavior, and examining the patient's body.[39] To these verbal and visual techniques, the 18th-century physician added procedures that were increasingly manual, such as performing percussion.

Medical theory and therapeutics in the 17th century continued to be strongly influenced by humoralism, formalized by Galen in the second century A.D. Basically, this tradition maintained that health resulted from a natural balance of bodily humours (yellow bile, blood, phlegm, and black bile) and their manifested qualities (cold, heat, moistness, and dryness). When not associated with injury due to physical trauma, disease was thought to result from an imbalance of the humours. To restore balance, physicians administered preparations of natural substances, either topically or internally. Feverish patients, for example, would be advised to follow a cooling regimen; they might also be bled to help eliminate excess heat produced by too much blood. Those suffering from chills and a runny nose would take preparations that warmed and dried. If a patient's stool was abnormally dark, the physician might conclude that there was an excess of black bile and prescribe an enema or an orally administered laxative. The followers of Galen thus advocated an active role for the physician in the management of disease.

During the 17th century continental influences and a renewed interest in Hippocratic doctrines affected both medical theory and practice in England. By 1600, the influence of Paracelsus (1493–1541) was increasing.[40] This revolutionary Swiss physician taught that diseases were real entities produced by specific causes. An advocate of mineral or chemical remedies, he was successful in convincing pharmacopoeia compilers to include mercury (for syphilis), lead, sulphur, iron, arsenic, copper sulfate, and potassium sulfate in the published drug lists. Apothecaries and apothecary-surgeons were certainly in a position to expand their functions by popularizing and taking advantage of this advice.

The most important 17th-century proponent of Hippocratic teaching was Dr. Thomas Sydenham (1624–1689), who based his medical practice, and instruction, on an eclectic combination of ancient and modern theories.[41] As an admirer of Hippocrates and believer in the healing power of nature, and *vis medicatrix naturae*, he believed that nature and experience, not theory, should be the healer's guide; this philosophy earned him the title of "The

English Hippocrates." His English physician contemporaries often ignored his advice, but colonial American practitioners at least paid him considerable lip service. With the availability of such advice from an Englishman, the physician's competitors were encouraged to broaden the level of their intervention in the management of illness or injury.

In spite of anatomical and physiological attacks on the underpinnings of humoralism during the 16th and 17th centuries, Galenic doctrines, or some variation of them, found acceptance well into the 19th century. But as Galen's authority came to be questioned on many points, various theories were proposed to replace it. Something a diversity of healers had encouraged since antiquity, a by-product of this competition for orthodoxy in medical theory seems to have been a willingness to innovate, to deviate from recommended therapy. For example, in a diary entry for 1602–1603, the writer records how Dr. Henry Gellibrand of provincial Kent expressed the uncertainties implicit in the therapy that he gave to "a minister, [who] was very sick. Gellibrand gave him a glyster [enema], and let him blood the same day, for a fever; his reason was, that not to let him blood had been very dangerous but to let him blood is doubtful, it may do good as well as harm."[42]

Although medical thought tended toward uniformity in both rural and urban areas, similarities in practice may have been offset by variations in degree. Given the information we now have, we cannot be certain that rural medical practice was less heroic in its use of bleeding and purging than that found in the large towns and cities.[43]

NEW ENGLAND

The North American mainland colonies were the foremost locations for English settlement, followed, not far behind, by Ireland.[44] The period of actual settlement spanned more than a century: from Virginia in 1607 to Georgia in the 1730s. With few exceptions the basic problems of settlement were quickly overcome, and the inhabitants rapidly increased their numbers. From a few thousand sprinkled along 1,000 miles of coastline in 1630, the population grew to 250,000 by 1700; and at the beginning of the American War for Independence in 1776, some two and one-half million people resided there, nearly one-third the number of residents in England and Wales. Not until the time of the first U.S. Census in 1790, did about 3.3 percent of the population live in communities with 8,000 or more inhabitants.[45] To be sure, there were cities in America that surpassed or rivaled the provincial centers in England, the three largest being Philadelphia (42,000 in 1790), New York (21,000), and Boston (18,000). Yet, in 1730, Boston, with 13,000 inhabitants, had led Philadelphia and New York, which each had about 8,500. Throughout the colonial period, New England attracted the most homogenous group of immigrants, who emigrated largely from counties in the west country and East Anglia.

Writing about the growth of New England, which contained about 25 percent of the U.S. population in 1790, is in effect like writing about the expansion of Massachusetts. Under the direction of the Puritan elect, some 20,000 provincial English took part in the Great Migration to Boston between 1630 and 1643.[46] Not represented by a cross section of English society, New England was "levelled" at the start. Most of the immigrants belonged to the socioeconomic group called the "common sort," which included farmers, laborers, artisans, and merchants, as well as a variety of medical personnel, ministers, holders of small political offices, and lawyers. Some of these immigrants had attended a university (mostly Cambridge), but most still acquired their expertise through occupational inheritance or by serving an apprenticeship. As congregational or presbyterian dissenters from the Church of England, they preferred local autonomy with the church and town meeting serving as loci of community action.

The Puritan venture to create a city upon a hill, a Holy Commonwealth, ran into difficulty when more immigrants arrived than could be safely located near the colony's center of power, Boston.[47] Immigrants moved out to the north, south, and west, creating the new colonies of Rhode Island, Connecticut, and New Hampshire.[48] Despite being perhaps the most densely populated region in North America, New England remained overwhelmingly rural, with most of its inhabitants residing in small villages and towns that were often named after provincial English towns (Boston, Cambridge, Ipswich) or counties (Norfolk, Suffolk, Essex). In Massachusetts alone, 81 towns existed in 1700, and by 1790 another 172 had been incor-

porated. Given these circumstances, in many ways Boston was to Massachusetts, or to New England, what Norwich was to Norfolk or East Anglia. During the 17th century immigration accounted for most of New England's population growth, but later increases from births became more important. During the 18th century, the population of Massachusetts doubled every 37.4 years, about 10 years longer than it took in the previous century.[49]

Not many colonial communities have received systematic demographic attention, but we know that in Andover, Massachusetts, life expectancy for male children born during the 17th and 18th centuries exceeded that of English male babies by more than 10 years. The high growth rate of Massachusetts also depended upon the ratio of males to females, early marriages, and the generally youthful age of the initial population. Although usually characterized by the dominance of families, both new and established, New England also attracted thousands of indentured servants, who ranged in age from the late teens to the early thirties. In England, conditions forced the postponement of marriage until the man was about 27 and the woman 25, but in Massachusetts the comparable figures were about 25 for men and 22 for women.[50] All of these factors help to account for the dramatic growth in population. But we must also consider the role played by disease.

Patterns of Disease

Accompanying immigrants to New England, and awaiting them upon their arrival, were various diseases. Investors and boat operators in search of quick profit and settlement leaders eager to secure needed manpower encouraged overcrowding on vessels during trans-Atlantic crossings, which in the 17th century lasted two months or more. Navigational improvements eventually reduced travel time to about six weeks, but the growing number of ships and passengers also increased the possibility of disease transfer, particularly those with epidemic potential in a rapidly expanding and mobile population. Although ship-board diets improved over time, scurvy frequently broke out, and poor dietary practices before, during, and after sailing for America probably lowered resistance to many diseases. Carving out an existence on the frontier also involved risk to life and health. During the 18th century, improved dietary habits and housing, as well as changes in the distribution of population, reduced the incidence of some diseases, but improved lifestyles also gave rise to medical problems common among prosperous people in England, particularly gout.

The most important epidemic diseases brought to the colonies were smallpox, diphtheria, scarlet fever, and measles. Yet, intestinal problems and respiratory ailments, such as pneumonia and pulmonary tuberculosis, were undoubtedly the major causes of high mortality. It is difficult, however, to determine the cause of death with accuracy for that period because of the ambiguous manner in which doctors recorded the medical problems of their patients.

While relatively free from the epidemic and endemic attacks of yellow fever, malaria, and typhus that plagued settlements to the south, New England was visited by most of the epidemic diseases of the times.[51] The first encounter with smallpox, imported along with the settlers to John Winthrop's Massachusetts Bay Colony, occurred in Boston during the 1630s. Originating in the British Isles, the West Indies, or in Canada, smallpox epidemics ravaged Massachusetts and most of New England in 1648, 1666, 1677, 1689, 1702, 1721, 1731, 1751, 1764, and in the 1770s. During the celebrated epidemic of 1721, during which inoculation was first practiced in America, over half of Boston's 10,670 inhabitants became infected and over 800 died. Death rates from smallpox seldom reached 14 percent in America, which was much better than the 18 to 40 percent that obtained in England during the period. Helping to reduce the severity of smallpox epidemics were effective quarantine measures, isolation facilities, and controlled inoculation campaigns. Smallpox, like dysentery and typhus, often followed the movements of armies, and the colonists waged many campaigns against the French, the Indians, and, of course, the British during the 18th century.

Measles, smallpox, and dysentery struck all age groups; diphtheria and scarlet fever, primarily children. First appearing in Boston in 1657, measles continued to be reported for the duration of the colonial period. Apparently the disease more often proved fatal in America and affected a larger number of adults than in England, where frequent epidemics conferred immunity upon those who survived it in childhood. The most destructive measles epidemic of colonial times, the New England epidemic of 1713, claimed over 100 deaths. Late in the 18th century, however, the disease seems to have become endemic; by 1800, it was primarily a disease of childhood.

Contemporaries called diphtheria by many names: cyanache, squinancy, quincy, angina, canker, bladders, rattles, hives, throat ail, and throat distemper. Diphtheria was also frequently confused with scarlet fever, which first appeared on the London Bills of Mortality in 1703, and which arrived in New England the year before. In 1735–36, scarlet fever and diphtheria struck simultaneously (for the former, largely in Boston), resulting in the most destructive epidemic of any childhood disease in American history. Whereas scarlet fever claimed over 100 deaths from the estimated 4,000 persons infected, diphtheria in New Hampshire alone took 1,500 from a population of 20,000.

Increased population density during the 18th century may also have added to the number of those without immunities, thereby assuring contagious diseases many victims. The incidence of epidemic diseases no doubt produced a widespread sense of vulnerability, regardless of available therapies and variations in mortality rates. Community leaders, concerned about economic disruption and loss of manpower, consulted with local medical practitioners and worked with them to develop public health policies.[52]

Although both New England and the parent country shared many disease experiences, survival rates in America were usually higher. Thus New England was generally a more healthful place in which to live.

The Origins and Training of Early New England Doctors

As late as 1800, neither in New England nor elsewhere in the new nation did there exist effective measures to regulate medical education and practice.[53] This situation prevailed both because of the duration of the initial period of settlement and because of the development of local practices in the physically isolated colonies. Although institutional and legislative responses to unregulated practice differed from place to place, the experience of New England, particularly Massachusetts, is well documented and serves as a useful example of this phenomenon.[54]

Of the nearly 1,600 doctors who practiced medicine in Massachusetts between the founding of Plymouth in 1620 and the election of Thomas Jefferson to the presidency in 1800, fewer than 100 were immigrants.[55] Generally the immigrant practitioners came from Great Britain rather than from continental Europe; at least 68 arrived from the British Isles, compared with fewer than 10 each from France, Germany, and the Netherlands. About one quarter of them received some form of documented training prior to their departure for America: four served apprenticeships, four earned M.D.s, eight attended universities, and, during the 18th century, 13 gained experience in the armed services of Britain or France. Regardless of their titles and functions in their homelands, in the colonies, where there were no regulations, they became "doctors" and combined the functions of Old World guilds.

Until the middle of the 18th century, there are few recorded objections in New England to this free use of the title "doctor." This practice seems to have originated even before the end of the Great Migration to Massachusetts in the 1640s, which included medical practitioners from both London and the provinces. Of the 54 English medical immigrants, at least 34 arrived before 1700; most of those from other areas came during the 18th century. The experiences of four of the 24 Englishmen migrating before 1650 reveal the conditions in which "doctor" became a common title for male medical practitioners.[56]

In the log books of the *Mayflower*, which landed at Plymouth in 1620, we find the name Samuel Fuller. A native of Redenhall, Norfolk, where he was born in 1580, Fuller joined the separatists bound for Leiden, the Netherlands, in 1609. It is believed that he studied some medicine in London and may have served an apprenticeship or attended medical lectures in Leiden, where he worked at least part-time as a silk weaver. The ship's log lists Fuller variously as physician, surgeon, and doctor. After his arrival in New England, he was summoned as "Dr. Fuller" by the leaders of Salem and Boston to go north of Plymouth to assist the diseased inhabitants. His use of phlebotomy to treat scurvy indicates that he did not confine his duties, as a physician, to diagnosing and prescribing. He died of smallpox in 1633.

Another *Mayflower* passenger, the surgeon to the ship's crew, was Giles Heale, who had served an apprenticeship and was licensed to practice by the Barber-Surgeons Company of London, which required trans-oceanic passenger vessels to have surgeons.[57] Although his skills were desperately needed in Plymouth, Heale remained only four months before returning to England.

Giles Firmin and his father, an apothecary from Suffolk County, arrived in Boston in 1632. The younger Firmin had attended Cambridge University and then served an apprenticeship with Dr. John Clark in London. For over a decade he practiced in Boston and nearby Ipswich. Responding to the shortage of experienced practitioners, Firmin presented what was perhaps the first series of medical and anatomical lectures in the colonies. At least one of those attending the lectures served an apprenticeship with him. Firmin's suggestion to provide cadavers for dissection before groups of apprentices and practitioners seems to have gained legislative endorsement after his departure for England in 1644 to become vicar of Shalford in Essex. In 1646, the Massachusetts General Court endorsed a proposal that would allow the body of an executed malefactor to be dissected by such groups.

The only M.D. to arrive before 1650 was Robert Child of Norfleet, Kent, the only one of the 814 English country practitioners mentioned earlier known to have visited Massachusetts. In Boston by 1638, he seems to have been more interested in checking up on his business interests than in practicing medicine because local authorities reproached him for neglecting his calling. The Massachusetts Bay government in Boston expelled him for professing unorthodox religious beliefs. Two years later that same government approved legislation that addressed the need for more skilled practitioners to protect residents from the dangerous advice and therapies of the self-proclaimed doctor.

Enacted in 1649, the law applied to "Chirurgeons, Midwives, Physitians, and others."[58] Its origins are distinguishable in the provincial English background of the immigrants, including medical practitioners. By endorsing the introduction of the apprenticeship system and by seeking an examination for competency to practice administered by local practitioners and laymen, the law reflected the customs and practices of rural England. By encouraging competency in more than one type of practice, it further replicated the functional realities of medical organization and practice in provincial England. The blurring of medical functions already evident in England posed no problem in New England, where the "doctor" became a kind of general practitioner.

Immigrant medical practitioners constitute 31 percent of all 17th-century doctors in Massachusetts; in the next century they account for only 3.9 percent. Because of the void created by the absence of men like Heale and Firmin and because the 1649 law proved unenforceable, persons claiming medical skills whatever assumed the title "Doctor." In Massachusetts, between 1700 and 1794, 861 practitioners (62.8 percent of 1,370 identified), styled themselves as "doctors."[59] Following the initial period of immigration and settlement, then, medical practice in New England came to be dominated by native-born doctors, most of whom had received no formal training.

For about half of the colonial period the ratio of doctors to population remained stable.[60] In 1650, with 50 doctors serving a population of some 50,000 in Massachusetts, the ratio was 1:1,000. Fifty years later the ratio remained unchanged. During the 18th century, however, the ratio steadily increased to 1:569 in 1750 and to a high of 1:417 in 1780. This was much higher than in the new nation as a whole, estimates for which vary from 1:600 to 1:800.[61] At the time of independence, Massachusetts contained about 10 percent of the population of the United States, but approximately 20 percent of its 3,500 doctors. Over a third of its medical practitioners had served apprenticeships, compared to a tenth for the nation as a whole.[62]

Because medicine was rarely a full-time endeavor in colonial America (or England), doctors, like other settlers, moved out of Boston in search of clientele, land, and security. Just as the influx of immigrants into Boston during the 17th century had created a demand for medical skills, so, too, did the migration of settlers into Worcester, Hampshire, and Berkshire counties throughout the 18th century. There were few large towns at any time in colonial Massachusetts, and those with fewer than 1,000 residents usually had higher ratios of doctors to population (1:557) than did larger communities (1:950). In both England and New England, then, sparsely populated areas seem to have served as residences, both temporary and long-term, for more healers than large towns, where life expectancy was shorter. Although it is doubtful that variations in medical practice were primarily responsible for the differences in life expectancy between rural and urban areas, some doctors expressed concern that the low level of medical training was threatening the health of Massachusetts' inhabitants.

The educational background of New England doctors was as diverse, if not as intensive, as that of English practitioners. Prospective New England doctors did not need to attend college, enroll in hospital lectures, possess an M.D. degree, or even serve an apprenticeship. A few, however, such as Thomas Bulfinch, Sr., received extensive training.

Upon completing an apprenticeship with Dr. Zabdiel Boylston, a respected Boston physician, surgeon, and apothecary-shop owner, Bulfinch sailed for Europe.[63] During the winter of 1718 he observed and practiced the latest surgical techniques at St. Thomas Hospital in London, where he became the first American pupil of the celebrated William Cheselden. Later, Bulfinch crossed the channel and achieved another American first by studying anatomy and surgery with Jean Louis Petit at the Paris Charité Hospital. Returning in 1721 to his native Boston, where his brother, Adino, owned an apothecary shop, he soon married Judith Coleman, the daughter of a prominent minister. Later their son, Thomas, Jr., also chose a career in medicine.

After graduating from Harvard College in 1746, young Bulfinch served a three-year apprenticeship with his father. Encouraged by his father's success, he went to England in 1754 to round out his medical studies. In London he met the king's personal physician and "walked" the famous teaching hospitals—Middlesex, St. Thomas, and St. Bartholomew. Eventually he traveled to Scotland, where he attended the chemistry lectures of Dr. William Cullen and in 1757 received an M.D. degree from the University of Edinburgh. Bulfinch, Jr., subsequently returned to London, to assume a hospital post as consulting physician, but the death of his father forced him instead to sail home to settle his father's estate and take over his large practice. Eventually he took on pupils and apprentices of his own. Given the absence of effective measures to restrict medical practice in New England, young Bulfinch's extensive training was most atypical.

New England's first medical school, created at Harvard College, did not begin accepting pupils until 1783. Founded in 1636, Harvard sought above all to train ministers but also to keep students at home and to help maintain local autonomy. There was little support for creating professional schools of medicine or law until late in the 18th century because skills in both fields had traditionally been acquired through the apprenticeship system. Although medical degrees were available in Philadelphia and New York as early as the 1760s, few Yankees went south—or anywhere else—to study.[64] Only about 50 students from Massachusetts went abroad to study medicine and fewer than 20 of those received M.D.s.[65] Between 1749 and 1800, over 100 Americans earned M.D. degrees at Edinburgh, including 49 Virginians, 15 Pennsylvanians, 10 New Yorkers, but only four natives of Massachusetts.[66] One eminent New Englander offered an explanation for this provincial, stay-at-home behavior. Benjamin Waterhouse, who had attended lectures at Edinburgh and taken his M.D. at Leiden (1780), once remarked that Americans need not travel to Europe in order to "treat the disorders of their neighbors."[67] Thus most New Englanders who desired medical training took advantage of local opportunities, particularly apprenticeships.

Used in England until the middle of the 18th century, when formal academic training replaced it, the apprenticeship remained popular in New England into the 19th century. Before 1700, only about 140 men, and an undetermined number of women, practiced medicine in Massachusetts, and only about 20 percent of this number were apprentice-trained, most having no formal training at all. This contrasts with the situation in England, where, between 1603 and 1643, four out of five country practitioners had some formal training. Of the 1,370 doctors practicing in Massachusetts between 1700 and 1794, however, at least 488 or 36.6 percent had served apprenticeships.[68]

Although apothecaries and surgeons in England were required to serve apprenticeships lasting up to seven years, the available evidence for Massachu-

setts suggests that the average apprenticeship there lasted just over one year, five years being the longest. The following apprenticeship agreement, drawn up in Massachusetts in 1736, may be typical of such documents:

> Articles of Agreement Indented and made . . . Between Zabdiel Boylstone of Boston in the County of Suffolk Practitioner in Physick & Surgery of the one part, and Joseph Lemmon of Charlestown in the County of Middlesex, Esqr. on the other part. . . .
>
> Imprimis - The said Zabdiel Boylstone Doth Covenant and agree for himself to teach and Instruct the said Joseph Lemmon Junr. in the Arts, Mysterys and Businesses of Physick & Surgery during the term of two years . . . and also to find and provide for him good sufficient and suitable Dyet and lodging during the said two years. . . .
>
> In Consideration whereof the said Joseph Lemmon for himself his Executors and adminrs. doth hereby Covenant and agree to and with the said Zabdiel Boylstone to pay him two hundred pounds in full Satisfaction for his Sons dyet and lodging and for the Instruction which the said Boylstone shall give him in the said Mysterys of Physick and Surgery during the term of two years ending in March 1737. . . . [69]

Occasionally, a student would finish an apprenticeship with one mentor and then continue his studies under the guidance of another; this is what John Perkins did before he visited London in 1734.[70] Some scholars have suggested that the apprenticeship system weakened during the 18th century, but in Massachusetts it grew stronger. During the period 1751 to 1790, 45.5 percent of all doctors starting practice were apprentice-trained, an increase of 28 percent over the pre-1750 period.[71] Obviously, self-taught "doctors" were most typical of the period, but their shadowlike appearance in the historical records prevents our making firm generalizations about them.

Among the most easily identified of the 1,370 Massachusetts practitioners during the 18th century are the 399 (29.9 percent) who attended college, and especially the 360 of this number who received a B.A. degree.[72] The collegiate curriculum of the day offered little that would directly benefit a medical practitioner, and a bachelor's degree was not an essential prerequisite for the M.D. candidate in either Scotland, Leiden, or America. Of the

Massachusetts doctors who attended college, most (86.1 percent) went to Harvard, with the remainder coming from Yale (8.6 percent), and from American or European colleges and universities (5.3 percent).

Prior to 1700, less than 20 percent of Massachusetts' doctors attended a college or university, but during the 18th century collegiate preparation became increasingly common. Between 1701 and 1770, college graduates constitute a per decade average of 33.9 percent of all those starting to practice in each period. After 1750, Yale and Harvard graduates increasingly chose to become doctors or lawyers rather than ministers.

Custom dictated that about three years after receiving his bachelor's degree a graduate would return to his alma mater to obtain an M.A. degree by submitting and defending a thesis, classical in style and theoretical in substance. To some proud New Englanders, the M.A. degree represented the epitome of colonial education—the final polish for young gentlemen. In reality, the degree was little more than a formality. The subject of the thesis did not always indicate a career preference; 257 (64.4 percent) of the Harvard graduates who became doctors received M.A.s, yet only 76 (29.6 percent) elected to defend medical or physiological subjects.[73] Since even an M.A. degree did not prepare one to practice medicine, a number of prospective doctors continued their studies abroad.

In 1700, John Cutler of Boston became the first of at least 50 medical students from Massachusetts to study abroad. Of these travelers 35 (70 percent) had attended college, and 42 (84 percent), including many of the collegians, had served medical apprenticeships. Of the 43 who went to Britain, 38 spent some time in London, and at least 12 attended lectures at one of the Scottish medical schools. Virtually all the Massachusetts doctors who earned an M.D. degree abroad also studied midwifery, anatomy, and surgery in a Paris or London hospital—either before or after graduating from medical school. That so many studied midwifery and surgery suggests that these New Englanders, like their provincial English and Scottish counterparts, had no intention of practicing medicine in the manner of London physicians.

Before 1700, 22 percent of all Massachusetts doctors came from medical families of two or more generations; during the 18th century this percentage dropped to 15.7 percent, which suggests that medicine was attracting new names to the ranks. Yet, throughout the colonial period recruitment from medical families accounted for a substantial number of practitioners.[74] Between 1700 and 1794 nearly 20 percent (271) of all Massachusetts doctors either came from or married into medical families. There were at least 215 two-generation families and 52 of three or more generations. All sons in multigenerational families served apprenticeships, customarily with their fathers, but only one out of four attended college, and even fewer, such as Thomas Bulfinch, Jr., traveled abroad or received an M.D.

The tendency of multigenerational medical families to reside in towns of fewer than 1,000 inhabitants contributed to the high ratio of doctors to population in these small communities. Apparently confident of future business and interested in making money as mentors, many medical fathers trained paying apprentices from outside the family as well as their own sons. For example, Dr. Thomas Bulfinch, Sr., not only trained his son, Thomas, Jr., but Benjamin Stockbridge, Jr., and his son, Charles. Bulfinch, like other mentors, consented to train new doctors, and although the Stock-

Table 1
Percentage of College Graduates Entering
the Professions

	1701–45	1778–92
Harvard		
Medicine	13.4	16.9
Law	5.6	33.3
Divinity	36.0	25.6
Yale		
Medicine	6.8	10.4
Law	6.8	30.9
Divinity	50.0	27.7

Source: Eric H. Christianson, "Individuals in the healing arts and the emergence of a medical community in Massachusetts, 1700–1794: a collective biography" (unpubl. Ph.D. diss., Univ. of Southern California, 1976), pp. 88–89.

bridges did not live in Boston, he was, nevertheless, producing potential competitors in the medical marketplace.

Some evidence suggests that doctors without family connections in the profession may have resented the dominance of multigenerational medical families in some communities. Virtually all of the medical societies in Massachusetts drew their members from the ranks of the apprentice-trained doctors, especially those who had also graduated from college or had gone to Europe or Britain; but only 25 percent of all eligible medical-family members ever joined a county or state medical society.

As an agency for medical education, the multigenerational medical family seems to have assumed a greater importance in New England than in England, although the evidence is limited. While academic training beyond, or instead of, the apprenticeship first appeared in England, by the second half of the 18th century medical society organizers were promoting similar reforms in New England. For in Massachusetts, where only one out of three was even apprentice-trained, the ordinary practitioner dominated.

Medical Organizations

Designed to meet a variety of needs, medical societies were created earlier and in greater numbers in Massachusetts than elsewhere in British America, or even, except for the guilds of London and Edinburgh, in Britain.[75] At least 17 medical societies were either proposed or created in Massachusetts during the 18th century, 14 appearing after 1750. Their goals ranged from proposing minimum requirements for practice, improving the profession's public image, and facilitating self-improvement, to licensing as the only criterion for practice. Although the longevity of some remains uncertain, the locations of all are known; Suffolk County (Boston) supported six, but the majority arose in the country. In terms of the educational background of their members and their stated objectives, a few of the most important groups—notably the Boston Medical Societies of 1735 and 1780 and the Massachusetts Medical Society—resembled London's guilds, without their functional distinctions. Granted the similarities in both composition and function to the provincial and student societies of

England, most of New England's medical societies were created in response to local problems or the appearance of other societies in the area rather than as a conscious imitation of organizations in England.

Most of the Massachusetts societies proposed more training for medical practitioners; midwifery regulation was not on their agendas. The first explicit call for training beyond the apprenticeship came from the group organized in 1755 by William Jepson, who had recently completed an apprenticeship with Dr. Silvester Gardiner of Boston.[76] In the by-laws of his society, composed of former apprentices interested in improving their knowledge and skills, Jepson enumerated the "many disadvantages attending a separate way of study." Although he had learned much as an apprentice, he realized that there was more to be learned than an apprenticeship could offer. Sharing books, they would first cover anatomy, then surgery, midwifery, physick, and then "proceed to the Other Sciences." Other medical societies, primarily concerned with limiting medical practice to those who could demonstrate adequate training or who could pass a licensing examination, faced the problem of what to do with the hundreds of men already practicing who had received no training at all.

Boston's epidemics of smallpox in 1721 and scarlet fever in 1735 served as major stimuli for the creation of medical societies. The Club of Physicians (1721), the Physicall Club (1726), and the Boston Medical Society (1735) were all created by Dr. William Douglass, a Scotsman with an M.D. from Utrecht.[77] Douglass possessed the only M.D. in Massachusetts at the time, but he was not the only Boston practitioner with European training. For membership in all three of his groups, he selected only men who had graduated from college, served an apprenticeship, or studied in Europe. The 1721 group, though not primarily interested in regulating general medical practice, did oppose inoculating patients against smallpox without quarantining them. When the epidemic passed, the group disbanded. In 1726, Douglass started the Physicall Club with essentially the same personnel, who met occasionally to discuss medical philosophy and therapeutics. When scarlet fever struck Boston in 1735, those same doctors assisted one another in combatting the epidemic; encouraged by this coop-

Table 2
Medical Societies in Massachusetts, 1721–1794

Date	Name	Purposes
1721–22	Club of Physicians (Boston)	oppose inoculation
1726–35	Physicall Club (Boston)	self-improvement
1735–54	Boston Medical Society	self-improvement; combat scarlet fever
1755	Jepson's Proposal (Boston)	self-improvement after apprenticeship
1765	Association of Doctors (Middlesex County)	self-improvement; upgrade public image
1766	Ames' Proposal (Middlesex County)	eliminate competition from quacks
1768	Sociable Club (Middlesex County)	self-improvement; eliminate quacks
1771*	Martimercurian (Middlesex County)	self-improvement
1771*	The Spunkers (Middlesex County)	body snatching
1772*	Club of Generous Undertakers (Middlesex County)	body snatching
1780–	Boston Medical Society	repression of quackery by licensing & membership; proposed mandatory lectures at Harvard
1780	The Confederacy of Physicians (Middlesex County)	self-improvement
1781–	Massachusetts Medical Society	regulation by membership & licensing
1787–	Berkshire County Medical Association	self-improvement; regulation by membership & cooperation
1789–	Middlesex County Medical Association	self-improvement; regulation by membership & cooperation
1791–	Bristol County Medical Society	self-improvement; regulation by membership & cooperation
1794–	Worcester County Medical Association	self-improvement; regulation by membership & cooperation

*denotes Harvard College student societies

Sources: Eric H. Christianson, "'To Check the Growth of Imposters': the role of Massachusetts medical societies in preserving the apprenticeship system, 1721–1794" (unpubl. manuscript), and "'The Confederacy of Physicians': an historical oversight?" *J. Hist. Med.*, 1977, *32:* 73–78.

erative effort, Douglass organized the first Boston Medical Society.

In 1738, as president of the society, Douglass proposed that qualified practitioners and innocent patients be protected from "Shoemakers, Weavers and Almanack-Makers, with their Virtuous Consorts, who have laid aside the proper Business of their lives to turn Quacks."[78] He advised the General Assembly to enact legislation "so that No person shall be allowed to practice Physick within the limits of this Province" before passing an examination administered by "such regular, approved and learned Physicians and Surgeons as the Honourable Court shall see [fit] to appoint." His provision to include both physicians and surgeons as examiners of candidates for the practice of "Physick" indicates the blurring in New England of the medical functions associated with the London guilds. In view of the great numbers of self-trained practitioners, Douglass advocated a simple solution: regardless of training, anyone who could pass an examination would be judged competent to practice. His proposal paralleled the provisions of the 1649 Massachusetts law, the 1518 charter of the Royal College of Physicians of London, and the 1511 statute of Henry VIII that regulated provincial English practice. Douglass wanted to vest the regulation of medical practice in the hands of a few examining practitioners. The assembly, however, refused to grant a Boston society jurisdiction over medical practice throughout the province. For the duration of the society's existence (it seems to have fallen apart after Douglass' death in 1752), it functioned essentially as an organization for self-improvement, presenting lectures, dissections, and operations to audiences composed of members, interested apprentices, and laymen. This first Boston Medical Society raised issues that would not be forgotten. Throughout the remainder of the 18th century, medical-society leaders continued to explore avenues that might eliminate the self-taught.[79]

Most of the medical societies created between 1765 and 1794 were located outside of Boston. The tactics employed by these groups to combat the increase of untrained doctors differed markedly from those originating in Boston. The three efforts that Dr. Nathaniel Ames, Jr., made to organize a medical society in Middlesex County in the 1760s illustrate one approach to the problem. In language reminiscent of Douglass' earlier observations, Ames wrote of men "having Poor stomachs to return to the stall or plough from whence them came; some of them commence QUACKS and call themselves Doctors."[80] In the spring of 1765, the presence of several self-taught practitioners in the area around Dedham, where he lived, prompted Ames and 23 other doctors to form a medical society. A letter signed by "Graph Iatroon" (Greek for "writing of physicians") explained that there had been

> some time on foot a proposal forming medical societies or Associations of Doctors . . . for the more speedy improvement of our young Physicians . . . to get the Profession upon a more respectable footing in the Country by suppressing this Herd of Empericks who have heaped such intolerable contempt on the Epithet *Country Practitioner.*[81]

The contents of the letter were "carefully sealed and superscribed lest a telltale Wife or Child divulge that which must be as secret as Masonry till some Societies are established." Ames and his friends thought that it was imperative to avoid "degrading each other . . . before the patient or people by consulting with untrained men." The discord produced by such actions was "highly detrimental to the Profession" and provided "great advantage to the ignorant and designing"; in fact, it was the "chief Root from whence these very Empiricks spring." Although the existence of this society remained secret, Ames openly pushed for the regulation of medical practice.

Utilizing his popular Almanack to reach a wider audience, Ames, together with several colleagues, some of whom had recently returned from studying in Europe, continued the war on quackery. As founder of the Sociable Club (1768), formed to reduce bickering within the ranks of the trained doctors and to "keep up the Honor of the Profession of Physick," Ames declared that

> Titles are marks of honest men and Wise,
> The Fool or Knave that wears a title lies.[82]

He advocated warning trained colleagues and unsuspecting patients of the presence of quacks by publicly identifying them. Little came of this particular campaign, but renewed concern soon produced a more ambitious effort. In 1768, Ames circulated a copy of John Morgan's treatise on the need for American medical schools, published in

Philadelphia in 1765, in which he warned the quack, the "remorseless foe to mankind," to hold "thy exterminating hand."[83] Morgan himself had helped to found a medical school in Philadelphia in 1765, and Ames lamented that Massachusetts had allowed a "neighboring province so far to get the start of us in the regulation of this noble science, which of all others, most needs the protection of civil authority." To prevent medical practice from remaining open to "every ignorant drone that assumes the title of doctor," he submitted a bill to the House of Representatives in 1768 for "regulating the Practice of Physick." At the time, however, the legislature was preoccupied with Samuel Adams' circular letter opposing the Townsend Duties. This is the last known attempt to regulate the practice of medicine until 1780.

The War for Independence during the late 1770s stimulated many of those who served in the medical corps to reevaluate the need for medical regulation.[84] Among these reformers was John Warren of Boston. While at Harvard College in the early seventies, he started three student societies because the school offered no medical lectures. Even after completing his apprenticeship, Warren, like William Jepson and Nathaniel Ames, retained doubts about the limitations of such training. So did his Boston colleague Williams Smibert (M.D., Edinburgh), who observed that American medicine would never become a science as long as students were "taught to believe compleat [sic] medical knowledge is to be acquired in a few months under the tutelage" of someone who himself had not received training beyond the apprenticeship level.[85]

In 1780, Warren and 12 colleagues formed the second Boston Medical Society, demanding for membership not only a completed apprenticeship or its equivalent, but a medical or bachelor's degree as well. Members of the society believed that the existing apprenticeship could not accommodate the tremendous increase of information about medical discoveries and surgical techniques; that Harvard College would have to assume some responsibility for educating doctors by offering a course of lectures; and that quackery, which the legislature failed to suppress, could only be combatted by exclusive societies. Toward this end, Warren urged patients not to patronize self-taught practitioners.

The following year, Warren's group, now 14 mem-

bers strong, attempted to convince the state legislature to incorporate them as the Massachusetts Medical Society. But the legislature, which included many representatives from western counties, where distrust of Boston power ran high, insisted that the society expand its membership to 70 before receiving a charter.[86] So, from the more than 600 doctors in the state, about one-third of whom possessed apprenticeship training or more, the society selected an additional 56 members. Although the legislature granted the society the privilege of examining and "licensing" practitioners, it failed to provide an effective means of enforcing the society's standards. According to Warren, this legislation permitted the society to enable "the people at large (who might otherwise be incapable of fully discerning the qualification of candidates for practice), to distinguish the persons upon whom they may rely."[87] In other words, the profession had received permission to regulate itself, if it could.

In some ways the efforts of Warren and his Boston colleagues resemble that of Thomas Linacre and his five associates who, in 1518, convinced King Henry VIII to incorporate them as the Royal College of Physicians of London. Yet, the Boston society was much more ambitious; it sought control over the entire state, whereas the London physicians had sought merely to monopolize the practice of medicine in London, not in the provinces. In 1785, acknowledging its unrepresentative membership throughout the state, the Boston-based society invited correspondence with groups that may have already been in existence, and encouraged the formation of affiliated county and district medical societies that would recommend candidates for examination and licensing. Apparently, informal associations of practitioners already existed in some locales, but the first counties to organize societies were Berkshire (1787), Middlesex (1789), Bristol (1791), and Worcester (1794). Given the delayed response time (two to nine years) to the 1785 communication issued by the Massachusetts Medical Society, it seems unlikely that county medical society members perceived that their right to practice was threatened legally by the existence of the Boston-based group. They may, however, have sought to protect themselves from competition presented by other practitioners in their communities.

The by-laws of these local voluntary associations indicate a willingness to include any practitioner who wished to improve himself and his profession. These organizations did not state explicitly that women were ineligible, but the available evidence reports the absence of female practitioners in the membership lists. The more inclusive nature of these county societies is reflected in the ratio of members to population. In contrast to the Massachusetts Medical Society with a ratio in 1781 of 1:5,657, Berkshire had 1:1,219, Middlesex 1:2,487, and Worcester 1:1,234.[88]

County societies were created to "eliminate the manifold inconveniences" that resulted from the "want of a regular and uniform method of educating pupils in physick, especially in the country." Yet, they, too, indirectly sought control over practice; for example, the Berkshire County Medical Association stipulated that if "Any person residing within the limits of this county [Berkshire], and pretending to practice physic and shall refuse . . . to become a member by attending the meetings and subscribing to the rules, . . . shall be treated with entire neglect by all that are members."[89] This open membership policy stands in contrast to the original intention of the state society to require all members to have a college education and an apprenticeship, "as it is in the mind of most of the Gentlemen of this town [Boston] never to vote for one that has not had one."[90] But the first president of the state society, Dr. Edward A. Holyoke, who possessed a bachelor's degree and apprenticeship training, did not believe that a college education should be a prerequisite for membership, nor did he "think it advisable to enjoin an Attendance on the medical lectures at Harvard," which were first offered in 1782, because it would be a "great inconvenience to many" living outside Boston and Cambridge.[91]

In some ways the regulatory ambitions of the Harvard Medical School, created in 1783, and the Massachusetts Medical Society, founded two years before, clashed.[92] Harvard produced M.D.s, but the society claimed the right to examine and certify them along with other, degreeless, practitioners. A similar situation obtained in London, where the Royal College of Physicians certified the graduates of medical schools in Scotland and England.

At the close of the 18th century, despite the appearance of new institutions, the regulation of medical practice differed little from what it had been in the 17th century. Although the "Great design" of the state medical society had been "to check the Growth of Imposters and lay a foundation for Improvement in Medical Knowledge," in practice, complained one member, the "institution of the Massachusetts Medical Society has not in the least degree prevented the Increase of Empirics."[93]

Boston doctors failed to centralize medical authority in Massachusetts, just as London practitioners failed to control medical practice in the provinces. But unlike the English guilds, which at least partially succeeded in regulating medical practice in London, the medical societies of Massachusetts had little to show for their efforts. In many ways the organization of medicine in Massachusetts, and in all of New England, resembled arrangements neither in London nor in the English provinces. In Massachusetts, where there were few degree-holding practitioners, the level of training was low and titles were employed indiscriminately; in London, or even in the provinces, "physicians" possessed M.D.s, regardless of their function. During the 18th century, however, just as the structure of medical practice in provincial England began more closely to resemble that in London, so too did the structure of practice in rural Massachusetts begin to bear similarities to that in Boston. The trend already evident in England toward formal academic training and certification for practice was also slowly gathering momentum in New England.

Medical Practice and Therapeutics

Although England had supported hospitals since the Middle Ages, few survived the Reformation. The beginnings of the voluntary hospital movement in the second quarter of the 18th century soon had parallels in the British colonies, beginning with the Philadelphia Hospital, which opened in 1751. Hospitals did not appear in New England, however, until the 19th century.[94] Several factors contributed to this delay. Only large towns and cities could afford to maintain a hospital, and during the colonial period New England remained predominantly rural. Besides, most immigrants came from provincial England and thus, unlike London residents, were unaccustomed to the presence of hospitals. Families and municipalities assumed responsibility for medical charity.[95] Finally, the apprenticeship system of training doctors encouraged practice in the homes of their patients.

The only "hospitals" in New England before

1800 were the almshouses, pesthouses, and poor-houses found in the larger communities, such as Boston; the temporary facilities used for quarantine and inoculation during smallpox epidemics; and the military hospitals established during the several colonial wars.[96] Many of these institutions contracted with doctors on a year-to-year basis; at least 118 Massachusetts practitioners supplemented their income in this way. Sometimes the awarding of commissions to care for the indigent involved political considerations. In 1773, for example, Dr. Joseph Warren, a Harvard graduate with apprenticeship training, was replaced as physician to the Boston almshouse by a man of more "conservative politics," Dr. Samuel Danforth, who possessed similar medical credentials.[97]

Between 1764 and 1780, 73 percent of the 92 inoculators in Massachusetts worked in specially prepared isolation facilities rather than in the homes of their patients. During the War for Independence, Dr. Nathaniel Ames, Jr., a college graduate with apprenticeship training, earned a handsome fee for inoculating patients at the smallpox hospital in Marblehead, north of Boston.[98] Such public services were usually provided by the medical elite, who at least had served apprenticeships.[99]

Responsibility for medical care in New England, as in the mother country, often devolved upon the sick themselves or their families, who picked up the rudiments of medicine from a variety of sources, including self-help manuals, newspapers, almanacks, and even medical texts.[100] When families could not treat their own medical problems, they called in a neighbor or sought the advice of local persons reputed to have medical expertise: grocers, booksellers, midwives, nurses, bone setters, and ministers, as well as apothecaries, surgeons, and physicians.[101] Unfortunately, we know little about the activities of most of these practitioners.

Contemporary records suggest that male doctors did not challenge the monopoly of female midwives until after 1750. The most familiar names in the annals of colonial midwifery are those associated with the settlement of Boston in the 1630s: Anne Hutchinson and Jane Hawkins, who were exiled from the town for theological reasons, and Margaret Jones, who was executed for witchcraft in 1648.[102] The extent of some practices is indicated by the fact that Elizabeth Phillips, a London-certified midwife, delivered over 3,000 babies between 1719 and 1761.[103] Although most midwives worked as private practitioners in the homes of their clients, some were paid by local communities to serve prospective mothers who were unwed or who lived in the poorhouse.

During the early decades of the 18th century, Boston supported at least a dozen apothecary shops. Under the sign of the Unicorn & Mortar, Dr. Silvester Gardiner conducted what was perhaps the largest wholesale drug business in New England until the 1770s: importing herbs and chemicals, preparing medicines, and selling medical and surgical books and apparatus.[104] Some Boston doctors, like their London counterparts, may have relied upon apothecaries to prepare their prescriptions; but in most of the settlements around Massachusetts, as in provincial England, medical practitioners either prepared their own prescriptions or had their apprentices do it. A survey of 167 inventories of 18th-century doctors' estates reveals that over two-thirds contained a substantial quantity of herbs, chemicals, and instruments necessary for compounding prescriptions.[105] These estates also included London pharmacopoeias (standard lists of ingredients of commonly prescribed preparations).

The estate inventories also document that virtually every doctor owned a set of basic surgical instruments: lancets for bloodletting or opening abscesses and scalpels for removing superficial abnormalities. Most colonial practitioners, however, seem to have left excisions of large fleshy tumors, amputations of extremities, and lithotomies to medical men having the requisite anatomical and surgical training. Dr. Gardiner's success at performing lithotomies indicates that he was just as capable, if not as fast, as his mentor, William Cheselden in London.[106] Gardiner, Bulfinch, Sr., Zabdiel Boylston, and James Lloyd, among others in the Boston area, brought home from England and Europe the most advanced surgical techniques, and they, in turn, taught these to their many students who practiced in the colony.

Regarding his practice in Boston during the first half of the 18th century, the European-trained William Douglass confided that he could

live handsomely here by the incomes of my practice, and save some small matter . . . here we have a great trade and many strangers with whom my business chiefly consists. I have a practice here among four sorts of People. . . .

His four types of patients were (1) families that paid him an annual fee for his services, (2) occasional patients in need of immediate care, (3) poorhouse or "free" patients, and (4) native New Englanders from whom he had difficulty collecting fees.[107] The mobility of New Englanders produced many "strangers" not only in Boston but throughout the state. Douglass' account of his practice in Boston, though in some respects unique to large seacoast towns, was in many ways descriptive of medical practice throughout the region, and even in provincial England, although his reference to poorhouse patients differentiates his practice from the majority of his contemporaries. Douglass, who practiced primarily as a physician, did not prepare his own prescriptions, but he did assist in some surgical cases, thus demonstrating that even the best-trained doctors of New England combined the functions of Old World guilds.

Generalizations about therapeutics in New England before 1800 have often been based on William Douglass' description. Medical practice, he alleged, "was very uniform, bleeding, vomiting, blistering, purging, anodyne, etc., if the illness continued there was repetendi, and finally murderandi." Also, he claimed, his fellow New Englanders followed the English physician Sydenham "too much in giving paregoricks, after catharticks, which is playing fast and loose." According to Douglass, the most commonly used drugs, taken in "unbelievable quantities," were calomel (mercurous chloride), opium preparations (laudanum, paregoric), ipecac, Jesuit's bark (cinchona bark), and snake root.[108] The evidence in recent studies, however, fails to support his contention that most New England practitioners over-prescribed these drugs and that they dealt in "quackish medicines."

Analyses of over 7,000 patient visits made by five apprentice-trained New England doctors—three of whom practiced in New Hampshire and two in the Boston area—provide a radically different picture of colonial medical practice.[109] The ledger books of these doctors, covering the years 1770 to 1795, show that bleeding was prescribed for only about 7 percent of their patients. In 10 percent of their non-surgical cases, these doctors gave no drugs at all, only advice about diet and regimen. The most commonly prescribed medications, in order of frequency, were liquid laudanum, paregoric elixir, and ipecac (all 1 percent or less), cinchona bark (2.48 percent to 11 percent), calomel (less than 3 percent), and snakeroot and anodyne balsam (less than 4 percent). Although the native plants of New England were, in the opinion of one recent scholar, "insufficient to furnish most of the medical profession's therapeutic needs," all but one of the doctors used the "emetic weed" (*Lobelia inflata*), one of the "few native plants that found major use in early American medicine."[110] Of the 100 most commonly used ingredients, 68 were botanical, 26 chemical or mineral, and 6 were derived from animal products. Nearly half of these substances were imported; many of the remainder came from native plants, but they constituted "only a small proportion of all the drugs administered."[111] All five doctors, regardless of their location, tended to follow uniform practices. Thus it seems that the heroic therapies described by Douglass were neither widely nor commonly administered.

When Dr. John Perkins visited London in 1734, he discovered great similarities between medical practice in the English city and that of Boston, and when, in 1777, he committed his recollection of that visit to paper, he evidently saw little reason to modify his original impressions. The actual practice of medicine, whether in treating gout, worms, dysentery, or broken bones, seems to have differed little between the colonies and the English provinces, or even London. New England medical practitioners admired and followed Sydenham's Hippocratic teachings to advise the patient of proper diet and fresh air, and to recognize the *vis medicatrix naturae* (the healing powers of nature). Experience-oriented as they were, these doctors did not produce new medical theories to replace those of the Old World, although some evidence suggests that they may have been more flexible and less harsh in their therapeutics than European authorities recommended. The successful New England experiments with controlled smallpox inoculation, which antedated similar developments in England by years, serves as the most obvious example of this tendency.

CONCLUSION

The differences between medical organizations, practice, and therapeutics in Old and New England during most of the 17th and 18th centuries cannot be attributed solely to climatic variations between the two regions. In 1623, Governor Edward Winslow, Jr., of Plymouth, observed:

> I can scarce distinguish New England from Old England in respect of heat and cold. . . . Some object because our plantation lieth in the latitude of 42 degrees it must needs be much hotter. I confess I cannot give the reason of the contrary; only experience teacheth us that if it do exceed England, it is so little as must require better judgements to discern it.[112]

Emphasizing the similarities that he observed at the time, William Wood wrote in 1639 that "onions and whatever grows well in England" also flourished in New England soil.[113]

The absence of guild restrictions, together with the presence of rapid demographic change, did not produce a unique form of medical thought and action among New Englanders. The institutional and intellectual environment of the New World was not sufficiently different from that of the Old World to produce more than variations in degree. The blurring of medical functions was widespread in provincial England as well as in America; both regions were moving in the same direction but at different rates of speed. English controversies over functions and titles, such as "physician," were similar in many respects to the debates in Massachusetts concerning the training appropriate for the title of "doctor." In their attempts to modify or eliminate medical traditions, reformers in both areas failed to reach a consensus regarding the necessity of academic and practical training. As the historian Michael Kammen has observed, the professions in England were not as highly developed as we have assumed, "nor were they so primitive in provincial America as some had suspected."[114]

NOTES

I wish to thank Professor Harold J. Cook, Department of the History of Medicine, University of Wisconsin–Madison, for reading a draft of this essay and making valuable critical remarks. His book *The Decline of the Old Medical Regime in Stuart London* (Ithaca, N.Y.: Cornell Univ. Press, 1986) is a major contribution to our understanding of 17th-century medicine in England. I also wish to express appreciation to Darlene Mickey and Dottie Leathers for preparing the typescript.

1 In particular, Daniel J. Boorstin, *The Americans: The Colonial Experience* (New York: Vintage, 1958). A brief, penetrating analysis of the historiography of such comparative efforts is Hugh Kearney, "The problem of perspective in the history of colonial America," in K. R. Andrews, N. P. Canny, and P. E. H. Hair, eds., *The Westward Enterprise: English Activities in Ireland, the Atlantic, and America, 1480–1650* (Detroit: Wayne State Univ. Press, 1979), pp. 290–302. The quotation is from John Perkins, "Memoirs of the life writings and opinions of John Perkins physician lately of Boston. begun March. 1777. and continued to 1778," manuscript memoirs, American Antiquarian Society, Worcester, Mass. (with permission of the Society), p. 2.

2 D. B. Quinn, *England and the Discovery of America, 1481–1620* (New York: Knopf, 1974).

3 E. A. Wrigley and R. S. Schofield, *The Population History of England, 1541–1871* (Cambridge, Mass.: Harvard Univ. Press, 1981).

4 H. C. Darby, ed., *A New Historical Geography of England* (Cambridge: Cambridge Univ. Press, 1973), pp. 293–298, 381, 459.

5 Thomas H. Breen and Stephen Foster, "The way to the New World: the character of early Massachusetts immigration," *William and Mary Quarterly*, 3rd ser., 1973, *30*: 189–222.

6 Ernest Caulfield, "Some common diseases of colonial children," Colonial Society of Massachusetts, *Transactions*, 1942–1946, *25*: 4–65, quotation on p. 36.

7 Charles Creighton, *A History of Epidemics in Great Britain*, 2 vols., 2nd ed. (New York: Barnes and Noble, 1965), I: 646–692; William H. McNeill, *Plagues and Peoples* (Garden City, N.Y.: Anchor Books, 1976), p. 152; Frederick W. Cartwright, *Disease and History* (New York: Crowell, 1972), pp. 121–122, 132–133.

8 Quoted in Owen H. and Sara D. Wangensteen, *The Rise of Surgery from Empiric Craft to Scientific Discipline* (Minneapolis: Univ. of Minnesota Press, 1978), p. 65.

9 Cartwright, *Disease and History*, p. 132; and John Duffy, *Epidemics in Colonial America* (Baton Rouge: Louisiana State Univ. Press, 1953), p. 165.

10 J. Worth Estes, *Hall Jackson and the Purple Foxglove: Medical Practice and Research in Revolutionary America, 1760–1820* (Hanover: Univ. Press of New England, 1979), p. 131.

11 On medical institutions, practices, and personnel, see, e.g., Sir George Clark, *A History of the Royal College of Physicians of London* (Oxford: Clarendon Press, 1964); Zachary Cope, *The Royal College of Surgeons of England* (Springfield, Ill.: Charles C. Thomas, 1959); Josephine A. Dolan, *History of Nursing*, 12th ed. (Philadelphia: W. B. Saunders, 1969); F. N. L. Poynter, ed., *The Evolution of Medical Practice in Britain* (London: Pitman Medical Publishing Co., 1961); John L. Thornton, ed., *James R. Aveling's English Midwives, 1872* (London: Hugh K. Elliot Ltd., 1967); and E. Ashworth Underwood, ed., *A History of the Worshipful Society of Apothecaries of London*, 2 vols. (London: Oxford Univ. Press, 1963), I: 1617–1815.

12 Quoted in John A. Raach, "Five early seventeenth century English country physicians," *J. Hist. Med.*, 1965, *20*: 213.

13 Margaret Pelling, "Occupational diversity: barber-surgeons and the trades of Norwich, 1550–1640, *Bull. Hist. Med.*, 1982, *56*: 484–511, esp. p. 489; and Pelling and Charles Webster, "Medical practitioners," in Charles Webster, ed., *Health, Medicine and Mortality in the Sixteenth Century* (Cambridge: Cambridge Univ. Press, 1979), pp. 165–235, esp. pp. 165–168.

14 Surgical developments for the period are discussed in William N. Boog Watson, "Four monopolies and the surgeons of London and Edinburgh," *J. Hist. Med.*, 1970, *25*: 311–322; Lloyd G. Stevenson, "A note on the relation of military service to licensing in the history of British surgery," *Bull. Hist. Med.*, 1953, *27*: 420–427; Richard Hardaway Meade, *An Introduction to the History of General Surgery* (Philadelphia: W. B. Saunders, 1968); and Wangensteen and Wangensteen, *The Rise of Surgery*.

15 Pelling and Webster, "Medical practitioners," pp. 182–184.

16 The changing prospects for apothecaries and for other types of practitioners in urban and rural areas are discussed in Sir Humphrey Rolleston, "History of medicine in the City of London," *Ann. Med. Hist.*, 1941, *3*: 1–17; Pelling, "Occupational diversity"; Pelling and Webster, "Medical practitioners"; John R. Guy, "The episcopal licensing of physicians, surgeons and midwives," *Bull. Hist. Med.*, 1982, *56*: 528–542; Thomas R. Forbes, "Apprentices in trouble: the training of surgeons and apothecaries," *Yale J. Biol. & Med.*, 1979, *52*: 227–237; J. J. Keevil, "The seventeenth century English medical background," *Bull. Hist. Med.*, 1957, *31*: 408–24;

and R. S. Roberts, "The personnel and practice of medicine in Tudor and Stuart England," *Med. Hist.*, 1962, *6*: 363–382; and 1964, *8*: 217–234.

17 Lester S. King, *The Medical World of the Eighteenth Century* (Chicago: Univ. of Chicago Press, 1958), pp. 18–29; and Rolleston, "Medicine in London," pp. 4–6.

18 See, e.g., A. J. Rook, "Cambridge medical students at Leyden," *Med. Hist.*, 1973, *17*: 256–265; and R. W. Innes-Smith, *English-Speaking Students of Medicine at the University of Leyden* (Edinburgh: Oliver and Boyd, 1932).

19 Gweneth Whitteridge and Veronica Stokes, *A Brief History of the Hospital of Saint Bartholomew* (London: Brown, Knight & Truscott Ltd., 1961); Pelling and Webster, "Medical practitioners," pp. 180–181; and Susan C. Lawrence, "'Desirous of improvements in medicine': pupils and practitioners in the medical societies at Guy's and St. Bartholomew's hospitals, 1795–1815," *Bull. Hist. Med.*, 1985, *59*: 89–104.

20 Pelling, "Occupational diversity," pp. 489–490; and Pelling and Webster, "Medical practitioners," pp. 165, 233–235.

21 Joseph F. Kett, "Provincial medical practice in England, 1730–1815," *J. Hist. Med.*, 1964, *19*: 17–29.

22 *Ibid.*, p. 27.

23 Guy, "Episcopal licensing," pp. 529–537.

24 Kett, "Provincial practice," pp. 26–29; and Ivan Waddington, "The struggle to reform the Royal College of Physicians, 1767–1771: a sociological analysis," *Med. Hist.*, 1973, *17*: 107–126.

25 Pelling and Webster, "Medical practitioners," pp. 225–227.

26 *Ibid.*, p. 215.

27 Pelling, "Occupational diversity," p. 508; and Leslie G. Matthews, "Licensed mountebanks in Britain," *J. Hist. Med.*, 1964, *19*: 30–45.

28 Pelling, "Occupational diversity," pp. 498, 507–508.

29 Pelling and Webster, "Medical practitioners," p. 188.

30 The data in this section are derived from John A. Raach, *A Directory of English Country Physicians, 1603–1643* (London: Dawson's of Pall Mall, 1962).

31 Pelling and Webster, "Medical practitioners," p. 232.

32 *Ibid.*, p. 235. Later, in "Occupational diversity," p. 508, Pelling would increase the ratio to 1:200, suggesting also that the "true ratio may well be higher."

33 The basic sources here are A. Batty Shaw, "The oldest medical societies in Great Britain," *Med. Hist.*, 1968, *12*: 232–244, and Lawrence, "Desirous of improvements."

34 Lawrence, "Desirous of improvements," p. 91 n. 12.

35 *Ibid.*, p. 89.

36 *Ibid.*, p. 101.

37   Walter Radcliffe, "The Colchester Medical Society, 1774," *Med. Hist.*, 1976, *20*: 394–401.

38   Owsei Temkin, *Galenism: Rise and Decline of a Medical Philosophy* (Ithaca, N.Y.: Cornell Univ. Press, 1973), and Wesley D. Smith, *The Hippocratic Tradition* (Ithaca, N.Y.: Cornell Univ. Press, 1979).

39   Stanley Joel Reiser, *Medicine and the Reign of Technology* (Cambridge: Cambridge Univ. Press, 1978), pp. 1–22.

40   Walter Pagel, *Paracelsus: An Introduction to Philosophical Medicine in the Era of the Renaissance* (New York: S. Karger, 1958); and Webster, "Alchemical and Paracelsian medicine," in *Health, Medicine, and Mortality*, pp. 301–334.

41   Reiser, *Reign of Technology*, pp. 8–10; and Lester S. King, *The Growth of Medical Thought* (Chicago: Univ. of Chicago Press, 1963; Midway Reprint, 1973), p. 19.

42   Raach, "Country physicians," p. 218.

43   Because of the great variety of medical assistance available in London and in provincial cities and towns, no definitive generalization can be made at this time (Pelling and Webster, "Medical practitioners," pp. 233–234). Major surgical intervention may have been an exception. Patients requiring extensive surgical treatment may have been encouraged to avoid the traveling lithotomist and to seek instead the practitioners in larger towns, like Norwich. The emergence of one provincial town as a regional center for risky operations is detailed in A. Batty Shaw, "The Norwich School of Lithotomy," *Med. Hist.*, 1970, *14*: 228–259.

44   Nicholas Canny, "The permissive frontier: the problem of social control in English settlements in Ireland and Virginia, 1550–1650," pp. 17–44, and Karl S. Bottigheimer, "Kingdom and colony: Ireland in the westward enterprise, 1536–1660," pp. 45–64, in Andrews, Canny, and Hair, eds., *Westward Enterprise*.

45   Evarts B. Greene and Virginia D. Harrington, *American Population before the Federal Census of 1790* (New York: Columbia Univ. Press, 1932); and U.S. Bureau of the Census, *Historical Statistics of the United States: Colonial Times to 1957* (Washington, D.C.: U.S. Government Printing Office, 1961).

46   Ralph J. Crandall, "New England's second Great Migration: the first three generations of settlement, 1630–1700," *The New England Historical and Genealogical Register*, 1975, *129*: 347–360; and Breen and Foster, "Early Massachusetts immigration," pp. 189–222.

47   Darrett B. Rutman, *Winthrop's Boston: A Portrait of a Puritan Town* (Chapel Hill: Univ. of North Carolina Press, 1965); Edmund S. Morgan, *The Puritan Dilemma: The Story of John Winthrop* (Boston: Little, Brown, 1958); and Crandall, "Second Great Migration," p. 359.

48   Morgan, *Puritan Dilemma*, pp. 120–133; and Sumner Chilton Powell, *A Puritan Village* (Middleton: Wesleyan Univ. Press, 1963).

49   Eric H. Christianson, "The emergence of medical communities in Massachusetts, 1700–1794: the demographic factors," *Bull. Hist. Med.*, 1980, *54*: 64–77, esp. p. 66.

50   John Demos, *A Little Commonwealth: Family Life in Plymouth Colony* (New York: Oxford Univ. Press, 1970); and Philip J. Greven, *Four Generations: Population, Land, and Family in Colonial Andover, Massachusetts* (New York: Norton, 1970).

51   This section is based on Duffy, *Colonial Epidemics*; Caulfield, "Common diseases"; John B. Blake, *Public Health in the Town of Boston, 1630–1822* (Cambridge, Mass.: Harvard Univ. Press, 1959); James H. Cassedy, "Meteorology and medicine in colonial America: beginnings of the experimental approach," *J. Hist. Med.*, 1969, *24*: 193–204; James H. Cassedy, "Church record-keeping and public health in early New England," in Philip Cash, Eric H. Christianson, and J. Worth Estes, eds., *Medicine in Colonial Massachusetts, 1620–1820*, Publications of the Colonial Society of Massachusetts, v. 57 (distributed by the Univ. Press of Virginia, 1980), pp. 249–262; Rose S. Lockwood, "Birth, illness and death in eighteenth-century New England," *J. Soc. Hist.*, 1978, *12*: 111–128; and Otho T. Beall, Jr., and Richard H. Shryock, *Cotton Mather: First Significant Figure in American Medicine* (Baltimore: Johns Hopkins Press, 1954).

52   Medical practitioners and community leaders were often the same individuals, whose medical advice reflected local political alignments. Dennis Don Melchert, "Experimenting on the neighbors: inoculation of smallpox in Boston in the context of eighteenth century medicine" (unpubl. Ph.D. diss., Univ. of Iowa, 1973).

53   For purposes of comparison, Whitfield J. Bell, Jr., "A portrait of the colonial physician," *Bull. Hist. Med.*, 1970, *44*: 497–517; Wyndham B. Blanton, *Medicine in Virginia in the Eighteenth Century* (Richmond: Garrett and Massie, 1931); David L. Cowan, *Medicine and Health in New Jersey: A History* (New Brunswick, N.J.: Rutgers Univ. Press, 1964); Maurice Bear Gordon, *Aesculapius Comes to the Colonies: The Story of the Early Days of Medicine in the Thirteen Original Colonies* (Ventor, N.J.: Ventor Publishers, 1949); Jonathan Harris, "The rise of medical science in New York, 1720–1820" (unpubl. Ph.D. diss., New York Univ., 1971); Joseph F. Kett, *The*

Formation of the American Medical Profession: The Role of Institutions, 1780–1860 (New Haven, Conn.: Yale Univ. Press, 1968); William Frederick Norwood, *Medical Education in the United States before the Civil War* (Philadelphia: Univ. of Pennsylvania Press, 1944); Francis R. Packard, *The History of Medicine in the United States*, 2 vols. (New York: Hafner, 1963); Byron Stookey, *A History of Colonial Medical Education in the Province of New York, with Its Subsequent Development, 1767–1830* (Springfield, Ill.: Charles C. Thomas, 1962); Joseph M. Toner, *Contributions to the Annals of Medical Progress and Medical Education in the United States before and during the War of Independence* (Washington, D.C.: U.S. Government Printing Office, 1874); and Joseph Ivor Waring, *A History of Medicine in South Carolina, 1670–1825* (Charleston: South Carolina Medical Association, 1964).

54 Christianson, "The medical practitioners of Massachusetts 1630–1800: patterns of change and continuity," in Cash, Christianson, and Estes, eds., *Medicine in Colonial Massachusetts*, pp. 49–67, esp. p. 57; and C. H. Brock and Eric H. Christianson, "Appendix: a biographical register of men and women from and immigrants to Massachusetts between 1620 and 1800 who received some medical training in Europe," in Cash, Christianson, and Estes, eds., *Medicine in Colonial Massachusetts*, pp. 117–143.

55 Christianson, "Medical practitioners," pp. 52–54, 65–66; Samuel A. Green, *History of Medicine in Massachusetts* (Boston: A. Williams and Company, 1881), p. 15; and Henry R. Viets, *A Brief History of Medicine in Massachusetts* (Boston: Houghton Mifflin, 1930), pp. 11–12.

56 The information about these four men is from my unpublished manuscript "The tachygraphy of Dr. Jasper Gunn (1606–1671)"; and Malcolm Sydney Beinfield, "The early New England doctor: an adaptation to a provincial environment," *Yale J. Biol. & Med.*, 1942–1943, *15*: 99–132, 271–288.

57 Kett, "Provincial England."

58 The act is quoted in Richard H. Shryock, *Medical Licensing in America, 1650–1965* (Baltimore: Johns Hopkins Press, 1967), p. vii.

59 Christianson, "Medical practitioners," pp. 52–54. On the underestimation of practitioners by earlier studies, see Christianson, "The historiography of early American medicine," in Cash, Christianson, and Estes, eds., *Medicine in Colonial Massachusetts*, pp. 20–25. The estimations in the present study are derived primarily from a systematic analysis of the vital records (births, deaths, and marriages), and histories of some 300 Massachusetts towns and eleven counties. Women practitioners and midwives were not generally reported in vital records;

studies of court records may identify many of them, and nurses as well. Future research will undoubtedly identify more practitioners that will produce higher ratios.

60 *Ibid.*, pp. 54–55.

61 Toner, *Annals of Progress*, pp. 105–106; and Shryock, *Medicine and Society*, p. 12.

62 Kett, *Institutions*, p. 170, and Christianson, "Medical communities," pp. 69–70.

63 The material on the Bulfinches is based on Edward Jacob Forster, "A sketch of the medical profession of Suffolk County," in *The Professional and Industrial History of Suffolk County, Massachusetts in Three Volumes* (Boston: Boston History Company, 1881), III: 259; Francis R. Packard, "Cheselden's American pupils," *Ann. Med. Hist.*, 1937, *9*: 533–548, esp. p. 536; Edgar M. Bick, "French influences in early American medicine and surgery," *Journal of the Mt. Sinai Hospital*, 1957, *24*: 499–509; Rolleston, "Medicine in London," pp. 11–16; Charles Coury, "L'Hôtel Dieu de Paris, un des plus anciens hopitaux d'Europe," *Medizin historisches Journal*, 1967, *2*: 269–316; Clifford K. Shipton, *Sibley's Harvard Graduates . . .*, 17 vols. to date (Boston: Massachusetts Historical Society, 1933–), XII: 16–23; Dr. William Cullen's manuscript list of "Students in the College of Chemistry, 1755–1765" (Univ. of Edinburgh Library); Lewis, "American graduates"; and Christianson, "Individuals in the healing arts and the emergence of a medical community in Massachusetts, 1700–1792: A collective biography" (unpubl. Ph.D. diss., Univ. of Southern California, 1976), pp. 84–85.

64 Whitfield J. Bell, Jr., "Medicine in Boston and Philadelphia: comparisons and contrasts, 1750–1820," and "Appendix: New England students at the Pennsylvania Hospital from the Revolution to 1820," in Cash, Christianson, and Estes, eds., *Medicine in Colonial Massachusetts*, pp. 159–183.

65 Brock and Christianson, "Biographical register"; and Christianson, "Medical practitioners," pp. 56–57.

66 Lewis, "American graduates," pp. 159–165.

67 William Pepper, *The Medical Side of Benjamin Franklin* (Philadelphia: W. J. Campbell, 1911), pp. 41–44. Like John Perkins, Waterhouse did not find much difference in the practice of general medicine in Europe, England, or by then, the U.S.

68 Christianson, "Medical practitioners," p. 57.

69 BMS misc. papers, with the permission of the Francis A. Countway Library of Medicine, Boston. This agreement is also quoted in Henry R. Viets, "The medical education of James Lloyd in colonial America," *Yale J. Biol. & Med.*, 1958, *31*: 1–13, esp.

pp. 6–7. Other agreements are cited and discussed in Genevieve Miller, "Medical apprenticeship in the American colonies," *Ciba Symposia,* 1947, *8*: 502–510. For a discussion of medical pupils who paid a flat fee rather than served as apprentices, see Christianson, "Medical practitioners," p. 52 n. 8.

70 Perkins, "Life writings and opinions," pp. 2, 3, 71. He studied with Dr. William Davis in 1718 and Dr. Francis Archibald in 1721, both of Boston.

71 Christianson, "Medical communities," p. 69, and Christianson, "Individuals in the healing arts," pp. 114–116.

72 This section is based on Christianson, "Individuals in the healing arts," pp. 85–116; Christianson, "Medical communities," p. 69; and Christianson, "Medical practitioners," pp. 57–58.

73 Edward J. Young, "Subjects for master's degrees in Harvard College from 1655 to 1791," *Proceedings, Massachusetts Historical Society,* 1880, *18*: 119–151; and Christianson, "Individuals in the healing arts," pp. 88–92.

74 This section is based on my unpublished manuscript "'In Case he inclines to follow a Doctor's Calling': the role of the medical family in the training of early American doctors"; and Christianson, "Individuals in the healing arts," p. 73.

75 Radcliffe, "Colchester Medical Society"; Shaw, "Oldest medical societies"; and Christianson, "Individuals in the healing arts," pp. 155–188.

76 Uncatalogued Silvester Gardiner Papers, courtesy of the Francis A. Countway Library of Medicine, Boston; and Eric H. Christianson, "The colonial surgeon's rise to prominence: Dr. Silvester Gardiner (1707–1786), and the practice of lithotomy in New England," *The New England Historical and Genealogical Register,* 1982, *136*: 104–114.

77 The standard account is George H. Weaver, "Life and writings of William Douglass, M.D. (1691–1752)," *Bull. Soc. Med. Hist. Chicago,* 1921, *9*: 229–259.

78 *The Boston Weekly News-Letter,* Dec. 29, 1737–Jan. 5, 1738.

79 G. B. Warden, "The medical profession in colonial Boston," in Cash, Christianson, and Estes, eds., *Medicine in Colonial Massachusetts,* pp. 145–159, argues that without a legal monopoly, medical societies were hopeless as agencies of control.

80 "An elegy on the death of the late Dr. Ames," in Nathaniel Ames, *An Astronomical Diary; or Almanack, for . . . 1765* (Boston: Draper, Edes & Gill, Green & Russell, and Fleet, 1764), n.p.

81 Quoted in Walter A. Burrage, *A History of the Massachusetts Medical Society with Brief Biographies of the Founders and Chief Officers, 1781–1922* (privately printed, 1923), pp. 3–7.

82 Christianson, "Individuals in the healing arts," pp. 168–173.

83 Shipton, *Sibley's Harvard Graduates,* XV: 3–15; Sarah Breck Baker, "Extracts from the Ames Diary," *Dedham Historical Register,* 1890–1903, *1*: 9, *2*: 24, 59, 60, 148, 150; John Morgan, *A Discourse upon the Institution of Medical Schools in America* (Philadelphia, 1765), p. 24. For a discussion of Morgan's ideas, see Toby Gelfand, "The origins of a modern concept of medical specialization: John Morgan's *Discourse* of 1765," *Bull. Hist. Med.,* 1976, *50*: 511–535.

84 See, e.g., Philip Cash, *Medical Men at the Siege of Boston, April, 1775–April, 1776* (Philadelphia: American Philosophical Society, 1973).

85 Quoted in W. B. McDaniel, II, "A letter from Dr. Williams Smibert of Boston, to his former fellow-student at Edinburgh, Dr. John Morgan, of Philadelphia, written February 14, 1769," *Ann. Med. Hist.,* 1939, *1*: 194–196.

86 Burrage, *Massachusetts Medical Society,* pp. 16, 68–84, 323–349.

87 *Ibid.,* pp. 181–182.

88 Christianson, "Medical communities," p. 75.

89 Manuscript Records of the Berkshire District Medical Society, 1785–1864, Berkshire Athenaeum, Pittsfield, p. 4.

90 Nathaniel Walker Appleton to Edward A. Holyoke, 9 June 1782, Holyoke Family Papers, MSS49, Essex Institute, Box 16, Folder 2.

91 Edward A. Holyoke to Nathaniel Walker Appleton, 18 April 1789, *ibid.,* Folder 3.

92 Burrage, *Massachusetts Medical Society,* chs. 2, 3, 10; Kett, *Institutions,* 15; and Thomas F. Harrington, *The Harvard Medical School: A History, Narrative and Documentary,* 3 vols. (New York: Lewis, 1905), I.

93 Ebenezer Hunt to Dr. Samuel Danforth, 13 Oct. 1789 (BMS b. 75.1 F76); and Israel Atherton to Nathaniel Appleton Walker, 20 Oct. 1789 (BMS b. 75.1 F723), courtesy of the Francis A. Countway Library of Medicine, Boston.

94 Bell, "Medicine in Boston and Philadelphia," pp. 163–164.

95 Christianson, "Medical practitioners," p. 52; and Douglas Lamar Jones, "Charity, medical charity, and dependency in eighteenth-century Essex County, Massachusetts," in Cash, Christianson, and Estes, *Medicine in Massachusetts,* pp. 199–213. Other informative works discussing medical charity are Shryock, *Medicine and Society,* pp. 104–107; and Shaw, "The Norwich School of Lithotomy," pp. 223–226.

96 Shipton, *Sibley's Harvard Graduates,* XIII: 380–389.

97 *Ibid.,* XII: 512–527; and John H. Cary, *Joseph Warren: Physician, Politician, Patriot* (Urbana: Univ. of Illinois Press, 1961), p. 31.

98 Shipton, *Sibley's Harvard Graduates*, XV: 3–15.

99 Christianson, "Medical communities," p. 70; and Christianson, "Individuals in the healing arts," pp. 118, 131, 151.

100 George E. Gifford, Jr., "Botanic remedies in colonial Massachusetts, 1620–1820," in Cash, Christianson, and Estes, eds., *Medicine in Colonial Massachusetts*, pp. 263–288; Wayland D. Hand, *Magical Medicine: The Folkloric Component of Medicine in the Folk Belief, Customs, and Rituals of the Peoples of Europe and America* (Berkeley: Univ. of California Press, 1980); Guenter B. Risse, Ronald L. Numbers, and Judith Walzer Leavitt, eds., *Medicine without Doctors: Home Health Care in American History* (New York: Science History Publications, 1977); Otho T. Beall, Jr., "Aristotle's MASTERPIECE in America: a landmark in the folklore of medicine," *William and Mary Quarterly*, 3rd ser., 1963, *20*: 207–222; and Christianson, "The description and treatment of the fly-blown ear: an aspect of science and medicine in colonial America," *The Melsheimer Entomological Series*, 1979, *26*: 1–12.

101 Few would doubt that, from the earliest settlements, local ministers also engaged in medical activities in the absence of trained medical personnel.

102 Green, *Massachusetts*, pp. 2–32; Viets, *Brief History*, pp. 14–40; and Jones, "Charity," pp. 207–208.

103 Green, *Massachusetts*, p. 55.

104 Christianson, "Lithotomy in New England," p. 107.

105 Christianson, "Individuals in the healing arts," pp. 143–145, 150.

106 Christianson, "Lithotomy in New England," pp. 106, 108, 110.

107 Douglass to Dr. Cadwallader Colden, 20 Feb. 1720/1, in "Letters from Dr. William Douglass to Cadwallader Colden of New York," *Collections, Massachusetts Historical Society*, 1854, 2: 164–189, esp. pp. 165–166.

108 The quotations from Douglass are found in Duffy, *Epidemics*, p. 8. Douglass' description is endorsed by Shryock, *Medicine and Society*, pp. 17, 19; Bell, "A portrait of the colonial physician;" and Malcolm Sydney Beinfield, "The early New England doctor."

109 Estes, *Hall Jackson;* and J. Worth Estes, "Therapeutic practice in colonial New England," in Cash, Christianson, and Estes, eds., *Medicine in Colonial Massachusetts*, pp. 289–383. Dr. Estes has extended this uniformity of practice into the 19th century in his article "Naval medicine in the Age of Sail: the voyage of the *New York*, 1802–1803," *Bull. Hist. Med.*, 1982, *56*: 238–253.

110 Estes, "Therapeutic practice," pp. 338, 344.

111 *Ibid.*, pp. 344–345.

112 Quoted in John A. Goodwin, *The Pilgrim Republic* (Boston: Houghton Mifflin, 1920), p. 581.

113 William Wood, *New England's Prospect . . .* , 3rd ed. (London, 1639; Boston: Fleet, Green & Russell, 1764), p. 15. These observations have been confirmed by J. I. Falconer, *History of Agriculture in the Northern United States, 1620–1860* (1925; rpt. Clifton, N.J.: Augustus M. Kelley Publishers, 1973), pp. 16, 99.

114 Michael Kammen, *People of Paradox: An Inquiry Concerning the Origins of American Civilization* (New York: Vintage, 1973), pp. 28–29.

4

Martha Moore Ballard and the
Medical Challenge to Midwifery

LAUREL THATCHER ULRICH

Some time in the afternoon or evening of October 9, 1794, David Sewall of Hallowell sent word to Martha Ballard that his wife was "unwell." There was nothing unusual in the call or in the casual entry in Mrs. Ballard's diary, "I was there all night." In her 16 years as a midwife, Martha Ballard had lost many nights' sleep attending a woman in labor. This birth would be unusual, however. On October 10, in Mrs. Sewall's chamber, a minor drama in the history of American midwifery unfolded. "They were intimidated," wrote Mrs. Ballard, "& Calld Dr. Page who gave my patient 20 drops of Laudanum which put her into such a stupor her pains (which were regular & promising) in a manner stopt till near night when she pukt & they returned & shee was delivered at 7 hour Evening of a son her first Born."[1]

The three elements in this story—the patient's "intimidation," the doctor's employment of laudanum, and the midwife's annoyance—fit perfectly into the larger history of childbearing in late 18th-century America. Recent studies have concluded that the transition from traditional midwifery to medical obstetrics began in the northern United States between 1760 and 1820 and that it was a consequence both of new medical technology and of changes in the attitudes of women. Midwives had always been taught to call doctors in medical emergencies, but beginning in the 1760s in urban cen-

ters like Philadelphia, New York, and Boston, doctors trained in Edinburgh and London began officiating at normal births, employing forceps to hasten delivery and administering opiates to relieve pain. "By the second decade of the nineteenth century," Jane Donegan has concluded, "northern urban women of means, convinced that the superior training of doctors equipped with instruments meant safer and shorter parturition, hardly ever employed a midwife."[2]

Although traditional midwifery would persist among immigrants and in isolated rural communities to the end of the 19th century, the physicians who founded America's first medical societies and colleges were successful in associating medical delivery with scientific progress. By 1820, an influential Boston physician could pronounce the exclusion of women from the practice of obstetrics one of "the first and happiest fruits of improved medical education in America." He argued that a revival of female practice would threaten the safety of Boston's women. "Heretofore, where midwifery has been in the hands of women they have practiced among the poorer and lower classes," he wrote, "the richer and better informed preferring to employ physicians."[3]

Women's historians have argued that male doctors promoted "science" at serious cost to women. Midwives were not only deprived of their occupation but were also shut out of the new medical education. Childbearing women were also hurt as birth became a medical event to be managed by interventionist attendants. Giving up the comforting circle of female support, mothers faced an increasing threat of infection and the possibility of overdoses of anesthesia or damage from forceps, as medical education failed to keep pace with the

LAUREL THATCHER ULRICH is James Duncan Phillips Professor of Early American History, and Professor of Women's Studies, Harvard University, Cambridge, Massachusetts.

Reprinted with permission from *Maine in the Early Republic: From Revolution to Statehood*, edited by Charles E. Clark, James S. Leamon, and Karen Bowden, © 1988 University Press of New England, pp. 165–183.

expansion of the profession. "Only after 1940," Judith Walzer Leavitt has argued, "did medicine begin to achieve a record of safety commensurate with the promises it had held out to women centuries earlier."[4]

Martha Ballard's account of Hannah Sewall's delivery seems to summarize, then, the key themes in the transformation of childbirth in the late 18th century: A traditional midwife patiently waiting for the operations of nature was upstaged by an interventionist physician whose opiates promised a frightened young woman relief from pain. A closer look at the participants in this drama reinforces that impression. At 59, Martha Ballard came to Mrs. Sewall's chamber with the authority of her own motherhood as well as with the specialized skills she had acquired in her 16 years as a midwife. Benjamin Page, on the other hand, was barely 24, still unmarried, fresh from his apprenticeship with Dr. Thomas Kittredge of Andover, Massachusetts. Yet he brought the promise of the new "scientific" obstetrics. In Dr. Kittredge's library he had probably read the works of Dr. William Smellie, the British physician who popularized the new medical obstetrics. Under Kittredge's supervision he may have used the improved forceps Smellie designed.[5]

Hannah Sewall, too, fits the model. Though only 20, she might well have classified herself among the "richer and better informed" of Hallowell women. Married less than a year, she had come to the Kennebec valley from the coastal town of York where her father, Nathaniel Barrell, and her grandfather, Jonathan Sayard, were prominent merchants. Her older sister, Sally (later known as "Madame Wood"), became Maine's first novelist. Although another sister, Ruth, the wife of Moses Sewall of Hallowell, had already delivered four children with the assistance of Martha Ballard, Hannah may have come to the town sympathetic to doctors if not already predisposed toward medical delivery. Another sister, Olive, had recently married Dr. Samuel Emerson of Kennebunk; her younger brother later married the daughter of Dr. Job Lyman, whose obstetrical forceps are now owned by the Old York Historical Society.[6]

All the elements characterizing the transition to medical obstetrics seem to have been in place by 1794 on the Maine frontier. Yet Hannah Sewall's delivery was only one among 48 that Martha Ballard performed in that year, and only one of the 797 deliveries she recorded in her diary between January 1785 and May 1812.[7] Seen in this context, Dr. Page's intrusion was a minor annoyance, a bungling effort set right by nature. There was no question in Martha Ballard's mind that she knew more about delivering babies than Ben Page, whom she once described as "that poor unfortunate man in the practice."[8] Her practice was increasing, not decreasing, in 1794. Only after 1800, when age and frequent illness undermined her strength and when a move to a new farm took her farther away from the river and from major roads, did she sharply curtail her practice.

Martha Ballard's diary gives us a surer sense of the differences between "male" and "female" obstetrics than earlier studies based largely on prescriptive literature and on the lives of medical leaders. In late 18th-century Hallowell, doctors were not as scientific or midwives as ignorant as older accounts would suggest, nor was there as much friction between the two specialties as more recent literature might lead us to believe. For the most part, medical obstetrics and midwifery coexisted peacefully. Yet there were important differences between men and women practitioners. That doctors were both better paid and less experienced than midwives created a troublesome discrepancy between the promise and the practice of the new medical obstetrics. Ironically, it was not "science" that undermined traditional midwifery but a new appreciation by doctors of the "ordinary" births that had long been managed by women.

Born in Oxford, Massachusetts, in 1735, Martha Moore Ballard emigrated to the Kennebec River country in 1777 with her husband, Ephraim, and five children. The Ballards spent the rest of their lives in Maine, living in four different locations in what would later be the towns of Hallowell and Augusta. Ephraim Ballard was a miller and surveyor. Although by no means a wealthy man, he was a respected citizen who served for a time as selectman of Hallowell.[9] Martha Ballard was a dutiful and productive housewife, who raised pigs and poultry, cultivated a large garden, and produced both woolen and linen cloth in addition to practicing midwifery and physic. There is no indication that she had anything more than a primary education, though her younger brother, Jonathan, graduated from Harvard College in 1761 and eventually became the minister of the First Church of Rochester,

Massachusetts.[10] She later noted that she delivered her first baby in 1778, just after coming to Maine. Although she may have kept some sort of record of deliveries from that point on, her diary begins in 1785, the year she turned 50. It closes in May 1812, a month before her death at the age of 77.[11]

The focus of this chapter is on the first 15 years of the diary, 1785–1800, the period of Martha Ballard's most active years as a midwife. In 1785, Hallowell was still a frontier town, but it was growing rapidly. In 1797, the section of the town near Fort Western separated from Hallowell, becoming the town of Augusta. The population of Hallowell had been 1,199 in 1790; by 1800, the combined population of the two towns was 2,680.[12]

The categories *doctor* and *midwife* inadequately describe the diversity of medical practice in this 18th-century town. Hallowell's "doctors" ranged from the eminently visible Dr. Daniel Cony, Justice of the Peace, Representative to the General Court, and founding President of the Kennebec Medical Association, to the anonymous "negro woman doctor" who appears two or three times in the diary. Not until after 1796, when Benjamin Vaughan arrived in Hallowell with an M.D. from Edinburgh, did the area have a college-trained physician.[13] Although most male physicians had served apprenticeships, it would be a serious mistake to read back into this period 20th-century notions of "medical science."

Doctor Cony, for example, had trained with Dr. Samuel Curtis of Marlborough, Massachusetts, a Harvard graduate, who later practiced in New Hampshire where he published "A Valuable Collection of Recipes" that included, in addition to homespun medical remedies, formulas for removing grease spots, bedbugs, and fleas, and for bleaching straw, mending china, and making red hair black.[14] His work is a striking example of the practicality and disdain for theory that characterized New England medicine, even of the "professional" variety, in the late 18th century.[15] In his *American Medical Biography*, published in 1828, Thomas Thacher praised Benjamin Page for his treatment of spotted fever between 1810 and 1816 and acknowledged the "Moral rectitude and public virtue" of Daniel Cony, but he concluded that the District of Maine "possessed little claim to the merit of contributing to the improvement of medical science." Clearly, though there were the beginnings of medical organization in Hallowell before 1800 (Dr. Cony was a member of the Massachusetts Medical Society as early as 1787), there was very little "science," even by contemporary standards.[16]

It is equally inappropriate to distinguish between midwives and doctors, as some writers do, by saying that one group has been concerned with a natural process (birth) and the other with a medical event (illness). Like most early American midwives, Martha Ballard not only delivered babies but also treated the sick.[17] She seems to have specialized in diseases of women and children, though she also treated grown men, especially those suffering from burns, rashes, or frostbite. Mrs. Ballard's patients came to her house seeking salves, pills, syrups, ointments, or simply advice. Forty to 70 times a year she went to them, spending a few hours or several days administering "clisters" (enemas), dressing burns, or bathing inflamed throats. She did not pull teeth, set bones, or let blood (though once she drew blood from a cat and applied it to a man who was suffering from shingles). She cut infant tongues, lanced abscessed breasts, and composed remedies for intestinal worms and the itch. She also dressed and laid out the dead. Martha Ballard was both a midwife and a doctor; significantly, her most frequent term for labor was "illness."[18]

The diary mentions six women—Mrs. Fletcher, Mrs. Hinkley, Mrs. Ingraham, Mrs. Clark, Mrs. Cox, and Mrs. Winslow—often enough or in such a context as to suggest that they also were midwives. Whether they practiced physic is difficult to say, although it is perfectly clear that several local physicians practiced obstetrics. Doctors Cony, Williams, and Hubbard, who were working in Hallowell or nearby towns by 1785, performed deliveries. In fact, Dr. Cony's only known literary contribution to the Massachusetts Medical Society, a one-page paper submitted in 1787, described "a circumstance which I had never before met with" in a delivery that he himself had performed in August of that year. Dr. Page, who arrived in Hallowell in 1791, and Dr. Parker of Pittston, who appears in the diary at about the same time, also delivered babies. Dr. Samuel Colman did not, nor did Dr. Steven Barton of Vassalboro, Martha Ballard's brother-in-law.[19]

Yet Martha Ballard's diary demonstrates that relations between midwives and physicians were more cooperative than competitive. Mrs. Ballard delivered Dr. Colman's children and even attended the doctor himself during an illness; she borrowed medicines from Daniel Cony and he from her; and on various occasions she summoned Drs. Hubbard, Williams, and Cony to her own patients. For the most part, her remedies seem to have been compatible with theirs. When her daughter, Hannah, became delirious about 10 days after delivering, Mrs. Ballard sent for Dr. Cony, who simply, as she said, "approved of what I had done—advised me to continue my medisin till it had opperation."[20] In general, her records support the conclusion of Richard Brown that "learned" and "folk" medicine in Massachusetts "were part of the same medical spectrum and overlapped considerably."[21]

After three men were injured when a fieldpiece misfired during militia training on May 31, 1792, Mrs. Ballard was summoned to bathe and dress the burns. Although Doctor Colman also attended the men during the next few days, it was Martha Ballard who "made an ointment" and applied it to Samuel Johnson's "soars" on June 2.[22] Some of the services Mrs. Ballard performed for the sick might today be defined as nursing, but in very few cases does she seem to have operated under a doctor's direction. She prepared her own medicines and apparently determined the best method of treatment. Furthermore, there are many references in the diary to "nurses," or specialists who seem to have been younger than midwives. Martha Ballard usually left her own patients about two hours after delivery in the care of an "after nurse."[23]

In addition to their joint commitment to what Martha Ballard would have called "pukes" and "purges," doctors and midwives had something else in common: the part-time nature of their work. As housewives, the midwives had heavy burdens at home including responsibilities for gardening, food processing, and animal care, but the doctors also divided their interests, none so strikingly as Dr. Steven Barton, who was a carpenter as well as a physician. In June 1774, for example, he charged Jonathan Ballard for "Visits & attendance" to his sick child and then for making a coffin.[24] Daniel Cony, Hallowell's most distinguished physician, was a judge, politician, agriculturist, and land speculator. His varied interests were typical of gentlemen physicians of the period. In 1808 he apologized to the President of the Massachusetts Medical Society for failing to convene a group of doctors in the District of Maine, explaining that "The dispersed situation of your Committee joined with their various avocations has prevented a meeting."[25]

Nor did Hallowell's practitioners claim a certain set of patients as their own. In troublesome cases, nearly everyone with any expertise—and some without—offered advice. When Martha Ballard's niece, Parthena Pitts, was suffering from a prolonged illness, she got up one morning about an hour after sunrise, as her aunt reported it, and "went out & milkt the last milk from the Cow into her mouth & swallowed it." This peculiar remedy had been "recommended as very Beneficial by Mr. Amos Page."[26] An 18th-century patient might summon more than one doctor or midwife at once and then employ whomever she chose. Mrs. Parker did not think it amiss to borrow Mrs. Ballard's horse "to go and see the negro woman doctor." Nor did a sense of loyalty to Dr. Colman, who had earlier treated him, prevent Calvin Edson from summoning Dr. Williams as well, or from applying to Martha Ballard for salve.[27] Some of this overlapping may have originated with the healers themselves—one practitioner might understandably consult with another—but even allowing for that, the territorial boundaries seem to have been very loose.

Perhaps the best evidence of Martha Ballard's cooperative relationship with male doctors is her attendance at autopsies. She observed at least three "dissections" between 1794 and 1801, carefully recording the results in her diary. Although none of these cases was obstetrical, two of the subjects were women and the third a small child. At Nabby Andrews' autopsy in September of 1800, Mrs. Ballard reported that 12 doctors and three midwives were present, though she did not give names.[28] That the women were there at all argues that medical practice in frontier Maine was still relatively open; neither professional exclusiveness nor womanly delicacy barred Hallowell's midwives from what was after all an important educational experience. That they were outnumbered four to one by male doctors testifies to the localized nature of their specialty as much as to their minority position in the medical world. The 12 doctors had no doubt gathered from miles around.

Thus Martha Ballard's diary modifies the picture of late 18th-century midwifery presented in secondary accounts. In Hallowell, relations between doctors and midwives were less antagonistic and the two specialties less separate than we might have

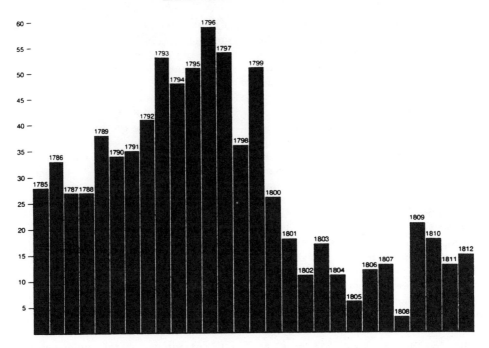

Fig. 1: Martha Ballard's obstetrical practice (1785–1812)—number of deliveries

supposed. What, then, of the encounter between Martha Ballard and Benjamin Page at the delivery of Hannah Sewall?

Though the Sewall incident was atypical, it does suggest broader differences between men and women practitioners during this period. Recall that in 1794, Benjamin Page was 24 years old, unmarried, fresh from his apprenticeship with Dr. Kittredge of Andover, Massachusetts. Martha Ballard was 59, a mother and grandmother many times over, and at the height of her obstetrical career. The contrast, though stark, was repeated in important ways in the medical community as a whole. Where midwives in 18th-century Hallowell claimed the authority of experience, physicians carried the aura of learning; where midwives depended on the support of a local community of women, doctors affiliated with a wider society of professionals; where midwives primarily turned to their own gardens and kitchens for remedies, doctors had recourse to instruments and to imported drugs with Latin designations.

We must not assume, however, that the practical training of midwives was necessarily less effective or their local reputations less powerful than the learning or the professional affiliations of the physicians. Martha Ballard's diary gives striking evidence to the contrary. By the time of her death in 1812, she had performed 981 deliveries, 797 of which are listed in the diary. The habit of counting deliveries seems to have been common to midwives. In June 1790, Mrs. James Marsh from nearby Vassalboro, called on Martha Moore Ballard. "The Old Lady informs me Shee has extracted 756 children in the coars of her practice," Mrs. Ballard wrote in her diary.[29] Martha Ballard's statistical habit makes it relatively easy to reconstruct her 35-year career. Retrospective entries show that she delivered about 30 babies a year between 1778 and 1785. Figure 1 shows yearly totals for the years 1785–1812. Note the expansion in her career between 1792 and 1800, the very years during which Benjamin Page was attempting to establish his practice.

At the end of the last entry for 1789, Martha Ballard tabulated total Hallowell births and deaths for the year 1785–1790. Although she does not give the source of these figures, they allow us to estimate the scope of her practice in relation to that of her competitors. If we assume that all the births recorded in the diary occurred in Hallowell unless marked otherwise, Mrs. Ballard delivered 64 percent of the babies born in the town during this six-year period.[30] There is no reason to suggest that her share of Hallowell births for the next 10 years was any less. Had a single practitioner delivered the 36 percent of Hallowell babies not accounted for in her records, Martha Ballard's eminence would still be unquestionable, yet these births were divided among five physicians and at least six other midwives, as the diary clearly shows.

Many of Martha Ballard's references to her competitors appear as "news" entries in her diary, that is, bits of information received by hearsay. After her daily entry for July 19, 1787, for instance, she turned her page perpendicularly and wrote: "Mrs. Church was dilivered of a son last Tusday marn at 1 Clock & 20 m Doct Coney operator." Below that she added, "Jerymy Baddoks wife the same Night of a dafter, old Mrs. Fletcher performed the ofice of a midwife for her." Other entries come from more direct experience, as on April 26, 1785, when she wrote, "I was Calld at 2 O'clok to Nathan Tylors wife in Travil found her delivrd of a Dafter Doct Williams operator." Apparently some women, afraid of being caught without any help at all, called more than one attendant.

In 32 of the 481 births to which Martha Ballard was called between 1785 and 1796, some other person performed the delivery, usually because she did not arrive in time. In 12 cases the other attendant was a midwife, in four a physician; in the remaining 16, the identity of the assistant is not given. This set of cases may give us a rough estimate of the proportion of deliveries performed by physicians in the period. If we apply the same ratio to all Hallowell births, we might conclude that doctors performed 9 percent of the deliveries in the town and other midwives the remaining 27 percent. Of course, it is entirely possible that those families who did not summon Mrs. Ballard may have called a physician more frequently than those who did summon her. Still, by any numerical measurement, midwifery prevailed over medical delivery in late 18th-century Hallowell.

Yet the seemingly casual appearances of Drs. Cony, Williams, Parker, and Hubbard at the bedsides of Hallowell women, like the more dramatic incursion of Benjamin Page, undermined a traditional distribution of responsibility. Midwives had

always been taught to summon a doctor in an emergency. Dismembering a dead fetus was a necessary skill for a surgeon; knowing when to let blood or to prescribe drugs was a proper role for a physician. In normal deliveries, the quite different talents of a midwife were enough.[31] That even a few Hallowell families could summon *both* a doctor and a midwife to a *normal* delivery, employing whichever attendant arrived first, suggests a remarkable change.

We must consider here the different backgrounds of doctors and midwives. Although medical reformers lamented the haphazard training of New England physicians, most male doctors could make some small claim to "learning." At the least they had mastered prescribed texts and spent a year or two as an apprentice. Dr. Page had attended Phillips-Exeter Academy before apprenticing with Dr. Thomas Kittredge. A hundred years earlier, of course, a literary education would have been enough; many New England ministers doubled as physicians.[32]

For midwives, giving birth themselves was an essential part of the training process; assisting at other women's deliveries was another. A traditional midwife was simply the most skilled member of an assemblage of female neighbors who assisted at each birth. After one delivery, for example, Mrs. Ballard wrote, "my comp[anions] were Old Lady Cox, Pitts, Sister Barton, Moody, Soal, & Witherel." At the Abial Herington house in June 1796, she noted, "there were 22 in number slept under that roof the night."[33] A doctor might be 24 years old and unmarried, like Benjamin Page, but only in middle age, and usually only after her own childbearing years were over, could a woman acquire the full stature of a midwife.

Physicians can be located in town records and on tax lists (as well as in Martha Ballard's diary) by the title *Doctor,* which originally was simply a designation for a man of eminent learning. No woman, whether a midwife or a practitioner of physic, can be discovered by title. Hallowell's mysterious black healer might be referred to as a "doctor" or "doctoress," but she is never called *Doctor* Black. If midwives had any sort of distinguishing label in this period, it was probably the word *old*. Mrs. Ballard referred to "Old Mrs. Fletcher," "Old Mrs. Ingraham," and "Old Lady Cox," and when Mrs. Marsh came to visit, she wrote of her as "The Old Lady."[34]

In traditional society such terms connoted respect, a respect for expertise acquired by experience rather than through systematic study.

Martha Ballard respected the formal training of Hallowell's physicians. She had obviously been taught, like other midwives, to call a doctor in an emergency. On November 11, 1785, when she arrived at Henry Babcok's house too late for the delivery and found the patient "greatly ingered by some mishap," though Mrs. Smith, who had delivered the baby, did "not allow that shee was sencible of it," Mrs. Ballard summoned Dr. Williams who "prescribed remedies." The inexperienced midwife had undoubtedly caused the injury. Though Martha Ballard felt competent to "inquire into the Cause," she did not attempt to correct it without a physician.

Only once in 797 births, however, did she herself feel incapable of handling a delivery. She described the delivery of Mrs. Prescott on May 19, 1792 in the following manner:

> Her Case was Lingering till 7 pm I removd dificulties & waited for natures opperations till then, when she was more severely atackt with obstructions which alarmed me much I desired Doct Hubard might be sent for which request was compiled with but by Divine assistance I performed the oppration, which was blisst with the preservation of the lives off mother and infant the life of the latter I dispard of for some time.

In the margin, she wrote: "the most perelous sien [scene] I Ever past thro in the Cours of my practice blessed be God for his goodness." Her ability to negotiate this "perelous sien" without Doctor Hubbard's assistance may have given Martha Ballard renewed faith in her own abilities, although characteristically she gave the credit to God.

In her obstetrical practice, Martha Ballard was obviously quite capable of managing alone. Still, the unspoken assumption that a man's knowledge was in some sense superior to a woman's experience colored her more ordinary relations with the men. She may have considered Ben Page incompetent, yet most of the time when in the actual presence of a physician she seems to have deferred. In May 1792, for example, she was called to Pittston to see the wife of Peter Grant and wrote, "They had called Doctor Parker before I arived and he seemed to chuse to perform the opperation which took place at 1 h 11 am." Her use of the verb *choose* is telling. On another occasion she and Dr. Page both appeared at a delivery. "I Extracted the child." she reported. "He Chose to close the Loin."[35] There is a silent acknowledgment here of the place of women in the traditional medical hierarchy.

Ironically in at least one respect—the keeping of records—Martha Ballard was a more methodical and progressive practitioner than most male doctors of her time. While medical leaders in early America urged the keeping of journals and commonplace books, few country physicians seem to have done so. Besides Martha Ballard's diary, the only surviving medical document from 18th-century Hallowell is Daniel Cony's one-page letter to the Massachusetts Medical Society. In some respects, Martha Ballard was herself a product of the medical enlightenment of the 18th century. Her willingness to attend autopsies, her meticulous recording of medical and obstetrical detail, her concern with vital records, and her commitment to "facts" in general, all suggest habits of mind not far removed from the best doctors of her age.

Her descriptions are often annoyingly formulaic ("she was not so well as Could be wisht") and her notation of methods exasperatingly obscure ("used means"); yet taken as a whole her records tower above those of most of her male contemporaries.[36] Where one Maine physician wrote, "delivered a dead child" and another scrawled "parturition," Martha Ballard routinely gave the approximate length of labor, the condition of the mother and infant, and sometimes even the weight of the newborn baby. Most entries read simply "safe delivered" or "left mother and infant cleverly," but others indicate specific symptoms and results. While we might have hoped for a more precise description of the "perelous sien" that prompted Martha Ballard to call Dr. Hubbard in May 1792, there is surely no more revealing description of medical obstetrics in the Maine literature than her account of Benjamin Page's administration of laudanum. Had there been a women's equivalent of the Massachusetts Medical Society, this event would surely have merited a one-page paper.

There was no professional association of midwives, of course. Women like Martha Ballard derived their authority from the larger community of women rather than from an exclusive society of professionals. This difference between doctors and midwives was reflected in their fees. Although there are no surviving accounts from Hallowell physicians for the period between 1785–1800, records from other Maine physicians suggest that doctors may have charged from two to three times as much as midwives.[37] Martha Ballard must have been pleased, therefore, when her more affluent patients treated her as they might have treated a doctor. About 10 days after the encounter with Dr. Page at the delivery of Hannah Sewall, Mrs. Ballard was called to the home of Chandler Robbins, a Harvard graduate and new resident of the town. "Mrs. Robbins Linguerd till 4 h pm when her illness came on," she wrote, adding, "Doct Parker was Calld but shee did not wish to see him when he Came & he returned home." Then she noted without comment that Mr. Chandler had given her 18 shillings, three times her usual fee.[38] In Hallowell, the "richer and better informed" might still prefer a midwife.

Considering Martha Ballard's eminence, we can only wonder whether there really was a medical challenge to midwifery in her lifetime. Although the sources clearly indicate the actions that in urban centers supposedly eliminated women from the practice of obstetrics–the development of medical societies, the appearance of physicians at normal deliveries, and the use of instruments and drugs–the results were hardly impressive. Martha Ballard respected the supposedly more specialized skills of the doctors, but she seldom required their help. As an accoucheur, she was more experienced than the doctors, and she was cheaper. Nor did the presence of male physicians have anything to do with her gradual withdrawal from practice after 1800. What we would like to know, of course, is who took up the work she eventually laid down.

Unfortunately, there is no evidence in the diary that she trained a successor, which in itself is a suggestive omission, although any of the many (mostly anonymous) women attendants at Hallowell births may have been preparing informally to succeed her. That the interaction between doctors and midwives continued into the 19th century is clear, however. In a diary entry for March 27, 1812, just a few months before her death, Mrs. Ballard wrote:

I was called at 10 h am Edwd Savage to go and see his wife who was in labor I had a fall on my way but not much hurt found the patient had called two midwives and Doct Ellis before she saw me I found her mind was for Doct Cony he was called and as Providence would have she called on me to assist her I performed the case.

The entry is ambiguous. The fact that Mrs. Savage called three midwives and two doctors shows that traditional midwifery survived into the second decade of the 19th century. Yet the form of the entry, the anonymity of the women, and the patient's apparent wavering between Martha Ballard and Daniel Cony suggests that the medical challenge was real.

Without additional evidence we cannot be certain of conditions in Hallowell after the death of Martha Ballard. Her diary allows us to see, with a clarity not provided in other sources, the relative strengths of traditional and medical obstetrics and the probable course of change. Clearly, there would be little potential for a gentleman physician to increase his obstetrical practice as long as he charged as much for delivering one baby as a midwife did for delivering three. In 18th-century terms, however, the medical consequences of this arrangement were serious, even if every midwife were as sober, experienced, and successful as Martha Moore Ballard.

As long as medical science was itself seen as rather static, something to be acquired primarily through mastery of the writings of ancient authorities, and as long as midwifery was considered a manual art best sustained and transmitted within the community of women, the book learning of the doctors and the practical learning of the midwives could complement one another. In the medical enlightenment of the 18th century, medical science became more experimental, more progressive. As a consequence, the lore of the midwives, passed from one generation of women to another, became suspect, while the active practice of obstetrics became more appealing. Historians have given considerable attention to the first consequence but little to the second.

The earliest collection of papers of the Massachusetts Medical Society provides a striking instance of medical disdain for traditional midwifery. In an account of an incident that had occurred sometime in the 1760s, Dr. Edward Augustus Holyoke of Salem told of being called from his bed to

visit a newly delivered woman who was dying because the midwife had used such force in extracting the afterbirth that she had disengaged the
uterus itself. "But what better can be expected
from an Operator utterly destitute of all Knowledge of the Figure, Situation, & Anatomy of the
Parts," he concluded.[39] We recall here the "perelous
sien" reported by Martha Ballard in which an inexperienced midwife unknowingly damaged her
patient.

This situation is a classic confrontation between
medical and traditional obstetrics, as seen from
the medical side. But suppose the situation were
reversed, and the doctor rather than the midwife were the incompetent practitioner. Suppose
that wide experience in managing ordinary births
rather than formal training in anatomy were the
decisive factor. Suppose, in fact, that a young physician fresh from his apprenticeship met an experienced and confident midwife at the bedside of a
laboring woman. What then?

We need not go into the diary of Martha Ballard
or to secondary accounts by feminist historians to
find the answer. Dr. William Smellie's *Collection of
Preternatural Cases and Observations in Midwifery* will
do quite well. Doctor Smellie, the British physician
whose improvements in obstetrical forceps supposedly paved the way for an interventionist and male-
dominated practice of obstetrics actually showed
genuine respect for the skills of traditional midwives. His works, which Benjamin Page might well
have read in the library of his mentor, Dr. Thomas
Kittredge, even included an account of a successful caesarean operation performed by an illiterate
Scots midwife named Mary Donally.[40] Although
Smellie included an occasional anecdote about the
stereotypical loquacious and ignorant midwife, far
more often he criticized half-educated physicians
who foolishly intervened in normal births. In a
number of these accounts, midwives come across as
more effective practitioners than physicians.[41] Obviously, what Smellie hoped to encourage was the
integration of the practical skills of the midwives,
who had long managed normal births, and the
more specialized training of doctors, who had now
acquired improved instruments for use in difficult
deliveries. He did not put it in those terms, of
course, but the direction of his argument is obvious: Midwives had skills that doctors needed.

The Massachusetts physician who in 1820 wrote
*Remarks on the Employment of Females as Practitioners
in Midwifery* understood the problem perfectly. "A
man must be a universal practitioner in midwifery,
before he is qualified for a practitioner in difficult
cases," he wrote. That is, a person unfamiliar with
normal deliveries could not manage preternatural
ones. If midwifery were reintroduced among the
upper classes in Boston, he argued, there would be
an inevitable decline in the quality of emergency
obstetrics because physicians would be denied the
day-to-day experience that good medicine required. That midwives might learn to manage difficult as well as normal deliveries was to him simply
unthinkable. No woman could pass through the
dissecting room and the hospital without losing
"those moral qualities of character, which are essential to the office."[42] We can only wonder what he
would have thought of Martha Ballard and the
other midwives who attended autopsies in late
18th-century Hallowell.

Seen in this light, Benjamin Page's effort to administer laudanum at the delivery of Hannah Sewall is hardly the ominous foreshadowing of later
medical triumphs that it seemed at first glance.
Rather, it is yet another example, to be added to
those of Smellie, of the limited obstetrical experience of 18th-century physicians. As long as Martha
Ballard continued to dominate the practice of
obstetrics in the region, it is difficult to imagine
Benjamin Page or any other doctor acquiring the
practical expertise that the new medical obstetrics
demanded. That was exactly the point: It was not
Dr. Page's competence but his incompetence that
made it essential for him to practice midwifery.

The history of midwifery in 19th-century Maine
is yet to be written, but the general direction of
change can be glimpsed in a document published
at Norridgewock in March of 1823 by an association of Somerset County physicians. Following the
example of Boston doctors of a generation earlier,
these men established fees for various treatments,
from "Extirpating tumors" to performing an "Operation for Hare-Lip," each man agreeing not to
"under value his own services" nor to "undermine
the practice of others" by charging less than the
minimum prescribed. The doctors also pledged
not to visit a patient previously treated by another
physician "unless it shall be the frank and unbiased
wish of the party calling him to dismiss the other
from further attendance." Significantly, their rates

included $4 for attendance "in ordinary Obstetric cases" and $8 "In cases where a Midwife has been first employed."[43]

The Somerset County agreement nearly defines the structural changes that transformed the medical world of Martha Ballard into the medical world with which most of us are familiar. These doctors cared about territorial boundaries in a way that she would have found puzzling. They were also bent on destroying the old system that had allowed midwives to perform "in ordinary cases" while calling doctors in an emergency. By charging twice as much for backup calls as for deliveries, they put extraordinary pressure on midwives as well as on their clients. The expected outcome, we may be sure, was that parturient women would call a doctor first.

Whether the Somerset physicians were successful in their reforms we do not know, yet the long-term direction of change is clear. The expansion of medical obstetrics was part of a larger process through which medicine changed from a learned specialty to a full-time profession.[44] For women the consequences of that shift cannot be overemphasized. Martha Ballard had sustained her practice while simultaneously running a household, supervising textile production, and rearing her youngest children. In her century such a pattern of part-time specialization was not unusual—for men or women. By the middle of the 19th century, this condition was no longer so. Doctors were doctors; midwives were also housewives. Ironically, then, the medical enlightenment of the 18th century, in teaching physicians to value ordinary midwifery, eventually guaranteed the exclusion of women from its practice. Still denied "learning," women no longer had the advantage of "experience."

None of this could have been foreseen, of course, on that October day in 1794 when Martha Ballard encountered Dr. Page at the delivery of Hannah Sewall. Having delivered 40 babies already that year and having made dozens of medical calls, she had reason to feel superior to young Ben Page, "that poor unfortunate man in the practice."

NOTES

For helpful comments on an earlier version of this chapter, I would like to thank Dr. Marcella Sorg of the University of Maine at Orono, Professor Sarah McMahon of Bowdoin College, and Professor John Demos and members of the American social history seminar at Brandeis University.

1   Diary of Martha Moore Ballard, 1785–1812 (hereafter MMB) in Maine State Library, Augusta, Me. (hereafter MeSL). An 18th-century English doctor suggested 10–20 drops of laudanum as the appropriate dosage for "weak women" and 12–24 drops for "weak men & midling women." See J. Worth Estes, "John Jones's Mysteries of Opium Reveal'd (1701): key to historical opiates." *J. Hist. Med.*, Apr. 1979, *34*: 202.

2   Jane B. Donegan, *Women & Men Midwives: Medicine, Morality, and Misogyny in Early America* (Westport, Conn., and London: Greenwood Press, 1978), p. 141. See also Catherine M. Scholten, "On the importance of the obstetrick art: changing customs of childbirth in America, 1769 to 1825," *William & Mary Quarterly*, July 1977, *34*: 426–445.

3   Judy Barrett Litoff, *American Midwives: 1860 to the Present* (Westport, Conn., and London: Greenwood Press, 1978), p. 26; A Physician, *Remarks on the employment of females as practitioners in midwifery* (Boston: Cummings and Hilliard, 1820), pp. 12, 21. Some historians attribute this work to William Channing, the first professor of obstetrics at Harvard Medical School; the Countway Medical Library, Harvard Medical School, Boston (hereafter CML), also lists it under the name of John Ware.

4   Judith Walzer Leavitt, "'Science' enters the birthing room: obstetrics in America since the eighteenth century," *J. Am. Hist.*, 1983, *70*: 281–304.

5   *Collections of the Maine Historical Society*, 2nd ser., 1898, *9*: 428 (hereafter *Coll. MeHS*); Mabel T. Kittredge, *The Kittredge Family in America* (Rutland, Vt.: Tuttle, 1936), p. 34.

6   *Coll. MeHS*, 2nd ser., 1896, *7*: 439–440; Doris R. Marston, "A lady of Maine: Sally Sayward Barrell Keating Wood, 1759–1855" (M.A. thesis, Univ. of New Hampshire, 1970), pp. 84, 110, 207; conversation with Dr. Eldridge Pendleton, Director, Old York Historical Society. Marston says that Dr. Hall Jackson of Portsmouth delivered Sally Keating's first child.

7   Charles Elventon Nash included an abridgement of Martha Ballard's diary in *The History of Augusta*, a book printed in 1904 but not bound and published until 1961. The Nash abridgement, heavily biased toward genealogy, includes only about one-third of the original, though it does suggest the sort of material in it. Except for brief reference to Nash's version in Nancy F. Cott, *The Bonds of Womanhood: 'Woman's*

Sphere' in New England, 1780–1835 (New Haven and London: Yale Univ. Press, 1977), pp. 19, 29, and in Richard W. Wertz and Dorothy C. Wertz, *Lying-in: A History of Childbirth in America* (New York: Schocken Books, 1977), pp. 9–12, 18, 20, the diary has been unused by historians. [See subsequent book by Laurel Thatcher Ulrich: *A Midwife's Tale: The Life of Martha Ballard Based on Her Diary, 1785–1812* (New York: Alfred A. Knopf, 1990).]

8 MMB, June 14, 1798. In this case, the child Dr. Page delivered was born dead.

9 *Vital Records of Oxford, Massachusetts* (Worcester, Mass.: Franklin P. Rice, 1905), pp. 13, 14, 268; MMB, Oct. 14, 1797; Nash, *History of Augusta,* pp. 235, 301; James W. North, *The History of Augusta, Maine* (Augusta, Me.: Clapp & North, 1870; reprinted, Somersworth, N.H.: New England History Press, 1981), p. 819. Ephraim Ballard was in the top 20 percent of taxpayers of the middle parish of Hallowell in 1794, although his estate was less than 10 percent that of James Howard, the largest taxpayer. See Invoice of rateable property, MeSL.

10 John L. Sibley, *Biographical Sketches of Graduates of Harvard University* and Clifford K. Shipton, *Biographical Sketches of those who attended Harvard College,* 17 vols. (Boston: Massachusetts Historical Society, 1873–1975), vol. 15, pp. 80–82 (hereafter *Sibley's Harvard Graduates*).

11 MMB, January 15, 1796. Because the diary begins January 1, 1785, with rather cursory entries and gradually takes on a more complex form over the next two or three years, I have assumed that it is complete. There may have been earlier lists of births, however, or even some sort of record kept in an almanac.

12 North, *History of Augusta,* p. 301; *Heads of Families at the First Census of the United States Taken in the year 1790: Maine* (Washington, D.C.: U.S. Government Printing Office, 1908), p. 9; Federal Census, 1800, microfilm.

13 North, *History of Augusta,* pp. 388–389, 836–837. I have not yet been able to identify the "black healer." There were very few nonwhite persons in Hallowell in this period. Martha Ballard occasionally mentions "Beulah" or "black Hitty," but she never mentions the "doctoress" by name.

14 *Ibid.,* pp. 170, 201; *Sibley's Harvard Graduates,* vol. 16, pp. 352–354; vol. 10, pp. 282–284; Samuel Curtis, *A valuable collection of recipes . . .* (Amherst, N.H.: Elijah Mansur, printer, 1819), pp. 52–54.

15 Eric H. Christianson, "The medical practitioners of Massachusetts, 1630–1800: patterns of change and continuity," *Pubs. Col. Soc. Mass.* 1980, *57:* 56–67; Joseph Kett, *The Formation of the American Medical Profession: The Role of Institutions, 1760–1860* (New Haven and London: Yale Univ. Press, 1968), pp. 9–14.

16 James Thacher, *American Medical Biography* (Boston: Richardson & Lord, 1828), p. 45. North, *History of Augusta,* 170–172; "Documents illustrative of the early history of the Massachusetts Medical Society," II: 22, MS, CML; *Massachusetts Register and United States Calendar* (Boston: Manning & Lorring, 1804), p. 43.

17 Douglas Lamar Jones notes the importance of midwives as healers in "Charity, medical charity, and dependency in eighteenth-century Essex County, Massachusetts," *Pubs. Col. Soc. Mass.,* 1980, *57:* 207–208.

18 MMB, e.g., burns, Sept. 17, 1786; bruises, Oct. 8, 1785; frostbite, Nov. 28, 1786; salves, Nov. 19, 1786; pills, Jan. 17, 1792; syrups, June 5, 1794; ointments, Feb. 18, 1787; clisters, May 25, 1791; dressings, Jan. 28, 1791; bathing throats, Sept. 13, 1787; bleeding cat, Oct. 13, 1786; cutting tongue, Feb. 1, 1786; lancing breast, Aug. 1, 1788; worms, Aug. 3, 1802; itch, Mar. 27, 1786; laying out dead, June 9, 1788; labor as illness, Mar. 31, 1790, Feb. 21, 1791. She also refers to woman in labor as "unwell," e.g., Apr. 12, 1791.

19 MMB, Dr. Cony, July 19, 1787, May 19, 1792, Nov. 4, 1802; Dr. Williams, Apr. 26, 1787; Dr. Hubbard, Aug. 23, 1794; Daniel Cony, "An extraordinary case in midwifery," read April 1788, in "Documents illustrative of the early history of the Massachusetts Medical Society," II: 22, MS, CML; MMB, Dr. Page, Nov. 17, 1793, Oct. 9, 1794, July 5, 1798, July 27, 1799, July 10, 1796; Dr. Parker, Oct. 22, 1794, May 3, 1792.

20 MMB, Oct. 30, 1791, Dec. 13, 1785, Sept. 18, 1786, May 19, 1792, Dec. 10, 1787, Mar. 26, 1787, July 25, 1785, Oct. 27, 1795.

21 Richard D. Brown, "The healing arts in colonial and revolutionary Massachusetts: the context for scientific medicine," *Pubs. Col. Soc. Mass.* 1980, *57;* 41–42.

22 MMB, May 31, June 1, June 2, 1792.

23 *Ibid.,* e.g., Sept. 23, 1790, Jan. 24, 1794.

24 Stephen Barton Account Book, MeSL.

25 Daniel Cony to the President and Counsellors of the Massachusetts Medical Society convened in Boston, June 1808, CML.

26 MMB, July 23, 1794.

27 *Ibid.,* Nov. 9, Nov. 15, 1793, Aug. 12, Nov. 10, 1785.

28 *Ibid.,* Sept. 2, 1794, Sept. 16, 1800, Feb. 4, 1801.

29 *Ibid.,* June 3, 1790.

30 Additional genealogical research to identify her patients may reveal more out-of-town births, although a cross-check of 1790 births with the federal census for that year suggests otherwise. The only births for which names could not be found in the census were either clearly identified as out of town (usually Winthrop) or listed as having occurred in the home of a Hallowell resident. For example, "Magr Stickney's" child was born "at Jacksons."

31 For a typical injunction to send for a doctor in an emergency, see William Salmon [Aristotle], *The Experienced Midwife* (Philadelphia, 1799), p. 39.

32 *Coll. MeHS*, 2nd ser., 1893, *4*: 428; Brown, "Healing arts," pp. 37–40; Laurel Thatcher Ulrich, *Good Wives: Image and Reality in the Lives of Women in Northern New England, 1650–1750* (New York: Alfred A. Knopf, 1982), pp. 132–134.

33 MMB, Mar. 5, 1801, June 15, 1796.

34 *Ibid.*, e.g., June 4, 1785, June 18, 1786, May 31, 1785, July 19, 1787.

35 *Ibid.*, May 3, 1792, Nov. 17, 1793.

36 As the frontispiece for a prospective history of Maine medicine, Dr. Jeremiah Barker of Gorham quoted Benjamin Rush's injunction to good record keeping but he concluded that few Maine doctors had taken his advice. "Diseases in the District of Maine," MS, MeHS. In Chapter 2 he praised Dr. Cony and other Lincoln County physicians but lamented their failure to keep "records of extraordinary cases, which occurred in their extensive practice." I have examined, in addition to the Barker notes, the account books of John Swett, York, Maine, 1775–1790 and Josiah Gilman, York, Maine, 1803–1813, Old York Historical Society; and Joseph and George Osgood Accounts, Andover, Massachusetts, 1770–1805 and Daniel Peirce Accounts, Kittery, Maine, 1762–1801, CML.

37 When Martha Ballard was charging six shillings, the physicians cited above were charging from 12 shillings to a pound. Dr. Peirce of Kittery, who performed only about six deliveries in 1792, charged extra for long labors or false alarms: Peirce Account Book, pp. 4, 14, 18, 19, 24, 36, 42, 54, 72, 228. A fragmentary account shows that between 1804 and 1811, Dr. Page charged one patient six dollars for each of three deliveries, although in 1796 Dr. Moses Appleton of Waterville charged his landlord only two dollars for delivering his wife; William Mathew debtor to Dr. Benjamin Page, Feb. 29, 1804–Apr. 4,

1814, Benjamin Vaughan Pps., MeHS; Diary of Moses Appleton, Sept. 5, 1796, Waterville Historical Society, Waterville, Me. (hereafter WHS).

38 MMB, Oct. 22, 1794.

39 Edward Augustus Holyoke, "Account of an inverted uterus," read July 1, 1784, in "Documents illustrative of the early history of the Massachusetts Medical Society," III, MS, CML.

40 William Smellie, *A Collection of Cases Preternatural and Observations in Midwifery*, 3rd ed. (London, 1764), vol. 3, pp. 420–421.

41 *Ibid.*, pp. 533–534. A handwritten transcription of lectures by a London physician in the papers of Dr. Moses Appleton of Waterville supports this argument. The English doctor argued that earlier doctors were dangerously interventionist because "all cases were difficult and so bad before they were called in that every one was frightened. . . . But if they had understood midwifery and what great things nature was capable of doing they would . . . seldom have had occasion to act in that manner." (Moses Appleton's notes from Dr. William Hunter's Lectures, 1793, MS, WHS.) I am grateful to Stephanie Hart of Colby College for calling this document and the Somerset physicians' agreement cited below to my attention.

42 *Remarks*, pp. 14, 7, and *passim*.

43 Articles adopted by Somerset County Physicians, Mar. 23, 1823, MS, WHS.

44 That early efforts at professionalization were strenuously resisted by various nonacademic movements in the 19th century is clear. See William G. Rothstein, *American Physicians in the Nineteenth Century* (Baltimore: Johns Hopkins Univ. Press, 1972), pp. 128–138; Kett, *The Formation of the American Medical Profession*, pp. 101–107. On the complexities of professionalism for women, see Virginia G. Drachman, *Hospital with a Heart: Women Doctors and the Paradox of Separatism at the New England Hospital, 1862–1969* (Ithaca, N.Y.: Cornell Univ. Press, 1984).

THERAPEUTICS

Therapeutics, the means of treating illness, is the core of medical practice. Prior to the 20th century, with the introduction of such "wonder drugs" as antibiotics, described by John Parascandola, regular physicians possessed only a handful of specific remedies, one of the most valuable of which was quinine, used to treat malaria and other fevers. In the absence of specific therapeutic agents, early 19th-century physicians focused their efforts on restoring the body to its "natural" state. To accomplish this, they often bled or blistered their patients, purged them with calomel, or induced vomiting with tartar emetic—measures that came to symbolize the regular practice of medicine. This "heroic therapy" dated back to antiquity, but it became especially prominent in the years after a yellow fever epidemic in 1793, when the Philadelphia physician Benjamin Rush bled and purged his patients back to health.

The therapeutic revolution described by John Harley Warner occurred in the middle third of the 19th century, when regular physicians turned increasingly from depletive to supportive therapies. Among the many factors contributing to this change were criticisms from irregular practitioners, who ridiculed the "heroic" practices of their rivals. These sectarian physicians, many of whom possessed M.D. degrees, offered the sick a range of alternatives to regular medicine, from the infinitesimal doses of the homeopaths to the water cures of the hydropaths. By the close of the century these sectarians, primarily homeopaths and eclectics

(who relied largely on botanic remedies), constituted an estimated 16 to 24 percent of all American practitioners. By this time, however, their methods of healing often differed little in practice from those of regular physicians, and in the 20th century the old sectarians faded away, leaving the irregular field to such newcomers as osteopaths, chiropractors, and Christian Scientists.

Therapeutic intervention did not always come at the hands of a physician, either orthodox or heterodox. Throughout American history many people have chosen to treat themselves, while others have sought help from faith healers, quacks, and assorted personnel on the fringes of the healing arts. Self-treatment has long been practiced by persons lacking access to physicians, wanting to save money, or simply embarrassed by the prospect of discussing delicate problems. In times of need they have often turned for advice to a do-it-yourself manual, a friend, or perhaps a patent-medicine peddler. Not infrequently, they have supplemented a physician's prescription with over-the-counter drugs or concoctions of their own making.

So-called quacks offered still other varieties of therapy, usually secret and sometimes expensive. During the middle decades of the 19th century, when licensing laws were nonexistent and anybody could be a "doctor," these pretenders to medical knowledge flourished openly. And even today, quackery remains a thriving business, annually milking hapless citizens of millions of dollars for cures deemed worthless by modern science.

5

From Specificity to Universalism in Medical Therapeutics: Transformation in the 19th-Century United States

JOHN HARLEY WARNER

Medical therapeutic explanation and practice in the United States were altered fundamentally over the course of the 19th century.[1] Traditional practices, founded upon assumptions about disease shared by doctor and patient and oriented toward visibly altering the symptoms of sick individuals, began to be supplanted by therapeutic strategies grounded in experimental science that objectified disease while minimizing the differences among patients. Concurrently, the principles that ordered proper therapeutic behavior began to change. Through the mid-19th century therapeutics was governed by the principle of specificity, the notion that treatment had to be matched to the idiosyncratic characteristics of individual patients and to the physical, social, and epidemiological peculiarities of their environments. During the final third of the century, a new therapeutic ideal, defined by an allegiance to knowledge produced and validated by experimental science and characterized by universalized therapeutic and diagnostic categories, was clearly in ascendancy.

The 19th-century transformation of therapeutics can be represented by a number of discrete shifts—in science, from clinical observation to laboratory experimentation; in practice, from heroic drugging that modified the whole system to therapeutic minimalism and narrowly targeted therapies; in

JOHN HARLEY WARNER is Professor of the History of Medicine, Yale University, New Haven, Connecticut.

Reprinted with permission from *History of Therapy: Proceedings of the 10th International Symposium on the Comparative History of Medicine—East and West*, edited by Yosio Kawakita, Shizu Sakai, and Yasuo Otsuka (Tokyo: Ishiyaku EuroAmerica, 1990), pp. 193–224.

the paradigm of health, from a natural balance to a normal state; and in epistemology, full circle from rationalism to empiricism and back to rationalism. Each of these changes merits close scrutiny. But my focus here is a wider one. All these permutations, I want to suggest, were linked to a broader transformation in therapeutic principle, from *specificity* (or individualism) to *universalism*. This shift underlay the changing ways physicians thought about their therapeutic knowledge, formulated plans to advance it, and sought to act at the bedside. It marked a fundamental transfiguration in clinical cognition, the emergence of a new conceptual structure. By the 1880s, when I have elected to end my study, this change was hardly complete; yet the beginnings of the sovereignty of *universalism* and demise of *specificity* were clearly evident.

MEDICAL THERAPEUTICS AND THE PRINCIPLE OF SPECIFICITY

Throughout the early and mid-19th century, American physicians viewed disease as essentially a systemic imbalance. The multitude of competing theories of pathogenesis shared the underlying assumption that for therapeutic purposes, the primary characteristic of sickness was the body's excessive excitement or enfeeblement. Treatment was premised upon an image of the body as an interconnected whole, and therapies addressed to individual symptoms usually modified the system's imbalance as well. The fundamental objective of therapy was to restore the *natural balance*, which defined the healthy condition. This was to be accomplished by depleting or lowering the overexcited patient, and by stimulating or elevating the patient enfeebled and exhausted by disease.[2]

Until the 1840s, however, physicians believed that nearly all of the prevailing diseases were over-stimulating, tipping the patient's vital balance to a dangerously overexcited condition. Ordinarily, therefore, treatment lowered the morbidly animated patient down to a healthy, natural state, often by aggressive depletive therapy. Such heroic depletion typically was brought about by extracting blood by opening a vein with the lancet (venesection) or drawing blood by suction from scarified skin (wet cupping); producing drastic catharsis by mercury-containing drugs (such as calomel) or other cathartics; causing debilitating vomiting by drugs such as tartar emetic; and prescribing a low diet.

Heroic depletive practices gave clear evidence that they worked. No one could doubt that the violent purging and debility that followed a large dose of calomel were the sequelae of giving the drug. Similarly, when a patient with a hard fast pulse, high temperature, and delirium became calmer—physiologically lower—after a large volume of blood had been let, the therapy's effect was undeniable. Confirmation of efficacy was generally rapid, and the physician, patient, and patient's family could all witness and appreciate the simple fact that the treatment was working. This was important, for it not only reaffirmed the validity of the treatment and the therapeutic system of which it was a part, but also demonstrated that the physician was in control.

Physicians' belief that most diseased conditions were overstimulating and required lowering therapy does not mean that they treated their patients by rote. What constituted the natural state, the desired endpoint of therapeutic effort, was highly individualistic. The natural condition was defined not for a population but for the individual, and was molded by such individuating factors as gender, family background, diet, occupation, and moral status. Moreover, what was natural for a person during one season, in one physical and social environment, and at one age changed as that person grew older, altered his or her social position, or moved to a different locale. The physician who knew best what was natural for the patient, and who therefore was the most capable of restoring the system to a natural balance when disease dis-

rupted it, was thought to be a practitioner well acquainted with the patient's personal history and with the peculiarities of the locality and its diseases. Therapeutic depletion could be produced in a variety of ways, and it was an important exercise of the physician's judgment to determine the best drugs, dosages, and regimen to accomplish it on the basis of the symptoms he observed and what he knew about the patient's background and environment.

This notion that treatment had to be fitted to the particular patient's idiosyncratic natural or healthy state was one expression of the *principle of specificity*, a central dogma of the orthodox therapeutic belief system.[3] Specificity demanded an individualized match between medical therapy and the specific characteristics of a patient and his or her social and physical environments. Treatment was to be sensitively gauged not to a disease entity, but to such attributes of the patient as ethnicity, age, socioeconomic position, and habits, and of place as climate, topography, population density, and social structure. "All remedies are relative agents, that is they only can act curatively by a judicious application to the individual case in hand," one physician noted in 1848. "*Individualism*, not *universalism*, attaches therefore to all our therapeutic measures."[4] Remarkably durable, the principle of specificity reigned virtually uncontested throughout the first two-thirds of the 19th century.

The idea of specificity was built into most of the rationalistic, speculative systems of medical practice early 19th-century medicine inherited from the Enlightenment. Benjamin Rush, for example, who had constructed the most influential American system of medicine, held that political and social organization, the physical environment, and health were linked, making disease and therapy functions of both place and culture. "Diseases of warm & cold climates require different treatment," a student listening in 1809 to Rush's directives wrote in his notebook. "The season of the year should be attended to. Epidemics differ. . . . The dress & moral habits should be attended to. The inhabitants of Egypt require stronger purges than other persons. The nation of which a person is a native should be attended to."[5] In principle, treatment was carefully adjusted to circumstances.

Nevertheless, as a reaction against rationalistic systems grew from the 1820s, critics often charged that the chief evil of systematic practice was that it followed fixed rules in therapy and gave insufficient attention to individuating factors. Between about 1820 and 1860, literally hundreds of American physicians studied in the hospitals of Paris, and many of them returned firmly committed to the ideal of therapeutic empiricism they acquired there. They identified therapeutic rationalism as a source of evil responsible for many of the medical profession's ills, and mounted a vigorous campaign against it. A reorientation from rationalism to empiricism, they believed, would elevate the profession's clinical and social power. Their ardent clinical empiricism and anti-rationalism, which encouraged them to denounce mechanical rules in therapy and stress close observation of individual patients, further secured the central standing of the principle of specificity in orthodox thought.[6]

Medical therapeutic knowledge, because of its domination by the principle of specificity, occupied an epistemological status fundamentally distinct from that of the basic medical sciences at least until the 1860s. In contrast to such branches of medicine as anatomy, physiology, and chemistry, and the mechanical manipulations of surgical and dental treatments, all of which were universalized in their generation and validation, medical therapeutics was specific to patient and place. Constructing universally applicable rules of practice was wrongheaded, for therapeutic knowledge was only valid in a context that closely approximated (in type of patient and environment) the one in which it had been produced. "Idiosyncrasy, or the peculiarities of the individual, are as anomalous and impossible to reduce to rule and measure, as the passage of the clouds," one physician asserted. "What is true of one place may not be true of another."[7] By stressing the specificity, not universalism, of therapeutic knowledge, physicians plainly expressed their conviction that knowledge pertinent for certain places or individuals could be inappropriate for others, and could direct practice that was dangerous both to the patient and to the physician's reputation.

It is important to emphasize that at least until the 1860s, the principle of specificity rendered disease-specific treatment, which disregarded idiosyncrasies of patient and place, to be professionally illegitimate. "Do not," a medical student copied his professor's warning in 1830, "prescribe the remedy for the name of the disease. For it may be essentially changed by climate, season, idiosyncrasies & other

causes. And while we bleed in this section of the United States the practitioners to the southward, never open a vain [vein], the disease there never demanding it."[8] The idea of matching therapies to diseases was inconsistent with prevailing theories of pathogenesis. Physicians did increasingly recognize the existence of specific diseases, but they simultaneously believed that a host of environmental influences could nudge one disease into another, and that a single disease could take on a variety of forms. Disease entities were not fixed but fluid. Further, the actions of remedies were also inconstant, modified by myriad influences. Two patients with the identical disease could require opposite treatments. This was precisely the point Harvard medical professor John Ware made in admonishing his students to distinguish between "a pathological and a therapeutical diagnosis." The name that pathological diagnosis assigned to a patient's disease was not a trustworthy guide in therapy, he urged, for "cases of which the pathological character is precisely the same may require a treatment diametrically opposite."[9] Diagnosis was of only secondary importance in determining appropriate treatment, for naming a disease at most suggested therapeutic possibilities and pointed to courses the disease might follow that could be anticipated therapeutically.

The principle of specificity's hegemony placed severe strictures upon the production and transmittability of therapeutic knowledge. It meant that knowledge sound within one context might be unsuitable if applied within another. For example, it was not at all clear that the findings of therapeutic research in urban hospitals were applicable to the vast majority of sick Americans. Physicians commonly believed that more energetic lowering treatment was demanded for country than for city residents. The environment of a city such as London or Philadelphia debilitated its inhabitants, a medical teacher in a small town told his students. "A particular mode of living in those large cities," he claimed, meant that "their diseases are of a nervous, rather than sanguineous character," requiring elevating therapy opposite to the debilitating treatment appropriate in the country. "The treatment of the same disease, Erysipelas for instance, in one of these large towns and here, would be totally different. In one you would stimulate, in the other deplete."[10] Further, because the beds of

American hospitals were peopled largely by the laboring-class poor and immigrants, populations often enfeebled by their social circumstances, therapeutic knowledge gained from hospital study might be misleading if transferred to a middle-class, native-born, private patient. "Let a student of medicine from the country take a course of instruction in the hospitals of one of our large cities and then let him go home and commence practice among the yeomanry of his native town," one physician asserted in 1846, "and he will soon find that he has a set of very different patients[,] and different class of diseases[,] and that [they] require a very different course of treatment from those he found in the hospitals."[11]

Just as specificity dictated that practice in hospitals might be inappropriate for patients treated outside their wards, so too treatment valid in one part of the country might be misleading in another. Regional variations in climate, topography, settlement patterns, and vegetation demanded distinctively northern, southern, and western therapeutics. In the South, for example, solar heat purportedly altered both the constitutions of the inhabitants and the actions of remedies, sustaining the proposition that in that region purgatives like calomel were called for more often than in the Northeast and depletion by venesection less so. Further, the South had a large population of black slaves that the North lacked, and the belief that this patient group required treatment different from that of white patients meant that a separate body of therapeutic knowledge was essential for southern practice. Such reasoning formed the conceptual core of a strident movement urging distinctively southern medical institutions, especially schools and literature, for transmitting knowledge matched to the region's medical needs. "Our practice here," one southern physician claimed, voicing a widely shared belief, "is entirely different from that taught in northern institutions, and by northern writers."[12]

The principle of specificity similarly informed American therapeutic nationalism. This was the notion that the epidemiological, meteorological, social, and demographic differences between the Old World and the New meant that European therapy was unsuited to many American circumstances. In general, physicians maintained that American constitutions and diseases were inherently more energetic than those of Europeans, and that therefore Americans required more forcefully depletive treatment when sick. The same disease could require opposite treatments in Europe and America. Typically, one medical student in Kentucky recorded in his class notebook that inflammation is "modified by a variety of circumstances. . . . Location or situation modifies [it]. In *Paris* we would stimulate by Porter, Bark &c—In Kentucky we would bleed & Purge."[13] European therapeutic knowledge was not merely irrelevant in managing American patients, but it could be actively harmful by directing the converse of the appropriate treatment, and this made Americans suspicious of European therapeutic precepts.

In establishing the limitations specificity placed upon using in one context therapeutic knowledge gained in another, physicians took care to point out that this did not draw into question the universality of certain other categories of medical knowledge. While practical therapeutic precept fell under the sway of the principle of specificity, the basic medical sciences and underlying principles of medicine did not. The latter were universal, independent of environment and valid everywhere. Thus one Southerner studying in the northern city of Philadelphia confidently claimed that "as for learning the theoretical part of Medicine and studying those branches on which we must found our practice (as Anatomy & Physiology) Philadelphia presents as many advantages as any other place in the united states." But he recognized the superiority of region-specific education in therapeutic practice, and applauded a fellow Southerner's decision to attend a medical school in the South, commenting that "the practice there will correspond better with that of our country."[14]

Yet, as long as specificity rather than universalism characterized what physicians regarded as valid therapeutic knowledge, much wisdom about treatment was necessarily tied to the place it was generated and used. In essence, it was local knowledge. It was up to the physician practicing in a particular place, on a particular type of patient, to develop the therapeutic knowledge suited to his own idiosyncratic needs. Unlike universalized knowledge, which could be reported and built upon through shared national and international publications, practical medical precepts could not be easily imported. Indeed, in knowledge about practice, one physician claimed, "there is no choice between a foreign supply and home production. Our medical literature cannot be *manufactured* for us abroad."[15]

THERAPEUTIC CHANGE

While relatively little of the evidence that could support generalizations about 19th-century American physicians' actual prescribing habits has been closely studied, what has been analyzed tends to suggest that the variations in treatment predicated by the principle of specificity were often carried out in practice. It is clear, for instance, that as therapeutic principle called for, the hospitalized sick tended to receive less aggressive therapy than their counterparts in private practice. Regional variations in treatment, corresponding to those predicted by principle, are also evident in practice. For example, venesection tended to be used more aggressively in the Mid-Atlantic region than it was in either New England or the Midwest; but in both of the latter regions practitioners resorted to the lancet more often than their brethren in the South did.[16]

More striking than regional differences, however, was the fact that practice during this period was changing throughout the United States. From the second decade of the century physicians' reliance on heroic depletive therapies gradually diminished. This change was driven by a variety of factors, including growing faith in the healing power of nature; French skeptical empiricism; the use of the numerical method in therapeutic evaluation; and patient demand for milder treatment. Important too was competition from such unorthodox practitioners as homeopathists (who prescribed greatly diluted doses of drugs) and their attacks upon heroically used orthodox therapies as murderous. Through the 1860s orthodox physicians continued to rely on the same armamentarium, by and large, but gave heroic therapies less frequently and in less aggressive dosages. The percentages of visits in American physicians' private practices in which venesection was employed, for example, tended to cluster in the 1820s and 1830s around four to eight; in the 1840s, three to five; and in the 1850s and thereafter, between a few per-

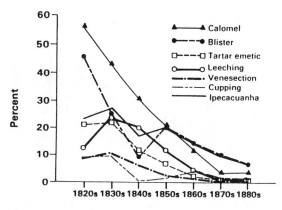

Fig. 1: Massachusetts General Hospital, Boston. Percentage of cases in which selected therapies identified with heroic depletive therapies were used. (Reproduced with permission from John Harley Warner, *The Therapeutic Perspective* [Cambridge, Mass.: Harvard Univ. Press, 1986].)

centage points and none. Some few practitioners continued to venesect patients occasionally, albeit rarely, at least into the 1880s.[17]

Figure 1 illustrates graphically the declining frequency with which the principal therapies identified with depletion were prescribed at one American institution, the Massachusetts General Hospital in Boston, between the 1820s and 1880s.[18] These treatments were not abruptly abandoned, but were gradually prescribed less often. Beneath the smooth contours of this overall pattern there was, of course, considerable fluctuation, and the decline was not always a steady one. Nevertheless, the overall tendency was toward prescribing depletive therapies less frequently. Moreover, when such treatments were prescribed, they tended to be given in increasingly milder doses. The mean dose of calomel given at the Commercial Hospital of Cincinnati in the 1830s, for example, was 21.7 grains, but diminished to 10.5 grains by the 1850s and 3.0 grains by the 1880s.[19]

As therapeutic depletion declined, reliance upon its opposite—therapeutic stimulation—rose. Especially from the early 1850s physicians came to believe that most diseases were no longer overstimulating, but instead had become debilitating. To restore a natural balance it was often necessary to elevate the patient's system, forcing it up to its natural level. The turn to stimulants at the bedside is illus-

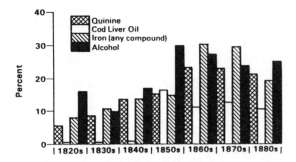

Fig. 2: Massachusetts General Hospital, Boston. Percentage of cases in which selected supportive and stimulant therapies were prescribed. (Reproduced with permission from John Harley Warner, *The Therapeutic Perspective* [Cambridge, Mass.: Harvard Univ. Press, 1986].)

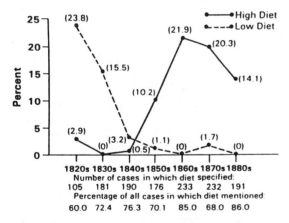

Fig. 3: Massachusetts General Hospital, Boston. Percentage of cases in which either elevating or lowering diet was prescribed. (Reproduced with permission from John Harley Warner, *The Therapeutic Perspective* [Cambridge, Mass.: Harvard Univ. Press, 1986].)

trated in Figure 2 by the rising use of such therapies as alcohol, iron compounds, and quinine at the Massachusetts General Hospital. Similarly, a depletive dietary regimen was replaced by a supportive one, as Figure 3 shows.

In certain respects, the shift from depleting to stimulating treatment marked a complete reversal in practice. But a shared objective guided the uses of both stimulants and depletants, namely, to restore the patient's system to its natural harmony. Elevating treatment often sought to accomplish this in the same way lowering therapy had, by righting

imbalance. Sometimes the use of stimulants became every bit as heroic as the extremes of heroic depletion had been, as in one American Civil War hospital where two-thirds of the patients received between three and eight ounces of whiskey per day over a period of many days.[20] Increasingly, however, as physicians came to doubt their ability to actively cure disease by medical art and transferred their faith to the healing power of nature, stimulant therapy did not seek to cure by force. Instead, stimulants sustained the patient's vital energy at a natural level while nature's healing powers proceeded.[21] Nothing in this challenged the reign of specificity. But the 1860s did witness the general eclipse of heroic therapy, understood as either an aggressively lowering or elevating plan.

Most physicians who appraised the state of therapeutics in the 1860s saw progress in the demolition of the rationalistic systems that had once dominated practice; in the move from heroic excesses toward therapeutic moderation; and in the mounting recognition of nature's cardinal role in healing and the practitioner's secondary function in assisting it. Yet they regarded it as *negative* progress, good but not constructive. Therapeutics had been improved principally by pruning past errors and excesses, not by nurturing the growth of anything new and vital. "In the use of medicine, the knowledge attained has been negative," one practitioner typically wrote. "Physicians have ascertained what medicine cannot do, rather than what it can do."[22] From the early 1860s, as physicians came to believe that the benefit of such a negative plan of therapeutic improvement had reached its limits, a tinge of despair colored therapeutic rhetoric.

The pivot upon which physicians' distress turned was the domination of therapeutics by the principle of specificity and the restraints it imposed. The perceived progress and growing certainty of such universal basic sciences as physiology and chemistry presented a disturbing contrast to the stagnancy and inevitable uncertainty of therapeutics. Because of the seemingly limitless variability of what constituted valid therapeutic knowledge for different peoples and places, there could be no fixed, universally applicable rules in therapeutic practice, no hope for therapeutic certainty. "In surgery, in anatomy, in physiology and in chemistry, there is a large degree of certainty," one Ohio physician affirmed in 1877. "But in the application of remedies to diseases, uncertainties, exceptions, disputes and defeats are interminable."[23] So long as physicians subscribed to the principle of specificity, the troubling disparity between the basic sciences and therapeutics would continue to worsen. Some practitioners came to believe that the bondage of therapeutics to specificity was humiliating proof of professional limitations, and warranted therapeutic pessimism.

EXPERIMENTAL THERAPEUTICS AND THE DISSIPATION OF GLOOM

By the mid-1860s, most physicians foresaw only dismal prospects if they continued to pursue therapeutic change in the same channel. They overwhelmingly agreed that since the century's start therapeutics had been improved by tearing down, and now redirected their energies toward building up. Programmatic schemes to escape the therapeutic gloom that had befallen the profession proliferated. Some advocated selective revival of remedies earlier in vogue, while others urged physicians to emphasize hygienic more than drug therapy. There were efforts to improve practice through empirical observation of remedies' effects, and some even turned their attention from healing individuals to the promise of state preventive medicine.[24]

Of all the designs for therapeutics put on the market, the one that was to remold medical enterprise most forcefully held that knowledge produced by physiological experimentation should become the new foundation for medical therapeutics. This was the plan for "experimental therapeutics"—what Americans usually called "physiological therapeutics"—that the French physiologist Claude Bernard gave its canonical form in his *Introduction to the Study of Experimental Medicine* (1865). According to the "physiological method," experimentation in the laboratory would elucidate physiological processes in health and disease as well as the actions of remedies. On the basis of this knowledge, the practitioner would know how processes going on in the patient's body deviated from normal and what adjustments would be required to correct them. By understanding the specific physiological alterations various therapies induced, he could then select the specific treatment that would precisely alter the deviant process and restore physiological normalcy.[25] While enthusiasm for the medical value of experimental physiology was by no means new to Americans in the 1860s, the crucial new element that emerged in that decade was the idea that knowledge produced in the physiology laboratory could direct, not just explain, therapeutic behavior.[26]

Physiological therapeutics sought to elevate therapeutic knowledge to a fundamentally new epistemological category. Basing therapeutics largely upon reasoning from physiological experimentation in the laboratory, not exclusively upon empirical observation at the bedside, meant that rationalism would regain its sovereignty over therapeutics. And with the new rationalism's ascendancy would come the restoration of those attributes of therapeutic knowledge linked to rationalism that the American empiricists had toiled so hard since the 1820s to banish. Many physicians believed that the reconstruction of therapeutics upon the foundation of experimental science would uplift therapeutics to the universalized level of the basic sciences. It would supplant the limitations of therapeutic specificity with the prospect of universalism, fixed laws, systems, and even an approach to certainty. By offering a pathway out of sterile clinical empiricism, the new rationalism would thwart prevailing skepticism and bring about a therapeutic renaissance.

To its enthusiasts, the new rationalism of experimental therapeutics, by informing universalized therapeutic laws based on experimental physiology, dispelled the causes of pessimism about the future. "We should not sit down in the ashes and mourn over uncertainties and doubts," one physician asserted in 1872, arguing that the physiological method justified therapeutic optimism.[27] Physiology would lead to boundless therapeutic progress, unhampered by the principle of specificity. Therapeutic skepticism may have been justified as long as the value of drugs could be assessed only on the basis of clinical experience, a Boston physician proposed. But now, he proclaimed in 1871, "as we are beginning to [understand], the real unquestioned effects of certain medicinal agents on the parts affected, a ray of light falls upon the healing art, and the practice of medicine begins to be something more than the observation of cases." He concluded that "the treatment of symptoms must yield, and be followed by the scientific appliance of remedies whose effect can be predicted to the removal of diseases."[28]

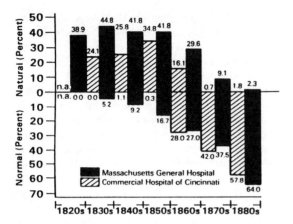

Fig. 4: Percentage of case histories in which terms "Natural" and "Normal" appear. (Reproduced with permission from John Harley Warner, *The Therapeutic Persective* [Cambridge, Mass.: Harvard Univ. Press, 1986].)

The program for physiological therapeutics was a formal, in some ways radical, articulation of an ongoing shift at the level of daily clinical conceptualization and behavior. By the start of the 1860s, the physician's way of looking at the objectives of treatment had begun to change. The *natural* began to be replaced by the *normal* as the paradigm of bodily order and health. Physicians came to think about their patients' disorders less as systemic imbalances in the body's natural harmony, and more as complexes of discrete signs and symptoms that could be analyzed, separated and measured in isolation. The hallmark of the new way of thinking about the goals of treatment was the reduction of signs of physiological order and disorder to objectively measured, quantified indices of normality.

The shift in clinical cognition is reflected by an important transformation in medical vocabulary. The term *natural* was replaced by the term *normal* in clinical discourse, mirroring the change in the way physicians thought about disorder and, hence, in what they believed was to be altered by treatment. This permutation, evident in all the conventional records of medical rhetoric, is most vividly exhibited in hospital case histories, which offer sustained documentation of clinical language over a long duration of time and which were produced with a reasonably consistent purpose and format. Figure 4 shows the changing frequencies with which each of the words *natural* and *normal* ap-

peared between the 1820s and 1880s in a sample of about 4,000 medical case histories taken from two American hospitals. The period of most marked change occurred between about 1860 and the mid-1870s, precisely the period during which the shift in medical thinking was progressing most rapidly.[29]

Commensurate with this reorientation, physicians, in looking for guides to therapeutic intervention, gradually turned away from judging a patient's well-being by comparison with that individual's natural state of heath. They began, instead, to weigh specific indicators in the patient against criteria of health expressed as norms for a population and as the universalized norms defined by laboratory science. Therapeutic attention from the 1860s focused increasingly on discrete physiological processes as meaningful in themselves, not merely as indicators of systemic disarray. In practice, physicians became preoccupied with bringing disordered processes back into line with fixed norms. During the 1860s and 1870s, quantification and graphical representation of such variables as respiration rate and temperature, and quantitative assay of, for example, the chemical composition of urine, became common methods of tracking a disease's course. Treatment, in turn, was explicitly aimed at making these measurements match figures that represented normal standards.[30]

Changes in etiological theory reflected and reinforced these shifts. Physicians began to think more in terms of specific disease entities and disease-specific causation, and less in terms of general destabilizing forces that unbalanced the body's natural equilibrium. Accompanying a more rigid ontology of diseases—in which disease was something abnormal but discrete and identifiable—was the idea that treatment could restore health not just by righting imbalance, but by specifically manipulating abnormal processes and even etiological agents.[31]

The increasingly reductionist perspective on patients which this cognitive restructuring entailed is apparent in clinical case records. Medical case histories became much shorter as physicians gave less attention to the individuating factors of social background and constitutional idiosyncrasy that shaped a particular patient's natural state of health. At the same time, physicians relied more and more on instrumental and chemical analyses to describe the patient's condition. Quantified signs and symptoms emblemized the newer approach. In many case histories daily entries consisted principally of temperatures and pulse rates, and these, along with respiration rates and urinalysis reports, stood out as the most visually striking features of case records by the 1880s. Much of the narrative description that in earlier decades sought to characterize the patient's destabilization from a natural condition was dropped from record-keeping conventions. Case records were more uniform and regimented, less individualized and discursive; contained fewer words, but more numbers; were more concise, but less able to evoke images of sick individuals.[32]

The most striking clinical changes which the ideal of physiological therapeutics effected through the 1880s were in attitude. As the therapeutic conception of disease shifted—from a disturbance of the natural balance to deviation from the normal state—physicians came to favor descriptions of therapeutic action that referred to discrete physiological processes. Equilibrium, of course, remained a common model of health, and in practice stimulant and supportive remedies remained among the most common treatments. But the physiological therapeutist found it more satisfying to conceptually break down the body into discrete units or systems, assess the function of each, and severally minister to them therapeutically. In general, physicians used fewer drugs but aimed them more narrowly at manipulating specific physiological processes. In the 1870s and 1880s, the salicylates and injections of morphia mitigated pain; chloral hydrate and the bromides produced sleep; antipyretics normalized temperature; and aconite slowed the pulse. The mean number of treatments prescribed per patient in hospitals diminished.[33] Instead of using a large number of therapies to restore the natural state of the body as a whole, physicians tended to prescribe a few remedies narrowly targeted to make specific physiological modifications.[34]

While only a minority of physicians took up the extreme position for experimental science, Americans widely adopted elements of it. The universalization in principle of therapeutic knowledge was expressed as a tendency away from therapeutic individualism toward more uniform practice, evident in the convergence of prescribing habits in the various regions of the country. Substantial regional variations in the employment of therapies such as venesection and calomel subsided from the 1860s, and the use of newer drugs never developed strong regional patterns sustained by the principle of specificity. To be sure, epidemiological differentials meant that some broad variations in drug usage endured, notably the more extensive use of quinine in the malarious regions of the South and West. Still, a move to standardization—in the strict sense, normalization—is clearly displayed by the practice of hospitals located in the various regions, where the differences that persisted, in dietary prescriptions, for example, tended to stem more from economic realities than from medical therapy.[35]

FROM SPECIFICITY TO UNIVERSALISM

While a shift to experimental science was certainly not completed between the 1860s and 1880s, during these decades its more devout disciples sketched out a detailed, expansive blueprint for how medicine should be reconstructed. They described how the proposition that experimental science should govern practice ought to transform the physician's professional identity; the doctor-patient relationship; the professional meanings of the hospital, medical education, and relations with sectarian practitioners; and the very definition of medical orthodoxy. In a narrowly therapeutic context, the implications of experimental science's ascent were most clearly expressed by the shift from specificity to universalism, and the move from empiricism to rationalism that underlay it.

In grounding medical therapeutics upon experimental science, and thereby elevating it to the universalism of the basic sciences, the advocates of experimental therapeutics sought to lift it out of the separate epistemological category that had excluded therapeutics from the basic sciences' rapid progress. Therapeutic advancement would be accelerated, stirred from its stagnation, by linking it to progress in physiology. Moreover, in principle, universal rules would make therapeutics a simple, certain endeavor. "I believe in time," one medical student enthusiastic about the promise of the laboratory wrote to his mother, "that the Physician[']s task will be almost as easy and irresponsible as that of the Engineer on Rail Roads or Pilot on a Boat."[36] If normal physiological processes were known and the patient's departure from normalcy determined, and if the physiological actions of drugs were defined, then in principle the patient's physiological deviance could be engineered back to normal by selecting the drug that would precisely effect the needed adjustment. The postulated identity of the physiological action of drugs with their therapeutic action promised something akin to experimental determinism in therapeutics. "While the notions of the actions and uses of drugs engendered by experience and observation are constantly changing, the deductions of experiment have the same value as the same methods in other experimental science," the advocate of physiological therapeutics Roberts Bartholow asserted in 1880, extolling the promise of universalism. "To this end we should direct our best efforts, and rest satisfied with no less certainty than that which belongs to the exact sciences, until we have attained to such a degree of perfection that, the disease being given, the remedy follows."[37]

One liberating effect anticipated from therapeutic universalism was some measure of freedom from the preoccupation with patient individuality that the principle of specificity had demanded. Experimental therapeutics matched therapy to a particular physiological process more than to the individual patient's constitution and background. Accordingly, individuating characteristics forfeited their earlier significance. The experimental therapeutist's scrutiny was closely centered on a physiological process (often elucidated by vivisectional research) rather than on a sick individual, and in some respects it made relatively little difference whether that process was going on in an Irish immigrant or a laboratory dog; therefore, whether the physician was prescribing for an immigrant or native-born patient, a distinction once seen as important, mattered very little indeed. "In seeking a remedy," one physician asserted in 1874, "it is logically legitimate only to consider the physiological requirement."[38] Knowing the normal state that the universalized standards of experimental science defined became relatively more important than knowing the natural condition that described health for an individual patient.

Therapeutic decision making was also freed to a large extent from domination by the modifying peculiarities of different social and physical environments. Prior to the 1860s, orthodox practitioners had upheld discriminating attention to environmental and patient variability as a mark of the model physician. But the emerging medical ideology assured practitioners that would they but submit their therapeutic decisions to arbitration by the universalized criteria of experimental science, correct treatment would ensue (so long as their scientific acumen was sound enough).

In practice, of course, the plan to completely universalize therapeutics has never been fulfilled. In the late 20th century, both prudent and invidious discriminations among patients on the basis of such factors as age, gender, stamina, ethnicity, socioeconomic class, and residence remain a part of what determines an individual's therapeutic management: treatment for a young, vigorous woman might be inappropriate for an enfeebled, elderly man, and a rural patient in the Third World might receive substantially different care from a wealthy urban resident of an industrialized nation. But it was a different type of medical reasoning than was prevalent during the first two-thirds of the 19th century, when therapeutic knowledge was bound by the principle of specificity to typologies of patients and locales. With the ascendancy of experimental therapeutics, therapeutic regionalism, nationalism, classism, moralism, and racism were stripped of much of their earlier justification. Such discriminations did not become unimportant, but they lost their prior significance as cardinal determinants of therapeutic propriety, and attention to them was no longer a salient sign of correct professional behavior. The free application of knowledge from one context to another—moving across national, regional, and ethnic boundaries—gained unprecedented epistemological and professional legitimacy.

Thus, the emergence of the new medical ethos increased, not decreased, the importance of the hospital as an authoritative source of medical therapeutic knowledge. However, in making a case for the practical potential of laboratory science, the proselytizers of experimental research often used rhetoric that polarized the laboratory and clinic. In a limited political sense, it is true that these two medical arenas were in competition for both believers and resources in the pursuit of therapeutic truth. But it would be a mistake to accept this politically motivated rhetoric of a power struggle at face value. The program for therapeutic progress that experimental science informed accorded clinical observation an important place within it, especially in refining the natural history of diseases and in testing therapies. More than this, it called for the epistemological changes that were necessary in order to legitimize the hospital as the heart of therapeutic research.

In the early and mid-19th century, the principle of specificity had been the primary wellspring of skepticism regarding the pertinence of therapeutic knowledge generated in hospitals to patients outside their walls. The weakening of specificity and strengthening allegiance to universalism after the 1860s substantially removed hospital knowledge from its singular status. Just as it became increasingly legitimate to freely transport therapeutic knowledge produced in one country or region to another, so too it appeared to physicians more and more reasonable to take therapeutic knowledge based on the observation of hospital patients and apply it in private practice. Thus, at the same time that reductionist laboratory science was beginning to secure its claim as one source of therapeutic authority, the hospital also gained a fundamentally new kind of authority as a source of therapeutic knowledge of wide applicability. The laboratory did not supplant the clinic; rather, they concomitantly found in the ideals of experimental science new supports for their claims to therapeutic legitimacy.[39]

This transformation carried with it a significantly altered attitude toward the hospital as a research institution in therapeutics. The rise of Parisian hospital medicine early in the century, though it instigated clinical research along many lines, had failed to make American physicians look unreservedly to the hospital for therapeutic instruction. Faith in the principle of specificity, reinforced by a dependence upon empirical observation, had rendered the hospital's therapeutic dictums suspect for application elsewhere. It was the epistemological transformation experimental science effected in medical therapeutics that validated the hospital as a leading source of therapeutic knowledge relevant for general medical practice.[40]

THE APPEAL OF EXPERIMENTAL SCIENCE

Looking backward from the perspective of the late 20th century, the very real positive benefit that the experimental laboratory eventually brought to medical therapeutics is unmistakable. But the simple fact remains that those Americans who between the 1860s and 1880s stridently insisted that experimental science should be the foundation of therapeutics did so *before* laboratory science had yielded any substantial proof of its power to make physicians better therapeutists. Starting in the 1860s, the proselytizers of experimental science pledged marvelous therapeutic products on the laboratory's behalf, but there was as yet scant evidence that science could deliver the goods in terms of a demonstrably elevated power to cure. To be sure, over the next couple of decades some new therapies arguably generated in part by experimental science—notably chloral hydrate and the salicylates—did enter practice. But they were few, and certainly failed to give clinicians the measure of therapeutic control the more sanguine spokesmen for the laboratory continued to promise. However adamantly those physicians who had given over their hearts and minds to experimental science foretold the coming revolution in therapeutics, their faith was sustained more by the promise of future relevance than by demonstrated results. During these decades they never found the compelling evidence they wanted to uphold their bold assertions. That the anticipated therapeutic millennium did not have its advent in their lifetimes was a source of acute disappointment and frustration.

Nevertheless, between the 1860s and 1880s the beginnings of a shift in therapeutic thinking informed by experimental science is evident not only among the laboratory's zealots but also among many of those who regarded its ideals and products with reserve or even suspicion. And, it is clear that this therapeutic change was but one part of a broader project to construct a new medical ethos upon the foundation of experimental science. But to say simply that therapeutics took part in this transformation, and to explicate how it proceeded, is to describe the change in therapeutics without explaining it. The question remains: If American physicians had little hard evidence of laboratory science's therapeutic fruitfulness, why did they choose to embrace it? What motivated their turn to experimental science?

Traditional historiography tended to answer this question by suggesting that because experimental laboratory science offered medicine a better grounding than it had hitherto, physicians, understandably, moved in its direction. The spokesmen for experimental therapeutics were simply exceptionally perceptive physicians who recognized the correct path to progress. The fact that the laboratory did not rapidly generate the kinds of therapies that eventually emerged in the 20th century did not particularly matter, since physicians somehow knew that they were on the right track. A modified version of the same reasoning points out that the biological sciences in general were moving from a natural historical to an experimental model, and medicine, as a science, necessarily followed this lead. Especially during the past two decades, however, historians have criticized such explanations as deterministic and presentistic, and pointed out that ultimately they beg the question.[41]

Instead, historians have recently suggested that American physicians embraced experimental science because they perceived it as a powerful vehicle for professional uplift. They saw it as a source of authority they could exploit to elevate the profession's standing in American society and the individual physician's prestige. The turn to experimental science was in its essence one strategy (and a brilliantly successful one) in medical professionalization, and is explained by physicians' socioeconomic aspirations. It was the cultural, not technical value of science that appealed to physicians, who used it as an ideological platform for enhancing the esteem of their profession.[42]

What these explanations share is the notion that, therapeutically, physicians gained nothing of immediate value by taking up experimental science. Yet I want to suggest that this was not the case. It is certainly true that at least until the mid-1890s, the laboratory contributed only very modestly to the practitioner's actual power to cure. However, in turning to experimental science, physicians also shed their commitment to the principle of specificity and bound themselves to a belief in therapeutic universalism with all its possibilities. Increased therapeutic efficacy ensuing from this shift might come only at some vaguely defined time in the future. Indeed, a physician in the 1860s could not know if it would ever come at all. But faith in experimental therapeutics *immediately* granted physicians belief in the possibility of rapid therapeutic advancement, a basis for dispelling therapeutic despair, and a structure and plan for orienting their own careers with optimism and energy. Even if the

early converts to experimental therapeutics did in some respects make a blind leap of faith, it rewarded them with confidence, purpose, and hope, however unproven its bedside utility. The shift in principle from therapeutic specificity to universalism provided a catalyst for dissipating the therapeutic gloom to which the profession had fallen prey, and this in itself constituted a potent incentive to take up experimental therapeutics and its values.

To be sure, some American physicians were more inclined than others to take this step. Those who spoke for physiological therapeutics typically shared several characteristics. They tended to be younger than the proponents of therapeutic empiricism in the tradition of the Paris clinical school against whom they aligned themselves. They also tended to be among the intellectual elite, and to be actively seeking professional distinction and advancement at the time their vocal advocacy of the physiological method began. Further, they tended to look to Germany more than to France for medical instruction, and many had studied in German laboratories and clinics. They clearly regarded German laboratory medicine as the harbinger of medicine's future, and denigrated French clinical medicine. Although they, themselves, were clinicians, they sought to relate laboratory work to their own activity at the bedside.

There was one overriding factor which divided the advocates of physiological therapeutics from their opponents: those physicians who embraced the new rationalism had not fought the empiricist battle against rationalistic systems of practice and heroic excesses that had so preoccupied intellectually alert American physicians from the 1820s through the 1850s. While the major division between those who supported and opposed the idea of physiological therapeutics was indeed generational, to regard this as a simple opposition of old conservatives to young progressives would be to miss a more important point. The central intellectual experience in the professional careers of many older physician-intellectuals had been overturning the rationalism that sustained medical systems, and this was an experience their younger colleagues lacked. Older physicians were not able to espouse physiological therapeutics and its underlying rationalism because they were committed to something else—a hard-won therapeutic empiricism—which they saw as being at odds with it. Younger physicians who lacked the zealot's allegiance to empiricism were more inclined to take empiricism for granted and condemn its shortfalls. By the 1860s virtually all American physician-intellectuals shared the perception that therapeutics was growing increasingly stagnant, and many held clinical empiricism and the principle of specificity responsible. And yet not all thinking physicians were emotionally or epistemologically prepared to turn to experimental science, rationalism, and therapeutic universalism as an alternative.

NOTES

This paper was supported in part by NSF Grant SES-8107609; NIH Grant 03910-01, -02 from the National Library of Medicine; a research award from an Arthur Vining Davis Foundation grant to the Department of Social Medicine and Health Policy, Harvard Medical School; a NATO Postdoctoral Fellowship from the National Science Foundation; and a research fellowship from the Wellcome Trust.

1 I refer in this essay only to medical therapeutics, not to the mechanical aspects of surgery, dentistry, or obstetrical treatment. These branches occupied an epistemological category different from that of medical therapeutics at least through the 1860s, and followed a different pattern of change in physicians' thinking.

2 On therapeutic thought in the 19th-century United States, see Alex Berman, "The heroic approach in 19th-century therapeutics," *Bulletin of the American Society of Hospital Pharmacists*, 1954, *11*: 320–327; Charles E. Rosenberg, "The therapeutic revolution: medicine, meaning, and social change in nineteenth-century America," *Perspect. Biol. & Med.*, 1977, *20*: 485–506; William G. Rothstein, *American Physicians in the Nineteenth Century: From Sects to Science* (Baltimore and London: Johns Hopkins Univ. Press, 1972), pp. 41–62, 177–197; and John Harley Warner, "'The nature-trusting heresy': American physicians and the concept of the healing power of nature in the 1850's and 1860's," *Perspect. Am. Hist.*, 1977–1978, *11*: 291–324.

3 On the principle of specificity and its implications, see John Harley Warner, *The Therapeutic Perspective: Medical Practice, Knowledge, and Identity in America, 1820–1885* (Cambridge, Mass: Harvard Univ. Press, 1986), which places the shift from specificity to universalism in the context of social changes in American medicine and relates it to the transformation in the physician's professional identity.

4 [John P.] H[arrison], "Notices of empiricism," *Western Lancet and Hospital Reporter*, 1848, *8*: 122–124, quotation on p. 122; my emphasis.

5 John Austin, Notes Taken on Lectures Given by Benjamin Rush [University of Pennsylvania], 1809 (Historical Collection, Rudolph Matas Medical Library, Tulane University, New Orleans, Louisiana).

6 On American physicians' use of Parisian therapeutic empiricism, and on their broader relationship with French therapeutics, see John Harley Warner, "The selective transport of medical knowledge: antebellum American physicians and Parisian medical therapeutics," *Bull. Hist. Med.*, 1985, *59*: 213–231.

7 David Cheever, "The value and the fallacy of statistics in the observation of disease," *Boston Med. & Surg. J.*, 1860–1861, *63*: 449–456, 476–483, 497–503, 512–517, 535–541, quotations on pp. 483, 514.

8 Samuel Murphey, Notes Taken on Lectures Given by [Nathaniel] Chapman, [University of Pennsylvania], 1830 (University Archives, University of Pennsylvania, Philadelphia).

9 John Ware, "Success in the medical profession," *Boston Med. & Surg. J.*, 1850–1851, *43*: 496–504, 509–522, quotation on p. 501.

10 Anonymous, Notes Taken on Lectures Given by [Benjamin] Dudley, 1830 (Historical Collection, Rudolph Matas Medical Library).

11 Daniel D. Slauson, Notes Taken on Lectures of Prof. Webster, Geneva College, Introductory Lecture, 1846, Medical Daybook, 1846–1877 (Daniel D. Slauson Papers, Department of Archives and Manuscripts, Library, Louisiana State University, Baton Rouge).

12 Jas. C. Billingslea, "An appeal on behalf of Southern medical colleges and Southern medical literature." *Southern Med. & Surg. J.*, 1856, n.s. *12*: 398–402, quotation on p. 399. On the principle of specificity's role in arguments urging distinctively regional medical therapeutics, as well as the necessity for institutional separatism, see John Harley Warner, "The idea of Southern medical distinctiveness: medical knowledge and practice in the Old South," in Judith Walzer Leavitt and Ronald L. Numbers, eds., *Sickness and Health in America: Readings in the History of Medicine and Public Health*, 2nd ed. (Madison and London: Univ. of Wisconsin Press, 1985), pp. 53–70, and John Harley Warner, "A Southern medical reform: the meaning of the antebellum argument for Southern medical education," *Bull. Hist. Med.*, 1983, *57*: 364–381.

13 [Lawrence] Jefferson Trotti, Notebook, bound with catalog from Transylvania University, 1828 (Special Collections and Archives, Frances Carrick Thomas Library, Transylvania University, Lexington, Ken-

tucky). The place of the principle of specificity in sustaining therapeutic nationalism in the United States is analyzed in Ronald L. Numbers and John Harley Warner, "The maturation of American medical science," in Nathan Reingold and Marc Rothenberg, eds., *Scientific Colonialism, 1800–1930: A Cross-Cultural Comparison* (Washington, D.C.: Smithsonian Institution Press, 1987), pp. 191–214 (ch. 8, this book).

14 Quotations from James K. Nisbet to N[athaniel] E. McClelland, Philadelphia, letters of 26 Feb. 1830 and 18 Sept. 1830 respectively (McClelland Family Papers, Southern Historical Collection, University of North Carolina at Chapel Hill, Chapel Hill).

15 S. L. Grier, "The negro and his diseases," *New Orleans Med. & Surg. J.*, 1852–1853, *9*: 752–763, quotation on p. 763.

16 These generalizations are based upon a study of manuscript record books in which American physicians chronicled their private practices, and comparison with records of hospital practice; see Warner, *The Therapeutic Perspective*. Martin S. Pernick similarly found that the use of anesthesia in surgery correlated well with the typologies of patients and patients' needs established in medical principle (*A Calculus of Suffering: Pain, Professionalism, and Anesthesia in Nineteenth-Century America* [New York: Columbia Univ. Press, 1985]).

17 A partial list of the private practice ledgers upon which these statements are based is given in Warner, *The Therapeutic Perspective*.

18 Information about therapeutic behavior at the Massachusetts General Hospital was derived from a computer-aided analysis of the complete run of its medical case-history books for the years from 1823 to 1885 (384 volumes), deposited in Oliver Wendell Holmes Hall, Francis A. Countway Library of Medicine, Harvard Medical School, Boston, Massachusetts. The sample used in calculations represents 10% of the admissions by year to the hospital's male medical wards, with the exception that for the comparatively small Massachusetts General, 25 cases per year were sampled if fewer than 250 male patients were admitted to the medical wards during the course of the year. This yielded a sample of 1762 cases; the breakdown of cases sampled by decade was for the 1820s—175; 30s—250; 40s—249; 50s—251; 60s—274; 70s—341; 80s—222. Cases were evenly sampled from throughout each calendar year's admissions to reflect the actual seasonal distribution of admissions.

19 The data base for statements about the Commercial Hospital in Cincinnati, later called the Cincinnati General Hospital, was compiled from that hospital's medical case-history record books, deposited at the History of the Health Sciences Library and Museum,

University of Cincinnati. These records include 51 volumes of case histories from the male medical wards spanning the period from 1837 to 1881. There is a gap from 1856 to 1859, and statistics for the 1850s therefore refer only to the first half of the decade. The sample used in calculations represents 10% of the admission by year to the male medical wards, yielding a sample population of 2023 cases. By decade, the numbers of case histories sampled were for the 1830s—79; 40s—550; 50s—342; 60s—243; 70s—700: and 80s—109.

20 Prescription Book, 2nd North Carolina Military Hospital, Petersburg, Virginia, 18 August–25 September 1864 (Confederate Hospital Records, Southern Historical Collection); and see Confederate States of America, Hospital Prescription Book [Virginia], 1864–1865, labeled "General Prescription Book, No. 2" (Manuscripts Department, Alderman Library, University of Virginia, Charlottesville).

21 From the 1840s and especially 1850s, physicians also made palliatives more prominent in their prescriptions, motivated in part by the idea that reducing pain contributed to an environment in which nature's healing powers could operate unhampered.

22 John P. Spooner, "The different modes of treating disease, or the different action of medicine on the system in an abnormal state," *Boston Med. & Surg. J.*, 1862, *66*: 245–251, 266–274, 287–291, 328–333, 349–354, 373–378, quotation on p. 245.

23 T. L. Wright, "Transcendental medicine.—an enquiry concerning the manner of morbific and remedial agents independently of empiricism," *Cincinnati Lancet and Observer*, 1877, *38*: 1015–1034, 1150–1158, quotation on pp. 1016–1017.

24 Exemplary plans for physiological therapeutics include Roberts Bartholow, "The degree of certainty in therapeutics," *Transactions of the Medical and Chirurgical Faculty of the State of Maryland*, 1876: 33–50, and [Horatio C. Wood], "The principles of modern medicine," *Boston Med. & Surg. J.*, 1884, *10*: 597–598.

25 For example, see Henry Ingersoll Bowditch, *Preventive Medicine and the Physician of the Future* (Boston: Wright and Potter, 1874), and *idem, Public Hygiene in America* (Boston: Little, Brown, and Company, 1877).

26 The relationship between physiological theory and medical therapeutics at mid-century is analyzed, using the British context, in John Harley Warner, "Physiological theory and therapeutic explanation in the 1860s: the British debate on the medical use of alcohol," *Bull. Hist. Med.*, 1980, *54*: 235–257 and *idem*, "Therapeutic explanation and the Edinburgh bloodletting controversy: two perspectives on the medical meaning of science in the mid-nineteenth century," *Med. Hist.*, 1980, *24*: 241–258.

27 Roberts Bartholow, Introductory Lecture on Experi-

mental Therapeutics, quoted in "Opening of the schools," *Cincinnati Lancet and Observer,* 1872, *33:* 635–636, quotation on p. 636.

28 Joel Seaverns, "Recent advances in medicine and their influence on therapeutics. The annual address delivered before the Norfolk District Med. Society, May 10, 1871," *Boston Med. & Surg. J.,* 1871, *85:* 113–120, quotation on p. 118.

29 These language changes were not merely coincidental, and all of the patient characteristics described as natural in the case records early in the century were later described instead as normal.

30 Illustrating the strengthening impulse to quantify, quantified respiration was recorded at the Massachusetts General Hospital in (seriatim between the 1820s and 1880s) 4.0%, 12.4%, 12.1%, 16.7%, 22.3%, 22.6%, and 78.8% of the cases. Quantified temperature, never noted before the 1860s, was given in 9.9% of the cases in that decade, 42.2% in the 1870s, and 82.0% in the 1880s. (For source, see note 18.)

31 On the history of the notion of disease as a specific entity, see Owsei Temkin, "The scientific approach to disease: specific entity and individual sickness," in A. C. Crombie, ed., *Scientific Change: Historical Studies in the Intellectual, Social and Technical Conditions for Scientific Discovery and Technical Invention from Antiquity to the Present* (New York: Basic Books, 1963), pp. 629–647; and Knud Faber, "Nosography in modern internal medicine," *Ann. Med. Hist.,* 1922, *4:* 1–63. A useful introduction to etiological thinking during this period is Charles E. Rosenberg, "The cause of cholera: aspects of etiological thought in 19th-century America," *Bull. Hist. Med.,* 1960, *34:* 331–354.

32 This description is founded principally upon the records described in notes 18 and 19.

33 Whereas at the Massachusetts General Hospital the mean number of treatments given per patient from the 1820s to the 1860s fluctuated by decade between 7.4 and 8.7, it dropped in the 1870s to 6.5 and in the 1880s to 5.2. The mean number of treatments given to patients at the Commercial Hospital, always lower than at the Massachusetts General, was reasonably constant from the 1830s to the 1850s, and thereafter steadily fell from 5.9 treatments in the 1850s to only 2.5 by the 1880s. (For source, see notes 18 and 19.)

34 See tables giving the percentages of cases at the Commercial Hospital and Massachusetts General Hospital in which selected drugs were prescribed, in Warner, *The Therapeutic Perspective.*

35 The move to uniformity can be seen in the hospital records described in notes 18 and 19, as well as in, for example, records of the Charity Hospital of New Orleans (Department of Archives and Manuscripts, Library, Louisiana State University) and the New York Hospital (Medical Archives, The New York Hospital-Cornell Medical Center, New York).

36 E. B. Thompson to Mary Eleanor Thompson, New Orleans, 30 Dec. 1866 (Benson-Thompson Family Papers, Manuscript Collection, William R. Perkins Library, Duke University, Durham, North Carolina).

37 Roberts Bartholow, *On the Antagonism between Medicines and between Remedies and Disease Being the Cartwright Lectures for the Year 1880* (New York: D. Appleton and Company, 1881), p. 114.

38 E. W. Gray, "The relation of physiology to the practice of medicine," *Tr. A.M.A.,* 1874, *25:* 153–166, quotation on p. 159.

39 The best recent histories of the emergence of the modern hospital in America have failed to notice this point; for example, see David Rosner, *A Once Charitable Enterprise: Hospitals and Health Care in Brooklyn and New York, 1885–1915* (Cambridge: Cambridge Univ. Press, 1982), and Morris J. Vogel, *The Invention of the Modern Hospital: Boston, 1870–1930* (Chicago and London: Univ. of Chicago Press, 1980).

40 For a more detailed account of support for physiological therapeutics and its perceived implications in America, as well of opposition to it, see Warner, *The Therapeutic Perspective.*

41 The shifting historiography of the place of science in American medicine is analyzed, and the polemics that differing views have produced are reviewed, in John Harley Warner, "Science in medicine," in Sally Gregory Kohlstedt and Margaret Rossiter, eds., *Historical Writing on American Science: Perspectives and Prospects, Osiris,* 1985, n.s. *1:* 37–58.

42 For example, see E. Richard Brown, *Rockefeller Medicine Men: Medicine and Capitalism in America* (Berkeley and Los Angeles: Univ. of California Press, 1979); Gerald L. Geison, "Divided we stand: physiologists and clinicians in the American context," in Morris J. Vogel and Charles E. Rosenberg, eds., *The Therapeutic Revolution: Essays in the Social History of American Medicine* (Philadelphia: Univ. of Pennsylvania Press, 1979), pp. 67–90, (ch. 7, this book); and S. E. D. Shortt, "Physicians, science, and status: issues in the professionalization of Anglo-American medicine in the nineteenth century," *Med. Hist.,* 1983, *27:* 51–68.

6

The Introduction of
Antibiotics into Therapeutics

JOHN PARASCANDOLA

Although the antibiotics only entered medical practice in a significant way less than half a century ago, these so-called "wonder drugs" have already earned a secure place in the history of therapeutics. In fact, scholars discussing the history of penicillin and other antibiotics have tended towards hyperbole in assessing the role of these drugs in therapy. Thus, Gwyn Macfarlane has claimed that "penicillin therapy is probably the greatest single medical advance of all time."[1] James Whorton has called the antibiotics "the most spectacular therapeutic advance of medical history."[2] And Felix Marti-Ibañez has stated that in medical history "there is perhaps no other event as revolutionary as the discovery of antibiotics."[3]

While other writers on the history of medicine might dispute the choice of antibiotics as the most significant medical advance in all of history, it would probably be difficult to identify a class of therapeutic agents which has had a greater impact on human health. It therefore seems appropriate in a symposium on the history of therapeutics to devote a paper to an examination of the development of these drugs and their introduction into therapeutics. In this paper, I shall be using the term "antibiotic," coined in the modern sense by Selman Waksman in 1942, to mean a chemical substance produced by a microorganism which destroys or inhibits the growth of other microorganisms.[4]

John Parascandola is the Public Health Service Historian, U.S. Public Health Service, Rockville, Maryland.

Reprinted with permission from *History of Therapy: Proceedings of the 10th International Symposium on the Comparative History of Medicine—East and West*, edited by Yosio Kawakita, Shizu Sakai, and Yasuo Otsuka (Tokyo: Ishiyaku EuroAmerica, 1990), pp. 261–281.

EARLY WORK ON ANTIBIOSIS

Although the antibiotics have only made their impact felt in therapeutics within the last four decades, the phenomenon of antibiosis itself has been known for more than a century. I shall leave aside here any speculations about whether certain folk remedies involving moldy bread and similar substances may have had therapeutic value due to the presence of antibiotics, and begin my story with scientific investigations of the phenomenon of microbial antagonism in the late 19th century. The first entry under antibiotics in the Garrison-Morton *Medical Bibliography* dates from 1876.[5] In a paper in the *Philosophical Transactions* that year, John Tyndall reported on the antagonism between certain bacteria and the *Penicillium* mold.[6] Tyndall was not the first to observe microbial antagonism, which was not uncommonly reported in the literature from the early 1870s on.[7]

Louis Pasteur may have been the first to clearly suggest that the antagonism between microbes might have therapeutic potential. In 1877, he noticed that urine, which is normally an excellent culture medium for the anthrax bacillus, would not support the growth of this organism if the urine were also inoculated with common aerobic bacteria. He interpreted this to mean that there was a struggle for existence between various kinds of microbes, just as between plants and higher animals, and that one microorganism might successfully prevent the multiplication of another. Pasteur went so far as to suggest that these facts perhaps justified the highest hopes for therapeutics.[8]

Several attempts were made over the next few decades to use "antibiosis," a term introduced by Vuillemin in 1889 to describe this phenomenon of "life against life,"[9] for therapeutic purposes, albeit without much success. Probably the antibiotic substance receiving the most extensive trial in this period was pyocyanase, an extract prepared from old cultures of the bacillus *Pseudomonas pyocyanea*. Pyocyanase was introduced by Emmerich and Löw in 1899 and seems to have received its most widespread use in the treatment of diphtheria, usually as a local spray. By 1914, however, pyocyanase had fallen into disuse. This was perhaps due to toxic side effects as well as to a greater interest in other therapeutic approaches such as antitoxin therapy and Ehrlich's synthetic chemotherapy.[10]

THE DISCOVERY OF PENICILLIN

When Alexander Fleming observed the antibacterial effects of a *Penicillium* mold in 1928, the phenomenon of antibiosis was thus already well known to microbiologists and medical researchers. The story of Fleming's discovery of penicillin is too well known for me to discuss in detail here, hence I shall just briefly review the circumstances surrounding the event. While investigating the staphylococcus bacterium at St. Mary's Hospital in London, one of Fleming's culture plates was accidentally contaminated by a *Penicillium* mold. Upon observing that no colonies of the bacterium were growing for some distance around the mold, he decided to further investigate the antibacterial properties of the mold. Fleming proceeded to culture the mold on the surface of a broth and to test the filtrate of the broth culture, which he called penicillin, for antibacterial activity. He found that on culture plates penicillin inhibited the growth of such pathogenic organisms as hemolytic streptococcus and pneumococcus. He also tested the toxicity of the substance towards animals and found it to be exceptionally low. In a paper published in 1929, he reported these results and suggested that penicillin might prove useful as an antiseptic for local application to infected areas. He apparently did not envision its use, however, as an internal or systemic chemotherapeutic agent.[11]

Some attempts were made to isolate the active agent of the mold, by Craddock and Ridley in Fleming's laboratory and by Raistrick, Lovell, and Clutterbuck at the London School of Tropical Medicine, but without success. There were even a few isolated attempts to use the mold juice clinically as an external agent in infections in the early 1930s (e.g., by Fleming himself and by C. G. Paine at the

Sheffield Royal Infirmary), but they did not lead anywhere.[12] Neither Fleming nor anyone else seems to have taken the step of testing the substance by injection into experimentally diseased animals. Penicillin thus remained essentially a laboratory curiosity for about a decade after the publication of Fleming's paper.

The introduction of penicillin into therapeutics was a result of the work of scientists at the William Dunn School of Pathology at Oxford University. Australian-born Howard Florey was appointed to the chair of pathology at Oxford in 1935. Ironically the subject which eventually led Florey and his colleagues to an interest in penicillin was another discovery of Alexander Fleming's—lysozyme. In 1922, while studying a culture of his own nasal secretion, Fleming discovered a new bacterium, *Micrococcus lysodeikticus*, which was readily lysed by a substance present in the nasal secretion. He later found that this same substance, which he called lysozyme, was also present in other bodily secretions such as tears and saliva, as well as in egg white and certain plant tissues. Although he was never able to isolate lysozyme, Fleming concluded that it must be an enzyme. It seemed likely to Fleming that this natural substance played a role in the defenses of the organism against bacteria, but since common pathogenic bacteria such as streptococci and staphylococci proved to be resistant to lysozyme, it did not arouse much interest in the medical community.[13]

Howard Florey became interested in lysozyme through his own work on mucus. Florey set out with a colleague in 1929 to test the hypothesis that mucus has an antibacterial action and that any such action was related to its lysozyme content. They examined various animal preparations and eventually had to conclude that the presence or absence of lysozyme seemed to have little to do with natural immunity. An appendix to their paper in the *British Journal of Experimental Pathology* in 1930 discussed inhibition of one bacterium by another. This appendix described the inhibition of the growth of one of their lysozyme test organisms in the vicinity of *E. coli* on a culture plate, and pointed out that this type of bacterial antagonism was a well-known phenomenon, citing a 1928 review article on the subject. No mention was made, however, of Fleming's 1929 paper on penicillin which was published in the same journal as their own paper.[14]

Florey's interest in lysozyme continued over the years. Soon after going to Oxford, Florey began to look around for a biochemist for his laboratory. At the recommendation of F. Gowland Hopkins, he hired a young Jewish refugee from Nazi Germany, Ernst Boris Chain. Once lysozyme was isolated in crystalline form by E. P. Abraham and Robert Robinson in 1937, Chain began to investigate its mode of action on bacteria. He and L. A. Epstein showed that lysozyme was a polysaccharidase and that it attacked a glucosamine in the bacterial cell wall.[15]

During the course of his review of the literature on lysozyme, Chain apparently became interested in the whole subject of antibacterial substances produced by microorganisms, a subject which, as we have seen, had earlier attracted Florey's attention. Among the papers which Chain came across in his literature search was Fleming's paper on penicillin. Penicillin attracted his attention because it was presumed to be an enzyme with lytic powers, such as lysozyme. Florey and Chain decided to investigate the antibacterial action of several products of microbial origin, originally focusing on penicillin, pyocyanase, and actinomycetin. The latter was a bacteriolytic filtrate from a species of actinomycetes which had been studied extensively by Gratia and Dath in the 1920s and by Welsch in the 1930s. Although they mentioned the therapeutic potential of these substances in grant applications at the time, both Florey and Chain later stated that it was a scientific interest in the problem rather than a therapeutic goal which initially motivated their researches.[16]

In 1939, when Florey and Chain were still in the early stages of this work, another development occurred which stimulated further interest in the therapeutic potential of antibiosis. Rene Dubos isolated an organism from soil, later identified as *Bacillus brevis*, which was capable of attacking gram positive microorganisms. From this bacillus he obtained a substance which he called tyrothricin, which was later shown to consist of two polypeptide components, gramicidin and tyrocidine. Unfortunately, these antibiotic substances proved to be too toxic for internal use in humans, although gramicidin found some use for topical applications. More importantly, Dubos' work encouraged research on antibiotics. One of those influenced to undertake work in this area was Dubos' former teacher, Selman Waksman, about whom I shall have more to say later.[17]

Meanwhile the Oxford group, which had come to include several other scientists such as Norman Heatley and E. P. Abraham, made significant progress in their work on penicillin. They managed to isolate a brown powder that exhibited such extraordinarily high antibacterial properties that they assumed it was relatively pure. Actually, penicillin was so much more effective as an antibacterial agent than any other drug known at the time, including the then recently introduced sulfonamides, that the Oxford team at first greatly underestimated its potency. The crude brown powder that they initially isolated was later shown to contain less than 1% penicillin.

Toxicity tests yielded no evidence that penicillin was harmful to the experimental animals used. Then in May of 1940 came the first test of penicillin's activity against pathogenic organisms *in vitro*. Eight mice were injected with a virulent strain of streptococci, with four left untreated as controls and four treated with penicillin. The results were sufficiently promising to convince Florey, Chain, and coworkers of penicillin's promise as a therapeutic agent. By the next morning, all four of the untreated mice were dead. All of the mice treated with penicillin were alive and three were in perfectly normal health, although the fourth died two days later.

In the early work at Oxford, the penicillin was produced by growing the mold on the surface of a broth in an Erlenmeyer flask, essentially as Fleming had done. The conical flasks took up a lot of space and were not available in sufficient numbers to produce the large amounts of penicillin that would be needed for further testing of the drug. It was felt that a flat vessel with a large surface area would be more advantageous, and the Oxford group began to experiment with various kinds of common objects such as dishes, trays, pans, and biscuit tins.

The old-style urinal or bed pan was found to be a very satisfactory vessel. It provided a relatively large surface area over a shallow fluid and had a side arm for inoculation and withdrawal of the fluid. But the demand for these in military hospitals made it impossible to obtain them in large quantities for penicillin production. Britain was being subjected to heavy bombing, and it was not easy to obtain many types of supplies. The Oxford laboratory managed to arrange, however, for a manufacturer to produce a similar type of container for use in the penicillin work, a rather flat rectangular ceramic vessel. These vessels could be stacked horizontally for greater economy of space, provided relatively more surface area than an Erlenmeyer flask, and had a side arm for ease of inoculation.[18]

As more penicillin was produced, further animal studies were carried out, and in August of 1940 the Oxford team published a brief account of their work on "Penicillin as a chemotherapeutic agent" in *The Lancet*.[19] This first publication on penicillin from Oxford attracted the attention of Alexander Fleming, who immediately contacted Florey and went to visit the laboratory to see what was being done with his "old penicillin."[20]

Eventually Florey and his colleagues were prepared to undertake their first trials of the drug in human patients. First, the drug was given to a volunteer dying of cancer, as a test of its toxicity in humans. The first patient to receive penicillin produced at Oxford in an effort to treat an infection was a policeman suffering a severe case of staphylococcal and streptococcal infection resulting from a cut on his face. He was near death when penicillin therapy was begun in February of 1941, and within a few days he was making a striking recovery. Unfortunately, the supply of available penicillin soon ran out, and the patient relapsed and died. Florey was disappointed, and he decided that it was necessary to accumulate a larger supply of the drug before treating future cases and to work with children, if possible, so that the dose required would be smaller.[21]

Actually, the policeman treated in Oxford was not the first human patient to receive the drug internally in an effort to treat an infection. Some four months earlier a patient at Presbyterian Hospital in New York had been treated for subacute bacterial endocarditis with penicillin, although without success. Martin Henry Dawson, a clinician at the College of Physicians and Surgeons of Columbia University, recognized on reading the first publication of the Oxford group on their animal experiments that penicillin had therapeutic potential in the treatment of streptococcal and pneumococcal infections. He obtained cultures of the penicillin-producing strain of *Penicillium notatum* from Chain in England and from Roger Reid of the Johns Hopkins Hospital, who had obtained the mold from Fleming several years earlier. Dawson and coworkers Gladys Hobby and Karl Meyer managed to ob-

tain penicillin extracts from Reid's sample with which they treated four patients with subacute bacterial endocarditis and eight patients with chronic staphylococcal blepharitis. Sufficient penicillin was not available for adequate therapy in the case of the former group of patients, but those with the staphylococcal eye infection responded well to the treatment. The Columbia researchers reported their results on penicillin at a meeting of the American Society for Clinical Investigation in May, 1941. Although this work was published only in abstract form, it was reported on in newspapers such as *The New York Times* which helped bring penicillin to the attention of Americans, including officials at Charles Pfizer and Company.[22]

Clinical trials in Oxford continued and a report on the first ten cases was published in *The Lancet* in August of 1941.[23] Although the results were encouraging, they were hardly statistically significant. Florey recognized that much greater quantities of penicillin would be needed to carry out the large-scale clinical trials that were necessary to establish the therapeutic value of penicillin.

DEVELOPMENT OF LARGE-SCALE PRODUCTION OF PENICILLIN

Unable to generate as much interest in penicillin as he had hoped to on the part of the pharmaceutical industry in wartime Britain, Florey traveled to the United States with Norman Heatley in July 1941 in an effort to stimulate American interest in the drug. One of the first places that they visited was the Northern Regional Research Laboratory of the United States Department of Agriculture in Peoria, Illinois. The Peoria group agreed to tackle the problem of trying to improve the yield of penicillin produced by the mold, and did manage, over a period of time, to increase penicillin production dramatically through a variety of innovations.

One avenue of attack undertaken in Peoria was to find strains of the *Penicillium* mold that produced greater quantities of penicillin than the original Fleming strain. Of all the samples of mold collected and tested, many of them sent in from great distances, it is ironic that the most productive strain discovered came from a moldy cantaloupe melon from a Peoria fruit stand. The "cantaloupe strain" was later further improved in academic laboratories by producing mutant strains through the use of chemicals or irradiation. The Peoria group also played a key role in the development of deep-tank fermentation, which was much more efficient than surface-culture fermentation of the mold.[24]

While Florey was in the United States, he visited several pharmaceutical companies and tried to interest them in penicillin, with some success. In fact, a few such firms had already begun work on the antibiotic before Florey arrived. For example, Randolph Major, director of research at Merck, developed an interest in penicillin. He was encouraged by microbiologist Selman Waksman of Rutgers University, who had become a consultant to the firm in 1938. Merck chemists were struggling to isolate penicillin at the time of Florey's visit.[25]

Perhaps more important was Florey's success in creating interest in the drug on the part of the Office of Scientific Research and Development (OSRD), recently established by the Federal government, and its Committee on Medical Research (CMR), chaired by University of Pennsylvania pharmacologist A. N. Richards. The OSRD and CMR set about to organize a concerted program of penicillin research involving the pooling of information and results. With the encouragement of the CMR, several pharmaceutical companies agreed to undertake collaborative efforts in penicillin research and development. Industrial cooperation also occurred in Britain where, in 1942, the Therapeutic Research Corporation of Great Britain was established by five pharmaceutical firms for the purpose of sharing research results. Later that year a sixth company joined the group, and a committee was established, at the request of the Ministry of Supply, to coordinate work on the production and purification of penicillin.

With the entry of the United States into the war, the American pharmaceutical industry became accountable to the War Production Board (WPB). The WPB soon established a formal penicillin program with Albert Elder as coordinator. The Board was concerned not just with research in industry, as the CMR had been, but also with productivity. Its penicillin program involved some 20 commercial firms as well as government and academic research laboratories. The Board also controlled the supply of penicillin in the United States with most of it going to the military. A prime goal of the government was to have adequate supplies of the drug available for the planned invasion of Europe, scheduled for the spring of 1944.[26]

Feelings of wartime patriotism greatly stimulated work on penicillin in Britain and the United States. The coordinator of the American penicillin program, for example, wrote to manufacturers in 1943, "You are urged to impress upon every worker in your plant that penicillin produced *today* will be saving the life of someone in a few days or curing the disease of someone now incapacitated. Put up slogans in your plant! Place notices in the pay envelopes! Create an enthusiasm for the job down to the lowest worker in your plant."[27]

Meanwhile penicillin was being tested clinically on a large scale for its effectiveness on a number of pathological conditions. The drug was shown to be effective in the treatment of a wide variety of infections, including various pneumococcal, streptococcal, staphylococcal and gonococcal infections. The United States Army established the value of penicillin in the treatment of surgical and wound infections. Clinical studies demonstrated its effectiveness against syphilis, and by 1944, it was the primary treatment for this disease in the armed forces of Britain and the United States.[28]

As publicity concerning this new "miracle drug" began to reach the public, the demand for penicillin increased. But supplies were at first very limited, and as noted earlier, priority was given to military use. In the United States, Dr. Chester S. Keefer of Boston, Chairman of the National Research Council's Committee on Chemotherapy, had the unenviable task of rationing supplies of the drug for civilian use. Keefer had to restrict the use of the drug to cases where it was fairly clear that penicillin would be helpful and where other methods of treatment had failed. Part of his job was also to collect detailed clinical information about the use of the drug so that a fuller understanding of its potential and its limitations could be developed. Not surprisingly, Keefer was besieged with pleas for penicillin.[29] As one newspaper account noted, "Many laymen—husbands, wives, parents, brothers, sisters, friends—beg Dr. Keefer for penicillin. In every case the petitioner is told to arrange that a full dossier on the patient's condition be sent in by the doctor in charge. When this is received, the decision is made on a medical, not an emotional basis."[30]

Fortunately, penicillin production began to increase dramatically by early 1944. Production of the drug in the United States jumped from 21 billion units in 1943 to 1,663 billion units in 1944.[31] The American government was eventually able to remove all restrictions on its availability, and as of March 15, 1945, penicillin was distributed through the usual channels and made available to the consumer in his or her corner pharmacy. By 1949, the annual production of penicillin in the United States was 133,229 billion units, and the price had dropped from 20 dollars per 100,000 units in 1943 to less than 10 cents.[32]

PENICILLIN OUTSIDE THE ANGLO-AMERICAN COMMUNITY

Knowledge of the therapeutic potential of penicillin soon began to spread to other countries, arousing interest in the production of the drug. Research on penicillin began at the Swiss Federal Institute of Technology in 1942, but did not yield spectacular results.[33] There were also efforts to produce the drug in Germany as early as 1942, but relatively little was accomplished during the war years.[34]

Work on penicillin was carried out in several German-occupied countries during the war. A Dutch fermentation firm, the Gistfabriek (or "yeast factory"), first learned of penicillin through British radio broadcasts and pamphlets dropped by British planes. The company then obtained the August 7, 1943, issue of *Klinische Wochenschrift* containing an article by Manfred Kiese which abstracted various English-language papers on penicillin published in the period 1940–1943. Papers on other antibiotic substances were also cited in this review. Gistfabriek began to produce small amounts of penicillin in 1944. In occupied France, penicillin was being produced at the Pasteur Institute and at the Rhone-Poulenc firm in 1943, with the monthly output reaching 765,000 units by December of 1944.[35] Rumors of the promise of penicillin reached occupied Czechoslovakia late in 1942, and work on the drug was begun at the Fragner Company. Enough penicillin was eventually produced to carry out clinical trials during the war.[36]

Research on penicillin was initiated in Japan in February, 1944, when the first meeting of the Peni-

cillin Committee was held at the Military Medical College in Tokyo. The Japanese first became aware of penicillin through the review article by Kiese in *Klinische Wochenschrift* mentioned earlier. A copy of this journal had been sent from Germany by submarine. A plan to organize a penicillin group, drafted by surgeon major Katsuhiko Inagaki and his colleagues, was approved by the War Minister. By the end of the war, the Japanese were producing relatively small amounts of penicillin and treating patients with it. In the post-war era, Japan has emerged as a major contributor to antibiotics research and production.[37]

Although a laboratory of antibiotics was established by the Ministry of Health in Moscow as early as 1942, penicillin does not seem to have received extensive attention in the Soviet Union during the war years. Dubos' work on tyrothricin would appear to have had more impact, and a new antibiotic, gramicidin S (Soviet gramicidin) was isolated from *Bacillus brevis* in the summer of 1942. The drug received widespread use for topical application to control local infections.[38] At the close of the war, a project was organized in the United States to raise funds to give the Soviet people a penicillin plant and research laboratory. The effort did not succeed, however, but fell prey to the cold-war mentality that soon developed between the two former allies.[39]

ANTIBIOTICS AFTER PENICILLIN

The tremendous success of penicillin as a systemic chemotherapeutic agent opened up the "era of antibiotics." Even before the end of the war, several other antibiotic substances had been discovered. The work of Dubos and the discovery of gramicidin S have already been mentioned above. One of the most active and productive wartime antibiotics research laboratories was that of Selman Waksman at Rutgers University. Trained as a soil microbiologist, Waksman had long been aware of the phenomenon of antibiosis. It was only after the work of his former student Dubos on tyrothricin, however, that he began a systematic search for antibiotic substances produced by soil microorganisms.

Waksman focused his researches on a group of microorganisms known as actinomycetes, which had interested him from early in his career. These organisms display some of the characteristics of bacteria and some of the characteristics of fungi. In retrospect this choice of organisms was a fortunate one, since they have turned out to be producers of some of the most important antibiotics introduced into therapy. In 1940, Waksman and his colleagues isolated their first antibiotic substance from actinomycetes, and called it actinomycin. Although it was active against a large number of bacteria, it was too toxic for use as a therapeutic agent. Streptothricin was isolated in 1942 and seemed at first to have therapeutic promise, but it was later shown to have a delayed toxicity in the animal body.

The real breakthrough in Waksman's laboratory came with the isolation of streptomycin in 1943. The name was derived from *Streptomyces*, the name assigned by Waksman to the particular group of actinomycetes from which the antibiotic was isolated. Streptomycin's greatest value proved to be in the treatment of tuberculosis. Because of the development of resistant strains of the tubercle bacillus and because of the drug's adverse side effects, it was later found that tuberculosis could be treated more effectively by using streptomycin in combination with another drug, such as para-aminosalicylic acid or isoniazid.[40]

Within a few years after the end of the war, three more important antibiotics had been added to the list of such drugs: chloramphenicol (1947), chlortetracycline (1948), and oxytetracycline (1950). These drugs were the first so-called "broad spectrum antibiotics," characterized by their ability to inhibit a variety of different types of bacteria (both gram-negative and gram-positive) and some ricksettsia. All were isolated from *Streptomyces* in the laboratories of American pharmaceutical companies. Their discovery was the result of large-scale screening programs aimed at identifying antibiotic substances produced by soil microorganisms. The discovery of oxytetracycline, for example, represented the culmination of an extremely intensive screening program involving more than 100,000 samples of soil from many different parts of the world.[41]

Over the past three decades hundreds more antibiotic substances have been discovered, although only a relatively small proportion of these have found significant therapeutic use. Among the more important of these therapeutically are erythromycin, tetracycline, the cephalosporins, and the semisynthetic penicillins. With very few exceptions,

these drugs are produced by microbial fermentation rather than by chemical synthesis. It may be noted in passing that most antibiotics have been named on a subjective basis, sometimes leading to interesting results. Antibiotics have been named, for example, after the discoverer's laboratory (nystatin, for the New York State Board of Health Laboratories), his wife (helenine), his secretary (vernamycin), and his mother-in-law (saramycetin). Bacitracin was named after a girl named Tracy, from whose wound the bacillus that produced it was isolated. One antibiotic was even named after a movie (rifomycin, from *Rififi*).[42]

IMPACT OF ANTIBIOTICS ON THERAPEUTICS

Antibiotics were found to be effective in the treatment of a host of infectious diseases: pneumococcal pneumonia, rheumatic fever, bacterial endocarditis, syphilis, gonorrhea, tuberculosis, meningococcal meningitis, diphtheria, typhus, Rocky Mountain spotted fever, etc. Within a decade after penicillin was first made freely available for civilian use in the United States in 1945, the antibiotics had become the most important class of drugs in the treatment of infectious disease. In 1948, antibiotics prescriptions accounted for only 1.5% of the total number written in the United States; by 1952, that figure had risen to 13.7%.[43]

A Federal Trade Commission report issued in 1958 attempted to analyze the effect of the introduction of antibiotics on American public health. Although admitting that antibiotic use was only one factor contributing to declining rates of incidence and mortality in many infectious diseases over the previous decade or so, the report pointed to evidence suggesting that the antibiotics made a significant contribution to this decline. In examining the occurrence of reportable diseases, for example, it was found that over the period 1946–1955 there was a 42% drop for diseases in which antibiotics were effective and only a 20% drop for diseases in which antibiotics were not regarded as effective. The Commission concluded that "it appears that the use of antibiotics, early diagnosis and other factors have limited the epidemic spread and thus the number of these diseases which occurred."[44]

More convincing evidence for the impact of antibiotics was provided in the Commission's analysis of mortality statistics for certain infectious diseases. The Commission examined mortality statistics for eight important diseases for which antibiotics offered effective therapy: tuberculosis, syphilis, dysentery, scarlet fever, diphtheria, whooping cough, meningococcal infections, and pneumonia. They found a 56.4% decrease in the total number of deaths for these diseases combined in the period 1945–1955. The decline for all other causes of death over this same period was only 8.1%. The figures were also dramatic for certain diseases where antibiotics offered a significant new therapy. For example, the decrease in tuberculosis mortality over this period was 75.2%, as compared with a decrease of 28.6% during the previous 10 years.[45]

In another 1958 publication, C. C. Dauer, Medical Adviser, National Office of Vital Health Statistics, United States Public Health Service, reported that mortality rates from infectious diseases had declined at a more rapid rate since sulfonamides and antibiotics had come into use as therapeutic agents. He estimated that 1.5 million lives had been saved over the previous 15 years as a result of this accelerated decline in mortality, recognizing, however, that other factors in addition to sulfonamide and antibiotic therapy contributed to the saving of these lives.[46]

Whatever the exact number of lives saved by the antibiotics since their introduction into therapeutics, it would be difficult to deny that they have played an important role in the declining mortality due to infectious disease. The past few decades have seen a significant change in the patterns of disease and death in industrialized nations. One study of changes in causes of deaths in Graz, Austria, for example, showed that deaths due to infections decreased 56% over the period 1930–1970. By contrast, deaths from malignancies over those same four decades increased 27% and deaths from degenerative diseases increased 44%.[47] It has been estimated that four out of five of the leading causes of deaths in children were microbially related at about the time that the sulfonamides and antibiotics were first introduced, whereas 20 years later four out of five were non-microbial.[48]

Antibiotics have also expanded what surgeons can do by helping to control infections. Walsh McDermott has argued that "there are several areas of surgery today that would hardly be possible were it not for the antimicrobial technology. The arena of cardiac surgery, organ transplantation and the management of severe burns can serve as examples."[49]

The introduction of antibiotics also served as a stimulus to the growth of the pharmaceutical industry and to the expansion of research in this industry in countries such as the United States and Japan. The total number of pounds of antibiotics produced by American pharmaceutical firms increased from 240,332 in 1948 to 3,081,373 in 1956. Twenty-nine new antibiotic substances were introduced from 1949 through 1956 by these firms, although all of them did not remain in production. Typically these products were protected by patents and, consequently, were produced exclusively by one company. The ethical drug industry in the United States became much more dependent on the development of patentable new drugs for generating profits in the post–World War II period, with antibiotics leading the way at first. By 1956, sales of antibiotics represented 17% to 39% of the total sales of major producers such as Lilly, Parke Davis, and Pfizer.[50]

On the negative side, concern has been expressed about the misuse and overuse of antibiotics. Evidence about the adverse side effects of antibiotics began to surface, and it was recognized that frequent exposure to antibiotic drugs could increase the proportion of antibiotic-resistant strains in a given population of microorganisms. Antibiotics were frequently prescribed in conditions where they were not effective, such as virus-induced respiratory tract infections. These negative developments also had a positive side, in that they helped to increase awareness of the problem of iatrogenic disease and to stimulate efforts toward a more rational therapeutics.[51]

The evidence suggests, however, that the "golden age" of antibiotics is over. In 1979, Philip Paterson pointed out that all of the major classes of clinically significant antibiotics had been discovered by 1959, and that subsequently introduced antibiotics largely represented molecular rearrangements or semisynthetic modifications of previously discovered drugs. In addition, he noted, strains of pathogenic microorganisms which were resistant to antibiotics seemed to be appearing more often. In Paterson's view, the "bug-drug" perspective that had come to dominate therapeutics in the antibiotic era had deflected much attention from other promising lines of research, such as host immune defense systems and the development of new vaccines. He felt, however, that the balance was shifting as the limitations of antibiotic therapy were increasingly realized.[52]

Antibiotics will no doubt continue to play a significant role in therapeutics, and new antibiotics will probably continue to be discovered, although probably not at the same pace and with the same dramatic results as in the "golden age" of the 1940s and 1950s. As with other new drug discoveries, however, there was too much of a tendency at first to see antibiotic therapy as a panacea. In time, we have begun to recognize the limitations of these "wonder drugs." Historians a generation or two in the future will be able to look back on the "antibiotic era" with greater perspective than we can do today, and to more fully assess the role of antibiotics in 20th-century therapeutics.

NOTES

I would like to acknowledge the assistance of Margaret Kaiser, reference librarian in the History of Medicine Division, National Library of Medicine, in locating some of the source materials used in preparing this paper. I would also like to thank John Swann and my fellow participants at the 10th International Symposium on the Comparative History of Medicine—East and West for their helpful comments on earlier drafts of this paper.

1 Gwyn Macfarlane, *Alexander Fleming: The Man and the Myth* (Cambridge, Mass.: Harvard Univ. Press, 1984), p. 268.

2 James Whorton, "'Antibiotic abandon': the resurgence of therapeutic rationalism," in J. Parascandola, ed., *The History of Antibiotics: A Symposium* (Madison, Wis.: American Institute of the History of Pharmacy, 1980), p. 126.

3 Felix Marti-Ibañez, *Men, Molds, and History* (New York: MD Publications, 1958), p. 1.

4 Selman Waksman, "History of the word 'antibiotic,'" *J. Hist. Med.*, 1973, *28*: 284–286.

5 Leslie T. Morton, *A Medical Bibliography (Garrison and Morton): An Annotated Check-List of Texts Illustrating the History of Medicine*, 4th ed. (Aldershot: Gower, 1983), p. 248.

6 John Tyndall, "The optical deportment of the atmosphere in relation to the phenomenon of putrefaction and infection," *Philosophical Transactions of the Royal Society, London*, 1876, *166*: 27–74.

7 For discussions of early observations of microbial antagonism, see J. Brunel, "Antibiosis from Pasteur to Fleming," *J. Hist. Med.*, 1951, *6*: 287–301; J. K. Crellin, "Antibiosis in the 19th century," in Parascandola, ed., *Antibiotics*, pp. 5–13; H. W. Florey, E. Chain, N. G. Heatley, M. A. Jennings, A. G. Sanders, E. P. Abraham, and M. E. Florey, *Antibiotics*, vol. 1 (London: Oxford Univ. Press, 1949), pp. 1–73.

8 Louis Pasteur and Jules F. Joubert, "Charbon et septicémie," *Compte rendu hebdomadaire des séances de l'Académie des sciences (Paris)*, 1877, *85*: 101–115.

9 Paul Vuillemin, "Antibioses et symbiose," *Compte rendu de l'Association française pour l'avancement des sciences*, Pt. 2, 1889 (publ. 1890), *18*: 525–543.

10 On pyocyanase, see Florey et al., *Antibiotics*, vol. 1, pp. 19–26; Rudolf Emmerich and Oscar Löw, "Bacteriolytische Enzyme als Ursache der erworbenen Immunität und die Heilung von Infectionskrankheiten durch dieselben," *Zeitschrift für Hygiene und Infectionskrankheiten*, 1889, *31*: 1–65.

11 Alexander Fleming, "On the antibacterial action of cultures of a penicillium, with special reference to their use in the isolation of *B. influenzae*," *Brit. J. Exper. Path.*, 1929, *10*: 226–236. See also Macfarlane, *Fleming*, pp. 117–138.

12 Gladys Hobby, *Penicillin: Meeting the Challenge* (New Haven, Conn.: Yale Univ. Press, 1985), pp. 48–50. For a discussion of the work of C. G. Paine, whose claim to have used penicillin therapeutically at this time has been substantiated by documented clinical notes, see Milton Wainwright and Harold T. Swan, "C. G. Paine and the earliest surviving clinical records of penicillin therapy," *Med. Hist.*, 1986, *30*: 42–56. I am grateful to Vivian Nutton for calling my attention to this paper.

13 Macfarlane, *Fleming*, pp. 98–111; Gwyn Macfarlane, *Howard Florey: The Making of a Great Scientist* (Oxford: Oxford Univ. Press, 1979), pp. 177–179.

14 N. E. Goldsworthy and H. W. Florey, "Some properties of mucus, with special reference to its antibacterial functions," *Brit. J. Exper. Path.*, 1930, *9*: 192–208; Macfarlane, *Florey*, pp. 179–182.

15 L. A. Epstein and E. B. Chain, "Some observations on the preparation and properties of the substrate of lysozyme," *Brit. J. Exper. Path.*, 1940, *21*: 339–355; Macfarlane, *Florey*, pp. 276–279.

16 Macfarlane, *Florey*, pp. 279–286; Florey et al., *Antibiotics*, vol. 1, pp. 50–52; Trevor Williams, *Howard Florey: Penicillin and After* (Oxford: Oxford Univ. Press, 1984), pp. 86–92.

17 Rene Dubos, "Studies on a bactericidal agent ex-

tracted from a soil bacillus. I. Preparation of the agent. Its *in vitro* activity," *J. Exper. Med.,* 1939, *70:* 1–10; Rene Dubos, "Studies on a bactericidal agent extracted from a soil bacillus. II. Protective effect of the bactericidal agent against experimental pneumococcal infections in man," *J. Exper. Med.,* 1939, *70:* 11–17; Florey et al., *Antibiotics,* vol. I, pp. 422–442.

18 This discussion of the efforts to culture and isolate penicillin at Oxford is based on Florey, et al., *Antibiotics,* vol. II, pp. 635–647; H. W. Florey and E. P. Abraham, "The work on penicillin at Oxford," *J. Hist. Med.,* 1951, *6:* 302–317; E. Chain, "A short history of the penicillin discovery from Fleming's early observations in 1929 to the present time," in Parascandola, ed., *Antibiotics,* pp. 15–29.

19 E. Chain, H. W. Florey, A. D. Gardner, N. G. Heatley, M. A. Jennings, J. Om-Ewing, and A. G. Sanders, "Penicillin as a chemotherapeutic agent," *Lancet,* 1940, *2:* 226–228.

20 Macfarlane, *Florey,* p. 323.

21 *Ibid.,* pp. 330–331; Lennard Bickel, *Rise Up to Life: A Biography of Howard Walter Florey Who Gave Penicillin to the World* (London: Angus and Robertson, 1972), pp. 121–123.

22 Bickel, *Rise Up,* pp. 124–129; Hobby, *Penicillin,* pp. 69–79; "'Giant' germicide yielded by mold," *N.Y. Times,* May 6, 1941, p. 23. I am indebted to my colleague Peter Hirtle and to Donald Shay of the Center for the History of Microbiology, University of Maryland, Baltimore-County for a copy of the *N.Y. Times* article.

23 E. P. Abraham, E. Chain, C. M. Fletcher, A. D. Gardner, N. G. Heatley, M. A. Jennings, and H. W. Florey, "Further observations on penicillin," *Lancet,* 1941, *2:* 177–188.

24 On the Peoria research, see Robert Coghill, "The development of penicillin strains," in Albert Elder, ed., *The History of Penicillin Production* (New York: American Institute of Chemical Engineers, 1970), pp. 15–21; Emerson Lyons, "Deep-tank fermentation," *ibid.,* pp. 33–36; Hobby, *Penicillin,* pp. 94–103.

25 W. H. Helfand, H. B. Woodruff, K. M. H. Coleman, and D. L. Cowen, "Wartime industrial development of penicillin in the United States," in Parascandola, ed., *Antibiotics,* pp. 31–33.

26 On the CMR and WPB involvement with penicillin, see *ibid.,* pp. 38–50; Albert Elder, "The role of the government in the penicillin program," in Elder, ed., *Penicillin Production,* pp. 3–11; A. N. Richards, "Production of penicillin in the United States (1941–1946)," *Nature,* 1964, *201:* 441–445; John Swann, "The search for synthetic penicillin during World War II," *British Journal for the History of Science,* 1983, *16:* 154–190.

27 Elder, "Role of the government," p. 7.

28 On the use of penicillin in treating various diseases, see Hobby, *Penicillin,* pp. 115–124, 141–170; Harry Dowling, *Fighting Infection: Conquests of the Twentieth Century* (Cambridge, Mass.: Harvard Univ. Press), pp. 136–157.

29 On Keefer and the allocation of penicillin, see Dowling, *Infection,* pp. 132–133; Hobby, *Penicillin,* pp. 141–143.

30 J. G. Rogers, *N.Y. Herald Tribune,* Oct. 17, 1943.

31 Hobby, *Penicillin,* p. 197.

32 George Urdang, "The antibiotics and pharmacy," *J. Hist. Med.,* 1951, *6:* 388–405. The figures are from p. 403.

33 L. Ettlinger, "Wartime research on penicillin in Switzerland and antibiotic screening," in Parascandola, ed., *Antibiotics,* pp. 57–67.

34 Hobby, *Penicillin,* pp. 207–208.

35 *Ibid.,* pp. 202–206.

36 Letter from K. Wiesner to David Perlman, Nov. 29, 1976 (copy in possession of author).

37 Yukimasa Yagisawa, "Early history of antibiotics in Japan," in Parascandola, ed., *Antibiotics,* pp. 69–90.

38 G. F. Gause, "Gramicidin S and early antibiotic research in the Soviet Union," in Parascandola, ed., *Antibiotics,* pp. 91–95.

39 Patricia Spain Ward, "Antibiotics and international relations at the close of World War II," in Parascandola, ed., *Antibiotics,* pp. 101–112.

40 Hubert Lechevalier, "The search for antibiotics at Rutgers University," in Parascandola, ed., *Antibiotics,* pp. 113–123; Dowling, *Infection,* pp. 158–173.

41 On the "broad-spectrum" antibiotics, see Dowling, *Infection,* pp. 174–192.

42 On the naming of antibiotics, see David Perlman, *Antibiotics* (Chicago: Rand McNally, 1970), p. 10.

43 Federal Trade Commission, *Economic Report on Antibiotics Manufacture, June 1958* (Washington, D.C.: United States Government Printing Office, 1958), p. 270.

44 *Ibid.,* p. 277.

45 *Ibid.,* p. 279.

46 C. C. Dauer, "A demographic analysis of recent changes in mortality, morbidity, and age group distribution in our population," in Iago Galdston, ed., *The Impact of Antibiotics on Medicine and Society* (New York: International Universities Press, 1958), pp. 98–120.

47 Broda Barnes, Max Ratzenhofer, and Richard Gisi, "The role of natural consequences in the changing death patterns," *Journal of the American Geriatrics Society,* 1974, *22:* 176–179.

48 Walsh McDermott (with David Rogers), "Social ramifications of control of microbial disease," *Johns Hopkins Medical Journal,* 1982, *151:* 302–312. The figures are from p. 308.

49 *Ibid.*, p. 308.
50 See Federal Trade Commission, *Economic Report,* pp. 65–109, 199–224 for information on antibiotics production and sales.
51 See Whorton,"'Antibiotic abandon,'" pp. 125–136.
52 Philip Paterson, "Infectious diseases—Into the 1980's and beyond," *J. Infect. Dis.,* 1979, *140*: 125–126. See also Guenter Risse and John Parascandola, "From bug and drug to human host: the control of infectious disease," unpublished paper delivered at a symposium on "Science in medicine, 1906–1981," Emory University, Atlanta, 1981.

THE SCIENCE OF MEDICINE

Before the late 19th century American physicians earned their medically related living by treating patients or, on occasion, teaching students. No jobs existed for aspiring medical researchers. As Ronald L. Numbers and John Harley Warner point out, American physicians tended to value wealth above knowledge and to pride themselves more on their skill as practitioners of the art of medicine than on their contributions to the science of medicine. Not surprisingly, some of their most original advances came in the practical fields of surgery and dentistry.

Science and medicine were not, however, as divorced as some historians have implied in describing the period before the rise of laboratory science. Medical practices often reflected the scientific theories of the time, such as the pathogenic role of miasmas and the equation between health and keeping the bodily fluids in equilibrium. From the colonial period onward at least some American physicians engaged in the study of medical botany and medical geography, hoping thereby to discover new medicinal plants or a correlation between environment and disease. And, as Warner has recently pointed out, physicians learned about the natural history of diseases at the bedsides of their patients.

With the reform of medical education at the turn of the century came a growing emphasis on experimental science. Rather than assuming the inevitability of this relationship, Gerald L. Geison provocatively asks why this was so. What did experimental physiology, for example, contribute to the task of healing? Or, more generally, how did scientific theory relate to the day-to-day clinical activities of physicians? Because the practice of cardiology and surgery benefited so little from what was being discovered in physiological laboratories, Geison argues that medical reformers embraced science more for its cultural and educational value than for its clinical utility. Like Latin in an earlier age, science in the late 19th century came to symbolize the professional standing of physicians.

As medical professors freed themselves from a reliance on student fees in the late 19th century, they grew increasingly hospitable to basic research. The scientifically oriented medical schools of the 20th century often devoted as much effort to discovering new medical knowledge in the laboratory and clinic as to disseminating it in the classroom. Private foundations, pharmaceutical companies, and government agencies also became patrons of medical science. After World War II the federally funded National Institutes of Health (NIH) picked up the lion's share of the tab for basic research. By the mid-1990s the NIH budget was approaching $12 billion a year. According to the political scientist Stephen P. Strickland, this largesse resulted in part as compensation for Congress's failure to create a national health-care system. Federal legislators could demonstrate their concern for the health of Americans by supporting medical research while at the same time opposing national health insurance.

7

Divided We Stand: Physiologists and Clinicians in the American Context

GERALD L. GEISON

> When you enter my wards your first duty is to forget all your physiology. Physiology is an experimental science—and a very good thing no doubt in its proper place. Medicine is not a science, but an empirical art.
>
> *Samuel Gee, 1888*[1]

INTRODUCTION

By the mid-19th century, European physiologists had largely won their campaign to secure the independence of their subject from medical anatomy. They had achieved this emancipation by self-consciously adopting an experimental approach toward the study of vital processes. They exploited for their own purposes recent advances in physics and chemistry, and they especially emphasized the value of vivisection experiments in the investigation of animal function. Henceforth they could claim to belong to a separate discipline no longer to be regarded as a mere "handmaiden" of medicine.[2]

A generation later, English and American physiologists had begun to enjoy a similar sense of independence. Yet the prospects of the new discipline remained closely bound up with the destinies of medicine and medical education. If physiology found its most receptive home in universities (occasionally even in philosophical faculties rather than medical schools), its audience nonetheless consisted overwhelmingly of intending physicians. Without those premedical and medical students, and without the resources that came to it by virtue

of its association with medicine, the newly "independent" discipline would have withered on the vine. However distasteful it may have been for some research physiologists to admit it, they remained essentially parasitic on the larger medical enterprise from which they had emerged.

But perhaps medicine was itself becoming dependent upon its new disciplinary offspring. Perhaps, to modify my earlier metaphor, it was less the case that physiology remained parasitic on medicine than that the two had entered into a symbiotic relationship. Many physiologists certainly believed (or hoped) so. The benefits that medicine derived from this symbiosis have traditionally been described (by physiologists and medical historians alike) in terms of the presumed utility of physiological theories, techniques, and instruments for the medical problems faced by practicing doctors. The word "presumed" is used advisedly, for these traditional allusions to the medical utility of physiology are disappointingly brief and vague.[3] They tend to take for granted the point at issue. No one, to my knowledge, has yet made a sustained effort to identify the specific ways in which experimental physiology has contributed to the healing task.

Some readers may feel that the contributions of the discipline to medical practice were (and are) so obvious as to require neither detailed elaboration nor systematic defense. Yet repeated assertions as to the medical value of experimental physiology have failed to still a remarkably persistent stream

GERALD L. GEISON is Professor of History, Princeton University, Princeton, New Jersey.

Reprinted with permission from *The Therapeutic Revolution: Essays in the Social History of American Medicine,* edited by Morris J. Vogel and Charles E. Rosenberg (Philadelphia: University of Pennsylvania Press, 1979), pp. 67–90.

of skepticism toward the discipline on the part of practicing doctors. The evidence for this skepticism is admittedly somewhat fragmentary, and must usually be surmised from the remarks of physiologists themselves. But since medical history, like other history, reflects the views of academic elites, one may wonder whether there have not been thousands of fellow travelers in spirit for every busy doctor whose skepticism has found its way into print. Moreover, it is the persistence of this attitude, rather than its extent, that is perhaps more striking and more in need of explanation.

Although this essay focuses on the division between physiologists and clinicians in the American context, it is important to recognize that there was nothing uniquely American about the situation. We do not have to do here with some trivial manifestation of the alleged pragmatism of American society. As the discipline of physiology took root in Europe, so, too, did the skepticism of clinicians toward it. When the great French physiologist Claude Bernard began to lecture on the medical significance of experimental physiology, he lamented the number of physicians who believed that "physiology can be of no practical use in medicine," that it was "but a science *de luxe* which could well be dispensed with."[4] It was presumably in an effort to change this attitude that Bernard wrote his famous *Introduction to the Study of Experimental Medicine* (1865). He offered there his vision of a new "scientific" medicine as the way out of the therapeutic uncertainty and nihilism of that era. Obviously irritated by the "false opinion" that medicine was not a science but a mysterious art, Bernard went on to dispute the long-standing claim that the best physiologists are the worst doctors, the "most awkward when action is necessary at the patient's bedside." To Bernard it seemed obvious that "solid instruction in physiology . . . , the most scientific part of medicine," was precisely the one thing that physicians most needed.[5]

Skeptical doctors may have been bemused to compare Bernard's prescription with that of his contemporary, the Prussian pathologist Rudolf Virchow. For Virchow, writing in the 1840s and 1850s, the royal road to medical certainty lay not through ordinary physiology ("a 'respectable' science but thus far a very incomplete one"), but rather through Virchow's own specialty of "pathological physiology." Unlike ordinary physiology, pathological physiology recognized that even a complete knowledge of drug action under normal conditions would be inadequate for understanding the therapeutic effects of drugs under pathological conditions. Moreover, pathological physiology "does not stand before the gates of medicine but lives in its mansion":

> [It] receives its questions in part from pathological anatomy, in part from practical medicine; it derives its answers partly from observation by the sickbed, to this extent being a division of the clinic, and partly from animal experiment.[6]

For all of that, however, the similarity between Bernard's campaign and Virchow's was more striking than their differences in matters of detail. Both would have medicine built upon basic science, and for both (as Virchow put it in 1847) "experiment is the final and highest court."[7] From Virchow himself, we gain some sense of the reception his program got from medical men. A surgeon named Schuh, whom Virchow had accused of failing to grasp the real significance of the new scientific knowledge and techniques, responded by saying that he was no more competent than anyone else to undertake the task Virchow had in view. As Virchow reported it, Schuh said that "he gladly left to others the saccharine practice of dreaming and the enjoyment of infallibility; meanwhile he, as a practical surgeon, stood on the same field of observation as his forebears had for centuries."[8]

One can, of course, look upon such reactions as merely inevitable in an era when the new or emerging disciplines of experimental biology had yet to establish their relevance and value for medical practice. For physiology, as I have suggested elsewhere, the yoke of utility was especially burdensome. At least as late as the 1870s, even the most aggressive spokesmen for the discipline found it difficult to think of any physiological discovery that had made a significant, direct impact on the art of healing. In his *Introduction*, Bernard could only point rather weakly toward the experimental eradication of the "itch" (scabies). In 1874, Michael Foster, founder of what was to become the great Cambridge School of Physiology, tried to justify vivisection by claiming that it had already played an important role in the advance of the healing task. But he could offer only two specific examples, both of them dubious: allegedly experimental advances in methods of ligature had improved the treatment of aneurysms, and Claude Bernard's work on the glycogenic function of the liver provided the only available source of illumination into the problem of diabetes, which unfortunately remained outside the therapeutic pale.[9]

What Bernard and Foster really sought to convey was the medical promise of experimental physiology, but their efforts won only guarded and partial acceptance from medical men. Before the 1870s, precisely because its therapeutic value was considered dubious, experimental physiology (and laboratory science in general) found no place in the English medical curriculum, which was shaped predominantly by the pragmatic demands of the London hospital schools. Foster's colleague and compatriot, John Scott Burdon Sanderson, faced the problem candidly and directly in 1872, when he told the British Association for the Advancement of Science that the revival of English physiology depended above all on overcoming "that practical tendency of the national mind which leads us Englishmen to underrate or depreciate any kind of knowledge which does not minister directly to personal comfort or advantage."[10]

Burdon Sanderson's address came on the eve of the exciting work of Koch and Pasteur in bacteriology and immunology. Obviously encouraged by these dramatic developments in neighboring fields, some physiologists (and "pathological physiologists") now found it easier to speak confidently about the medical utility of their research. As early as 1877, in fact, Virchow could write as follows:

> It is no longer necessary today to write that scientific medicine is also the best foundation for medical practice. It is sufficient to point out how completely even the external character of medical practice has changed in the last thirty years. Scientific methods have been introduced everywhere into practice. The diagnosis and prognosis of the physician are based on the experience of the pathological anatomist and the physiologist. Therapeutic doctrine has become biological and thereby experimental science. Concepts of healing processes are no longer separated from those of physiological regulatory processes. Even surgical prac-

tice has been altered to its foundations, not by the empiricism of war, but in a much more radical manner by means of a completely theoretically constructed therapy.[11]

Nonetheless, there is evidence to suggest that many clinicians continued long afterward to doubt the utility of all this newfangled science. The epigraph with which this essay begins reminds us that as late as 1888 an academic clinician could advise his ward clerks that their "first duty" was to "forget all [their] physiology."[12] Especially now, when the scientific basis of modern medicine is taken so much for granted, the long-standing split between doctors and research physiologists seems worthy of more systematic attention than it has hitherto received.

In what follows, the persistence of this division after 1870 is explored with specific reference to the American context. The latter half of the paper very briefly surveys several factors that may help to explain the phenomenon, including the possibility that the skeptical physicians may actually have had some justification for their doubts about the medical utility of laboratory physiology.

But two cautionary and qualifying remarks should be recorded now. First, this is an exploratory effort, not yet buttressed by the sort of extensive evidence we shall surely need to do full justice to the issues it raises. Second, while every effort has been made to focus on the attitudes of practicing doctors toward physiology per se, these attitudes almost invariably reflect a very similar posture toward experimental science in general. Indeed, a much longer essay (or book) might well be written on the persistent skepticism of many ordinary doctors toward experimental science from its beginnings to the present day.

THE PERSISTENCE OF THE DIVISION IN THE UNITED STATES AFTER 1870

Like his colleagues in Europe, Henry Newall Martin, first professor of biology at the new Johns Hopkins University (from 1876 to 1893), insisted that physiology "should be cultivated as a pure science absolutely independent of any so-called practical affiliation." Yet he could scarcely ignore the fact that his discipline's raison d'être "in the mind of even the educated public rested on its relation to medical instruction." Martin therefore worked hard to establish rapport with the medical profession, at one point inviting local physicians to his course of physiological demonstrations.[13]

But in Baltimore, as elsewhere, experimental physiology must have seemed very far removed from the immediate problems of practicing doctors. There was as yet, in the 1870s and 1880s, precious little evidence of its pragmatic value, and no obvious reason to suspect that the situation might soon change. In a sense, American doctors probably agreed with Martin that experimental physiology, if it were to be pursued at all, must indeed be pursued independently of pragmatic medical concerns.

That assessment was clearly shared by those most directly responsible for medical education. For while many medical students might want or need some exposure to human physiology, few indeed would perceive any special need for the sort of experimental training emphasized by the emerging band of professional investigators in the discipline. Insofar as these students did represent a natural (or even captive) audience for physiology, their primary need was for oral (rather than laboratory) instruction in the settled points of functional human anatomy. Thus, even as medical schools did begin increasingly to employ professional, laboratory-trained physiologists, they did not really expect their students to repeat the most salient features of the training those physiologists had themselves received.

Even at Harvard Medical School, which had established an assistant professorship in physiology as early as 1870 and had provided a laboratory for the first incumbent (H. P. Bowditch), two decades passed before students were required to take a course in laboratory physiology. By then, laboratory courses had been established at two other major medical schools (the University of Michigan, in 1887, and Columbia University, in 1891), but both courses were elective, rather than required. The course at Columbia did not become required of medical students until 1902.[14] Partly, no doubt, because it was so expensive to provide, laboratory physiology found precious little place in the American medical curriculum before 1900. The process by which it finally did become a regular part of

medical education has gone virtually unexplored, though it seems likely that laboratory physiology was one of the chief beneficiaries of that more general reform movement in medical education associated with the activities of the American Medical Association's Council on Medical Education (established in 1904) and with the famous Flexner Report of 1910.[15]

Even then, the AMA's classification of the discipline suggests some continued uncertainty about its precise relationship to medicine. From 1901 until at least the 1930s, physiology was placed in the AMA's section on "Pathology and Physiology," having previously been relegated to sections on "Medical Jurisprudence, Hygiene and Physiology" (1847–73), "Practical Medicine, Materia Medica and Physiology" (1874–91), and "Physiology and Dietetics" (1892–1900).[16] Professional physiologists doubtless preferred their new union with the pathologists to those earlier sectional associations, but they may also have wondered whether their own research interests would now be subordinated to the interests of pathologists, "pathological physiologists," or bacteriologists. Perhaps that concern helped to motivate the provision, adopted at the outset, that a physiologist would serve as chairman of the new section every third year.[17]

Implicit, at any rate, in the curricular neglect of laboratory physiology was the assumption that practicing doctors had little or no need of it. Occasionally, that assumption found explicit verbal expression as well. In the 1890s, when the American Physiological Society almost collapsed from lack of interest, even some spokesmen for "scientific medicine" were heard to say that "physiology had done all it could for medicine."[18] In the 1902 edition of his spectacularly popular *Book on the Physician Himself, and Things that Concern His Reputation and Success*, D. W. Cathell claimed that the new scientific knowledge might actually be damaging to his primary readership, ordinary general practitioners. He warned his readers not to be "biased too quickly or too strongly in favor of new theories based on physiological, microscopical, chemical or other experiments, especially when offered by the unbalanced to establish their abstract conclusions or preconceived notions." To submit too readily to the appeal of theoretical science "may impair your practical tendency, give your mind a wrong bias and almost surely make your usefulness as a practicing physician diminish." Scientific curiosity was all well and good for those "scholars and scientists" who did not depend upon practice for their "bread and butter." But the "first question for you, as a practitioner, seeking additional and better tools, to ask yourself in everything of this kind is 'What is its use to me?'"[19] Very different in tone, but not remarkably different in its conclusions about the direct utility of experimental physiology, was the complaint of clinician S. J. Meltzer, in 1904, that internal medicine had received little benefit from physiology "because this science is keeping aloof from medicine and its problems."[20]

As laboratory physiology became an established part of the medical curriculum, the number of skeptics perhaps declined, but they certainly did not disappear. From Meltzer through Rufus Cole and Alfred Cohn, in the 1920s and 1930s, to Frank McLean, in the 1960s, one can trace the theme that physiology—if it had once been useful for clinical medicine—was becoming less so every day. That, at any rate, was part of the justification these men offered for their efforts to establish and extend a new independent discipline of experimental clinical medicine. They hoped that this new discipline would reduce some of the obvious distance between the laboratory and the ward.[21]

In a way, the frequency with which physiologists and academic clinicians continued to insist upon the need for closer interaction between the two fields is itself a striking indication of the persistence of their separation. Physiologist W. H. Sewall might insist, in 1923, that "today . . . every physician recognizes that he is likely to understand his sick man in proportion as he apprehends 'clinical physiology,'"[22] but that same year, one such "clinical physiologist," A. B. Luckhardt, lamented the continuing split between the clinician and the laboratory worker. "Although both groups are intensely interested in the progress of medicine," he wrote, "each group, curiously enough, views the work of the other either with a silent disregard or more often with disdain or openly expressed contempt."[23] Five years later, physiologist C. J. Wiggers thought he detected a few encouraging signs of increased cooperation between clinicians and physiologists, but he could still only hope that they would soon be "walking arm in arm," rather than "making mere gestures of shaking hands across the street."[24]

As a matter of fact, most such appeals for greater

cooperation between the clinician and the laboratory worker probably understated the extent and depth of the split. The problem went beyond the mere indifference or skepticism of clinicians toward laboratory science. Some clinicians, including a few of the most celebrated, continued to echo D. W. Cathell's concern that laboratory training might actually *damage* the practitioner's ability to treat patients effectively. Although my first two examples of his sentiment will be drawn from statements by great English clinicians, there is no doubt that American counterparts could be found. Sir Archibald Garrod, renowned for his work on the "inborn errors of metabolism," wrote in 1911 of his belief that "laboratory findings are little less fallible than clinical inferences, and . . . in some cases they actually mislead."[25] In 1919, the eminent cardiologist Sir James Mackenzie expressed the same reservation far more harshly in his book, *The Future of Medicine*. "Laboratory training," he wrote, "*unfits* a man for his work as a physician, for the reason that, not only does the laboratory man fail to educate his senses, but he puts so much trust in his mechanical methods that he never recognizes their limitations and he fails to see that there are other methods which are essential to the interpretation of disease."[26] A decade later the American clinician Alfred Cohn, who generally adopted a more nuanced and less hostile tone, did nonetheless insist that "the history of medicine since the Renaissance has shown plentifully that whenever the approach to an understanding of disease is made by scholars trained primarily in other pursuits of knowledge [including physiology] . . . the result, so far as understanding disease is concerned, is disappointing and sometimes grotesque."[27] As recently as 1967, in his book *Clinical Judgment*, American cardiologist Alvan Feinstein produced a perceptive and sometimes eloquent statement of the position that reliance upon "scientific" medicine—simplistic animal experiments, elegant physiological theories, and elaborate diagnostic instruments—can actually distort the clinician's judgment when he encounters disease, in all its complexity, in real human beings. For Feinstein, not so incidentally, the persistence of the split between experimental biology and clinical medicine is reflected in the continued division of medical journals and conferences into separate scientific and clinical sections, which seem to have little to do with each other.[28]

To be sure, the major aim of Feinstein's book is to lay the foundations for his own version of a truly scientific clinical medicine. But the "science" upon which he would have clinicians build is mathematical logic. More specifically, he advocates the use of Boolean algebra and Venn diagrams as the basis of a new and subtler form of disease "taxonomy." Whatever its merits, Feinstein's version of scientific medicine is clearly a world apart from Bernard's (or Virchow's) experimental "determinism," and it is the skepticism of clinicians toward that experimental vision of medicine that we shall now seek to explain.

THE ROLE OF ECONOMIC AND TEMPERAMENTAL FACTORS IN THE DIVISION

We cannot begin to understand the persistence of the division between doctors and research physiologists unless we emphatically reject one of Virchow's more exasperated admonitions. "Whether someone is or is not a practitioner *ex professo* has little to do with the matter," he insisted. "If only people would finally stop finding points of disagreement in the personal characteristics and external circumstances of investigators."[29] What Virchow would have us do is to ignore an important part of reality—as if unanimity of opinion or convergence of interest could be expected from individuals and groups with deeply different "personal characteristics" and "external circumstances."

In the case at hand, differences in "personal characteristics" were decidedly reinforced by wide differences in "external circumstances"—both in the nature of the tasks performed and in the structure of the respective reward systems. It is one thing to publish papers; it is quite another to treat patients. And surely no reader of this volume need be told that eminent American doctors (and, increasingly since the 1930s, ordinary doctors too) have always enjoyed higher social status and higher incomes than eminent research scientists. Even in the first decade of this century, the difference was already obvious. In 1909, when S. J. Meltzer sought to entice a group of graduating physicians into his proposed new discipline of experimental clinical medicine, he felt the need to appeal to the example of the German university system. There, he claimed, scientific research was more highly val-

ued than medical practice, and the character of youth was not formed by "sport and the habits of millionaires' sons." There, unlike the United States, "the worth of the individual is not measured exclusively by a gold standard." Should anyone in his audience remain unclear, Meltzer warned them that medical practice was "a bewitching graveyard in which many a brain has been buried alive with no other compensation than a gilded tombstone."[30]

Meltzer's ally, that ubiquitous layman Abraham Flexner, also had good reason to respect the power of the "gold standard." Throughout the 1910s and 1920s, his efforts to create "full-time" salaried clinical chairs ran into serious difficulties, initially even at John Hopkins—perhaps partly because American physicians rather oddly found the very concept of a salary in some way unprofessional, but surely also because the proposed amount of the salary (initially, $10,000) was so much less than the income to which leading clinicians had become accustomed from their practices.[31]

From the mid-1940s on, the income differential between research physiologists and practicing doctors became so striking that the American Physiological Society considered it a serious obstacle to the recruitment of high-quality personnel. According to the data collected in the society's remarkable self-survey of the 1950s, the average net income of physicians in 1940 was $4,400 compared to $3,700 for physiologists. A mere five years later this modest gap had widened to $11,000 for physicians and $4,625 for physiologists, while by 1952 it was $14,080 for physicians and $6,360 for physiologists. The absolute differential is surely much larger now [i.e., 1976]—with physicians earning a tremendous range of incomes, with a median of perhaps $65,000, while the salaries of employed physiologists cluster within a much narrower range, around $30,000.[32] Already in the 1950s, the American Physiological Society was more concerned about the loss of "brainy people" to medical practice than Meltzer had once been about the "loss to clinical medicine of the brainy men who now devote their energies to the pure sciences."[33] In the academic market of the 1970s, a cynical (or honest) physiologist might warn potential aspirants to the field that "research is a bewitching graveyard in which many a body has been buried alive with no other compensation than an occasional published paper."

Yet throughout the period of widening income differentials, an appreciable (if declining) number of students proved willing to make economic sacrifices for the sake of physiological research. And as late as the 1950s, at least, most physiologists expressed general satisfaction with their chosen career.[34] Once captured by research, few physiologists with M.D.s (from Claude Bernard on) forsook it for the medical practice they could have entered. For most of them, the obvious and growing financial appeal of a medical career could not divert them from their personal inclination; for them, an unfavorable difference in "external circumstances" could not overwhelm a more profound difference in "personal characteristics." It did, however, help to increase the distance separating them from practicing doctors.

By now the available sociological and psychological data seem sufficient to establish the existence of significant differences in background, personality, and values between those intending to be physicians and other members of their age cohort, including those who undertake careers in scientific research.[35] Indeed, the so-called "Two Cultures" gap between humanists and scientists may be as nothing compared to that which separates pragmatic, action-oriented, client-dependent professionals (including physicians) from those with essentially scholarly sensibilities (including research physiologists).[36]

Long before any data were systematically compiled, experimental scientists and practicing physicians were aware of deep differences between them. Such major 19th-century physiologists as Claude Bernard, Carl Ludwig, and Emil du Bois Reymond, although sometimes at odds over issues within science itself, were united in that mixture of disdain and grudging envy with which they regarded practicing doctors. Vallery-Radot, in his *Life of Pasteur,* captures some of this ambivalence in a presumably apocryphal, but nonetheless revealing exchange said to have been initiated by Bernard, who, unlike Pasteur, did at least possess an M.D. degree. Vallery-Radot has Claude Bernard ask Pasteur, "with a smile under which many feelings were hidden, 'Have you ever noticed that when a doctor enters a room, he always looks as if he was going to say, "I have just been saving a fellow man"?'"[37]

The physiologists of more recent times have continued to feel, and occasionally to express, that sense of ambivalence. In 1945, for example, the Harvard physiologist Walter Cannon wrote as follows:

> My father's wish that I might become a physician was . . . never realized. Instead of engaging in practice I engaged in teaching medical students. This was what my predecessor, Dr. [H. P.] Bowditch had done. He told the tale of a conversation between one of his children and a little companion. The companion asked, "Has your father many patients?" and the answer was, "He has no patients." "What! A doctor and no patients?" Thereupon the apologetic answer, "Oh, no, he is one of those doctors who don't know anything!" Possibly the children of other physiologists suffer from the same sense of inferiority. One of my daughters, on being informed proudly by a little friend that *her* father was a doctor, remarked somewhat sadly, "*My* father is only a father."[38]

If that is a joke, it is a deeply symbolic one, and we may wonder whether Cannon (like Bowditch before him) is not indulging the familiar parental habit of speaking through their children.

That is not to say that Cannon himself actually felt an outright "sense of inferiority" vis-à-vis doctors; the rest of his autobiography suggests otherwise. Rather, it is as if he felt the need to call attention to the high (but ephemeral) social status of individual doctors in order to distinguish the very different and basically intellectual motivations that lay behind his calling. That self-identification with intellectual goals, which also permeates most other autobiographical accounts by research physiologists, helps us to understand the existence and persistence of the split between them and practicing doctors. If physicians have often looked upon research physiologists as remote "dreamers," or worse, the physiologists have tended to regard doctors as mere "technicians" who cannot or will not appreciate the value of basic scientific research or of scholarship in general. The result, quite obviously, has been to increase whatever distance between them might have been accounted for by any actual disjunction between physiological knowledge and medical practice.

In the following section, these psychological (or temperamental) differences between physiologists and physicians are invoked to help us understand the response of physiologists to the fragmentation of physiology as a discipline. Because of this fragmentation, physiologists faced the charge that their discipline had become both intellectually sterile and progressively less relevant to medical science. The physiologists, it seems, found the former charge more worrisome than the latter.

THE FRAGMENTATION OF PHYSIOLOGY: THE PRIORITY OF THE INTELLECTUAL CHALLENGE

Quite steadily, from about 1900 onward, physiology lost its privileged place in the world of experimental biology. Just as physiology had once declared its independence from medicine and medical anatomy, so new fields and specialties now seemed to declare their own independence from physiology. Actually, we know very little as yet about the emergence of these new disciplines. It is certainly premature, and probably misleading, to claim that the new disciplines evolved directly and simply out of physiology. But that was the way many physiologists saw it. For them, a sense of fragmentation resulted from the creation of independent societies by specialists who had begun to feel some dissatisfaction with the American Physiological Society, founded in 1887. In the early years of the 20th century, there appeared in rapid order the Society for Experimental Biology and Medicine (1903), the American Society for Biological Chemists (1906), the American Society for Pharmacology and Experimental Therapeutics (1908), and the American Society of Experimental Pathology (1913).[39] As early as 1911, the increasing fragmentation of what had once seemed to be physiology inspired the following remark from physiologist Henry Sewall. Ever ready to deploy (or mix) metaphors, Sewall voiced his impression that:

> the course of evolution has ordered it that whereas physiology was then [in the 1870s] the dependent runt of the medical family, it is today the eldest son in a stable system of primogeniture. As with a noble jewel, whose beauty depends upon the cutting, we may name one facet Pathology, another Pharmacology, another Bio-chemistry, another Psychology, and so on, the jewel remains and ever will be Physiology.[40]

Other physiologists, then and later, expressed similar sentiments with remarkable frequency, and their common use of biological metaphors in which physiology remains the "trunk" or "mother-stem" of the vigorous new disciplinary branches does not entirely mask an underlying concern. Along with a growing sense of estrangement from the increasingly intricate methods, techniques, and instruments of modern experimental biology or "biophysics," some "classical" medical physiologists began to wonder whether physiology any longer existed as a discrete intellectual entity at all. The conceptual doubt was reinforced by recruiting difficulties, for the number of new Ph.D.s in physiology actually declined in the 1940s—not only absolutely, but also (and more importantly) in proportion to new Ph.D.s in other biological disciplines.[41]

Certainly, by then, if not before, the situation had provoked something of an internal crisis in the field. In the early 1950s, the American Physiological Society, funded by the new National Science Foundation, undertook a remarkably complete and revealing survey of the discipline. The survey addressed itself in part to the question of how physiology could be defined, and in one striking aside, survey director R. W. Gerard carried Sewall's evolutionary metaphor to one of its possible conclusions. For if Sewall, in 1932, could still see physiology as the "eldest son in a stable system of primogeniture," Gerard raised the possibility that its new disciplinary offshoots might be better adapted to the age, while physiology itself faced the prospect of extinction, of becoming "a fossil on musty library shelves."[42]

The threat of extinction came from at least two directions simultaneously. For the new disciplinary subspecies not only included fields which seemed intellectually more exciting than physiology—biochemistry, cytology, genetics, cellular physiology, and biophysics—but also others which seemed (at least on the surface) of more immediate utility to medicine—notably bacteriology or microbiology, pharmacology, nutrition, immunology, and endocrinology. And so, as the fragmentation became increasingly obvious, physiologists had to decide which of these two developments they considered more threatening.

On the whole, despite the once aggressively utilitarian rhetoric of spokesmen like Bernard, physiologists apparently found the charge of intellectual sterility more immediately damaging than that of medical irrelevance. They certainly noticed both aspects of the challenge, but they responded to the latter with rather considerable restraint. Often enough, they referred to the value that physiology might derive from a closer interaction with clinical medicine (rather than the other way around), and they rarely exaggerated the direct medical utility of their work—even on so profoundly practical a problem as wound shock, which occupied a truly staggering proportion of American physiologists and their *Journal* after the nation entered World War I.[43]

Insofar as physiologists did confront the clinical challenge, they seemed most concerned about such "academic" clinicians as Meltzer, Cole, Cohn, and McLean. The challenge these critics posed was at once intellectual and institutional. For Meltzer and his three spiritual descendants, all of whom had significant ties to Abraham Flexner and the Rockefeller Foundation, took the almost paradoxical position that the best solution to the medical remoteness of physiology was to create yet another entirely new independent discipline, "experimental clinical medicine." In the more or less typical language of biomedical reformers, Meltzer claimed that this new discipline would keep more firmly in touch with medical problems. Yet he also insisted that it absolutely required independence from actual medical practice—more so, even, than such "ancillary" sciences as physiology.[44]

Physiologists, already concerned about the fragmentation of their subject into new academic fields, began to wonder aloud what would be left of the discipline if it now faced a frontal assault from the proposed new discipline of "experimental medicine." When physiologist C. A. Lovatt Evans expressed open concern about these "Meltzerian" proposals, in 1928, Alfred Cohn probably did little to reduce that concern when he responded as follows:

> Although physiology has made itself independent, Professor Evans still harbors fears. He fears to cut the guiding strings of the alma mater [medicine], lest physiology lack nourishment. And like many . . . children, he fears lest the ancient mother be too feeble intellectually and too powerless, having reared and weaned her children, to be able to continue to order and to develop her own house. But

the situation is just this: having learned as it were and indicated to her many offspring how they might best set up houses of their own, medicine is at length free to cultivate her own garden.[45]

At this point, it seems to me, the rank-and-file practitioner might have been excused for casting a smile at the medico-academic elite. So the long campaign of research physiologists to achieve independence from medicine was to end, after all, in nothing more than an opposing effort by "academic" clinicians to reclaim part of the lost territory! To the ordinary doctor, the situation was doubtless reminiscent of other comical but irrelevant disputes among academics. In addressing themselves above all to the challenge posed by fellow academics, the physiologists were responding to the intellectual inclinations that had attracted them to research in the first place. But ordinary doctors can only have felt still further estrangement from such physiologists as Nobel Laureate Otto Loewi, who actually suggested in 1954 that the discipline might never have entered its "crisis" if only its practitioners had pursued problems of even more "remote usefulness for medicine."[46]

AN HERETICAL SUGGESTION: COULD THE SKEPTICAL CLINICIANS HAVE A POINT?

Medical historians have not entirely ignored the skepticism of practicing doctors toward experimental physiology (or experimental science in general), and doubtless most would agree that economic and temperamental factors contributed importantly to it. But perhaps the most common "explanation" for the phenomenon—essentially an echo of Claude Bernard's position—is that one could not expect "short-sighted" clinicians to appreciate the benefits that medicine would "soon" reap from the triumph of experimental science. On this view, older clinicians, who had not been trained in the new scientific methods, would be especially unwilling to acknowledge their dependence on experimental science and especially likely to persist in the delusion that medicine was essentially an art. These older clinicians were also most likely to feel vulnerable to any economic threat posed by the rising breed of "scientific" and presumably more effective physicians. And so, out of ignorance and self-interest, they would naturally dispute the medical utility of laboratory science.[47]

At some point, however, this sort of explanation can no longer suffice. Surely by about 1920, when most doctors would have been exposed to laboratory science, its value should have been so obvious as to overwhelm their skepticism, except perhaps for a few quacks and cranks. From the evidence already presented, it should be clear that skeptics nonetheless remained, including at least a few leading clinicians. If such an eminent clinician as Feinstein can continue even today to express reservations about the medical value of laboratory science, medical historians ought to be willing at least to reconsider the matter. Insofar as the skeptics have focused on physiology per se, there is perhaps particular reason not to dismiss their views out of hand.

In 1959, the Downstate Medical Center of the State University of New York sponsored an ambitious symposium on "the historical development of physiological thought," resulting in a book of that title. The book contains 16 essays, many of which remain valuable studies, but only one directly confronts the relationship between physiology and medical practice. That essay, by Owsei Temkin, begins as follows: "If we were to awaken a man from sleep by shouting into his ear: 'medicine depends upon basic scientific thought,' he might unthinkingly say 'amen!'" With typical perspicacity, Temkin goes on to show how thoroughly complicated the issue looks when the immediate response is replaced by a thoughtful one. Indeed, the thoughtful response soon threatens to dissolve into utter confusion because of the difficulty of even defining such terms as "physiology," "science," "medicine," or "health." In the end, Temkin seems to suggest, medicine does indeed depend partly upon basic scientific thought, but often of a sort that research physiologists might fail to recognize as "scientific" at all.[48] At the same conference (though his paper was published separately), Paul Cranefield briefly pondered the far more specific question of whether the preceding half-century of "infinitely delicate and beautiful studies of microscopic physiology" had influenced medical practice. His rather rueful answer: no, "by and large they have not."[49]

In other cases, too, in which it has seemed obvious that experimental physiology must have made vital contributions to medical practice, it may prove instructive and sometimes surprising to examine the situation critically. Consider, for much too brief a moment, the clinical field in which the pragmatic value of physiology might seem to be most obvious—cardiology. This is, moreover, a field of direct concern to the average office-based practitioner, and one in which physiological research has been accorded a quite immediate and specific role. As C. J. Wiggers put it in 1951, referring to the "clinical applications" of modern circulatory physiology:

> The physiological interpretations of graphic recordings of the pulse enabled [Sir James] Mackenzie to give physiological interpretations to many of the common [cardiological] irregularities. Additional experiments by [Joseph] Erlanger and [Arthur] Cushny elucidated the phenomena of heart block and atrial fibrillation, and their discoveries soon proved valuable in clinical diagnoses. The foremost step in the decade was the invention in 1903 of the string galvanometer by Einthoven and his prompt application of this tool to the study of physiological and clinical problems . . . while the discerning mind of Thomas Lewis immediately envisaged the great strides that could be made in the field of clinical cardiology through correlation of electrographic phenomena in patients and experimental animals.[50]

Upon closer inspection, however, the situation begins to look rather more problematic than Wiggers's summary suggests. To be sure, electrocardiography did emerge from an important tradition in purely "academic" electrophysiology. But can one trace any major innovations in therapeutic cardiology to this new physiological knowledge? At the turn of the century, physiological experiments on dogs led Arthur Cushny to the diagnosis of auricular fibrillation for the clinical condition previously known as *delerium cordis*. In the course of this research, Cushny complained bitterly that clinicians had failed to keep abreast of the recent avalanche of knowledge in the physiology of the mammalian heart. Yet his own further research, and that of others, produced no dramatic departures in therapy, but mainly provided a new rationale for the use of the digitalis compounds that had already been introduced "empirically" for the treatment of such cases.[51]

Astonishingly, some leading clinicians even disputed the diagnostic value of the new electrophysiology and electrocardiography. Among them was Sir James Mackenzie himself, who figured prominently in the clinical developments to which Wiggers refers, not only through his own contributions, but also as sometime collaborator with Arthur Cushny and as mentor of Sir Thomas Lewis. It was Mackenzie, we should recall, who warned in 1919 that laboratory training could actually distort the clinician's judgment and "unfit" him for his work as a physician. Mackenzie, moreover, practiced what he preached, eventually abandoning even the primitive polygraph he had used in his early cardiological studies for the more direct and more traditional methods of clinical observation.[52] Recall, too, that Feinstein, who to some degree shares Mackenzie's skepticism toward laboratory science, is also a cardiologist.

Thus, even in the apparently straightforward case of cardiology, the medical utility of physiological research is open to some qualification, at the least. Surgery represents a second major clinical field in which experimental physiology has seemed to play an important and immediate role. Quite obviously, future surgeons can enhance their surgical skills by performing animal experiments, even if the immediate object is physiological. Moreover, some leading surgical mentors, notably Owen Wangensteen of the University of Minnesota, certainly did, and do, stress the value to surgeons of a thorough training in physiology. Indeed, one branch of modern "total surgery" even bears the name "physiologic surgery." In the 1964 edition of a leading American surgical textbook, this branch of surgery is said to have as its aim the alteration of "normal" but deleterious bodily function for the general good of the patient and to have its theoretical roots in Walter Cannon's concept of homeostasis.[53] Cannon himself was considerably more restrained about the medical utility of his physiological work: "It is said," he wrote in 1945, "that our researches on the bodily effects of emotions have been helpful because they give the doctor pertinent information in explaining to his nervous patients the reasons for their functional disorders."[54] Perhaps Cannon was being excessively modest here, and perhaps his work has had an immense impact on medical and surgical practice. Yet it is not immediately obvious exactly how and in what sense his physiological insights should have led to

any significant reorientation of medical or surgical practice.

But let us not carry our own skepticism too far. Let it be clear that it would require infinitely more research to examine the wealth of possible contributions experimental physiology may in fact have made to clinical medicine or to health. At least on the surface, it seems obvious that experimental physiology must have made important contributions to the so-called "replacement therapies," of which insulin for diabetes, thyroid extract for myxodema, and liver extract for pernicious anemia serve as classic examples. It also seems clear that experimental physiologists must have contributed importantly to our understanding of such medically significant aspects of blood chemistry as pH levels, electrolyte balance, and incompatible blood types.[55] The existing literature provides a host of other possible examples, thus far mostly in the form of schematic lists, which clearly deserve more extensive inquiry.[56]

In the course of this more extensive research, a number of distinctions should be kept very firmly in mind. Different sorts of medical practitioners would doubtless come to very different sorts of conclusions about the utility of particular investigations in experimental physiology. What is useful to hospital-based specialists, for example, is liable to be very different from what is useful to office-based pediatricians, internists, and general practitioners. It is also vital to distinguish between the therapeutic, diagnostic, and preventive aspects of medicine with respect to the possible utility of experimental physiology. And whatever conclusions one may reach about the direct medical utility of laboratory physiology, one may well wish to consider separately its cultural value and its proper role in medical education. Let me conclude with some preliminary conclusions growing out of this last distinction.

REFLECTIONS ON THE GENERAL ROLE OF LABORATORY SCIENCE IN MEDICAL EDUCATION

Throughout much of this paper, I have insisted upon the persistent skepticism of practicing doctors toward the medical utility of experimental physiology. In the section immediately above, I have raised the question whether that skepticism may not have had some justification for at least some sorts of physicians. Yet Abraham Flexner, in his famous report of 1910 on American medical education, called physiology "the central discipline of the medical school."[57] It has retained an important place in the American medical curriculum ever since. For some readers, that will surely serve as prima facie evidence of its value to medical practice. Unless laboratory physiology is medically useful, why did it acquire and why does it retain such an important role in medical training?

To address this question, which might seem to dismiss out of hand any need to reexamine the relationship between experimental physiology and medical practice, we must give at least passing attention here to the more general role of laboratory science in medical education. What especially needs to be recognized is that doctors have gained at least one important benefit from the study of experimental science, quite apart from whatever direct medical utility it has, and quite apart from any indirect contribution it may make to medically effective modes of thinking.[58] For the experimental sciences, like Latin in an earlier era, have given medicine a new and now culturally compelling basis for consolidating its status as an autonomous "learned profession," with all of the corporate and material advantages that such status implies.

It was Flexner, once again, who asserted that the "possession of certain portions of many sciences arranged and organized with a distinct practical purpose in view . . . makes [medicine] a 'profession.'"[59] Underwritten by the Carnegie Foundation for the Advancement of Teaching, supported in crucial ways by the AMA's Council on Medical Education, and dependent for much of its influence on Flexner's skillful use of Rockefeller money, the Flexner Report accelerated the process by which the supply of American doctors was greatly reduced, as the expense, duration, and quality of their education was greatly increased.[60] The eventual result was to elevate the American medical profession to the remarkable position it enjoys today. Along the way, the basic sciences came to occupy a newly central role in the medical curriculum. And in a society that reveres (even as it fears) science, doctors clearly owe some of their status to the Flexnerian perception that they are in some sense "scientists."

This is not to claim that the medical profession, either in its "prescientific" or "scientific" phases,

has consciously and cynically exploited the available cultural resources (whether polite Latin learning or experimental science) for its own self-serving ends. Even if the medical profession, like other professions, has sometimes acted in a patently self-interested way, there is no *necessary* conflict between professional self-interest, responsible medical care, and the humanitarian ideals traditionally espoused by physicians. And even if proper examination should reveal that our "scientific" medicine has contributed only marginally, if at all, to any measurable improvement in health, it does not automatically follow (as Ivan Illich assumes) that our medical system is misconceived or unduly expensive.[61] Perhaps the most striking feature of the recent wave of controversy over the cost and quality of American medicine is the tendency of both critics and apologists (but especially "academic" critics) to forget that sick people do not think in statistical terms and that medicine serves vital functions not yet captured in "vital statistics"—among them, prevention or relief of pain and morbidity, preservation or restoration of physical or social function, making sense of illness, and the dispensing of peace of mind and general human support.[62]

But if we do not yet really know the extent to which physicians influence ordinary vital statistics, we have even less sense of whether or how much their scientific training enhances their ability to perform their less dramatic "supportive" functions. The slim available evidence does suggest—contrary to a segment of popular opinion—that scientific orientation and training are not inimical to the humanitarian aspects of the physician's role.[63] But it remains to be established that physicians so trained perform *more* effectively in their supportive tasks, and there is perhaps particular reason to wonder whether such primary-care physicians as pediatricians and internists are being appropriately trained to handle the medical problems that they actually face.[64] Especially in view of the high cost of training "scientific" doctors, we deserve rather better evidence that the expense is justified.

For Flexner, who had graduated from Johns Hopkins during its golden early years as the American embodiment of "Germanic" research ideals, it was a self-evident proposition that an expansion of laboratory training in the medical curriculum would result in improved health for Americans. Thus he noted in passing—as if establishing a causal connection between two parallel phenomena—that "the century which has developed medical laboratories [roughly, 1810–1910] has seen the death-rate reduced by one-half and the average expectation of life increased by ten or twelve years."[65] It is precisely this alleged connection between laboratory training and health that requires critical scrutiny. And surely it will not do for medical historians to assume in advance the health benefits of experimental science or to ignore entirely the other functions that laboratory training plays in medical education.

NOTES

This is a considerably revised version of the paper I delivered at the conference "Two Hundred Years of American Medicine," in Philadelphia, 2–4 December 1976. As originally presented, the essay included a section that argued for the existence of an "Anglo-American" style of physiology. I now expect to develop that theme in a separate publication. Moreover, since giving my paper, I have profited from the criticisms of James Secord, Robert Bernstein, and an anonymous referee for the University of Pennsylvania Press, and (especially bibliographically) from reading an as yet unpublished essay on the "social and intellectual location of physiology in America" by David Bearman. In the notes that follow, I have tried to indicate my more specific debts to Mr. Bearman.

1 Quoted by K. D. Keele, *The Evolution of Clinical Methods in Medicine* (London: Pitman Med. Pub., 1963), p. 105.

2 On the emergence of physiology as an independent discipline, see Joseph Schiller, "Physiology's struggle for independence in the first half of the nineteenth century," *History of Science,* 1968, *7*: 64–89; and John E. Lesch, "The origins of experimental physiology and pharmacology in France, 1790–1820: Bichat and Magendie" (Ph.D. diss., Princeton Univ., 1977).

3 A perhaps typical, if not classic, example of the genre is Carl J. Wiggers, "The interrelations of physiology and internal medicine," *J.A.M.A.,* 1928, *91*: 270–274, where a long but utterly schematic list is given of the "physiologic researches . . . of immediate interest to the clinician."

4 Rene Vallery-Radot, *The Life of Pasteur,* trans. Mrs. R. L. Devonshire (New York: Garden City Publ. Co., n.d.), p. 226.

5 Claude Bernard, *An Introduction to the Study of Experimental Medicine,* trans. H. C. Greene (New York: Dover 1957), pp. 203–205.

6 Rudolf Virchow, *Disease, Life and Man,* trans. L. J. Rather (New York: Collier, 1962), pp. 50, 66, 76; indented quotation on p. 50.

7 *Ibid.,* p. 50.

8 *Ibid.,* p. 91. This was almost certainly Franz Schuh, one of the most "scientific" surgeons in Europe. See Erna Lesky, *The Vienna Medical School of the 19th Century,* trans. L. Williams and I. S. Levij (Baltimore: Johns Hopkins Univ. Press, 1976), pp. 168–173. If so, one can only imagine the response of less academic surgeons.

9 See G. L. Geison, "Social and institutional factors in the stagnancy of English physiology, 1840–1870," *Bull. Hist. Med.,* 1972, *46*: 30–58, esp. pp. 41–42.

10 *Ibid.,* p. 55.

11 Virchow, *Disease,* p. 163.

12 See Keele, *Evolution,* p. 105.

13 See Henry Sewall, "The beginnings of physiological research in America," *Science,* 1923, *58*: 187–195, esp. pp. 194–195.

14 Walter J. Meek, "The beginnings of American physiology," *Ann. Med. Hist.,* 1928, *10*: 111–125, esp. pp. 122–124.

15 See text below, section 6, and the sources cited in n. 60.

16 Arthur H. Sanford, "The role of the clinical pathologist," *J.A.M.A.,* 1930, *95*: 1465–1467.

17 *Ibid.,* p. 1466.

18 See Frederick W. Ellis, "Henry Pickering Bowditch and the development of the Harvard Laboratory of Physiology," *New Eng. J. Med.,* 1938, *219*: 819. I owe this reference to David Bearman.

19 As quoted by William G. Rothstein, *American Physicians in the Nineteenth Century: From Sects to Science* (Baltimore: Johns Hopkins Univ. Press, 1972), pp. 265–266.

20 S. J. Meltzer, "The science of clinical medicine: what it ought to be and the men to uphold it," *J.A.M.A.,* 1909, *53*: 508–512, quotation on p. 510.

21 See Franklin C. McLean, "Physiology and medicine: a transition period," [1960], in *The Excitement and Fascination of Science* (Palo Alto, Calif.: Annual Reviews, Inc., 1965), pp. 317–332, esp. pp. 317–319.

22 Sewall, "Beginnings," p. 190.

23 Arno B. Luckhardt, "The progress of medicine: a plea for the concerted efforts of the clinician and the laboratory worker," *J.A.M.A.,* 1923, *81*: 347–349.

24 Wiggers, "Interrelations," p. 274.

25 A. E. Garrod, "The laboratory and the ward," in *Contributions to Medical and Biological Research: Dedi-*

cated to Sir William Osler (New York: P. B. Hoeber, 1919), pp. 59–69, quotation on p. 63. I owe this reference to David Bearman.

26 Quoted by Keele, Evolution, p. 105.

27 Alfred E. Cohn, "Medicine and science," Journal of Philosophy, 1928, 25: 403–416, quotation on p. 409.

28 Alvan R. Feinstein, Clinical Judgment (Baltimore: Johns Hopkins Univ. Press, 1967), esp. introduction.

29 Virchow, Disease, p. 163.

30 Meltzer, "Science," pp. 511–512.

31 See Donald Cousar, "The establishment of full-time clinical chairs at Johns Hopkins," unpublished essay for Junior Independent Work, Princeton Univ., 1975. Cf. Abraham Flexner, Universities: American, English, German (London: Oxford Univ. Press, 1930), pp. 85–96.

32 For the 1940 to 1952 data, see R. W. Gerard, Mirror to Physiology: A Self-Survey of Physiological Science (Washington, D.C.: American Physiological Society, 1958), p. 66. I owe the estimate of current physician income to my colleague, medical economist Uwe Reinhardt. The estimate for current physiologists' salaries is merely a rough guess based on the salary of chairmen of physiology departments at medical schools ($35,000–$40,000).

33 Meltzer, "Science," p. 511; and Gerard, Mirror, passim.

34 See Gerard, Mirror, ch. 6.

35 See, inter alia, Rashi Fein and Gerald I. Weber, Financing Medical Education (New York: McGraw Hill, 1971); James A. Knight, Medical Student: Doctor in the Making (New York: Appleton-Century-Crofts, 1973); and the flood of papers on the topic in virtually any issue of the Journal of Medical Education.

36 Perhaps useful in this connection is Eliot Freidson's distinction between consulting and other professions. See Freidson, Profession of Medicine (New York: Dodd, Mead, 1970). Cf. also Norman W. Storer, The Social System of Science (New York: Holt, Rinehart, 1966), esp. pp. 91–97.

37 See Vallery-Radot, Life of Pasteur, pp. 225–226.

38 Walter B. Cannon, The Way of an Investigator (New York: Norton, 1945), p. 21.

39 See Sewall, "Beginnings," p. 190.

40 Henry Sewall, "Henry Newell Martin . . . ," Johns Hopkins Hosp. Bull., 1911, 22: 327–333, quotation on p. 328.

41 See Gerard, Mirror, esp. p. 39.

42 Ibid., p. 249.

43 See, e.g., Carl J. Wiggers, "Physiology from 1900–1920: incidents, accidents, and advances," [1951], in Excitement and Fascination of Science, pp. 547–566, who says (on p. 558) that his haemodynamic studies "tended to confuse rather than clarify" the problems of shock. More generally, see the other essays in Excitement.

44 Meltzer, "Science," p. 508.

45 Quoted by McLean, "Physiology," p. 319.

46 Otto Loewi, "Reflections on the study of physiology," [1954], in Excitement, pp. 269–278, quotation on p. 276.

47 This attitude is at least implicit in Rothstein, American Physicians, and perhaps also in R. H. Shryock's classic, The Development of Modern Medicine (New York: Knopf, 1947).

48 Owsei Temkin, "The dependence of medicine upon basic scientific thought," in The Historical Development of Physiological Thought, ed. Chandler McC. Brooks and Paul Cranefield (New York: Hafner, 1959), pp. 5–21.

49 Paul F. Cranefield, "Microscopic physiology since 1908," Bull. Hist. Med., 1959, 33: 263–275, quotation on p. 274.

50 Wiggers, "Physiology," 553–554.

51 See G. L. Geison, "Arthur Cushny," Dictionary of Scientific Biography (New York: Scribner's, 1978), vol. 15, pp. 99–104, esp. the primary sources cited on p. 104.

52 See Thomas M. Durant, The Days of Our Years: A Short History of Medicine and the American College of Physicians (brochure, American College of Physicians, Chicago, 1965), p. 14.

53 Loyal Davis, ed., Christopher's Textbook of Surgery, 8th ed. (Philadelphia: W. B. Saunders Co., 1964), pp. 1–3, 19–20.

54 Cannon, Way, pp. 213–214.

55 I am indebted to Robert Bernstein for emphasizing to me the examples from studies of blood chemistry. It may well be significant that Rufus Cole, who was otherwise somewhat skeptical about the utilitarian claims of physiologists, did concede the medical value of their research in these areas. See Rufus Cole, "Progress of medicine during the past twenty-five years as exemplified by the Harvey Society Lectures," Science, 1930, 71: 619–627.

56 Cole, ibid., offers a few other examples that may be all the more deserving of investigation since they come from a clinician and moderate critic of physiologists' claims for medical utility. For an expansive list of such claims, I refer the reader again to Wiggers, "Interrelations."

57 Abraham Flexner, Medical Education in the United States and Canada (New York: Carnegie Foundation, 1910), p. 63.

58 I mean to suggest here, though none of my sources develops the idea, that laboratory training may encourage a receptivity to novelty. That claim is sometimes made on behalf of liberal studies in general.

59 Flexner, Medical Education, p. 58.

60 See Robert Hudson, "Abraham Flexner in perspective: American medical education, 1865–1910," Bull. Hist. Med., 1972, 46: 545–561 (ch. 12, this book); and

the remarkable series of annual reports by the AMA Council on Medical Education that began to appear in the *Journal of the American Medical Association* from 1904.

61 Ivan Illich, *Medical Nemesis* (London: Marion Boyars, 1975).

62 Cf. Walsh McDermott, "Evaluating the physician and his technology," in "Doing better and feeling worse: health in the United States," ed. John Knowles, *Daedalus*, 1977, *106*: 135–157.

63 See Earl R. Babbie, *Science and Morality in Medicine* (Berkeley: Univ. of California Press, 1970).

64 Cf. Kerr L. White et al., "The ecology of medical care," *New Eng. J. Med.*, 1961, *265*: 885–892.

65 Flexner, *Medical Education*, p. 62.

8

The Maturation of American Medical Science

RONALD L. NUMBERS AND JOHN HARLEY WARNER

"What does the world yet owe to American physicians or surgeons?" an essayist for the *Edinburgh Review* asked contemptuously in 1820.[1] Offended Americans, unable to claim any medical heroes of their own, responded defensively by arguing that they excelled in the practice of medicine. Nevertheless, they resented their obviously dependent status in the medical sciences and yearned for the respect that would come with scientific achievement. It is time, proclaimed one patriotic surgeon in 1856, "to declare ourselves free and independent of our transatlantic brethren, as we did eighty years ago declare ourselves free and independent of the British crown."[2] In an effort to spur the medical community into action, the *Philadelphia Journal of the Medical and Physical Sciences* for several years prominently displayed the insulting Edinburgh query on its title page.

The American struggle for medical independence lasted for nearly a century after the 1820 incident. As Table 1 indicates in a crude, quantitative way, the United States continued to lag behind Europe in contributing to the medical sciences until late in the 19th century. However, the Americans overtook the English in the 1880s, the French in the 1890s, and the Germans in the 1910s. By 1920 they led the world in medical research.[3]

It is important in tracing the maturation of American medical science to distinguish between the history of the so-called basic medical sciences (e.g., anatomy, physiology, biochemistry, pathol-

ogy, and pharmacology) and the development of the clinical sciences, particularly therapeutics. The former reached maturity in the years between 1890 and 1920, when the United States created a self-sustaining institutional base for medical research. Medical therapeutics followed a much different course, achieving maturity during the last third of the 19th century, when American clinicians abandoned their insistence on a distinctively "American" practice in favor of therapies based on the principle of medical universalism. It is also important to bear in mind that even during the period when American physicians failed to keep pace with their European colleagues in using hospitals and laboratories for medical research, they not uncommonly engaged in armchair theorizing about the functions of the human body or investigated empirically the relationship between climate and

Table 1
Numbers of Discoveries in the Medical Sciences, by Nation, 1820–1919

	U.S.A.	England	France	Germany
1820–29	1	12	26	12
1830–39	4	20	18	25
1840–49	6	14	13	28
1850–59	7	12	11	32
1860–69	5	5	10	33
1870–79	5	7	7	37
1880–89	18	12	19	74
1890–99	26	13	18	44
1900–09	28	18	13	61
1910–19	40	13	8	20

Source: Joseph Ben-David, "Scientific productivity and academic organization in nineteenth century medicine," *American Sociological Review,* 1960, *25*: 830. Based on Fielding H. Garrison's "Chronology of medicine and public hygiene."

RONALD L. NUMBERS is William Coleman Professor of the History of Science and Medicine, University of Wisconsin–Madison.

JOHN HARLEY WARNER is Professor of the History of Medicine, Yale University, New Haven, Connecticut.

Reprinted from *Scientific Colonialism: A Cross-Cultural Comparison,* edited by Nathan Reingold and Marc Rothenberg (Washington, D.C.: Smithsonian Institution Press, 1987), pp. 191–214, by permission of the publisher. Copyright © 1987.

disease. Regarded retrospectively, such activities may have contributed little to the advancement of medical science, but they were scientific nonetheless.

THE BASIC SCIENCES

Nineteenth-century American physicians tended to attribute their meager scientific output to the relative immaturity of their country. "In the great family of enlightened nations, we are the last born," explained one doctor. "In our youth we must be sustained."[4] The United States may indeed have been a youth among nations, but, as Richard H. Shryock long ago pointed out, it lacked neither the population nor the wealth to support scientific research. By 1860 the United States claimed eight cities with populations in excess of 150,000; its per capita income exceeded that of any European nation; its industry led the world in mechanization.[5] Thus whatever the reasons for the country's failure to contribute to medical science, they did not stem from either poverty or the absence of an urban culture.

Some American physicians, embarrassed by being "the mere recipients" of European knowledge, blamed the scientific inactivity of their colleagues on the availability of foreign literature and feelings of national inferiority. Harvard's Oliver Wendell Holmes, for example, discerned a "fatal influence" to the growth of indigenous science emanating from the indolence created by the "fairest fruits of British genius and research [being] shaken into the lap of the American student."[6] The much-discussed American custom of pirating foreign works and selling them well below the cost of the original editions, which continued until the United States recognized international copyrights late in the century, encouraged this parasitical tendency.[7]

American reliance on foreign works reflected what some contemporaries diagnosed as "a morbid feeling of inferiority to our transatlantic brethren," a paralyzing fear that the humble efforts of Americans would elicit nothing but scorn from the scientific capitals of Europe.[8] Professional leaders repeatedly chastised American physicians for being slaves to foreign authority and urged them "to interrogate nature and experience more, and European opinions less," but little progress resulted.[9]

Although a proclivity for borrowing and a sense of intellectual inferiority may have contributed indirectly to America's poor record in the medical sciences, a far more basic cause was the commercial system of medical education that prevailed in the United States. Before the 19th century medical schools had traditionally stressed the dissemination rather than the production of scientific knowledge; they had frequently provided institutional homes for medical scientists, but had tended to leave organized research up to individual initiative or to scientific academies. During the first half of the nineteenth century this arrangement changed as educational reformers turned European medical schools, particularly German ones, into patrons of laboratory-based medical research. At the University of Berlin, for example, state-paid professors were expected by 1810 to conduct research as well as to teach.[10] In America most medical schools, even those nominally affiliated with a college or university, remained proprietary institutions, run for prestige and profit by ill-equipped local practitioners, many of whom could not have qualified for matriculation as medical students in Europe. Unlike most European governments, which regulated and supported medical education, the state legislatures in America granted charters virtually upon request — an estimated 457 by 1910 — and allowed schools to set their own standards.[11] Such legislative liberality may have provided the expanding nation with an ample supply of medical practitioners, but it did little to promote medical science.

American medical schools derived their income almost solely from student fees, which the professors divided among themselves. This scheme virtually guaranteed mediocrity, since high standards would inevitably have reduced the number of fee-paying customers. Because medical schools generally required less of their matriculants than liberal arts colleges, they often enrolled "*the leavings of all the other professions.*" Most medical students never attended college, and some barely knew how to read and write. The college boys who did go into medicine, complained one educator, were often those "who, from various causes — ill-health, poor scholarship, bad conduct and general discouragement — fall by the wayside and after one or two years of study, leave college without a diploma."[12] In view of such conditions, it is not surprising that contemporary critics frequently identified inadequate preliminary education as the highest barrier to the cultivation of medical science in America. "Our physicians and other professional men have genius enough," observed a Boston medical journal in 1833; "their defect is in mental discipline, which was not acquired during their preparatory studies in such a degree as to make the daily acquisition of knowledge, and the habitual exercise of the mental powers, become a primary object of pursuit, and a principal source of their highest enjoyment."[13]

Students not only entered medical school ill prepared for a scientific career, they left in the same condition. In contrast to the leading European schools, which at mid-century required attendance for four years and devoted from 37 weeks (e.g., Edinburgh and Paris) to 41 weeks a year (e.g., Berlin and Pavia) to lectures, the medical school of the University of Pennsylvania, one of America's best, required only 25 weeks a year for two years; and most American schools offered annual terms of only 16 weeks. To make matters worse, American medical students until the last quarter of the century customarily repeated the same courses during their second year that they had taken during their first.[14] In 8 to 12 months of formal training they were expected to learn anatomy, physiology, chemistry, and medical botany. Instruction in the basic sciences, except for anatomy, consisted of didactic lectures. As the historian John B. Blake has pointed out, "Until late in the century, anatomy was traditionally the only laboratory course in medical school. It was, however, generally taught simply for its practical value, chiefly for surgery, and, unlike the other medical sciences, gross human anatomy had very limited potential for stimulating original research."[15] American medical students may have picked up the vocabulary of science, but unlike German students, for example, they had little opportunity to learn its methodology.

Although the quality of American medical schools and their graduates varied greatly, by and large they produced craftsmen, not scholars. "In Europe an educated physician is presumed to be an accomplished *belles lettres* and professional scholar," noted the American Medical Association's Committee on Medical Education in 1863. "In this country . . . a doctor has no special prominence,

and, because a graduate, is not therefore regarded as educated or learned."[16] (It should be noted, however, that both American physicians and foreign visitors occasionally observed that individual physicians enjoyed higher social standing in the United States than in Europe.)[17] Given their cultural environment and training, American physicians understandably valued practice above science, wealth over scholarly reputation. William Beaumont, the frontier physician-physiologist, observed these traits during a visit to New York in 1833. "The professional gentlemen of this City have quite too much personal, political and commercial business on hand to permit them to turn their attention to animal and physiological chemistry, whose high honours and rewards to them are to be the results," he wrote to a friend. "Their curiosity once gratified, they are silent and aloof from the subject."[18]

The American obsession with practice and "getting ahead" deterred even scientifically inclined physicians from engaging in research. "You will lose a patient for every experiment you make in the laboratory," one medical professor warned a student contemplating a scientific career.[19] This attitude helps to explain why the nearly 700 American physicians who studied in Paris between 1820 and 1861 failed to establish a research tradition in America. An eminent American physician, upon hearing that his Paris-trained son wished to devote several years to clinical research before entering practice, explained to his son's French mentor why he could not approve of such plans. "In this country," he wrote, "his course would have been so singular, as in a measure to separate him from other men. We are a business doing people. We are new. We have, as it were, but just landed on these uncultivated shores; there is a vast deal to be done; and he who will not be doing, must be set down as a drone."[20]

American independence in the basic medical sciences did not come until the nation's medical schools freed themselves from dependence on student fees and acquired endowments sufficiently large to allow them to raise standards for admission and provide professors with the time and facilities to undertake scientific work. Since almost no medical professorships generated sufficient income from fees to provide a decent living, American professors customarily supported themselves by practicing medicine. In fact, observed one young physician, "a professorship in a medical college is generally sought as an advertisement in acquiring practice, rather than as an opportunity for study and investigation."[21] In 1878 the president of the American Medical Association contrasted conditions in the United States, where "the names of those who have made undeniable and valuable additions to the common stock" of medical knowledge could be counted on the fingers of two hands, with those in Europe, where scientists were supported by the hundreds. Americans, he said, must recognize

> that pure science, while it is a mine of wealth to the state, cannot remunerate the investigator; that it cannot live upon itself; that those who consecrate themselves to the pursuit of it must isolate themselves from the money-getting world around them; must be relieved from all care and anxiety as to their daily bread; and must be supplied with every necessary appliance while with concentrated thought and patient toil they seek to penetrate as it were with a diamond drill the flinty barriers which separate the known from the unknown. This is particularly true of those engaged in biological research . . . .[22]

Experience in Europe demonstrated that medical schools could provide a home for science, but as long as American institutions remained primarily business enterprises, they stood little chance of attracting the necessary governmental and philanthropic assistance. The "peculiar commercial organization of medical colleges," explained John D. Rockefeller's chief philanthropic advisor, accounted for the reluctance of the wealthy to support medicine "while other departments of science, astronomy, chemistry, physics, etc., had been endowed very generously."[23] It also helped to explain the preference of American millionaires for theology over medicine. In 1890 American seminaries claimed 171 endowed chairs compared with only 5 for medical schools, and none of the latter was adequate to pay even one professor's salary. The combined capital funds of all medical schools amounted to less than a quarter million dollars, approximately 1/48 of what theological schools, with half the number of students, possessed. This disparity, grumbled one jealous physician, existed despite the fact that Edward Jenner's discovery of smallpox vaccination "saves the community more

dollars in one year than all the endowments of all the theological schools in all time."[24]

The absence of salaries and laboratories that endowments could have provided influenced not only individual careers but the general pattern of activity in the biomedical sciences. Americans, noted one physician, displayed "a bias toward systematizing and utilizing the already existing knowledge, rather than the exploration of yet untravelled routes of investigation."[25] Those with scholarly inclinations often channeled their energies into writing financially remunerative textbooks and reference works rather than conducting basic research. John Call Dalton, for example, who in the late 1840s studied in Paris with the French physiologist Claude Bernard, achieved his greatest fame as a teacher and author of texts, not as a researcher. An admiring colleague commented on this unfortunate outcome:

> This eminent physiologist is by mental constitution evidently qualified to hold the position in America which in Paris is occupied by M. Bernard. He should be exploring the dark and untravelled regions of physiology instead of leading undergraduates along its beaten track; his pen should be occupied in tracing new provinces of thought added by his genius to the ever-spreading map of discovered biological science, instead of writing text-books for students. . . . his own proclivities would lead him to produce original monographs, circumstances confine him to the systematic routine of writing a college text-book.[26]

Practicing physicians—and virtually all biomedical scientists in America until the last quarter of the century did practice medicine—found little time for systematic scientific investigation. As one medical journal pointed out, "A man, fatigued with the details of practice, and whose time is never at his own disposal, can rarely do more than keep himself acquainted with the existing condition of medical science . . . ."[27] The experience of S. Weir Mitchell, a Philadelphian who studied the physiological effects of snake venom, illustrates the difficulties facing those who combined research and practice. "It was my habit," he wrote, "to get through work at three or four o'clock; to leave my servant at home with orders to come for me if I was wanted, and then to remain in the laboratory all the evening, sometimes up to one in the morning, a slight meal being brought me from a neighboring inn."[28] As more than one investigator discovered, such self-financed research could also be expensive. William Beaumont, whose experiments on digestion in the 1820s and 1830s won international acclaim, calculated his out-of-pocket expenses at over $3,000, and he suffered less than most because his position as a salaried surgeon in the United States Army provided a steady income and considerable free time.[29]

When the first Americans began returning from German laboratories in the 1870s, they, too, experienced difficulty in finding full-time employment as scientists. One of the earliest returnees, Henry Pickering Bowditch, who in 1871 established the first laboratory for experimental physiology in the country, was able to devote full time to research and teaching only because family money supplemented his Harvard salary.[30] T. Mitchell Prudden and William H. Welch, America's pioneer pathologists, were not so fortunate. Upon returning to the United States in 1878 both reluctantly practiced medicine for a period before finding institutional homes where they could continue their research. In a letter to his sister, Welch described the frustration he experienced in trying to launch his career as a professional scientist:

> I sometimes feel rather blue when I look ahead and see that I am not going to be able to realize my aspirations in life. . . . I am not going to have any opportunity for carrying out as I would like the studies and investigations for which I have a taste. There is no opportunity in this country, and it seems improbable that there ever will be.
>
> I was often asked in Germany how it is that no scientific work in medicine is done in this country, how it is that many good men who do well in Germany and show evident talent there are never heard of and never do any good work when they come back here. The answer is that there is no opportunity for, no appreciation of, no demand for that kind of work here. In Germany on the other hand every encouragement is held out to young men with taste for science.

All these evils, he continued, derived from the fact that "the condition of medical education here is simply horrible."[31]

But even as Welch penned these words, the reformation of American medical education was beginning. Although the proliferation of substandard schools continued unabated, the best institutions were grading and lengthening their curricula to three years, requiring evidence of preliminary education, and, led by the Harvard Medical College, abandoning proprietary status to become an integral part of a university.[32] The dramatic growth of laboratory-based medical science in the latter half of the century encouraged such reforms, as did the German training of approximately 15,000 Americans between 1870 and 1914. Although only a minority of the total specialized in the basic sciences, men like Bowditch, Prudden, and Welch succeeded in transplanting the research laboratory to American soil.[33] When it became apparent that student fees alone could not support such expensive facilities, medical schools began trading proprietary autonomy for the financial security of a university connection.

A further prod to educational reform came from the state legislatures, each of which passed some kind of medical licensing act between the mid-1870s and 1900. The state licensing boards influenced medical education in two ways. First, most of them required candidates to hold a diploma from a reputable medical school, that is, one requiring evidence of preliminary education and, in some cases, offering a three-year course of study, a six-month term, and clinical and laboratory instruction. This forced any school hoping to compete for students to upgrade its curriculum, at least superficially. Second, many states, especially during the late 1880s and 1890s, revised their statutes to require all candidates, even those holding medical degrees, to pass an examination. Although some of the weaker schools quickly learned how to coach students to pass these tests, graduates from strong institutions had a much better chance of passing. Medical commercialism, observed Abraham Flexner, thus "ceased to pay."[34]

No single event contributed more to the reformation of American medical education than the opening in 1893 of the Johns Hopkins School of Medicine under the leadership of Welch. At a time when, according to Welch, no American medical school required a preliminary education equal to "that necessary for entrance into the freshman class of a respectable college," the Hopkins faculty, at the insistence of its patron, demanded a bachelor's degree.[35] Modestly following the Hopkins example, more than 20 schools by 1910 raised their entrance requirements to two years of college.[36] As

Robert E. Kohler has pointed out, this reform, more than any other, "stretched the financial resources of the proprietary school beyond the breaking point. . . . higher entrance requirements disrupted the established market relation with high schools, diminished the pool of qualified applicants, and resulted in a drastic plunge in enrollment. Medical schools could not survive on fees."[37]

Blessed with a large endowment, Johns Hopkins became the nation's first real center for medical science. In addition to creating chairs in anatomy, physiology, pathology, and pharmacology, it provided their occupants—recruited nationally—with well-equipped laboratories and salaries sufficient to free them from the burdens of practice. Before long Hopkins students were spreading across the land, similarly transforming other medical schools. "It is no exaggeration to say that the few teachers who manned these [Hopkins] departments . . . revolutionized within a single decade the status of anatomy, physiology and pathology in America," reported a national body in 1915.[38] Welch, who in 1878 despaired of ever finding employment as a pathologist, was able less than a quarter century later to write:

> Today, pathology is everywhere recognized as a subject of fundamental importance in medical education, and is represented in our best medical schools by a full professorship. At least a dozen good pathological laboratories, equipped not only for teaching, but also for research, have been founded; many of our best hospitals have established clinical and pathological laboratories; fellowships and assistantships afford opportunity for the thorough training and advancement of those who wish to follow pathology as their career . . . and as a result of all these activities the contributions to pathology from our American laboratories take rank with those from the best European ones.[39]

By the turn of the century the medical schools at Harvard, Pennsylvania, Chicago, and Michigan had joined Hopkins as major medical research centers, but the nation still lacked an institution comparable to the Koch Institute in Berlin or the Pasteur Institute in Paris. However, in 1901 the United States Congress provided funds for a national Hygienic Laboratory to investigate infectious and contagious diseases, and, more important, John D. Rockefeller, the oil magnate, donated the

first of millions of dollars to create an institute that would become "the crown of medical research in this country."[40] The Rockefeller Institute for Medical Research not only freed its staff from practicing medicine but from teaching as well, allowing them to devote their entire lives to medical science. This environment brought the United States its first Nobel Prize in medicine—awarded to the French-born Alexis Carrel in 1912—and helped to reverse the flow of medical science and scientists from west to east. Its success soon inspired the creation of other American institutes for medical research and provided a model for the Kaiser Wilhelm Gesellschaft, which opened in Berlin in 1911.[41] By 1920 snide Europeans no longer asked what the world owed to American physicians and surgeons.

THE CLINICAL SCIENCES

In clinical medicine, which involved diagnosing and treating diseases, the process of maturation did not always parallel the transition from colonial dependence to independence that characterized the development of the basic sciences. Although Americans admired the superior clinical facilities of Europe, they commonly believed that singular circumstances in the United States demanded uniquely American responses to disease and made European knowledge suspect and in certain respects irrelevant. Thus, in areas like medical therapeutics American physicians never established a traditional colonial relationship, and they achieved maturity not by declaring independence, but by abdicating it.

The development of diagnostic tools and knowledge about morbid natural history flourished in the great hospitals of 19th-century Europe. Particularly in Paris, easy access to large numbers of diseased bodies made possible systematic clinical observation of disease processes, pathoanatomical correlation of these clinical patterns at autopsy, and statistical portraits of diseases based upon such studies. From the early 1820s, hundreds of American physicians were drawn to study in Paris both by its clinical facilities and by an environment conducive to medical research. "Merely to have breathed a concentrated scientific atmosphere like that of Paris," one physician studying in the French hospitals wrote home to Boston, "must have an effect on any one who has lived where stupidity is tolerated, where mediocrity is applauded, and where excellence is defied."[42]

Americans, however, lacked comparable institutions until late in the century, and physicians who studied in France found little opportunity at home to apply what they had learned abroad. William Wood Gerhard, who in the 1830s successfully employed Parisian methods to distinguish between typhoid and typhus fever, lamented the conditions at his hospital in Philadelphia. "I regret much the slender materials I possess and the difficulties wh[ich] seem inseparable from observation in this country," he wrote. Despite his inferior facilities for research, he optimistically expected that American physicians would place greater faith in his modest statistical studies on *American* patients than in conclusions drawn from manyfold more Parisians. Many of his countrymen nevertheless remained sceptical of his conclusions, which contradicted European opinion, until Sir William Jenner confirmed them in the 1840s.[43] American contributions to differential diagnosis and the natural history of diseases increased as the century progressed and large hospitals became common in American cities.[44]

It was in surgery that American clinical medicine attained its highest level in the eyes of both Americans and Europeans during the 19th century. Attributing their surgical skills to native mechanical ingenuity and frontier resourcefulness, Americans celebrated such pioneering work as Ephraim McDowell's 1809 operation in Kentucky for an ovarian cyst and J. Marion Sims's operation for vesicovaginal fistula, which he perfected on slave women while practicing in a small Georgia town during the 1840s. The first successful application of ether anesthesia for surgery, at the Massachusetts General Hospital in Boston in 1846, greatly inflated the American medical ego and convinced Americans that they no longer need apologize for their medical backwardness. American medical men "may not dive so deeply into abstract sciences, or linger there so long as in the old and somewhat *senile* establishments of Europe," declared one American surgeon, "but . . . as skillful operators, and practical men, they are the equals to any in the world, and second to none whatsoever."[45] Although a few exceptional achievements like the application of ether may not have warranted such

pride, surgery—and the similarly mechanical field of dentistry—did represent the most accomplished branches of clinical medicine in 19th-century America. The mechanical nature of dental and surgical therapeutics made them largely immune to arguments of American particularism, which were prominent in discussions of medical therapeutics.

The beliefs that therapeutic knowledge gained from experience with European patients and diseases might not be suitable for American practice and that therapeutic principles, unlike the tenets of the basic sciences, might not apply to all environments had deep roots in American medical thought. Although based in part on cultural nationalism, this conviction derived chiefly from the pivotal importance American physicians assigned to specificity in treating patients with different backgrounds and in different settings. Prevailing therapeutic theory stressed the necessity of tailoring therapy to the patient's age, gender, ethnicity, and habits, as well as to climate, topography, and population density.[46]

This commitment to specificity suggested, for example, that the therapeutic needs of the immigrant poor differed from those of native-born patients and, consequently, that information gained by observing pauperized Irish immigrants in a large urban hospital might be deceptive as a guide to treating middle-class private patients or even the hospitalized native-born. Thus physicians at the Commercial Hospital of Cincinnati in the mid-19th century prescribed rest for many patients "just from Ireland via New Orleans" while treating many of the other inmates with full depletive regimens of bleeding and purging.[47] A clinical lecturer at the Massachusetts General Hospital identified a phthisical woman to his students as "one of the cases of broken down health so often met with among her class";[48] another physician at that hospital, reflecting the consequences of this sort of class-specific constitution, noted that although copious bleeding and purging were appropriate for hale constitutions, hospital practice provided few opportunities to employ these remedies.[49] The notion was widespread that although active depletion might not be tolerated by degenerate urban dwellers who lived sedentary lives in vitiated surroundings, the robust farmer required a forcefully depletive therapeutic strategy.[50]

The stress on specificity fostered the notion that, because different regions of the country required distinctive therapeutic practices, physicians should be educated where they intended to practice. For example, the peculiar features of the South — its characteristic diseases, large population of Negroes, and warm climate — all argued for southern students studying medicine at southern medical schools. "Anatomy, Physiology, General Therapeutics, and Chemistry, may be studied to perfection in the Capitols of Europe and the United States," explained one proponent of this view; but, he warned, before practicing in the South such students would either have to unlearn the practical precepts their northern teachers had taught them or fail miserably in their efforts to heal the sick. In his opinion, an ill southerner would be

> better in the hands of some Planter or overseer who had long resided in this region, and who was perfectly familiar with the disease, than he would be in the hands of the ablest Physician of London or Paris, who had never practiced beyond their precincts, and who would be guided in his treatment solely by the general principles of Medicine.[51]

Just as northern therapeutic practices could be inappropriate for southerners, so too European practices might be invalid or even dangerous for Americans. Charles Caldwell, a medical professor in Louisville, Kentucky, returned from a European tour in the early 1840s to warn his classes that "the climate of London and Paris were entirely different from our own; the diet and habits of the people altogether different; and that these with other circumstances so modified the constitutions of the people and the character of the diseases, as to make the latter totally different from the diseases of this country." He expressed the common American suspicion of therapeutic knowledge generated in European clinics:

> The Hospitals of those great cities were very extensive and filled with persons laboring under great varieties of diseases; but they were from the very dregs of society[,] a class whose constitutions have been depraved by intemperance and want, and modified by vice, habit and climate until they possess no analogy in constitution or disease to any class in our own country. From this class or this kind of cases is the student of medi-

cine to derive his knowledge and experience in visiting the Hospitals of London & Paris.

Caldwell concluded that European "constitutions and diseases are so modified and so totally different from those in our country, the knowledge of Pathology and Therapeutics to be gained by visiting these hospitals can be of but little advantage to the practice of Medicine in the United States."[52]

Moreover, physicians generally believed that climate modified the influence of remedial agents on the body just as it influenced disease actions. One physician who held this view cited as his evidence "the different aspects of hyosciamus in England and Italy; of nitrite of silver in Naples and England; of the eau medicinale in Russia and France; [and] the vastly different effects of mercury in different climates."[53] If American practitioners remained "satisfied with the imbecility of European practice," they would, according to the estimate of one Boston physician, "undoubtedly lose a third or half of our patients."[54]

Americans criticized not only specific European therapies but also more general therapeutic philosophies. The therapeutic scepticism characteristic of the Paris clinical school, which argued for discarding any treatments whose clinical value had not been established by empirical observation, found an even more extreme expression in Vienna during the 1840s as therapeutic nihilism, that is, the complete rejection of therapeutic intervention in certain cases.[55] American physicians, who regarded active intervention as a crucial element in professional identity and legitimacy, found such inactive postures impractical and perhaps immoral. "The temporalizing course pursued by the French renders their therapeutics often inefficient," argued a Cincinnati practitioner, referring to the French inclination to leave the patient's cure to the healing power of nature.

> In anatomy, physiology, and pathology, they stand unrivaled; but beyond this they seem scarcely to look. Having made a *diagnosis*, the next most important matter is to prove its correctness; and as this can only be verified in the *dead body*, more enthusiasm is manifested in a post mortem examination than in the administration of medicine to cure disease. *The triumph of these physicians is in the dead-room.*[56]

As this quotation suggests, adherence to therapeutic localism did not imply a belief in the relativity of all medical knowledge. Among the medical sciences, therapeutics was largely exceptional. Although medical therapeutic knowledge did not function equally in all environments, the tenets of such basic sciences as chemistry and anatomy were universally applicable. American physicians also admitted the possibility that European therapeutics might have some applicability in the United States, but they insisted that each therapy be validated independently for the American market. Most believed, however, that therapeutic knowledge grounded upon American experience held far more promise than knowledge of foreign provenance.

In proclaiming their distrust of European clinical knowledge, American physicians assumed the burden of investigating American diseases and cures. A Kentucky student emphasized in his medical thesis the broad gap that existed between the clinical principles set forward by European medical writers and the requirements of American circumstances. Writing in the 1830s, he suggested that this

> imposes on us the greater necessity of observing for ourselves, and of culling, from among the useless rubbish of their productions, something worthy of an extensive and enlightened nation. Thus, although it has been vauntingly asked by one of their writers, "what does medicine owe to America," her sons may yet explore her wilds, and collect the materials, to rear upon the ruins of Eastern speculation, an edifice both complete and durable.[57]

The American environment provided both the opportunity and the responsibility for the reconstruction of medicine to meet American needs.

The program for medical research this emphasis on the American environment implied was avowedly localistic, drawing from a region knowledge to be applied within that region. Among the most active areas of research in early- and mid-19th-century American medicine was a species of natural historical investigation that linked together meteorological, epidemiological, and therapeutic observations. Perhaps the most original medical theses were of this genre. While most theses were merely derivative exercises, many a student elected to write an original essay based on his own investigations of the topography, climate, diseases, and

therapeutic practices of his home county.[58] Daniel Drake's massive treatise *On the Principal Diseases of the Interior Valley of North America* (1850–1854) was, in many respects, only a singularly ambitious expression of the same endeavor.[59] Studies of prevailing diseases, weather conditions, and appropriate treatments also thrived in the discussions of local medical societies, whose meetings were otherwise thin in scientific content.[60] Although American physicians did not excel in those branches of medical science that held universal interest or applications, they did actively conduct research in a sort of environmentally oriented, clinical natural history that was of considerable local import.

American allegiance to therapeutic localism and knowledge gained from direct clinical observation could be seriously challenged only during the last third of the 19th century, when a new therapeutic epistemology took its grounding in experimental laboratory science. Growing interest in this way of generating medical knowledge was both reflected in and fostered by the return from Germany of American physicians eager to exploit laboratory science as a means of transforming medical practice and elevating the status of the profession. Central to their program of reform was the idea that the laboratory was a legitimate arbiter of therapeutic knowledge.

From the early 1870s a number of prominent American physicians began arguing forcefully that the laboratory should join the bedside as an appropriate locus for the generation and validation of therapeutic knowledge. The ensuing clash between the advocates of the laboratory and the defenders of empirical clinical observation did not pit science against art, but entailed two largely incommensurable conceptions of the proper boundaries of therapeutic epistemology. For example, in the mid-1870s Alfred Stillé, a Philadelphia practitioner committed to clinical observation and environmental specificity, argued that the intrusion of laboratory science into the realm of therapeutics was presumptuous and destructive:

> The domain of therapeutics is, at the present day, continually trespassed upon by pathology, physiology, and chemistry. Not content with their legitimate province of revealing the changes produced by disease and by medicinal substances in the organism, they presume to dictate what remedies shall be applied, and in what doses and com-

binations. Their theories are brilliant, attractive and specious . . . . When submitted to the touchstone of experience, they prove to be only counterfeits. They will neither secure the safety of the patient nor afford satisfaction to the physician.[61]

Roberts Bartholow, an American enthusiast for the therapeutic promise of laboratory physiology, denounced Stillé's views to the members of a Baltimore medical society as "reactionary." "Modern physiology," he asserted,

> has rendered experimental therapeutics possible, and has opened an almost boundless field which is being diligently cultivated . . . . It is obvious that no science of therapeutics can be created out of empirical facts. We are not now in a condition to reject all the contributions to therapeutics made by the empirical method, but a thorough examination of them must be undertaken by the help of the physiological method.[62]

Bartholow and like-minded physicians rejected clinical observation as the principal way of gaining therapeutic knowledge—and as insufficient for the creation of a science of therapeutics—but they did not rule out the clinic as a source of therapeutic progress; rather, they advocated a new role for clinical observation in testing laboratory-generated therapeutic principles and practices.

The ascendance of this new view fundamentally altered the relationship between American and European therapeutics. Experimental science investigated disease processes and the practices that altered them. A tacit assumption that animated the rising vogue of vivisectional research during this period was that some fundamental tenets of physiological and therapeutic knowledge could be transferred profitably from the lower animals to man; medical scrutiny focused upon physiological processes, and it was to a certain extent irrelevant whether these processes occurred in a laboratory animal or an Irish immigrant. In this context, the heretofore crucial differentiae between northerner and southerner, immigrant and native, and American and European grew small indeed. The new experimental science, gradually taken up during the next few decades, prescribed in principle standardized treatments for diseased bodies, and considerations based on national variations (other than incidental ones) became stigmata of inferior medical practice.

Recognition of the therapeutic relevance of knowledge generated in the laboratory—abstracted from both the patient and the patient's environment—meant that therapeutic knowledge could be transferred freely between Europe and America. Thus, at the same time that American physicians acquired their own institutions for clinical research, they also freed themselves from their commitment to a distinctive "American" practice. Maturity in this context implied international reciprocity grounded upon an allegiance to medical universalism, not national independence.

NOTES

This paper was prepared for a conference on "Scientific Colonialism, 1800–1930," held at the University of Melbourne, Australia, May 25–29, 1981. Numbers is primarily responsible for the section on the basic medical sciences; Warner, the section on the clinical sciences. The former wishes to thank Mark Shale for his research assistance. The latter gratefully acknowledges the support of NSF Grant SES-8107609, which supported part of the research for this paper.

1 [Sydney Smith], Review of *Statistical Annals of the United States of America*, by Adam Seybert, *Edinburgh Rev.*, 1820, *33*: 79.

2 S. D. Gross, "Report on the causes which impede the progress of American medical literature," *Tr. A.M.A.*, 1856, *9*: 348. On American excellence in medical practice, see, e.g., the Prospectus, *Philadelphia J. Med. & Phys. Sci.*, 1820, *1*: ix.

3 By the late 1870s Americans were publishing more articles on surgery, obstetrics and gynecology, and diseases of the nervous system than any other nationality; see Mary E. Corning and Martin M. Cummings, "Biomedical communications," in *Advances in American Medicine: Essays at the Bicentennial*, ed. John Z. Bowers and Elizabeth F. Purcell, 2 vols. (New York: Josiah Macy, Jr. Foundation, 1976), II, 731–733.

4 A. B. Palmer, "Report of the Committee on Medical Literature," *Tr. A.M.A.*, 1858, *11*: 231.

5 Richard H. Shryock, *American Medical Research: Past and Present* (New York: Commonwealth Fund, 1947), p. 28; Robert William Fogel and Stanley L. Enger-

man, *Time on the Cross: The Economics of American Negro Slavery*, 2 vols. (Boston: Little, Brown, 1974), I, 248–250; Thomas C. Cochran, *Frontiers of Change: Early Industrialism in America* (New York: Oxford University Press, 1981), p. 114.

6   Oliver Wendell Holmes and others, "Report of the Committee on Literature," *Tr. A.M.A.*, 1848, *1*: 286–287. The phrase about being "mere recipients" appears in Samuel Jackson and others, "Report of the special committee appointed to prepare 'A statement of the facts and arguments which may be adduced in favour of the prolongation of the course of medical lectures to six months,'" *ibid.*, 1849, *2*: 365.

7   See, e.g., Gross, "Report," pp. 344–346; and Alfred Stillé and others, "Report of the Committee on Medical Literature," *Tr. A.M.A.*, 1850, *3*: 181. For British reaction to this practice, see "Report on the progress of midwifery and the diseases of women and children," *Half-Yearly Abstract of the Medical Sciences*, July–Dec. 1855 (No. 22): 208.

8   Thomas Reyburn, "Report of the standing committee on medical literature," *Tr. A.M.A.*, 1851, *4*: 493.

9   Usher Parsons, "Address," *ibid.*, 1854, *7*: 48–49. For references to American physicians being "slaves" and "toadies," see S. D. Gross, "On the results of surgical operations in malignant diseases," *ibid.*, 1853, *6*: 157; and Gross, letter to the editor, *Med. Rec.*, 1868, *4*: 191. For a counteropinion, see the *Am. J. Med. Sci.*, n.s., 1857, *33*: 389–390.

10   Hans H. Simmer, "Principles and problems of medical undergraduate education in Germany during the nineteenth and early twentieth centuries," in *The History of Medical Education*, ed. C. D. O'Malley (Berkeley and Los Angeles: Univ. of California Press, 1970), p. 189; Theodor Billroth, *The Medical Sciences in the German Universities: A Study in the History of Civilization* (New York: Macmillan Co., 1924), p. 27. French medical schools only belatedly supported laboratory-based science; see Erwin H. Ackerknecht, *Medicine at the Paris Hospital, 1794–1848* (Baltimore: Johns Hopkins Press, 1967), pp. 123–126.

11   Abraham Flexner, *Medical Education in the United States and Canada* (New York: Carnegie Foundation, 1910), p. 6; Alfred Stillé, "Address," *Tr. A.M.A.*, 1871, *22*: 83; William O. Baldwin, "Address," *ibid.*, 1869, *20*: 75. On the teaching of the various medical sciences, see Ronald L. Numbers, ed., *The Education of American Physicians: Historical Essays* (Berkeley and Los Angeles: Univ. of California Press, 1980).

12   "American vs. European medical science again," *Med. Rec.*, 1868, *4*: 182–183.

13   "Medical improvement.— No. 1," *Boston Med. & Surg.*

J., 1833, *9*: 92. On inadequate preliminary education, see, e.g., Stillé and others, "Report," p. 173; and N. S. Davis, "Report of the Committee on Medical Literature," *Tr. A.M.A.*, 1853, *6*: 125.

14   F. Cambell Stewart and others, "Report of the Committee on Medical Education," *Tr. A.M.A.*, 1849, *2*: 280. See also E. Giddings, "Report of the Committee on Medical Education," *ibid.*, 1871, *22*: 137.

15   John B. Blake, "Anatomy," in *The Education of American Physicians*, pp. 39–40.

16   Charles Alfred Lee, "Report of the Committee on Medical Education," *Tr. A.M.A.*, 1863, *14*: 84.

17   See, e.g., Stewart and others, "Report," p. 344; "American surgery," *Boston Med. & Surg. J.*, 1875, *92*: 21.

18   Ronald L. Numbers and William J. Orr, Jr., "William Beaumont's reception at home and abroad," *Isis*, 1981, *72*: 598.

19   S. Weir Mitchell, "Memoir of John Call Dalton, 1825–1889," National Academy of Sciences, *Biographical Memoirs*, 1895, *3*: 181. See also "The scarcity of working medical men in America," *Med. Rec.*, 1867, *2*: 277; and John S. Billings, "Literature and institutions," in *A Century of American Medicine, 1776–1876* (Philadelphia: H. C. Lea, 1876), pp. 363–364. Allegiance to medical practice helped to kill the short-lived Philadelphia Biological Society; see Bonnie Ellen Blustein, "The Philadelphia Biological Society, 1857–61: a failed experiment?" *J. Hist. Med.*, 1980, *35*: 188–202.

20   James Jackson, *A Memoir of James Jackson, Jr., M.D., with Extracts from His Letters to His Father; and Medical Cases, Collected by Him* (Boston: I. R. Butts, 1835), p. 55. On Americans in Paris, see Russell M. Jones, "American doctors and the Parisian medical world, 1830–1840," *Bull. Hist. Med.*, 1973, *47*: 40–65, 177–204.

21   Simon Flexner and James Thomas Flexner, *William Henry Welch and the Heroic Age of American Medicine* (New York: Viking Press, 1941), p. 85.

22   T. G. Richardson, "Address," *Tr. A.M.A.*, 1878, *29*: 96–97.

23   George W. Corner, *A History of the Rockefeller Institute, 1901–1953: Origins and Growth* (New York: Rockefeller Institute Press, 1964), p. 579.

24   Shryock, *American Medical Research*, p. 49. On support for theological and medical schools, see Flexner and Flexner, *William Henry Welch*, p. 237.

25   Henry F. Campbell, "Report of the Committee on Medical Literature," *Tr. A.M.A.*, 1860, *13*: 773.

26   *Ibid.*, pp. 774–775.

27   "American medicine," *Philadelphia J. Med. & Phys. Sci.*, 1824, *9*: 405.

28   Edward C. Atwater, "'Squeezing Mother Nature': experimental physiology in the United States be-

fore 1870," *Bull. Hist. Med.*, 1978, *52*: 330. Atwater emphasizes the importance of financial support for the progress of physiology.

29   Numbers and Orr, "William Beaumont's reception," p. 596. In this instance, Beaumont generously padded his expense account. On self-supporting science, see also the *Autobiography of Samuel D. Gross, M.D.*, 2 vols. (Philadelphia: George Barrie, 1887), I, 96–97.

30   W. Bruce Fye, "Henry Pickering Bowditch: a case study of the Harvard physiologist and his impact on the professionalization of physiology in America" (M.A. thesis, Johns Hopkins Univ., 1978), p. 78.

31   Flexner and Flexner, *William Henry Welch*, pp. 112–113; *Biographical Sketches and Letters of T. Mitchell Prudden, M.D.* (New Haven: Yale Univ. Press, 1927), p. 32.

32   See Martin Kaufman, *American Medical Education: The Formative Years, 1765–1910* (Westport, Conn.: Greenwood Press, 1976); and Robert P. Hudson, "Abraham Flexner in perspective: American medical education, 1865–1910," *Bull. Hist. Med.,* 1972, *56*: 545–561.

33   Thomas Neville Bonner states that "German study, especially in the basic sciences, was probably the most important factor in explaining the remarkable progress in medical studies in this country after 1870"; *American Doctors and German Universities: A Chapter in International Intellectual Relations, 1870–1914* (Lincoln: Univ. of Nebraska Press, 1963), p. 137. Robert G. Frank, Jr., Louise H. Marshall, and H. W. Magoun identify study in Germany as the "essential ingredient" in the maturation of the neurosciences; "The neurosciences," in *Advances in American Medicine*, p. 557.

34   Flexner, *Medical Education*, p. 11; Martin Kaufman, "American medical education," in *The Education of American Physicians*, p. 19.

35   Flexner and Flexner, *William Henry Welch*, pp. 219, 222–223.

36   Flexner, *Medical Education*, p. 28.

37   Robert E. Kohler, "Medical reform and biomedical science: biochemistry—a case study," in *The Therapeutic Revolution: Essays in the Social History of American Medicine*, ed. Morris J. Vogel and Charles Rosenberg (Philadelphia: Univ. of Pennsylvania Press, 1979), p. 32.

38   Richard H. Shryock, *The Unique Influence of the Johns Hopkins University on American Medicine* (Copenhagen: Ejnar Munksgaard, 1953), p. 22. See also Edward C. Atwater, "A modest but good institution . . . and besides there is Mr. Eastman," in *To Each His Farthest Star: University of Rochester Medical Center, 1925–1975* (Rochester: Univ. of Rochester Medical Center, 1975), p. 6.

39   Flexner and Flexner, *William Henry Welch*, pp. 266–267.

40   Kohler, "Medical reform," p. 53; Corner, *Rockefeller Institute*, p. 149; A. Hunter Dupree, *Science in the Federal Government: A History of Policies and Activities to 1940* (Cambridge, Mass.: Harvard Univ. Press, 1957), pp. 267–268.

41   Corner, *Rockefeller Institute*, pp. 76, 150–151; Shryock, *American Medical Research*, p. 93.

42   Oliver Wendell Holmes to his parents, Paris, Aug. 13, 1833, reprinted in John T. Morse, Jr., *Life and Letters of Oliver Wendell Holmes*, 2 vols. (Cambridge, Mass.: Riverside Press, 1896), I, 108–109.

43   William Wood Gerhard to James Jackson, Jan. 1, 1835, James Jackson Papers, Francis A. Countway Library of Medicine, Boston; Dale C. Smith, "Gerhard's distinction between typhoid and typhus and its reception in America, 1833–1860," *Bull. Hist. Med.,* 1980, *54*: 368–385. See also Ackerknecht, *Medicine at the Paris Hospital*; and Jones, "American doctors and the Parisian medical world."

44   Phyllis Allen Richmond, "The nineteenth-century American physician as a research scientist," in *History of American Medicine: A Symposium*, ed. Felix Marti-Ibañez (New York: MD Publications, 1959), pp. 142–155. On American hospitals, see Morris J. Vogel, *The Invention of the Modern Hospital: Boston, 1870–1930* (Chicago: Univ. of Chicago Press, 1980).

45   Valentine Mott, quoted in Courtney R. Hall, "The rise of professional surgery in the United States, 1800–1865," *Bull. Hist. Med.*, 1952, *26*: 234. American surgical excellence is discussed in "American vs. European medical science," *Med. Rec.,* 1869, *4*: 133–134; S. D. Gross, "American vs. European medical science," *ibid.*, pp. 189–191; and John Eric Erichsen, "Impressions of American surgery," *Lancet*, Nov. 21, 1874: 717–720.

46   For a particularly useful analysis of 19th-century American medical therapeutics, see Charles E. Rosenberg, "The therapeutic revolution: medicine, meaning, and social change in nineteenth-century America," in *The Therapeutic Revolution*, pp. 3–25.

47   Casebooks for Medical Ward Female, May 30, 1848–Mar. 7, 1850, Cincinnati General Hospital Archives, History of the Health Sciences Library and Museum, University of Cincinnati Medical Center, Cincinnati.

48   John Ware, Clinical Lectures, 1830, John Ware Papers, Francis A. Countway Library of Medicine, Boston.

49   George Cheyne Shattuck, Diary Notes on Patients, Vol. II, entry for Dec. 12, 1832, Francis A. Countway Library of Medicine, Boston.

50   "Effects of breathing impure air," *Boston Med. & Surg. J.,* 1832, *6*: 14; Northern Medical Association

of Philadelphia, "Discussion on bloodletting," *Med. & Surg. Rep.*, n.s. 1859, *3*: 271–274, 495–500, 515–521; 1860, *4*: 34–39, 486–497, 517, 518.

51 "Introductory address," *New Orleans Med. J.*, 1844, *1*: ii–iii. See also Jas. C. Billingslea, "An appeal on behalf of southern medical colleges and southern medical literature," *Southern Med. & Surg. J.*, 2nd series, 1856, *12*: 398–402; and on this idea, see John Duffy, "A note on ante-bellum southern nationalism and medical practice," *J. Southern Hist.*, 1968, *34*: 266–276, and John Harley Warner, "The idea of southern medical distinctiveness: medical knowledge and practice in the Old South," in *Science and Medicine in the Old South*, ed. Ronald L. Numbers and Todd L. Savitt (Baton Rouge: Louisiana State Univ. Press, 1989), pp. 179–205.

52 Courtney J. Clark, Notes on the Medical Lectures of Charles Caldwell, Medical Institute of Louisville, Kentucky, 1841–1842, Courtney J. Clark Papers, Manuscripts Department, Duke University Library, Durham, North Carolina.

53 Edward H. Barton, *Introductory Lecture on the Climate and Salubrity of New-Orleans and Its Suitability for a Medical School* (New-Orleans: E. Johns and Co., 1835), p. 17.

54 Celsus, "Treatment demanded by malignant diseases," *Boston Med. & Surg. J.*, 1832, *6*: 141; "Public medical information," *ibid.*, p. 336.

55 Ackerknecht, *Medicine at the Paris Hospital*, pp. 129–138; Erna Lesky, *The Vienna Medical School in the Nineteenth Century*, trans. L. Williams and I. S. Levij (Baltimore: Johns Hopkins Univ. Press, 1976). I do not suggest by this that Viennese nihilism was fully derived from Parisian scepticism; see Erna Lesky, "Von den Ursprüngen des therapeutischen Nihilismus," *Sudhoffs Archiv für Geschichte der Medizin und der Naturwissenschaft*, 1960, *44*: 1–20.

56 Review of "Lectures on the theory and practice of physic.—by William Stokes . . . and John Bell . . . ," *Western Lancet*, 1842-43, *1*: 354–357. On American attitudes toward the healing power of nature and its associations with therapeutic scepticism, see John Harley Warner, "'The nature-trusting heresy': American physicians and the concept of the healing power of nature in the 1850's and 1860's," *Perspect. Am. Hist.*, 1977-78, *11*: 291–324. See also "Andral's medical clinic," *Western Lancet*, 1843-44, *2*: 148; John P. Harrison, "On the certainty and uncertainty of medicine," *ibid.*, 1844-45, *3*: 118; and "Modern practice of medicine," *Boston Med. & Surg. J.*, 1835, *12*: 351-352.

57 William Wood, "An inaugural dissertation on the causes of epidemics" (M.D. thesis, Medical Department of Transylvania Univ., 1834), Special Collections and Archives, Transylvania University, Lexington, Kentucky.

58 Typical of such theses are Robert H. Hanna, "An inaugural dissertation on the medical topography and epidemic diseases of Wilson County Kentucky" (M.D. thesis, Medical Department of Transylvania Univ., 1835), Special Collections and Archives, Transylvania University, Lexington, Kentucky; and Thomas Hunter, "A dissertation on the topography of south Alabama and the diseases incident to its climate" (M.D. thesis, Medical College of the State of South Carolina, 1843), Waring Historical Library, Medical University of South Carolina, Charleston.

59 Daniel Drake, *A Systematic Treatise, Historical, Etiological, and Practical, on the Principal Diseases of the Interior Valley of North America*, 2 vols. (Vol. I, Cincinnati: Winthrop B. Smith and Company, 1850; Vol. II, ed. S. Hanbury Smith and Francis B. Smith, Philadelphia: Lippincott Crombe and Co., 1854).

60 The proceedings of one such local society are recorded in the Minutes of the Union District Medical Association, Vol. I, 1867–1880, Walter Havinghurst Special Collections Library, Miami University, Oxford, Ohio.

61 Alfred Stillé, *Therapeutics and Materia Medica*, 2 vols. (Philadelphia: Henry C. Lea, 1874), I, 31. Stillé discusses the influence of such factors as climate, season, and occupation on the actions of medicines on pp. 33, 90–94. The pairing of this and the following quotation is suggested by Alex Berman, "The impact of the nineteenth century botanico-medical movement on American pharmacy and medicine" (Ph.D. dissertation, Univ. of Wisconsin, 1954), pp. 36–37.

62 Roberts Bartholow, *Annual Oration on the Degree of Certainty in Therapeutics* (Baltimore, 1876), pp. 12–14. For an assessment of the relationship between experimental physiology and therapeutics in the mid-19th century, see John Harley Warner, "Physiological theory and therapeutic explanation in the 1860s: the British debate on the medical use of alcohol," *Bull. Hist. Med.*, 1980, *54*: 235–257; and *idem*, "Therapeutic explanation and the Edinburgh bloodletting controversy: two perspectives on the medical meaning of science in the mid-nineteenth century," *Med. Hist.*, 1980, *24*: 241–258.

THE ART OF MEDICINE

Historians of medical practice have focused primarily on the scientific accomplishments and forward-looking activities of physicians. But there has always been another side to medicine: the one-on-one interactions with sick people that formed, and still form today, the daily, repetitive activities that are the bread and butter of medical practice. It is in this intimate human connection that physicians have earned their keep and made their biggest contributions. There is no greater satisfaction, some physicians would say, than that gained from giving good patient care and being appreciated for it.

Most physicians in the past did not hope to make new scientific breakthroughs in their work. They lived and practiced in rural areas, small towns, and large cities where they visited their patients in their homes or, increasingly in the 20th century, in their offices or outpatient departments of hospitals. They were part of the communities in which they practiced, and their medical interactions were part of their larger social relationships. Physicians lived "complexly" with their patients, and the boundaries between medicine, family, and community life were blurred. In their patients' homes and in their own home offices, as Judith Walzer Leavitt relates, the domestic and social context of medical practice often dictated actual medical procedures.

Good bedside medicine required more than up-to-date medical knowledge. In addition to learning the latest techniques, physicians needed to know how most effectively to interact with patients. Diagnosing a disease and prescribing a medication were only the first steps toward effecting a cure; patients also needed to be convinced to accept the diagnosis and take the medicine once they were home. Success in this "art" of medicine often depended on personal communication skills and social adeptness.

But the art of medicine went even further. Physicians needed to keep case records that could help them remember treatment events over long periods of time or consult with their colleagues when they sought advice or published their successes. Steven M. Stowe analyzes how mid-19th-century southern physicians crafted and used individual medical case narratives as more than a practical tool to record things they had to remember. The case narrative, the most widespread form of medical writing in the 19th century, was story-telling. Narratives gave historical meaning—through timing, character, and plot—to medical practice, reconfiguring the wishes and fears of patients as well as the accomplishments and failures of their healers. As one mid-century physician wrote, the intensity of personal experience can illustrate more than "a volume of theorizing." The narratives provided physicians with means to sort and value their experiences, and they provide us with detailed and complex notions of the importance of daily medical work, of the art of medicine.

9

"A Worrying Profession": The Domestic Environment of Medical Practice in Mid-19th-Century America

JUDITH WALZER LEAVITT

One hot August evening in 1854, after returning home from a 16-mile journey on horseback to visit a sick patient, William H. Brisbane, M.D., wrote in his diary, "[I] feel tired to night from my medical business. It is a worrying profession indeed."[1] He did not refer to problems of diagnosis and treatment. He was confident of his ability with the technical parts of his practice and only rarely called for advice from his colleagues or voiced anxiety that his treatment was not working. He kept up with his reading of medical journals—most important to him seemed to be the *American Journal of Medical Science*—and was ready to act when his patients called. What concerned Brisbane and filled his days and the pages of his 33-volume journal were all the other things that accompanied him in his day's work and, to his mind, colored his reactions about his practice of medicine: the weather, his horse, the condition of the roads or the river, his children, his wife, the house and farm, his own health. He had not seen many patients in the week preceding this remark—and none had posed a difficult challenge—but he had seen the hired man break Brisbane's wagon to pieces, his son Benjamin leave home in pursuit of his unhappy wife, his own wife suffer from boils under her arms, and his son William complain of not feeling well. The following week, Brisbane himself felt so ill "with something

like cholera" that he thought it "prudent" to sign his will and "make proper arrangements about [his] business."[2] The temperature continued to hover in the 90s. He dismissed his hired man and hired another. He spent only one day that week "reading, botanizing & practicing."[3] His time had been full of worrisome things, indeed, but the actual practice of medicine occupied a relatively small portion of the concerns he found worth recording in his diary. All through his voluminous journal, Brisbane viewed his medical practice in conjunction with other activities having to do with his family, his house, and his farm. These personal factors that we often label extraneous to medicine deeply affected Brisbane's perceptions of his practice.

In this article, I explore some of the nonmedical aspects of life that provided the context in which much mid-century medicine was practiced. I think we miss attaining a full understanding of medicine unless we study physicians in the family and home contexts in which they carried out their practice. When and where Brisbane received his medical training, the therapeutic tradition in which he viewed himself, his professional connections and loyalties, all—obviously—are important to understanding what his example can teach us about mid-19th-century medical practice. But those are not the factors that William Brisbane himself identified as most influential day in and day out as he lived the life of a small-town and rural physician.[4]

In the daily details, incessant repetitiveness, and domestic base of medical life, male physicians in the mid-19th century—especially those with small-town and rural practices—followed closely the experiences of female healers, midwives such as Martha Ballard, whose life and context has been

JUDITH WALZER LEAVITT is Professor of the History of Medicine, History of Science, and Women's Studies, and Associate Dean for Faculty, University of Wisconsin–Madison.

From the *Bulletin of the History of Medicine*, 1995, 69: 1–29. Copyright © 1995 by the Johns Hopkins University Press. Reprinted by permission of the Johns Hopkins University Press.

so wonderfully described by Laurel Ulrich.[5] In this sense, medicine crossed gender lines and allowed men to experience—in fact, demanded that they experience—many of the same problems and rewards that female "social" healers knew intimately.

Most mid-19th-century male physicians practiced their profession in their own homes and in the homes of their patients. Their experiences were circumscribed and defined within these domestic settings, which we have previously (and mistakenly, I think) associated only with women. As Charles Rosenberg reminded us 28 years ago (and since), medical practice can only be understood within "the institutional context in which [medical ideas] are elaborated."[6] Just as the urban environment and the hospital, for those physicians and patients who inhabited them, ordered the practice of medicine, as Rosenberg demonstrated,[7] so too did the household for those practitioners who worked largely within its perimeters.[8]

I came to understand the importance the domestic environment might have on the people who practiced within patients' and physicians' homes from the recent work of women's historians, who have demonstrated how women's lives—including career women's lives—were shaped in their homes, and in the tensions between the public and private spheres.[9] Women's household responsibilities sometimes interfered with and always complicated their public activities. But not just women lived and worked within domestic spaces. Men, too, occupied homes and shared family concerns. Medical men, particularly, lived, worked, and had responsibilities within the households of America. Physicians, by virtue of spending their days within the domestic sphere—relying on its rhythms and smooth functioning—were an important component of what we have previously labeled domestic, female, and private. They daily crossed the line between public and private, between what might be viewed as male breadwinning activities and what we have typically seen as private female occupations.

Of course men continued to carry with them cultural norms of masculinity, just as women physicians later in the century exhibited strong notions of appropriate female behavior when they shaped their medical practices, as we have learned from Regina Morantz-Sanchez and others. Gendered distinctions existed within the home as well as outside of it, and we can expect therefore to see how male

gender expectations continued to provide essential boundaries within which physicians lived and worked.[10] Sewing and daytime visiting, for example, remained solidly within the female sphere. But some lines that we previously thought separated men's and women's experiences may have been quite permeable. Brisbane carried out various domestic chores on a regular basis, as we will see. Nancy Tomes, Steven Stowe, Joan Brumberg, and others have demonstrated that medicine is an excellent subject through which to explore gender boundaries because it exposes intimate human interactions.[11]

Those of you who are familiar with Laurel Ulrich's book *A Midwife's Tale* will recognize that my argument builds upon hers, but takes it in a different direction and applies her lessons to the practices of male physicians.[12] In her study, Ulrich connected women's domestic medical practice to the rest of their lives. We learned that what women healers did in the sick room was an extension of their traditional roles in the garden, in the parlor, and in the domestic economy. Midwife Martha Ballard did not leave behind her central worries about her own family when she went to attend others. Her garden, containing food for her family and herbs for her practice, was testament to her integration of the two parts of her life. The social networks extending from her roles as wife and mother opened the paths to her healing practice. In church, she sat next to her family, her friends, and her patients. If she was unable to find domestic helpers at home, it was difficult for her to get away to her patients. If a woman's labor was protracted, attending her interrupted the normal flow of Ballard family meals and chores. In Martha Ballard's story we come to understand the melding of the public and the private, and the interdependence of the two. And we see, also, that the weather, the horses, the condition of the roads and the river all impinged greatly on Ballard's ability to care for her patients.

I contend here that the men who practiced medicine were also rooted in family and domestic life. Men crossed into what we have thought of as women's domain with the practice of medicine when they entered their patients' homes and when their own home life surrounded them as they carried on the business of medicine. Just as Martha Ballard could not divorce her life and experience as a wife, mother, and woman from her practice of attending the sick, William Brisbane could not divorce his life and experience as a husband, father, and provider from his medical practice, as he, too, crossed the line between public and private. Mid-19th-century gendered distinctions were blurred in the home-based practice of medicine, whether by men or women, and in the very blurring of the social divisions we better understand medicine itself.[13]

In order to illustrate the importance of the familial environment for male physicians, I examine the lives and work of two small-town and rural-based mid-19th-century Wisconsin physicians: William H. Brisbane and Horace B. Willard. Through the daily journals of these two physicians we can learn their major preoccupations and worries and see that their personal lives intertwined with their medical practice.

Brisbane and Willard were fellow members of the Wisconsin State Medical Society, although I have been unable to determine whether they knew each other personally. They were both migrants from the East to Wisconsin. They were both ardent antislavery advocates and political activists in the state. Both men were medical school graduates, and their medical practices were comparable, although Brisbane spent more of his time with vaccination and tooth pulling than did Willard. Both Willard and Brisbane repeatedly scorned homeopathy and celebrated the tenets of regular medical theory and practice. I do not believe either man to have been unique or extraordinary as a rural practitioner, and, acknowledging the limitation of studying only two physicians, I nonetheless believe that their particular stories reveal certain general realities about the practice of rural and small-town medicine, in line with how other historians—recently, Jan Coombs, Paul Berman, J. Worth Estes, Edward Atwater, and Jacalyn Duffin—have described it.[14] Brisbane and Willard may or may not be singular in their personal journalizing (I think most school-trained physicians probably kept some written records), but we can definitely credit their descendants and archivists for preserving their jottings.

Brisbane was the older of the two.[15] Born in Beaufort, South Carolina, in 1806, of Scotch extraction, he was the son of a plantation and slave owner. He was ordained as a Baptist minister in 1830, served a medical preceptorship with Henry Rutledge Frost, M.D., and graduated from the Medical College of the State of South Carolina in Charleston in 1837.[16] When he himself inherited slaves, he emancipated them all and accompanied them to safe settlement in Ohio. He continued throughout his life to speak against slavery.[17] He lived subsequently in New Jersey and Pennsylvania, edited an antislavery paper, the *American Citizen*, farmed land back in Ohio, and then moved to Wisconsin in 1853. Settling in the small town of Arena, he farmed, operated a ferry across the Wisconsin River, and continued informally to preach. He also occasionally served as legislative clerk, town assessor, and school district supervisor. During the Civil War he took a chaplaincy with the Eleventh Wisconsin Volunteers, and after the war lived briefly again in South Carolina before returning to Wisconsin. He married Anna Lawton when he was 18 years old, and she, with seemingly decreasing eagerness, followed him through his geographical and career meanderings, bearing 10 children, of whom only four came with the family to Wisconsin and three survived their teenage years. Brisbane died in Arena in 1878, at 72 years of age.[18]

Arena is located on the southern bank of the Wisconsin River, about 30 miles west of Madison. Founded in 1848, the year of statehood, it stood on land that had been a trading location for local Indian groups. Its white settlers ran a store, a tavern, a lumber business, and a steam mill. When the Milwaukee & St. Paul Railroad chose to run its line through Arena in the 1850s, the town's future as somewhat more than a farming community was secured. Brisbane was Arena's first physician, and at the time of his death there were two others to carry on the work of serving the population that had grown from slightly over 400 when he arrived to about 2,000 at the time of his death.[19]

Brisbane's medical practice can be recovered through a careful reading of his diary, although no account books survive. During the six-month period January through June 1856, Brisbane prescribed for and visited patients a total of 267 times, a monthly average of 44.5 patients. The large majority of his patient contacts, almost 200 of his 267, were visits to their homes.[20] He visited patients' homes in Arena and in nearby Blue Mounds, Helena, Kalezonia, Dover, Smithfield, Wyoming, and Hogarth. A high proportion of his patient interactions involved women and children: during the same six-month period, Brisbane prescribed 52 times for women and children and 17 times for men; he visited men only 42 times, while his women and child patients numbered 156. He attended 9 women in childbirth. The practice was never sufficient to support his family, and Brisbane's farm was essential to their livelihood. Money problems followed him throughout his career.

Horace Birney Willard was born in Oswego County, New York, in 1825, the fifth of six children born to a stonemason and his wife.[21] He taught public school before apprenticing himself to Dr. William B. Coye of Gilbertsville, New York. A college benefactor allowed him to attend Geneva Medical College, from which he graduated in 1849, along with classmate Elizabeth Blackwell.[22] The young Willard and his bride Elizabeth Vickery settled in Aztalan, Wisconsin, where Horace set up his first medical practice. The Willards had one child. In 1856, Willard moved his practice to the larger community of Watertown, and the following year joined a practice in Lake Mills.[23] Willard served on the county board of supervisors as well as one term in the state legislature. He gave up the practice of medicine in 1866—some say because of failing health—and moved to the largest town in Jefferson County, Fort Atkinson, where he lived out his life as a businessman. H. B. Willard died in Fort Atkinson in 1900 at the age of 75.

Aztalan, the town in which Willard first sought his fortune, had been settled in 1836 on the site of ruins of a stockaded Indian village. It vied with Madison to become the capital of the territory in 1839 (losing by one vote), and was a thriving community of almost 600 settlers when the Willards arrived.[24] On the banks of the Crawfish River, it was Jefferson County's leading business center in the 1850s when Willard became the third physician to settle there. The population reached 1,000 by 1860, but the town's growth was capped when the railroad bypassed the community. Willard saw patients in Aztalan and traveled the county to Milford, Johnson Creek, Farmington, Oakland, Lake Mills, and Watertown. He, too, found medical practice insufficient to maintain his family and constantly fretted about patients who did not settle their accounts with him.

A sense of Willard's medical practice can be reconstructed from his account books, diary, and obstetrical and medical case books. Unfortunately,

these sources overlap for only short periods of time. Willard's practice was considerably busier than was Brisbane's, but Willard, too, suffered the ups and downs of uncertain patient loads throughout his years of practice. During the six-month period January through June 1852, Willard prescribed for and visited patients a total of 438 times, or a monthly average of 73, but in April of that year he served only 28 patients. The majority of Willard's contacts were visits to his patients' homes: of the 438 contacts, 316 were home visits. Again in a similar pattern to Brisbane, a high proportion of Willard's home visits involved women and children: during this same six-month period, Willard visited women and children 236 times compared with only 80 times for his adult male patients. He attended seven women in childbirth in this period, and his overall confinement record accounted for approximately 3.6 percent of his total patient visits.

The diaries of these two physicians allow historians to get to know them as people and to see their medical practices in the full context of their lives, in the way biographers might. Biographers of women in recent years have demonstrated the importance of both the private and public parts of their lives.[25] Biographers of men, however, have traditionally not placed very much importance on the domestic side of their subjects' lives, and provide only brief references to wives and children. They do not usually attribute public significance to the swirl of domestic arrangements.[26] I am here positing that 19th-century rural and small-town physicians showed a steady concern with familial issues and were themselves enmeshed within this sphere as well as able to separate themselves from it. Paying attention to the daily concerns these two men revealed in their diaries forces us to acknowledge the importance to them of the domestic world of the family. We see male physicians not only as public persons, representatives of the learned profession of medicine, but also as family-oriented men whose personal satisfaction was closely caught up with their wives' activities and their children's achievements and whose family obligations helped define their mental preoccupations.

What these diaries suggest is not just that these two men were good, caring people, but that their familial orientation was part of what it meant to be a physician in small rural communities in this period. At the same time as professional, school-based, male, learned medicine became available in isolated small communities such as Arena and Aztalan, Wisconsin, with these Eastern-trained physicians, the domestic base that we have previously associated with women and midwives similarly infused the physicians' lives. Brisbane and Willard were able to bring the learning and prescriptions of their teachers to these communities. But they did this only within the context of their own family rhythms, and they wove together their domestic experiences with their perceptions of their own medical practices. Calomel and ipecac were always with them; so too were their personal concerns about their integration in the community, the weather, the condition of their horses and wagons, their brother's or son's indiscretions, their daughter's love letters, their wife's mental state, their daughter's Latin lessons, the landings of the ferryboats, the productivity of their land, and even the new parlor wallpaper. Furthermore, within the homes of their patients, they encountered the familiar group setting of family and friends within which they negotiated their bedside behavior.[27] Their medicine was in part structured by the dual domestic settings in which they lived and practiced. Their lives as physicians were intimately linked with their lives as husbands and fathers. In familiar small-town environments physicians spoke both the discourse of medicine and the language of the home.

These diarists repeatedly connected all the parts of their lives, the medical alongside the domestic, indicating that they saw the parts as an integrated whole. We historians must acknowledge this intimate linking and see as they saw that failure or problems in one realm leaked into another, that successes in one made the other side look rosier. William Brisbane's medical business became "worrying" when all that surrounded it was not in order. His domestic life sometimes made visiting a patient seem hopelessly difficult.

I want to push this one step further, one that is admittedly more speculative. Medicine was the kind of profession that gave positive reinforcement to the domestic skills men learned from being at home and in homes. In part because the majority of their patients were women, men's routine participation in family matters sharpened their sensitivity and perhaps appreciation for domestic tasks, which made them better able to relate to the pressures in their patients' lives and made them more

attentive practitioners. It was a two-way street. Being a good husband and father made a man a better doctor; being a better doctor reverberated back home and made a man more aware of his own domestic arrangements. When a man practiced medicine, he was—and had to be—attuned to family concerns and an active participant in domestic affairs.

I will present evidence underscoring my thesis in two directions. First, using Brisbane as my main example, I demonstrate the intimate link between medicine and family, as he crossed what we usually think of as a gendered line in his active involvement with the daily concerns of his wife and children. Second, using Willard's diary in particular, I provide examples of how the domestic environment influenced his therapeutic decisions.

I turn first to a portion from William Brisbane's diary that illustrates his daily combination of activities:

Friday Jany 11th 1856. The Lyceum met this evening at the school room. Went to Kalezonia to see a sick Irishman & brought [back] a load of corn.

Saturday Jany 12th/56. Arena, Wis. Paid my taxes. Wife & I supped at old Mr. Hatch's.

Sunday, January 13th/56. Arena, Wis. I preached from the words, "Thus it becometh us to fulfill all righteousness." Went in the afternoon to Kalezonia to visit a patient. Will & Addie went with me. We supped at Mr. Bartletts.

Monday, Jany 14th/56. Arena, Wis. Attended to our school district matters. Prepared deed, &c. Had corn & wheat prepared for Mill.

Tuesday Jany 15th/56. Arena, Wis. Weather has moderated. Lost the sixth one of my cattle for this winter four by the cold & two by drowning. Have given Wm the Blacksmith's shop on condition he works in it two years. Sent corn & wheat to Mill but had to leave it there to take its turn.

Wednesday Jany 16th/56. Arena, Wis. Sent Wm to Kalezonia for corn. He brought the 5th load from there. Mrs. Bartlett dined with us.

Thursday, Jany 17th/1856. Arena, Wis. Sent two wagons for corn . . . our girl Mary Smith having left us yesterday morning my wife has been quite in a fret about it. She left because she was dissatisfied about her wages. Wife was unwilling to give her more than $1.00 a week. We had prayer meeting this evening at Mr. Brown's.

Friday, Jany 18th/56. Arena, Wis. Mrs. B. & I went to Helena and dined at Mr. A. Bernard's. Wm hauled with the horses 32 baskets of corn from the field at Kalezonia & Fred 62 baskets with the oxen. Visited [patient] Antony Haley. Our children went to a party at Ashmore's this evening.

Saturday, Jany 19th 1856. Arena, Wis. I feel much displeased that my children should have staid at the party until 2 & 3 o'clock this morning & engaged in dancing also. Got breakfast between 9 & 10 o'clock. Wm brought the wheat & buckwheat flour from the Mill. Had some wood hauled up. Got a letter from Dr. Robert.

Sunday, Jany 20th, 1856. Arena, Wis. I was called up at 3 o'clock A.M. & had a cold ride to James Young's; whose wife I delivered this evening of her first child, a daughter.[28]

In this 10-day excerpt, Brisbane recorded seeing only four patients, well below his average. From the entries, we get a sense of how he fit medicine around his other duties and how the various parts interacted. In general, when called to attend a sick patient, he gave the medical visit priority over his other commitments unless he was sick himself. When lumbago kept him at home in February 1856, he sent prescriptions to the three sick members of the Brodie family, but wrote in his diary, "I could not go to see them, on account of the pain in my back."[29] The diary entry tells us how he integrated his visits to patients with other activities. After visiting a sick man in Kalezonia, he brought back a load of corn from his farm. Another time, he took two of his children with him on a patient visit and, together with them, had dinner at a friend's house. He combined another patient call with a social evening with his wife, leaving his son and hired hand to haul the corn without him. Brisbane did not compartmentalize medicine; it was part of everything and everything was part of it.

Many other things beckoned him when patients did not. During this particular time, getting the corn to the mill seemed most important.[30] Brisbane preached at the Sunday church service and attended a midweek prayer meeting. Throughout his life, he continued his pastoral duties. He also participated in the intellectual life of the community through the Lyceum discussion group, which he missed only when it conflicted with his patient visits. He and his wife enjoyed visiting with neighbors and friends, some of whom were connected to his medical practice and others who seemed separate from it. They dined in the homes of others and reciprocated with entertaining in their own home.

Brisbane's chastisement of his children for their partying echoed a constant family theme during these years.[31] On 23 March 1855, his 14-year-old daughter Addie noted in her diary, "Father did not like our staying [out late]. I think I shall not try it again."[32] William was involved in raising his children, and his interactions with them extended far beyond acting as disciplinarian. They frequently accompanied him on his rounds, he oversaw their education and other extracurricular activities, and he worried about their behavior and their future.

Brisbane's daughter Addie frequently wrote in her diary about her father and noted how her activities were scheduled around his. For example: "We had no meeting to-day as father had to go to see a patient"; or "Father took Belle and myself to Mr. Adams' this morning on his way to Hayworth to see a patient."[33] Addie followed her father's medical work, writing, for example, "Mrs. Brodie was very sick last night and sent for father about the middle of the night. She is better now. She has the cholic."[34] Addie also provided some insight into the roles of doctors' wives when she wrote, "Mother went this morning to see Mrs. Rockwell, as father was not here; she had a daughter."[35] Brisbane's medical practice was a family affair.[36]

Brisbane noted trouble with domestic help at some length in the excerpt above and throughout other parts of his diary. The closeness of his observations about the subject of hired help and his wife's feelings about them was no doubt related to the fact that he was in and out of the home during the day and he was of necessity concerned with the efficiency of its operations. We do not see what historians have assumed was the normative 19th-century domestic sexual division of labor in the Brisbane family. William Brisbane participated along with his wife in his children's education, giving his daughter frequent Latin lessons. He delighted in the new parlor wallpaper, glad the family at last had a "tolerably decent looking room" in which to entertain.[37] In the winter months he was the first to rise and light the fires so that the house would be warm when his wife and children arose. He often made his own breakfast and sometimes that of his children while his wife slept. His public life was complicated when things did not run smoothly at home, and he paid close attention to domestic organization. He persuaded his wife to renegotiate with Mary Smith and she came back to work at higher wages, but that was not the end of it. In June 1856, Brisbane wrote, "Wife had a blow up this morning with Catherine & Christian which resulted in my settling up with them and their departure from my service."[38] The next day he paid off another worker, noting, "He spoke very passionately & insultingly to Mrs. Brisbane because she tried to show him the unreasonableness of his demand for pay. . . . He was so insolent I had to drive him off the place."[39] In these instances, William was in the house during the disturbances and acutely aware of his wife's feelings.

There are two issues of significance here. The first concerns Brisbane's financial problems. One of the reasons Anna Brisbane tried to protect domestic finances was that her husband did not get a good return on his medical fees. In March 1856, after attending a woman in a protracted labor and finally delivering "a fine large girl," Brisbane returned home to find Mary Smith demanding her wages. He wrote in his diary, "I was sorry not to have the money for her. . . . It is so hard to make [medical] collections that I am left without the means of paying my hired people."[40] Another time, again following an obstetrics case, he wrote, "I sent William out to day to collect medical bills but he did not get a cent. So it goes, & I am very much in want of money."[41] Brisbane practiced medicine at a time and in a place when the profession was economically vulnerable and few could make a living solely from the practice of medicine. Anna Brisbane had cause to try to protect the family purse.

A second problem concerns his wife Anna's mental health. In 1854, when their son Benjamin married a woman they did not know, William wrote that "Benjamin's [letter informing them of the event] fills me with anxiety."[42] But Anna Brisbane had a much stronger reaction: she "takes on at a grievous rate about it. . . . [And] she says she would rather Benjamin be dead, and that she is going to write to him not to come about her any more."[43] Brisbane at first attributed Anna's strong reactions to her excessive worries about "saving money & property," although he thought her emotional disposition instead of helping "has hindered . . . our prosperity."[44] But later he realized, "She is becoming more and more fretful, so as to make herself and others unhappy. . . . If I make any remark to her upon the effect of her action or words upon other minds, she immediately charges me in an

extraordinary manner with always finding fault with her."[45]

Again mulling over the same subject, Brisbane wrote, "The Lord save me from the disaffection of my loved Anna in our advancing years. What does her coldness towards me for some time past mean? I cannot feel that she has any ocasion [sic] for it in my conduct to her, and cannot but fear there is some unfortunate condition of her mind, perhaps arising from what is termed the critical period of woman's life."[46] When his wife left home to accompany their children Addie and William to school in Ohio months later, however, these difficult interactions quickly faded from Brisbane's mind and he missed her. When he did not hear from her, he left his farm and medical work to follow her, because "I cannot stand it longer, I must see [my] wife. It is out of the question for me to enjoy life without her."[47] Upon their reunion he wrote, "I feel truly happy in the loving & endearing embraces & kisses of my precious wife. The cloud seems to have passed from her mind & she is her loving self again."[48] Maybe he was right to attribute her mood swings to the menopause; then again, perhaps in their happy cross-country reunion, Anna was responding to what was an extraordinary show of his own affection for her. Whichever it was, they returned together to Arena, where Brisbane was able to resume medical and farm work with fewer domestic interruptions.

This example does more than uncover a family drama. For months at a time and over a period of years, Brisbane carried his concern about his wife's health around with him as he saw patients and attended to the farm work. He left his patients to pursue Anna to Ohio to settle family affairs. In this instance, he scheduled his medical business around his familial concerns. Feminist biographers have suggested that women (more than men) "make decisions and sacrifices on the basis of personal needs."[49] Susan Ware wrote that "Whom you share your bed with and how you pay the bills do have an impact on events beyond the household." Ware believes the insight, which might apply to male subjects, too, "more salient" for women subjects, who "almost inevitably had to make decisions and sacrifices that had potentially profound effects on their personal lives. In charting a woman's public achievements," Ware concluded, "we need to pay special attention to both the benefits and the costs of such personal choices."[50] William Brisbane demonstrated that such experiences are definitely salient to understanding the lives and careers of men as well as women, and that a possible gender difference—if one exists on this point—needs to be more carefully delineated.

This example of the interacting influence of family affairs and medical obligations was consistent with medicine but did not extend to Brisbane during the period he pursued other employment. A comparison of two time periods proves the point. In 1856, still practicing medicine, Brisbane wrote long sections of his diary about his daughter Addie and a possible romantic entanglement of hers of which he did not approve. For example, he wrote, "A letter our daughter got from George W. Ashmore . . . has been a subject of much anxiety to me. He makes love to her; & I exceedingly deprecate anything that will tend to keep her from getting a complete education."[51] The source of the letter to Addie was the son of a local Arena family; they were part of Brisbane's church and lyceum group and patients in Brisbane's medical practice. Brisbane lamented, "I apprehend [this incident] will do my family no essential good."[52]

The following year Brisbane assumed a job as clerk of the state legislature, became the town assessor in Arena, and then a justice of the peace. This period of his diary contains very few references to family matters and none to Addie's love letters and relationships, although they were ongoing, as detailed in her diary. It seems that when he temporarily left medicine he also abandoned his intimate connections to the daily occupations of his wife and children, which suggests that medicine fostered particular sensitivity to domestic concerns as well as a need to influence them. Government service, on the other hand, did not seem to require such close knowledge of or involvement with domestic matters.[53]

The diary quotation illustrates an additional point about the importance of weather and road conditions to medical work. Brisbane was obsessed with watching the mercury on his thermometers, which he placed on all four sides of his home. Many days he rode 30 miles or more in the course of making medical visits, and expressed pleasure in the beautiful or cool weather. But more frequently the weather made his travels difficult and uncomfortable, or even impossible, and sometimes he refused

to see patients rather than brave the cold temperatures. One February evening he wrote, "The mercury this morning was 22 [degrees] below zero at 8 o'clock; at zero the most of the Day & tonight it is at 8½ o'clock 10 [degrees] below zero on the North side of the house & 14 [degrees below] on the west side."[54]

Summer could be even worse than winter. On 10 June 1854, Brisbane wrote, "Went to the farm; & paid two medical visits. The mosquitoes are terrible."[55] The next day he again noted, "The Mosquitoes are so severe that I am apprehensive we cannot do anything here. The family are very much dissatisfied. I never saw the like of them in my life. It is very disheartening indeed."[56] When the mercury soared in early July to 96 degrees "in the shadiest part of the house" and he had to keep up both medical visits and farm work, Brisbane despaired.[57] The "excessively oppressive" weather continued over the month as his family voiced their eagerness to move away from Arena.[58] The weather exacerbated all the other problems with his son and his son's wife, the hired man, and his wife's health. It was this confluence of familial concerns, problems with the hired help, and unbearable heat that brought Brisbane, upon the occasion of his next patient visit, to call medicine a "worrying profession."[59]

Horace B. Willard's diary provides evidence of the second point I am making in this article—that medical practice itself was influenced by its location within the physicians' or patients' homes. Willard moved to Aztalan in 1849 and established a viable medical practice rather quickly. He billed for $474 during his first year, and almost doubled that amount his second and third years ($731 and $736, respectively), making his income comparable to other medical graduates who practiced in rural areas.[60] But, like Brisbane and countless others, he had difficulty collecting his fees. Over the parts of his career for which we have complete data, he collected a maximum of 24 percent of his billed fees, and much of this in kind rather than in cash. His experience of 15 August 1857 was all too common: "Yesterday went in the 7 1/2 train to Oconomowoc and thence walked up to Summit to see see [sic] a man who is owing me, but could get no money, and got back to the Depot and returned by the 5 1/2 train."[61]

Financial problems accounted for an important

part of Willard's worries, but his diary dwells on others equally momentous in his daily reckoning. In Willard's diary, as with Brisbane's, we see evidence of close interdependence between home life and medical practice. He missed few opportunities to write about his wife and daughter and their importance to him, often waxing romantic: "Freely do I admit that my hearts affections centre in those two beings, parts and parcels of myself."[62] When his family was away from home, his medical duties seemed heavier. When his family was together, he wrote: "How many charms are thrown around the word *home*. The happy private family circle. The ever sacred hearth stone."[63] When his family was around him his medical work seemed less onerous. "O how much of this life's happiness is wrapped up in the dear ones we love," he remarked.[64] One time after a group of house guests left them, Willard wrote, "I am alone with my wife and baby—O happy solitude! Visited one or two patients. The day has been warm and pleasant." All parts of his life were in order.[65]

Like Brisbane, Willard was plagued by the harsh Wisconsin weather and the geography of his practice. In April 1851, the five miles he had to travel to attend a woman's delivery prevented him from being there before the baby arrived, and thus from charging a fee. In February 1857, the bad weather—in this case rain—made his practice uncomfortable in the extreme. On the 24th, he rode five miles in the pouring rain to see a patient. But by the next morning, when he was again called to nearby Farmington, the rain froze and made for rough traveling. He reached his patients nonetheless, and "Found the sick some better and the well ones some sicker."[66] The bad weather continued. In March he complained, "A month of *toil* and care, of fearful anxiety. Have had much to do. And much *mud* to do it in."[67] Trouble continued when in April, usually a time of moderating temperatures (even in Wisconsin), it snowed: "the snow falling fast. At first it fell still & it was quite warm—And I started for the country—visited my patient about 4 [7?] miles out but the wind came up and a cold North easter set in, and I was oblige[d] to face the storm which was as severe as I had experience[d] during this winter—Mud so deep that I was oblige[d] to walk my horse all the way. Such a 10th of April I have never seen before."[68] The next day he stayed at home, feeling the effects of his exertions. The

rest of the month and throughout May, Willard recorded, "Rain, Rain, drizzle, drizzle, and no cessation. No Sun Shine. No dry wood. No good roads."[69]

Lest Wisconsin get too negative a reputation from such notations, let me quote from the diary on one of Willard's better weather days, in which he made several professional calls: "It has been a beautiful day as it is a glorious night. There are but few who seem to appreciate the beauties of a moonlight night. Those thousand bright and laughing eyes that look so cheeringly upon us. The broad blue *immensity*—it seems like Heaven's shield thrown over us to protect . . . Such a night fills the soul with reverence and love . . . shows man his finer feelings and reveals his better nature. The time for thought—pure, lofty, sacred, meditations."[70] Surviving the bad conditions made the good ones that much sweeter.

Willard, of course, had more on his mind than the weather, as important as that could be on certain days. Unlike Brisbane, whose life outside the sick room integrated business, community, and family affairs, Willard, when not seeing patients, usually remained at home. He read copiously, including sentimental novels, biography, history, and philosophy, and he tried to learn German.[71] When his patients improved, he wrote happily, "Visited my patients in the woods. Found them doing well—& therefore I myself *feel* well."[72] But just as often, his precarious finances and his lack of community activities weighed him down. In April 1857, he wrote, "Have tried to read but could not bring my mind to bear upon the subject. Am somewhat melancholy and for what I am unable to say, unless it is my future prospects, which I must confess are not decidedly pleasant."[73] In May he lamented, "This is a *blue* day and this is a *bluer* fellow what is writing in this diary. The dark and gloomy feelings within accords well with the lowering clouds, & mist and rain & mud without. O! who will deliver me from this living death? Have . . . done no [medical] business."[74]

Despite Willard's depression over his financial prospects, he had many successes in the sick room. He demonstrated considerable medical and social skills with patients, often in ways that provide some vivid evidence of the impact of the domestic environment on medical practice. In his medical and obstetrical journals there are repeated examples that he adapted his medical therapeutics to the familial situation in which he met his patients. When visiting patients' homes, Willard usually encountered the patient's friends and family members, who, worried about their loved ones, offered advice about what Willard should do. Sometimes, under these conditions, Willard acquiesced. On at least one occasion he refused to do what his patient's friends wanted. In the communal domestic setting in which he saw his patients, at the least Willard needed to explain and justify his actions. He often wrote of his therapy in terms of its popularity with patients and their friends.

Willard attended Harvey Foster's wife through her fourth delivery in 1854 and noted that her labor was very slow. The parturient had "excruciating pains, but no progression. . . . Friends became much alarmed." Willard decided to try manual version, which, he happily reported, "I did with perfect success—though not without hard work—for it was with the greatest effort that I could force the child back to introduce my hand." He preferred manipulation over forceps because, "the idea of instruments is *horrible* to *friends & patients*." He concluded, "With such a *hand* as I am blessed, turning is much the better way."[75]

Reflecting on his obstetrical cases the following year, Willard estimated that one-quarter of them involved very slow progress in which the os did not dilate sufficiently. In these cases, "I have neither bled nor given tartar emetic or any other sickening medicine. I have no doubt but had I done so, labor would have been facilitated some. But whether my reputation would have been fairer is very questionable in my mind. The idea of being vomited at such a time is exceedingly repugnant to the patient and to friends."[76]

In these examples, we see clear evidence that Willard altered his medical practices specifically because of the group of friends gathered around the delivery bed in the homes he entered. Unlike physicians visiting patients in a hospital setting where the physicians might be able to carry out whatever therapy they thought necessary, Willard and others who practiced in patients' homes surrounded by patients' families and friends often had to moderate their medical therapy to these others' wishes and needs.

During the slowly progressing labor of another patient, Willard resisted this friendly pressure. At-

tending Mrs. Fobes in her third delivery, Willard recorded these signs and symptoms: irregular contractions, increasing numbness of the hands, dimming of vision, and slurring of speech. Mrs. Fobes' friends "urged" Willard to bleed her, but he "refused[.] Administered tinct. Assafoetida and removed symptoms in a degree." Then, showing sensitivity to the female dynamics in the birthing room, Willard steered events in another direction: "I learned about this time [from her friends] that one of the female attendants had not been expected to be present, and that patient had a little antipathy towards her. Although she was the oldest one present, and the mother of 13 children, I sent her away. Nervous symptoms entirely subsided. Pains became more regular . . ." But the contractions remained ineffective, and in time the older woman returned. With her, "the nervous symptoms also returned in aggravated form. . . . Gave her laudanum, which alleviated the symptoms."[77] In this ultimately successful delivery (of an eight-pound girl), Willard demonstrated helpful bedside interactions and keyed his activity to the social situation in the room. Able to resist the parturient's friends' therapeutic advice, he nonetheless responded sensitively to the interpersonal dynamics in the birthing room.[78]

In his obstetrical cases particularly, Willard effectively reached across a line that separated men's and women's spheres of experience to demonstrate impressive interpersonal skills that clearly helped his medical practice. In his medical cases, too, Willard repeatedly noted the presence of friends, their opinions about therapeutic matters, and their influence on medical care. For example, his treatment of a man suffering from ulceration of the lower portion of the bowel was interrupted when "Friends became alarmed" and insisted on the consultation of another doctor.[79] Indeed, physicians practicing within the homes of male and female patients needed to be comfortable negotiating their treatment plans within the domestic environment. They developed interpersonal familial skills through the kinds of experiences I have detailed in this article, and their domestic and medical worlds intertwined and supported each other.

In the experiences of Brisbane and Willard we see the intimate linkage between domestic concerns and medicine. During the period when medicine was practiced predominantly at home—in the patients' homes and in the physicians'—it was influenced by the domestic world that swirled around it. Placing physicians within the home environment allows us to see them as healers who interacted with and who influenced and were influenced by the people and events within the domestic sphere. Just as the hospital influenced and shaped the medicine that was practiced within it, the home, too, made its presence felt. Recognition of the domestic does not replace our understanding of how medical theory, public events, and other institutions also shaped medicine; rather, it adds another layer to the composite picture.[80]

Studies of rural and small-town general practitioners today illustrate some of the same themes I have developed in this article. In the words of family practitioner John Frey, rural family doctors today "live complexly" with their patients, and the boundaries between medicine, family, and community life are "blurred." Much of medical practice takes place in informal interactions in which the family and social group know all about their friends' medical problems and eagerly contribute their perspectives. A physician's full community involvement, family members, and activities all become essential to defining medical lives and practice.[81]

In the middle of the 19th century, when most of American medicine was rural and small-town based, most physicians followed a similar pattern and were intimately linked to their families' and their neighbors' lives. Mid-19th-century male practitioners woke up and went to sleep in their homes; they took their meals in their homes and worked within the same walls. They bore responsibility for many domestic tasks and were involved in the daily comings and goings of their families. They carried out their patient care within the domestic environment of their neighbors' and fellow church members' homes. As biographers of women have long realized, it makes sense to recognize that they "struggled to merge public with private demands."[82] An important theme in women's biography has been the tension between the public and the private. But it is not gender alone that made the private domestic sphere influential. What has sometimes been taken as attributable to gender— even in a somewhat essentialist way—has in fact been situational, constructed by contexts and affecting both sexes. The private parts of women's lives indeed took time and added stresses to the burdens of making a public life. I have suggested here that the same can be said for mid-19th-century American male physicians.

NOTES

This is a revised version of the Fielding H. Garrison Lecture presented at the 67th Annual Meeting of the American Association for the History of Medicine in New York, 29 April 1994. I wish to express my gratitude for the honor to the committee that selected me, chaired by Caroline Hannaway, and to the entire membership of the Association.

I am grateful for the research assistance and wisdom of Evelyn Fine, Sarah Leavitt, Jennifer Munger, and Lian Partlow, and for the discussions, readings, and criticisms of John Frey, Susan Stanford Friedman, Linda Gordon, Lewis A. Leavitt, R. David Myers, and Ronald L. Numbers.

1 William H. Brisbane diary, 28 Aug. 1854, State Historical Society of Wisconsin Archives (hereafter SHSW). There are four boxes and 50 volumes in the William Henry Brisbane Papers, covering the period 1829–1913. Volumes 1–33 are Brisbane's journal and diary. Volumes 42–46 and 50 contain Addie Brisbane's diary (volumes 44 and 45 under the name Adeline Brisbane Reed). I am very grateful to Lian Partlow for her research assistance with this article, par-

ticularly for calling this collection to my attention.

2 William Brisbane diary, 30 Aug. 1854.

3 William Brisbane diary, 29 Aug. 1854.

4 In this article I refer to combined small-town and rural medical practices because I describe both. Brisbane and Willard lived in small towns, but their medical practices encompassed visiting patients living in small towns (their own or nearby) and in rural areas.

5 Laurel Thatcher Ulrich, *A Midwife's Tale: The Life of Martha Ballard, Based on Her Diary, 1785–1812* (New York: Alfred A. Knopf, 1990).

6 Charles Rosenberg, "The practice of medicine in New York a century ago," *Bull. Hist. Med.,* 1967, *41:* 253.

7 On the importance of the hospital to shaping medical practice, see Charles E. Rosenberg, *The Care of Strangers: The Rise of America's Hospital System* (New York: Basic Books, 1987); John Harley Warner, *The Therapeutic Perspective: Medical Practice, Knowledge, and Identity in America, 1820–1885* (Cambridge: Harvard Univ. Press, 1986); Morris J. Vogel, *The Invention of the Modern Hospital: Boston, 1870–1930* (Chicago:

Univ. of Chicago Press, 1980); David Rosner, *A Once Charitable Enterprise: Hospitals and Health Care in Brooklyn and New York, 1885–1915* (Cambridge: Cambridge Univ. Press, 1982); Virginia G. Drachman, *Hospital with a Heart: Women Doctors and the Paradox of Separatism at the New England Hospital, 1862–1969* (Ithaca, N.Y.: Cornell Univ. Press, 1984); Rosemary Stevens, *In Sickness and in Wealth: American Hospitals in the Twentieth Century* (New York: Basic Books, 1989); Martin S. Pernick, *A Calculus of Suffering: Pain, Professionalism, and Anesthesia in Nineteenth-Century America* (New York: Columbia Univ. Press, 1985); and Vanessa Northington Gamble, *Making a Place for Ourselves: The Black Hospital Movement, 1920–1945* (New York: Oxford Univ. Press, 1995).

8   In his excellent study of colonial medicine, J. Worth Estes makes this point about the importance of community context, although he does not discuss home life or articulate the situation of physicians in gendered terms. J. Worth Estes and David M. Goodman, *The Changing Humors of Portsmouth: The Medical Biography of an American Town, 1623–1983* (Boston: Francis A. Countway Library of Medicine, 1986). On rural medicine in the 19th century, see Jan Coombs, "Rural medical practice in the 1880s: a view from central Wisconsin," *Bull. Hist. Med.*, 1990, *64*: 35–62; Jacalyn Duffin, "A rural practice in nineteenth-century Ontario: the continuing medical eduation of James Miles Langstaff," *Canadian Bulletin of Medical History*, 1988, *5*: 3–28; Paul Berman, "The practice of obstetrics in rural New England, 1800–1860" (Paper presented at the annual meeting of the American Association for the History of Medicine, Louisville, Kentucky, 15 May 1993); Walter Vanast, "Constancy, continuity, and curiosity: a reassessment of the 'low estate' of nineteenth-century American doctors" (*ibid.*, 15 May 1993); and various articles in *Wisconsin Medicine: Historical Perspectives*, ed. Ronald L. Numbers and Judith Walzer Leavitt (Madison: Univ. of Wisconsin Press, 1981).

9   For some early examples see Nancy F. Cott, *The Bonds of Womanhood: "Woman's Sphere" in New England, 1780–1835* (New Haven: Yale Univ. Press, 1977); Barbara Welter, *Dimity Convictions: The American Woman in the Nineteenth Century* (Athens: Ohio Univ. Press, 1976); and some of the essays in Mary S. Hartman and Lois W. Banner, eds., *Clio's Consciousness Raised: New Perspectives on the History of Women* (New York: Harper & Row, 1974). For a more recent perspective on this issue, see Nancy Grey Osterud, *Bonds of Community: The Lives of Farm Women in Nineteenth-Century New York* (Ithaca, N.Y.: Cornell Univ. Press, 1991). African-American historians, too, showed that community culture defined some basic parameters of the lives of black Americans. See, for example, Earl Lewis, *In*

Their Own Interests: Race, Class, and Power in Twentieth-Century Norfolk, Virginia (Berkeley: Univ. of California Press, 1991). For the perspective of labor historians, see Ava Baron, ed., *Work Engendered: Toward a New History of American Labor* (Ithaca, N.Y.: Cornell Univ. Press, 1991). Nancy Tomes, in "The private side of public health: sanitary science, domestic hygiene, and the germ theory, 1870–1900," *Bull. Hist. Med.*, 1990, *64*: 509–539 (ch. 33, this book), demonstrates how medical historians can benefit from inclusion of the domestic.

10 Regina Morantz-Sanchez, *Sympathy and Science: Women Physicians in American Medicine* (New York: Oxford Univ. Press, 1985). In this regard, see also Alice Kessler-Harris, "Treating the male as 'other': redefining the parameters of labor history," *Labor Hist.*, 1993, *34*: 190–204. Kessler-Harris defines gender as "the socially shaped cluster of attributes, expectations, and behaviors assigned to different sexes," a definition I find very helpful (p. 191). I am very grateful to Linda Gordon for calling my attention to this article. See also E. Anthony Rotundo, *American Manhood: Transformations in Masculinity from the Revolution to the Modern Era* (New York: Basic Books, 1993).

11 See, for example, Tomes, "Private side," and two other essays by Nancy Tomes: "The wages of dirt were death: women and domestic hygiene, 1880–1930" (Paper presented at the annual meeting of the Organization of American Historians, Louisville, Kentucky, 1991); and "Spreading the germ theory: sanitary science, household bacteriology, and the home economics movement, 1880–1930," in *Rethinking Women and Home Economics in the Twentieth Century*, ed. Sarah Stage and Virginia Vincenti (Ithaca, N.Y.: Cornell Univ. Press, 1995); Steven M. Stowe, "Obstetrics and the work of doctoring in the mid-nineteenth-century American South," *Bull. Hist. Med.*, 1990, *64*: 540–566; Joan Jacobs Brumberg, *Fasting Girls: The Emergence of Anorexia Nervosa as a Modern Disease* (Cambridge: Harvard Univ. Press, 1988), and her "Something happens to girls: menarche and the emergence of the modern American hygienic imperative," *Journal of the History of Sexuality*, 1993, *4*: 99–127.

12 The vagaries of labeling physicians in this period are well articulated in Eric H. Christianson, "The medical practitioners of Massachusetts, 1630–1800: patterns of change and continuity," *Medicine in Colonial Massachusetts, 1620–1820* (Boston: Colonial Society of Massachusetts, 1980), pp. 49–67. See also Christianson's "Medicine in New England," in *Medicine in the New World: New Spain, New France, and New England*, ed. Ronald L. Numbers (Knoxville: Univ. of Tennessee Press, 1987), pp. 101–153 (ch. 3, this book).

13 Rural and small-town medical practices were simi-

larly influenced by the physician's place in the community, but distinctions between them cannot be systematically followed in this essay.

14 In addition to the sources listed in n. 7, see Jacalyn Duffin, *Langstaff: A Nineteenth-Century Medical Life* (Toronto: Univ. of Toronto Press, 1993); Edward C. Atwater, "The physicians of Rochester, N.Y., 1860–1910: a study in professional history," *Bull. Hist. Med.,* 1977, *51*: 93–106, and Atwater's "Medical personalities: Samuel Beach Bradley, M.D., 1796–1880: a rural practitioner," *N.Y. St. J. Med.,* 1976, *76*: 1883–1888. See also James H. Cassedy, "A country doctor in Connecticut medicine and politics: William L. Higgins, 1867–1951," *Connecticut Medicine,* 1992, *56*: 363–369.

15 Brisbane's biographical information is taken from the *History of Iowa County, Wisconsin* (Chicago: Western Historical, 1881), pp. 930–931, and the *Wisconsin State Journal,* 12 and 13 April 1878. See also the information filed with Brisbane's papers at the State Historical Society of Wisconsin Archives.

16 Information about Henry Rutledge Frost (1795–1866) is found in Joseph Ioor Waring, *A History of Medicine in South Carolina, 1825–1900* (Charleston: South Carolina Medical Association, 1967), pp. 230–232. On the medical college, see William Frederick Norwood, *Medical Education in the United States before the Civil War* (New York: Arno Press, 1971), pp. 251–258; and Kenneth M. Lynch, *Medical Schooling in South Carolina, 1823–1969* (Columbia, S.C.: R. L. Bryan, 1970). Brisbane's graduation date was confirmed by Elizabeth Young Newsom, curator at the Waring Historical Library of the Medical University of South Carolina, who kindly sent a copy of the title page of Brisbane's medical thesis and a list of students from the medical school catalog. Brisbane mentions visiting with his preceptor Frost at an American Medical Association meeting in his diary entry of 6 May 1856.

    I do not know how many physicians of Brisbane's generation combined the ministry with medicine, which had been a fairly popular option at an earlier time.

17 Brisbane's antislavery activity is recounted in Blake McNulty, "William Henry Brisbane: South Carolina slaveholder and abolitionist" (paper presented to the Citadel Conference on the South, Charleston, S.C., April 1981; typescript available from the Waring Historical Library, Charleston, S.C.).

18 A particularly helpful view of Brisbane's Arena years is found in Betty Cass, "Mysteries of ancient Arena house solved: amazing Dr. Brisbane diary gives up secrets," *Capital Times* [Madison, Wisconsin], 29 June 1963, pp. 4, 8.

19 On Arena, see *History of Iowa County, Wisconsin* (Chicago: Western Historical, 1881); George and Robert M. Crawford, eds., *Memoirs of Iowa County, Wisconsin, From the Earliest Historical Times Down to the Present* (Northwestern Historical Association, 1913); *Iowa County Heritage, Vol. 1: The Census Records* (Dodgeville, Wis.: Fieldhouse Foundation, 1967). See also various maps in the State Historical Society of Wisconsin Archives.

20 The number of patients in his practice compared, for example, to the first year of practice of James Miles Langstaff, the rural Toronto community practitioner who began his practice in the 1850s, thoroughly studied by Jacalyn Duffin. But Langstaff went on to a much busier practice, making close to 2,000 visits in a comparable six-month period by the 1870s. See Duffin, in "A Rural Practice" and *Langstaff.*

21 Biograhical information on H. B. Willard is from "Death of Dr. H. B. Willard," *Jefferson County Union,* 6 April 1900, p. 7; "Koshkonong Township," in *The History of Jefferson County, Wisconsin* (Chicago: Western Historical, 1879), pp. 699–700; *Biographical Sketches of Old Settlers and Prominent People of Wisconsin* (Waterloo, Wis.: Huffman & Hyer, 1899), *1*: 68–74; and from [Helmut Kneis], "H. B. Willard—A Life in Pictures," *Fort Atkinson Hist. Soc. Newsletter,* 1988, *4*: 3–6. I am grateful to Jennifer Munger, Sarah Leavitt, and Helmut Kneis for their help in locating this material. Willard's obstetrical and medical journals and his diary are located in the State Historical Society of Wisconsin Archives and his account books are at the Fort Atkinson Historical Society. Eve Fine first alerted me to Willard's obstetrics journal, and I remain very grateful to her.

22 I would like to thank Eric Luft at the Rare Books and Special Collections, State University of New York Health Science Center Library, Syracuse, who provided three different lists of the 1849 graduating class of Geneva Medical College from the alumni records, on which the names Horace Birney Willard and Elizabeth Blackwell are found.

23 Willard's medical practice has been briefly discussed by recent historians. See Guenter B. Risse, "From horse and buggy to automobile and telephone: medical practice in Wisconsin, 1848–1930," in *Wisconsin Medicine: Historical Perspectives,* ed. Ronald L. Numbers and Judith Walzer Leavitt (Madison: Univ. of Wisconsin Press, 1981), p. 27, and Judith Walzer Leavitt, *Brought to Bed: Childbearing in America, 1750–1950* (New York: Oxford Univ. Press, 1986), p. 48, both of which view Willard in the tradition of heroic and activist medicine.

24 Hannah Swart, *Koshkonong Country: A History of Jefferson County, Wisconsin* (Fort Atkinson, Wis.: W. D. Hoard & Sons, 1975), pp. 161–170.

25 See Sara Alpern, Joyce Antler, Elisabeth Israels Perry, and Ingrid Winther Scobie, eds., *The Challenge*

of *Feminist Biography: Writing the Lives of Modern American Women* (Urbana: Univ. of Illinois Press, 1992).

26 I pose the issue in these terms in part because of a seminar discussion with W. Bruce Fye concerning how he thought about family issues in his work on the history of key American physiologists (graduate seminar, University of Wisconsin History of Medicine Department, Spring 1989). See W. Bruce Fye, *The Development of American Physiology: Scientific Medicine in the Nineteenth Century* (Baltimore: Johns Hopkins Univ. Press, 1987).

27 This is an extension of the argument I presented about obstetrics in *Brought to Bed*.

28 William Brisbane diary, 11–20 Jan. 1856.

29 *Ibid.*, 5 Feb. 1856.

30 Because he had to wait for his flour to be processed, which happened to him repeatedly, Brisbane wrote a booster pamphlet, "A Word to Immigrants and All Who are Seeking a Western Home" (1858; in the William H. Brisbane Papers, SHSW), in which he sent word East that Arena needed another mill and that a settler involved in the business would thrive. He also offered personally to help such a settler.

31 Brisbane regularly listened to Addie's Latin lessons and had long talks with her about her education. He wrote frequently about his children's behavior, worrying repeatedly about William's future prospects because he did not show enough responsibility. The reference in this excerpt to Brisbane's giving William the blacksmith shop with conditions reflects this concern. Even Benjamin, the most responsible child, caused Brisbane and his wife anxiety. Brisbane's active participation in his children's lives is an important part of the story of how his deep domestic involvement affected his medical practice.

32 Addie Brisbane diary, 23 Mar. 1855.

33 *Ibid.*, 21 Jan. 1855 and 17 Feb. 1855.

34 *Ibid.*, 25 June 1855.

35 *Ibid.*, 15 Sept. 1855.

36 See Addie Brisbane diary, *passim*. On 6 Feb. 1855, she noted, "Father & Dr. H were away all day yesterday & a good part of last night attending Mrs. Jane Rockwell, who is confined."

37 William Brisbane diary, 1 July 1856.

38 *Ibid.*, 11 June 1856.

39 *Ibid.*, 12 June 1856.

40 *Ibid.*, 13 Mar. 1856.

41 *Ibid.*, 5 Mar. 1856.

42 *Ibid.*, 3 Jan. 1854.

43 *Ibid.*

44 *Ibid.*

45 *Ibid.*, 5 Aug. 1856.

46 *Ibid.*, 24 Sept. 1856.

47 *Ibid.*, 4 Oct. 1856.

48 *Ibid.*, 7 Oct. 1856.

49 "Introduction," Alpern et al., *Challenge of Feminist Biography*, p. 9.

50 Susan Ware, "Unlocking the Porter-Dewson partnership: a challenge for the feminist biographer," in *ibid.*, p. 61.

51 William Brisbane diary, 11 Apr. 1856.

52 *Ibid.*, 11 Apr. 1856 and *passim*.

53 The ups and downs of Addie's relationship with George Ashmore can be followed in her diaries for 1856 and 1857 and compared with Brisbane's for the same dates. See, for example, Addie's Dec. 1857 preoccupation with George; no mention is made in her father's diary entries of the same dates.

54 William Brisbane diary, 3 Feb. 1856. See also, for example, the entry for 3 Dec. 1856.

55 *Ibid.*, 10 June 1854.

56 *Ibid.*, 11 June 1854.

57 *Ibid.*, 1 July 1854.

58 *Ibid.*, 28 July 1854 and 2 Aug. 1854.

59 *Ibid.*, 28 Aug. 1854.

60 See, for example, Jan Coombs, "Rural medical practice," pp. 48–49; and Risse, "From horse and buggy," pp. 28–29. See also Paul Starr, *The Social Transformation of American Medicine* (New York: Basic Books, 1989), pp. 64, 84. Starr cites Lemuel Shattuck's 1850 report that the average Massachusetts practitioner billed $800 and collected about $600 per year.

61 Horace Willard diary, 15 Aug. 1857.

62 *Ibid.*, 11 July 1856.

63 *Ibid.*, 1 Jan. 1858.

64 *Ibid.*, 25 July 1856.

65 *Ibid.*, 25 May 1857. An example of a time when home life and medical life did not mesh well came one rainy day when Willard, returning home after visiting a patient, found his house empty and locked, with no comfort to be found. He wrote, "Got quite wet—reached home at 9 in the evening. Raining briskly—found no one to receive me on my arrival. After turning in my horse to the stable—I tried to affect an entrance into my [illeg] home—succeeded in the course of time in raising [one] window through which I made my ingress after quite a struggle. Cold, wet and dirty—This was solitude indeed and I was led to exclaim, "Oh! Solitude where are the charms that others have seen in thy face" (*ibid.*, 27 May 1857; see also 8 July 1857).

In 1856, the year Willard left Aztalan and set up practice in Watertown, his personal life did not go well. There is some evidence that a scandal of sorts influenced his leaving Aztalan—he makes only veiled references to how it hurt his mother's heart—and for a while his wife took their daughter and went home to New York. Willard meanwhile traveled around Wisconsin, Iowa, and Minnesota looking for a new place to settle. He was miserable. Ultimately, he followed his wife to New York, where they were re-united and moved as an intact family to Watertown. Willard continued to visit patients in Aztalan, indicating that whatever had sent him from that town did not hinder his medical practice there.

66 *Ibid.*, 25 Feb. 1857.

67 *Ibid.*, 26 Mar. 1857.

68 *Ibid.*, 10 Apr. 1857.

69 *Ibid.*, 1 June 1857.

70 *Ibid.*, 5 June 1857.

71 Willard read, among other things, many sentimental "women's" novels, which greatly affected him (perhaps another example of crossing stereotypical gender lines). One was *The Newsboy*, by Elizabeth Oakes Prince Smith (New York: J. C. Derby, 1854): "full of pathos, reaching down deep into the heart, and touching the most tender cords. [The author] shows that a noble generous philanthropic heart can throb beneath poverty's poorest vesture, that *Manhood* is not measured by the purse, piety by the face, or humility by the attitude" (Horace Willard diary, 22 June 1856). Another was *The Lamplighter*, by Maria S. Cummins (London: Clarke, Beeton and Co., 1854): "Its authoress must have possessed an almost fathomless depth of soul—she reaches the reader's heart, she softens and elevates the spirit" (Horace Willard diary, 26 July 1856). He also read biography, and wrote at some length about his reactions to the autobiography of Frederick Douglass. Both *Newsboy* and *Lamplighter* are discussed in Nina Baym, *Woman's Fiction: A Guide to Novels by and about Women in America, 1820–1870* (Ithaca, N.Y.: Cornell Univ. Press, 1978). See also, Baym's introduction to a reprint edition of *The Lamplighter* by Maria Susanna Cummins (New Brunswick, N.J.: Rutgers Univ. Press, 1988).

72 Horace Willard diary, 9 Dec. 1856.

73 *Ibid.*, 29 Apr. 1857.

74 *Ibid.*, 28 May 1857. Willard's medical practice may have suffered from his lack of community involvement. Although he understood social activity to be connected to a successful practice, he did not find participation easy. When he first moved to Watertown to set up a practice there, he wrote, "During [the last few months] I have been moving and arranging &c. &c. and am now fairly settled in the city of Watertown. Have done some business in the line of my profession. But what is hardest to become reconciled to is I have lost my favorite horse Prince" (*Ibid.*, 6 Dec. 1856). Rather than get out and meet people and encourage business, Willard chose to wait for it to come to him: "I am in my office," he wrote one Sunday, "from choice. And many are I know wondering why I should stay at home from church. But I am in a degree disgusted, not with the preaching so much as with the incentives held out

for me to attend worship. 'You will be thought more of,' says one. 'It will be money in your pocket,' says another pious lady. 'You must attend the Presbyterian Church because all the "elite" go there' says a third. 'Everybody is despised that attends the Methodist Church & you will be if you go there,' says the fourth. And so it goes. Never once has the idea been carried by Christian or infidel that I should go to Church . . . for the good of my soul—for a spiritual feast. But the worldly gain or character is the whole great Motive power." That evening, perhaps perversely, Willard attended the Methodist meeting (*ibid.*, 7 Dec. 1856).

Willard seriously considered leaving medicine for law at one point. He wrote, "Was in [court] a few moments last evening to hear the pleadings of a case of Rape—The case was decided in favor of the Defendant. I never listen to a plea but I feel that I am out of my sphere in the Profession I chose—I would like to be a member of the Bar. I would like to plead for the innocent and the injured" (*ibid.*, 11 June 1857). A month later the idea was still with him: "today for the first time I entertain the idea of reading law in some office, and before it is known turn up a lawyer" (*ibid.*, 16 July 1857). The next day he set his mind to it: "I have determined to go into Wm. Dutcher's office & read law—And I am conscious that when the fact is known—'fool' will escape from many lips. But a knowledge of law will hurt no man. . . . *An honest lawyer need not starve*," he concluded (*ibid.*, 17 July 1857; emphasis in original). He took one law book from Dutcher's office, but thereafter the diary reveals no more about the law. In fact, he continued to practice medicine for a few more years after this discouraging time, but by 1866, 17 years after seeing his first patient, he gave it up. Willard gained material success in his new business and found a place in the county history books as a pillar of the community. He was buried with a tombstone that towered over those of its neighbors, second in height in the entire cemetery only to that of the governor, which indicated that Willard's reputation improved after he left behind the practice of medicine.

75 Horace Willard obstetrical journal, 29 June 1854.

76 *Ibid.*, 4 Feb. 1855. Willard voiced his concerns about obtaining patient confidence and worry about his reputation also on 26 Mar. 1857 and 29 Apr. 1857.

77 *Ibid.*, 7 May 1850.

78 Mrs Fobes's sister-in-law, Mrs. Ebinezer Fobes, engaged Willard about nine months later when she went into labor in January 1851, indicating that the family was pleased with his obstetrical services. When another of Willard's obstetrical patients (Mrs. Townsend) died, he indicated his sympathetic connection to his patients by writing, "The earth has

need to mourn for a brilliant star has fled her galaxy. A gentle spirit has left its loving friends and returned to its place of origin. . . . none saw her but to admire. I speak of the dead as she was. Amiable and kind in disposition; immaculate in her moral character; interesting as an acquaintance; frank and generous as a friend, and a companion confiding and affectionate" (Horace Willard medical journal, 22? Aug. 1850, p. 62). In the account of Mrs. Townsend's delivery in his obstetrical journal, Willard noted that following delivery the placenta was adherent: "[I]ntroduced my hand and separated it and put her to bed. No flooding." (Horace Willard obstetrical journal, 25 Aug. 1850.) It is possible that the process also introduced the infection that killed her. For other examples of male physicians fitting in with the women attending the sick, see the diary of Daniel Cameron (25 May 1853, SHSW Archives).

79 Horace Willard medical journal, 27 Oct. 1850. See also 29 Dec. 1850, 14 Oct. 1852, and 17 Apr. 1853.

80 The point that situation affects perceptions and functions is not new. It is what literary critics call *positionality* and philosophers call *standpoint theory*. See, for example, Sandra Harding, *Whose Science? Whose Knowledge? Thinking from Women's Lives* (Ithaca, N.Y.: Cornell Univ. Press, 1991), esp. pp. 119–133, and Linda Alcoff, "Cultural feminism versus poststructuralism: the identity crisis in feminist theory," in *Signs*, 1988, *13*: 405–436. Historians, too, have been sensitive to relative points of view, and some interesting recent work has focused on seeking out multiple perspectives on historical events. See, among medical historians, for example, Stowe, "Obstetrics and the work of doctoring," pp. 540–566, whose work is a real model in this area.

81 John Frey, "Taking care of neighbors: oral history of small town doctors," William Snow Miller lecture, 2 Nov. 1993 (Univ. of Wisconsin, Madison).

82 Joyce Antler, "Having it all, almost: confronting the legacy of Lucy Sprague Mitchell," in Alpern et al., *Challenge of Feminist Biography*, p. 107.

10

Seeing Themselves at Work: Physicians and the Case Narrative in the Mid-19th-Century American South

STEVEN M. STOWE

Writing for the *Transylvania Journal of Medicine* in 1836, Kentucky physician Lunsford Yandell told about nine of his most difficult cases of dysentery, frankly discussing his encounters with free patients and slaves, children and adults. "Cases of Dysentery, with Remarks" is what Dr. Yandell called his essay, meaning by "remarks" a sharp challenge to the approved therapy for this serious affliction. It was no small matter to make such a challenge. Because many therapies were only erratically effective, and because patients had many different kinds of healers to choose from, orthodox physicians thought twice about openly questioning their standard treatments. Disputes among physicians too easily spilled into public debate, raising the fears of already skeptical patients. But Yandell was convinced that the "great confidence" with which he had used the standard treatment was no longer merited. His bedside experience had shown it to be "utterly futile," with disastrous consequences.[1]

Although Lunsford Yandell happened to be editor of the *Transylvania Journal*, his essay might have found a home in any of the half-dozen other major medical publications that began to flourish in the South by the 1830s. It focused on a commonly diagnosed disease of great interest to southern physicians. And while arguing forcefully for reconsidering the established therapy, Yandell spoke in a modest, professional voice, acknowledging his own shortcomings. Most important, he built his essay on cases taken from his own practice. The story of

something gone wrong, and of the doctor's attempt to restore what had been lost, was the story physicians found uniquely compelling in its many variations, making the case narrative the most widespread form of medical writing in the mid-19th century. Yandell's was such a tale, and one with a hidden twist. He did not mention that Case Number 7, "W. Y. a boy [age] 6 years, of robust frame," who sickened and died with a suddenness unusual even in dysentery, was his own son Willie. Case 8 was his wife, Susan, and Case 9 a young female friend of the family, both of whom recovered but only after the approved therapy had failed them, too.[2]

In what follows, Dr. Yandell's essay—and his family tragedy—frame an exploration of the stories doctors told about their work. In one sense, of course, the case narrative was simply a practical tool; doctors used stories to record things they needed to remember. But because it was a story, not a list of facts or a numerical equation, a case narrative provided a literary opportunity as well. Narratives made otherwise odd or silent events into a "case," where meaning was created through timing, character, and plot. And each time a doctor revisited a narrative, he had the occasion to relive the drama of the case and, if he wished, to reconfigure the significance of his work. Different historical meanings also inhabit these narratives. Case stories might be read as brief ethnographies of sickness, for example, in which are recorded the wishes and fears of patients. Or they might be read as part of the history of changing notions of professionalism. Mid-19th-century professionals, including many physicians, rightly have been portrayed as troubled by questions concerning their identity and authority. The knowledge they claimed

STEVEN M. STOWE is Associate Professor of History, Indiana University, Bloomington, Indiana.

Reprinted with permission from the *American Historical Review*, 1996, *101*: 41–79. Notes have been revised for this edition.

seemed full of promise and power. And yet their claim to legitimacy still lacked broad popular support. After the 1860s, especially, many professionals responded to public skepticism (and perhaps their own self-doubt as well) by scaling down the realm of their knowledge so that they might better command it, at the same time redefining it to express their sense of an ever more complicated and thus specialized reality.[3]

Insofar as physicians' published case narratives also struggled with these issues, they may be read as evidence of the ideological crisis of mid-19th-century professionalism. But my primary intention here is to see them as comprising a different genre of writing, as work stories that record how a practitioner's sense of his worth grew from the pressing immediacy of his daily efforts. If we look at these narratives in this way, as ontological texts born of the work of doctoring, they may be seen to include important meanings not found in doctors' more self-conscious ideological attempts to redefine professionalism. In the comparatively roomy, common-sense work story are the patterns of gratification and discontent underlying physicians' slow and troubled shift toward a new, "modern" vision of their calling.

Indeed, this shift was a slow one in part because of the way doctors told their stories. Physicians, like many others in this era, were caught up in the excitement of a world open to empirical manipulation; the desire to observe, collect, and describe phenomena shaped fields of knowledge from botany to the Bible. But the storytelling power of medical narratives derived less from the information they reported than from the way they framed the meaning of medical work in terms of a doctor's own personal experience. Narratives were, most importantly, first-person tales, brief fragments of autobiography. By embracing the tensions in doctors' daily work in this markedly personal way, I will argue, narratives helped to shape a mentality that made broad changes in medicine difficult even as it allowed a doctor to see himself at work.[4]

One major tension of practice acknowledged by increasing numbers of American physicians in the South and elsewhere during the 1830s and 1840s concerned their inability to effect cures with any consistency or even to give relief. Although medicine's failures were scarcely something new, physicians in these years were nurturing new hopes for clinical success and professional dominance and thus were especially vexed by their persistent shortcomings. Although physicians were able to do some things to their own and their patients' satisfaction, doctors could not help but be dismayed by the numbers of people who either avoided orthodox remedies or, worse, got sicker despite submitting to everything the doctor wanted to do.[5]

The older historiography of Western medicine placed therapeutic failure into a comforting teleology: the 19th century was the last of the dark before the dawn of modern medical success. In this interpretation, the quotidian details of "premodern" medical practice often were skipped over or treated as curiosities of interest to specialists only. More recently, however, the work of Charles E. Rosenberg, Judith Walzer Leavitt, John Harley Warner, and many others has begun to restore the context of premise, routine, and knowledge that shaped 19th-century American practice. Physicians' training, institutions, ethics, and especially their social relations have become the setting for understanding the medical past. Instead of emphasizing how the 19th-century medical world fell short of modern achievements, this revisionist work turns to the more interesting question of how and why this world made sense to the people who lived within it.[6]

From this point of view, doctors' inability to restore sick people to unambiguous health is important because, among other things, it challenged doctors' professional self-esteem. M.D.s cherished the historical myths that elevated their healing tradition as the most learned and selfless in a marketplace crowded with unregulated competitors. Even physicians in remote plantation communities knew that their lineage stretched from the Roman god Aesculapius to the great Scottish physician and teacher William Cullen. Moreover, as medicine made intellectual claims to the useful knowledge and rigorous inquiry associated with an emerging new science by the 1830s, physicians had additional reason to feel superior to folk healers, whom they considered—at best—mere craftsmen. Thus the conscientious M.D. was all the more frustrated when his daily practice revealed that much of orthodox medicine was itself little more than a craft with few opportunities for reading, systematic problem-solving, and accurate prediction.[7]

This world of ordinary practice, confronting

physicians with the gap between medicine as a way of knowing and a way of working, was the world captured in the case narrative. "Much that we take as observations about 'reality,'" Kenneth Burke has suggested, "may be but the spinning out of possibilities implicit in our particular choice of words." So physicians made their realities by relying on the personal narrative as the chief way to explain and describe their problematic work. Especially after the 1820s, with the rise of medical periodicals, these stories combined two important ingredients: a teeming variety of subject matter given shape by a flexible, yet distinct, narrative form. With Yandell's essay as the threshold, the significance of the great variety of subject matter will be considered first; compared to the tightly framed discourse of modern medical research, a wide-open array of stories— energetic, vernacular, first-person—flowed from the work of 19th-century doctoring. This variety was brought under control by a particular narrative form: whatever the diversity of tales, physicians almost always spoke of patients' bodies and biographies, seeking a way to combine them in the name of healing.[8]

Then, returning to the death of Lunsford Yandell's son, I will suggest how case narratives traced an important boundary between doctors' professional and personal lives, excluding some meanings of medical work while including others. It is possible to do this in Yandell's case because there exists a second account of the family tragedy silently included in his journal essay. This other account is a series of letters, amounting to a kind of diary, written by Yandell to his parents-in-law during the illness of his son and wife. These letters antedate his published essay by only a few weeks, and the differences between his two stories—one professional, one familial—reveal the key role of the autobiographical narrative in shaping—and limiting—a practitioner's vision of his work.

Yandell's cases of dysentery are a good point at which to begin exploring the variety of subjects and voices in doctors' narratives because Lunsford Pitts Yandell was, in 1836, a physician whose career was moving along a path taken by many M.D.s during this century. At 31 years of age, Yandell already had been practicing for a dozen years, first in Murfreesboro, Tennessee, and then in Nashville, where his career had branched out into teaching and editing.

His professional ambitions were considerable, but they had not taken him away from a typical, community-based practice. Indeed, because there were no widespread institutional or professional alternatives to local practice, everything about most physicians' work tied them to their communities, intellectually as well as economically. A line between physicians and "lay" people existed in terms of training and knowledge, but in practice (to the chagrin of many doctors who desired professional exclusivity), the line often was vague and remarkably permeable.[9]

This social reality of medical work helps explain the accessible literary style of mid-19th-century case narratives. Although physician-authors sometimes went far into technical detail or wrote from within a cloud of theoretical language, most accounts of bedside practice were written in a plain, vernacular style—like Yandell's—that could be easily read by non-M.D.s as well as professional peers. This is not to say that narratives were artless; indeed, Yandell's essay has a calculated literary shape. He introduces his cases in a way that suggests how narratives persuaded and informed by keeping the cases—the heart of the essay—fully illuminated by the doctor's personal experience of them. Typically, Yandell begins his essay on an autobiographical scale and never departs from it:

> An experienced and eminent physician of this city [Lexington] lately assured me that he had not lost a single case of dysentery, exclusively under his own charge from first to last, during a practice of more than five-and-twenty years. The remedies upon which he chiefly relies are calomel and ipecac., in small doses. With these, he uses hot and stimulating applications externally, when the surface [of the skin] is cold, and confines his patients to warm drinks. I confess I have not been so fortunate in combatting this disease.[10]

With this opening, Yandell lets his readers know that his essay, like so many others, will be a therapeutic tale. It promises to be an informative story, doubtless yielding practical advice on particular medicines. But it is also clearly *his* tale, one fueled by his own experience of pursuing disease and by his sense of the intellectual context of therapeutic decision-making. Specifically, he frames his story in terms of the problematic power of medical authority, represented by the older physician and the approved therapy. Yandell at first seems to lean on his more experienced colleague, but this "eminent" man quickly becomes a foil in a more complicated story. Yandell may be the less experienced practitioner, but he is much more self-aware. Introducing an eminent (and complacent) colleague in order to shunt him aside, Yandell suggests he will tell a darker, perhaps confessional, story of therapeutic failure.

And, in fact, Yandell goes on to tell how he, too, had been content with the approved therapy for several years until he came up against cases of dysentery in which "every plan pursued was unavailing," and he watched his patients die. The cases he has chosen to write about would illustrate something his older colleague had not experienced: that there is abroad a dysentery "of a very different character, . . . especially when it attacks children." And here Yandell calls on his readers for "a fuller investigation" of the approved therapy, which he has discovered "was not the proper one" after all. His patients had worsened under it, improving only "when a different set of remedies was adopted." Not content simply to accept that every doctor loses some patients, Yandell pushes the personal point: when faced with therapeutic failure, "a conscientious man will ask himself, why?"[11]

Here, Yandell abruptly changes course in the way he introduces his cases. Calling into question an approved therapy by contrasting unfavorably an "eminent" older colleague with "a conscientious man" obviously himself, Yandell comes close to speaking evil of a professional brother, something frowned on by physicians. Therefore, in contrast to his intimate opening paragraph, he turns to a broad, uncontroversial truth: the potency of all diseases and therapies is influenced by the local climate. He invokes an exemplary historical authority, Thomas Sydenham, as witness. Not only did Sydenham have the brilliance to see the climatic influences in sickness, he had the courage to challenge the established way of doing things, and he "had the candor to confess that he generally lost the first patients he was called to attend when a new epidemic arose."[12]

By introducing his nine cases in this way, Yandell created an intellectual context that was both personal and yet far-reaching. Although the social stage for his nine cases was expansive (more than once, he describes dysentery as "epidemic"), the source of his reformist passion was lodged in the small but significant realm of an individual's work among his neighbors. Yandell appealed to physicians' curiosity, to their heroes, and to such overarching factors as climate. But all of these were focused by the image of the "conscientious man," whose daily work with the unknown revealed the particular, important truths learned only through practice: that respected colleagues and approved therapies can and do fail, that the tenets of good health are shockingly malleable, and that even known diseases change their identities unpredictably.

His narrative power is the power that inheres in autobiography; the story will be meaningful because the storyteller has witnessed what he will write about, and it has changed him. Although Yandell stops short of revealing that his wife and son are among the cases, he does drop a hint. Recalling the "feelings of mortification" he experienced when two of his first patients died, he admitted, "I have experienced the same feelings since in a more intense degree." In this allusion to his son's death, Yandell sought the attentive reader who would ask, *why* "more intense"? He imagined a reader familiar with the inevitable personal meaning of medical work—a meaning that could be at its core too devastating to reveal outright but too important to exclude.[13]

Of course, Yandell's hint at this deepest personal dimension of his story is just that, a pause for only a beat or two. Most of his readers doubtless did not linger over it. Instead, they moved into the nine cases to see what particular substance Yandell's stories offered. Substance came in different ways. Case-based essays usually offered practical advice of some kind, and physicians looked for such articles to clip and save in scrapbooks for later reference and sharing with colleagues. But essays such as Yandell's had a larger, epistemological dimension, too. In the modern medical journal, the privileged place in the creation of new knowledge belongs to articles based on institutionalized research occurring in venues far removed from the office call. In the mid-19th century, however, the bedside case from a doctor's rounds was most often the featured, "original" article. For Yandell and his contemporaries, the essence of medical science was at the bedside, and thus the individual practitioner's cases held the promise of broad, scientific advance that has since been relocated to the laboratory and teaching hospital.

A contrast with modern medical publications will sharpen this key distinction. An individual physician's clinical experience still has a place in medical journals. But it is a far more circumscribed place, sharply limited as a source of new knowledge. The format of the clinical-pathological conference (CPC) is a good example of how modern journals use physicians' clinical stories. In this kind of essay, an individual physician (or team of physicians) presents a particularly puzzling case to an audience of colleagues. The patient's problematic symptoms invariably are enumerated in terms of established medical categories: physical appearance, vital signs, family history of illness, and the recent events that led the patient to seek help. The patient's social context is largely invisible, except in generalized demographic terms of sex, age, and occupation. The typical CPC is most intent on sketching the patient through the medical inspection of the body, inside and out. The language is cool and technical, and the description a check list: "The head and neck were normal. The lungs were clear, and the heart, breasts, and abdomen were normal . . . Neurologic examination was negative." The meeting between patient and physician is noted in terms of the complaint; we do not see the meeting itself: "A 34-year-old woman was admitted to the hospital because of a dry cough." "A 39-year-old man was admitted to the hospital because of a fever." [14]

The scene shifts quickly to the laboratory, where the patient is reconfigured in terms of various measures of physiological processes, those revealed by blood and urine tests, for example. Here, the patient (or the sick body numerically abstracted in laboratory reports) is once again compared to the normal and to whether a given finding is negative or positive according to diagnostic indices. Specialists are introduced to comment on how they used pieces of data to rule out certain diagnoses. Usually, the presenting physicians conclude the CPC by advancing more than one possible diagnosis, keeping the clinical puzzle alive until the CPC wraps up with the pathologist's report revealing which diagnosis was correct.

These stories of doctors and patients have limited, if important, purposes in the realm of modern medicine. They efficiently describe accepted diagnostic method and impart the casuistry of modern medical problem-solving. They articulate the relation between bedside findings and laboratory tests, at the same time underscoring the essence of medical practice as an applied science—not excluding the intellectual excitement of clinical sleuthing. If the specialized, calibrated, and often highly quantified language drains the patient's suffering from the case, physicians, too, become personally remote, as every clinical event is reduced to what it can or cannot contribute to identifying the affliction. Physicians' language is streamlined and made muscular by jargon; their individual efforts are relentlessly yoked to those of others; their encounters with patients are made standard by a generic hospital setting that serves as a kind of signature of modern medical power. [15]

Moving back from the form of this modern professional genre to the case narratives of Lunsford Yandell's medical world is like stepping away from retrospection into the very swirl of events. The expressive vitality of 19th-century medical writing, not its failure to match modern scientific standards, is what should impress; narratives crowd against one another in a profusion of individual voices and plots. Invariably, the vernacular voice of one man's experience gave narratives their persuasive power. "March 16th, 1833, [I] was called before sunrise to visit a negro woman." "I took from her twelve ounces of blood." "I waited about fifteen minutes, when she had a severe convulsion." "I unloosed the bandage from her arm." Case stories are built around such images of individual doctors traveling, talking, deciding, acting. The work itself, not the diagnostic puzzle abstracted from it, supplied the purpose, narrative coherence, and dramatic energy to make these stories full of meaning. [16]

It is not surprising, therefore, to find so many published narratives taking the form of diary entries, as doctors sought to persuade or inform by reconstructing the familiar, time-bound immediacy of caring for the sick. Though retrospective, the diary-essay countered the foreshortening of time— the leap over events to reach the outcome—by insisting that readers see the significance of how each activity in a doctor's work paralleled each day—or hour—of a patient's illness. Such accounts of work were more than reflections on it; they became a part of the work itself. A conscientious physician created a set of bedside notes, which he shared or which were published still in notebook form, and then other doctors used these stories to discover meaning at their own patients' bedsides.[17]

The homological relation between text and work helps explain the strikingly wide variety of subject matter and plot in published narratives. The medical journal—like practice itself—was a wide field of triumph, catastrophe, and bathos, and physicians never tired of the compelling ways these things could combine in any given case. There was no substitute for writing about the full complexity of events. The timing of the doctor's arrival and his particular combination of experience and eagerness shared the spotlight with blind luck and sudden disaster. Much could be learned from each new configuration of events and actions around the bedside. While most published cases were written in a practical register, many stories have no *particular* advice or discovery to relate. Rather, the key ingredient in a story was what an individual doctor found personally "interesting" about a case. This plain but capacious word carried much merit and promise, shaping an intellectual world with the dramatic power of autobiography at the heart of every doctor's tale.[18]

Rita Charon has suggested that the often frustratingly different way modern physicians and patients tell stories of sickness, a difference underlying many of the ethical and legal disputes between them, is not a simple misunderstanding that can be fixed by brushing up everyone's "communications skills." The difference lies much deeper in the radically different work of doctor and patient. The modern physicians' effort, as in the CPC, to rule out diagnoses, to narrow down the case and trim the story, is at odds with the patient's desire to elaborate a story large enough to hold the terrible enormity of his or her illness. Following Charon, there is much about the 19th-century doctor's wide-ranging fascination with the various things that were "interesting" that resembles the modern patient's perspective more than the modern physician's. To some extent, that is, the work of doctoring imposed the daunting, world-transforming power of illness onto the doctor as well as the patient.[19]

Yandell's dysentery essay, seen in this context, is comparatively sophisticated in the way it linked case stories to the larger aim of reforming therapy. But in admitting his failures, and using them as testimony to his candor, Yandell was doing what many other physicians did in writing case stories: he was calling for help. Calls for help appeared in many published tales, with a doctor depicting himself as confronted by "a mystery which is beyond my comprehension." Such stories engaged readers precisely because the puzzle still was unsolved; no pathologist waited at the end to write the final paragraph revealing what "really" was wrong with the patient. A rural physician wrote in hopes of finding a colleague who had seen something similar. His essay was a kind of shout in the woods of everyday practice that medical journals helped amplify.[20]

Consider Dr. Henry A. Ramsay's case of Ben, a slave living on a plantation near Raysville, Georgia—a young, strong man who one day came down with a chill serious enough to alarm his owner. Ramsay arrived, looked Ben over, took note of his muscular body and good constitution. He also observed that Ben's abdomen was unusually hard to the touch; Ben said that he had been suffering pain from a blow he had received. Ramsay drew off a gallon of Ben's urine, gave him opium and calomel for inflammation, prescribed a warm bath, and left. Returning the next day, the doctor was shocked to learn that Ben had died during the night. Ramsay wondered whether a diseased spleen was somehow involved, but he did not speculate further on the cause of death. His purpose in writing was to call for help, to hope aloud that someone was wiser than he: "Are such cases common? Could such a case have been remedied?"[21]

A related kind of "interesting" story came from doctors who were moved to tell about intense or stressful cases they had experienced. Like the call for help, the story of a runaway challenge to a doctor's skill often did not have a specific discovery or advice to relate. Instead, meaning was in the drama of a man's immediate experience. F. A. Bates's 1849 story about the scarlet fever that swept through his community is such a tale, interesting because it was such a personal trial by fire. Bates wrote of his hurried meetings with neighbors, the humid and excessively hot weather that surely strengthened the disease, and everyone's fatigue. He wrote of the bodies of his patients, of settling at last on a therapeutic course of light diets and mucilaginous drinks. He told of the rapid pulses of the victims, the "alarming diarrhea," and wrote in summary,

"Every effort was made to arrest this condition, but without avail. The hot air bath, frictions, enemata of tr. opii, sinapisms, calomel and opium all failed." Similarly, the title of J. M. Hamilton's case story posed the agonizing question "How Long Shall We Wait?" to crush the skull of an undeliverable fetus in order to save the life of a woman in labor. But nowhere in the article did he answer—or even ask—this practical question. Instead, he simply told about the harrowing craniotomy he had performed one day at a neighbor's.[22]

The autobiographical texture of these narratives is just as tangible in accounts that have neither mystery nor trial by fire to relate but that speak instead of the strangeness of life opened to the physician's view. A doctor comes to a neighbor's door and finds something weird or astonishing. He *must* tell about it, and so there are stories of wonders: of women sickened by their husbands' bodily "exhalations"; a black woman with skin turning white; a child born to a dead woman at her wake; a man suffering a gunshot wound in 1840, and the bullet is not recovered until his death twenty years later, when it is found in the chambers of his heart; an unfortunate man who tried to cure his piles (as he said he read somewhere) by putting a greased half-pint whiskey flask up his rectum.[23]

There were moral stories to be told as well, tales of the timeless qualities of human nature encountered on neighborhood rounds. Physicians told of courageous patients, as in Samuel Leland's story of a woman giving birth unassisted and doing very well without his help. (She was "some pumpkins," Leland wrote admiringly.) Kentucky's A. A. Patteson praised a patient's willingness to endure a drawn-out death; the dying man deeply impressed Patteson with his "firm and cheering faith of a Christian." There were reformist stories, too, like Yandell's and like that of a Louisiana doctor who was fed up with the lack of skilled midwives and strongly recommended training more of them.[24]

The significance of the way southern doctors' stories, in particular, arose from bedside practice, giving rise to a jostling variety of tales, is further seen in the way narratives hooked into the social fabric of slavery. African-American patients appear interchangeably with whites in the various kinds of tales; physicians understood that whites had no corner on courage, suffering, or wonder. However,

physician storytellers were silent about certain details, suggesting that doctors were similarly selective in what they allowed themselves to see. For example, there is an almost complete absence of explicit stories about treating slaves for the whippings they suffered, a remarkable silence given what most doctors must have encountered. Yet there were stories exposing other threats to slaves' health. The hazards of daily field labor and the bad effects of a monotonous diet formed the heart of case stories that advised better food and more rest for workers. The racial prejudices of slave owners were not always accepted at face value. While some physicians wrote about the supposed "typical" slave who feigned illness in order to escape work, others warned owners that it was dangerous to assume that slaves' complaints were mostly fabricated.[25]

Perhaps most strikingly, this focus on the particular, and on the individual doctor's experience in treating slave patients, suggests that broad racist theories did not rigidly determine doctors' routine care, nor, in most cases, did doctors appear to have much interest in developing or extending racist thinking through their case narratives. Slaves' supposed biological tolerance for long hours in the hot sun, for instance, or the supposed natural "ease" of childbirth among slave women are only rarely articulated in narratives. Rather, slaves were written about as having the individual differences inherent in any case. The grander the theory, the less it appears to have shaped doctors' actual practice. And thus the attempts of a few physicians like Samuel Cartwright to argue in an eager and sometimes bizarre manner for the racial peculiarity of black pathology and physiology, for example, or Josiah Nott's expansive theory of separate racial origins simply were not useful in a livelihood where usefulness was the touchstone.[26]

Most ordinary physicians, then, relating or reading stories in a plain style, expecting knowledge to emerge from the great variety of stories, believed, like Dr. J. R. Freese, that even a single case drawn from the intensity of personal experience would "illustrate . . . better than a volume of theorizing." In one sense, of course, this is a hermeneutic approach with a sharply limited vision of the means and objectives of theory. Certainly, case-by-case thinking undercut not only racial generalizations but also other kinds of systematic thought that later in the century were widely adopted as part of a new medical science: establishing criteria for what counts as data, manipulating data under controlled conditions, looking for general patterns across several cases, and so on.[27]

The key point about these narratives, though, is not the obvious one—that they were "pre-modern." Rather, the point is to understand why these stories were so popular and profuse—what needs they satisfied and how. In the heat of practice, the vernacular, autobiographical tale was a work story that strengthened the worker by holding out the hope that the case might have meaning not revealed except in the telling. It gave the physician an opportunity to celebrate the many routes clinical success might take and gave him some resilience in the face of failure. Moreover, narratives acknowledged the shared work of medicine, supporting a kind of democracy of knowledge among physicians at a time when at least some country doctors were beginning to feel that their rural isolation might be an intellectual disadvantage. Within the confines of the narratives, any careful, observant man was the equal of any other if he could tell plainly what had happened to him and what he found or wanted. A story needed only to be lifelike to persuade. And, if told well enough, any case narrative might reveal larger implications for the betterment of medicine. If, as most physicians believed, the very nature of disease varied dramatically incident by incident, a single case—the next case—might supply a key piece in some broader medical puzzle that would place a doctor's individual patient into the wide field of medical science.[28]

This brings us close to the bedside, the place where sick people's stories first reached out to the doctor and where he first made them part of his own tale. Here, we can see in the jumbled variety of case stories two crucial narrative elements in common. These gave shape to the diversity of tales, tying their narrative urgency (which physicians believed was inherent in the patient's predicament and in the "course" of the disease itself) to the social world of doctors and patients (which physicians helped to create through their stories). All narratives spoke of sick people in terms of their bodies and in terms of what might be called their biographies—or that part of their lives exposed by malady. Joined together in a single narrative, these elements none-theless were two distinct ways the doctor gave expressive form to what happened in his work.

It is important to recall how physicians like Lunsford Yandell, working alone or in hastily arranged consultations, often were care-givers of the last resort in isolated country neighborhoods. The call went out to them only when the bleeding, fever, or pain overflowed the family's efforts. What had happened to the patient—or was still happening—had to be contained by hand and word lest it also overwhelm the doctor himself. In terms of the emotional energy demanded of the physician, the modern analogy is the emergency room, not the ordered quiet of a private-practice office. Walking into a house, the mid-19th-century doctor grasped, literally, the patient's distress and his own need to act by taking hold of the sick body. But this was not all. The sheer physical reality of sickness and treatment also had to be apprehended in terms of the sick person's life as the doctor knew it, or as he was led to discover it when he walked through the door. Putting the sick person's biography and body together was at the center of the doctor's overwhelming need to act, and it became the framework of his stories as well.

Doing this was no simple matter, as many confused or inconclusive narratives reveal. Running throughout the many kinds of stories is a tension doctors encountered as they struggled to shape body and biography into a single piece of medical work. It was a tension between the inside and the outside—whether of the body or the sick person's life—and physicians strove to use it to their advantage. With regard to the body, the relation of inside to outside appears when the doctor observed in order to touch, and he touched what was outside in order to imagine what lay within. In terms of the patient's biography, the physician's act of entering the sickroom, of making a place for himself at the bedside, inevitably was the act of an outsider seeking entrance to—and help from—the social world of the sick person he must now engage.[29]

Narratives often hang poised at the sickroom's threshold for the first few sentences, as the doctor took the measure of his patient in terms of sex, race, and "habit," an assessment of the patient's basic constitution and temperament. Taken together, these provided a context for whatever actions were to follow. And by introducing the patient in this objectifying way, the doctor paused briefly, taking his own measure and setting the stage for his physician-readers. In most narratives, this is the only point in the story where the patient is remote and seems merely to be awaiting the physician's work. Unlike modern patients who, in the language of CPCs, are said to "present" their symptoms, the 19th-century sick are comparatively passive, usually "found" (as in "I found the patient in bed"); each of Yandell's dysenteric patients, for example, "was attacked," or "was seized with the disease," or was found "under the disease." The disease, too, is present in the room, having its own story or "course," which the physician must discover.[30]

Neighbors and strangers, African-American or white, children and adults, were thus made to lie quietly at the beginning of the tale so the physician could move in close to the stilled body. The physician then engaged in a kind of physical portraiture, beholding the relevant body in its illness, though only from the outside. Alabama doctor Courtney J. Clark's description was typical: "Found his face flushed, skin hot, pulse 120 to 125—small, tense and corded, lips dry, tongue somewhat moist . . . Stools yellow and of moderate thickness. Tongue has a whitish coat, edges of natural color." Pulse had special importance, for of all the body's signs and properties, blood had the most far-reaching effects and sent the most insistent signals. To "take" a patient's pulse was to seize upon the essence of the closed, coded body. The heart's throb, so deep and reflexive as to be "absent" in the healthy person, had an almost preternatural presence in the sickroom. Physicians' words describing the pulse were so numerous that their evocative power spilled over the confines of technical usefulness: pulses were small, quick, tense, soft, hard, corded, bounding, thready, shallow, weak, undulating, jerking, full, resistant—and more. Not the working of the heart but the supply of blood was the object of this intense discrimination of touch and vocabulary. Physicians believed that most forms of disease affected the movement and amount of blood in the body; they witnessed bodies "congested" with thick and poisoned blood. The doctor, his senses collapsed into touch alone, cast nets of pulse words in ever-tighter weaves: when he felt a patient's full, tense, and bounding pulse shifting to a beat that was tense, resistant, and thready, he believed he learned something of the strength of both patient

and the foe disease. With his fingertips, through his words, the skillful physician worked to visualize the tell-tale flow of blood through its hidden map.[31]

Looking at a patient's stool was as revealing as the pulse, though in a different way. From ancient times, the examination of stool was the nearest a physician could come to seeing inside the body, and for 19th-century Americans, stool was more "present" to consciousness than pulse: even in health, most people paid a great deal of attention to their bowels. Doctors' terms for stool were less numerous than their pulse words, in part because stool, as a tangible product of the body, could be judged in a far greater number of ways—and it could be altered or created. "Stools inconsistent last night but still clay colored," noted Charles Hentz of a patient whom he was medicating in 1858 in order to effect a change in bowel movements. "A large one today with bile in it—consistent & stinking—rich bile in it—good." The color, consistency, odor, and amount of stool were the chief signs of a body altered by disease as well as symbols of a body taken in hand by the doctor. The importance of stool as a substance released by the body to the outside, to the waiting doctor, was reflected in the memorable images he used to capture its qualities. Working with patients whom he feared were dysenteric, Lunsford Yandell dreaded most of all to find the characteristic mix of blood and mucous in the stool. He was encouraged when one patient's bloody but clear stool called to mind "the washings of raw beef." But another patient's stool was marked by the much-feared mixture. It reminded Yandell of "strawberries and cream."[32]

By the 1840s, some rural southern doctors were adopting new ways of physically inspecting patients that took some of the focus away from blood and stool—systematically employing auscultation, palpation, and percussion, for example. However, despite calls for a new, standardized science of physical diagnosis, pulse and stool continued to define the body in most narratives. Indeed, examination of the body doubtless retained a particular importance in narratives because it was sanctioned by both the new clinical science and ancient tradition. But appeals to science or tradition did not supply the narrative energy behind body talk. The energy arose from a man's personal story of bedside labor; a tale of purpose and discovery in the craft of dismantling each body into its key elements by working them over with his own hands, senses, and words.[33]

This activity was a necessary prelude to working over the body with medicines. The essence of most stories was therapeutic, with the doctor's prescriptions giving final shape to the medical body. It is striking that medicines and disease shared an identity in doctors' stories. Both were subtle, mutable forces ranging through the bodies of the afflicted. Despite physicians' increasing skepticism of harsh, heroic therapy by the 1840s, orthodox medicine still aimed to discover the course of disease—its plot—and to overpower it. The extent to which this was a personal struggle for physicians is plainly seen in the heightened way they spoke about the conflict. Many physicians prescribed aggressive, "operative" medicines not simply because this was the orthodox thing to do; rather, prescribing such drugs was the doctor's personal answer to the challenge of disease. As Dr. W. R. Sharpe of North Carolina wrote of an outbreak of deadly brain fever in his community, the aggressiveness of the disease "aroused me, and caused me to adopt [a] very active treatment." Physicians had a sense of disease sweeping through settlements much like the weather, its harbinger and sometime ally. If they could not prevent sickness, they would do their best to match it. Therapy, not diagnosis, thus became the thrilling heart of the doctor's tale and the focus of his discriminating interest.[34]

In writing about the therapeutic encounter, then, physicians continued to write about the patient as a body, and of themselves as outsiders, but with a difference. The effective prescription was the doctor's way inside. Opened by medicine, a diseased body became less an opaque object and more like a domain through which the physician pursued disease. Because so many contingencies of practice could thwart healing—arriving too late, for example, or overlooking one crucial detail—doctors spoke warmly of their successful medicines' "noble effects." Beloved therapies sometimes reached into the hidden interior so far that they seemed to outstrip even science. "These means had a magical effect," Dr. George Grant wrote in awe of his fortunate prescription of laudanum and an abdominal bandage. Physicians' lovingly detailed stories of medicating the body show that orthodox practice, for all of its rigors, was not necessarily a meat-ax approach to healing. Though powerful,

orthodox medicines were perceived as anything but crude; like disease, they possessed an almost infinite subtlety and malleability. Disease constantly shifted its very identity in narratives, beginning, for example, as gonorrhea yet "assuming a typhoid character" before the doctor's very eyes. One disease might impersonate another, as in the well-known propensity of intermittent fever to appear "with great exactness as a phrensy, a pleurisy, hepatitis." But these deadly transformations in the body's interior were matched by the physician's drugs; medicines were the gallant mount he rode inside the body of his imagining.[35]

"Medical care begins when corporeal events achieve the status of words," a contemporary clinician has observed.[36] In 19th-century narratives, the body was dark until the doctor's words illuminated it from the outside in. Although bodies and stories were many, every story pursued the same tale: how the doctor, working from outside the body, encountered it, ripe with signs, and labored to make it responsive to his hands and drugs. These stories are not chiefly marked by the frustrations of being premodern; rather, they convey a physician's sense of personal challenge in discovering each body's physicality and the journey to be taken inside. Inevitably, then, these were autobiographical stories—and here, the rural context of most southern practice lent a sharp accent to the drama—of coming on the field of a patient's body, gazing from the outside, getting inside, chasing down disease.

White male doctors took up bodies in this way regardless of sex or race. Although most physicians doubtless accepted many of the typical, demeaning race- and gender-based stereotypes of slaves and women, the case narrative did little to strengthen these abstractions. In fact, narratives expressed a certain somatic equality at the bedside because, in practice, the physician desired access to all bodies equally, often under urgent conditions. Beginning with the body not only had obvious diagnostic aims, it also allowed the doctor to begin his labor (and later his narrative) in a quiet, controlled way. Focusing on the body established the bedside as his workplace, clearing a space among worried family and friends who also crowded close, anxious to assist in the work of definition and relief. One modern physician-author's newly discovered delight in writing full, subtle case stories "*combining* . . . physiology with a [patient's] personal experience" stems in part from the fact that medicine since the 1880s has tended to *separate* "personal experience" from physiological data.[37] But the 19th-century doctor was in a different predicament. Rather than coaxing forth his patients' personal lives, he was more likely to fear being swamped by them. He needed to hold the full flood of an individual's life at bay, at least for a while; in his close focus on the body, he did just that.

Yet the time came soon enough when the doctor had to turn from the body to the patient's biography—to the events and habits of the sick person that had led them both to the bedside moment. Because a southern physician's patients were usually his neighbors and friends, or the human property of neighbors and friends, they were known to him on a number of social levels. He saw them around town, at celebrations, in the fields; he prayed with them, traded stories with them, joked with their children. Then, sometimes with terrible swiftness, he saw them naked and broken, bleeding, in pain. Even when he did not wish to, he learned their secrets, their satisfactions and shame, and they learned something of his. While this work often was marked by a craft-like directness—just get the painful tooth out—it could turn dark and confounding, all eyes on the doctor who hesitated.

So just as he stood outside the body before moving in with words and drugs, the country physician may be seen as both outsider and insider in the social life of the sick. He was inside the community in a sense many modern physicians are not: he had few professional retreats and no way to establish a livelihood apart from living fully and each day in his community. His neighbors were his patrons and his judges; their satisfaction with his work mattered much more than that of his far-flung peers. Yet he inevitably appeared as an outsider, too. He had to be fetched from *his* place to the *other* place, and he needed help from those already around the bed of the sufferer. They had witnessed the seizure, the fall, the onset of labor, and he had not. The doctor needed to establish a work space by handling the body, but he could not give care without family and friends; he needed to hear what they had to say and make it a part of his work.

For these reasons, the physician drew on whatever he shared with his patient, and his effort to combine biography with body may be seen in another kind of body talk in case narratives: descriptions of the sick person in a rural vernacular that the doctor could rely on as meaningful in his patient's life as well as his own. Here, the afflicted body was not measured in terms of medicine's standards for pulse and stool but likened to the surrounding rural world. Examples are everywhere in the case stories: a Kentucky man had a tumor "sprouting out below the eye." Infected glands resembled "that decayed portion of honey comb usually called bee bread." Vomit was likened to "a mass of wet gunpowder," a woman's rigid cervix to "an ivory ring," and a splintered bone to a "broken stick of tough, straight-grained wood." Doctors easily folded such plain talk into their evaluation, relying on blunt, conversational terms for how patients felt: "right smart," "diminished," "reduced," "laboring under disease," "sinking," "taking a bad turn."[38]

In using the vernacular, physicians were not merely condescending to patients, as is often the case today when so much definitive power belongs to the physician. Instead, common language signaled how thoroughly a common world shaped the description of malady and suffering. Of course, the physician was no less the teller of his tale; the patient's biography as it entered into narratives should not be mistaken for the patient's own story. Yet once the doctor departed from a close description of the body, the plain, expansive nature of what he had to say took its power from the wider world both he and his patient knew. In this way, too, as the doctor ventured into his patient's biography, he was writing autobiography as well.

Descriptions of patients' lives were less ordered and predictable than descriptions of their bodies. For instance, the episode that led a person to summon the doctor (in modern jargon, the indispensable "history of the present illness") is not included in many narratives. Or else it is oddly placed (by modern standards) at the end of the story, where it gives some context but does not help build an argument supporting the doctor's subsequent course of action. Sometimes, the outcome of the patient's illness is recounted, but many times it is not. Maladies begin but do not end; people appear and disappear in unexplained ways that once again suggest how medical practice was shaped more by the drama of care-giving than by diagnostic goals. In short, accounts of patients' lives persuaded through a verisimilitude born of a doctor's personal experience, not because they adhered to a set of abstract standards divorced from the intensity of a doctor's bedside work.

Thus, like their maladies and their bodies, patients' lives, too, entered narratives in ways that expressed the rhythms of a physicians's work. Patients were identified by name, sex, and by household status or neighborhood, though rarely in terms of age (except in very general terms) or previous illness (unless directly related). Some identification of the patient was necessary in published case reports to show that the case was authentic; but too much might give the patient unwelcome publicity. Most southern doctors described their white patients with an initial (as in "Mrs. A.") and their slave patients by first name and owner's initial (as in "Sam, the property of Mr. C."). Another form, reminiscent of fiction, substituted asterisks for letters in a name (as in "Mrs. F****** of Natchez"), seeming almost to entice readers to guess the identity. These usages, reflecting the social biases of slave society, were anything but consistent, however; race was a factor but not always in predictable ways. Certain slave patients remained entirely anonymous, probably in some cases because the doctor did not either ask or recall their names. Yet many patients, slave and free, were introduced in ways that would make possible their identification, with precise references to where they lived, who they were related to, or some other marker. In still other stories, patients of both races and all classes simply were named outright. The variety revealed a reality of practice: the doctor knew certain patients better than others, and certain patients claimed his attention more than others. This was as it should be. To adopt a uniform mode of identifying a patient would be to choose a less lifelike form, lessening the power of the narrative.[39]

What counted as a "case" similarly derived from this personal register of medical work. Physicians did not equate a case with an individual patient. In Lunsford Yandell's essay on dysentery, for example, only six of the nine cases he narrates were individuals. The other three cases turn out to be inhabited by more than one person. Case 5, for instance, consists of four white children. Yandell introduces Case 4 as "a little girl, sister of the foregoing [case], about 6 years of age." These two children are white; but, in the second paragraph, the story suddenly opens up to reveal "several black children in the same family," whom Yandell also treated on succes-

sive visits. None is described individually. In another variation, the "negro boy, [age] 6 years" of Case 6 is the sole occupant of this story; Yandell describes the five days of caring for this slave child in more detail than he devotes to anyone in his essay, save for his disguised family members.[40]

Clearly, the "case" as a conceptual device took its form from the social setting of each call for help, not from the identity of a single patient. Shaped by the demands of rural practice, medicine was not fundamentally one-to-one, physician and patient. The doctor was called to a household that untied a diverse group of individuals. Under these conditions, and along with his sense of disease as dangerously fluid by nature, a doctor easily saw sickness traversing several bodies in one afternoon, or over a series of days, and thus conceptualized as a single case all of the sick people in this multiple challenge to his skills. Bedside medicine was not a matter of fitting people into a conceptual world of disease divorced from the doctor's immediate experience. It was a matter of encountering sickness in its individual worlds, and a moving narrative must recreate this reality.

The various ways in which patients were represented and grouped in case narratives inevitably exposed other conditions as well: how carefully or extensively the doctor took notes, for example, how well he had previously known (and liked) the patient, and how much time he found to write. These conditions, too, were fashioned into meaningful narratives not by a unified vision of science, nor by uniform training, but rather by the sanction given to autobiographical writing. A physician testified to his worthiness as a practitioner not in terms of clinical successes or biomedical discoveries but by writing convincing personal sketches of his summons to this or that household. Thus he mastered the various lives of his patients, and all of the turns and vagaries of disease, by telling a story of what happened to *him*. The absence of a rigid form for doing this is best seen not as a pre-modern sloppiness but rather as a literary roominess, a dramatic—and useful—way of seeing oneself at work.

Although this style misses by a wide mark the modern concern for patients' rights or autonomy, the doctor's personal vision nevertheless gave a defining clarity to the way lives were disrupted by injury or malady. Physicians' stories did not omit the fundamental power of illness to transform a life; in story after story, disease is compelling because it appears out of the blue to strike a person down. As Hannah, a slave, was "lifting a heavy piece of timber, she felt something give way in the lower part of the abdomen." "Mrs. L*****, . . . whilst stretching out a hank of cotton yarn, suddenly felt pain" in her arm, and now the "limb is painful and almost useless." Narratives are thickly populated with people like the Alabama merchant "sitting on a chair . . . his features greatly contracted, his countenance anxious," the young Virginian "scarcely able to walk about his chamber . . . sallow, haggard," and the woman Dr. J. S. Dyer recalled only as "the misshaped patient."[41]

Doctors' stories, opening onto the life of the sick person, also allowed family and friends to come forward, just as they did at the bedside. Here, too, the autobiographical style freely admitted the variety in doctor-patient encounters. Sometimes, there is an edginess in the physician's account of coming in as an outsider. "When I walked into the room she looked on me with an eye of distrust," W. A. Shands reported of a new patient. Some doctors, like William P. Hort in 1847, were frustrated in the use of a particular procedure because "the prejudices of the family . . . were very strong, and they would not tolerate it." Another practitioner urged the husband of a woman in labor to send for a consulting physician, but "the case being one of emergency, he declined calling in any other help" because it would take too much time. Though provoking great anxiety in the doctor, the husband's wish prevailed.[42]

Other stories, however, just as freely portray harmony, or at least a shared determination, between a household and the doctor, suggesting ways a physician might successfully join the bedside group to further his recommended treatment. One physician, for example, recounted his alarm at finding a six-year-old girl blue in the face and gasping for breath when he entered the house. Though highly agitated, the girl's mother told him to wait and watch—the seizure would soon pass. She proved to be correct. Impressed, the doctor continued to rely on the woman's observations, quoting her authoritatively when he wrote the case. A Virginia physician, struggling to stanch the bleeding of a woman in labor, was "staggered by a grave proposition from the horror stricken husband . . . that all unite in prayer for divine assistance." The doctor told the

man that he "would not object to his prayer, provided it was short," yet he himself ended up drawing strength from kneeling and praying with the family.[43]

In telling the lives of patients by telling his own experience, a doctor's narrative reveals, too, how thoroughly social was the practice of medicine. Because sickness did not isolate a person from familiar surroundings, and because these surroundings were seen as essential to understanding health and illness, a patient's personal relationships were implicated in all stages of discovery and treatment. Many doctors, like H. V. Wooten in an 1850 case of a youth with remittent fever, believed that they did not need to speak with seriously ill patients before treating them as long as family and friends joined in. Wooten received enough information about the delirious youth from neighbors and a landlady—all of whom knew the young man's comings and goings—to lead to a suitable course of therapy. At times, patients' own words break into the story with particular power. A white woman giving birth screams, "O Doctor, my womb is split, and I shall die!" Mary Ann, a slave, suffering in her labor, tells the doctor, "I want you to cut me open, and take out this little devil in my belly . . . for I am determined not to die." Such words burst from the circumstances of the narrative, just as they electrified the bedside moment. They represented unforgettable—and significant—points where the case pivoted dramatically from one attitude to another. They were moments the doctor could not ignore in the telling of the case, just as he could not ignore them when engaged in treatment.[44]

Both free and slave patients' lives are visible in these stories, just as their bodies are, with doctors similarly taking notes and recalling previous treatments for patients of both races. Although white doctors usually dismissed slaves' views of disease, African-American remedies—and folk remedies in general—often were treated with at least some respect; help came from strange quarters, as any working physician knew. At the very least, most doctors simply went along with the cornmeal or cabbage-leaf poultices, cobweb swabs, and hog-lard emetics. Folk remedies, Dr. R. S. Bailey reasoned, "are for the most part, harmless," and so he "acquiesced, not desiring to lessen [patients'] confidence in the means employed," nor, probably, their confidence in himself.[45]

In all of this, too, most rural doctors do not appear to have found much use for broad, systematic racial views. Again, this is not to say that physicians saw no racial differences, only that these differences seldom were seen either as determining the specific therapy or as explaining the results. Lunsford Yandell, for instance, stated that "dysentery is not so fatal a disease with Africans as among white men." Yet he prescribed the same therapy to patients of both races, and the concrete conditions of slave patients' lives, not racial theory, explained particular outcomes. In Yandell's Case 6, for instance, the central fact in the death of the black child was not race but slavery. The child fell victim to "the inattention of the overseer," who neglected to regulate the boy's medicine, leading to a fatal overdose.[46]

There is a particularly significant way in which slavery shaped doctors' stories of black patients that illuminates the entire issue of how a physician struggled to combine body with biography. Slave patients almost always seem alone in doctors' stories. Even slaves who are given names and voices usually have no one around them—no family or friends such as fill out the biographies of white patients. Owners sometimes lurk in the far background; but the white doctor and the black patient seem a lonely couple unlike any other in case stories. This blinkered vision of a whole class of patients suggests, once again, how narratives revealed, and shaped, what a physician was able to see and thus how he conceived of illness and caring.

Courtney J. Clark, for example, a conscientious Alabama physician who made careful bedside notes, described in 1850 his physical examination of an 18-year-old slave patient, Charles. Clark portrays Charles in a lively way that makes the young man engagingly human: the doctor knows Charles's medical past ("he has always been a very healthy boy"), reports Charles's self-assessments ("he felt better"), and gives his readers a vibrant, colloquial body ("his pulse was thumping away"). There is nothing in this style of seeing and writing that distinguishes Charles from Clark's white patients, except the absence of any social setting.[47]

This omission of the social group gives a strikingly different emphasis to the doctor's struggle to fit biography together with body. The most "interesting" stories—like the most compelling bedside experiences—struck a balance between body and biography, creating a significant tension within a narrative. This occurred in stories, as we have seen, because the doctor's vision of the body encompassed two awkwardly joined purposes. Descriptions of the body served not only to diagnose—to move inside—but also to hold at bay the clutter of the patient's life that rushed to meet the doctor as he came into the room. The doctor, in focusing on stool and pulse, got his footing amid the urgency of a person's pain and the family's anxiety. Yet, at the same time, a physician needed what he found in this rushing stream of another person's now-stricken life: information, cooperation, trust—stories from which to make his own and thus ride the tension between body and biography to a beneficial end.

The comparatively sparse biographies of slave patients, then, suggest how slavery crippled this basic level of the white doctor's work. The tension between body and biography was not as marked because the white doctor was, in a key way, not the *slave's* physician at all. The doctor was called by, and responsible to, the slave's owner and thus easily saw the slave's life as nesting within that of his master. There was no need to pursue the patient's biography very fully or far. Seen in this less complete way by white doctors, African Americans under slavery became, in narratives, simpler, more physical patients. They could never be as "interesting" as white patients because they did not compel the doctor to achieve the precarious balance of body and biography, thus investing his own story in theirs. For, as we have seen, the storytelling energy released by the doctor's work, and thus the meaning of caregiving itself, was in neither the practical details nor the theoretical speculations but in the way narratives tapped into his own autobiography. Uniting the struggle to know disease with the struggle to give care, and joining both success and failure in a single worthy effort, was the testifying, personal voice of the doctor.[48]

The favored place occupied by the case narrative in mid-19th-century medical literature thus expressed its peculiar breadth and flexibility as a text inscribing the essential realities of a doctor's daily work. This narrative style at times furthered certain "modern" habits of medical activity—focused observation, careful note-taking, reflective critique.

But it was the autobiographical heart of these narratives—bedside-born, diverse in motive and purpose, and, above all, fundamentally personal—that made them dear to physicians and thus importantly resistant to a cooler, more abstract style.

And yet, for all of the open, autobiographical energy in case narratives, there remained personal aspects of a doctor's experience that these stories did not touch, or touched only obliquely. Even as it enjoyed a privileged place, the older narrative style was not without limits; there were some things about the dangerous bedside that a physician did not tell. Things not said in narratives, though obviously difficult for historians to discover, suggest the extent to which a doctor created an image of himself at work that excluded realities his narrative could not resolve. Silences and omissions thus suggest important restraints on the narrative's autobiographical voice. These restraints, perhaps ironically, may be seen as harbingers of a new way of writing in which a doctor diminished his personality in order to reconfigure his bedside practice in terms of a science in which mastery of a more impersonal method was all that needed telling.

Lunsford Yandell's case stories of dysentery may be considered once again with this in mind. Although writing about his nine patients in the typical personal style, it will be recalled, Yandell concealed the most personal dimension of all: his family tie to Cases 7, 8, and 9.[49] By silently including his family among his cases—especially his young son Willie, who died—Yandell shaped his personal tragedy into a call for medical reform, combining in one story the collapse of both his medical world and his family's happiness. And he did something else, too: as he told about his work, he left out much about his feelings, and even as he related certain details of his treatment, he omitted others. The effect was to close off avenues of his personal experience and, significantly, efface the awful contingency that marked even the most focused bedside work.

Although Yandell began his essay by portraying himself as a modest practitioner, we have seen that his aim was bluntly to warn his profession of the failure of an orthodox therapy in the face of a new, virulent dysentery. He relied on his final three cases to drive home his discovery, beginning with six-year-old Willie. Transforming his son into "W. Y.," Yandell proceeds in the typical fashion. We see something of the boy's body and his biography. Though physically "robust," he became feverish on June 9—a "slight" fever. Nevertheless, Yandell induced vomiting, and "that evening [he] appeared well." On June 11, however, he had "a bad night" of it, vomiting to exhaustion. His limbs were cold despite the application of flannels soaked in hot spirits. "Thirst moderate," Yandell observes in crisp, note-taking style. "And drinks toast and slippery elm water, but asks for ice." Willie's pulse was rapid and his stools increasingly bloody and mucous-filled; he quickly became too weak to get out of bed. The case, which takes just under a page to relate, tumbles to a sudden end with "Great tossing . . . restlessness increasing. At 12 [midnight], becomes delirious. At half past 8 in the morning, dead." The sickness seemed "extraordinary," going beyond dysentery itself, as if the boy had fallen "into the collapse of cholera." Yandell underscored the shock he felt at the approved medicine's devastating lack of effect: "It is quite remarkable that these symptoms should have come on *while he was under the operation of calomel.*"[50]

Yandell's wife Susan (identified only as "the mother of this little boy") and their friend Matilda Cantrell ("Miss C.") fared much differently. Four days into Susan's worsening symptoms—five days after Willie's death—Yandell tried something new. Thirsty, the patient asked for ice, and he gave it to her, along with an "effervescing" drink consisting principally of bicarbonate of soda. The standard therapy discouraged giving cold water or ice to dysenteric patients. But the standard therapy was failing him once again, so Yandell broke with it and tried an "experiment" urged by the patient herself. "After eating the first lump of ice [she] felt refreshed" and slept well. In Yandell's recollection, "the relief afforded by the change of treatment was instantaneous." He treated Miss C. in a similar way, and she, too, recovered straightaway. The emotional heart of the tale may be heard at this point, beating just beneath Yandell's casebook style: he wonders whether ice and the soda drink might have helped "the little boy." While this "cannot now be determined, . . . it is impossible to repress a feeling of regret that the experiment was not made."[51]

In striking contrast, Yandell's grief, here compressed into a professional's "regret," bursts in full force from the letters he wrote to his wife's parents during the crisis. Intense anxiety pushes the letters along, as Yandell sees in the failure of his therapy a dreadful prognosis. His first letter, written on June 12 "under the most painful oppression," is a father's at the line between hope and despair. Willie is "in the most imminent danger . . . His death, in less than 24 hours, would not surprise me! . . . O Lord! have mercy on us." His next letter, the following morning, proclaims the loss even as it occurs: "Our dear little Wilson is still alive, but to all appearance, in the article of death! The trial to us is, indeed awful. Our hearts were bound up in the lovely boy. Half the charm of our existence goes with him." Then, in a postscript, "It is all over. Even while writing the lines within, he began to gasp for his last breath. He died at half past 8 in the morning—being 5 years, 11 months & one day old . . . I am unfit to add more—"[52]

But, in the two weeks following Willie's death, Yandell wrote a great deal more, daily letters that rode three powerful themes, which appear only dimly or not at all in his essay. Two themes he entirely elided from his published narratives—those of religious faith and family feeling. The third theme concerned the work of doctoring but in a way quite different from that in the brisk, problem-solving essay. His letters suggest the wearying and, above all, treacherous labor of care-giving. In omitting these themes when he wrote his essay, Yandell was relying on the narrative, as did other doctors, to hold at bay the life-swallowing enormity of dread disease. And it is also clear that his decision to reshape his story in the way invited by the case narrative had specific consequences for how he saw himself at work.

In his family letters, for example, religion gave Yandell a powerful language that reached out for an explanation large enough to match his loss. Writing as Willie lay dying in the next room, Yandell yearned for "the comforts of that hope which looks beyond the grave . . . Thank God! if we may not hold the beloved one with us, we may go to where he will shortly rest." He freely admitted to being overwhelmed by grief, gently adding, "I cannot say that Susan bears it well." During the next six days, he spoke in this religious idiom, combining hope for consolation, deep personal guilt, and something close to a sense of divine injustice. Guilt was foremost, but it was mixed with questions for God. Two days after Willie's death, Yandell feared that "having been so luke-warm a Christian," his

own lack of faith might have "caused this blow to fall upon our lovely child." Yet, invoking Job, he wrote bluntly—he calls it a confession—that religion was not helping him as he expected it would, as it should. In the same breath in which he says he will not "murmur" against God, he does murmur, confessing that he did not feel "resigned" to God's will as a Christian ought. Indeed, while feeling "obliged" to express resignation, he also feels "grief and shame" that God may have taken Willie in retribution for his father's tepid faith. Despite his own guilt, the sharp edge of Yandell's words is tilted toward a God who punished innocent children.[53]

This fierce religious craving diminished somewhat within the next few days as Yandell gradually realized that his faith had not foundered entirely. Although religious images continued to appear in his words, he quoted St. Paul's injunctions to persevere rather than Job's lament. He urged himself to see his loss in an earthly frame. "Time, I know, is the remedy," he wrote on June 21. "But how slow!" And five days later, he remarked, "Nature does not exert its usual recuperative power" over his and Susan's grief. Though "low spirited" still, it was to Nature, not God, that the doctor now turned.[54]

An even more consistent theme in Yandell's letters was family need and obligation, especially his desire to have his wife's parents join them. Yandell wrote to assure them that the dysentery was not epidemic, and thus not a threat to them, saying in more than one way "how much we need your society to comfort us." Love was shaded by doubt, however, perhaps building on old, hidden resentments. So great was his desire to see them that it seemed to overwhelm his sense of epistolary time: almost from the beginning, he wrote as if he might daily expect a reply even though Susan's parents would scarcely have had time to receive his letters. Nevertheless, Yandell began to wonder aloud why they had not yet arrived. He began to employ an imperative epistolary style indicating injured feelings. His use of the word "singular," for instance, connotes inexplicable and perhaps inappropriate oddness: "[W]e cannot help looking upon it as a little singular that we have not heard one word from any of you," he wrote on June 18. On the 22nd, he wrote his father-in-law even more pointedly, "I hope, my dear sir, that you will not disappoint us."[55]

But disappointed he was, and his sense of being neglected by family twined about his physician's sense of what good medical care required. "Susan, as may be supposed, is low spirited," Yandell wrote eight days after Willie's death. "How consolatory to her it would be ... to have the society of her mother." Two days later, he reminded them once again that it was their own daughter who so urgently desired their company, that it was her health at stake: "she bids me urge you ... to commence your journey without delay ... Impatience, you know, is almost inseparable from the sickbed."[56]

In erasing these themes of family and religion from his journal article, Yandell drew a sharp line between his work as a doctor and his conflicted feelings as a father and husband. That he would make this choice was not at all predictable in a medical world that favored the personal and vernacular. Yet the contrast between essay and letters makes it clear that even though the autobiographical narrative furthered a medical knowledge privileging—even celebrating—the personal context of work, it separated a physician's work from other kinds of self-awareness. In writing his essay, Yandell was able to place the medical vision of body and biography at the center of his intellectual and emotional horizon; he laid Willie to rest by bringing "W. Y." to life. In this way, he reduced his struggles with loss, family, and faith to a problem of practice and reform in his work. Moreover, in thus shying away from the deeper dimensions of the personal—in showing how the narrative in fact sharply delimited what counted as "personal"—Yandell distanced himself from the similar loss and sorrow he had witnessed many times in the homes of other frightened and grieving families. To the extent that he stepped away from the fullness of his own feelings, then, he removed himself from his local community while joining his professional one. His narrative suggests how such distancing was becoming a hallmark of professional practice.

Perhaps most intriguing of all the contrasts between his letters and his essay is the way Yandell diminished in his essay what was probably one of the most pervasive realities of routine care-giving: the particular way that medical work could outstrip and exhaust the physician, deceiving him into a misstep. As we have seen, practitioners' puzzlement and failure are not excluded from most case stories, nor are they from Yandell's. But his letters suggest a related dimension that narratives did not directly explore: the treacherous contingency at the heart of all that the physician tried to do with the bodies and biographies of sick people. Even a "conscientious man" discovered that doing his work meant mastering the shifting, dirty details of care-giving while, simultaneously, realizing that an insight or discovery might not follow from this mastery; insight might come from any quarter, at any time. The key was to be neither careless of detail nor stupefied by it, so that the saving insight might be recognized and seized.

In his article, Yandell spoke of his bedside activities candidly but in the typically brisk, imperative voice of a professional in full stride. Although he admitted to being baffled at certain points, his narrative nonetheless portrayed his work as something orchestrated: he "instructed' his patients; he "ordered" things to be done. But who was carrying out these orders? From his letters, it is clear that Yandell was doing most, if not all, of the tiring work himself. He did consult with a colleague, Dr. Benjamin Dudley; but with the possible exception of a slave assistant (he never mentions one directly), Yandell, like most rural doctors fully engaged in a difficult case, worked alone. He compounded, timed, and administered his medicines himself; he handled stool and pulse; he stayed up nights, sleeping only when his patients slept, acceding, as they did, to the rhythm of illness.[57]

The nature and course of the disease, too, are cast in a much more problematic light in Yandell's letters. In his essay, he styled the dysentery as a cleanly identifiable foe, epidemic in force. Yet, in his letters, Yandell was struck by the oddness of the affliction. "This disorder seems almost confined to my family," while the neighborhood as a whole was healthy. Moreover, in contrast to the passive, "found" patients in his narrative, the letters highlight the ways in which patients' feelings and perceptions shaped his care. Yandell reported in his article, for instance, that one of his wife's relapses resulted from an adverse reaction to some milk and mush she had eaten for breakfast. In his letter, however, he revealed that this assessment was Susan's own diagnosis, which he later adopted.[58]

Overall, the comparatively distant physician of the essay is shown to have been much swayed by Susan's despair and grief, to the extent that both of them became bound together by the illness, rather than by his ministrations. The power of illness dictated its own story. "We are, indeed, in great darkness," Yandell wrote his in-laws on June 21. "And, like dying men, catch at straws to keep our hearts from utterly sinking." His fragile efforts drove him to focus on himself rather than his patient. After more than 10 days of bedside anxiety, he craved release, aware that his exhaustion clouded his caregiving: "I want occupation. Books do not interest me, as formerly, and my duties in the sick chamber forbid my seeking employment without. Thus depressed, I do not afford that salutary excitement which flows from a bright countenance & cheerful manner."[59]

All of this bears, finally, on the fact that Yandell's narrative, like most other case stories, is a therapeutic tale. Here, too, his letters reveal that much of his experience slipped through the looser weave of his essay. A physician hoped his tale would be one of finding the means to lift the patient, in Yandell's sweet phrase, to the "rising ground" of convalescence. We can recall that his final two cases were this kind of story, relating the discovery of the ice and the soda drink that so relieved his wife and friend. In his letters, though, he writes a markedly different tale.[60]

The right therapy, we have seen, was itself a character in narratives, making itself known immediately in many stories, a shining ally through which the doctor made himself known, first in the body and then in the broader life of his patient. Yandell's soda and ice therapy has such a presence in his essay. Susan's symptoms had reached "an alarming height," he writes, but the "first piece of ice and soda-powder" gave her marked relief; the change was "instantaneous," its power as extraordinary as the virulence of the disease. In his letters, however, the soda drink is noted without fanfare, as merely making Susan "much more comfortable." This therapy followed others and seems, from Yandell's terseness, very likely to have its own successor in the course of treatment. The flat, diary-like letters have no heroes, only necessities, and they suggest the extent to which happenstance and experiment structured doctors' work in difficult cases, and how narratives later memorialized what was at first only dimly seen or even fortuitously discovered.[61]

But was the ice and soda an accidental find? Like so many doctors, Yandell does not say, in his essay, how he came to make a particular trial. Just as typically, he does not explore his understanding of the chemical or physiological transformations entailed

in the therapy. He simply remarks, almost as an afterthought, that "the [bicarbonate] salts had set up a different secretion" that carried the sick body to higher ground.[62] The bedside, not the pathology, remains at the center of his tale; the dramatic reversal of fortune is the best plot.

Yet we can also recall that, in his narrative, Yandell expressed "regret" that he had not given the ice and soda to "the boy." And among Yandell's letters are a few notes from Willie's last days that suggest the provenance of the new therapy. Yandell's discovery of the ice, like much else about medical work, is revealed as rooted in the awful contingencies that seethed just below the surface of all medical narratives. As he wrote his daily letters, Yandell recorded some of his son's last words and moments. "I'm sick so often," Willie said at one point, and, prefiguring Yandell's own judgment that the boy's affliction was more cruel than ordinary dysentery, "I expect I'll have smallpox next and then cholera—and that is worse than any, isn't it Pa?" Yandell told of how he carried Willie to the window to watch a parade of militiamen and of how "his last effort to rally was on Sunday, in the afternoon when he heard his puppy bark. My darling boy Wilson."[63]

Most terribly in retrospect, "he said Pa couldn't I have a little ice." But it seems the father, staying with the approved therapy, said no, even though Willie allowed it could be "a piece so small that you could hardly see it." The moment of therapeutic discovery thus presented itself, but the doctor did not grasp it. "Thirst moderate," Yandell says of "W. Y." afterwards, and he drops the barest hint: "[He] drinks . . . slippery elm water, but asks for ice." Later, his wife asked for ice, too. Thinking it over, watching Susan recover, and then writing his essay, Yandell buried this terrible turn in the case. In the light of his fatal adherence to orthodoxy, the essay's last paragraph, in which Yandell calmly notes the weather conditions surrounding those days in June, seems more than a conventional bow toward tradition. It seems an attempt to pull his story aloft, into a realm where illness was more like a grand force of nature and less like the messy force of circumstance around the bedside.[64]

Yandell's two ways of writing about Willie's death might be read to suggest many things about how a doctor, father, and husband chose to cope with medical failure and personal tragedy. Of importance here is the way the difference between his letters and his formal essay reveals some key limits to the personal case-narrative style and thus to the meaning of medical work. That is, even during the period of its greatest influence, the autobiographical voice diminished many of the most particular— yet far-reaching—aspects of everyday practice: the drudgery, the guilt, the accidents, the wavering sense of Providence. Yandell's writings suggest how and why physicians let this material fall out of their stories. It was not so much that giving expression to these aspects of practice was not "modern" but, rather, that the misgivings and confusion encountered every day were too painful, too intellectually exhausting to recount. To become so fully personal was to acknowledge too openly the terrible chaos

that lay just beneath one's daily efforts. Among other things, then, finding the right words for opportunities missed or chances not taken was too daunting a literary task, and thus doctors like Yandell limited the personal disclosure at the heart of seeing themselves at work. These limits would expand and toughen in the decades to come, as physicians came to see a different kind of professional authority to be gained by further loosening the tie between work and self. For these physicians, modernity would consist of doing what Yandell did through his case-narrative essay: become more remote, impersonal figures in their own stories, permitting the harrowing bedside contingencies of practice—the pieces of ice given or not given— silently to melt away.

NOTES

1 Lunsford P. Yandell, "Cases of dysentery, with remarks," *Transylvania J. Med.*, 1836, 9: 240–250, quotations on p. 242.

2 Yandell, "Cases of dysentery," p. 242. Medical journals in the United States grew phenomenally in number after 1820. See Myrl Ebert, "The rise and development of the American medical periodical, 1797–1850," *Bull. Med. Lib. Assn.*, 1952, 40: 243–276, and W. F. Bynum, Stephen Lock, and Roy Porter, eds., *Medical Journals and Medical Knowledge: Historical Essays* (London, 1992).

3 For a suggestive view of professionals as men searching for ways to define and exercise power, see Daniel H. Calhoun, *Professional Lives in America: Structure and Aspiration, 1750–1850* (Cambridge, Mass., 1965). Professionalization as reflecting shifts in general as well as particular knowledge is the view in Thomas L. Haskell, *The Emergence of Professional Social Science: The American Social Science Association and the Nineteenth-Century Crisis of Authority* (Urbana, 1977). See also Burton J. Bledstein, *The Culture of Professionalism: The Middle Class and the Development of Higher Education in America* (New York, 1976).

4 For interesting modern examples of physicians speaking in this personal register, see David Hilfiker, *Healing the Wounds: A Physician Looks at His Work* (New York, 1985); Melvin Konner, *Becoming a Doctor: A Journey of Initiation in Medical School* (New York, 1986); Perri Klass, *Other Women's Children* (New York, 1990). On the scientific urge in this period, see Robert V. Bruce, *The Launching of Modern American Science, 1846–1876* (New York, 1987); Theodore Dwight Bozeman, *Protestants in an Age of Science*

(Chapel Hill, N.C., 1977); Elizabeth B. Keeney, *The Botanizers: Amateur Scientists in Nineteenth-Century America* (Chapel Hill, N.C., 1992).

I do not make the "southernness" of my physicians a central argument here. Much of what I am exploring in terms of southern medical practice also obtained in the mid-19th-century North, though I do highlight the distinctly southern emphasis on race and slavery. For collections of essays that thoroughly survey the southern context for medicine, see Ronald L. Numbers and Todd L. Savitt, eds., *Science and Medicine in the Old South* (Baton Rouge, La., 1989); Todd L. Savitt and James Harvey Young, eds., *Disease and Distinctiveness in the American South* (Knoxville, Tenn., 1988). See also the pathbreaking work on medicine in the South, Todd L. Savitt, *Medicine and Slavery: The Diseases and Health Care of Blacks in Antebellum Virginia* (Urbana, Ill., 1978).

5 Orthodox, or "regular," physicians, the focus here, were those healers who held M.D. degrees. The U.S. Census counted about 18,500 people identifying themselves as physicians in the South in 1860; all but a handful were male. But even among the orthodox there was considerable variety in terms of knowledge and experience. Illuminating interpretations of the significance of therapeutic change during these years are in Charles E. Rosenberg, "The therapeutic revolution: medicine, meaning, and social change in nineteenth-century America," *Perspect. Biol. & Med.*, 1977, 20: 485–506, and John Harley Warner, *The Therapeutic Perspective: Medical Practice, Knowledge, and Identity in America, 1820–1885* (Cambridge, Mass., 1986).

Change was slow, however, especially in rural areas where the great majority of Americans lived. They still depended on the traditional blend of self-help with various kinds of local healers, including M.D.s, and, in the South, a large number of slave caregivers. For brief explorations of the practice of M.D.s in terms of the domestic setting for care, see Judith Walzer Leavitt, "'A worrying profession': the domestic environment of medical practice in mid-19th-century America," *Bull. Hist. Med.*, 1995, *69*: 1–29 (ch. 9, this book); Steven M. Stowe, "Writing sickness: a southern woman's diary of cares," in Susan Donaldson and Anne Goodwyn Jones, eds., *Haunted Bodies: Gender and Southern Texts* (Charlottesville, Va., forthcoming, 1996).

6   For recent work in the history of 19th-century American medicine that significantly shifts the historiographical focus to the social context of institutions, ideology, and practice, see Charles E. Rosenberg, *The Care of Strangers: The Rise of America's Hospital System* (New York, 1987); Judith Walzer Leavitt, *Brought to Bed: Childbearing in America, 1750–1950* (New York, 1986); Warner, *Therapeutic Perspective;* Martin M. Pernick, *A Calculus of Suffering: Pain, Professionalism, and Anesthesia in Nineteenth-Century America* (New York, 1985); Savitt, *Medicine and Slavery;* Laurel Thatcher Ulrich, *A Midwife's Tale: The Life of Martha Ballard Based on Her Diary, 1785–1812* (New York, 1990); Kenneth M. Ludmerer, *Learning to Heal: The Development of American Medical Education* (New York, 1985); Regina Morantz-Sanchez, *Sympathy and Science: Women Physicians in American Medicine* (New York, 1985); Nancy Tomes, *A Generous Confidence: Thomas Story Kirkbride and the Art of Asylum Keeping, 1840–1883* (Cambridge, Eng., 1984); Susan M. Reverby, *Ordered to Care: The Dilemma of American Nursing, 1850–1945* (Cambridge, Eng., 1987).

For overviews of the recent historiography of medicine, see Ronald L. Numbers, "The history of American medicine: a field in ferment," *Reviews in American History*, 1982, *10*: 245–263; David Rosner, "Tempest in a test tube: medical history and the historian," *Radical Hist. Rev.*, 1982, *26*: 166–171; Judith Walzer Leavitt, "Medicine in context: a review essay of the history of medicine," *Am. Hist. Rev.*, 1990, *95*: 1471–1484.

7   In formal training and through apprenticeships, young physicians were (and are) importantly inculcated with stories of their origins. See, for example, Genevieve Miller, "Medical history," in Ronald L. Numbers, ed., *The Education of American Physicians: Historical Essays* (Berkeley, Calif., 1980), pp. 290–308; Guenter B. Risse, "The role of medical history in the education of the humanist physician," *J. Med. Educ.*, 1975, *50*: 458–465.

On physicians' attempts to come to terms with the troubling gap between an ideal of science and the difficult realities of routine medical practice, see John Harley Warner, "Science, healing, and the physicians's identity: a problem of professional character in nineteenth-century America," *Clio Medica*, 1991, *22*: 65–88, and Warner, "Science in medicine," in Sally Gregory Kohlstedt and Margaret W. Rossiter, eds., *Historical Writing on American Science: Perspectives and Prospects* (Baltimore, 1985), pp. 37–58; see also W. F. Bynum, *Science and the Practice of Medicine in the Nineteenth Century* (Cambridge, Eng., 1994).

8   Kenneth Burke, "Terministic screens," in *Language as Symbolic Action: Essays on Life, Literature, and Method* (Berkeley, 1966), p. 46. Case narratives are an ancient form of medical writing, but one that each generation re-invents from the stuff of practice. Many of the physicians I have been studying comment on how case stories satisfyingly combined personal and professional motives. As one "country" doctor observed, he wrote "partly for future reference, and partly . . . [for] younger brethren." See "W.," "Contributions of a country doctor," *Va. Med. J.*, Jan. 1857, *8*: 5. For other commentary on the importance of case histories to practice, see O[tis] F. Manson, "On malarial pneumonia," *Medical Journal of North Carolina*, Aug. 1860, *3*: 467; Daniel Ryan Sartor, "On medical ethics," M.D. Thesis, Transylvania University, 1843, Transylvania University Medical Library, Lexington, Ky., p. 8; Lewis D. Ford, "Remarks on the pathology and treatment of intermittent and remittent fevers, with cases," *Southern Med. & Surg. J.*, 1836, *1*: 340; J. Graham Tull, "An address delivered before the [North Carolina] State Medical Society," *Medical Journal of North Carolina*, Aug. 1858, *1*: 29–30.

9   Lunsford Yandell (1805–1878) began his career in his early 20s as an apprentice to his doctor father in Murfreesboro. Yandell went on to earn his M.D. in 1825 from Transylvania University in Lexington, Ky., then the largest medical school in the South. After graduation, Yandell teamed with his father in caring for the Murfreesboro community before moving to Nashville, Tenn., in 1830. A year later, Yandell joined the faculty at Transylvania as a professor of chemistry. (Biographical information is drawn from materials in the Yandell family papers, The Filson Club, Louisville, Ky.) See also John D. Wright, *Transylvania: Tutor to the West* (Lexington, Ky., 1980), and Hampden Lawson, "The early medical schools of Kentucky," *Bull. Hist. Med.*, 1950, *24*: 168–175.

Dysentery, regarded as a dangerously acute disease with epidemic potential, was a frequent diagnosis among southern practitioners. In writing about dysentery, Yandell was assured of an interested audience even if he had not assigned himself the task of calling for reform. See David K. Patterson, "Disease environments of the antebellum South," in Numbers and Savitt, *Science and Medicine,* pp. 152–165; James O. Breeden, "Disease as a factor in southern distinctiveness," in Savitt and Young, *Disease and Distinctiveness,* pp. 1–28.

10 Yandell, "Cases of dysentery," p. 240.

11 *Ibid.,* pp. 240–241.

12 *Ibid.,* p. 241. On Thomas Sydenham (1624–1689) and theories of infection, see Charles-Edward Amory Winslow, *The Conquest of Epidemic Disease: A Chapter in the History of Ideas* (1943; rpt. edn., Madison, Wis., 1971); Harry F. Dowling, *Fighting Infection* (Cambridge, 1977).

13 Yandell, "Cases of dysentery," p. 242.

14 The Clinical-Pathological Conference as a pedagogical device in medical school arose along with modern medical education after 1920. Although the CPC no longer holds the central place in clinical teaching that it once did, it remains, in the view of two influential observers of medical education, "an invaluable way to teach the process of internal soliloquizing, which enables the diagnostician to examine his or her own reasoning."

Edmund D. Pellegrino and David C. Thomasma, *A Philosophical Basis Of Medical Practice: Toward a Philosophy and Ethic of the Healing Professions* (New York, 1981), p. 129. See also Ran Oren and Yaacov Matzner, "Clinical problem-solving," *New Eng. J. Med.,* 1994, *330* (No. 1): 48–50. The prestigious *New England Journal* regularly publishes "Weekly Clinico-pathological Exercises" as a feature called "Case Records of the Massachusetts General Hospital." The quotations here are from this feature: "Case 5-1994," *New Eng. J. Med.,* 1994, *330* (No. 5): 347; "Case 44-1993," *ibid.,* 1993, *329* (No. 19): 1411.

15 A classic text on diagnosis reasonably accessible to laypersons is Alvan R. Feinstein, *Clinical Judgment* (Baltimore, 1967). On the recent history of diagnostic specialism, see Stanley Joel Reiser, *Medicine and the Reign of Technology* (Cambridge, Eng., 1978).

16 Robert S. Bailey, "Three cases of puerperal convulsions," *Proceedings of the South Carolina Medical Association* (Charleston, 1852), pp. 144–147. On the different, often subordinate, place of diagnosis in 19th-century practice, see Warner, *Therapeutic Perspective,* esp. ch. 4; Rosenberg, *Care of Strangers,* esp. ch. 3; Paul B. Beeson and Russell C. Maulitz, "The inner history of internal medicine," in Maulitz and Diana E. Long, eds., *Grand Rounds: One Hundred Years of Internal Medicine* (Philadelphia, 1988), p. 33. See also Charles E. Rosenbeg, "Framing disease: illness, society, and history," in Rosenberg, *Explaining Epidemics and Other Studies in the History of Medicine* (Cambridge, Eng., 1992), pp. 293–304.

17 Yandell's essay typically relates many of his cases as diary accounts. For general discussion of case records as historical sources, see Guenter B. Risse and John Harley Warner, "Reconstructing clinical activities: patient records in medical history," *Social History of Medicine,* 1992, *5*: 183–205; John D. Stoeckle and J. Andrew Billings, "A history of history-taking: the medical interview," *Journal of General Internal Medicine,* 1987, *2*: 119–127; Stanley Joel Reiser, "Creating form out of mass: the development of the medical record," in Everett Mendelsohn, ed., *Transformation and Tradition in the Sciences: Essays in Honor of I. Bernard Cohen* (Cambridge, Eng., 1986), pp. 303–316; Ellen Dwyer, "Stories of epilepsy, 1880–1930," in Charles E. Rosenberg and Janet S. Golden, eds., *Framing Disease: Studies in Cultural History* (New Brunswick, N.J., 1992), pp. 248–272.

18 Of course, modern physicians also refer to "interesting" cases and diseases but in a way different from their 19th-century counterparts. Modern doctors tend to use the term to describe puzzles of diagnosis and treatment that they find intellectually stimulating—as challenges to scientific mastery. Doctors in the earlier century tended to apply the word more broadly and more in terms of an entire range of problems in hands-on care. Helpful in thinking about the ethical issues in seeing sick people as "interesting" are Terry Mizrahi, *Getting Rid of Patients: Contradictions in the Socialization of Physicians* (New Brunswick, N.J., 1986), and Pellegrino and Thomasma, *Philosophical Basis of Medical Practice,* esp. ch. 9.

The rise of medical ethics in recent years, as a particular field of inquiry and a concern in medical writing and teaching, has sparked interest in the place of stories in modern medicine. A sampling of the literature I have found helpful includes Arthur Kleinman, *The Illness Narratives: Suffering, Healing, and the Human Condition* (New York, 1988); Kathryn Montgomery Hunter, *Doctors' Stories: The Narrative Structure of Medical Knowledge* (Princeton, N.J., 1991); Howard Brody, *Stories of Sickness* (New Haven, Conn., 1987); Kathryn Montgomery Hunter, "'There was this one guy . . .': the uses of anecdotes in medicine," *Perspect. Biol. & Med.,* 1986, *29*: 619–630; Rita Charon, "Doctor-patient/reader-writer: learning to find the text," *Soundings,* 1989, *72*: 137–152; Larry R. Churchill and Sandra Churchill, "Storytelling in medical arenas: the art of self-determination," *Lit. & Med.,* 1982, *1*: 73–79.

19 See Rita Charon, "To build a case: medical histories as traditions in conflict," *Lit. & Med.,* 1992, *11*: 115–132. In the modern debate over the meaning of medical language, some observers concede that physicians' attempts to be technically precise compel them to reduce people's illnesses to abstract disease

types, making medicine cold and impersonal. But they go on to argue that this is a necessary stage in efficient diagnosis, pointing out that a patient can be comforted by knowing the name of his or her affliction, and by the fact that others share the same problem. See for example, Stephen J. Kunitz, "Classifications in medicine," in Maulitz and Long, *Grand Rounds*, pp. 279–296. Also taking a positive, though not uncritical, view of the effect of modern medical language are Howard Brody, *Stories of Sickness* and Thomas W. Laqueur, "Bodies, details, and the humanitarian narrative," in Lynn Hunt, ed., *The New Cultural History* (Berkeley, 1989), pp. 176–204. Other observers take the harsher view that the mystifying, transforming impact of technical language—indeed, its performative authority—often is used by doctors to override patients' values and perceptions as if they do not matter. See William J. Donnelly, "Righting the medical record: transforming chronicle into story," *Soundings*, 1989, *72*: 127–136; Lawrence B. McCullough, "Particularism in medicine," *Criticism*, 1990, *32*: 361–370; Nancy M. P. King and Ann Folwell Stanford, "Patient stories, doctor stories, and true stories: a cautionary reading," *Lit. & Med.*, 1992, *11*: 185–199.

20 W. L. Wood, "Case of labor," *Va. Med. J.*, 1857, *9*: 377.

21 Henry A. Ramsay, "Inflammation of the bladder, gangrene, death, autopsy," *Charleston Med. J. & Rev.*, 1850, *5*: 47.

22 F. A. Bates, "On the prevailing diseases of Dallas County [Alabama]," *Southern Med. Rep.*, 1849, *1*: 377; J. M. Hamilton, "How long shall we wait?" *Nashville J. Med. & Surg.*, 1856, *10*: 33–34. See also A. A. J. Riddell, "Epidemic bloody flux," *New Orleans Med. & Surg. J.*, 1850, *7*: 99.

(In Bates, "tr. opii" is a tincture of opium, probably laudanum, often given as a sedative or analgesic; a sinapism is a mustard plaster. Craniotomy, in Hamilton's account, is the procedure of delivering an impacted or otherwise "undeliverable" fetus—one considered a threat to the woman's life—by crushing its skull, thus facilitating its removal.)

23 The case of fatal exhalations is in C. E. Lavender, "Anthropo-toxicologia: cases, with remarks," *New Orleans Med. & Surg. J.*, 1848, *5*: 33–37; the black woman turning white is in John Knox, medical casebooks, vol. 1, Jan. 21, 1848, Special Collections, University of Kentucky, Lexington, Ky.; the posthumous child is in Benjamin F. Fessenden, "A case of puerperal apoplectic convulsions, with spontaneous expulsion of the foetus after death," *Medical Journal of North Carolina*, 1858, *1*: 16–17; the found bullet is in "Gun-shot wound of the heart," *Southern Med. & Surg. J.*, 1867, *21*: 520–521; the man and his pile

remedy is in T. M. Harris, "A contribution to the curiosities of medical experience," *Western J. Med. & Surg.*, 3rd ser., 1848, *2*: 281–283.

24 Samuel Leland, Diary, March 1, 1852, South Caroliniana Library, University of South Carolina, Columbia, S.C. (hereafter SCL); A. A. P[atteson], "Patterson's [*sic*] cases," *Transylvania Medical Journal*, 1850, *1*: 329–330. Charles A. Hentz regularly expressed his approval of patients who bravely withstood painful therapies; see Hentz, Medical Diary, November 3, 1858, Nov. 18, 1858, Feb. 2, 1859, Feb. 14, 1859, Hentz papers, Southern Historical Collection, University of North Carolina, Chapel Hill, N.C. The physician recommending training midwives is R. H. Day, "Obstetrical cases," *New Orleans Med. & Surg. J.*, 1847, *4*: 223–227.

25 Treatment of slaves is addressed typically in "W.," "Contributions." For examples of narratives charging slaves with feigning illness and other deceptions, see William A. Brown, "A curious case of malingering," *Medical Journal of North Carolina*, 1860, *3*: 375–378; T. S. Hopkins, "A remarkable case of feigned disease," *Charleston Med. J. & Rev.*, 1853, *8*: 173–176; W. L. Sutton, "A case of doubtful paternity," *Southern Med. & Surg. J.*, 1852, *8*: 760–763. For a doctor using a poor, white patient as an object of moral judgment, see Louis A. Dugas, "A clinical lecture upon some of the effects of intemperance: delivered at the Augusta [Georgia] City Hospital," *Southern Med. & Surg. J.*, 1858, *15*: 1–11.

26 Most white physicians probably believed in at least some physiological differences between the races; certainly most doctors believed slaves were especially vulnerable to some forms of disease. The point here is that these presumed differences do not seem to have led physicians to give slaves significantly inferior care in general. In fact, it appears that physicians examined blacks and whites in the same way, applying most of what they learned about the body and illness across racial lines. Samuel Cartwright, a physician practicing in Mississippi and Louisiana, has gained historical notice for his invention of slave "diseases" such as draeptomania, the "sickness" of runaway slaves. He once argued that he had discovered racial differences in skin tissue. But Cartwright, though a frequent contributor to medical journals, was hardly mainstream, and his ideas on race seem to have been viewed by many of his peers as questionable. See Samuel Cartwright, "The diseases and physical peculiarities of the Negro race," *Southern Med. Rep.*, 1850, *2*: 421–429, and the skeptical reply by the journal's editor Erasmus Darwin Fenner, "State Medical Society of Louisiana," *ibid.*, 294–298. For a suggestive discussion of Cartwright and no-

tions of African-American susceptibility to certain diseases, see Kenneth F. Kiple and Virginia Himmelsteib King, *Another Dimension to the Black Diaspora: Diet, Disease, and Racism* (Cambridge, Eng., 1981), ch. 12.

Josiah Nott's theory of the separate creation of the races is not apparent at all in the hundreds of case stories I have seen. Most ordinary practitioners were wary of theorizing in general, and, in any event, were wary of challenging religious orthodoxy which held to the single creation of mankind. On Nott and the debate, see the lucid account of Reginald Horsman, *Josiah Nott of Mobile: Southerner, Physician, and Racial Theorist* (Baton Rouge, La., 1987), esp. ch. 4. For typical, non-theoretical descriptions of African-American bodies, see James. E. Smith, "Cases illustrating the practice of medicine in the counties of Rusk and Panola [Texas]," *New Orleans Med. & Surg. J.,* 1857, *14*: 166–170, and C. H. Jordan, "Thoughts on Cachexia Africana or Negro consumption," *Transylvania Journal of Medicine and Associate Sciences,* 1832, *5*: 18–30.

27 J. R. Freese, "Healing art vs. the knife," *New Orleans Med. & Surg. J.,* 1836, *13*: 13.

28 On the conceptual satisfactions of narrative, see Hayden White, "The value of narrativity in the representation of reality," in White, *The Content of the Form* (Baltimore, 1987); Louis O. Mink, "Narrative form as a cognitive instrument," in Robert H. Canary and Henry Kozicki, eds., *The Writing of History: Literary Form and Historical Understanding* (Madison, Wis., 1978), esp. 144–147; David Carr, *Time, Narrative, and History* (Bloomington, Ind., 1986), esp. 68–69. On the persuasive power of stories' "lifelikeness," see Jerome Bruner, *Actual Minds, Possible Worlds* (Cambridge, Mass., 1986), pp. 10–12.

29 On the "pre-modern" body as found in the work of one practitioner, see Barbara Duden, *The Woman Beneath the Skin: A Doctor's Patients in Eighteenth-Century Germany* (Cambridge, Mass., 1991). On the reduction of a sick person's biography to his or her illness, see Mary Rawlinson, "Medicine's discourse and the practice of medicine," in Victor Kestenbaum, ed., *The Humanity of the Ill: Phenomenological Perspectives* (Knoxville, Tenn., 1982), pp. 69–85.

30 Yandell, "Cases of dysentery," pp. 242, 243, 245. Among the general characteristics of patients, age was rarely noted by 19th-century southern physicians—and even then only in terms of the extremes of infancy and old age. Mostly, the broad categories of "adult" and "child" sufficed, doubtless in part because many people did not know their exact chronological age. In contrast, a person's sex was invariably noted. As for race, "mulatto" is a frequent term used

by doctors, and the terms "yellow negro" and "black negro," denoting perceived shades of skin color, also enter into some descriptions. Most doctors doubtless made these discriminations for descriptive reasons alone, though some physicians felt that a person of mixed race had special health problems. For a sharp critique of what modern medicine chooses to emphasize in a patient's medical history, see William Frank Monroe, Warren Lee Holleman, and Marsha Cline Holleman, "'Is there a person in this case?'" *Lit. & Med.,* 1992, *11*: 45–63.

31 Courtney J. Clark, medical notebook, ca. 1844, William Perkins Library, Duke University, Durham, N.C., p. 7. Although certain folk therapies also focused on blood, the concern for its supply became an emblem of orthodox medicine, as did bloodletting. But as with some other allopathic tenets, this one is not considered wholly groundless even today. Malaria, for example, one of the South's most prevalent diseases, does in fact put stress on the body's production of blood. On changes in blood-letting, see Warner, *Therapeutic Perspective,* esp. ch. 7.

Most of the pulse-words used by southern doctors were Galenic in origin. Although such terms reached toward precision, they condensed into something more like poetry. For one practitioner's brave but elusive attempt to define pulse-words, see John Bernard Vandergriff, "Southern practice of medicine, surgery, obstetric, therapeutics, toxicology, and useful notes on various methods of treatment" (New Orleans, n.d. [*ca.* 1870s]), ms. in Matas Library, Tulane University School of Medicine, New Orleans, La., pp. 331–334.

On the healthy body as "absent" compared to the sick body, see the suggestive discussion in Drew Leder, *The Absent Body* (Chicago, 1990), esp. pp. 1–2, 39–45, 81.

32 Charles A. Hentz, medical diary, Nov. 27, 1858; Lunsford Yandell, "Cases of dysentery," pp. 246–247.

33 Auscultation is the act of listening to internal body sounds, either with a stethoscope or unmediated; palpation involves examining by touch, especially the abdomen; percussion is the act of tapping various parts of the body, listening and feeling for particular reactions.

By the 1840s, medical students in most of the larger schools were required to write a doctoral thesis, and the students who chose to write on physical diagnosis provide succinct accounts of how it was taught as a subject. See, for example, Benjamin Robinson Mitchell, "On diagnosis," M.D. Thesis, Transylvania University, 1844, Medical Library, Transylvania University, Lexington, Ky.; Pinckney A.

Williams, "Physical diagnosis," M.D. Thesis, University of Nashville, 1854, Special Collections, Vanderbilt Medical Center Library, Nashville, Tenn. These two archives, along with that of the Waring Historical Library, Medical University of South Carolina, Charleston, S.C., hold the largest collections of 19th-century medical theses, an invaluable source for suggesting how clinical practice was conceptualized and transmitted to students.

34   W. R. Sharpe, "Report on the diseases of Davie County [N.C.]," *Medical Journal of North Carolina,* Feb. 1860, *3*: 279–280. On the ideas linking medicines to disease, see Warner, *Therapeutic Perspective,* esp. ch. 3; Rosenberg, "Framing disease."

35   Day, "Obstetrical cases," p. 225; Geo[rge] R. Grant, "A case in which the placenta was retained thirteen days after delivery at full term," *Southern Med. & Surg. J.,* 1836, *1*: 196; Charles Chester, "Four cases of cerebro-spinal meningitis," *New Orleans Med. & Surg. J.,* 1847, *4*: 315; S. C. Farrar, "General report on the topography, meteorology and diseases of Jackson, the capital of Mississippi," *Southern Med. Rep.,* 1849, *1*: 358. For an example of what medical students typically learned about the subtlety of medicines and the small but crucial variations achieved by mixing or timing them differently, see the 21 categories of therapeutic "action," both "direct" and "indirect," in Philip Southall Blanton, "The modus operandi of therapeutic agents," M.D. Thesis, Medical Department of Randolph-Macon College, [1848], in the Armistead-Blanton-Wallace family papers, Virginia Historical Society, Richmond, Va.

36   Charon, "To build a case," p. 115.

37   Oliver Sacks, "Clinical tales," *Lit. & Med.,* 1986, *5*: 21. For the phenomenological aspect of 19th-century rural practice, see Jacalyn Duffin, "A rural practice in nineteenth-century Ontario: The continuing medical education of James Miles Langstaff," *Canadian Bulletin of Medical History,* 1988, *5*: 3–28; Steven M. Stowe, "Obstetrics and the work of doctoring in the mid-nineteenth-century American South," *Bull. Hist. Med.,* 1990, *64*: 540–566. Helpful in thinking about modern rural practice is Ruth Purtilo and James Sorrell, "The ethical dilemmas of a rural physician," *Hastings Cent. Rep.* August, 1986, *16*: 24–28.

38   The examples of rural imagery here are from A. A. P[atteson], "Patterson's cases," p. 329; Jordan, "Thoughts on Cachexia Africana," p. 25; John Terrill Lewis, "Remarkable case of hemorrhage," *Transylvania J. Med.,* 1830, *3*: 261; T. B. Camden, "'Missed labor,'" *New Orleans Journal of Medicine,* 1868, *21*: 152; Freese, "Healing art," p. 13. The reference to tissue looking like "bee bread" or honeycomb points to the fascinating way in which medical langauge derived both from a doctor's own observations of the natural world as well as from traditional usage among physicians. Here, Dr. Jordan's mention of "bee bread" seems to draw an analogy to honeycomb and bees in the southern countryside, as it may well have. Yet "bee bread," as a term referring to a certain kind of skin pustule, had origins at least as remote as the 13th century. Thus, some apparently immediate, "southern" terms may in fact have been traditional "technical" terms, passed down through the centuries by medical mentors. Or, in any given case, such language may well have been truly vernacular, with southern doctors in effect reinventing technical terms by returning to nature for their metaphors. Such cross-hatching of tradition and practice was basic to the intellectual texture of a doctor's work. I am indebted to Faye Getz for pointing out to me the medieval usage. For a helpful if sometimes quirky introduction to medical language, see John H. Dirckx, *The Language of Medicine: Its Evolution, Structure, and Dynamics,* 2nd ed. (New York, 1983).

39   It appears that when doctors used an initial for the patient's name, it was more often than not the correct initial, a usage that seems to have derived from note-taking shorthand as much as from a concern for confidentiality—a concern which was not highly developed before the 20th century. It is not clear whether physicians asked their patients for permission to tell about them; I suspect most did not. For instances of physicians naming patients outright, see Dr. M. S. Watkins, "Case of Mrs. Watkins' cure of recto-vaginal lacerations, by Dr. J. Marion Sims, of New-York: reported by M. S. Watkins, M.D., of Jackson, Miss., husband of the patient," *New Orleans Med. & Surg. J.,* 1855, *11*: 645–647; W. J. W. Kerr, "Letter from Mississippi," *Nashville J. Med. & Surg.,* 1869, n.s. *5*: 70.

40   Yandell, "Cases of dysentery," pp. 244–245, and *passim.*

41   R. L. Scruggs, "Clinical notes from private practice," *New Orleans Med. & Surg. J.,* 1852, *9*: 29; E. H. Macon, "On the diuretic virtues of azalea, or honeysuckle," *Southern Med. & Surg. J.,* 1837, *2*: 23; C[ourtney] J. Clark, "Remarks on the existence of typhoid fever in Alabama (Communicated to Dr. J. C. Harris, of Wetumpka)," *New Orleans Med. & Surg. J.,* 1850, *6*: 464; P[atteson], "Patterson's cases," pp. 332–333; J. S. Dyer, "Twins, with enormous quantity of liquor amnii," *Nashville J. Med. & Surg.,* 1852, *2*: 47.

42   W. A. Shands to William Anderson, July 22, 1880, William Anderson papers, SCL; William P. Hort, "An enquiry whether in the southern states there is a specific disease that can properly be called congestive fever; with cases and remarks," *New Orleans Med. &*

Surg. J., 1847, *4*: 61; Day "Obstetrical cases," p. 225. Two brief introductions to modern medical ethics which I have found useful in thinking about how physicians confront or avoid conflict with patients and their families are Alastair V. Campbell, *Moral Dilemmas in Medicine: A Coursebook in Ethics for Doctors and Nurses*, 2nd ed. (New York, 1975), esp. ch. 5; Richard Warner, *Morality in Medicine: An Introduction to Medical Ethics* (Sherman Oaks, Cal., 1980), esp. pp. 1–15.

43 The physician attending the 6-year-old girl is Scruggs, "Clinical notes," p. 31; the Virginian is L. Faulkner, "Cases from my note book and memory," *Virginia Med. J.*, 1856, *6*: 465.

44 Wooten's case is in H. V. Wooten, "On the topography and diseases of Lowndesboro' and its vicinity, during the year 1850," *Southern Med. Rep.*, 1850, *2*: 341; J. C. C. Blackburn, "A case of ruptured uterus," *Southern Med. & Surg. J.*, 1849, n.s. *5*: 73; James S. Lawton, "Case of rupture of the womb," *Charleston Med. J. & Rev.*, 1854, *9*: 183.

45 R. S. Bailey, "Essay on medical faith," *Transactions of the South Carolina Medical Association* (Charleston, 1856), p. 20. Bailey's article is an instance in which a white doctor specifically describes certain folk medicines as African-American in origin. For other cases in which physicians admit adopting, or at least tolerating, folk remedies, see Jno. W. Richardson, "Case of obstruction of the intestines," *Western J. Med. & Surg.*, 1848, 3rd ser., *2*: 209–210; Scruggs, "Clinical notes," p. 30; Archer, "Memoranda," p. 10. As might be expected, given the anonymous institutional setting, poor white hospital patients also were described without reference to family; see A. H. Cenas, "Obstetrical memoranda," *New Orleans Med. & Surg. J.*, 1847, *4*: 312–314. White physicians practicing in a plantation slave community known to them, however, would have to actively omit African-American friends and kin from narratives. On African-American traditional medicine, see Loudell F. Snow, *Walkin' Over Medicine* (Boulder, Colo., 1993).

46 Yandell, "Cases of dysentery," pp. 245–246. See also Savitt, *Medicine and Slavery*, pp. 10–17, 33–37.

47 Clark, "Remarks on the existence," pp. 465–466.

48 Clearly, most southern physicians were not much interested in raising questions about the nature of slave patients' consent to treatment. Some white doctors comment on reluctant slave patients, but in much the same way they take note of non-compliant free patients. Interestingly, doctors sometimes do mention slaves asserting one kind of control after medicine has failed: slaves' frequent refusal to hand over the bodies of deceased family members for autopsy (a refusal acceded to by some owners). There is a vast literature on ethical issues of medical consent. Helpful are Tom L. Beauchamp and James F. Childress, *Principles of Biomedical Ethics* (New York, 1979); George Annas et al., *Informed Consent to Human Experimentation: The Subject's Dilemma* (Cambridge, Mass., 1977). On medical experimentation and slaves, see Todd L. Savitt, "The use of blacks for medical experimentation and demonstration in the Old South," *J. Southern Hist.*, 1982, *45*: 331–348. For examples of physicians being refused bodies for autopsies, see Richard Jarrot, "Amputation for gangrene of the foot, successfully performed on a Negro, at the advanced age of one hundred and two years," *Charleston Med. J. & Rev.*, 1849, *4*: 301–305; William T. Briggs, "A case of traumatic tetanus—treated by inhalation of chloroform—result unsuccessful," *Nashville J. Med. & Surg.*, 1851, *1*: 30–38.

49 Yandell's letters to his wife's parents, it will be recalled, reveal his family ties to the three cases. Because he wrote daily letters without receiving a response, the letters have a diary-like quality, but they also take on the character of clinical notes.

50 Yandell, "Cases of dysentery," pp. 246–247. Although Yandell clearly believed that there was a well-established therapy for dysentery, it is notable that 19th-century medical textbook authors were relatively vague about the timing and details of various therapies; they often reported treatments instead of recommending them. See, for example, the much-used textbook by John Eberle, *A Treatise on the Practice of Medicine*, 2 vols. (Philadelphia, 1831).

51 Yandell, "Cases of dysentery," pp. 247–250.

52 Lunsford Yandell to Sarah Wendel, June 12, 1836; June 13, 1836. Yandell's letters are in the Yandell family papers, Filson Club. All citations are from this collection. At least some of the extant letters are the ones Yandell actually mailed to his wife's parents, Sarah and David Wendel, but others may be first drafts or copies that Yandell may have saved as part of his bedside notes.

53 Lunsford Yandell to Sarah Wendel, June 13, 1836; June 15, 1836.

54 Lunsford Yandell to David Wendel, June 21, 1836.

55 Lunsford Yandell to Sarah Wendel, June 15, 1836; June 18, 1836. Lunsford Yandell to David Wendel, June 22, 1836.

56 Lunsford Yandell to David Wendel, June 21, 1836; June 23, 1836.

57 The reference to Dr. Dudley is in Yandell's June 19 letter to Sarah Wendel.

58 Lunsford Yandell to David Wendel, June 21, 1836. Yandell records the mush and milk explanation in "Cases of dysentery," p. 248, and in his letter to David Wendel, June 23, 1836.

59 Lunsford Yandell to David Wendel, June 24, 1836.

60 Lunsford Yandell to David Wendel, June 23, 1836.

61 Yandell, "Cases of dysentery," p. 249; Lunsford Yandell to Sarah Wendel, June 19, 1836.

62 Yandell, "Cases of dysentery," p. 250.

63 Yandell recorded Willie's words on the reverse, blank side of a printed sheet announcing the publication of the *Transylvania Journal of Medicine,* which is among the series of letters to his in-laws in the Yandell papers, Filson Club.

64 Yandell, "Cases of dysentery," pp. 247, 250. The source of Willie's words is in n. 63 above.

EDUCATION

America's first medical school opened in Philadelphia in 1765. Prior to that time, colonials who wanted to become physicians either studied in Europe or, more commonly, served a brief apprenticeship with a local practitioner. Even after the advent of formal medical education the apprenticeship remained the primary means of obtaining clinical experience, supplementing the lectures available in schools.

By 1800 Columbia, Harvard, and Dartmouth also had established medical departments, and within a few decades medical schools were flooding the country. Most of the new schools were proprietary in nature, run by groups of doctors who cared as much about profits as pedagogy. The typical mid-century institution comprised five or six nonsalaried professors, usually local physicians, whose income depended on how many lecture tickets they could sell to students. Each term lasted only three or four months and consisted of perhaps half a dozen courses, ranging from anatomy to the theory and practice of medicine. To graduate with an M.D. degree, students attended the same lectures for two terms and—at least in theory—completed an apprenticeship.

Efforts to improve this dismal situation, inspired in part by the German experience, culminated in Abraham Flexner's 1910 expose, which hastened the closing of the most wretched schools. The institutions that survived into the 1920s offered a four-year, graded curriculum, usually divided between the basic sciences (e.g., anatomy, physiology, and biochemistry) and practical experience in the clinic or hospital. Unlike in times past, when reading and writing were virtually the only skills required for admission, students now had to attend college for at least two years before admission.

As medicine grew more and more complex in the 20th century, medical educators found it increasingly difficult to squeeze into four years all the clinical training a physician or surgeon would need to practice medicine effectively. Thus in the 1910s state licensing boards began requiring an additional year of internship in a hospital. But even this proved insufficient training for some specialties, and by the 1920s and 1930s a number of physicians were continuing to train for several years after the internship in postgraduate residency programs, specializing in fields such as ophthalmology, obstetrics, and orthopedic surgery. After World War II it became customary for all but general practitioners to study medicine for about eight years before engaging in independent practice.

Until shortly before the Civil War, American medical schools admitted only white males, but this barrier fell under the influence of the antebellum abolitionist and feminist movements. Elizabeth Blackwell in 1849 became the first American woman to earn a medical degree. Women made rapid strides in separate as well as coeducational medical schools during the late 19th century—only to fall back again in the 20th. Between 1910 and 1960, few medical schools had enrollments of more than 5 percent women, and they admitted an even smaller percentage of African-American and other minority students. But once again social ferment forced these schools to reevaluate their admission policies, and by the mid-1990s almost half of entering students were women and about one-eighth represented various minority groups.

11

The German Model of Training Physicians in the United States, 1870–1914: How Closely Was It Followed?

THOMAS NEVILLE BONNER

The history of American medical education is a field in ferment. No longer is it enough to contrast the backward conditions of American medical schools before 1870 with the remarkable changes brought by study in Germany, an awakened profession, and the famous Bulletin no. 4 of the Carnegie Foundation. New research, new questions, and new perspectives have broadened and deepened our understanding of the timing and importance of the forces that transformed the American system of training physicians. Ronald L. Numbers and others have stressed, for example, the crucial importance for medical education of the shift from passive to active learning; Kenneth M. Ludmerer has focused attention on the early and continuing role of an academic elite, assisted by universities and foundations, in forcing this shift to "learning by doing"; and William G. Rothstein has raised serious questions about the unskeptical approval by American educators of intensive laboratory and inductive teaching in training practitioners for the larger community.[1]

In this emerging view, it was an elite group of academic scientists, university leaders, and foundation executives who, for good or ill, gave both form and content to the modern system of professional medical education. Their reforms were made possible by the changing nature of American society in the last third of the 19th century: the acceleration of industrial reorganization, the growing demand for technical and scientific education, the accumulation of great wealth, which was used by some to promote university and medical education, and the hunger for native institutions to end the long dependence on Europe for scientific training. The views this elite held on medical education, and the scientific ideology that stirred them, according to all accounts, owed much to the example of the German university. What they saw in German medicine was a vision of a laboratory-based, all-conquering science that could sweep away all doubts before it. Many were zealous advocates of everything German in medicine and became almost apostles of the German scientific spirit. "A good many people," the admiring Henry Sewall wrote to John J. Abel in 1886, "are mere cattle in their power of appreciating European observation, but I have felt that it is only on the continent that one comes into the sanctuary of science."[2] At about the same time, Franklin P. Mall, also profoundly affected by what he saw in German laboratories, was writing his sister in almost religious terms: "It is a sin to seek pleasure only. . . . My aim [now] is to make scientific medicine a life work."[3] Fresh from the laboratories and workplaces of Germany, returning Americans sought to make a revolution in medical education. Not medical practice, but medical science as a career became their goal in life. To achieve that goal, they first had to create medical schools like those they had seen in Germany, where teacher and scientist were one. Tirelessly, they proselytized the strengths of the German system—the uniformly high standards, the excitement of doing original work, the freedom enjoyed by the scientist, the unity of research and teaching, the advanced de-

THOMAS NEVILLE BONNER is Distinguished Professor of History and Higher Education, Wayne State University, Detroit, Michigan.

From the *Bulletin of the History of Medicine*, 1990, *64*: 18–34. Copyright © 1990 by the Johns Hopkins University Press. Reprinted by permission of the Johns Hopkins University Press.

gree of specialization, and the superbly equipped laboratories that were led by men on the frontiers of medical science.[4]

University presidents, leaders of foundations, and spokesmen for organized medicine all came to share the enthusiasm for a new experimental science that would make of medicine a respected intellectual pursuit. Some university leaders had themselves studied or travelled in Germany; others were persuaded of German superiority in medicine by their faculty members who had studied abroad. President Charles Eliot of Harvard, after a period of hesitation, praised German methods of investigation and urged that scientific medicine be taught in the American university by means of laboratory work which would be "required of every student."[5] Henry Pritchett of the Carnegie Foundation, a zealous convert to the new academic medicine, wrote that American medical schools must "ground themselves in the fundamentals upon which medical science rests.[6] Arthur Dean Bevan, a leader of the American Medical Association's effort to raise standards of medical education in the early 20th century, told Abraham Flexner in 1912, "I believe that the best model we can follow in this country . . . is the medical departments of the German speaking universities."[7] A remarkable consensus thus developed among professional and public leaders on the value of experimental science, which together with the vast social changes sweeping over the country, would make possible extraordinary changes in American medical training. William Henry Welch voiced a widely acknowledged truth in 1907 when he said that the German medical sciences had "been the pervasive and dominant foreign influences of the last four decades."[8]

What was medical education actually like for the average German student who completed minimal requirements and then went into practice? By the late 19th century, the sole route to medical practice was through university education. Entrance to the university came only after graduation from a *Gymnasium*, a classically oriented preparatory school for boys (and later girls) aged 9 to 18, or, less frequently, from an *Oberrealschule* that taught science and modern languages. After receiving the *Abitur,* signifying successful graduation (only a small minority of all students did), the student went on to one of the 28 German-speaking universities of the Second Reich, Austria, or Switzerland. At the university, the future physician would spend the first two years in lectures and laboratory work in basic sciences such as chemistry, physics, biology, anatomy, and physiology. The last two years were devoted to medicine, surgery, and obstetrics. By the latter part of the century, it was normally compulsory to take courses in dissection, to complete laboratories in chemistry and physiology, and to attend clinical lectures and demonstrations in medicine, surgery, and obstetrics. Lectures remained a principal and dominant medium of teaching medicine. Examinations were held after the first two years and again on completing formal study. Not every student took the doctoral degree in Germany, but all practitioners were required to pass the demanding *Staatsexamen*, which might necessitate a fifth or sixth year of preparation. Those taking the M.D. were required also to write a thesis representing original research. Practical bedside and clinical experience was difficult to come by and most students were forced to find it during vacations or in the year following their university studies. Not until 1901 was a fifth-year internship mandated for all students throughout Germany.[9]

How closely in fact was this German model of training ordinary physicians actually followed in the United States? As in other fields of study, American medical educators borrowed selectively from the German experience. "The German model," writes Harold Perkin, "was most self-consciously admired and followed in the United States, with the least Germanic of results. Nothing could be further from the German state-controlled and -financed university than the buccaneering, free-market system of American higher education in the 19th century.[10] Certainly, the traditional American liberal arts college and the independent American medical school free of any state or university control were improbable soil in which to plant the seeds of the German university system. A number of features of the German system understandably never took hold in the American environment, despite efforts to transplant them to the United States. The use of private teachers or *Privatdozenten,* who had completed their advanced training and were licensed to teach students informally outside the official curriculum, for example, never found favor in the more egalitarian academic life of the United States. The tradition of students making payments to professors as well as to private teachers in Germany, which was becoming an ever larger source of professorial income by the 1880s, had no counterpart in the British-American practice of paying for instruction with a single tuition payment. This payment of *Kollegiengeld* by students made German full professors far more wary of collegial competition and more resistant to the introduction of new subdisciplines than were their colleagues in America, who had relatively equal status in their departments which, in Jonathan Harwood's phrase, could "grow incrementally, incorporating new research areas as necessary."[11] Likewise, the large degree of freedom given the German student in choosing electives outside the few required courses and in moving from university to university was never considered as an option by educators used to the paternal atmosphere of an American college. Similarly, the *Lehrfreiheit* enjoyed by German professors, whereby they exercised exclusive control over which courses were taught, never extended to America. Although some American professors trained in Germany lamented the academic lockstep of prescribed courses, rigid curricula, and frequent examinations in their British-influenced system, their protests made little headway in American universities. "How different," wrote Franklin P. Mall in 1905, "is the study of medicine in Europe from that of America! There freedom reigns and students wander from place to place being controlled only by a fairly rational system of examinations in case they wish to graduate. Weak students fall out, for there is no cram system to drive them onward; able students select great men as teachers and thereby develop themselves and become stronger."[12] Neither did American educators adopt the German practice of deferring major clinical experience to the year of internship after graduation; they adopted instead the British system of clinical clerkships in the last years of medical education, as made popular by William Osler at the Johns Hopkins Hospital. The arrangements for clinical teaching in the United States likewise remained true to older British and American models and did not emulate the German practice of making full-time university appointments in clinical medicine.[13]

The German model then was transformed in important ways as it made its way across the Atlantic. The loose American environment, the prevalence of British undergraduate organization, the tradition of egalitarianism, and the absence of external controls in American higher education discouraged some features of the highly structured German university system and encouraged others. The academic department and the graduate school, which developed as peculiarly American responses to the revolution in higher education, took over many of the functions in America of the German institute and its director.[14] The Ph.D. degree, which emerged, in Nathan Reingold's view, out of a misunderstanding of the German system, differed sharply from the German doctorate, which could be awarded after three years of study beyond the *Gymnasium* and a brief dissertation. The American Ph.D. came to more nearly resemble the lengthy preparation for *Habilitation,* or university teaching.[15] In medicine, American educators copied the form of the undergraduate laboratory and the emphasis throughout the future physician's studies on experimental science but continued also to emphasize the importance of practical training during the years in medical school. The structure of the American medical school from the 1870s onward, while rigid in comparison to the antebellum years, was far more flexible, more open to curricular innovation, and more tolerant of new specialties than the hierarchical German schools of medicine.[16]

The emerging American model of university medical education after 1870 differed from German medical training, too, in the emphasis it placed on laboratory instruction, and later, clinical experience for the average undergraduate medical student. Kenneth Ludmerer interprets this as a democratic extension of a thorough scientific and clinical education to all students rather than to an elite of selected students as in Germany and England. German students, of course, were also exposed to laboratory instruction. As William Coleman and his collaborators have stressed, the institute was justified originally as a means of providing practical experience to medical students in the use of apparatus and the conduct of experiments.[17] But more so than in Germany, the leading American schools, which became models for the others, insisted that all students have extensive personal experience in a number of scientific laboratories and in using a variety of instruments during the first two years of medical school. Some questioned at the time, and many have questioned since, the utility of the heavy American emphasis on scientific study and use of laboratory instruments to the thousands of medical students who sought only to prepare themselves to practice medicine. Critics insisted that educators were ignoring the important differences between scientific study and professional education. "The failure of medical schools to recognize the differences between liberal arts and professional training," writes William Rothstein, had a profound effect on removing medical education "from the hurly-burly of professional life."[18]

Where did the creators of the American model of university education find their principal inspiration? It hardly seems possible that they were greatly influenced by the actual teaching that existed for undergraduates in Germany after 1870. On the contrary, most of those who had studied in Germany were sharply critical of the heavy dependence on didactic lectures and clinical demonstrations in the normal German system of instruction. The famous laboratories in which Welch, Mall, and the others had worked had brought them into personal contact with few ordinary medical students. The problems of aspiring academics in Germany were also of little concern to these American students since their futures were not dependent on their generous hosts. Nor did they see, as Ben-David points out, "how different the departmental structure was from the combination of chair and institute which they admired and thought they were establishing in their own universities."[19] While the medical student population in Germany grew after 1850, the size of university faculties had not kept pace. By the time Americans were coming in large numbers in the 1870s, an explosion of new enrollment had changed drastically the relations between faculty and students. Where there had been 2,541 students of medicine in Germany in 1865, 7,644 had taken their places by 1885.[20] The German university was entering what one historian has called its period of "stagnation and national crisis" at the very time the flood of American students to Germany reached its peak.[21] As Charles E. McClelland notes, what was most remarkable about the spectacular burst of German research and publication after mid-century was "that it occurred without a dramatic increase in the number of fully salaried employees on the teaching staff."[22] Inevitably, the teaching of run-of-the-mill students in medicine suffered in comparison with research activity and the training of advanced students. Reli-

ance on *Privatdozenten* increased as the numbers of full professorial chairs grew only slowly. As early as 1864, there were half as many unsalaried *Privatdozenten* as there were professors.[23]

Lecture classes grew swiftly in size, while the bedside and laboratory opportunities of most students declined. The *Hauptvorlesung* or grand lecture was increasingly used by famous professors to enable them to teach the growing numbers of students. Friedrich Frerichs, for example, erected a lecture hall for more than 200 students in Berlin in 1859 and others followed his example.[24] More and more of the practical instruction in laboratory methods and physical diagnosis was left to assistants, often without proper supervision. Laboratory exercises (*Übungen*) grew larger and were frequently uncoordinated with the relevant lectures in a subject and could be taken in different semesters. They had become, in the words of Welch's biographer, Donald Fleming, "mechanical, listless, and routine."[25] Contacts between professors and medical students were fewer and more impersonal. As Hans Simmer has written, "The freedom of teaching [in Germany] was such that the professor was even permitted to give up the close contact with his students. Good contact existed earlier in the 19th century, but became less and less towards its end especially in the larger universities."[26] In the clinical lectures, few of the *praktikanten* (probationers) among the medical students now had more than a fleeting chance to observe and learn at the bedside.[27] "In a word," said the Munich internist Hugo von Ziemssen in 1874, "the clinical student sees the sick too little . . . he does not learn . . . how a diagnosis is made . . . [or] how a treatment is finally carried out."[28] And a few years later, Adolf von Strümpell complained that "students hear lectures from eight to ten hours a day. From morning to night their time is taken up with classes; they rush out of one lecture hall into another hearing a huge mass of facts and theories put forward."[29] This was the crisis facing undergraduate medical education in Germany during the years the Americans were transforming their own system.

What is particularly surprising, then, was the subsequent American attack on the didactic lecture, which was not only the staple of their own medical education but by that time of the German system as well. Even in Carl Ludwig's famous institute, a lecture hall was built to hold 259 students.[30] Without question, the American system of medical education in 1870 was even more deeply rooted in the didactic teaching of the lecture and textbook and provided less individual instruction in laboratory or clinic than did that of Germany or any country of Europe. Many of the founders of the Johns Hopkins School of Medicine, as well as those who brought reform to Harvard, Michigan, and Pennsylvania, were themselves educated in traditional American schools, where laboratories of the German type or clinical teaching of the London kind simply had not existed. If they had gone to Germany for postgraduate instruction, as most had, they must have observed, as did Welch and Mall, how little the undergraduate medical student, despite declared intentions and public pronouncements, shared in the scientific life of the laboratory and the clinic. The famous figures who went abroad in the 1870s and 1880s went chiefly to the institutes of leading scientists, were assigned places to work in small laboratories, and were given research projects of their own. They saw little of ordinary medical students and heard few lectures. For the most part, they ignored the gulf separating the advanced work of the institutes, with their cadres of devoted students, and "the old massive lectures for the hundreds of careerists, with their half-empty benches on the many days of *Korps* festivals."[31] Most missed the significance of the severely autocratic rule of the institute chiefs. Efforts at university reform in Germany foundered in these years on the conservatism of the *Ordinaria*, who resisted any change that would impinge on their often dual role as senior members of the faculty and directors of institutes. The institutes, in the words of Joseph Ben-David, often became "feudal fiefs" of the professor.[32]

The Americans returned home with a burning sense of mission to create a new type of medical career where opportunities for research, laboratory study, and teaching could be combined as in Germany. They understood little of the intricate structures, severe inequalities, and academic and state politics of the places they left behind. Scientific careers of the kind they sought were thinkable at first only within the tiny handful of universities then endeavoring to build a broad spectrum of programs for research and graduate study. But these universities, with the exception of Johns Hopkins, were built around a core of undergraduate teaching programs that had no real counterparts in the German university. Teaching had to be integrated with research in a different way than in the German institutes. Graduate programs grew up side-by-side or as an extension of undergraduate programs of liberal arts in America, while professional schools such as medicine were only gradually and incompletely brought into the university orbit. The new university professional schools, moreover, had no monopoly over medical and other professional education as in Germany, and were forced to compete with highly practical training schools for students. This meant that an American medical school, even an elite one, had to provide a far larger amount of practical professional training than its German counterpart.[33] A much broader range of quality clearly existed in the United States schools than in the tightly controlled German system. While in Germany, the pioneers who built the American system of educating physicians had seen large quantities of research being carried on in special laboratories, while the normal medical teaching of professors was accomplished by means of lectures and demonstrations to large classes. The laboratory work of run-of-the-mill students in the German institute took place in ever-larger laboratories supervised by busy assistants. After 1870, the state itself took increasing responsibility for the institutes and seminars in which research and advanced study important to the nation was carried on. These institutes and seminars became, in McClelland's phrase, "a sort of university within the university" serving well relatively few students and often bypassing the formal faculty structure of the university. In medicine alone, 173 new institutes were established in the 50 years after 1860. Institutes became the principal concern of state authorities involved with higher education. It was these specialized institutes that became "a second home" to advanced students, including the large cadre of influential Americans who worked in them.[34]

A similar division between normal teaching and advanced work in America was neither economically nor professionally feasible in the fragile university medical schools of the late 19th century. The first research efforts in the new American programs were of necessity joined much more closely to the teaching of small groups of regular medical students. At the same time American proponents of research were claiming this was also the case in the

German university. To a larger degree than in late 19th-century Germany, the expansion of medical staff and facilities had to be justified to public authorities and donors in terms of the effective teaching of medical students by the new methods. "In spite of the growing interest in science," writes a group of recent researchers, "most American universities did not [as yet] view research as an end in itself. . . . They argued to wealthy patrons, industrial donors, and state legislators that . . . laboratories would extend the university's primary mission, teaching."[35]

Tirelessly, the educational pioneers at Hopkins and elsewhere preached the pedagogical doctrine that only close personal teaching of the kind they had experienced as researchers in German laboratories was suited to the teaching of beginning medical students. They ignored altogether the popular didactic lectures of Germany, many of which were taught by the same men they praised so highly. Only the knowledge that comes to students from personal laboratory experience, said William H. Welch in a dozen public addresses, "is real and living, and not that which comes from mere observation of external appearances, or from reading or being told about things, or, still less, merely thinking about them."[36] "Knowledge lives in the laboratory," said Charles Minot of his teaching methods, "when it is dead, we bury it decently in a book."[37] At the University of Minnesota, C. M. Jackson followed the injunction "Never tell a student anything he can observe for himself; never draw a conclusion or solve a problem which he can be led to reason out for himself; and never do anything for him that he can do for himself."[38] Some American teachers, notably Franklin P. Mall, took an almost perverse delight in avoiding lectures or any kind of structured presentation. "The point in which he differed from the German system," wrote his student and biographer Florence Sabin, "was the time at which full intellectual responsibility was put upon the students, namely, at the very beginning of the medical course instead of toward the end."[39] By 1910, so great was the shift to personal laboratory training for the normal medical student that Abraham Flexner could declare in his famous report, "The student no longer merely watches, listens, memorizes; he *does*. His own activities in the laboratory and in the clinic are the main factors in his instruction and discipline."[40] American medical educators had succeeded in creating a curriculum that, in the words of Merriley Borell, "was distinctly American, emphasizing hands-on experience with research instruments at the introductory level."[41]

The wonder of it is that so radical and expensive a pedagogical doctrine could have been accepted at all in the medical world of the late 19th century. This method of education moved directly away from what was happening in the overcrowded medical schools of Germany. It sought unconsciously to re-create the more personal teaching environment of the German medical schools of several decades before. The builders of the university-type medical school in America, it is not unfair to say, did not simply copy the German system they saw; they transformed it and turned it upside down. They indeed created, as Ludmerer observes, a conceptual revolution in the way that medicine was learned by ordinary students. "With the introduction of laboratory and hospital work [for every student]," he writes, "students were expected to be active participants in their learning process, and the new goal of medical training was to foster critical thinking."[42] Many were conscious of the growing differences in medical teaching from that practiced in Germany. As early as 1876, John Shaw Billings wrote president Daniel Coit Gilman of the new Johns Hopkins University from Leipzig that he was disappointed in Ludwig's famous institute because it was so difficult for ordinary students to take advantage of the laboratories and instruments.[43] A decade and a half later, Charles Withington of Harvard described the ordinary German medical student as "the subject of a constant intellectual *gavage*, in which . . . the pabulum is of the best quality . . . but it is so thoroughly predigested that very little work is left for his own intellectual powers."[44] And William Osler, after seeing Ludwig's institute in 1884, remarked that the teaching in such an institute did not bring the renown that research did and "is apt to be neglected in the more seductive pursuit of the 'bauble reputation.'"[45] By 1912, Abraham Flexner was lecturing to professional audiences in Germany, Britain, and the rest of Europe of the new American model of training physicians. He contrasted it with the German system with its overemphasis on lecturing, its want of apparatus for the individual student, and its huge classes. The laboratory work done by German students, he said, was at best an "isolated group of experiments." Even the best clinical lectures, he told his readers, were "words, words, words." He even questioned whether medical education "on the grand scale" of Europe was possible any longer "without sacrifice of its scientific character."[46]

The massive emphasis on individual instruction in the laboratory and later in the clinic in America was possible, of course, only in medical classes that were small and well financed. As early as 1875, John Shaw Billings described a new type of medical school to the trustees of the Johns Hopkins University, in which only a small number of students would have superior opportunities to become teachers and investigators of medical science in America. It should be wealthy enough so that "being entirely independent of students, [it] can therefore afford to consult their welfare instead of their wishes." Quality must be the only aim; a medical diploma from Hopkins should mean not only that its recipient was a well-educated physician but "that he has learned to think and investigate for himself."[47] The "clinical advantages" needed to produce a scientifically-trained physician, Billings added two years later, "cannot be offered for the instruction of a large number of students."[48] The large lecture to hundreds of students that was so characteristic a feature of both American and German undergraduate education at this time must give way to much more personal experience in laboratory and clinic. It was both expensive and necessary. "Inductive teaching" became the watchword of the pedagogical revolution. In one of his New York classes, before even coming to Baltimore, Welch told his few advanced students:

> I shall make the leading features of the course the demonstration of fresh pathological specimens and the making of postmortem examinations. In doing this I must sacrifice . . . systematic didactic instruction. I am governed by the fact that it is possible for you to learn from books the didactic part, whereas nothing can replace the careful study of fresh specimens.[49]

In Baltimore, Mall wrote to his old master, Wilhelm His, "I am strongly tempted to attempt teaching human anatomy with but few lectures, but to carry most of them to the dissecting rooms. With a small number of students, this may, I think, be done."[50] In New Haven, Russell Chittenden cautioned that

the new science of physiological chemistry could be taught only through "personal instruction and personal experience in the laboratory" and that "instructors . . . must be provided, so that at the most not more than fifteen men shall be under the guidance of one teacher." Such an opportunity, he concluded, should be "open to every student of medicine."[51] The immense wealth invested in the leading medical schools of America would make possible an internal equality unknown in Europe.

It is clear that Chittenden, Welch, Billings, Mall, and the others were striking out in a new direction from that which they had seen followed in Germany. The German system, by sharply dividing laboratory research for professors and the training of advanced students from undergraduate teaching in crowded lecture halls, clinics, and laboratories, had made it possible to have side-by-side a "practitioner-type" medical school of good scientific quality and a small center of advanced study for research and laboratory work in the same university. At first, the American pioneers believed that a similar division of types of education might take place in the United States. "The Medical Schools in Boston, New York, Philadelphia, Baltimore, and the West," Billings told a Johns Hopkins audience in 1877, "are educating practitioners. . . . what I propose is, that the field shall be left perfectly clear to them, and that we undertake to do what they cannot do." In Billings's view, "there should be some institution in the country where superior facilities for, and incentives to, medical study and research can and will be provided."[52]

But the ambitious and influential leaders at Harvard, Yale, Pennsylvania, Michigan, and other universities with aspirations were not content to allow other schools to become the premier medical training centers of the nation. "Once a single institution had committed itself to research and graduate education," writes Ben-David, "the elite colleges had no choice but to follow suit."[53] They, too, were swept up in the limitless enthusiasm for the creation of careers in research and scientific teaching. Schools of the second and third rank likewise struggled to raise their standards and reputations. Not all, of course, were able or willing to follow the German-Hopkins example. The German model, which had been sharply recast in the elite schools, was even more seriously attenuated in the struggling majority of schools. Some schools disappeared quickly in the face of the heavy demands made by state and professional licensing bodies; others fought valiantly to preserve something of the older professional orientation of American medical schools; still others were able to survive by minimal but timely accretions to staff and facilities at critical points in their history. In Chicago, for example, the three oldest medical schools all became affiliated with universities in the 1890s; all made some provision for laboratory instruction by full-time teachers; and all were able to withstand Flexner's characterization of Chicago as "the plague spot of the country" in his 1910 report.[54] In Vermont, the state university took over the financial management of its weak medical school in 1908, then survived a devastating report on its quality by Flexner, and finally appointed enough full-time faculty in the basic sciences to keep its doors open.[55] Similarly, the medical school at Little Rock, Arkansas, which was organized as a proprietary institution in 1879, was forced into an unhappy merger with a second Little Rock school, won state financial support, and clung to life during the blizzard of medical school closings after 1910.[56] In all of these cases, the struggling colleges, representing the large majority of American medical schools, were at best only pale imitations of the German example. While they had moved closer to universities, had constructed minimal laboratory facilities for students, and made some full-time appointments, they were miles away from the German model in their lack of scientific spirit, disdain for original research, lack of provision for superior students, and instruction in clinical subjects by local practitioners.

The emerging university model of training physicians in the elite schools in America, as yet imperfectly realized, was clearly more individualized, more uniformly based in scientific study, more egalitarian, more flexible, and far more expensive than anything yet seen in Europe. It was an amalgam of advanced German medical training, British bedside and clinical traditions, and a homegrown philosophy of learning by doing. Its focus was on the way in which students must learn to comprehend the mass of new information that was pouring from the workplaces of scientific medicine. Several decades before Dewey, it put emphasis on the long tradition of practical learning in medicine and the need for a problem-solving approach to modern medical science.[57] Visiting Europeans were clearly aware that Americans were not simply copying the

German model. When Friedrich Müller visited the leading medical schools of America in 1907, for example, he was astonished, he said, to see the mountains of equipment for the teaching of ordinary medical students and the many practical analyses carried out by every student in the physiological laboratories. One university alone, he told his German listeners with amazement, used 4,000 frogs in a single year for teaching purposes! The results, he admitted, were far better than in Germany, where "the student only hears passively of experiments in the lecture and sees the curves on the blackboard."[58]

The developing American system depended on small enrollments, large amounts of instruments, and generous endowments or lavish state support. No German or European medical school of the turn of the century came close to expending so large an amount of money and material on the ordinary medical student. The new method of training physicians, Frank Billings told the American Medical Association in 1903, "involves an expense which is appalling when compared with the methods of teaching formerly practiced in all schools, and still adhered to in many schools." Its success, however, he told his hearers, was important to the whole profession, for a physician thus educated "would be, in truth, a member of a learned profession. From an educational point of view he would rank as an equal with the scholar in philosophy, law and theology."[59] Physicians and their teachers would at last have a firm and respected position in American society. Billings was a leader in the fight to put the influence of organized medicine on the side of the new system of teaching and, with the timely help of Abraham Flexner and the great foundations, the struggle was successful. Organized medicine thus made common cause with academic scientists and foundation officials in advancing the new education.

But was the highly personalized, scientifically inductive system of instruction begun at Hopkins the only or even the best way to train the ordinary practitioner of medicine? Could not the German and European distinction between the training given to the average professional student with the science-oriented advanced student have been made in America? Welch and Flexner argued repeatedly that it could not. There were not two kinds of medical schools, said Welch, "one designed for mediocre students to make practitioners, and the other for superior students to make investigators and teachers."[60] Then and since, some critics have charged that medical education took a wrong turn when it sought to make medical scientists of all practitioners. American educators, it seemed to some, had "out-Germaned" the Germans. This change made medical training much more exclusive and expensive, say the critics; it drove the practical medical school from the American educational scene; it opened a deep ravine between basic medical science and its clinical applications; it weakened the holistic approach of the clinically trained physician to patients; it stressed research at the expense of medical teaching; and it often barred the weaker and less powerful groups in society from a chance to practice medicine. Why must every student have intensive laboratory experience, not only in anatomy, but in physiology, biochemistry, and pharmacology? What did the student learn that made him a better physician? Was the physician thus educated a better problem-solver, more likely to look for the origins and not just the surface manifestations of disease? Many, of course, responded with an enthusiastic "yes," but the questions have continued to echo through the years and the empirical evidence remains shallow and inconclusive. "The priority most medical faculty members accord to research, patient care, and training of residents and graduate students," charges the most recent general study of medical education in 1984, "has militated against the education of medical students."[61]

But whether one agrees with the critics or not, the founders of scientific medicine in America made a spectacular revolution in the teaching of students. Their ambition went far beyond the system they admired in Germany. In creating a new kind of scientific career in this country, they simultaneously changed the fundamental pedagogy of all medical education. While the spirit of their teaching was that of the German investigative laboratory of the 1880s, it had much less to do with the kind of teaching received by the average German doctor. By the First World War, Abraham Flexner could chide the Germans themselves for their "mistaken pedagogy" in teaching medical students. "The most brilliant demonstration" or lecture, he wrote, of the typical way of teaching doctors, is "less educative than a more or less bungled experiment executed by the student with his own hands."[62] As late as 1930, the American Commission on Medical Education was still describing the German system in Flexner's terms as "distinctly a passive process" in which the student was expected to learn by listening and looking on.[63] The German system, as pictured in the 1930 report, was not greatly different from that of 1890: a sorting process in which a great many students entered medical school at a university, a considerable number later dropped out of their studies, most went on to the competent practice of medicine, and a few absorbed the scientific spirit permeating the faculty and entered careers in research, teaching, or advanced specialization. Neither British nor French medical educators were to follow the new American model in its heavy emphasis on scientific training. British medical teaching, according to the Commission, was still "primarily practical" in 1930 with the basic sciences ancillary to the clinical subjects, while French medical education, they wrote, "is essentially clinical training with a single objective, the production of doctors by practical methods."[64]

The introduction of the inductive, personal model of educating physicians in America, it now seems clear, owed much to the German research laboratory where so many influential Americans had worked as postgraduate students. In America, more than in Germany, this method of education was supported as crucial to ordinary medical students. It seemed the only feasible way in the 1880s and 1890s (as it had been earlier in Germany) to create scientific laboratories and scientific careers like those in Germany. "The object of the laboratory," wrote Franklin P. Mall, "is to teach students, to train investigators, and to investigate. Although [teaching] requires the greater portion of the instructor's time, its importance is by no means as great as the second and third."[65] A remarkably progressive theory of medical learning derived from this elite group's experiences as researchers or observers (as in the case of Billings) was developed to explain the importance of personalized inductive or laboratory teaching even for beginning students. The pedagogical revolution coincided with unparalleled successes in experimental medicine worldwide and a great burst of philanthropic support in the United States. By 1914, the pedagogical transformation was largely complete, the older practitioner-type medical school, still struggling, was on its way to extinction, and the organized profession itself had become the leading spokesman for the new type of medical teaching.

NOTES

An earlier version of this paper was presented at the 61st meeting of the American Association for the History of Medicine, 6 May 1988, in New Orleans, Louisiana. It profited from the discussions there and especially the detailed comments of Kenneth M. Ludmerer and Robert G. Frank, Jr.

1 Ronald L. Numbers, ed., *The Education of American Physicians: Historical Essays* (Berkeley and Los Angeles: Univ. of California Press, 1980); Kenneth M. Ludmerer, *Learning to Heal: The Development of American Medical Education* (New York: Basic Books, 1985); William G. Rothstein, *American Medical Schools and the Practice of Medicine: A History* (Oxford: Oxford Univ. Press, 1987).

2 Henry Sewall to John J. Abel, 17 Oct. 1886, John J. Abel Papers, Alan M. Chesney Archives, Johns Hopkins Medical Institutions, Baltimore, Maryland.

3 Mall is quoted by Florence R. Sabin, *Franklin Paine Mall: The Story of a Mind* (Baltimore: Johns Hopkins Press, 1934), p. 64.

4 For a full discussion of the influence of German study and travel on American physicians, see Thomas N. Bonner, *American Doctors and German Universities: A Chapter in International Intellectual Relations, 1870–1914* (Lincoln: Univ. of Nebraska Press, 1963).

5 John Harley Warner, "Physiology," in Numbers, ed., *Education of American Physicians,* p. 60.

6 Introduction to Abraham Flexner, *Medical Education in the United States and Canada: A Report to the Carnegie Foundation for the Advancement of Teaching,* Bulletin no. 4 (New York: Carnegie Foundation for the Advancement of Teaching, 1910), p. xiii.

7 Arthur Dean Bevan to Abraham Flexner, 9 Jan. 1912, Abraham Flexner Papers, Library of Congress, Washington, D.C.

8 William Henry Welch, "Some of the conditions which have influenced the development of American medicine, especially during the last century," reprinted in his *Papers and Addresses,* 3 vols. (Baltimore: Johns Hopkins Press, 1920), 3: 301.

9 For a general discussion of German medical education in this period, see Hans H. Simmer, "Principles and problems of medical undergraduate education in Germany during the nineteenth and early twentieth centuries," in *The History of Medical Education,* ed. Charles D. O'Malley (Berkeley and Los Angeles: Univ. of California Press, 1970), pp. 173–200. A good contemporary summary by an American is Charles F. Withington, "Medical teaching in Germany," *Boston Med. & Surg. J.,* 1893, *129:* 585–588.

10 Harold Perkin, "The historical perspective," in *Perspectives on Higher Education: Eight Disciplinary and Comparative Views,* ed. Burton R. Clark (Berkeley and Los Angeles: Univ. of California Press, 1984), p. 37.

11 Jonathan Harwood, "National styles in science: genetics in Germany and the United States between the world wars," *Isis,* 1987, *78:* 409.

12 Franklin P. Mall, "Wilhelm His: his relation to institutions of learning," *American Journal of Anatomy,* 1905, *4:* 141.

13 A full discussion of this divergence from German practice is in Thomas N. Bonner, "Friedrich von Müller and the growth of clinical science in America, 1902–1914" (paper read at the 28th International Congress of the History of Science, Hamburg, West Germany, 4 August 1989).

14 Nathan Reingold, "Graduate school and doctoral degree: European models and American realities," in *Scientific Colonialism: A Cross-Cultural Comparison,* ed. Nathan Reingold and Marc Rothenberg (Washington, D.C.: Smithsonian Institution, 1987), p. 133.

15 *Ibid.,* p. 130.

16 On the importance of the academic department and the graduate school in the United States, see Joseph Ben-David, "The universities and the growth of science in Germany and the United States," *Minerva,* 1969, 7: 6–9.

17 See Ludmerer, *Learning to Heal;* William Coleman and Frederic L. Holmes, "Introduction," p. 11; William Coleman, "Prussian pedagogy: Purkyně at Breslau, 1823–1839," p. 38; Arleen M. Tuchman, "From the lecture to the laboratory: the institutionalization of scientific medicine at the University of Heidelberg," p. 91; Timothy Lenoir, "Science for the clinic: science policy and the formation of Carl Ludwig's institute in Leipzig," in *The Investigative Enterprise: Experimental Physiology in Nineteenth-Century Medicine,* ed. William Coleman and Frederic L. Holmes (Berkeley and Los Angeles: Univ. of California Press, 1988), p. 173.

18 Rothstein, *American Medical Schools,* p. 331.

19 Ben-David, "Universities and the growth of science," pp. 7–8. Quotation on p. 8.

20 Konrad H. Jarausch, "The social transformation of the university: the case of Prussia, 1865–1914," *J. Soc. Hist.,* 1979, *12:* 611–613.

21 Harwood, "National styles in science," p. 391.

22 Charles E. McClelland, *State, Society, and University in Germany, 1700–1914* (Cambridge: Cambridge Univ. Press, 1980), p. 164.

23 *Ibid.,* p. 166.

24 Simmer, "Principles and problems of medical education in Germany," p. 188.

25 Donald Fleming, *William H. Welch and the Rise of Modern Medicine* (Boston, Mass.: Little, Brown, 1954), p. 110.

26 Simmer, "Principles and problems of medical education in Germany," p. 187.

27 Claudia Huerkamp, *Der Aufstieg der Ärzte im 19. Jahrhundert: vom Gelehrten Stand zum Professionellen Experten* (Göttingen: Vandenhoeck & Ruprecht, 1985), pp. 98–101.

28 Hugo von Ziemssen, "Ueber den klinischen Unterricht in Deutschland," *Deutsches Archiv für Klinische Medicin*, 1874, *13*: 13.

29 Adolf von Strümpell, *Über den medizinisch-klinischen Unterricht: Erfahrungen und Vorschläge* (Leipzig and Erlangen: Deichert, 1901), p. 11.

30 Lenoir, "Science for the clinic," p. 141.

31 McClelland, *State, Society, and University*, p. 287.

32 Ben-David, "Universities and the growth of science," p. 3.

33 Joseph Ben-David, *Centers of Learning: Britain, France, Germany, United States; An Essay* (New York: McGraw-Hill, 1977), pp. 62–65.

34 McClelland, *State, Society, and University*, pp. 79–85.

35 Merriley Borell, Deborah J. Coon, H. Hughes Evans, and Gail A. Hornstein, "Selective importation of the 'exact method': experimental physiology and psychology in the United States, 1860–1910" (paper presented at joint meeting of the History of Science Society (U.S.) and British Society for the History of Science in Manchester, England, 13 July 1988), p. 4.

36 See, for example, Welch, "The evolution of modern scientific laboratories," in his *Papers and Addresses*, 3: 200–211. Quotation on p. 208.

37 Quoted by Richard M. Pearce, "The experimental method: its influence on the teaching of medicine," in *Medical Research and Education*, ed. J. McKeen Cattell (New York: Science Press, 1913), p. 92.

38 C. M. Jackson, "On the improvement of medical teaching," p. 371.

39 Sabin, *Franklin Paine Mall*, p. 153.

40 Flexner, *Medical Education in the United States and Canada*, p. 53.

41 Merriley Borell, "Instruments and an independent physiology: the Harvard Physiological Laboratory, 1871–1906," in *Physiology in the American Context, 1850–1940*, ed. Gerald L. Geison (Bethesda, Md.: American Physiological Society, 1987), p. 295.

42 Ludmerer, *Learning to Heal*, p. 5.

43 W. Bruce Fye, *The Development of American Physiology: Scientific Medicine in the Nineteenth Century* (Baltimore: Johns Hopkins Univ. Press, 1987), p. 147.

44 Withington, "Medical teaching in Germany," p. 587.

45 Harvey Cushing, *The Life of Sir William Osler*, 2 vols. (Oxford: Clarendon Press, 1940), 1: 218.

46 Abraham Flexner, *Medical Education in Europe: A Report to the Carnegie Foundation for the Advancement of Teaching*, Bulletin no. 6 (New York: Carnegie Foundation for the Advancement of Teaching, 1912), pp. 81, 85, 171. See, too, *I Remember: The Autobiography of Abraham Flexner* (New York: Simon & Schuster, 1940), p. 170.

47 Johns Hopkins Hospital, *Five Essays Relating to the Construction, Organization, & Management of Hospitals* (New York: William Wood, 1875), pp. 3–5.

48 John Shaw Billings, "Two papers by John Shaw Billings on medical education," *Bulletin of the Institute of the History of Medicine*, 1938, *6*: 315.

49 Simon Flexner and James Thomas Flexner, *William Henry Welch and the Heroic Age of American Medicine* (New York: Viking Press, 1941), p. 122.

50 Sabin, *Franklin Paine Mall*, p. 135.

51 Russell H. Chittenden, "The importance of physiological chemistry as a part of medical education," *N.Y. Med. J.*, 1893, *58*: 373–374.

52 Billings, "Two papers," pp. 321, 337.

53 Ben-David, *Centers of Learning*, p. 61.

54 Thomas N. Bonner, *Medicine in Chicago, 1850–1950: A Chapter in the Social and Scientific Development of a City* (Madison, Wis.: American History Research Center, 1957), pp. 108–117.

55 Martin Kaufman, *The University of Vermont College of Medicine* (Hanover, N.H.: Univ. Press of New England, 1979), pp. 130–151.

56 W. David Baird, *Medical Education in Arkansas, 1879–1978* (Memphis, Tenn.: Memphis State Univ. Press, 1979), pp. 36–113.

57 Ludmerer, *Learning to Heal*, pp. 63–68.

58 Friedrich Müller, "Amerikanische Reiseeindrücke," *Münchener medizinische Wochenschrift*, 1907, *54*: 2389–2390.

59 Frank Billings, "Medical education in the United States," *J.A.M.A.*, 1903, *40*: 1273, 1275.

60 Flexner and Flexner, *William Henry Welch*, p. 105.

61 Association of American Medical Colleges, *Physicians for the Twenty-First Century: The GPEP Report* (Washington, D.C.: Association of American Medical Colleges, [1984]), p. xv.

62 Flexner, *Medical Education in Europe*, p. 84.

63 Commission on Medical Education, *Medical Education and Related Problems in Europe* (New Haven, Conn.: Commission on Medical Education, 1930), p. 67.

64 *Ibid.*, pp. 24, 80–81.

65 Franklin P. Mall, "The anatomical course and laboratory of the Johns Hopkins University," *Bulletin of the Johns Hopkins Hospital*, 1896, *7*: 85.

12

Abraham Flexner in Perspective: American Medical Education, 1865–1910

ROBERT P. HUDSON

The Flexner report of 1910[1] justifiably stands as a monument in the reform of American medical education. Yet there is much in the report itself to suggest that Flexner's contribution was not so much revolutionary as catalytic to an already evolving process. The present study reassesses Flexner's impact, but only peripherally. The principal aim rather is to sketch in some of the pertinent social and educational trends during the half century between the end of the Civil War and Flexner's genteel thunderbolt.[2]

The Civil War abated only temporarily the spread of proprietary medical schools. This spread, it should be remembered, was malignant less because it produced too many physicians and more because the ensuing competition for students diluted the already limited resources and eroded academic standards.[3] Plumpers for reform who had sounded their monotonous litanies in the quarter century after the founding of the American Medical Association were by 1870 generally pessimistic. They not only appeared to agree with John Shaw Billings that "it does not pay to give a $5000 education to a $5 boy,"[4] but by now they had despaired of doing anything about either education or the boy.

Still defeatism was not universal. A few observers thought they saw glimmers of improvement both in the quality of incoming students and in the educational process itself.[5] These disparate views can be made at least comprehensible. Part of the improvement was purely on paper. Medical school catalogues became increasingly imaginative as

ROBERT P. HUDSON is Professor Emeritus in the Department of the History and Philosophy of Medicine, University of Kansas Medical Center, Kansas City, Kansas.

From the *Bulletin of the History of Medicine*, 1972, 56: 545–561. Copyright © 1972 by the Johns Hopkins University Press. Reprinted by permission of the Johns Hopkins University Press.

competition for students grew ever more heated. The catalogues of around 1880 contained few stipulations which could be described as non-negotiable. Most medical colleges specified, for example, that applicants must have a high school diploma or its "equivalent." Flexner, with characteristic candor, disposed of the equivalent as "a device that concedes the necessity of a standard which it forthwith proceeds to evade."[6]

There can be no serious disagreement with Shryock's conclusion that American medical training remained "relatively inferior" during the two decades following the Civil War.[7] But it was a complex period, one in which entrenched mediocrity and rising reform coexisted. While new schools continued to pop up with the discouraging persistence of dandelions, worthier seeds were sprouting. These early improvements, though they frequently interlocked, can be considered under three headings: admission requirements, medical school curricula, and postgraduate training.

The 35 years before the turn of the century saw something of a revolution in higher education in America. The public, which previously saw little utility in education, began to change its mind. This new attitude was reflected in the state's perception that it had an obligation to make educational opportunities more accessible to the masses. The 1862 Morrill Act not only initiated the land grant university movement, but indirectly gave rise to a new institution, the public high school. These multiplied so rapidly that by 1892 their enrollment surpassed that of the academies.[8] Thus premedical education of improved quality became generally available during the period under consideration.[9] Even though compulsory education reportedly existed in only seven states by 1896,[10] the fact remains that the opportunity at least for better premedical preparation expanded rapidly in the post-bellum decades.

The question remains—did more students bound for medical studies during this time take advantage of their new opportunities? [11] A precise answer in numerical terms is not possible. There is evidence that students increasingly pursued premedical training of steadily improving quality, and that medical colleges upgraded their entrance requirements both on paper and in fact. But the case should not be overstated. The situation was characterized by a spectrum rather than by uniformity, and most students still received inferior premedical training. The Johns Hopkins required an academic degree from the outset (1893) but by 1906 the Council on Medical Education of the American Medical Association was still recommending only a minimal one year of college for medical school admission. [12]

Still trends slowly pushed toward higher standards. John Rauch, a champion of reform in Illinois, reported that the number of medical colleges exacting certain specified educational requirements for matriculation was 45 in 1882, 114 in 1886, 117 in 1889 and 124 in 1890. [13] Of the 155 schools Flexner surveyed, 16 required two years of college and 6, which demanded only one year, were scheduled to go to a two-year requirement in 1910. [14] Among these 16 schools, 1,850 students had satisfied the two-year requirement as of 1908–1909. [15] By way of balance it should be recalled that in 1910 some 50 medical colleges demanded for admission only a high school education or the much abused equivalent. [16]

One of the most significant features of the decades under study was that many aspiring physicians fashioned educational programs superior to those dictated by law or custom. This held true for collegiate preparation as well as medical and post-graduate training. Regarding premedical studies evidence of this is found in an analysis of the 1,513 physicians in the *Dictionary of American Medical Biography* (*D.A.M.B.*) who took their medical training in 19th-century America. A surprising 607 or some 40 percent of these men earned predoctoral degrees before beginning medical studies. Another 430 had education beyond grammar school but short of a degree. [17]

By definition men selected for inclusion in the *D.A.M.B.* are not representative of the general physician population. Still, if properly qualified, the figures have value. For one thing the 1,513 physicians may be more representative than they appear at first glance. In the 19th century America did not produce fifteen hundred physicians who distinguished themselves by scientific or literary contributions to medicine. Deciding just why a given physician was selected for inclusion in the *D.A.M.B.* was imprecise at best, but in this study 531 were included apparently because they achieved excellent reputations in medical practice and nothing more. These men must have differed little from many others who were not selected by editors Kelly and Burrage.

Two other features of the *D.A.M.B.* study apply to the question at hand. The average age at which the 607 men took their predoctoral degree increased from 19.5 years in the first decade to 21.1 in the last. There are several possible explanations for this, but the most reasonable is that as the century progressed these students undertook more college preparatory work or a longer collegiate program or both.

Finally in the *D.A.M.B.* the percentage of men earning a predoctoral degree increased steadily decade by decade. In the seventh decade the figure was 46.9 percent and by the tenth, 63.4 (that is 63.4 percent of all men completing medical training in the tenth decade had earned a predoctoral degree). In short both the quality of premedical education and the number of men pursuing predoctoral degrees increased as the century progressed. Putting aside the question of representativeness, the *D.A.M.B.* sample was itself improving with time.

The 45 years encompassed in the present study saw a persistent though phlegmatic upgrading of the formal medical curriculum as well. The process involved the teaching of laboratory medicine as well as the length and content of clinical training. Before the Civil War medical studies centered about a preceptorship (nominally three years in length, but steadily diluted in practice during our period) and two lecture courses of some three months duration. The lecture courses were ungraded and the second was identical to the first—tedium compounded. Earlier requirements for Latin and the M.D. thesis were by now defunct, and since students were assessed a diploma fee final examinations rarely took a scalp. Even the rule that medical graduates must be 21 years of age was widely ignored.[18]

Throughout the first four post-bellum decades basic science teaching, with few exceptions, remained the domain of practicing physicians. In the earlier decades of the 19th century these men generally had no formal basic training beyond their own medical school experience. This too began changing as the 20th century approached. Laboratory medicine (along with laboratory science generally) was scantily taught prior to the Civil War,[19] but increased thereafter with the period of rapid development beginning between 1890 and 1895.[20] During this time the number of nonmedical graduate students rose "with astonishing speed."[21] These men formed the pool from which specialized basic science teachers eventually came, but by Flexner's time the appointment of Ph.D. teachers in medical schools was still an innovation to be "watched with interest."[22] Widespread support of basic research in America was spurred by the impressive practical successes of German science. The movement was well under way by 1900 and was bound to contribute to the favorable reception accorded Flexner's recommendations.

With a few significant exceptions[23] clinical teaching prior to 1865 took place in the crowded formality of classrooms and amphitheaters. Riesman scored the principal defect of the method when he said of his own experience at Pennsylvania, "It really made no difference what ailed the patient; the professor could use him as text for almost any disease and we would be none the wiser."[24] Even in medical schools with affiliated hospitals the situation could be deceptive. Norwood could justly say of Philadelphia schools, "What appears to be a formidable array of clinical material was in reality not always made available for clinical teaching."[25]

But again improvement was under way. After the Civil War a growing number of medical colleges graded the curriculum[26] and extended its length. Those who opposed stretching the term argued that a six-month session was "preposterous, in violation of all laws of health, physical or mental. . . ."[27] As always in this situation it is difficult to sort out true convictions from the economic incentives that in fact dominated many actions by proprietary professors. In any event the transition was far from painless. The University of Pennsyl-

vania attempted a six-month course of study as early as mid-century, but had to abandon the experiment when competing schools refused to fall in line.[28]

Between 1800 and 1830 the lecture term generally remained at 13 weeks.[29] By 1882, 42 schools reportedly required courses of six months or more and by 1890 the figure was 76.[30] Improvement continued so that by 1910 practically all schools offered four years of graded instruction although the number of months required each year continued to vary.[31]

Despite the many defects of the American M.D. degree program at this time, a physician must have been better prepared for practice by earning a degree than by the simple process of self-ordination. Another bit of positive evidence appears in the *D.A.M.B.* study where the number of men earning medical degrees increased consistently in the three decades after the Civil War.[32]

For the more affluent student of this period two domestic postgraduates experiences were available — house staff training and postgraduate clinics. The latter came on the scene relatively late; the New York Polyclinic, for example, opened its doors in 1882.[33] A few of these clinics flourished, but generally they never attained the prestige of their European counterparts. To Flexner the earlier clinics were nothing more than "undergraduate repair shops,"[34] though this is less an indictment than Flexner may have intended, because a measure of repair was needed. Still the postgraduate clinic was not as important for our purposes as house staff training. The clinics rather quickly settled on imparting the technical skills of a specialty, and were not looked to as a source of broad clinical training.

Insofar as it helped remedy the deficiencies of formal clinical instruction the role of postgraduate hospital training[35] has been underestimated. The situation is inevitable if one thinks only in terms of formal internships and residencies as they exist today. In the pre-Flexnerian period house staff training was arranged more or less informally, and there is evidence that such personal arrangements were not rare. As early as 1914 an internship was made requisite to licensure in Pennsylvania,[36] and in that same year the American Medical Association published its first list of approved internships plus a list of 95 hospitals offering specialty training.[37] Even more striking is the report that a year or two earlier 70 percent of the nations' medical graduates elected an internship.[38] Of the 1,513 physicians in the *D.A.M.B.* sample 248 took some form of house staff training. In the first decade of the 19th century the figure was only 1.6 percent, for the seventh decade, 20.2 percent, and by the last decade, 43.9.[39]

In part the importance of house staff training in improving physical education during this time has been understressed because the movement arose spontaneously and informally with no visible external crusade such as that directed against the undergraduate curriculum by the American Medical Association and others from mid-century on. Hospitals were growing in numbers and practitioners needed resident physicians. At the same time medical graduates were increasingly aware of their inadequate clinical skills. The two needs met quietly in a symbiotic relationship which was well established and which had most of the characteristics of modern house staff training by the time Flexner came along.

The third postgraduate option open to American students was medical training abroad. Remembering that students were under no legal compulsion to undertake such a costly venture, the numbers involved are mildly staggering. Bonner estimated that some fifteen thousand visited Germany and Austria alone in the period 1870–1914.[40] In the *D.A.M.B.* study 372 (24.6 percent of the 1,513 subjects) went abroad within five years after completing their American medical training. Only 14.8 percent did so in the first decade, a figure that rose to 24.6 percent in the seventh decade and 36.6 in the tenth.[41]

It could be argued and correctly that this amazing European exodus merely underscored the inadequacy of clinical opportunities at home. Yet the larger significance is that they *did* go abroad, most of them presumably intent on improving their medical proficiency. Each wave of returning students evangelized a new group which in turn came back impressed with what medical training could and should be like. The influence of these European-trained men in abetting eventual reform is impossible to assess quantitatively, but undoubtedly it was greater than observers appreciated at the time. When the Johns Hopkins University opened a medical school along German lines, J. S. Billings feared it would be many years before the

school could hope to graduate a class of 25. The first class graduated only 15, but two years later the figure was 32.[42]

Field points out that 16 of the 28 founding members of the American Physiological Society (1887) were trained in Germany and four others had studied in England or France. The group included a number of men who later led the battle to reform medical education in this country, such names as H. P. Bowditch, W. H. Howell, H. N. Martin, C. S. Minot, William Osler, and William Welch.[43] It is reasonable to conclude that the fifteen thousand or so men who saw German medical education firsthand must have been a powerful, though never organized, influence supporting Flexner's efforts of 1910 and thereafter.

It was Welch who summarized the period by saying "The results were better than the system."[44] He was right, and in part the results were better than the system because an increasing number of students forsook the system to create a system of their own. In assessing the status of American medical education before Flexner it is important to look beyond the extant degree requirements and consider the finished product as well. While the majority continued to settle for the prevailing mediocrity, a growing number of young physicians around the turn of the 20th century voluntarily supplemented their formal medical education with undergraduate collegiate preparation, house staff training, and study abroad. During the same period certain social forces began to be felt, and these too contributed to the catalytic effect of the implacable little schoolmaster from Louisville. Among the external forces were organized medicine, state licensing boards, and the improbable educational experiment made possible by Baltimore businessman Johns Hopkins.

The two principal medical organizations involved were the American Medical Association (A.M.A.) and the Association of American Medical Colleges (A.A.M.C.). From its inception in 1846 to the end of the century the A.M.A. battled for reform with little visible effect. Lacking any legal bite the A.M.A. sought to work by moral suasion alone. A main reason for its failure is found in the structure of the A.M.A. itself. From the outset the A.M.A.'s architects faced an organizational dilemma. If physician-owners of proprietary schools were to be induced to join a national organization which strongly favored educational reform, they would have to be offered greater voter representation than their numbers would dictate. On the other hand, of course, giving the medical schools too large a voice would block any chance for reform. The founders attempted a compromise which was doomed from the beginning. Only about a third of the eligible medical schools were present at the A.M.A.'s organizational convention in 1846.[45] From 1847 to 1852 medical college representation fell from 59 to 39 while medical society membership increased from 178 to 226.[46] The rift grew so wide that in 1853 there was an attempt to eliminate medical college delegates altogether.[47] Thus a town-gown split was all but assured by the A.M.A.'s original constitution, though it is difficult to imagine any scheme that would have satisfied the contending factions. Perceiving that the A.M.A. was essentially powerless, the physician-professors refused to participate and thus were free to ignore the A.M.A.'s exhortations. The best that can be said for the A.M.A. before 1900 is that its monotonous editorial lamentations did keep the matter somewhere in the far reaches of the profession's conscience. So frustrated did matters become in reality that at one time the A.M.A. entertained such now-startling possibilities as federal legislation and a national medical college.[48]

In 1904 the A.M.A. formed a permanent Council on Medical Education and from that point on its voice began to be heard. Staffed originally with a group of outstanding men, the council periodically inspected medical schools and began a rating system of A (worthy), B, and C (hopeless). Their reports were broadcast and the ensuing publicity undoubtedly contributed to the closure or merger of 29 schools between 1906 and Flexner's report[49] four years later.

In 1908 the council helped enlist Carnegie Foundation support for Flexner's survey and made available to him data accumulated from the council's previous investigations.[50] Dr. N. P. Colwell, secretary of the council, accompanied Flexner on several inspection trips, and, although Colwell's reports had to be couched "cautiously and tactfully,"[51] his medical insights must have proved helpful to Flexner. When it came time to write, Flexner, with his independent status, was free to put down with brutal objectivity many of the findings Colwell had felt free only to mutter about.

During the years after 1846 the disabling internecine problems of the A.M.A. became apparent to a small number of reform-minded men in the ranks of medical education itself. Responding to an organizational call on June 2, 1876, representatives from 22 medical colleges met to consider the possibility of reform from within. The following year they organized as the American Medical College Association.[52] Their laudable idealism soon came to grief. In 1880 the A.M.C.A. voted to require three years of medical training with at least six months of each year in a "proper medical college." Within two years ten schools withdrew, including several of the better founding colleges, and the organization collapsed. A successful reorganization took place in 1890 and from then on the pace of reform picked up. By 1896 the renamed Association of American Medical Colleges could report that 55 of the nation's 155 medical colleges were "cooperating," which meant at least nominal adherence to the higher self-imposed standards of A.A.M.C. members.[53] This, it should be recalled, at a time when even catalogue adherence to higher standards placed a school at a real disadvantage in the fierce competition for students.

The overriding importance of the A.A.M.C.'s early years was that at last reform was stirring within the proprietary system itself. Educators themselves now admitted at least that a mess existed. Earlier A.A.M.C. standards, while not always met by member schools themselves, created something of a self-fulfilling prophecy. The fact that a number of schoolmen could now organize for reform finally cracked the solid front presented by proprietary professors for most of the century. As its power and prestige grew the A.A.M.C. demanded a series of higher standards from its member colleges, and in 1905 the Confederation of State Medical Examining and Licensing Boards adopted the A.A.M.C.'s standard curriculum. From this point on the story became one of increasingly effective cooperation between the A.A.M.C., the A.M.A., and a new force, state legislatures, intent upon revitalizing medical practice acts.

During the early 19th century licensure was under some measure of legal control in most states of the Union. Regulatory mechanisms variously involved medical societies, medical schools, and governing boards of universities. Around 1830 licensure laws began to be repealed and by 1850 legislative control of medical licensure was practically nonexistent.[54] The anarchy that characterized American medicine from 1830 to 1875 came about for a number of reasons which cannot be detailed here. In general, abandonment of medicine by the state paralleled the rise of medical sects and indeed in the judgment of Reginald Fitz the spread of Thomsonianism was "the first serious blow to the regulation by the State of the practice of medicine. . . ."[55]

To put the matter perhaps too simply, public dissatisfaction with the heroic therapy of regular physicians coupled with widespread confusion over the conflicting claims of some two dozen different medical sects led legislators to withdraw legal sanctions from all the contending parties. A nadir of sorts was reached in 1838 when Maryland made it legal for any citizen of that state to charge and be paid for medical services.[56]

In medical practice absolute freedom can corrupt as thoroughly as absolute power, and in truth the two become effectively synonymous. N. S. Davis, whose prescience becomes more remarkable with time, resolved at the first convention of the A.M.A. that licensing be placed in the hands of a single independent state board in each state.[57] Not surprisingly the resolution failed and by 1860 editorialists such as Stephen Smith had given up hope for state control and tossed the matter back to the A.M.A.[58] By 1875 the chaos could not be ignored any longer. The process of repeal reversed itself and states began writing new laws controlling licensure and practice. New York was in the vanguard, and the story there in many ways exemplifies the complex seesaw evolution of legislative attempts at controlling medical licensure.[59] By 1894 Fitz reported that all states except New Hampshire and his own Massachusetts had laws regulating medical practice.[60] Although these laws varied in their inherent worthiness as well as the energy with which they were prosecuted, Fitz concluded that "to them, more than any one cause, is due the difference which exists between the condition now [1894] and in 1870."[61]

The state licensing boards created by this new wave of legislation confederated and finally there existed a national organization invested with the power that for years the A.M.A. and A.A.M.C. could only sigh after. To Flexner's eye the state boards of 1910 left much to be desired—their continued loose interpretation of his *bête noire*, the "equivalent," for example—but he wisely perceived that the legal approach held the best hope for reform. "The state boards," he said, "are the instruments through which the reconstruction of medical education will be largely effected."[62]

Viewed in terms of the ground they plowed for Flexner, legislative reforms after 1870 were far more significant than their immediate success in eliminating diploma mills, quacks, and ignorant but sincere sectarians. The new laws reflected a wholly new national attitude. No longer was the scene dominated by fears of class legislation, by the pervasive attitude of *laissez-faire*, or Herbert Spencer's version of social Darwinism which made the patient responsible for his own folly when he chose his physician unwisely.[63] The confused lawlessness of 1830–1870 had by now convinced the populace and their lawmakers that some degree of state regulation was both proper and necessary.

Other factors contributed. Regular medicine emerged as genuinely superior to its sectarian competition, which was handicapped by a simplistic, usually unitarian, approach to disease causation and therapy. Whatever its faults regular medicine retained the capacity for change, a trait that allowed it to accommodate the remarkable advances of European medicine, albeit at times with some chauvinistic delays. The regulars could scarcely claim a jockey's role in medicine's late 19th-century emergence as a science, but at least they backed the right horse. Nationalistic elements remained, but by Flexner's time the profession and the public generally accepted the superiority of German science.

The next step was eminently reasonable. If the fruits of German medical science could be imported into this country, why not the educational system that underpinned it all? Indeed the process had begun as far back as 1876 in the hands of a few visionary men and a Baltimore merchant who was wise enough to avoid all restrictive adjectives when he endowed a new type of American university.[64] The Hopkins' story is too well known to need retelling here.[65] The fact that the Johns Hopkins medical school succeeded almost two decades before the Flexner report testifies that American attitudes toward medical education had changed drastically between 1865 and 1893. True, a number of proprietary schoolmen refused to see the message on the wall, but by 1910 the writing was plainly there.

A final force deserves mention, that of economics. The same free marketplace which originally encouraged the wild growth of medical schools had begun to take its toll. Supply now exceeded valid demand, and as Flexner wryly remarked, "Nothing has perhaps done more to complete the discredit of commercialism than the fact that it has ceased to pay. It is but a short step from an annual deficit to the conclusion that the whole thing is wrong anyway."[66]

Indeed commercialism had ceased to pay. Flexner found that during the single year prior to his report a dozen schools had collapsed and "many more are obviously gasping for breath."[67] Had this attrition rate prevailed consistently it would more than have accounted for the decrease in schools from a 1906 peak of 161 to the 100 or so surviving in 1915.[68]

Thus the tide of reform was running heavy by 1910. Flexner himself observed that "there is no denying that especially in the last fifteen years, substantial progress has been made."[69] Yet none of this need detract from Flexner's renowned contribution.[70] There is no doubt but that his meticu-

lously documented survey shocked the nation and hastened the closing of a number of marginal schools. But that was not his chief bequest. The sorry state of America's medical schools was no secret before 1910. His enduring legacy derived from what he accomplished quietly after the sensation created by his report had subsided. Largely due to Flexner's efforts John D. Rockefeller donated almost fifty million dollars to improve the nation's medical schools through the ministrations of the General Education Board. Within a few years a fine new school was established at the University of Rochester and major overhauls had been financed at schools such as Iowa, Cornell, Vanderbilt, and Washington at St. Louis. Nor did the tide stop there. As Flexner and members of the General Education Board anticipated, state rivalries erupted around the few schools selected for financial assistance.[71] Many of these states, when they failed to secure philanthropic help, took upon themselves the long-neglected fiscal responsibility for training their own physicians. Thus Flexner is properly remembered not so much for the fire he set as for his blueprint of the new structure which was to rise from the ashes.

NOTES

1   A. Flexner, *Medical Education in the United States and Canada* (New York: Carnegie Foundation, 1910).

2   Elements of the story appear in chapters by W. F. Norwood and J. Field in *The History of Medical Education*, ed. C. D. O'Malley, U.C.L.A. Forum Med. Sci. No. 12 (Los Angeles: Univ. of California Press, 1970), pp. 463–499 and 501–530.

3   N. S. Davis calculated that physicians and population had increased *pari passu* in the 35 years prior to 1876. *Contributions to the History of Medical Education and Medical Institutions in the United States of America 1776–1876* (Washington, D.C.: Government Printing Office, 1877), pp. 41–42. Davis makes no mention of men entering practice without degrees. Stern agreed when he estimated one physician for 572 persons in 1860 and one for 578 persons in 1900. B. J. Stern, *American Medical Practice in the Perspectives of a Century* (New York: Commonwealth Fund, 1945), p. 63. Stern is quoting the *Eighth Census* (Washington, D.C., 1860), pp. 670, 677. Flexner, *Medical Education*, p. 14, concurred in Stern's figures, but argued that the ratio was too great. Cur-

rent thought would tend to support Flexner, but transportation and distribution of physicians would demand a higher ratio for the last part of the 19th century.

4   F. H. Garrison, *John Shaw Billings: A Memoir* (New York: Putnam's, 1915), p. 256.

5   Samuel Chew, generally inclined to the rosy view, said, "A favorable change has already taken place. The number of well educated young men who engage in the study of medicine is every year increasing." *Lectures on Medical Education* (Philadelphia: Lindsay and Blakiston, 1864), p. 136. The bulk of contemporary editorial opinion of course was distinctly pessimistic.

6   Flexner, *Medical Education*, p. 30.

7   R. H. Shryock, "Public relations of the medical profession in Great Britain and the United States: 1600–1870," *Ann. Med. Hist.*, 1930, n.s. 2: 327.

8   T. R. Sizer, *Secondary Schools at the Turn of the Century* (New Haven: Yale Univ. Press, 1964), p. 4.

9   R. F. Butts summarized the trends in higher American education during the 19th century as follows:

growth of large elective curricula, broadening of the traditional B.A. degree, rise of the German ideal of the university, increase in teaching of science and technology, decline of the notion that intellectual discipline could be honed only on the classics, increasing secularism, greater personal freedom for students, and opening college to all classes including women. *A Cultural History of Education* (New York: McGraw-Hill, 1947), pp. 519–521.

Flexner, speaking of the South, which lagged behind the rest of the nation in this regard, described the scene as an educational "renaissance." *Medical Education*, p. 40. He reported some 300,000 students enrolled in public high schools in 1910 and about 120,000 (males) in colleges of the North and West as of 1908, excluding preparatory and professional students. *Ibid.*, p. 42.

10 F. V. N. Painter, *A History of Education* (New York: Appleton, 1898), p. 322.

11 On the negative side McIntire surveyed 222 New Jersey and Pennsylvania physicians in 1882 and found 178 with no collegiate preparation at all. Unfortunately he does not indicate how many of his sample took their medical training before the Civil War. C. McIntire, Jr., *The Percentage of College-Bred Men in the Medical Profession* (Philadelphia: American Academy of Medicine, 1883), pp. 3–6.

12 M. Fishbein, *A History of the American Medical Association: 1847–1947* (Philadelphia: Saunders, 1947), p. 243.

13 J. H. Rauch, *Report on Medical Education, Medical Colleges and the Regulation of the Practice of Medicine in the United States and Canada 1765–1890* (Springfield: Illinois State Board of Health, 1890), pp. iii–iv.

14 Flexner, *Medical Education*, p. 28.

15 *Ibid.*, p. 46.

16 *Ibid.*, p. 29.

17 H. A. Kelly and W. L. Burrage, eds., *Dictionary of American Medical Biography* (New York: Appleton, 1928). The study is by R. P. Hudson, "Patterns of medical education in nineteenth century America" (Unpublished essay for the M.A. degree, The Johns Hopkins University, 1966). The section dealing with premedical education is on pp. 23–34. This study will be referred to several times in the present paper, so a *caveat* is in order. The *D.A.M.B.* analysis is essentially statistical. This demanded the employment of certain arbitrary definitions and categories which are carefully spelled out in the essay. These should be consulted firsthand before any figures or conclusions from the study are used. The analysis included the reasons individuals were included in the *D.A.M.B.*, patterns of premedical, medical, and postgraduate education decade by decade, patterns of age, the influence of economic factors, effects of

father's occupation and marriage, and educational patterns of the various specialties.

18 W. F. Norwood *Medical Education in the United States Before the Civil War* (Philadelphia: Univ. of Pennsylvania Press, 1944), p. 406, states that the requirement for an attained age of 21 "seems to have been the only general regulation that was not grossly violated at some time during the decades covered by this survey." However the *D.A.M.B.* study revealed that 100 of the 1,513 men earned the M.D. degree before the age of 21. Hudson, "Patterns of medical education," p. 46. See also H. B. Shafer, *The American Medical Profession 1783 to 1850* (New York: Columbia Univ. Press, 1936), p. 82, and the implication of Daniel Drake's plea in *Practical Essay on Medical Education and the Medical Profession in the United States, 1832* (as reprinted in Baltimore by the Johns Hopkins Press, 1952), p. 57.

19 R. Hofstadter and C. D. Hardy, *The Development and Scope of Higher Education in the United States* (New York: Columbia Univ. Press, 1952), p. 20. The authors refer to W. P. Rogers, *Andrew D. White and the Modern University* (Ithaca: Cornell Univ. Press, 1942), pp. 23–29, and D. J. Struik, *Yankee Science in the Making* (Boston: Little Brown, 1948), chs. VI, X.

20 This is the estimate of R. M. Pearce, "An analysis of the medical group in Cattell's Thousand Leading Men of Science," *Science*, 1915, n.s. *42*: 264. Flexner dated the beginning of improved laboratory teaching at 1878, when Francis Delafield and William H. Welch opened clinical laboratories at the College of Physicians and Surgeons (N.Y.) and Bellevue Hospital Medical College respectively. *Medical Education*, p. 11.

21 Figures given are 198 in 1871, 2,382 in 1890 and 9,370 in 1910. Hofstadter and Hardy, *Development and Scope of Higher Education*, p. 64.

22 Flexner, *Medical Education*, p. 72.

23 Probably because of its French connections New Orleans apparently had a genuine go at bedside teaching as early as the 1850s. Speaking of his New Orleans School of Medicine in 1857, D. W. Brickell said flatly, "Six, out of ten, of us give *daily* bed-side instruction in the great Charity Hospital. . . ." *New Orleans Med. News & Hosp. Gaz.*, December 1857, *4*: 601. See also A. E. Fossier, "History of medical education, in New Orleans from its birth to the Civil War," *Ann Med. Hist.*, 1934, n.s. *6*: 432. For the French influence see R. Matas, *The Rudolph Matas History of Medicine in Louisiana*, ed. J. Duffy (Louisiana State Univ. Press, 1962), II, 237–268. Norwood appears to accept the New Orleans claim. *Medical Education . . . Before the Civil War*, p. 370.

Abraham Jacobi reportedly initiated bedside

pediatrics teaching in New York in 1862. *The American Pediatric Society 1888-1938* (Privately printed, 1938), p. 4. According to Shryock, Jacob Da Costa began bedside teaching at Jefferson around 1870. "Medicine in Philadelphia during the nineteenth century," *Bull. Soc. Med. Hist. Chicago,* 1948, *6*: 70. On the other hand, Osler maintained that as of 1890 there was "not a single medical clinic worth the name in the United States." W. Osler, "The coming of age of internal medicine in America," *International Clinics,* 1915, *4*: 3.

24   D. Riesman, "Clinical teaching in America with some remarks on early medical schools," *Tr. Coll. Physicians Phila.,* 1939, 3rd series, *7*: 109.

25   Norwood, *Medical Education . . . Before the Civil War,* p. 107.

26   Credit for the first graded curriculum is difficult to assign. To be answered the question must be put with precision. Who first advocated grading the curriculum? Where was it first attempted? Was it optional or required? Where did it first take hold? See F. C. Waite, "Advent of the graded curriculum in American medical colleges," *J. Assn. Am. Med. Colls.,* 1950, *25*: 315–322.

27   Quoted by Brickell, *New Orleans Med. News & Hosp. Gaz.,* December 1857, *4*: 599.

28   J. Carson, *A History of the Medical Department of the University of Pennsylvania* (Philadelphia: Lindsay and Blakiston, 1869), p. 172.

29   Shafer, *American Medical Profession,* p. 51.

30   Rauch, *Report on Medical Education,* p. iv.

31   Flexner, *Medical Education,* pp. 10–11.

32   In the *D.A.M.B.* sample (n. 17) subjects were categorized as degreed, nondegreed, and delayed, meaning the man earned a degree but only after he had begun medical practice. Forty-eight were delayed, and 68 nondegreed. The remaining 1,397 earned medical degrees before practising. In the first decade only 55.7 percent earned the M.D. Figures for the last four decades were 96.0, 99.4, 100, and 100 respectively. All figures excluded honorary and *ad eundem* degrees. Hudson, "Patterns of medical education," p. 53.

At the other extreme Norwood states that two-thirds to three-fourths of medical school matriculates did not remain to take the degree, implying that they entered practice without ever earning the M.D. *Medical Education . . . Before the Civil War,* p. 432. This estimate strikes me as too dismal. It was based on a "sampling of school circulars and catalogues and contemporary literature." The use of catalogues and circulars for this purpose is particularly precarious, since many students wisely took the second lecture course at a different school from the first. Such students would appear in successive catalogues as dropouts rather than transfers.

33   T. E. Keys, "Historical aspects of graduate medical education," *J. Med. Educ.,* 1955, *30*: 260. As Keys makes clear, this was not the first attempt at graduate medical education in America.

34   Flexner, *Medical Education,* p. 174. Flexner summarizes postgraduate schools as of 1910 on pp. 174–177.

35   House staff training is used here to cover all hospital-based clinical training taken after the medical degree. So used, the term must be approached with caution, because terminology describing hospital training shared the general educational confusion of the time. See, for example, *Graduate Medical Education* (Chicago: Univ. of Chicago Press, 1940), p. 98. The terms "resident" and "resident physician" have a tortuous history dating back at least 125 years. *Ibid.,* p. 97. The earliest hospital training programs centered around medical students who were called resident students. They were outgrowths of the indenturing practice such as that initiated in 1773 at Pennsylvania Hospital. Beginning in 1820 at City Almshouse of Philadelphia, house pupils were required to pay a fee and were dignified by the title of house physician or house surgeon. By 1823 a candidate for this position had to be a medical graduate. Norwood, *Medical Education . . . Before the Civil War,* p. 49. The following year the same held true of Pennsylvania Hospital, and a similar evolution occurred elsewhere. T. Morton and F. Woodbury, *The History of the Pennsylvania Hospital 1751–1895* (Philadelphia: Times Printing House, 1895), p. 480.

For most of the 19th century "interns" were medical students, the equivalent of the early hospital pupils. To illustrate, in 1880 the Board of Trustees of Charity Hospital in New Orleans decided that interns must have one year of medical study before appointment. A. E. Fossier, "The Charity Hospital of Louisiana," reprinted from *New Orleans Med. & Surg. J.,* May–October 1923, *75–76*: 38. As applicants for the position of intern gradually were required to be physicians, hospital pupils came to be known as externs, and in 1892 we find the two working side by side "without any official friction." *Ibid.,* p. 43. The Philadelphia and New Orleans examples are not offered as precise for other geographical situations, but rather to provide illustrations of the confusion that followed the evolution of hospital training positions.

The term "internship" reportedly first appeared about the time of the Civil War. *Graduate Medical Education,* p. 31. The word gained distinction only slowly and did not achieve its modern meaning until about 1914. From the end of the Civil War to the end of the century, however, "intern" continued to be used interchangeably with "resident" and "resident physician." Today's highly standardized residency program was patterned on the German system and formalized by William Halsted at the Johns Hopkins at the beginning of the 20th century. See W. S. Halsted, "The training of the surgeon." *Johns Hopkins Hosp. Bull.*, 1904, *15*: 271 ff. The current use of the word "resident" followed.

36 Fishbein, *A History of the American Medical Association,* p. 899.

37 *Ibid.,* p. 899–900.

38 *Ibid.,* p. 899. The source of this figure is not given and it is difficult to know if the year referred to is 1912 or 1913.

39 Hudson, "Patterns of medical education," p. 108.

40 T. N. Bonner, *American Doctors and German Universities* (Lincoln: Univ. of Nebraska Press, 1963), p. 23.

41 Hudson, "Patterns of medical education," p. 88. The five-year limitation was imposed in an attempt to eliminate the large number of physicians who toured Europe after years of successful practice.

42 Flexner, *Medical Education,* p. 46.

43 O'Malley, *History of Medical Education,* p. 504. Field apparently surveyed the founders' biographies in W. J. Meek, W. H. Howell, and C. W. Greene, *History of the American Physiological Society Semicentennial 1887–1937* (Baltimore, 1938).

44 Flexner, *Medical Education,* p. 10, quotes Welch, "Development of American medicine," *Columbia Univ. Quart. Suppl.,* December 1907.

45 N. S. Davis, *History of the American Medical Association* (Philadelphia: Lippincott, 1855), p. 37.

46 *Ibid.,* p. 117.

47 *Ibid.,* p. 115.

48 Dr. William Baldwin, president of the newly formed Association of American Medical Editors, suggested federal legislation in an address to the A.M.A. in 1869. Fishbein, *A History of the American Medical Association,* p. 79. A resolution supporting the establishment of a national school of medicine was introduced at the 1870 meeting of the A.M.A. (*Tr. A.M.A.,* 1870, *21*: 37–39), and the idea was revived in his presidential address of 1876 by J. Marion Sims. *Ibid.,* 1876, *27*: 94–95.

49 See O'Malley, *History of Medical Education,* pp. 507–508. Of 140 schools visited by the Council on Medical Education and reported to the A.A.M.C. in 1910, 68 were rated A, 38 B, and 34 C.D.F. Smiley, "History of the Association of American Medical Colleges, 1876–1956," *J. Med. Educ.,* 1957, *32*: 520.

50 H. S. Pritchett in Flexner, *Medical Education,* p. viii.

51 A. Flexner, *An Autobiography* (New York: Simon and Schuster, 1960), p. 74.

52 This story can be found in Smiley, "History of the Association of American Medical Colleges."

53 In the earlier years the A.A.M.C. had difficulty mounting a vigorous leadership role. Their higher standards usually were adopted only after a number of schools had led the way. In 1900 premedical requirements were upgraded, but the A.A.M.C. did not insist that member colleges require three years of collegiate preparation until 1952. *Ibid.,* p. 523. By 1904 66 member schools reported a required four-year medical curriculum with terms ranging from six to nine months. *Ibid.,* p. 518.

54 R. H. Fitz, "The legislative control of medical practice," *Med. Comm. Mass. Med. Soc.,* 1894, *16*: 306–307. Fitz's account is perhaps the best to come from the contemporary scene. To his mind, laws regulating medical practice were unpopular because (1) they were considered class legislation, (2) legislators suspected the motives of regular physicians, (3) the populace believed every person had a right to choose his own medical attendant, and (4) regular physicians themselves objected, out of a belief that the very act of regulating irregulars would exert a protective influence upon them. *Ibid.,* pp. 282 ff.

55 *Ibid.,* pp. 301–306. Fitz saw Thomsonianism as paving the way for homeopathy, "which proved to be the more effectual agent in annulling the licensing of physicians."

56 *Ibid.,* p. 306.

57 Davis, *History of the American Medical Association,* p. 35.

58 "What shall be our title?" *Am. Med. Times,* July 28, 1860, p. 63. The editorial is unsigned, but in a personal communication Gert Brieger expressed his conviction that Smith was the author.

59 J. B. Bardo, "A history of the legal regulation of medical practice in New York State," *Bull. N.Y. Acad. Med.,* 1967, *43*: 924–949. Bardo's narrative is revealing in that it shows the difficulties encountered in efforts at regulating licensure when neither the state, the medical schools, nor the profession had clearcut legal backing. In 1684 New York law prohibited medical practice without the "consent of those skillful" in it. In 1760 specified magistrates were empowered to examine and license, and in 1797 the magistrates could license by endorsing a preceptor's certificate. An 1806 law made the medical profession responsible through state and local medical societies. Three years later the University of the State of New York could authorize colleges to issue the M.D. degree, which was then a license to practice as well. By 1872 the Regents could appoint a board of examiners, which again made licensure a state function. In 1880 medical societies were divested of any licensing function, and by 1890 the M.D. degree no longer conferred automatic licen-

sure, which was now solely in the Regents' hands. To insure professional input, Boards of Medical Examiners were created, one each for regular practitioners, homeopaths, and eclectics. In 1893 existing laws were brought together as part of a Public Health Law and in 1907 the three separate boards were made one. At no point in this long evolution did enforcement of a law necessarily follow its enactment.

60   Fitz, "Legislative control," pp. 294, 313.

61   *Ibid.*, p. 315.

62   Flexner, *Medical Education,* pp. 167–173.

63   Spencer is quoted as saying, "Unpitying as it looks, it is best to let the foolish man suffer the appointed penalty of his foolishness. For the pain—he must bear it as well as he can; for the experience—he must treasure it up and act more rationally in the future." *Social Statics,* 1851, p. 373, quoted in Fitz, "Legislative control," p. 284.

64   A. Flexner, *Daniel Coit Gilman: Creator of the American Type of University* (New York: Harcourt, Brace, 1946), p. 35.

65   *Ibid.*, and especially R. H. Shryock, *The Unique Influence of the Johns Hopkins University on American Medicine* (Copenhagen: Munksgaard, 1953).

66   Flexner, *Medical Education,* p. 11.

67   *Ibid.*

68   The figures are from *The General Education Board* (New York: General Education Board, 1915), p. 161.

69   Flexner, *Medical Education,* p. 10.

70   In some current circles it has become fashionable to criticize Flexner for the nationwide adoption of the Hopkins' plan—for what has come to be called the tyranny of the four-year lockstep curriculum. This is akin to holding Galen responsible for the 1500 years his doctrines influenced medical thought after his death. Worse, it ignores Flexner's own unequivocal caveat, "In the course of the next thirty years needs will develop of which we here take no account. As we cannot foretell them, we shall not endeavor to meet them." *Medical Education,* p. 143.

71   This story is told in Flexner, *An Autobiography.* See esp. pp. 178 ff.

THE ALLIED HEALTH PROFESSIONS

For well over a century physicians have sought to monopolize, or at least to control, health care in America. Yet despite the acquisition of unparalleled professional power, they have, as Ronald L. Numbers points out, repeatedly been forced to settle for considerably less control than they desired. In addition to the various sectarian healers, whom regular physicians hoped to suppress, there were dentists, midwives, optometrists, podiatrists, pharmacists, and psychologists competing for slices of the health-care pie. The division of labor among these diverse occupational groups seems at times to defy logic. Why, for example, did the treatment of most of the body evolve into medical specialties controlled by physicians, while the care of the teeth and feet fell to practitioners outside the medical profession? History provides some clues.

During the second quarter of the 19th century a handful of physician-dentists sought to turn their field into a medical specialty, like surgery or ophthalmology. But their suggestion that dentistry be included in the medical curriculum met with the response that it was too mechanical. After all, dentists filled and extracted teeth; they did not heal them. Rebuffed by their erstwhile colleagues, these dentists about 1840 founded their own society, started their own journal, and opened their own school, the Baltimore College of Dental Surgery. Within a few decades American dentistry was preeminent in the world, far surpassing American medicine in international repute.

Podiatry, also known of chiropody, developed from a similar quasi-medical background. When doctors specializing in the care and treatment of the feet urged medical schools at the turn of the 20th century to assume responsibility for training in this area, deans turned them away with the comment that corns and bunions were too trivial to warrant such attention. So with the medical profession's blessing, podiatrists, like dentists a half century earlier, set up their own institutions.

Pharmacists and optometrists trace their roots not primarily to medicine but to the businesses of drug selling and "spec" peddling. Both professions worked out reasonably harmonious divisions of labor with physicians, distinctions ultimately sanctioned by law. Pharmacists prepared and dispensed drugs, but left the prescribing to physicians. Optometrists tested for corrective lenses, but referred treatment of the eyes to ophthalmologists. In recent years, however, these distinctions have become increasingly blurred as pharmacists and optometrists have sought, often successfully, to expand the range of their clinical activities.

The female-dominated fields of midwifery and nursing developed along different historical lines. As Charlotte G. Borst shows, the demise of midwifery in the 20th century resulted more from the failure of midwives to organize themselves professionally than from the machinations of physicians. Besides, it was not as easy for midwives and obstetricians to divide their work as it had been for, say, optometrists and ophthalmologists. Nurses, in contrast, flourished numerically by submitting to the authority of physicians.

13

The "Connecting Link": The Case for the Woman Doctor in 19th-Century America

REGINA MORANTZ-SANCHEZ

Little more than a decade before the Civil War, Elizabeth Blackwell, daughter of a prominent reform-minded family, achieved notoriety by becoming the first woman to earn a medical diploma in the United States. Soon after she received her degree in 1849, Geneva Medical College, a small regular institution in upstate New York, closed its doors to women. Undaunted, women continued to seek medical training. Within two years, three female students at the eclectic Central Medical College in Syracuse — the first coeducational medical school in the country — gained medical licenses.

In Philadelphia a group of Quakers led by Dr. Joseph Longshore pledged themselves to teach women medicine and established the Woman's (originally Female) Medical College of Pennsylvania in 1850. The following year eight women graduated in the first class. A Boston school, founded originally by Samuel Gregory in 1848 to train women as midwives, gained a Massachusetts charter in 1856 as the New England Female Medical College. Here Marie Zakrzewska, former medical associate of Elizabeth Blackwell, early female graduate of the Cleveland Medical College, and the influential founder of the New England Hospital for Women and Children, came to teach in 1859.

Meanwhile, in New York the Homeopathic New York Medical College for Women, established in 1863 by still another early graduate, Clemence Lozier, enjoyed such success that by the end of the decade it had matriculated approximately 100 women. Five years later Elizabeth Blackwell, now joined in practice by her sister Emily, also a graduate of Cleveland, opened the Women's Medical College of the New York Infirmary.

By 1870 a handful of medical schools, both orthodox and sectarian, accepted women on a regular basis. In fact, over 300 women had already graduated, though mostly from sectarian institutions. Several of these women founded dispensaries in New York, Boston, and Philadelphia to offer needed clinical training to the increasing numbers of female colleagues. Of course, opportunities for women in medicine remained circumscribed, for throughout the century the majority of institutions barred them from attendance. Nevertheless, the ranks of female physicians grew: by 1900 their numbers reached an estimated 7,387.[1]

Women entered the medical profession as part of a broader 19th-century movement toward self-determination in which all reformist women, from conservative social feminists to radical suffrage advocates, played a significant part. Like many members of their sex, women doctors sought to examine and redefine the concept of womanhood to fit the changing demands of a complex, industrializing society. Furthermore, the move to educate women in medicine grew out of the antebellum health reform crusade. Abolitionists, peace advocates, temperance reformers, and women's righters participated in the drive to improve the nation's health, and middle-class women were particularly active in health and dietary reform in these years.[2]

Indeed, medicine attracted more women votaries in the 19th century than any other profession except teaching. Female physicians took seriously their role in health education. For many, medical training appeared the logical outcome of their prior interest in health issues. Borrowing arguments

REGINA MORANTZ-SANCHEZ is Professor of History at the University of Michigan, Ann Arbor, Michigan.

from the health reformers, they emphasized the importance of giving information on hygiene and physiology to all women. Women's new and central role in family life made systematic knowledge imperative. Female physicians, as teachers and clinicians concerned with such issues, would be essential to improving the health and nurturing the expertise of all women.

Although they remained a small minority, women doctors were conspicuous because they violated 19th-century norms of female behavior. Consequently they became the focus of a vigorous controversy over women's proper role in and relationship to public and private health. Whereas amateur female instructors of physiology could be dismissed as objects of public ridicule, professionally trained women doctors were entirely another matter. By the end of the 1860s protests against them mounted from within the profession, forcing them and their supporters to refine and elaborate an ideology to defend their cause.

The arguments with which they chose to justify themselves revealed women physicians both as ideological innovators and as daughters of their century. Sharing this role with other women of their generation who sought an expansion of their activities, their ideas fell well within the mainstream of feminist thinking. Their ideology served a dual purpose. As self-explanation it enabled them to convey to the world what they hoped to accomplish and why. By their use of ideas they attempted to place themselves in cultural and historical perspective. Yet their ideology also exerted a powerful influence on the way they perceived reality, shaping that perception and giving it meaning within the context of Victorian values.

This essay examines the means by which women physicians justified their role in the 19th-century health revolution. Because their arguments deviated only slightly from those of others favoring a more active place for women in society, such an investigation illustrates how similar groups of 19th-century women came to grips with their culture. The reasoning of women physicians was often brilliant and effective, and their practical work extremely important. Yet their ties to 19th-century values also impaired their vision. Only a handful of them understood the ways in which a self-limiting conception of their potential contribution hampered their progress within the profession.

I

The opponents of medical education for women deserve attention first because their objections fixed the context of the debate. Women doctors remained sensitive to criticism, realizing that professional opposition reached far beyond ideology in its implications, sharply curtailing their opportunities to study and practice.

Most doctors were traditionalists who shared the same ambivalence towards women's role that dominated 19th-century American culture. Placing women on a pedestal and cementing them firmly within the confines of the home, they justified an emotional preference for sequestered women by making them the moral guardians of society and the repositories of virtue. Fearing that women who sought professional training would avoid their childrearing responsibilities, they reminded their colleagues in overworked metaphors that "the hand that rocks the cradle rules the world." Woman, argued a spokesman, held "to her bosom the embryo race, the pledge of mutual love." Her mission was not the pursuit of science, but "to rear the offspring and ever fan the flame of piety, patriotism and love upon the sacred altar of her home."[3]

Rational legitimation of the female role often veiled less rational preferences: the home represented for 19th-century Americans a refuge from an immoral and often brutalizing world. A woman who dared to move beyond her sphere was "a monstrosity," an "intellectual and moral hermaphrodite." Nevertheless, insisted Dr. Paul de Lacy Baker, women controlled society, government, and civilization through the "home influence." Home was

> the place of rest and refuge for man, weary and worn by manual labor, or exhausted by care and mental toil. Thither he turns him from the trials and dangers, the temptations and seductions, the embarrassments and failures of life, to the one spot beneath all the skies where hope and comfort come out to meet him and drive back the demons of despair that pursue him from the outside world. There the sweet enchantress that rules and cheers his home supports his sinking spirits, reanimates his self-respect, confirms his manly resolves and sustains his personal honor.[4]

While revering the purity and repose of the home, doctors, like other Victorians, feared the animal in man and dwelt on the significance of female moral superiority in curbing man's most brutal instincts. Woman's venturing out into the world would bode ill for civilization, for women kept men respectable. In copying men, they ran the risk of demoralizing both sexes. Men, confessed Dr. J. S. Weatherly to his colleagues, were "little less than brutes," and "where men are bestialized, women suffer untold wrongs." Woman's great strength and safety, he concluded, was in the institution of marriage, and "everything she does to lessen men's respect and love for her, weakens it, and makes her rights more precarious; for without the home influence which marriage brings, men will become selfish and brutal; and then away go women's rights."[5]

Traditionalists also worried that teaching women the mysteries of the human body would affront female modesty. "Improper exposures" would destroy the delicacy and refinement that constituted women's primary charms. Men recoiled at the thought of exposing young women to the "blood and agony" of the dissecting room, where "ghastly" rituals were performed.[6]

Despite the popularity of this defense of female delicacy, traditionalists compromised their case when they admitted that women's ready sympathy made them excellent nurses. Praising Florence Nightingale's achievements in the Crimea, they credited them primarily to her ignorance of scientific medicine. Medical education, they argued, would surely have hardened her heart, leaving her bereft of softness and empathy.[7]

Supporters of female education quickly discovered, however, that respect for Victorian delicacy could work in their favor. Was the mother who nursed her family at the bedside ever shielded from the indelicacies of the human body? they asked. Furthermore, if the issue was female modesty, then why should men — even medical men — ever be allowed to treat women? As the use of pelvic examinations became part of ordinary practice, male physicians posed a greater threat to feminine delicacy than women practitioners. Indeed, many supporters of medical training for women were conservative champions of female modesty.[8] Though the arguments of these supporters were extremely persuasive within the context of Victorian values, traditionalist physicians continued throughout the century to object to training women on the grounds of Victorian delicacy.

Some male physicians alleged other unsuitable character traits against women besides their innocence. Many agreed that Nature had limited the capacity of women's intellect. They were impulsive and irrational, unable to do mathematics, and deficient in judgment and courage. Their passivity of mind and weakness of body left them powerless to practice surgery. And if these disadvantages were not enough, there remained the enigmatic side of the female temperament. Dependent, "nervous," and "excitable," women, "as all medical men know," were subject to hysteria over which they had no control. "Hysteria," regretted J. S. Weatherly, M.D., "is second Nature to them."[9]

Because at least some women physicians in the 19th century received inferior training, doctors often mistook poor preparation for innate ignorance. On this point critics were most disingenuous. The issue became particularly controversial as local and national medical societies began to debate the admission of female members in the 1870s and 1880s. Though circular, the reasoning seemed incontrovertible — at least for a time. First-rate regular medical schools barred women from attendance. Meanwhile opponents unfairly stigmatized the women's medical colleges as either irregular or of poor quality. While denying them access to the kind of education they could approve, critics held their alleged inferior training against them, often refusing to consult with women physicians and ostracizing male practitioners who did. The final irony was that a fair number of medical women received excellent training in the 19th century. A comparative study of curriculum and clinical offerings in several 19th-century medical schools suggests that the women who earned their diplomas from the New York Infirmary and the Woman's Medical College of Pennsylvania were exposed to a vigorous, demanding, and comparatively progressive course of study.[10] Self-conscious of their need for proper preparation, other women sought postgraduate training in Europe.

Probably much of the grumbling over inferior training arose out of disgust for the multiplication of proprietary schools in the 1830s and 1840s and the resulting sharp increase in the number of practitioners. Some physicians complained of the possibility of increased economic competition if women were admitted to study. Such concern dissipated toward the end of the century when women doctors appeared to segregate themselves in certain specialties. What remains most striking about this objection, however, is that it took for granted not only women's ability to practice medicine, but also their ready acceptance by the public.[11]

Rejecting many of the preceding arguments, an increasingly influential group of medical men preferred instead to take a more scientific posture. Rallying around Harvard professor E. H. Clarke's book *Sex in Education: A Fair Chance for Girls*, published in 1873, they based their case against women entirely on the predominating negative influence of their biology. When they chose to emphasize the debilitating and still mysterious effects of menstruation, traditionalists were indeed effective. Physicians knew little about the influence of women's periodicity, and the culture treated menstruation as a disease.[12] Reasoning that only rest could help women counteract the weakness resulting from the loss of blood, complete bedrest was commonly prescribed. Thus even if opponents appeared willing to concede women's intellectual equality—and many were prepared to do so—women's biological disabilities seemed insurmountable.[13] Since menstruation incapacitated women for a week out of every month, could they ever be depended on in medical emergencies?

Female physicians helped to dispel doubts about the effects of menstruation in their own professional lives. A few investigated the problem scientifically. Outraged by the influence of E. H. Clarke's book, the feminist community in Boston cast about for a woman doctor with the proper credentials to call its thesis into question. In 1874 women gained a public forum when Harvard announced that the topic for its celebrated Boylston Essay would be the effects of menstruation on women. Writing to Dr. Mary Putnam Jacobi in 1874, C. Alice Baker urged her to take up the "good work," and "win credit for all women, while winning for yourself the Boylston Medical Prize for 1876." Jacobi met the challenge; and her pioneering essay, "The Question of Rest for Women During Menstruation," which, to the opposition's chagrin, won the esteemed prize, was exemplary in its modern statistical methodology. Her conclusion, that there existed "nothing in the nature of menstruation to imply the necessity, or even the desirability, of rest for women whose nutrition is really normal," directly challenged conservative medical opinion.[14] Despite widespread respect for her paper, it failed to convince many physicians, and the debate continued into the 20th century.

In the realm of emotion, of course, apologists for women's sphere needed to be neither scientific nor consistent; they offered arguments for female inferiority, vulnerability, and dependence alongside claims for their moral superiority and responsibility. After all it was they who molded the life of future generations and spent tireless hours tending the sick long after the physician had retired from his daily rounds. Nineteenth-century society never did come to terms with this Manichean image of women, at least not until female physicians and other social feminists exploited the contradictions in the ideal and forced a reconciliation of the two extremes. In the process they created a new image for their sex and a broader definition of woman's sphere.

II

Women physicians argued within the context of the shared values of 19th-century culture. They and their male supporters took seriously the idea of their moral superiority and their ability as natural healers, embracing 19th-century definitions of feminine qualities intact. With intelligence and skill they made Victorian ideology yield to them and used these notions to justify seeking medical education.

One example of their subtle use of the value structure has already been noted: their reversal of the "delicacy" argument. Defending the "propriety of entrusting feminine life to feminine hands," they denounced exclusive male attendance to women's ailments as an outrage against female modesty pernicious to the "social and moral welfare of society."[15] Elizabeth Blackwell called it an "unnatural and monstrous arrangement" that women had "no resort but to men, in those diseases peculiar to themselves." Were the "methods of modern medicine quietly received and passively submitted to," she claimed, "it would indicate a terrible deficiency in some of the most important elements of womanly character." Here medical science clashed with morality, in the demand that to preserve morality women be taught science.[16]

Female physicians did not quarrel with the concept of separate spheres for men and women, though they meant something quite different from their opponents when they spoke of "woman's sphere." They sought an altogether novel definition of "femaleness." Examining the ethical implications of scientific methods for medicine and society, they claimed for women the task of integrating Science and Morality.

A glance at the titles of the popular health manuals of the period reveals the pervasiveness of the concept of Woman as Healer.[17] In response to these widespread assumptions, traditionalists had glorified women's abilities as nurses, while denying them the right to become physicians. Supporters of medical training for women were outraged by such logic. "Is not Woman man's Superior?" asked Dr. J. P. Chesney of Missouri. "It is an idea extremely paradoxical," he continued, "to suppose that woman, the fairest and best of God's handiwork, and practical medicine, a calling little less sacred than the holy ministry itself, should, when united, become a loathsome abomination . . . from which virtue must stand widely aloof." If the traditionalists' own logic were applied consistently, he noted, "men would long ago have been banished from obstetrics."[18]

Zealously cataloguing women's past contributions to healing, supporters of female medical education argued that women needed the tools of modern medicine. Women *would* attend in the sickroom, and instinct and sympathy were not sufficient as advances in medical therapeutics rendered folk medicine ineffective. "It has begun to occur to people," observed Dr. Emmeline Cleveland of the Woman's Medical College of Pennsylvania in 1859, "that perhaps the fullest performance of her own home duties" required of woman "a more extended and systematic education . . . especially in those departments of science and literature which have practical bearing upon the lives and health of the community."[19]

Women doctors and their allies frequently echoed Victorian sentimentality over womanly attributes. Like their opponents they constantly connected womanhood with the guardianship of home and children. Women were morally superior to men, claimed Elizabeth Blackwell, because of the "spiritual power of maternity." The true physician, male or female, she argued, "must possess the essential qualities of maternity."[20]

Medical training was necessary for scientific motherhood. Ignorance of their own bodies and poor training in child management were taking their toll on American mothers and offspring alike. "What higher trust could be dedicated to the wife and mother," asked Dr. Joseph Longshore in his introductory lecture to the first class at the Woman's Medical College of Pennsylvania, "than guardianship of the health of the household?" His colleague Emmeline Cleveland, a brilliant gynecological surgeon, affirmed the necessity of giving to all women knowledge of the human body. She reminded her students that their high vocation was "as nature's appointed guardians of childhood and youth," that as mothers and teachers they would "become the conservators of public health and in an eminent degree responsible for the physical and moral evils which afflict society."[21]

Medical women like Cleveland intended to play a central role in the elevation of their sisters. As science was brought to bear on domestic life, women physicians would become the "connecting link" between the science of the medical profession and the everyday life of women.[22] To accomplish this purpose, each of the female medical schools offered courses in physiology and hygiene to nonmatriculants, who were often mothers and teachers hoping to gain knowledge in health education.

When critics charged that medical training was wasted on women who would eventually marry and have children, female physicians responded by pointing out that medical knowledge was important for any woman, even if the skills acquired would not be used to practice. Competence in medicine made women better mothers. Others were bolder, displaying the temerity that helped to expand society's conception of woman's role: they saw no necessary conflict between motherhood and general practice. This recognition of the possibility of combining marriage and career marked a radical departure from 19th-century thinking. "A woman can love and respect her family just as much if not more," asserted Dr. Georgiana Glenn, "when she feels that she is supporting herself and adding to their comfort and happiness." Dr. Mary Putnam Jacobi agreed. Conceding that marriage complicated professional life, she nevertheless felt that "the increased vigor and vitality accruing to health women from the bearing and possession of children, a good deal more than compensates for the difficulties involved in caring for them, when professional duties replace the more usual ones, of sewing, cooking, etc."[23]

The social Darwinism of the post-Civil War period enabled traditionalists to reiterate their prejudices with the finality of scientific truth. Little changed in the controversy over female physicians besides the language of the debate. Scientific rationalism predominated by the 1880s, with evolution and eugenics mustered in defense of both sides. Critics of women doctors made pessimistic pronouncements that female higher education would be biologically destructive of the race.[24] Female physicians countered with measured optimism. They depicted themselves as living examples of the transition to higher life forms. Yet they also remained suspicious of prevailing trends, especially the frivolous pursuits of leisured women. Warning that the increased leisure accruing from technological advances demanded that women be given noble work to do, they urged society to check the notorious aimlessness of the civilized woman's life. Women's boredom was notorious: "For one case of breakdown from overwork among women," quipped Dr. Ruffin Coleman, "there are a score from ennui and sheer inanation from doing nothing."[25]

Along with most American scientists of the period, physicians accepted the neo-Lamarckian concept of the inheritance of acquired characteristics. Women doctors drew the logical object lesson: if mental as well as physical characteristics were inherited, the race would steadily improve only if women could uplift themselves. Their arguments remained a warning as well as a prophecy: Hold back your women—your mothers—and you retard the race.[26]

Medical women also insisted that they had special contributions to make to the profession. Feminization could enhance the practice of medicine, which concentrated on the eradication of suffering. Association with female colleagues would "exert a beneficial influence on the male." Combining the best of masculine and feminine attributes should raise medical practice to its highest level.

Occasionally supporters carried the implications of this reasoning even further. Female physicians expected to challenge heroic therapeutics directly. As the "handmaids of nature," women would place greater value on the "natural system of curing diseases . . . in contradistinction to the pharmaceu-

tical." They would promote a "generally milder and less energetic mode of practice." "The Past," claimed the health reformer Mary Gove Nichols, "with the lancet, and poision, and operative surgery, did not insult woman by asking her to become a physician; and the Past has not asked her to become a hangman, general or jailer. We may well excuse all believers in Allopathy, if they judge woman unfit for the profession."

Many women physicians did spurn heroic medicine. The husband of Hannah Longshore, the first female physician to establish a practice in Philadelphia, recorded the following in a biographical sketch of his wife:

> The Woman's Medical College claimed to be an entirely regular or old school institution and its faculty had a testimony to bear against homeopathy and eclecticism or in any irregularity of its graduates from the established old school practice. But many of its alumni [sic] discovered that the growing aversion to large doses of strong and disagreeable medicine among the more liberal and progressive elements in society and that many intelligent woman had become tinctured with the heresy of Homeopathy and gave a preference to the physician who would prescribe or administer their milder and pleasant remedies, and especially for the children who would take their medicines voluntarily. This discovery led the woman doctor to an investigation of their remedies and theories of therapeutics and to partial adoption of their remedies and methods of treatment. This conformity to the demands for mild remedies gave the women doctors access to many families whose views were in accord with the reform movements that recognized the growing interest in enlarging the sphere of woman. The woman doctors who saw that the door was opening for this reform of regular practice and prepared themselves accordingly were the first to get into successful business.

Marie Zakrzewska, who, as founder of the New England Hospital for Women and Children in Boston, earned the respect of even the most stubborn members of the male opposition, also remained skeptical of heroic dosing. In a letter to Elizabeth Blackwell she confessed that her whole success in practice was based on the cautious use of medicine, "often used as Placebos in infinitesimal forms." "In fact," she wrote, "I have the reputation among my large clientele, men, women, and children as giving hardly any medicine but teaching people how to keep well without it. This subject is a large theme and I am thankful from the innermost of my emotions . . . that nobody has ever been injured, if not relieved by my prescriptions."[27]

In keeping with this interest in prevention rather than cure, friends claimed that women physicians would become zealous advocates of public health and social morality. Emmeline Cleveland noted that women were naturally altruistic, while Elizabeth Blackwell expected her female colleagues to provide the "onward impulse" in seeing to it that human beings were "well born, well nourished, and well educated." Dr. Sarah Adamson Dolley urged women doctors to bring to the profession their "moral power." "Educated medical women," wrote Dr. Eliza Mosher, of the University of Michigan, "touch humanity in a manner different from men; by virtue of their womanhood, their interest in children, in girls and young women, both moral and otherwise, in homes and in society." Most of their male supporters agreed. Dr. James J. Walsh admitted that men did not recognize their social duties as readily as women. "Therefore," he confessed, "I have always welcomed the coming into the medical profession of that leaven of tender humanity that women represent."[28]

III

It is possible to claim that women doctors made their arguments, not necessarily because they believed them, but because they sensed the pervasiveness of traditional ideas about woman's sphere. Their rhetoric, as well as that of other feminists, has been interpreted as a calculated and well-planned offensive designed to challenge 19th-century assumptions with familiar ideas. Yet women physicians sincerely believed that they were different from men and that they had their own special contribution to make to society. Of course many of them realized the expediency of their case, especially because it did not wholly challenge traditional assumptions. Yet the desire to educate women in medicine remained part of 19th-century social feminism, a movement with roots in antebellum reform and nurtured in an atmosphere of perfectionist concern with the revival of moral values. Female physicians were not the only group of women to seek a broadening of women's role by expanding the notion of separate spheres.

Certainly many women chose a career in medicine for more private reasons than the belief in their own special female abilities. Many probably saw medical practice as a lucrative means of self support. But whatever their personal motives, such women belonged to a movement that justified itself in larger terms, and they gained their self-image from the social context in which they acted. After wishing the New York Infirmary graduating class of 1899 financial success — "we are always glad to hear of a woman's making money" — Dean Emily Blackwell urged her students to remember that "there are other kinds of success that . . . we hope you will always consider far higher prizes." These, she continued, were "the consciousness of doing good work in your own line, of being of use to others, of exerting an influence for right in all social and professional questions." Readily conceding that her students "doubtless all entered upon medical study from individual motives," she hoped that they had learned "that the work of every woman physician, her character and influence, her success or failure, tells upon all, and helps or hinders those who work around her or come after her."

Forty years earlier, a younger Emily Blackwell confided similar sentiments to her diary when she thanked God that she was only 25 and not yet too old to commence a life's labor full of "great deeds." Newly decided on a medical career, she prayed that at life's close God would grant her the ability to look back on a "woman's work done for thee and my fellows." Opportunities were then appearing for women to live a "heroic life," and Emily desperately wished to avail herself of them.[29]

Victorian ideology, then, was not entirely repressive to women. On the contrary, courageous individuals hoping to move out of the home could use the ideology for their own ends, and in a way consistent with their acceptance of prevailing values. The glorification of woman's moral power was their most potent weapon. Taking refuge in the concept of female superiority, they provided future generations of women with an imperfect but essential stepping stone on the route to sexual equality.

While both sides in the debate claimed to be seeking moral progress and civilization's advancement, female physicians, in their commitment to using women's abilities systematically and scientifically, diverged fundamentally from their conservative opponents. Women doctors hoped to reform society by feminizing it, a task which required the professionalization of "womanhood." Acknowledging that their goals required a broader interpretation of woman's sphere, they felt this a small price to pay for a morally righteous and civilized America.

Nineteenth-century women doctors never drifted out of touch with the mainstream of Victorian ideology. As proponents of the gradual expansion of women's role, they perceived gradual change to be the only kind the public would tolerate. Slowly they succeeded in creating a positive image for the female physician. A minority proved that wives and mothers could handle a professional career. Their inevitable interaction with male colleagues eventually convinced many critics that women could be competent doctors and still maintain their femininity.

Female physicians confined themselves to feminine specialties—obstetrics and gynecology in the 19th century, pediatrics, public health, teaching, and counseling later on. Such specialization was not due solely to resistance from male professionals, although women doctors occasionally blamed discrimination. Women practitioners gravitated to these specialties because they were conscious of their "special" abilities. They concerned themselves with the health problems of women and children because they hoped to raise the moral tone of society through the improvement of family life.

Doubtless for some unmarried women physicians, the practice of obstetrics or pediatrics gave them vicarious fulfillment from an intimate involvement in the primary events of the female life cycle, while freeing them from traditional Victorian marriage. Recently, historians have explored the female support systems that existed throughout the 19th century.[30] Many women physicians gravitated to feminine specialties—in fact, to medicine itself—out of a desire to perpetuate these support systems. Mothers, they argued, would more readily discuss their problems with women doctors because male physicians did not have "the patience to deal with the anxieties, ignorance, and frequent terrors of women who are often overwhelmed by some misunderstood condition in themselves or in their children." During the sen-

sitive period of gestation, a woman needed someone whom she could approach "without reserve," and from whom she could get "a woman's sympathy." "Being women as well as physicians," acknowledged Dr. Florence B. Sherbon, president of the Iowa State Society of Medical Women, "we share with our sex in the actual and potential motherhood of the race. Being women we make common cause with all women. . . . And being women and mothers, our first and closest and dearest interest is the child. . . ."[31]

Such arguments naturally had limitations. Confining themselves to women's concerns circumscribed women physicians' professional influence. A few even willingly advocated an informal curtailment of their medical role, hoping to gain support by taking themselves out of competition with men.[32] But the more perceptive disdained this approach. Such women converted early to the modern and empirical world of professional medicine, and their first love was science. Often uneasy in the moralistic world of their medical sisters, they exhibited a toughness and clarity of vision which set them apart from those women who used medicine primarily as a moral platform. Physicians like Mary Putnam Jacobi and Marie Zakrzewska insisted from the beginning that medical women need be of superior mettle. Fearing that specialization in diseases of women and children would mean a loss of grounding in general medicine, they warned that women would be justly relegated to the position of second-class professionals. Eventually their performance even within their specialty would become second-rate. If women were to succeed in medicine, they asserted, they had to be thoroughly trained.[33] Despite their predictions, specializing remained popular throughout the 19th century and into the next because it provided advantages in blunting the resentment of male colleagues.

The emphasis female physicians placed on the mother's child-rearing responsibility also reflected and reinforced social values which would present future problems. As more married women entered the work force, acute ideological conflicts arose to hamper the development of alternative methods of child-rearing. When Charlotte Perkins Gilman proposed to professionalize child care at the turn of the century, she was virtually ignored. Even more revealing is the intensity of the conflict over child care centers that emerged within the federal government during World War II, a time when married women were entering the work force in large numbers and government facilities were a palpable necessity.[34]

While stressing women's peculiar adaptability to medicine, women doctors perpetuated an exaggerated concept of womanhood, rendering their arguments inapplicable to other, less obviously "feminine" pursuits. Dr. Frances Emily White attributed women's great success in medicine compared with other professions to a "peculiar fitness" for the work and the lack elsewhere "of equal opportunities for the exercise of those qualities that have become specialized in women."[35] Pursuits like teaching and nursing fit the pattern well, but law did not. Though women lawyers chose to justify their interests in a similar fashion, it was harder for them to claim that their work was an extension of women's natural sphere, and, indeed, few women preferred law to medicine in the 19th and early 20th centuries. All the reasons for this disparity remain complex, but the "natural sphere" argument did exhibit vexing limitations as women moved out of the home and into the world.[36]

A few perceptive individuals struggled uncomfortably with the implications of such reasoning. The journalist Helen Watterson, for example, denounced the woman movement's emphasis on "woman's qualities." Mary Putnam Jacobi quipped that "recently emancipated people are always bores, until they themselves have forgotten all about their emancipation." And Marie Zakrzewska frowned on women who chose medicine out of female "sympathy." The only motives the profession permitted its votaries, she maintained, were "an inborn taste and talent for medicine, and an earnest desire and love of scientific investigation."[37]

The brilliant Jacobi sensed the psychological disadvantages that hampered women physicians from attaining equal status within the profession. Society was still against them, impairing both their own confidence and that of other women in them. Because society refused to judge medical women by their achievements, women doctors were in danger of setting lower standards for themselves. In the 19th century any woman who ventured beyond the domestic role was considered an anomaly. Jacobi insistently urged her students to measure themselves by the highest standards of professional excellence. Mediocre women doctors, she warned, would doom their cause.

Jacobi, forever impatient with the deficiencies of the women's medical schools, tolerated no compromises in quality because of their disadvantaged position. Her tough-mindedness remained a source of inspiration to those who occasionally lost sight of the larger goal. Still the courage and conviction of the early generations of medical women are beyond reproach. In the face of strenuous opposition they pressed for the right of women to study medicine as part of a redefinition of woman's role in a reformed industrial society. Although their goals remained circumscribed by a particular cultural vision, these pioneers effected real changes in their own lives and in the lives of many women around them. They did so without *radical* ideological innovation. Paving the way for the future by their deeds, they failed to understand that true equality between the sexes could not come if women remained confined to a "sphere," no matter how expansively it was defined. Nineteenth-century women doctors passed on a large legacy of unfinished business to their daughters and granddaughters. They left future generations of women to struggle with the implications of Mary Putnam Jacobi's prophetic warning, "if you cannot learn to act without masters, you evidently will never become the real equals of those who do."[38]

NOTES

1 The last statistic is taken from a survey done by W. C. Hunt, statistician, of Washington, D.C., reprinted in H. Scott Turner, "History of women in medicine," *Los Angeles J. Eclectic Med.*, 1905, 2: 125. The number of male physicians in the United States in 1900 is in dispute. Census records set the figure at 132,000, the A.M.A. at 119,749. See U.S. Bureau of the Census, *Historical Statistics of the United States: Colonial Times to 1970* (Washington, D.C., 1975), Part 1, pp. 75–76.

2 See William B. Walker, "The health reform movement in the United States, 1830–1870" (Ph.D. dissertation, Johns Hopkins Univ., 1955); John Blake, "Health Reform," in *The Rise of Adventism: Religion*

and Society in Mid-Nineteenth-Century America, ed. Edwin S. Gaustad (New York, 1975), pp. 30–49. For health reform and women see Regina Markell Morantz, "Nineteenth century health reform and women: a program of self-help," in *Medicine Without Doctors: Home Health Care in American History,* ed. G. Risse, R. Numbers, and J. Leavitt (New York, 1977), pp. 73–93, and "Making women modern: middle class women and health reform in nineteenth century America," *J. Soc. Hist.,* 1977, *10:* 490–507.

3   W. W. Parker, M.D., "Women's place in the Christian world: superior morally, inferior mentally to man—not qualified for medicine or law—the contrariety and harmony of the sexes," *Tr. Med. Soc. St. Va.,* 1892, pp. 86–107.

4   Paul de Lacy Baker, "Shall women be admitted into the medical profession?" *Tr. Med. Assn. St. Ala.,* 1880, *33:* 191–206. See also Julien Picot, "Shall women practice medicine?" *North Carolina Med. J.,* 1885, *16:* 10–21; N. Williams, "A discussion on female physicians," read before the Clay, Lysander and Schroeppel (N.Y.) Medical Association, *Boston Med. & Surg. J.,* 1850, *43:* 69–75; J. F. Ziegler, "Women's sphere," Presidential address to the Medical Society of Pennsylvania, *Tr. Med. Soc. St. Penn.,* 1882, *14:* 25–38; Joseph Spaeth, "The study of medicine by women," *Richmond & Louisville Med. J.,* 1873, *16:* 40–56; *Men and Women Medical Students and the Woman Movement* (Philadelphia, 1869), *passim.*

5   J. S. Weatherly, "Woman: her rights and her wrongs," *Tr. Med. Assn. St. Ala.,* 1872, *24:* 63–80. For a reversal of the argument, defending female medical education as a step *up* from primitive brutality, see Edwin Fussell, *Valedictory Address to the Students of the Female Medical College of Pennsylvania* (Philadelphia, 1857), pp. 5–6.

6   Reynell Coates, *Introductory Lecture to the Class of the Female Medical College of Pennsylvania* (Philadelphia, 1861), pp. 3–4. See also articles cited in n. 4 and n. 5. "Female physicians," *Boston Med. & Surg. J.,* 1856, *54:* 169–174, and "Female practitioners of medicine," *Boston Med. & Surg. J.,* 1867, *76:* 272–274.

7   Medical women, of course, claimed just the reverse. See Weatherly, "Woman: her rights," p. 76; Sophia Jex-Blake, *Medical Women: Two Essays* (Edinburgh, 1872), p. 36; J. P. Chesney, "Woman as a physician," *Richmond & Louisville Med. J.,* 1871, *11:* 6.

8   Samuel Gregory, who founded the New England Female Medical College, was one of those concerned primarily with the improprieties of male accoucheurs.

9   Weatherly, "Woman: her rights," p. 75.

10 See Martin Kaufman, "The admission of women to 19th-century American medical societies," *Bull.*

Hist. Med., 1976, *50:* 251–260. The assessment I give here of the quality of medical education at the women's medical colleges is based on my own preliminary research. Although the majority of male physicians opposed the medical education of women, female physicians had many prominent male supporters who were loyal and enthusiastic. These included Henry I. Bowditch and James Chadwick in Boston; Steven Smith and Abraham Jacobi in New York; Hiram Corson, Henry Hartshorne, and later Alfred Stillé in Philadelphia; and William Byford and I. N. Danforth in Chicago.

11 *Boston Med. & Surg. J.,* 1884, *111:* 90, 1849, *40:* 505; 1873, *89:* 23. "The practice of midwifery by females—by one of the class," *Boston Med. & Surg. J.,* 1849, *41:* 59–61; Coates, *Introductory Lecture,* pp. 3–4; D. W. Graham, "The demand for medically educated women," *J.A.M.A.,* 1886, *6:* 479.

12 Carroll Smith-Rosenberg, "Puberty to menopause: the cycle of femininity," in *Clio's Consciousness Raised,* ed. M. Hartman and L. Banner (New York, 1974), pp. 1–22. The debate over premenstrual tension and the related effects of woman's cycle on her psyche still goes on in medical circles. See K. J. and R. J. Lennane, "Alleged psychogenic disorders in women—a possible manifestation of sexual prejudice," *New Eng. J. Med.,* Feb. 8, 1973, *290:* 288–292.

13 "Female physicians," p. 169; Horatio Storer, "The fitness of women to practice medicine," *J. Gynec. Soc. Boston,* 1870, *2:* 266–267; "Female practitioners of medicine," pp. 272–274; E. H. Clarke, *Sex In Education: Or a Fair Chance for the Girls* (Boston, 1873); Lawrence Irwell, "The competition of the sexes and its results," read before the American Association for the Advancement of Science, August 1896, *Am. Medico-Surg. Bull.,* 1896, *10:* 316–320.

14 Mary Putnam Jacobi, *The Question of Rest for Women During Menstruation* (New York, 1877), p. 227. See also C. Alice Baker to Mary Putnam Jacobi, Nov. 7, 1874, Jacobi MSS, Schlesinger Library, Radcliffe. For studies by other women doctors see Elizabeth R. Thelberg, physician at Vassar, "College education as a factor in the physical life of women," Alumnae Association of the Woman's Medical College of Pennsylvania, *Transactions,* 1899, pp. 73–87; Elizabeth C. Underhill, resident physician at Mt. Holyoke, "The effect of college life on the health of women students," *Woman's Med. J.,* 1913, *22:* 31–33; Clelia Duel Mosher, "Normal menstruation and some factors modifying it," *Johns Hopkins Hosp. Bull.,* 1901, *12:* 178–179.

15 George Gregory, *Medical Morals* (New York, 1852), pp. 5, 20; Rev. William Hosmer, *Appeal to Husbands and Wives in Favor of Female Physicians* (New York,

1853), pp. 3–24; Thomas Ewell, *Letters to Ladies Detailing Important Information Concerning Themselves and Infants* (Philadelphia, 1817), pp. 23–27.

16 Elizabeth Blackwell, *Address on the Medical Education of Women* (New York, 1856), pp. 8–9.

17 For example, G. Fenning, *Every Mother's Book: or the Child's Best Doctor, Being a Complete Course of Directions for the Medical Management of Mothers and Children* (New York, n.d.), or D. Wark, *The Practical Home Doctor for Women* (New York, 1882).

18 Chesney, "Woman as a physician," p. 4.

19 Emmeline Cleveland, *Introductory Lecture to the Class of the Female Medical College of Pennsylvania* (Philadelphia, 1859), p. 7.

20 Elizabeth Blackwell, "Criticism of Gronlund's Cooperative Commonwealth; Chapter X–Woman," given before the Fellowship of New Life, n.d., pp. 9–10; *The Influence of Women in the Profession of Medicine* (London, 1889), p. 11; "Anatomy," Lecture Notes, n.d., all in Blackwell MSS, Library of Congress.

21 Joseph Longshore, *Introductory Lecture to the Class of the Female Medical College of Pennsylvania* (Philadelphia, 1859), p. 11; Joseph Longshore, *The Practical Importance of Female Medical Education* (Philadelphia, 1853), p. 6; Female Medical College of Pennsylvania, "Appeal to the Corporators," Medical College of Pennsylvania Archives [hereafter cited as M.C.P.]; Ann Preston, *Valedictory Address to the Graduating Class of the Female Medical College of Pennsylvania* (Philadelphia, 1858), pp. 9–10.

22 Elizabeth and Emily Blackwell, *Medicine as a Profession for Women* (New York, 1860), pp. 8–9; Richard C. Cabot, "Women in medicine," *J.A.M.A.,* 1915, *65:* 9–17; Cora B. Lattia, "Public health education among women," *N.Y. St. J. Med.,* 1913, *13:* 12–17; "Appeal to the Corporators."

23 Georgiana Glenn, "Are women as capable of becoming physicians as men," *The Clinic,* 1875, *9:* 243–245; Jacobi, "Inaugural Address," Women's Medical College of the New York Infirmary, *Chicago Med. J. & Exam.,* 1881, *42:* 580.

24 A. Lapthorn Smith, "Higher education of women and race suicide," *Popular Science Monthly,* 1905, *66:* 466–473; A. Lapthorn Smith, "What civilization is doing for the human female," *Tr. Southern Surg. & Gynec. Assn.,* 1889, *2:* 352–360; F. W. Van Dyke, "Higher education as the cause of physical decay in women," *Med. Rec.,* 1905, *67:* 296–298; William Goodell, R. Gaillard Thomas, M. Allen Starr, J. J. Putnam, "Symposium on the co-education of the sexes," *Med. News, N.Y.,* 1889, *55:* 667–672; J. T. Clegg, "Some of the ailments of woman due to her higher development in the scale of evolution," *Texas Hlth. J.,* 1890–91, *3:* 57–59; A. J. C. Skene, *Educa-*

tion and Culture as Related to the Health and Diseases of Women (Detroit, 1889), p. 39.

25 Elizabeth and Emily Blackwell, *Medicine as a Profession,* p. 4; Henry Hartshorne, *Valedictory Address to the Graduating Class of the Woman's Medical College of Pennsylvania* (Philadelphia, 1872), pp. 13–14; Louise Fiske-Byron, "Woman and nature," *N.Y. Med. J.,* 1887, *66:* 627–628. For Ruffin Coleman's remark see "Woman's relation to higher education and the professions as viewed from physiological and other stand points," *Tr. Med. Assn. St. Ala.,* 1889, *42:* 233–247.

26 Emily Blackwell made this argument even before social Darwinism came into vogue: "Mankind," she confided to her diary, "will never be what they should be until women are nobler." Diary, June 4, 1852, p. 83, Blackwell MSS, Library of Congress. See also *J. Hyg. & Herald Hlth.,* 1894, *44:* 236; J. G. Kiernan, "Mental advance in woman and race suicide," *Alienist and Neurologist,* 1910, *30:* 594–599; Elizabeth Blackwell, *Pioneer Work in Opening the Medical Profession to Women* (New York, 1895), p. 253.

27 Samuel Gregory, "Female physicians," *The Living Age,* 1862, *73:* 243–249; Mary Gove Nichols, "Woman the physician," *Water-Cure J.,* 1851, *12:* 3; Thomas Longshore's manuscript is in the M.C.P. Archives, p. 106. See also Marie Zakrzewska to Elizabeth Blackwell, March 21, 1891, Blackwell MSS, Schlesinger Library.

28 Blackwell, *Pioneer Work,* p. 253; Emmeline Cleveland, *Valedictory Address to the Graduating Class of the Woman's Medical College of Pennsylvania* (Philadelphia, 1874), p. 3; Eliza Mosher, "The value of organization—what it has done for women," *Woman's Med. J.,* 1916, *26:* 1–4; James J. Walsh, "Women in the medical world," *N.Y. Med. J.,* 1912, *96:* 1324–1328.

29 Emily Blackwell, "Address at the Thirty-first Annual Commencement of the Women's Medical College of the New York Infirmary," May 25, 1899, printed in *Final Catalogue* (New York, 1899). Also Emily Blackwell, Diary, October 1851, p. 47. Rarely until after 1900 does one come across the argument that medicine is enriching from the standpoint of personal development. It is primarily society which is to benefit; individuals gain *because* they are aiding society.

30 Carroll Smith-Rosenberg, "The female world of love and ritual: women and sexuality in nineteenth century America," *Signs,* Autumn 1975, *1:* 1–29.

31 Margaret Vaupel Clark, "Medical women's contribution to the education of mothers," *Woman's Med. J.,* 1915, *25:* 126–128; Glenn, "Are women as capable," pp. 243–245; "The woman physician and her special obligation and opportunity," *Woman's Med.*

J., 1915, *25*: 74–79. Numerous women doctors gave public health lectures.

32   For the use of the argument see J. Stainbeck Wilson, "Female medical education," *Southern Med. & Surg. J.,* 1854, *10*: 1–17; Emmeline Cleveland, *Valedictory Address to the Graduating Class of the Female Medical College of Pennsylvania* (Philadelphia, 1858), p. 10; Clark, "Medical women's contribution," pp. 126–128; Harriet Williams, "Women in medicine," *Texas Med. News,* 1903, *12*: 613–615.

33   See Jacobi, "Inaugural address," pp. 561–585.

34   Elizabeth Blackwell, for example, opposed public child care. Her belief was that parents as well as children suffered: "Neither can parental responsibility be wisely devolved upon the creche, the nursery, the school and the workshop. . . . No social arrangements must ever be allowed to destroy the essential education which is given to the parent by the children." "Criticism of Gronlund," p. 11. For government ambivalence see William Chafe, *The American Woman* (New York, 1972), pp. 159–172.

35   Frances Emily White, "The American medical woman," *Med. News, N.Y.,* 1895, *67*: 123–128.

36   An interesting example of how conservative this type of argument can be is provided by the situation in India and Pakistan. There "purdah," the seclusion of women, still exists. Yet similar arguments for training a core of professional women to administer to an exclusively female clientele are extremely popular. See Hanna Papenek, "Purdah in Pakistan: seclusion and modern occupations for women," *Journal of Marriage and the Family,* August 1971, pp. 517–530.

37   Helen Watterson, "Woman's excitement over 'Woman'," *Forum,* 1893, *16*: 75–85; Mary Putnam Jacobi, "An address delivered at the Commencement of the Women's Medical College of the New York Infirmary, May 30, 1883," *Arch. Med.,* 1883, *10*: 59–71; Marie Zakrzewska, *Introductory Lecture Before the New England Medical College* (Boston, 1859), pp. 3–26; C. L. Franklin, "Women and medicine," *The Nation,* 1891, *52*: 131.

38   Jacobi, "Commencement address," p. 70.

14

The Fall and Rise of the
American Medical Profession

RONALD L. NUMBERS

Midway through the 19th century American medicine lay in a shambles. Addressing the first annual meeting of the American Medical Association in 1848, President Nathaniel Chapman expressed the dejection felt by many physicians. "The profession to which we belong," he lamented, "once venerated on account of its antiquity—its varied and profound science—its elegant literature—its polite accomplishments—its virtues—has become corrupt and degenerate, to the forfeiture of its social position, and with it, of the homage it formerly received spontaneously and universally."[1] The golden age Chapman recalled may have been only an old man's fantasy, but American physicians had unquestionably fallen on hard times. For centuries, at least since the late Middle Ages, Western societies had commonly recognized medicine, along with divinity and law, as one of the prototypical professions, possessing an esoteric body of knowledge, requiring extensive training, and being entitled to exclusive rights protected by law.[2] Although many healers practiced outside the law and failed to acquire professional status, those who earned M.D. degrees or served a protracted apprenticeship formed an occupational elite. By the mid-19th century, however, medicine for many Americans had degenerated into little more than a trade, open to all who wished to try their hand at healing. "Any one, male or female, learned or ignorant, an honest man or a knave, can assume the name of a physician, and 'practice' upon any one, to cure or to kill, as either

may happen, without accountability," wrote one observer in 1850. "'It's a free country!'"[3]

Within a century American physicians and their allies effected a revolution. By 1950, even earlier, medicine had emerged as the strongest and most influential profession in America. Aspiring physicians competed intensely to enter the guild and studied for years to acquire the proper credentials and requisite knowledge. Powerful institutions guarded its interests and regulated its behavior. Laws in every state protected its rights. Indeed, as one lawyer noted enviously, no interest group in the country enjoyed "more freedom from formal control than organized medicine."[4]

THE FALL

During the early 19th century American physicians viewed their situation with optimism—and rightfully so. Although the country abounded with self-appointed healers and poorly trained doctors, professionally ambitious physicians seemed to be improving their position. With the opening of a medical school at the College of Philadelphia in 1765, Americans could, without going abroad, supplement their apprenticeships with formal lectures and acquire medical degrees. By 1830 the United States boasted 22 such institutions. By this time, also, 13 states had awarded physicians special privileges, usually in the form of laws giving medical societies exclusive rights to license practitioners. Few of these statutes granted licensed practitioners more than the right to sue for unpaid bills; yet they did provide the medical elite with a modicum of social recognition.[5]

In the decades after 1830 several factors converged to undermine the progress physicians

RONALD L. NUMBERS is William Coleman Professor of the History of Science and Medicine, University of Wisconsin–Madison.

From *The Professions in American History*, edited by Nathan O. Hatch (Notre Dame, Ind.: University of Notre Dame Press, 1988), pp. 51–72. Courtesy of the University of Notre Dame Press.

seemed to be making: an epidemic of "medical school mania" broke out, the profession fractured into rival sects, and physicians lost the legal standing they had just acquired. Of course, physicians were not the only professionals to suffer a loss of status at this time. As one historian has pointed out, the period from 1830 to 1880 witnessed the general "disestablishment and humbling of the professions" in America.[6] Nevertheless, the particulars of the medical story are unique.

Encouraged by revised state laws that accepted a medical school diploma as the equivalent of a license from an approved medical society, physicians who fancied themselves as professors began flooding the country with low-grade institutions. Already by 1830 the nation's 22 medical schools exceeded by several times the number in European countries of comparable size, and during the next three decades the quantity more than doubled. To make matters worse, most of these institutions, even those nominally affiliated with a college or university, were run for prestige and profit by ill-equipped local practitioners, many of whom could not have matriculated as students in the best European schools. Because of this commercial orientation, requirements for admission—where they existed at all—remained low and graduation often depended more on a student's ability to pay than on his competence. Some medical students could barely read and write, and most had never sat in a college classroom. The best arts students aspired to careers in divinity or law and looked upon medicine as a last resort. "It is very well understood among college boys," wrote one dispirited physician, "that after a man has failed in scholarship, failed in writing, failed in speaking, failed in every purpose for which he entered college; after he has dropped down from class to class; after he has been kicked out of college, there is *one* unfailing city of refuge—the profession of medicine."[7]

Until the last quarter of the century students typically obtained their M.D. degrees by successfully completing eight months of formal training, the last half of which merely duplicated the first. During this brief period they were expected to learn not only the basic sciences like anatomy, physiology, and chemistry, but to acquire clinical skills as well. This became especially true as more and more schools dropped a previous apprenticeship as an entrance requirement. It was no won-

der, commented the medical reformer Nathan Smith Davis in 1845, that 99 out of every 100 American physicians were poorly educated:

> With no *practical* knowledge of chemistry and botany; with but a smattering of anatomy and physiology, hastily caught during a sixteen weeks' attendance on the anatomical theater of a medical college; with still less of real pathology; they enter the profession having mastered just enough of the details of practice to give them the requisite *self-assurance* for commanding the confidence of the public; but without either an adequate fund of knowledge or that degree of mental discipline and habits of patient study which will enable them ever to supply their defects.[8]

Such practitioners, charged another contemporary, tended to pursue "medicine as a trade instead of a profession, [and to] study the science of patient-getting to the neglect of the science of patient-curing."[9]

A second development that undermined the status of medicine was its fragmentation into competing sects, each offering an alternative to the so-called heroic therapy of regular physicians. During the early 19th century practitioners gained considerable notoriety for copius bleeding and the seemingly routine prescribing of calomel, a mercury-based cathartic that often produced harmful and discomforting side effects. Such practices gave rise to a host of critics, among the most vocal and visible of which were the homeopaths and the various botanics, such as the Thomsonians and eclectics.[10]

The pioneer among American sectarians was a New Hampshire farmer, Samuel Thomson, who learned much of his botanic lore from a local female herbalist. Early in his healing career he became convinced that the cause of all disease was cold and that restoring the body's natural heat was the only cure. This he accomplished by steaming, peppering, and puking his patients, relying heavily on lobelia, a botanical emetic long used by American Indians. Not one to ignore the commercial possibilities of his discovery, Thomson in 1806 began selling "Family Rights" to his practice, for which he obtained a patent in 1813. During the 1820s and 1830s his agents fanned out from New England through the southern and western states, urging self-reliant Americans to become their own physicians. By 1840 approximately 100,000 Family Rights had been sold, and Thomson estimated that about 3 million individuals had adopted his system. In states as diverse as Ohio and Mississippi as many as one-half of the citizens reportedly cured themselves the Thomsonian way.

Unlike many sectarians who simply wanted the public to exchange one kind of physician for another, Thomson saw no need for professional healers of any kind. It was high time, he declared, for the common man to throw off the oppressive yoke of priests, lawyers, and physicians and assume his rightful place in a truly democratic society. Contrary to his wishes, his disciples in the 1830s began opening medical schools and competing head-to-head with regular physicians. By the 1840s, however, the Thomsonians were rapidly losing ground to a rival botanical sect, the eclectics, who achieved great popularity in the Midwest, where, in Cincinnati, they operated one of the nation's largest medical schools. Eclectic physicians and surgeons denounced blood-letting and calomel-dosing, but, except for preferring vegetable to mineral remedies, their practice differed little from that of the regulars.

Homeopathy originated in the mind of a regularly educated German physician, Samuel Hahnemann, who during the last decade of the 18th century constructed a novel system of healing based in large part on the healing power of nature and two fundamental principles: the law of similars and the law of infinitesimals. According to the first law, diseases are cured by medicines having the property of producing in healthy persons symptoms similar to those of the disease. An individual suffering from fever, for example, would be treated with a drug known to increase the pulse rate of a person in health. Hahnemann's second law held that medicines are more efficacious the smaller the dose, even down to one-millionth of a gram. Though regular practitioners — or allopaths, as Hahnemann dubbed them — ridiculed this theory, many patients flourished under the mild homeopathic therapy, and they seldom suffered harmful side effects.

Following its appearance in the United States in 1825, homeopathy quickly grew into the nation's largest medical sect, replete with its own medical schools, societies, and journals. Its practitioners, many of whom defected from the regular ranks, were often well educated and well received by the middle and upper classes. By 1860 there were an estimated 2,400 homeopaths nationally, compared with about 60,000 regular physicians, only a third of whom possessed M.D. degrees.[11]

The acrimonious debates between sectarians and regulars — to say nothing of bitter quarrels among the regulars themselves over the efficacy of drug therapy and bleeding — created an atmosphere in which it became fashionable, wrote one editor, "to speak of the Medical Profession as a body of jealous, quarrelsome men, whose chief delight is in the annoyance and ridicule of each other."[12] Although many individual physicians continued to enjoy the respect of their communities, the reputation of the profession as a whole does seem to have declined. Certainly public confidence in medicine was not bolstered when physicians like Harvard's Oliver Wendell Holmes suggested that, with few exceptions, "if the whole materia medica, *as now used*, could be sunk to the bottom of the sea, it would be all the better for mankind — and all the worse for the fishes."[13]

Indicative of the public's low esteem for the medical profession during the 1830s and 1840s was its stripping doctors of their privileged legal status. Although it is difficult to see how early-19th-century licensing laws deterred anyone from practicing medicine, Thomsonians especially felt persecuted and led the fight to repeal the so-called Black Laws. Apparently they were not without popular support, because at one point during the legislative debate in New York Samuel Thomson's son John arrived in the chamber, dramatically pushing a wheelbarrow full of petitions. Swayed by such displays of antimonopolistic sentiment, state legislatures one by one repealed the offending statutes, until by 1850 only two states retained laws in any way restricting the practice of medicine. Society, observed one dejected physician, had thrown the medical profession overboard, "to sink or swim as it can, without even a rope by which to sustain itself."[14]

Unhampered by legal prejudice, healers of every stripe assumed the title "doctor" and hung out their shingles. According to one 1850 survey, of the 201 practitioners in eastern Tennessee, 35 held M.D.'s from regular medical schools, 42 had taken only one course of lectures, 27 identified themselves as botanics, while the rest — nearly half of the total — had picked up their knowledge of medicine from casual reading.[15] With so many marginally qualified practitioners crowding the field, even medical school graduates often found it impossible to live on their medical income alone. It was high time, concluded professional leaders, to launch a counteroffensive — to protect the public health as well as their own pocketbooks.

EARLY REFORMS

Abandoned to their own devices by the states, medical reformers first explored what they could accomplish through self-regulation. As the editor of the *New York Journal of Medicine* wrote in 1845,

> The profession in almost every state in the union is now left, so far as legal enactments are concerned, to take care of itself; to make its own rules, and adopt its own standard of excellence. For if we cannot make rules or laws which will banish ignorance, stupidity, and empiricism, we can, at least, fix our *own* standard of qualification, and thereby say who *we* will recognize as *our* associates.[16]

To accomplish this end, members of the State Medical Society of New York in 1846 convened a national convention to explore ways of reducing internal discord, isolating sectarians, and, above all, elevating the quality of medical education. Although some medical professors denounced the movement as an "aristocratic" attempt to cripple medical schools, nearly 100 delegates from 16 states showed up.[17] The following year this group launched the American Medical Association, the first national society of regular physicians.

One of the top items on the A.M.A.'s agenda was to draw up a code of ethics, outlining not only the duties of the "professional" physician toward patients and peers, but also the obligations of patients to obey their physicians and avoid quacks. Although the code dealt more with etiquette than ethics, it did establish guidelines governing economic behavior, a source of much contention. In addition to banning advertising, the selling of secret nostrums, and other practices associated with quackery, it recommended the adoption of fee bills in each community to regulate the price of medical services. These schedules of minimum fees were to be binding in all cases except those involving the indigent, brother physicians, and certain public duties. American doctors commonly acknowl-

edged an obligation to put their patients' welfare above personal gain, but they also recognized, as one practitioner explained, that "a physician must train himself to be a professional man when treating patients, and a business man when collecting his bills."[18]

To segregate sectarians and to assist the public in identifying regular physicians, the code stipulated that no A.M.A. member could consult with anyone "whose practice is based on an exclusive dogma." Like the entire code, this provision remained advisory until 1855, when it became mandatory.[19]

No subject attracted greater attention among A.M.A. members than medical education, the improvement of which, they hoped, would simultaneously produce better physicians and reduce the number of potential competitors. At its first meeting delegates adopted standards requiring medical schools to offer anatomic dissections and clinical instruction, to extend their terms from four to six months, and to require entering students to possess "a good English education, a knowledge of Natural Philosophy, and the elementary Mathematical Sciences, including Geometry and Algebra, and such an acquaintance, at least, with the Latin and Greek languages as will enable them to appreciate the technical language of medicine, and read and write prescriptions." These standards were so unrealistically high that, in the opinion of one scholar, rigid enforcement "would have closed down practically every medical school in the country, and would have depleted the ranks of formally educated physicians in a few years."[20] Fortunately for medical educators, the A.M.A. represented only a small percentage of American physicians and remained too weak and ineffective throughout the 19th century to reform much of anything.

Medical education did, nevertheless, improve significantly during the latter half of the century. Although the proliferation of substandard schools continued unabated, the best institutions lengthened their curricula to three years, offered a new set of courses each year, required some evidence of preliminary education, and, led by the Harvard Medical College, abandoned proprietary status to become an integral part of a university. The dramatic growth of laboratory-based medical science in the latter half of the century encouraged such reforms, as did the German training of approximately 15,000 American physicians between 1870 and 1914.[21]

No event symbolized the reformation of American medical education more than the opening in 1893 of the Johns Hopkins School of Medicine under the leadership of the German-trained pathologist William H. Welch. At a time when, according to Welch, no American medical school required a preliminary education equal to "that necessary for entrance into the freshman class of a respectable college," the Hopkins faculty, at the insistence of its patron, demanded a bachelor's degree. Modestly following the Hopkins example, more than 20 schools by 1910 raised their entrance requirements to two years of college. This reform, Robert E. Kohler has argued, had far-reaching effects: it "stretched the financial resources of the proprietary school beyond the breaking point. . . . higher entrance requirements disrupted the established market relation with high schools, diminished the pool of qualified applicants, and resulted in a drastic plunge in enrollments. Medical schools could not survive on fees."[22]

A further prod to educational reform came not from within the profession itself, but from the states, every one of which passed some kind of medical licensing act between the mid-1870s and 1900. Although physicians generally led the crusade to restrict the practice of medicine, they were not without external support. As society came to rely on physicians to certify births and deaths, to control infectious diseases, and to commit the insane, it became increasingly apparent that licensing served a public as well as a professional function. In 1888 the Supreme Court in a landmark decision, *Dent v. West Virginia*, upheld the authority of the state medical examining board to deprive a poorly trained eclectic physician of his right to practice. In the opinion of Justice Stephen J. Field, no one had "the right to practice medicine without having the necessary qualifications of learning and skill," and no group had greater competency to judge these qualifications than well-trained physicians.[23] Thus society granted the medical profession one of its most cherished goals: the authority to exclude practitioners deemed unworthy. The fact that other professions won protective legislation at the same time suggests that the physicians' achievement resulted more from a change in so-

cial policy than from a recognition of the improved state of medical science, impressive though it may have been.

The state licensing boards influenced medical education in two ways. First, most of them required candidates to hold a diploma from a reputable medical school, that is, one requiring evidence of preliminary education and, perhaps, offering a three-year course of study, a six-month term, and clinical and laboratory instruction. This forced any school hoping to compete for students to upgrade its curriculum, at least superficially. Second, many states, especially during the late 1880s and 1890s, revised their laws to require all candidates, even those holding medical degrees, to pass an examination. Although some of the weaker schools quickly learned how to coach students to pass these tests, graduates from strong institutions had a much better chance of passing. A shallow medical education no longer paid.[24] The success of the licensing laws convinced professional leaders of the great advantage of legal sanctions over moral suasion in reforming medical education. "It is our opinion," declared the editor of the *Journal of the American Medical Association* in 1895, "that notwithstanding the far-reaching influence of the medical press, and the support of the various medical societies of the country, medical legislation alone caused the healthy reform. It is also apparent that the enforcement of efficient medical legislation in a few States will do more in destroying the dangerous work of the low grade college than all other factors combined."[25]

Medical practice acts not only set standards for licensing physicians, but defined the very practice of medicine. The Nebraska act, for example, stipulated that anyone "who shall operate on, profess to heal, or prescribe for, or otherwise treat any physical or mental ailment of another" was practicing medicine and thus subject to the provisions of the law. For various political reasons, the states often granted exceptions. Many laws specifically exempted dentists, midwives, medical students, and persons who gave emergency aid; and a few states, especially in New England, provided immunity for Christian Scientists and others who engaged in mental healing.[26] However, physicians, whose strategy called for the elimination or subordination of competitors, preferred to define the practice of medicine as broadly as possible. As the sociologist Eliot Freidson has argued, "an essential feature of a useful concept of profession is the possession of something of a monopoly over the exercise of its work."[27] American physicians would have agreed.

Since the appearance of rival sects in the first half of the century, regulars had sought to isolate and discredit them. Nevertheless, well-trained sectarians, especially homeopaths and eclectics, had prospered. During the latter part of the century it became increasingly difficult to distinguish between orthodox and heterodox practice, or to argue that one system was more efficacious than another. In the 1880s, for example, one life insurance company concluded that homeopathy was just as effective as allopathy in saving lives, while a comparative study at Cook County Hospital in Chicago showed the latter to have only a slight edge in mortality rates. Given the training, therapeutic success, and numerical strength of the sectarians, regulars found it impossible to legislate them out of existence; in fact, in at least 20 states they sat on the same licensing boards with homeopaths or eclectics—at a time when the A.M.A. still banned professional intercourse. In only three or four instances did they obtain what they most desired: a single licensing board composed exclusively of regulars.[28]

Although orthodox physicians liked to describe their efforts to suppress sectarians in humanitarian terms, evidence suggests that opposition stemmed as much from fear of competition as from a desire to safeguard the public. After all, as one homeopath perceptively noted, regulars made few attempts to police therapy among themselves:

> If you inform the people that you treat those who come to you according to Similia, so far as drugging goes, you are anathema with the "regular," but if you get inside his fold, you can use any old treatment you please—be it an "electro-therapeutist," a man of "suggestion," or of "serums," calomel, bleeding, anything, and be a "regular physician." Curious, isn't it? Looks as though the real thing at issue was the "recognition of the union" rather than the "welfare of the public."[29]

In addition to battling sectarians, the regular medical profession zealously fought to subordinate and control allied health personnel. As the number of trained nurses increased after the Civil War, physicians expressed concern that these women would attempt to expand their role and presume to act as physicians. Thus doctors attempted to limit the amount of theoretical training nurses received and insisted that nurses strictly obey their orders. As the *Boston Medical and Surgical Journal* explained, the physician's relationship to the nursing staff was to be "like that of the captain to his ship." To win acceptance and approval from the medical profession, nurses themselves went out of their way to reassure the doctors. For example, in her influential *Nursing Ethics* (1900) Isabel Hampton emphasized discipline as the key to success: "The head nurse and her staff should stand to receive the visiting physician, and from the moment of his entrance until his departure, the attending nurses should show themselves alert, attentive, courteous, like soldiers on duty."[30] Such an attitude posed no threat either to pocketbooks or egos.

Physicians experienced much greater difficulty trying to subdue male pharmacists, who not only sold medicines prescribed by doctors but frequently diagnosed minor ailments and suggested remedies. Pharmacists, warned one physician in the early 1880s, had "so industriously and energetically wedged themselves between the 'dear public' and the professional province of the physician" that they threatened to take over the practice of medicine. Incensed doctors denounced this intrusion as dangerous and illegal and sought to revise medical practice acts to ban practices such as over-the-counter prescribing. Their efforts, however, generally failed, and in at least one state (North Dakota) the pharmacists retaliated by securing passage of a law barring physicians from dispensing medicines. "If physicians wish to prevent encroachment on their domain," warned one pharmacist, "they should avoid invading others' property."[31]

Dentistry, which required mechanical skills more than medical knowledge, was one of the few healing activities that physicians attempted neither to crush nor to control. When dentists in the late 1830s tried to win a place for their specialty in the medical school curriculum, one physician quipped "that the author of such a plan, to be consistent, should not fill teeth but treat the constitutional causes of dental caries by bleeding, purging, and leeching his patients."[32] Thus rebuffed, a group of physician-dentists in 1840 established the Baltimore College of Dental Surgery, the first dental school in the country. Although some dentists fretted for decades about their identity—"If we are not medical specialists we are a set of carpenters," said one—the majority, believing that "dentistry was altogether too large to be made the tail end of the kite of medical practice," quietly created a set of institutions paralleling those of the medical profession: societies, schools offering D.D.S. degrees, and separate dental licensing boards. "A professional wall is now being built up," explained one dental leader, "and we hope it will be built so high and strong, that none can scale or break it down, but that all who enter will be compelled to do so through the legitimate and well-guarded gateways." Except for legal loopholes that often allowed physicians to practice dentistry, dentists succeeded by the end of the century in building their professional wall. They failed, however, to win equal standing among physicians, who regarded the filling and pulling of teeth as minor matters. "If dentists are ambitious to be considered medical specialists, they must undergo a general medical education," said one physician condescendingly. "An individual dentist who has taken the medical degree may assuredly be received as a brother practitioner, but a simple D.D.S. never."[33]

In spite of substantial improvements in medical education and the passage of licensing laws, the medical profession at the end of the century still contained, according to one knowledgeable physician, "a vast number of incompetents, large numbers of moral degenerates, and crowds of pure tradesmen." By 1900 there were 151 medical schools, and even the worst institutions sometimes managed to prepare their graduates to pass ineffectual licensing examinations. In one notorious case, the weakest of Chicago's 14 schools—"a school with no entrance requirement, no laboratory teaching, no hospital connections"—outperformed its 13 rivals on the state boards. The medical practice acts, complained one contemporary, allowed all but "the most flagrant quacks and charlatans from carrying on their business unmolested."[34] Thus after a half-century of reform much remained on the medical profession's agenda for elevating its status: the elimination of inferior medical schools, the enactment of stricter licensing laws, and the creation of a powerful national body to represent the interests of physicians.

THE REFORMATION COMPLETED

The professional leaders of American medicine faced the 20th century determined to complete the reformation they had begun, that is, to reduce the quantity and increase the quality of medical practitioners. Although the ratio of physicians to patients had scarcely changed during the previous 50 years—in fact, it had actually improved from 1:568 in 1850 to 1:576 in 1900—American doctors continued to view the overcrowding of their profession by poorly trained physicians as their greatest problem. Such individuals, argued the reformers, not only provided inadequate and sometimes dangerous care, but also depressed physician income. Well into the 20th century, most American physicians earned less than $2,000 a year. In 1914, for example, less than 60 percent of Wisconsin's approximately 2,800 practitioners earned enough even to pay income taxes; and of those who paid, the average income was only $1,488—well below that of bankers, manufacturers, and lawyers, though more than twice what professors earned.[35]

The profession's first order of business was to create a united front. Since its founding in 1847, the A.M.A. had remained virtually impotent; it had, noted its president sadly in 1901, "exerted relatively little influence on legislation, either state or national." Only about 7 percent of the country's physicians had joined the association, and independent state and county societies often operated at cross purposes to the will of the national body. In response to this situation, an A.M.A. committee in 1901 recommended a complete reorganization: the welding of local, state, and national units into one representative society that would "foster scientific medicine and . . . make the medical profession a power in the social and political life of the republic." Henceforth, membership in a local (generally county) society would automatically carry membership in the state and national organizations. This plan, approved at the 1901 annual session, produced immediate results. Most state societies fell quickly into line, and membership in the A.M.A. multiplied over sevenfold within five years. For the first time American physicians possessed an organization large enough and strong enough to further their interests effectively.[36]

Like its 19th-century parent, the reorganized A.M.A. fought to control access to the profession by tightening the requirements for medical education and licensure. In 1904 it created a Council on Medical Education, which soon began inspecting and grading medical schools. A few years later the council cooperated with the Carnegie Foundation in producing Abraham Flexner's famous report on *Medical Education in the United States and Canada* (1910). This muckraking exposé described conditions — often abysmal — at each of the country's 155 medical schools. The adverse publicity generated by the A.M.A.'s inspections and the Flexner report, together with continuing pressure from licensing bodies and the growing expense of providing laboratory and clinical instruction, forced many institutions to shut down — and finally brought a halt to the overproduction of unqualified physicians. Between 1910 and 1920 the number of medical schools declined from 155 to 85, and it continued falling for the next two decades. By the late 1910s the total number of physicians in the United States was dropping for perhaps the first time in history. The schools that survived this winnowing were, by 1930, generally requiring a bachelor's degree for admission and offering rigorous scientific and clinical training. Unlike the improvements of the 19th century, which had little to do with the A.M.A., these changes often resulted from the A.M.A.'s cozy relationship with the state licensing boards, which delegated to the A.M.A. (sometimes jointly with the American Association of Medical Colleges) the privilege of deciding which schools merited approval for the licensing of their graduates. By this means, and by accrediting hospitals for the training of interns, organized medicine gained considerable control over the education and supply of physicians.[37]

The medical profession enjoyed much less success in its efforts to monopolize the practice of medicine by outlawing rivals and controlling allies. As we have seen, during the last quarter of the 19th century regular physicians often united reluctantly with their old nemeses, the eclectics and homeopaths, to win passage of state licensing laws. In 1903 the A.M.A. took additional steps toward unity by deleting its ethical ban against consulting with irregulars and by welcoming as members eclectics and homeopaths willing to forsake sectarian dogma for scientific truth. This latter act proved to be the kiss of death for eclectic and homeopathic organizations, which, though still numerically strong, were now struggling to survive. Weakened by internal discord, defecting members, and the lack of state-supported medical schools, they soon ceased to be a factor in American medicine.

The demise of eclectics and homeopaths did not, however, eliminate sectarian competition. During the late 19th century Christian Science, osteopathy, and chiropractic made their appearance, and by the early 20th century these new "cults," as the medical establishment insisted on calling them, began threatening the therapeutic consensus based on scientific medicine as well as the economic goals of physicians. Despite an intensive and protracted campaign by the medical profession to have these sects declared illegal, all three won the legal right to practice their form of healing. The nature of their victory varied from state to state, but their experience in Wisconsin illustrates the various means they used to thwart the medical profession's monopolistic designs.

Like their spiritual leader, Mary Baker Eddy, Christian Science practitioners denied the existence of disease and the need for physicians. Instead of prescribing drugs, they relied on prayer and verbal persuasion to cure individuals who imagined themselves to be ill. Christian Scientists began practicing in Wisconsin in the 1880s, but physicians could do little to stop them before the state passed a medical practice act in 1897. Following enactment of this bill, authorities in Milwaukee arrested two Christian Science practitioners for violating the new law. At their trial the defendants argued that they were not guilty of practicing medicine "because they never administered drugs, never performed surgery, never manipulated the body or even touched their patients." Nevertheless, the court convicted them — only to be overruled by a higher court, which agreed that Christian Science had little in common with medicine. Thereafter, Christian Scientists in Wisconsin faced only the insults of physicians, not the threat of arrest. In some other states, as we have noted, legislators specifically exempted them in defining medical practice.

Osteopathy was founded by a Missouri physician, Andrew Taylor Still, who turned against regular medicine after drug therapy failed to save the lives of his children. Convinced that the human brain functioned as "God's drug store," he attributed all disease to obstructions inhibiting the flow of blood and nervous fluid, which he sought to cure using manual manipulation, particularly of the spine. The first osteopaths arrived in Wisconsin in the 1890s, and by 1900, 19 had located in the state. Charged in that year with illegal practice under the 1897 statute, the osteopaths lost their first court case. Seeking relief, they petitioned the legislature to create a separate licensing board for osteopathy. Regular physicians lobbied instead for the addition of an osteopath to the existing licensing board, believing that the "requirements are so high it is safe to say that but few, if any, osteopaths will ever be able to meet them." The regulars got their wish, but discovered to their chagrin that the underrated osteopaths routinely qualified for licenses to practice. Elsewhere, in over half the states, regulars suffered even greater humiliation, as legislators ignored their pleas and set up separate boards composed only of D.O.'s.

Chiropractic, the brainchild of an Iowa grocer, Daniel David Palmer, explained disease in terms of dislocations of the spine, which allegedly impeded the circulation of nervous fluid. Like osteopaths, with whom they were frequently confused, chiropractors relied therapeutically on adjustments of the spinal column. When the first of them moved to Wisconsin in the early 1900s, they landed in jail for practicing *osteopathy* without a license. Undeterred, they continued to practice illegally until 1915, when the legislature granted them immunity to work as unlicensed practitioners. Ten years later it disregarded the will of the medical profession and, like legislatures in many other states, voted to create a separate chiropractic board of examiners.[38]

In the early 1930s one study of medical care in America reported that "the efforts of the medical profession to prevent legal recognition of the chiropractors have met with almost universal defeat." In fact, by this time nearly a quarter of American healers were Christian Scientists, osteopaths, chiropractors, or irregulars of some stripe. It was clear, concluded the same study, that "in the United States the legislative regulation of the healing art is not accomplishing its acknowledged purpose," that is, creating a monopoly for regular physicians.[39]

Medical doctors encountered equal difficulty keeping assorted other health-care professionals from intruding on what they regarded as their rightful domain. Although they actually assisted podiatrists in achieving their independent status —on the grounds that corn-cutting like tooth-pulling was too trivial to control—they fought continually to limit the activities of such interlopers as optometrists, psychologists, and midwives, who competed directly with physicians specializing in ophthalmology, psychiatry, and obstetrics.[40] The medical profession hoped to restrict the practice of such individuals by defining medicine comprehensively; however, unsympathetic judges and legislators time and again sided with its opponents, as can be seen in the history of relations between physicians and optometrists.

Until the latter part of the 19th century physicians generally left the dispensing of spectacles to itinerant peddlers and other businessmen, who fitted eyeglasses on a trial-and-error basis. But after the 1860s, when a Dutch opthalmologist showed how various visual disorders could be treated by prescribing glasses, physicians became increasingly active in the field—and resentful of opticians who continued to test for lenses. The issue came to a head in 1892, when a couple of New York ophthalmologists warned an optician named Charles F. Prentice that he was violating the state's medical practice act by performing such tests. If he wished to prescribe lenses rather than simply sell them, they argued, "he must first get a degree of M.D. and then a license to practice in this State." Prentice responded to this veiled threat by organizing a state optical society and appealing to the legislature to pass a special practice act for opticians.

During the ensuing debate it became clear that opticians (or optometrists, as they came to be called) sought a status equivalent to dentists, who functioned independently of the medical profession, whereas physicians wanted to relegate optometrists to a position analogous to pharmacists, who simply filled physicians' prescriptions. "The optician must be forbidden to prescribe spectacles," declared one medical journal, "and it should be as illegal for the optician to prescribe glasses as it is for the pharmacist to prescribe morphine or arsenic. Both the optical instrument and the drug are medical agents, and only a physician is fitted to judge of the propriety of their use." Retorted Prentice, "A lens is not a pill."[41]

Despite medical harassment, including occasional arrests for violating medical practice acts, optometrists eventually acquired full legal recognition. During the two decades between 1901 and 1921 they won judicial decisions declaring optometry to be outside the sphere of medical practice, and every state in the Union passed laws setting up examining boards for optometrists. These laws usually barred them from using drugs and, to their great irritation, allowed physicians to refract eyes and prescribe glasses; nevertheless, such statutes represented a clear victory for the optometrists and another humbling defeat for the medical profession, which continued into the 1950s to shun optometrists and to debate whether optometry was a cult or merely a medical sect.[42]

THE PROFESSION AND THE PUBLIC

A century after the founding of the A.M.A. the American medical profession could look with pride on its various accomplishments. Although its efforts to monopolize the practice of medicine had fallen far short of its goals, it had, through its alliance with the law, eliminated overcrowding in the field, greatly reduced the number of incompetents practicing medicine, and acquired so much prestige and power that medicine became the envy of other professions. By the mid-20th century physicians had become the most admired professionals in the land, and, benefiting especially from the growth of health insurance, had passed bankers and lawyers to become the nation's highest paid workers.[43] Despite mounting criticism of the medical profession during the past couple of decades, physicians have, by and large, retained their elevated position in American society.[44] And despite the increasing intrusion of insurance companies, government agencies, and various allied health professionals into the medical domain, the gains physicians made during the first half of the

20th century have, to a great extent, remained intact.

Medical apologists have long argued that professional advancement brought corresponding gains to the public. "There is nothing for the benefit of medicine unless it is for the benefit of the people," declared one medical society official. "The two interests are identical." In recent years, however, critics of the medical profession have increasingly questioned such assumptions, arguing instead that the reforms we have described "centralized, bureaucratized, modernized and expanded medicine and medical education in the interests of physicians' own professional needs and with little regard for the needs of the public."[45]

The truth, I believe, lies somewhere between these two extremes. On the one hand, there can be little doubt that physicians benefited handsomely from their efforts to regulate and monopolize the practice of medicine. It is equally apparent that the elevation of the profession, in conjunction with other factors, drove up the cost of medical care, created a shortage of American-trained doctors, and damaged the chances for the poor and minorities to pursue careers in medicine.[46] On the other hand, only the most prejudiced observer would argue that the public did not gain as well. Curative medicine may have contributed little to the dramatic reduction in mortality during the past century, but physicians using preventive and ameliorative measures did significantly improve the quality and length of life in America.[47] And although the profession continues to harbor its share of scoundrels, patients today enter doctors' offices with much less cause for fear—and much more hope of being helped—than did their grandparents and great-grandparents. The interests of the profession and the public may not be identical, but neither are they antithetical.

NOTES

1 N. S. Davis, *History of the American Medical Association* (Philadelphia: Lippincott, Brambo and Co., 1855), p. 56.

2 See, e.g., Vern L. Bullough, *The Development of Medicine as a Profession: The Contribution of the Medieval University to Modern Knowledge* (Basel: S. Karger, 1966).

3 [Lemuel Shattuck], *Report of the Sanitary Commission*

of Massachusetts (Boston: Dutton & Wentworth, 1850), p. 58.

4 "The American Medical Association: power, purpose, and politics in organized medicine," *Yale Law J.*, 1954, *63*: 1018.

5 See, e.g., Joseph F. Kett, *The Formation of the American Medical Profession: The Role of Institutions, 1780–1860* (New Haven: Yale Univ. Press, 1968); and

William G. Rothstein, *American Physicians in the Nineteenth Century: From Sects to Science* (Baltimore: Johns Hopkins Univ. Press, 1972).

6   Samuel Haber, "The professions and higher education in America: a historical view," in *Higher Education and the Labor Market*, ed. Margaret S. Gordon (New York: McGraw-Hill Book Co., 1974), p. 251. The "mania" quotation comes from James H. Cassedy, "Why self-help? Americans alone with their diseases, 1800–1850," in *Medicine without Doctors: Home Health Care in American History*, ed. Guenter B. Risse, Ronald L. Numbers, and Judith W. Leavitt (New York: Science History Pubs., 1977), p. 35.

7   "American vs. European medical science again," *Med. Rec.*, 1869, *4*: 183. See also Martin Kaufman, "American medical education," in *The Education of American Physicians: Historical Essays*, ed. Ronald L. Numbers (Berkeley and Los Angeles: Univ. of California Press, 1980), pp. 7–28.

8   Harris L. Coulter, *Divided Legacy: A History of the Schism in Medical Thought*, 3 vols. (Washington, D.C.: McGrath Publishing Co., 1973), III, 143.

9   Worthington Hooker, *Physician and Patient; or, A Practical View of the Mutual Duties, Relations and Interests on the Medical Profession and the Community* (New York: Baker and Scribner, 1849), p. viii.

10   These paragraphs on sectarian medicine are extracted from Ronald L. Numbers, "Do-it-yourself the sectarian way," in *Medicine without Doctors*, pp. 49–72.

11   Kett, *Formation of the American Medical Profession*, p. 186.

12   Richard Harrison Shryock, *Medicine in America: Historical Essays* (Baltimore: Johns Hopkins Press, 1966), p. 151.

13   Rothstein, *American Physicians*, p. 178. For a positive view of medical prestige, see Barbara G. Rosenkrantz, "The search for professional order in 19th century American medicine," *Proc. Int. Cong. Hist. Sci.* 1974, *14* (No. 4): 113–124; see also ch. 16, this book.

14   Kett, *Formation of the American Medical Profession*, pp. 13, 165.

15   Richard Harrison Shryock, *Medical Licensing in America, 1650–1965* (Baltimore: Johns Hopkins Press, 1967), pp. 31–32.

16   Coulter, *Divided Legacy*, p. 182.

17   Davis, *History*, pp. 30–32.

18   J. J. Taylor, *The Physician as a Business Man; or, How to Obtain the Best Financial Results in the Practice of Medicine* (Philadelphia: Medical World, 1981), p. 94; "Code of medical ethics," *Tr. A.M.A.*, 1857, *10*: 607–620.

19   "Code of medical ethics," p. 614.

20   Rothstein, *American Physicians*, p. 120.

21   Robert P. Hudson, "Abraham Flexner in perspective: American medical education, 1865–1910," *Bull. Hist. Med.*, 1972, *56*: 545–561 (ch. 10, this book); Thomas Neville Bonner, *American Doctors and German Universities: A Chapter in International Intellectual Relations, 1870–1914* (Lincoln: Univ. of Nebraska Press, 1963).

22   Simon Flexner and James Thomas Flexner, *William Henry Welch and the Heroic Age of American Medicine* (New York: Viking Press, 1941), pp. 219, 222–223; Robert E. Kohler, "Medical reform and biomedical science: biochemistry—a case study," in *The Therapeutic Revolution: Essays in the Social History of American Medicine*, ed. Morris J. Vogel and Charles Rosenberg (Philadelphia: Univ. of Pennsylvania Press, 1979), p. 32.

23   Haber, "The professions," p. 260.

24   Kaufman, "American medical education," p. 19.

25   Donald E. Konold, *A History of American Medical Ethics, 1847–1912* (Madison: State Historical Society of Wisconsin, 1962), p. 30.

26   Alexander Wilder, *History of Medicine* (New Sharon, Maine: New England Eclectic Publishing Co., 1899), pp. 776–835.

27   Eliot Freidson, *Profession of Medicine: A Study of the Sociology of Applied Knowledge* (New York: Dodd, Mead & Co., 1970), p. 21.

28   John S. Haller, Jr., *American Medicine in Transition, 1840–1910* (Urbana: Univ. of Illinois Press, 1981), pp. 126, 266; Rothstein, *American Physicians*, pp. 308–309.

29   Coulter, *Divided Legacy*, pp. 433–434. For confirmation of this opinion by a regular physician, see Hooker, *Physician and Patient*, p. 255.

30   Philip A. Kalisch and Beatrice J. Kalisch, *The Advance of American Nursing* (Boston: Little, Brown, 1978), pp. 150–153; Mary Roth Walsh, *"Doctors Wanted: No Women Need Apply": Sexual Barriers in the Medical Profession, 1835–1975* (New Haven: Yale Univ. Press, 1977), p. 142; Janet Wilson James, "Isabel Hampton and the professionalization of nursing in the 1890s," in *The Therapeutic Revolution*, p. 221.

31   Haller, *American Medicine in Transition*, pp. 268–272; James G. Burrow, *Organized Medicine in the Progressive Era: The Move toward Monopoly* (Baltimore: Johns Hopkins Univ. Press, 1977), p. 114.

32   L. Laszlo Schwartz, "The historical relations of American dentistry and medicine," *Bull. Hist. Med.*, 1954, *28*: 545.

33   Robert W. McCluggage, *A History of the American Dental Association: A Century of Health Service* (Chicago: American Dental Association, 1959), pp. 155, 169–171. See also William J. Gies, *Dental Education in the United States and Canada* (New York: Carnegie Foundation, 1926), pp. 39–40.

34 Shryock, *Medical Licensing*, p. 60; Abraham Flexner, *Medical Education in the United States and Canada* (New York: Carnegie Foundation, 1910), p. 170; Rothstein, *American Physicians*, p. 310.

35 Ronald L. Numbers, *Almost Persuaded: American Physicians and Compulsory Health Insurance, 1912–1920* (Baltimore: Johns Hopkins Univ. Press, 1978), pp. 4, 9; Committee on Social Insurance, *Statistics Regarding the Medical Profession*, Social Insurance Series Pamphlet No. 7 (Chicago: American Medical Association, [1917]), p. 123.

36 Numbers, *Almost Persuaded*, pp. 27–28.

37 *Ibid.*, pp. 4, 113; Kaufman, "American medical education," p. 20; Shryock, *Medical Licensing*, p. 63; "The American Medical Association," pp. 969–970.

38 Elizabeth Barnaby Keeney, Susan Eyrich Lederer, and Edmond P. Minihan, "Sectarians and scientists: alternatives to orthodox medicine," in *Wisconsin Medicine: Historical Perspectives*, ed. Ronald L. Numbers and Judith Walzer Leavitt (Madison: Univ. of Wisconsin Press, 1981), pp. 59–68; Louis S. Reed, *The Healing Cults: A Study of Sectarian Medical Practice — Its Extent, Causes, and Control* (Chicago: Univ. of Chicago Press, 1932). Regarding osteopathy, see Norman Gevitz, *The D.O.'s: Osteopathic Medicine in America* (Baltimore: Johns Hopkins Univ. Press, 1982); and Erwin A. Blackstone, "The A.M.A. and the osteopaths: a study of the power of organized medicine," *Antitrust Bull.*, 1977, *22*: 405–440.

39 Reed, *The Healing Cults*, pp. 1, 121.

40 See, e.g., Maurice J. Lewi, "Medicine and podiatry in New York State," *N.Y. St. J. Med.*, 1954, *54*: 536–540; John C. Burnham, "The struggle between physicians and paramedical personnel in American psychiatry, 1917–41," *J. Hist. Med.*, 1974, *29*: 93–106; Frances E. Kobrin, "The American midwife controversy: a crisis of professionalization," *Bull. Hist. Med.*, 1966, *40*: 350–363 (ch. 14, this book); and Judy Barrett Litoff, *American Midwives: 1860 to the Present* (Westport, Conn.: Greenwood Press, 1978).

41 Charles F. Prentice, *Legalized Optometry and the Memoirs of Its Founder* (Seattle: Casperin Fletcher Press, 1926), quotations on pp. 27, 55; James R. Gregg, *American Optometric Association: A History* (St. Louis: American Optometric Association, 1972), p. 47.

42 Gregg, *American Optometric Association*, pp. 56–58, 79–80; Louis S. Reed, *Midwives, Chiropodists, and Optometrists: Their Place in Medical Care* (Chicago: University of Chicago Press, 1932), pp. 36–60; *1846–1958 Digest of Official Actions: American Medical Association* (Chicago: American Medical Association, 1959), pp. 536–540.

43 Robert W. Hodge, Paul M. Siegel, and Peter H. Rossi, "Occupational prestige in the United States, 1925–63," *Am. J. Sociol.*, 1964, *70*: 290; Andrea Tyree and Billy G. Smith, "Occupational hierarchy in the United States: 1789–1969," *Social Forces*, 1978, *56*: 887; Ronald L. Numbers, "The third party: health insurance in America," in *The Therapeutic Revolution*, pp. 192–193 (ch. 17, this book).

44 John C. Burnham, "American medicine's golden age: what happened to it?" *Science*, 1982, *215*: 1474–1479 (ch. 18, this book).

45 Ronald L. Numbers, "Public protection and self-interest: medical societies in Wisconsin," in *Wisconsin Medicine*, p. 96; Gerald E. Markowitz and David Karl Rosner, "Doctors in crisis: a study of the use of medical education reform to establish modern professional elitism in medicine," *Am. Quart.*, 1973, *25*: 107.

46 See, e.g., Herbert M. Morais, *The History of the Negro in Medicine* (New York: Publishers Co., 1967); and Walsh, *"Doctors Wanted,"* pp. 192–194. Although the percentage of women in medical schools declined during the first half of the 20th century, Walsh argues (p. 237) that they were not driven out of the profession by licensing laws and educational requirements. Rather, she attributes the decline to overt discrimination against women.

47 On the relationship between medicine and mortality in America, see the introduction to this book.

15

The Training and Practice of Midwives: A Wisconsin Study

CHARLOTTE G. BORST

On 1 February 1887, Caroline Kueny, 35 years old and the mother of five children, began her six-month course at the Milwaukee School of Midwifery. Attending daily lectures and 16 confinements during the course of her instruction, Kueny graduated after passing her final examinations to the satisfaction of the school's two physician examiners. She immediately registered with the Milwaukee Health Department and began attending births. Over the next 15 years she worked as a midwife in Milwaukee and had several more children of her own.[1]

While Caroline Kueny was obtaining her midwifery skills at school, Dora Larson, living in the small, rural village of Woodford, Wisconsin, became a midwife in a more traditional way. Thirty-seven years old and the mother of eight children, Larson learned midwifery techniques from assisting her midwife mother, by helping local physicians attending confinement cases, and from her own childbirth experiences.[2] She noted on her 1910 state license application that her midwife credentials included being the mother of eight children and the grandmother of four.[3]

In 1910, when Kueny and Larson applied for state licenses from the Wisconsin Board of Medical Examiners, many other applicants had experiences similar to theirs. Although the women who applied for licenses had become midwives in many ways—demonstrating the problem of defining the "Ameri-

can midwife"—patterns in the manner of their training and in the tempos of their practice are discernible. In addition, Kueny's and Larson's status as mothers and grandmothers typified most of Wisconsin's midwives: almost every midwife practicing in the state had children. Indeed, it seemed that motherhood was almost a prerequisite for the job.

Historical studies of American midwives generally have employed the term *midwife* to mean any non-physician, usually female, attending a childbirth. The focus of most scholarly studies—the elimination of midwives by a hostile medical profession—explains why many historians have not analyzed the differences among these female practitioners.[4] By evaluating midwives only within the scope of the "midwife question" of the early 20th century, historians have been forced to see these female practitioners through the eyes of the doctors engaged in the debate. That is, most of their sources of information about midwives are articles from medical journals or papers delivered at conferences controlled by physicians.

Frances E. Kobrin's pioneering article suggested that midwifery was the victim of a crisis of professionalism within the medical profession over the role of the new specialty of obstetrics.[5] However, I will argue in this article that American midwifery also faced its *own* professional crisis. Like physicians and nurses, schooled midwives began to replace their untrained or apprentice-trained counterparts at the turn of the 20th century. Unlike these other practitioners, however, midwives did not become professionalized as they became more educated; instead, they became increasingly subservient to physicians, their autonomy decreased, and, ultimately, they ceased functioning in the United States.

The evolution of midwifery training is not the

CHARLOTTE G. BORST is Associate Professor of History and Executive Director of the Historical Collection, University of Alabama at Birmingham.

From the *Bulletin of the History of Medicine*, 1988, *62*: 606–627. Copyright © 1988 by the Johns Hopkins University Press. Reprinted by permission of the Johns Hopkins University Press.

sole explanation for midwives' professional problems at the beginning of the 20th century. Most
midwives, however they learned their trade, were
married women who practiced within the limited
confines of a discrete ethnic and classbased community. Thus, any study of the demise of midwives
in the United States must incorporate an analysis
of the geographic, class, and ethnic dimensions of
the midwives' practices, a consideration of their
gender and marital status, and an examination of
the results of changes in their training.

The best way to understand this crisis of professionalization is to examine the cultural context of
midwife practice by focusing on a particular geographic location. This article examines midwifery in Wisconsin.

Wisconsin was fairly typical in its political stance
toward midwives. Beginning informally in Milwaukee in the 1870s and culminating with the 1909
Midwife Registration Act administered by the State
Medical Practice Board, physicians in the state
helped to determine which midwives could deliver
babies and the conditions under which they could
work.[6] Thus, Wisconsin did not ban these female
attendants outright when there were still women
who wished to practice, but neither did the state
leave the practice of midwifery unregulated.[7]

Sociologists and historians have identified an occupation's movement toward professionalization
with the "standardization allowed by a common and
clearly defined basis of training. . . . [This training]
is, in fact, the main support of a professional subculture."[8] Instruction, Magali Sarfatti Larson
points out, historically proceeds from the relative
informality of apprenticeships to formal education
based on the standardization and the codification
of knowledge.[9] While the rise of a profession is tied
intimately to the development of formal training, a
profession also needs a distinct body of knowledge
to claim exclusively. To then achieve a monopoly of
practice, the profession needs to control both the
"production of knowledge and the production of
producers," preferably within one institution.[10] In
summary, the professional practitioner must have
mastered a body of knowledge unique to the field
within a formal setting and then have the autonomy to decide when and under what circumstances
to apply this knowledge.

By the first decades of the 20th century, leaders

in both medicine and nursing recognized clearly
that better training fulfilled their professions' objective need for more knowledge and their subjective desire to improve the status of practitioners.
As one historian has written, "The aim and end of
reform . . . [was] professionalization."[11] Physician
and nursing leaders understood professionalization to mean credentials based on education and
licensing standards that would enable practitioners
to assert the value to human life of their specialized
knowledge and their special skills. At the same time,
professionalization was a movement to strengthen
status and to consolidate authority, which allowed
professionals to seek broad social prerogatives.[12]

This paper asks the same questions about midwives that medical historians have posed for physicians and nurses: Did midwife training follow the
classic path toward professionalization—from casual learning to apprenticeships to school training?
Were midwives able to claim an exclusive body of
knowledge? Were these practitioners then able to
control their own institutions and to determine
who could practice midwifery?

From state and county records, I identified 893
midwives practicing in Wisconsin between 1870
and 1920. Defined by the sources of their training, these practitioners may be divided into three
groups: (1) the neighbor-women, usually native-
born women who learned their techniques from
giving birth themselves and from assisting at the
deliveries of relatives and a few close friends; (2)
the apprentice-trained midwives, typically native-
born, who learned their craft from other, older,
midwives or from physicians; (3) school-educated
midwives, usually first- or second-generation immigrants, instructed in formal settings through both
didactic and practical work. Though all three kinds
of midwives could be found practicing in Wisconsin
during much of the 50-year period of my study,
there was a distinct evolution from neighbor-
women to apprentice-trained practitioners to school-
trained practitioners.

NEIGHBOR-WOMEN

Distinguished from the other midwives by their
completely informal training, neighbor-women
were the most "traditional" practitioners in my
study.[13] Their experiences are exemplified by Susan Washburn, a 60-year-old Millston, Wisconsin,

woman who wrote in 1909, "I would state that my [midwifery] education has been limited; has more in a careing [*sic*] for those that are going to be confined in my own neighborhood. I have followed this about 30 years with good results my experience has been much in midwifery: I never lost a case."[14]

Some neighbor-women did not even consider themselves "midwives," for besides their own confinements, their birthing experiences were limited to a few births in their families or among their friends. A 1919 Children's Bureau report of rural problems illustrated how blurred the distinction was between merely helping out one's neighbor and assuming a title conveying authority. Reviewing maternity care in rural America, the report revealed that "many country women are cared for entirely by neighbors who may or may not have acquired some skill from experience." Some women, the report added, were seen "by midwives."[15]

Some neighbor-women midwives developed more than the basic skills out of sheer necessity— many lived in isolated settlements where doctors or trained midwives were scarce and where even rudimentary skills were better than none. A study of childbearing conditions by the Children's Bureau in Marathon County, Wisconsin, portrayed the extreme isolation farm women faced in many rural areas. Only 5 percent of the county's roads were paved; the rest were dirt roads, impassable during much of the year. Families in this area, including childbearing women, were often forced to rely totally on nearby neighbors.[16]

The neighbor-women's skills were probably the most heterogeneous of any of the three midwife groups. However, these women did share several characteristics: their training, which I have described as informal or self-taught, their age, and where they lived.

Thirty-nine of 396 women who registered with the state of Wisconsin after 1909 reported training like Susan Washburn's. These 39 women shared two other characteristics: their mean age of 54 was older than the mean ages of other types of midwives,[17] and they were overwhelmingly rural residents—only one neighbor-woman who registered with the state lived in Milwaukee, the state's most urban area.[18]

An analysis of the 588 women who registered with the Milwaukee Health Department between 1877 and 1907 reinforced the state license data

correlating neighbor-women midwives with rural residence: over the 30-year period of the Milwaukee registry, only 21 women reported having no training.[19]

Characterized by the female-centered networks within which they operated and by their self-acquired knowledge, neighbor-women midwives were also the most autonomous midwives—the conditions under which they practiced permitted them complete freedom. But even in isolated rural areas, families turned to these practitioners only when they could not find a trained midwife. It is not surprising, therefore, that as frontier areas of Wisconsin became more settled and more accessible, trained midwives replaced neighbor-women. By the beginning of the 20th century, this family-based practitioner was found only in limited numbers, and in the most rural parts of the state.

APPRENTICE-TRAINED MIDWIVES

Throughout American history, many midwives, like many physicians, received their training by apprenticing themselves to older and more experienced practitioners. Indeed, in the United States until the last decades of this 19th century, almost all trained midwives learned their trade in this way.

Anne Nowak of Lublin (in north-central Wisconsin) reported her training in terms that would be familiar to many medical historians: "I have not taken up the course in any school but have practiced with a professional midwife until I was fully-capable of doing the work alone."[20]

Education at the side of a preceptor midwife was not the only form of apprenticeship. Many 19th-century midwives did learn their craft from older, more experienced midwives, but many more female birth attendants apprenticed themselves with local physicians. Indeed, the most striking feature of apprentice-trained midwives in Milwaukee was that virtually all of them were trained by physicians.

Historians who have analyzed physician debates over the "midwife question" of the late 19th and early 20th centuries have emphasized the hostility or at least the indifference most doctors felt toward these female practitioners. My analysis of Wisconsin data, however, indicates that midwives and physicians sometimes worked together in America's birthing rooms.[21]

Of the 49 apprentice-trained women who registered in 1909, 35 (or 71 percent) reported that their instruction came from physicians. Mary Greeley, a practitioner in western Wisconsin, was typical, declaring "[I] have been practicing midwifery for 30 years. Usually under the direction of physicians of River Falls, Wis. and Ellsworth, Wis."[22]

The timing of this kind of instruction in the midwife's career varied. Some women received their initial training from physicians and then went out to deliver babies, as Ellen Stiefvator, practicing in northern Wisconsin, related: "I practiced with Dr. Rauls Gadsden [identified by her as an Alabama physician] 30 yrs. ago and have practiced ever since: [the] last 18 years in Merrill, Wis."[23]

This cooperation between midwives and physicians in rural areas is not entirely surprising, as the 1919 Children's Bureau survey of childbirth attendants in northern Wisconsin pointed out. In one village, for example, the authors of the survey found that local doctors had ceded all of the obstetrical work to Mrs. M. (a local midwife). The doctors referred their maternity cases willingly, telling the survey authors, "We don't like that kind of work and have always more or less turned it over to [her]."[24]

A close interaction between midwives and doctors was not limited to rural maternity practice, however. Although historian Jane Pacht Brickman has concluded that "medical animus against the midwife focused on the urban practitioner," in Milwaukee, many physicians trained midwives.[25] While some of these physicians were European-educated émigrés and may have been continuing a traditional association common to their ethnic groups, other Milwaukee doctors who worked closely with midwives had trained at American medical schools. Furthermore, most of these American physicians were well regarded; they were not marginal practitioners at the fringes of the Milwaukee medical community.[26]

Apprenticeships for these women lasted from several months to a year, and ranged from what seemed to be private lectures to a combination of didactic and practical work. Frances Jahnz, for example, had "Studied with Dr. F. S. Wasielewski for (2) two months, [in 1901] then took up ten (10) practical cases of confinement in his presence according to his instructions."[27] Her entire course of instruction, she stated, lasted one year. Mary Browikowski had "received instructions in the art of

midwifery for fully five months from Dr. F. M. Hinz, M.D."[28] Dr. Hinz, a homeopathic physician trained in Illinois, offered seven months of "private instruction" to another Milwaukee midwife.[29] Like Dr. Hinz, other Milwaukee physicians lectured to midwives, offering courses ranging in length from 6 to 12 months.[30] One physician continued his mentor relationship with his pupil midwife after training her. Mary Holub received "practical training under Dr. Andrew Munro, attended confinement cases with him and studied under him. Have certificate from him and recommendation to practice. Attended cases with him for a period of seven years."[31]

Apprentice training, however, was not a major route to the profession for Milwaukee midwives. Overall, of the 588 women who registered with the Milwaukee Health Department between 1877 and 1907, only 51 (8.6 percent) reported an apprentice education. Analyzed year by year, the Health Department data also reveal the declining popularity of apprenticeships, from 15.7 percent of the registrants in the early years to 6.6 percent in the last years of the registry. The state licenses, covering a somewhat later period (1909–15), also demonstrate the decline in the 20th century in apprentice training, especially for urban midwives: of 189 Milwaukee midwives who registered with the state, only 12 (6.3 percent) were apprentice educated.

SCHOOL-TRAINED MIDWIVES

Though many European school-educated midwives were among the large numbers of immigrants flooding Milwaukee and the rest of the state in the late 19th century, the most compelling explanation for the decline in apprentice training for midwives in Wisconsin lies with the establishment of two schools for midwives in Milwaukee.

Organized by physicians together with some local midwives, these schools offered Milwaukee and many other Wisconsin women a potentially more rationalized course of instruction in the theory and the practice of midwifery than they could obtain by themselves or in tandem with a physician or experienced midwife. Many Milwaukee midwives received their training at these schools: between 1890 and 1907, over half of the Milwaukee Health Department registrants reported having obtained a local school education.

But did the Milwaukee and other American midwifery schools organized at the end of the 19th century offer formal training that resulted from a "codification of knowledge" in midwifery? Did the standardized training become "the main support of a professional subculture" for midwives? A brief examination of these schools' curricula, personnel, and philosophy demonstrates that regardless of the new skills they offered their students, the schools were not, nor were they intended to be, a step in promoting the professionalization of midwifery.

American midwifery schools developed from and found their roots in the immigrant community. It was not surprising, therefore, to find two schools based on the European model in Milwaukee. They were founded in the period of the heaviest immigration from Europe to America. Milwaukee in particular experienced an explosive growth in its population and much of its growth was related to immigration from the Old World: between 1880 and 1910, the city was tied for first place among all American cities in its percentage of foreign stock. By 1910, the total foreign stock—total foreign-born plus natives with at least one foreign-born parent—was 78.6 percent of the total population. Of the foreign-born, almost half of them came from Germany.[32] Partly as a consequence of this immense immigrant population, the city had a large number of practicing midwives. As a *Milwaukee (Wisconsin) Sentinel* article pointed out in 1889, "No city in the country has as many practicing midwives as Milwaukee, it is said, at least in proportion to population. There are only a comparatively few more physicians than midwives, there being 171 of the former and 115 of the latter."[33]

European midwives and physicians emigrating to the Untied States brought their ideas about training with them. They were proud of their education and sometimes quite critical of unlicensed, unskilled American midwives. As Agnes Pradzinska, a Milwaukee midwife, wrote to the *American Midwife,*

I wish to express my concern that the so-called butcher-women be prohibited. I have already complained many times to the local health department, but I always receive the same answer: "We will look into the matter." But they are [still] looking into it. The births should all be registered in the state of Wisconsin, such female bunglers don't do it

though. Among [female bunglers] is one, who calls herself a midwife and has hung out a shield with the inscription "midwife and birth assistant." But she can neither read nor write. . . . [She] also knows fortune-telling, because when she performs a delivery, she tells the mother how many more children she will have; even better yet is that she also knows how many boys and girls she will have.[34]

The Milwaukee School of Midwifery, founded in 1879 by a German-born and -trained midwife, and the Wisconsin College of Midwifery, established in 1885 by a German immigrant woman trained in America, shared many of the characteristics of other American midwifery schools at the turn of the century. Established by immigrant midwives in conjunction with immigrant physicians and their sympathetic American colleagues, both schools drew their students from Milwaukee's foreign-born community. Like many American midwifery schools, the Milwaukee School of Midwifery offered its course in German and in English. Both schools, like their American and European counterparts, relied heavily on the prestige of local, usually young, physicians for some instruction and for the approval of the final examination.[35]

Finally, both institutions, established as profit-making ventures, encountered many of the problems that plagued proprietary medical schools of the era: poor facilities, a dearth of instructional materials, and course work consisting of didactic lectures with limited clinical instruction. For example, a graduate of the Milwaukee School of Midwifery described her training as a course of "6 months—about 6 lectures per month. . . . [I] assisted Precepteress Mrs. Wilhelmine Stein at her hospital as well as at [her] private confinement cases."[36] Mrs. Stein's hospital was in her home, having probably fewer than 10 beds. The Wisconsin College of Midwifery offered a seemingly more rigorous course. Two graduates reported having taken a five-month course, with lectures in anatomy, biology, and physiology daily for two hours.[37] Students at this college gained practical experience by attending obstetrical clinics at least twice weekly, though many graduates reported that much of their entire four- to six-month term was spent working at the hospital. By 1912, the school's term had expanded to eight months, and students attended at least 25 confinements.[38]

These midwifery schools, despite their faults, probably offered more clinical instruction in obstetrics than many of the medical schools surveyed by J. Whitridge Williams in 1912.[39] However, both institutions suffered from a problem unique to the field of midwifery. Unlike nursing or medicine, midwifery could claim neither a distinct body of knowledge nor control over the production of its own practitioners. Milwaukee's two schools, like other midwifery schools in America and in Europe, needed physician support to establish and maintain respectability.

One way to maintain physician support was to downplay midwives' roles as teachers. For example, advertisements for the Wisconsin College of Midwifery in Milwaukee's city directory show how Mary Klaes's central role as founder and teacher was subsumed by her need to attract doctors to her school. Listing the physician president's name first and in bold print, the text of the advertisement urged other doctors to bring their cases to the school's hospital, with the promise that "trained nurses" were "always in attendance."[40] By 1901, the school referred to itself as "St. Mary's Sanitarium," having the "best physicians in charge." Though midwives were still being trained in the hospital at that date, the city directory advertisement made no mention of the fact.[41]

The problem the Milwaukee schools experienced in holding on to their "best physicians" was not unusual, and reflected a lack of autonomy by midwives at even the most prestigious institutions. The Bellevue Hospital School in New York City, for example, generally considered the best American midwifery school, denied any teaching role to midwives. Local doctors taught all of the courses.

The midwifery schools' curriculum reinforced and institutionalized their students' dependent role by emphasizing that good midwives should attend only normal, uncomplicated births. A physician associated with the Bellevue School explained this philosophy:

> Not the least advantage of our primitive attempt to educate the midwife at the Bellevue School is the thorough teaching of each candidate for graduation her limitations. . . . One important fact is instilled into the brain of each midwife, and that is the knowledge of her own limitations—the knowledge of what not to do, and when to seek the aid of a practising physician.[42]

In another study by American physicians of midwifery schools in Scandinavia, the writer noted approvingly that the students "are taught to revere the physician, and they are distinctly shown their limitations."[43] Although Milwaukee physicians left no written records of their feelings about the midwifery schools they were associated with, it is reasonable to assume, given their standing in the medical community, that the physicians instructed their students the same way the instructors at the Bellevue and Scandinavian schools did theirs.

Midwifery in the United States in the late 19th century faced a "crisis of professionalism" that related directly to the natural evolution of the occupation. At each stage of the developmental process, midwives' practice became less autonomous and more circumscribed. But a limitation on midwife practice that did not diminish as the period unfolded related directly to the nature of the practitioner herself: midwives in Wisconsin, as midwives everywhere else, were almost always mature married women who had children themselves. Most midwives, therefore, were limited by family responsibilities to practicing part time.

Like many other working women in their communities, midwives left their homes in order to earn money to help support their families. Their work-force participation as married women placed them in a unique category of working women, however. Historians studying women's work have emphasized the rarity of married women participating in the work force in the late 19th and early 20th centuries. Leslie Woodcock Tentler found that only 5.6 percent of all married women were reported at work in 1900, though she admits that the census data probably underenumerated wives working part time in the home.[44] Married women who did work outside the home, she points out, chose jobs with relatively short or flexible hours or part-time employment, and they often tried to stay within the immediate neighborhood.[45]

Eugene Declercq and Neal Devitt have shown that early 20th-century midwives were predominantly mature women of diverse ethnic backgrounds.[46] However, neither of these historians has

Table 1
Distribution by County and by Census of
398 Wisconsin Midwives

County of Residence	Number	Year of Census	Number
Dane	33	1870 federal	6
Milwaukee	301	1880 federal	77
Trempealeau	26	1900 federal	191
Price	38	1905 state	38
		1910 federal	86

Sources: U.S. Census Office, 9th, 10th, and 12th Censuses, 1870, 1880, 1900, *Population Schedules: Wisconsin* (Washington, D.C.: Census Office, 1870, n.d., 1901); U.S. Bureau of the Census, 13th Census, 1910, *Population Schedules: Wisconsin* (Washington, D.C.: Government Printing Office, 1913), National Archives microfilm publications; Wisconsin Dept. of State, *Wisconsin State Census, 1905: Population Schedules*, Wisconsin Secretary of State's Office (Madison: State Historical Society of Wisconsin, n.d.), Microfilm.

considered how these women's demographic characteristics might have influenced the practice of midwifery. In my investigation of some of Wisconsin's midwives, I analyzed the effects of age, marital status, immigrant background, and social class on the occupation and professionalization of midwifery in America.

Women training for or practicing midwifery in Wisconsin were usually immigrant, working-class, married women. Midwives' age, marital status, ethnicity, and social class, and the number and the ages of their children, conceivably influenced the patterns of their practice. In order systematically to study midwives as working women, I linked practitioners who identified themselves as midwives to the descriptions of their families in the federal or state manuscript census schedules.[47] From four Wisconsin counties that were chosen to sample an urban-to-rural-to-frontier continuum in a 50-year period (1870 to 1920), 893 different midwives were identified.[48] Of these practitioners, 398 were then linked to the 1870, 1880, 1900, or 1910 federal census, or to the 1905 state census (see Table 1). The census data were then analyzed to determine basic information about midwife families and about the practitioners themselves.[49]

To couple the data describing midwives with information about their patterns of practice, I sampled birth certificates in the same four counties every five years between 1870 and 1920.[50] The data sets contain 28,924 cases for all four counties and provided information on the patterns of practice of 893 midwives and 1,149 physicians.[51]

Neal Devitt's assertion that midwives were predominantly elderly women and that "advanced age" played a role in the demise of midwifery fails to acknowledge the fact that midwifery has always been an occupation for mature, married women. As historian Jane Donegan relates, "The acceptable midwife was the respectable older woman who carried herself correctly."[52] In fact, young, single women often were not allowed to become midwives. Dr. Marie Zakrzewska, the famous director of the New England Hospital for Women and Children, trained as a midwife in one of the most prestigious schools in Berlin. Even with influential support, however, her first application for admission was refused specifically because she was not married and because, at age 18, she was considered too young to be a midwife.[53]

Wisconsin's midwives, on the whole, were middle-aged but not elderly women. Two data sets provided information on the ages of these practitioners. The mean age of the women who registered with the state after 1909 was 48.1 years. Many women from this group were even younger—the numerous school-educated practitioners were a significantly more youthful group.[54] The census study, which identified practitioners earlier in their careers, confirmed the findings in the state file. In the census study, the 398 midwives had a mean age of 43.9, almost five years younger than the average age of the state licentiates.

The census study provided crucial information on marital status. These data emphasized the close relationship between the occupation of midwifery and the attendant's family situation. Of the 394 women whose marital status could be determined, 316 (80.2 percent) were married women living with their husbands. Another 59 women (15.0 percent) were widowed, and 11 midwives were divorcées. Thus, only 8 women (2.0 percent) were single. In Wisconsin, at least, midwifery was most definitely practiced by married women.

Like the midwives in Declercq's Lawrence, Massachusetts, study, almost all of the midwives identified in Wisconsin were literate. Of the 359 mid-wives in the census study for whom literacy could be ascertained, 343 (95.5 percent) were described as being able to read or write (not always in English). Only 16 women in this sample were not literate.[55]

Midwifery, like the prevalent 19th-century custom of keeping boarders and lodgers, was ideally suited in many ways as an occupation for mature, married women. Though midwives worked outside of their own homes, they delivered babies in other women's homes, remaining protected from the shop floor and the effects of working there, popularly accepted as deleterious. Like women who kept boarders, midwives could adjust their job schedule to meet the needs of their families, taking fewer cases if they needed the time at home, and taking more cases if they needed the extra money. Indeed, when viewed in the context of home-based wage earning, midwifery may have replaced keeping boarders. Only 31 midwife families in the census study (7.8 percent) also had boarders present. Instead, most midwife families were nuclear—287 midwives (72.1 percent) lived solely with their husbands and children.[56]

Many Wisconsin midwives were responsible for their own minor children even as they were on call to deliver other women's babies. Of the midwives in the census study, 323 (81.2 percent) had at least one child living at home with them, and most women had more. The mean number of children was three. The range was quite narrow, however, as most practitioners (77 percent) had four or fewer children at home.[57] Married, divorced, and widowed women did not have significantly different numbers of children.[58]

These children were not preschoolers. Midwives, unlike other mothers working outside of the home, did not work until their children were of school age. Most working-class mothers who earned money outside of the home, on the other hand, *stopped* working when their oldest child reached the age of nine or ten. Thus, the mean age of midwives' oldest children, 16.0, and the mean age of the youngest, 8.1, deviated considerably from Lynn Y. Weiner's findings for children of working mothers.[59]

To measure the effect of the number and the ages of midwives' children on the midwives' patterns of practice, data from the vital statistics were aggregated and merged with data from the census study. In a multivariant analysis of 274 midwives who were found in both the census study and in the birth records, no relationship was found between the number of babies a midwife delivered and the number of children she herself had, the age of her oldest child, or the age of her youngest child. A very weak relationship did exist between the size of a midwife's practice and her age.[60]

Like the age distribution of midwives' children, the distribution of midwives' husbands' occupations did not fit the usual pattern for married women workers. The census sample provided data on 314 husbands of midwives. Unlike the husbands of other married women wage earners, who were employed in very low status jobs, most husbands of midwives were skilled artisans (41 percent, n = 131). The next largest occupational groups were much smaller: 17.2 percent (54) of spouses of midwives worked on farms or in the lumber trade, and 16.9 percent (53) worked as laborers. Most midwives, therefore, were the wives of men who earned a comfortable working-class income.[61]

The data on patterns of practice provide no direct evidence of a midwife's economic incentive. When data on the size and the pace of practice were linked to the midwives' demographic characteristics found in the census study, tests of significance showed no connection between the total number of births a midwife attended and her husband's occupation. Thus, poorer practitioners were not more active than those who might be better off. Midwives married to day laborers were not significantly more active practitioners than midwives married to men having more secure occupations like carpentry or clerking.

While midwife families may not have been as desperately poor as the families of women working in factories, the status of their homes demonstrates their precarious financial position and provides some evidence of a midwife's economic incentive. Of the 320 midwife families whose housing status could be verified, 47 percent rented their homes (151 families), 29 percent held mortgages (92), and only 24 percent (77) owned their homes outright. As Stephan Thernstrom's work shows, home ownership in the 19th century was the result of a very slow and expensive process.[62] About half of the

Table 2
Home Ownership in Artisan and
Laborer Midwife Families

Occupation	Renters	Mortgage Holders	Owners
Artisan			
Number	61.0	34.0	10.0
Expected number	47.1	31.8	25.2
Laborer			
Number	13.0	13.0	13.0
Expected number	18.4	12.4	9.8

Sources: U.S. Census Office, 9th, 10th, and 12th Censuses, 1870, 1880, 1900, *Population Schedules: Wisconsin* (Washington, D.C.: Census Office, 1870, n.d., 1901); U.S. Bureau of the Census, 13th Census, 1910, *Population Schedules: Wisconsin* (Washington, D.C.: Government Printing Office, 1913), National Archives microfilm publications; Wisconsin Dept. of State, *Wisconsin State Census, 1905: Population Schedules,* Wisconsin Secretary of State's Office (Madison: State Historical Society of Wisconsin, n.d.), Microfilm.
Note: A chi-square analysis of midwives' husbands' occupations by home status was significant at the 0.0001 level. The artisan and laborer groups provided most of the deviations from the expected numbers.

midwife families in my study were involved in the process of home ownership, a figure consistent with Thernstrom's findings for working-class families in Newburyport, Massachusetts.[63]

Although a simple analysis of the size of a midwife's practice showed no statistical connection with the status of her family's house,[64] a more complex examination of home ownership by midwife artisan and laborer families revealed one potential use of a midwife's income. One would perhaps expect artisan families predominantly to own their homes and laborers mostly to rent, but in actuality many artisan families rented and not many of them owned houses outright. On the other hand, fewer laborer families rented and more of them had mortgaged or fully owned homes than would be expected (see Table 2). These surprising findings for artisan midwife families indicate that some of them lived near the edge of economic hardship. Thus, for those families who did not own their homes, the practice of midwifery may have been an important secondary source of income or a supplement to be saved for a house.

While the Wisconsin data provided no clear-cut economic rationale for midwifery practice, they demonstrated strongly the relationship between midwives' immigrant status and their chosen occupation. Most of the midwives in the Wisconsin census sample were first-generation immigrants; all but 10 had parents born in Europe. The majority of these women, not surprisingly, came from Germany (52 percent). Indeed, the German-speaking midwives, including those born in Germany, the Austro-Hungarian empire, and Switzerland, constituted 72 percent of all of the foreign-born women in the sample. The next largest groups of foreign-born attendants were Poles (13.3 percent) and Scandinavians (12.7 percent). An analysis of the 274 midwives who appeared both in the census study and as birth attendants in the vital-record study strengthened the evidence for a link between an immigrant background and the practice of midwifery. The 166 German and Swiss midwives delivered an average of 28 babies each, both the Austrian and the Polish practitioners averaged 22 births, and the American-born midwives attended an average of only 11.8 births. Scandinavian-born birth attendants, living mostly in rural counties, were called to a mean of only 7.6 births, the smallest average number of deliveries of any group (see Table 3).

Historians of midwifery in the United States have long noted the large percentage of midwives who were immigrant women. Frances Kobrin describes turn-of-the-century midwives as being employed primarily by the foreign-born and their children.[65] Jane Pacht Brickman writes, "Wherever immigrants settled, the midwife flourished. . . . The midwife, almost always foreign-born and living in the community, lay in the buffer that immigrant groups maintained against an already overwhelming cultural shock."[66] Almost none of the midwives Eugene Declercq found practicing in early 20th-century Lawrence, Massachusetts, were born in the United States: "Like the population they served, they came predominantly from Southern and Eastern Europe."[67]

Contemporary observers of these birth attendants documented their prevalence and popularity

Table 3
Distribution of Births by Ethnic Group

Group	Mean Number of Deliveries	95 Percent Confidence Levels
German	28.01	23.11 to 32.91
Swiss	28.01	23.11 to 32.91
Scandinavian	7.61	2.37 to 12.85
Austrian-Hungarian	22.05	8.47 to 35.62
Polish-Russian	22.45	14.45 to 30.45
American	11.83	2.05 to 21.62

Sources: Milwaukee, Price, Trempealeau, Dane Counties [Wisconsin], *Births,* Vital Records Office, Wisconsin Dept. of Health and Human Services, Madison. Also U.S. Census Office, 9th, 10th, and 12th Censuses, 1870, 1880, 1900, *Population Schedules: Wisconsin* (Washington, D.C.: Census Office, 1870, n.d., 1901); U.S. Bureau of the Census, 13th Census, 1910, *Population Schedules: Wisconsin* (Washington, D.C.: Government Printing Office, 1913), National Archives microfilm publications; Wisconsin Dept. of State, *Wisconsin State Census, 1905: Population Schedules,* Wisconsin Secretary of State's Office (Madison: State Historical Society of Wisconsin, n.d.), Microfilm.
Note: A one-way analysis of variance was very significant when attendants' total number of deliveries was tested with their own birthplace. The f-ratio was 2.86, with a probability of 0.0103. The German and Swiss groups also showed the least variability around the mean as measured by the 95 percent confidence levels about the mean. Other groups had much wider-ranging confidence levels.

in immigrant communities all over the United States. As Michael Davis noted in his study of immigrant health,

> The immigrant mother has rarely been accustomed to a man doctor at the time of confinement. She and her friends have used the midwife, who, in most European countries, is a woman of some standing, trained, and in many countries, carefully supervised. The midwife is the most important single element in the general question of the care of immigrant mothers.[68]

Grace Abbott's 1917 work on immigrant communities in Chicago also recounted immigrant women's preference for midwives at their lyings-in,[69] and Joseph B. DeLee, a virulent opponent of midwives, noted approvingly in 1915 that midwives

seemed to be "slowly disappearing in America," and were only found in "crowded communities of foreigners."[70]

"Communities of foreigners" were not restricted to crowded urban areas, however. A 1924 study of Minnesota midwives found that only 13.7 percent of them were born in America; 86.3 percent (101 of 117 midwives found by the survey) were foreign-born. Twenty-five percent of the practitioners in this sample were from Scandinavian countries; 17.1 percent were German.[71] The Children's Bureau study of maternity conditions in Marathon County, Wisconsin, found that the large number of German and Polish settlers clung "to their foreign customs and habits of thought and to a certain extent their languages, making the district as a whole distinctly foreign in its atmosphere."[72] Midwife attendance at childbirth was one of the "foreign customs" these women clung to. Twenty-four midwives, all members of the German or Polish communities, practiced in the county, though only a few were reported to be earning their living as midwives. Most female attendants "went out" as an accommodation to their neighbors. Of all of the midwives in the region, only six practitioners delivered more than 10 babies per year, and only two oversaw the births of more than 30 babies.[73]

As the Children's Bureau study suggests, midwives practiced not only in cities but also in rural areas where there were significant communities of immigrants. In my census study of Wisconsin midwives, American-born midwives were a distinct minority in each of the four counties. Tests of significance showed distinct ethic groupings in each county, with Scandinavians predominantly located in the three rural areas and Polish and Russian midwives exclusively based in Milwaukee. Very few American-born midwives practiced anywhere in Wisconsin. In Milwaukee County, only 7.6 percent of 301 birth attendants were born in America, and the numbers of American-born midwives in the other counties were so small that statistical analysis was not feasible.[74] In Milwaukee, the overwhelming majority of practitioners, 185 (61.5 percent), had emigrated from Germany. In Dane County, German-born midwives formed the largest group (n = 15, 45.5 percent), but Scandinavian attendants were almost as numerous (n = 10, 30.3 percent). In the other two rural counties in the study, Scandinavian women were the dominant ethnic group. In Trempealeau County, 65.4 percent (n = 17) of the attendants were Scandinavian, and in Price County, 34.2 percent (n = 13) were born in Norway, Sweden, or Denmark. Five midwives in Price County were natives of Finland.

Summarizing the practice of midwifery in the early 20th century on the lower East Side of New York City, historian Elizabeth Ewen described it as an occupation that was "an acceptable means for a married woman to contribute to her family economy."[75] This article has demonstrated that midwifery certainly was an occupation practiced by mature, married women. But the evidence from the analysis of Wisconsin's midwives also demonstrates that these women were not motivated solely by economic necessity. Though many Wisconsin practitioners undoubtedly contributed significant amounts of money to their family's economy, they practiced midwifery because it was an acceptable occupation in the *immigrant* community. Both historians and contemporary observers of European settlers in the United States found that these settlers had high regard for midwives. Midwives respected birth traditions of immigrant families and, as female birth attendants, they did not offend religious families' sense of moral propriety.[76]

Midwifery, unlike nursing or medicine, was practiced according to established customs. Women could adapt the size of their practice to the needs of their families and themselves. Unlike other wage-earning women, midwives did not have to punch a time clock or listen to a boss. However, despite having some very highly developed skills and the autonomy to set their own hours, midwives had more in common with housewives who kept boarders and lodgers than they did with nurses and female physicians. In these two branches of the healing arts, practitioners worked full time, accepting people from many diverse classes and ethnic groups as patients. Midwives, despite their education, practiced part time, birthing babies in their own immigrant neighborhoods. While the trained midwife may have delivered more babies and at-tended births on a more regular basis than the neighbor-woman, midwifery for both kinds of female practitioner was an occupation that was fitted in around the needs of the family. Given this priority within the context of women's lives, midwifery could not develop into a profession.

The demise of American midwives can best be understood through an analysis of the practitioners themselves. Many historians have emphasized the physician debates concerning midwives at the beginning of the 20th century. This work on Wisconsin's midwives adds a new dimension to these authors' analyses, one that is important to complete the story: that dimension is an understanding of the process of change within this quintessentially female occupation. Midwives, I have argued, faced the same problems as nurses and physicians at the turn of the century: practitioners needed to build a professional ethos based on standardized education, licensure, and self-governance. Nursing and medical leaders of the early 20th century gave direction to and helped shape their professions' goals. But there were no positions as leaders for midwives to assume. The presumptive leaders—trained midwives—relied on others to decide their qualifications and to license them for practice. In addition, the personal characteristics of the practitioners and the characteristics of their practice prevented the emergence of a united profession.

The demise of midwifery at the beginning of the 20th century in this country was due to a number of complex and interrelated factors, including, but not limited to, physician opposition to these birth attendants. In an era when Americans increasingly insisted on professionals in many aspects of their lives, the evidence on Wisconsin's midwives shows that midwifery remained a traditional female occupation. Failing to achieve any professional goals or identity, and lacking any outside professional support, midwives faced the inevitable: their replacement in the birthing room by the self-consciously professional obstetrician.

NOTES

Revised version of a paper read at the 59th annual meeting of the American Association for the History of Medicine, Rochester, New York, 1 May 1986. I wish to thank Judith Walzer Leavitt for her patient reading of and generous comments on several drafts of this paper. The Alice E. Smith Fellowship from the Wisconsin State Historical Society, and grants from the University of Wisconsin Graduate School and the

Maurice Richardson Fund for the History of Medicine, helped defray the cost of the data entry and computer analysis.

1 Information about Kueny's schooling was found in her midwife license application, Midwife File, State Board of Medical Examiners, Archives Division, Wisconsin State Historical Society, Madison, Wisconsin (hereafter Midwife File). Information concerning her family life came from the 1900 federal manuscript census for Milwaukee.

2 License application of Dora Larson, Midwife File.

3 Ibid.

4 For example, Frances E. Kobrin carefully delineates the characteristics of the various physician groups who were debating the "midwife problem" at the beginning of the 20th century. However, she only identifies the midwives as sharing "race, nationality, and language with their customers" (p. 351). See Kobrin, "The American midwife controversy: a crisis of professionalization," Bull. Hist. Med., 1966, 40: 350–363.

   Judy Barrett Litoff's sweeping study of American midwives also focuses primarily on the physician debates of the early 20th century. While her evidence demonstrates that many varieties of midwives practiced in different areas of the country, she allows no regional distinctions for the "American midwife," grouping together such diverse women as Southern black grannies, Northern, native-born white neighbor-women, and immigrant midwives trained in the European hospital schools (Litoff, American Midwives, 1860 to the Present [Westport, Conn.: Greenwood Press, 1978], pp. 53, 60). Despite the disparate nature of these groups of female practitioners, Litoff argues that there was no clear distinction between the official midwife and the neighbor (p. 27). Litoff acknowledges the difficulties in drawing a composite picture of early 20th-century midwives, but this problem relates directly to her sources. Because she did not use data directly focusing on the midwives themselves, such as midwife licenses or birth registration certificates, she could not provide detailed information about individual midwives and their practices.

5 Kobrin, "American midwife controversy," p. 350.

6 The Milwaukee Health Department began requiring midwife registration in the 1870s (see "Dr. Wight's wisdom," Milwaukee (Wisconsin) Sentinel, 20 July 1878). In 1909, the state of Wisconsin began requiring all midwives to have licenses. Women already in practice were allowed to continue, regardless of their training. But new practitioners were required to have a diploma from a "reputable school of midwifery," defined as being "connected to a reputable hospital or sanitarium, [offering] a twelve month course in the science of midwifery and experience in twelve confinements." Candidates were also required

to pass an exam relating to the clinical practice of midwifery. Wisconsin Statutes, 1909, ch. 528, s 1435f-12.

7 In 1894, Massachusetts was the first state to forbid midwifery practice by refusing to license midwives and by recognizing attendance at childbirth as medical practice. Thus, midwives delivering babies were subject to prosecution for practicing medicine without a license. Midwives continued to deliver babies in Massachusetts, however, often under the protection of a local physician who would sign the birth certificate. Contemporary critics of this approach to midwife regulation noted this discrepancy. See J. M. Baldy, "Is the midwife a necessity?" American Journal of Obstetrics, 1916, 73: 399–400. A recent historical study analyzed this phenomenon: Eugene Declercq, "The nature and style of practice of immigrant midwives in early twentieth-century Massachusetts," J. Soc. Hist., 1985, 19: 113–129. Many Southern states completely ignored their midwife practitioners in the early 20th century. Midwives in these states, mainly black "grannies," did not have to hide their practice, but the state had no records about them.

8 Magali Sarfatti Larson, The Rise of Professionalism: A Sociological Analysis (Berkeley and Los Angeles: Univ. of California Press, 1977), p. 45.

9 Ibid., pp. 44–45. Also Paul Starr, The Social Transformation of American Medicine (New York: Basic Books, 1982), p. 81.

10 Larson, Rise of Professionalism, p. 17.

11 Barbara Melosh, "The Physician's Hand": Work, Culture, and Conflict in American Nursing (Philadelphia: Temple Univ. Press, 1982), p. 4.

12 Ibid., p. 17. For a longer discussion, see Eliot Freidson, Profession of Medicine: A Study of the Sociology of Applied Knowledge (New York: Dodd, Mead, 1970).

13 Judith Walzer Leavitt has defined traditional childbirth as a birth that remained within a female-centered network. The attendant was a midwife, trained or not. (Leavitt, Brought to Bed: Childbearing in America, 1750–1850 [Oxford and New York: Oxford Univ. Press, 1986], pp. 73–74.) As I try to show in this article, the definition of who was a midwife in the 19th and early 20th centuries did not remain static. However, despite this occupation's move toward school training, its practitioners and its philosophy toward birth remained tied to gender and cultural norms that were not part of a "modern" professional outlook. For the best interpretation of the historical meaning of modernization, see Thomas Bender, Community and Social Change in America (New Brunswick, N.J.: Rutgers Univ. Press, 1978).

14 License application of Susan Washburn, Midwife File.

15 Elizabeth G. Fox, "Rural problems," Children's Bureau

Conference, Children's Bureau Publication no. 60 (Washington, D.C.: Government Printing Office, 1919), p. 187.

16 Florence B. Sherbon and Elizabeth Moore, *Maternity and Infant Care in Two Rural Counties in Wisconsin,* Children's Bureau Publication no. 46 (Washington, D.C.: Government Printing Office, 1919), pp. 20–22. Many areas of this county were served by minor roads, which the study noted were impassable by automobile and difficult even for wagons. Many parts of the county had no mail delivery, which required that some people send 12 miles for their mail.

17 Compared with an overall mean age of 47 for all of the registrants and a mean age of 45 for the school-trained group.

18 All of the licenses (396) were examined and coded for computer analysis. Using SPSS to analyze the data, t-tests to evaluate differences in mean age were run between the different groups. The t-test comparing the mean age of the school-trained versus no training was highly significant, with a t-value of ($-$ 4.56) and a significance level (2-tailed) = .00001. To investigate the link between region and type of midwife, the 52 counties represented by the registrants were grouped into six regions. When these six regions were cross-tabulated with kind of training, the result was highly significant: $\chi^2 = -102.33$, with 15 df, the significance level = .00001. Particular individual cells contributed to this finding: the western and central regions had three times more than the expected number of midwives reporting "no training," and the city of Milwaukee, where 20 untrained midwives would be expected, reported only one.

19 "Physicians' Register," City of Milwaukee Health Department, Archives Section, Milwaukee Public Library. (The title of this registry is a misnomer, for the registry contains registration data on both physicians and midwives.) There were 658 entries between June 1877 and October 1907, but many of the women who registered in 1906 and 1907 had previously registered. The registry, then, provided information on 588 different women.

A more detailed check of the data suggests that the 21 women who reported "no training" did not represent a larger contingent of unregistered, infrequent practitioners, perhaps singled out for prosecution. Dividing the registry into five-year periods, I found one or two neighbor-women midwives in Milwaukee signing up with the health department every year.

The figures (from the "Physicians' Register") were as follows:

	1877–79	1880–89	1890–99	1900–1907
Total registered:	46	171	227	144
"No training"	6	0	3	12

20 License application of Anne Nowak, Midwife File.
21 Leavitt finds that physicians called to a birth in the 19th century were forced to work within the limits imposed by "'women's domain of the birthing room," which often included a midwife and the parturient woman's friends (*Brought to Bed,* p. 100).
22 License application of Mary Greeley, Midwife File.
23 License application of Ellen Stiefvator, Midwife File.
24 Sherbon and Moore, *Maternity and Infant Care,* pp. 32–33.
25 Jane Pacht Brickman, "Public health midwives and nurses, 1880–1930," in *Nursing History: New Perspectives, New Possibilities,* ed. Ellen Condliffe Lagemann (New York: Teachers College Press, 1983), p. 70.
26 License applications for both the city of Milwaukee and the state of Wisconsin required two physician references. Each woman's references were coded, and the names of physicians in selected counties were checked against their own license applications (Physician Licenses, Board of Medical Examiners, Archives Division, Wisconsin State Historical Society). For physicians' activities within the medical community, I consulted Louis F. Frank, *The Medical History of Milwaukee, 1834–1914* (Milwaukee: Germania Publishing, 1915).
27 License application of Frances Jahnz, Midwife File.
28 License application of Mary Browikowski, Midwife File.
29 Frank, *Medical History of Milwaukee,* p. 255. Hinz practiced in Milwaukee from 1875 to 1903.
30 State license applicants reported four other Milwaukee physicians who offered lectures. Some of these doctors were instrumental later in the several midwifery schools in Milwaukee. (License applications of Kondaneygli Sythowski, Helena F. Mueller, Fredericka Wirth, and Mary Dudek, Midwife File.)
31 License application of Mary Holub, Midwife File.
32 Roger Simon, "The expansion of an industrial city, Milwaukee: 1880–1910" (Ph.D. diss., University of Wisconsin-Madison, 1971), p. 82.
33 "Figures that don't lie," *Milwaukee (Wisconsin) Sentinel,* 23 June 1889.
34 Agnes Pradzinska, "Letter to the editor" in the German-language section of the *American Midwife* [St. Louis, 1896], 2(6): 40–41. My thanks to Rebecca Bohling for her expert help in translating this letter for me.
35 These schools exerted a statewide influence: the Milwaukee registry listed 84 graduates of these schools who gave addresses outside the city, perhaps implying that some students came to Milwaukee specifically for their education. The State License file also showed the wide influence of the schools on Wisconsin's midwives: of the 121 women in the state file who reported their school was based in Milwaukee, 30 gave addresses outside of the city.

36   License application of Sophia Kegel, Midwife File.

37   Licenses of Caroline Fuss and Maria Timm, Midwife File.

38   Two 1913 graduates of the school, Stefonia Staigwillo and Gertrude Turzynski, noted on their state license applications that they had attended the "full course" of the Wisconsin College of Midwifery. The full course was described as an eight-month term from September to May with hospital time and attendance at 25 cases. (License applications, Midwife File.)

39   J. Whitridge Williams, "Medical education and the midwife problem in the United States," *J.A.M.A.,* 6 Jan. 1912, *58*:1–7. Williams found the general attitude toward obstetric teaching to be "a very dark spot in our system of medical education" (p. 1).

40   *Wright's Milwaukee City Directory,* 1893, *26*: 1178, advertisement under "Lying-In Hospitals." Midwives were never mentioned.

41   *Ibid.,* 1901, *34*: 1262, advertisement under "Hospitals."

42   J. Clifton Edgar, "Why the midwife?" *American Journal of Obstetrics,* 1918, *78*: 249.

43   George W. Kosmak, "Results of supervised midwife practice in certain European countries," *J.A.M.A.,* 10 Dec. 1927, *89*: 2009.

44   Leslie Woodcock Tentler, *Wage-Earning Women: Industrial Word and Family Life in the United States, 1900–1930* (Oxford and New York: Oxford Univ. Press, 1979), p. 137.

45   *Ibid.,* p. 142.

46   Declercq, "Nature and style," p. 118, and Neal Devitt, "The statistical case for elimination of the midwife: fact versus prejudice, 1890–1935" (pt. 1), *Women and Health,* 1979, *41*: 82.

47   Practitioners were identified from the Milwaukee Health Department licenses, the licenses filed with the state of Wisconsin, advertisements in city directories, and the names of women who signed birth records as the attendant.

48   The four counties were Milwaukee, Dane, Trempealeau, and Price. Milwaukee, of course, was the urban center for my study: Madison, Wisconsin's capital city, is located in Dane County, which nevertheless was largely rural in the late 19th and early 20th centuries. Trempealeau County was chosen both because it was the site of Merle Curti's famous study (*The Making of an American Community: A Case Study of Democracy in a Frontier County,* with the assistance of Robert Daniel et al. [Stanford, Calif.: Stanford Univ. Press, 1959]) and because, like Dane County, it was a fine example of a place with a settled, rural population. Price County, in northern Wisconsin, was a frontier county at the turn of the 20th century. The major industry in this county was lumbering, but a few farmers had settled on the clear-cut land.

49   Within the four counties, I identified 893 different midwives in the 50-year period of my study (1870–1920), and I was able to link 398 of these practitioners to the census (44.5 percent). I attempted to link practitioners to the census nearest to the date when they first began to practice, because I wanted to capture the full effects of family life on midwifery practice.

     Note that the large number of practitioners linked to the 1900 census reflects the absence of the 1890 manuscript census schedules. The unavailability of this schedule (it was burned in a Washington, D.C., fire near the turn of the century), coupled with the high mobility of midwives' artisan families, was the reason more midwives were not found in the census.

50   To obtain an estimate of the frequency of practice and to maintain a reasonable sample size, every birth was coded for a given three-month period in Milwaukee County, every birth for six months in a sample year for Trempealeau County and Price County after 1905, and every birth for the entire year for Dane County and Price County between 1880 and 1905.

51   The birth data are contained in two data sets. Milwaukee, Price, and Trempealeau counties were coded together in one set, with a total of 24,302 cases. Dane County data were coded alone, having a total of 4,622 cases. For each case, the following variables were coded: date of birth, including day, month, and year; place of the birth, including county, city, whether home or at a hospital; parents' home address; the occupation of the father; the birthplace of the mother and the father; the mother's parity; the attendant at the birth (if noted, otherwise a code for parents or other was included); the status of the attendant (midwife, physician, or other). Each attendant was given a unique five-digit number that reflected his or her county and the year he or she first appeared in any record of birth attendants. The data were analyzed using various statistical subroutines of SPSS.

52   Jane B. Donegan, "'Safe delivered,' but by whom? Midwives and men-midwives in early America," in *Women and Health in America: Historical Readings,* ed. Judith Walzer Leavitt (Madison: Univ. of Wisconsin Press, 1984), p. 304.

53   Agnes C. Vietor, ed., *A Woman's Quest: The Life of Marie E. Zakrzewska, M.D.* (New York: D. Appleton, 1924), p. 39. Zakrzewska was 18 years old when she first sought admission to the Midwifery School of Berlin. Finally, by the entreaties of a powerful physician friend to the King of Prussia, she was admitted to the school (pp. 39–42).

54   The mean age of all the respondents was 48.05, with a standard deviation of 12.37 (Midwife License File,

State Board of Medical Examiners, Archives Division, Wisconsin State Historical Society). When age was cross-tabulated with kind of training, the chi-square was 43.88, highly significant at the .00001 level.

55 Although the number of illiterate midwives was small, I tested for significance for birthplace and county. When literacy was cross-tabulated with county, the chi-square was significant (15.82, significant at .001). Half of the 16 illiterate midwives lived in Milwaukee County, three lived in Dane County, four in Trempealeau, and one in Price. The Dane and Trempealeau figures account for the significance, as fewer illiterate women were expected. Birthplace was also significant, with a chi-square of 22.9, significant at 0.0008. The Scandinavian group had more illiterate women than expected. For 39 midwives, literacy was either not reported or unknown.

56 The percentage of families with boarders is significantly low when midwife families are compared with families sharing similar ethnic and class backgrounds and age distributions. Modell and Hareven's study of New England families, for example, found that as high a proportion as 40 percent of immigrant, working-class families where the head of the household and his wife were in their forties had boarders. John Modell and Tamara Hareven, "Urbanization and the malleable households: an examination of boarding and lodging in American families," *Journal of Marriage and the Family*, 1973, *35*: 472. The implications of boarding for women's work in the home are discussed on p. 473.

57 Information on children was obtained for 387 women. The mean number of children was 2.99, with a standard deviation of 2.41. Number of children ranged from 1 to 13; 77 percent of midwives had 4 or fewer children, 48 percent had 2 or fewer.

58 An analysis of variance showed no significant differences between groups. The f-ratio was .99 and the f-probability equalled .37.

59 Lynn Y. Weiner, *From Working Girl to Working Mother: The Female Labor Force in the United States, 1820–1920* (Chapel Hill: Univ. of North Carolina Press, 1985), p. 85. The median age of oldest children of all midwives was 15, with a modal (most frequent number) age of 9. However, the range of ages was quite broad, from 1 to 54 years.

60 In order to analyze the number of births attended by each practitioner or group of practitioners, the births file was sorted and aggregated by county, attendant number, and attendant status. The resulting file had information on 274 midwives who had also been found in the census data. A new variable "NBABES" captured the total number of babies delivered by each midwife. Matching the midwives in this file to the census data by linking the attendant number, I performed a stepwise multiple regression, with number of babies as the dependent variable. The number of children, age of the oldest, age of the youngest, and the midwife's own age were entered as independent variables. Only age of the midwife had an F-value significant enough to enter into the regression equation ($f = 7.101$, significance = .0082). The resultant R^2 equalled 0.02194, showing age only very weakly explaining the total number of babies delivered.

61 To code for husbands' occupation, I used the categories defined by the federal census of 1940. The categories are roughly hierarchical, except for the two categories in the middle defining farming and mining. To approximate "white-collar" versus "blue-collar" occupations, I grouped the professional, managerial, clerical, and sales groups together, and the transport, artisan, laborer, and service categories together. Only 39 midwife husbands fell into the "white-collar" group (12.4 percent), while 201 husbands (64.0 percent) were clearly "blue-collar." The farm husbands, all located in the three rural counties, were hard to categorize, as farm property values varied tremendously.

62 Amortizing mortgages, common to today's home buyers, were rare. Working-class families were required, therefore, to save a large sum of money to pay off the principal at the end of the loan period, even as they paid the interest in semi-annual payments. See Stephan Thernstrom, *Poverty and Progress: Social Mobility in a Nineteenth-Century City* (New York: Atheneum, 1974), pp. 121–122, 129–130.

63 *Ibid.*, pp. 120–121.

64 One way ANOVAs were used to test whether the size of a midwife's practice was related to her socioeconomic status. For husbands' occupation, the f-ratio was 1.3700, with 7 d.f., and an f-probability of .2202. For home status, the f-ratio was 1.7649, with 2 d.f., and an f-probability of .1736.

65 Kobrin, "American midwife controversy," p. 350.

66 Brickman, "Public health, midwives, and nurses," p. 70.

67 Declercq, "Nature and style," p. 115. Table 2 (p. 117) indicates that all of the "major midwives" in Lawrence were born in Europe.

68 Michael M. Davis, "Birth rates and maternity customs," in his *Immigrant Health and the Community* (New York: Harper and Brothers, 1921), p. 195.

69 Grace Abbott, *The Immigrant and the Community* (New York: Century, 1917). Abbott declared that "sometimes the traditions and the prejudices of our immigrant population must be consulted in determining an important health policy. An admirable illustration

of this necessity may be found in the question of what should be our state and city policy with regard to the practice of midwifery" (pp. 145–146).

70   Joseph B. DeLee, "Progress towards ideal obstetrics," *American Association for the Study and Prevention of Infant Mortality*, 1915, *6*: 119.

71   E. C. Hartley and Ruth E. Boynton, "A survey of the midwife situation in Minnesota," *Minnesota Medicine*, 1924, 7: 439–446.

72   Sherbon and Moore, *Maternity and Infant Care*, p. 10.

73   *Ibid.*, pp. 30–32.

74   A chi-square analysis of county by birthplace was highly significant; the chi-square = 207.949, significant at 0.00001. The three rural counties had many more Scandinavian midwives than would be expected, and Milwaukee City had many fewer. (Mil- waukee had only 4 Scandinavian midwives; 37.1 were expected.) Trempealeau and Price counties also had many fewer German birth attendants than would be expected. (Trempealeau had no German midwives; 13.9 were expected. Price had 8 instead of 20.3.)

75   Elizabeth Ewen, *Immigrant Women in the Land of Dollars: Life and Culture on the Lower East Side, 1890–1925* (New York: Monthly Review Press, 1985), p. 133.

76   Several historians have noted the almost outright ban on male doctors in immigrant women's birthing rooms. See Brickman, "Public health, midwives, and nurses," p. 70; Kobrin, "American midwife controversy," p. 321; Ewen, *Immigrant Women in the Land of Dollars*, pp. 70–71.

16

"Neither for the Drawing Room nor for the Kitchen": Private Duty Nursing in Boston, 1873–1920

SUSAN REVERBY

"Neither for the drawing room nor for the kitchen" was how nursing leader Isabel Hampton Robb succinctly captured the ambiguous position of the private duty trained nurse working in a patient's home in 1900.[1] While this nurse's social standing in a household was uncertain, her particular position in the health care system was becoming clearer. At the turn of the century most Americans when ill, even seriously, took to bed at home, not in a hospital. If it appeared necessary and/or could be afforded, at the bedside stood a hired nurse. The omnipresent, harried staff nurse employed by a hospital is a figure whose vintage dates only from the World War II years. Until then, the majority of nurses we would now refer to as registered nurses worked in private duty as the employees of patients; and until the mid-1920s, most of them did this work in patients' homes.[2] This article explores the conditions and practices of this historically critical form of care giving and women's work.

To uncover the private duty world, this essay draws upon nurses' written memoirs, letters, and texts, journal articles, reports of nursing meetings, census data, and a sample of 539 nurses taken from the records of the 4,550 nurses who registered for private duty work at the Boston Medical Library (BML) Directory for Nurses, the first major registry for nurses in this country. By the last quarter of the nineteenth century, directories or registries, run by hospitals, medical and nursing societies, or private businesses, were organized to keep lists of available nurses and to match them, for a fee, to the needs of inquiring physicians or patients. Between 1880 and 1914, the BML Directory pro-

vided work for the majority of Boston's private duty nurses.[3]

The ubiquitous female ministering to the needs of the sick was not always a relative or an altruistic neighbor; women who were paid to nurse made their appearance in the colonies in the seventeenth century. Such "nurses" gained their label through self-proclamation coupled to some kind of experience of caring for the ill in their own families, domestic service, or hospital work.[4] In 1873, the first nursing schools were linked to hospitals and created to train women to nurse. By 1900, neither titles nor job functions clearly differentiated the "old-style" nurses from the new graduates. The label "nurse" was applied to a graduate of a two- or three-year training school, a nursing school dropout, an experienced worker with no formal education, a person who had taken a few lessons at the YWCA, or an attendant who had worked in a hospital. Lack of uniformity characterized even those who received "training" as the education was frequently haphazard and unstandardized. A nursing leader lamented in 1893 that the title "'trained nurse' may mean then anything, everything, or next to nothing."[5] The nursing associations struggled to have licensing laws enacted to differentiate the trained or graduate nurse, as she was labeled, from her many untrained competitors. But even the strongest of these generally weak registration laws could not limit which "nurses" sought work in the private duty market.

Sheer numbers were part of the problem. There were fewer than 500 graduate nurses in the United States in 1890. The numbers rose 634 percent by 1900; another 136 percent between 1900 and 1910. In Boston, the total number of nurses, both trained and nontrained, rose nearly 400 percent between

SUSAN REVERBY is Luella LaMer Professor for Women's Studies, Wellesley College, Wellesley, Massachusetts.

Reprinted from *Women and Health in America*, edited by Judith Walzer Leavitt (Madison: University of Wisconsin Press, 1984), pp. 454–466.

1880 and 1905.[6] By the turn of the century, the supply of nurses began to outpace the demand for their services, particularly in the urban centers. The overcrowding in nursing was a function of the limited fields of employment open to the increasing numbers of women seeking work, exacerbated by the deliberate policies of the hospital training schools.

Many hospitals, once the idea of training women to nurse was institutionalized, began to staff their wards almost entirely with nursing students and untrained nursing workers. Control over nursing education, and who was admitted to the training, was often the subject of pitched battles between nursing superintendents and hospital administrators, trustees and physicians. Under pressure from the hospitals to staff the wards as well as educate the students, nursing superintendents frequently sacrificed educational goals to the necessity of getting the work done. In many schools a two-track system developed. A few students were encouraged upon graduation to become head nurses or nursing superintendents in the hospitals while the majority were shunted into the increasingly crowded and undifferentiated private duty field.[7]

Private duty work quickly took on a peculiar, ambiguous status in the nursing and medical world. A medical student completing his training went into private *practice;* the nursing student, however, went into private *duty.* The physician was expected to apply his skills in independent action. The nurse, even without the control of the hospital and medical hierarchy, was still supposed to be submissive to higher authority and morally obligated to her work. In private duty, a nurse was working for a *doctor's* patients. Although employed by a family, she was primarily dependent on the physician to define what she did and to help her get work.

While devoted care to one private patient would seem to be the ultimate expression of a nurse's skill, the nursing superintendents expressed grave concern about the dangers of private duty and feared their students would find the work "exhausting." The exhaustion that worried them was both physical and spiritual, a loss of sheer strength and moral fiber. It was frequently assumed that a nurse could last only ten years in private duty, her collapse owing as much to the danger of "moral laxity" as to the physical labor.[8] Warnings were is-

sued about the danger of nursing single men in hotels (no respectable woman should) or the tempting advances of a patient's unscrupulous husband (to be spurned at all cost), along with admonitions to get enough rest. These warnings reflected less a fear of the vanquishing of the nurse's physical virginity than of her loss of spiritual virginity—the collapse of her moral purity, her gentleness, humanity, sympathy and tact because of the long hours and strain inherent in the work.

In private duty a nurse provided an array of services from the purely domestic to skilled nursing care. As might any domestic servant, mother, or wife, she often had to be an entertaining companion, an imaginative cook, and a competent laundress. She also had to monitor vital signs; insert catheters; prep and assist at operations and deliveries; provide cold packs, baths, and massages; even carry and decide when to give such medications as morphine. Alone with a patient, twenty-four hours a day, often weeks at a time, she crossed the ambiguous line between nursing and medical care with regularity, if trepidation. Freed from the hierarchy and controls of the hospital setting, she could practice her best skills with relative autonomy, or make terrible mistakes without supervision.[9]

The stress on the graduate nurse in private duty was due to the pressure to prove the role of the trained nurse; the ambiguity of her place in the household structure; the difficulty of the work; and the overcrowding in the field. In the 1880s and 1890s, the trained nurse was still a new creation whose necessity had to be proven to both physicians and families. Although much of what she did seemingly could be done and was done by nontrained nurses, domestic servants, or female relatives, she was somehow through her personality, bearing, and character to present herself as a new and vital creation, necessary to patient survival, worthy of being paid a high wage. Many physicians were not convinced that this kind of nurse was necessary for their patients and perceived hospital-trained nurses as a threat. Physician preference, patient income, and the nature of the illness often determined what kind of nurse was hired, and for how long.

Working in a patient's home at the time of the illness created a number of different stresses for the nurse and the family. In the patient's home, the nurse was an individual confronting a family social system. In the hospital, the patient was the lone individual, subject to a set of defined rules and a structured hierarchy. A private duty nurse warned that hard work, with many patients, "under some circumstances [may] demand much less wear and tear on the nervous system than that consequent upon the supervision of her own solitary self while engaged in nursing one patient in the bosom of that patient's family."[10]

The nurse also had to learn to make do in the home without all the equipment, supplies, and paraphernalia which even then marked hospital care. The author of one of the many manuals on "how to be a good private duty nurse" recounted the story of an overzealous private duty nurse who "thought she was distinguishing herself by extreme neatness, used to put thirty-five sheets in the wash in a week. She defeated her own end, for the laundress, naturally thought this a folly, and smoothed out those that looked clean, without washing them."[11]

The most obvious contrast between hospital and home-based care was the lack of structure in private duty. A system and schedule for performing duties was the sine qua non of the hospital training schools. But as one private duty nurse cautioned, "indeed a too loyal adherence to one certain system may prove a huge stumbling block in the way of success."[12] The very work rhythms of the home and hospital differed. In private duty, a nurse's success depended on her ability to reset her work to the demands and whims of the patient and family, not in a premeasured routine set to a rigid schedule. Private duty nursing consisted of the performance of a series of tasks whose order was determined by the ups and downs of the patient's illness and the needs of the family, as much as a farmer's work depended upon the weather and seasons. The patient and the family, not the work as in the hospital, had to be the center of the nurse's attentions.

The difficulty of private duty work was compounded by the contrast between the class of patients in the hospitals and that of the families who hired private nurses. In the hospital, the nurse confronted a patient population of primarily working-class men and women. In the home, however, most nurses were working for families whose class position was usually higher than their own. In a sample taken of the families who hired nurses through the BML Directory, more than 50 percent of the male family heads were either lawyers, owners of companies, or merchants. The others were skilled or white-collar workers and professionals. While the nurse could as easily be called to the home of a skilled bricklayer in a Dorchester triple-decker as to the bedside of a tea merchant in an elegant townhouse on Beacon Hill, few private duty nurses were hired to work in the tenements and boardinghouses to care for the majority of Boston's populace.[13]

Once in a patient's home, there were few guidelines to govern social relations for either nurses or families. Nurses, as a character in a nursing novel explained, were "always afraid of being asked to do too much. They're always afraid of being treated like ordinary servants."[14] Nurses were told during training that there was nothing which was beneath their dignity to do, but once in a patient's home a nurse had to draw the line and decide for herself what was a reasonable demand. If a nurse washed out a baby's clothes in one home because there was no laundress, should she be expected to do so in another home where such a domestic servant was employed? Should a family be subject to a nurse's wrath because she was asked to eat at a second table or in a kitchen with the servants?

The unspoken rules of class conduct between employers and servants were continually violated by private duty nurses. Clashes were inevitable between a family's expectation of servantlike behavior and the nurse's need to assert her standing above that of servants and to establish her autonomy. Patients' objections to the nurses echoed those made to servant girls unfamiliar with the furnishings of a crowded bourgeois Victorian home. Nurses were faulted for their clumsiness with precious objects, lack of knowledge of how to handle exotic pieces of furniture, and willingness to use expensive items carelessly.

At a time of illness and stress, small mistakes in social conduct by the nurse became magnified and compounded by the fears of death and disease which pervaded a household. "The families want to know 'what are the rules,'" a Philadelphia physician said. Yet, as one nursing superintendent noted, "there are no definite rules to be observed."[15] A nurse's inability to correctly judge the unwritten rules could be costly. In 1892, a Philadelphia nurse angrily reported:

> If by any chance a nurse gains the ill will of her first few patients, her career is ended. She is not told anything of this, simply waits in her boarding house until her last dollar is gone, . . . in suspense . . . and wondering why a "case don't come."[16]

The relationship of the nurse to the household's other servants was one of the problems of social conduct which caused the biggest difficulty. The dilemma centered on where the nurse would take her meals. Nurses often insisted (to make sure they were treated as ladies, not servants) that they be served their meals in the dining room with the family rather than in the kitchen with the servants. This demand was usually opposed by patients used to treating the nurse as a servant, and uncomfortable about sharing their dinner conversations with a stranger from another class. To lessen the conflict on this question, the BML Directory, for example, asked both trained and nontrained nurses on their application forms whether or not they would take their meals in the kitchen with the servants. In the sample from these records, only 40 percent of the trained nurses were willing to eat with the servants, as opposed to 74 percent of those without formal nurses' training. Of those not willing to eat with the servants, 82 percent were trained nurses, and less than 20 percent were the nontrained nurses.

Managing relationships to the family's servants required tact as the nurse had to be understanding but distant, above but not superior. A graduating nursing class at the Boston Training School for Nurses at the Massachusetts General Hospital was told by a physician: "Never assume an air of superiority when dealing with the servants; but on the other hand, never be too familiar with them. At best they recognize your superior position unwillingly, therefore do all you can to conciliate them."[17] The countless stories of the overbearing dictatorial nurse who left the household in an emotional and physical uproar suggest that finding a path to conciliation was not always easy. In desperation, families often turned to the more expensive private

room in the hospital or called for only an "old-style" nontrained nurse since in both situations the social relationships were clearer.

The letters and work records in the BML Directory suggest that the differences and difficulties in private duty existed between the trained and nontrained nurses as well. Class, age, and marital status often differentiated the two groups. Trained nurses, especially graduates of the larger and more elite schools, were likely to be middle- to lower-middle class in origin, while the nontrained nurses drew more heavily from women from the working class. However, graduates of the more numerous smaller training schools were more likely to be working class, compounding the difficulties of differentiating them on class grounds from the nontrained women. Except for large numbers of Irish and English Canadians from the Maritime Provinces, trained nurses in Boston were overwhelmingly native born. They were also much younger than the nontrained women: the graduates' average age was twenty-nine, while the nontrained average age was thirty-six. While the nature of nursing work demanded that its practitioners not be encumbered by family responsibilities, the nontrained nurses were more likely to be widowed or divorced, while 92 percent of the trained nurses had not yet married.

Few of the nurses of either type shared households with male relatives. As did other women workers, they crowded into the boardinghouse districts of Boston's South and West Ends. Many of the nontrained nurses were women already in Boston, who took up nursing as an extension of familial or domestic servant duties, while the graduates were more likely to be single women who had come to Boston or nearby towns to do their training and then stayed on in the city to work. Trained nurses, because they were younger and more often single, were twice as likely as the nontrained to leave Boston.[18]

Turnover figures had a morbid side as nursing was a dangerous occupation. Death reaped a much higher proportion of nurses than women of comparable ages in the Boston population. The BML Directory had a death rate of 16.36 per thousand, a rate greater than that for any adult age group, except those over fifty, in the general population. Even here, probably because of age and class, the nontrained nurses were twice as likely to die as the graduates.[19]

The work experiences and career patterns of the nurses varied considerably. The nontrained women had about two years more experience in nursing than the trained nurses, but both groups remained in the BML Directory on average for eight years. This figure does not measure their work commitment since dropping out of the Directory's records did not mean either giving up nursing or some other form of employment. Some nurses worked during their entire lifetimes, regardless of marital status or training. Others dropped out into related fields: to sell surgical equipment, to operate rest homes, or to work in training schools and hospitals. Some changed fields completely. Still others dropped out at marriage and never returned, while some came back to nursing when their children were older or widowhood necessitated employment.

While these nurses differed in age, morbidity, marital status, experience, and training, they competed in the same overcrowded labor market for the same jobs. It was certainly not clear to the physicians or patients who hired the nurses what skill really differentiated these women. Wage rates, specialities, and the seriousness of the patient's illness, however, were all factors in determining who was hired.

Wage rates clearly differentiated the nurses, as the graduates tended to charge the patients almost five to ten dollars a week more than the nongraduates. From the 1880s till the mid-1890s graduates received fifteen to eighteen dollars a week; by the late nineties they were commanding twenty to twenty-five dollars. But a graduate nurse just out of school and a graduate nurse with ten years experience were paid the same. Neither trained nor nontrained nurses were rewarded for their experience with a higher wage. There was clearly a "customary" wage established for each group, although there was a slightly greater dispersion in the fees asked for by the nongraduates.[20]

Nurses were allowed in most registries to state their preferences on cases. Comments such as "only surgical cases" or "no obstetrical cases" cover the application forms. Trained nurses specialized more and were less willing to care for postpartum patients, presumably because such work almost always guaranteed they would be asked to do more household labor. Nontrained nurses, in contrast, more willingly took what work they could get.[21]

The graduate nurses clearly felt their livelihood was always threatened by the nongraduates. There were frequent complaints that the nongraduates were receiving most of the work or that the directories did not apportion the work equitably.[22] But at least through the BML Directory it appears the nongraduates were the ones suffering discrimination. The graduates received consistently two or three times as much work as the nongraduates and about 20 percent more than they should have, given equal distribution.[23] Miss E. L. Blanchard, a nontrained nurse, bitter over her lack of work, wrote to the physician in charge of the BML Directory in 1895: "I feel it is a . . . way of pushing one out, for opening the way for those younger, and . . . in the training school for a few months. . . . Thus . . . experience goes for nothing." Another nontrained nurse explained the dilemma by pointing out that her costs were too high for the poor and too low for the rich who would, she believed, rather employ a graduate.[24]

Thus it appears that in the 1880s and early 1890s, the trained and nontrained nurses in Boston were competing directly with one another. By the late 1890s, the competition was caused by the enormous number of graduates entering the labor market. By 1915, in Massachusetts as a whole, the trained nurses outnumbered the nontrained two to one.[25]

Competition then developed between the graduates of the different training schools. Physicians at the BML Directory told graduates of the smaller schools that their training was not equal to that of nurses from the more prestigious hospitals.[26] Graduates of the more elite schools demanded and received higher wages than those from the smaller schools and more often demanded a wage differential to nurse male and contagious patients. The graduates of the larger schools also received slightly more cases, although the reason for this may have been that they were better known by local physicians.

Class background as much as training differences was probably the reason for this division among the graduate nurses. Graduate nurses from Long Island Hospital, the site of the city's asylum, were told they probably would not receive cases if they charged the graduate nurses' going rate.[27] One of the physicians in charge of the BML Directory bluntly wrote:

> It is preposterous to put the Long Island Hospital nurses on the same plane as those of the Mass. and City Hospitals. Certainly our patrons would not accept it. When we consider the class from which they come, their lack of education, etc. . . .the answer would seem to be obvious. The Long Island nurses are worth say from $10–15 and I could not with any feeling of fairness send them to first-class families and serious cases with a supply of the others on hand.[28]

Nursing employment in private duty for every type of nurse was sporadic, seasonal, and uncertain. The average graduate received only 3.2 cases a year from the BML Directory, the nongraduates 2.3; this multiplies out to be a work time, on average, of thirteen and nine weeks a year, respectively. In contrast, a Cleveland nurse reported that in her eleven years of private duty work from 1895 to 1906 she worked thirty-three to thirty-five weeks each year. The only time she was employed fully was the fifteen months she spent on a "luxury" case as a companion, as much as a nurse, for one wealthy patient. A comparison to other women workers in Massachusetts in 1890, however, shows nurses and midwives as sixth in unemployment frequency out of a list of twenty-five other women's occupations.[29] A nurse, trained or not, did make more money *when* she was employed than most other working women.[30] But the wait between cases was often so long that the extra wage could not make up for the unemployment.

The expense of the nurse also limited her calls. As early as 1888 nurses were aware that even what they called the "breadwinning middle class" could not afford their services. But because their own need for work made nurses feel they could not have an official sliding scale for patients, a nurse suggested that those who could not afford them should be "either taken to the hospital or put on the list of the visiting nurses."[31]

Nursing work was also seasonal, and the nurses, as much as any industrial workers, recognized that there were "slack" times. There was definitely a higher call for nurses in the midwinter months of January through March and a slowdown between May and July caused perhaps as much by the exodus of the rich from the city during the summer as a difference in disease incidence. Similarly the demand for nurses dropped during the economic depressions.[32]

One nurse echoed a common theme while reproaching the Directory in 1895: "if I had depended entirely on the Directory for a living, of course I should have starved long ago."[33] But registering with a number of directories was still no guarantee of work. A nurse could and often did spend weeks at a time waiting at her boarding-house for cases from any source. While the nursing leadership bemoaned the nurses' increasingly "mercenary spirit," a San Francisco nurse asserted that they could not afford to be "angels of mercy."[34] Worry and anxiety over where they would get their next case dominated the thinking of many nurses. Poignantly, a California nurse warned Boston nurses not to come to the West Coast and of the danger of "the gradual fading away of resources, courage, hope—too often self-respect—and sometimes suicide. The papers here suppress all that."[35]

One solution to the difficulty of finding work was for a nurse to "attach" herself to a physician and hope he would send her all his cases. Nurses had to make the rounds of physicians' offices to announce their presence and then wait to be called. Contacts made during training were critical to a nurse launching her career. If the nurse ventured to a new city, finding work took even more time. But nurses also competed for the patient dollar with the doctors since it was not just nurses at the turn of the century who were in oversupply.[36] A New York nurse complained in 1897 that a physician had her fired from a case and then began to make more visits to the patient himself.[37] There was the danger for physicians who had taken short apprenticeships or correspondence courses that a nurse with several years of hospital training might in fact know more scientific medicine.[38] The intensity of the economic competition makes more comprehensible the constant ideological stress by physicians on the need for the nurse to know her place and remain "loyal." But physicians held the master key which opened the employment door. An informal blacklist of sorts also circulated among physicians and directories of both dropouts and troublesome nurses.[39] Despite the fact that nurses were usually employed by the patients, in actuality they had to behave as if the patient's physician was their boss.[40]

Despite the uncertainty and limits on their autonomy, nurses fashioned a variety of *individual* means for surviving, reshaping, and enjoying the work. The stories of patients who took their nurses to Europe on trips, married them, left them large sums of money, or employed them for decades, however unusual and idiosyncratic, suggest that, as with servants and governesses, the step up in nursing could mean marriage or life as the family retainer. Nurses also found ways to create alliances in the household. Reexamining the figures on eating with the servants makes clear that nearly 40 percent of the graduate nurses and 74 percent of the nontrained nurses were willing to eat in the kitchen. Nurses quickly learned that a recalcitrant servant could make their lives miserable. Letters rebuking nurses for being too friendly with the servants and gossiping with them about the patient's illness and the household life suggest that congenial working relationships frequently developed.

Nurses never lacked resourcefulness in asserting some kind of control of their work. Especially on long convalescent cases, they would sometimes leave patients to go home, to go to the theater, or to visit friends or family. Some nurses had other businesses and conducted their other work while on a case. If the job was really miserable there were ways to be relieved. One nurse was disciplined for having another nurse call and lie for her, saying an aunt was ill and she was urgently needed at home. The nurse, when discovered, said she left the case because the patient was "poor" and conditions of work were terrible. Other nurses took to sleeping while on the job, or abandoning patients, or refusing to take cases they had agreed to.[41]

These attempts to control the work were perceived by the public and many nursing superintendents, however, as counter to devotion to duty. Nurses were, after all, expected to be more like mothers than workers, forever on call, cheerful and devoted. In a 1904 editorial entitled "The Path of Duty," for example, the *American Journal of Nursing* chastised nurses for refusing to take cases: "Such failure to meet our highest obligations, such violation of our common standards of right and duty, cannot be too sternly censured. The women who permit themselves to conduct their professional work in this manner are in this, at least, wrong through and through."[42]

Different kinds of organized efforts were made to introduce some rationality into the private duty labor market and to distribute the work more evenly. Those in charge of the BML Directory, for example, tried to influence the wage rate, often to the detriment of all the nurses. With nurses of equal skill, the Directory registrars would send out the nurse with the lowest fees.[43] Lavinia Dock, the outspoken socialist and feminist nursing leader, counseled nurses: "We must not undersell; that is treachery to fellow workers."[44] But Anna Maxwell, the nursing superintendent of Presbyterian Hospital in New York wrote the Directory in Boston in 1899 to ask if there would be work if more graduate nurses lowered their wage rates.[45] In *principle*, graduates tried to keep to an agreed-upon wage rate. But it was not uncommon for a nurse to lower her rates when necessary, or, more often, to overcharge the patient on items such as carriage fares and laundry bills as well as on her fees.[46]

The Directory officials also attempted to rationalize the system by grading the nurses by experience and training. This was done on a haphazard basis since they had to rely upon the nurse's self-reporting and incomplete patient and physician recommendations for their assessments. Registry officials remained uncertain, however, about their legal right to discipline nurses. All they ultimately could do was to refuse to put the nurse on their lists, but she could go elsewhere.[47]

Much of the nursing concern over the lack of work focused upon the registries. "Most graduates," a nurse reported, "do not feel that they are fairly treated by the Directory, but are afraid to complain for fear that it will be visited upon them."[48] In an attempt to equally distribute employment, when a nurse reported off a case, her name was supposed to be placed at the bottom of the rotation list. But, nurses charged, the registries played favorites, did not always follow this system, and could not control whether or not a nurse received a "good" case which guaranteed employment for a length of time in a decent home.[49]

Aware of the discontent over the registries, the extent to which both physicians' groups and commercial agencies were profiting from such services, and the necessity to control distribution, some nurses began to advocate the organization of nursing-controlled, centralized, and officially sponsored registries in each city. In Boston, for example, the BML Directory was closed as graduate nurses began to register with the nurses' officially sponsored Suffolk County Nurses' Directory. One nurse commented, however, "It is not so much a share in the government of directories, as a share in the work given out by them that is asked by the majority of nurses. . . . Each one should have a share."[50] There was no guarantee, however, that a nursing-controlled registry would mean any more work. In fact, nurses admitted that the commercial agencies often allowed the more experienced trained nurses to charge more, thus making them more attractive than the official registries, which set one rate.[51] The official agencies often enrolled only graduates from the "better schools," did not provide nontrained nurses, and were thus less able than the commercial registries to meet the varied community demand for nurses. Hospital and alumnae registries often served only to provide the institution's own graduates with positions as private duty "specials" in their hospital.

Despite the overcrowding, private duty work continued to absorb the majority of nursing graduates because there was very little else they could do to remain in nursing. Some private duty experience was considered essential for every nurse, but status in nursing quickly accrued to those in more "executive" positions within hospitals or public health nursing agencies. Private duty nursing, seen even within nursing as often no more than domestic service or a mother's work, gave those who did it little status. Furthermore, because there was no supervision, women with the weakest skills could hide in private duty. With the growth of hospital-based care in the early 1900s and the decline in the debilitating sicknesses which required more long-term nursing care, the status decline of private duty nursing increased. By the late 1920s, a major study of nursing's dilemmas could repeat the aphorism "Every nurse ought to do some private duty, but no good nurse ought to stay in the field more than a few years."[52]

Aware of the overcrowding as early as the 1890s, Lavinia Dock counseled nurses to specialize "by branching into auxiliary lines of work not strictly nursing, yet which can be better done by one having the training of nurses." Among Dock's suggestions were heading various departments in hospitals, becoming a dietician or pharmacist, directing nurseries, old people's homes, social service, settlement work, medicine and massage, or returning to nurse in small towns and the countryside. Her suggestions entailed the nurse's specializing *outside* of private duty work itself.[53] Lucy Drown, the superintendent of nurses at Boston City Hospital, was blunter in her advice. Sharing her concern over the lack of work with the physician in charge of the BML Directory she wrote:

> [there is] . . . less survival of the fittest [in nursing] than in some other walks of life. My advice to these young women would be that if they cannot make a place for themselves as nurses, to go back to the work they left when they came into the hospitals [for training], for they belong to the working class, and maintain themselves as teachers, stenographers, dressmakers, etc.[54]

But in fact, most stayed in nursing, waiting for cases, trying to find different kinds of nursing work, "making do" in an increasingly untenable form of employment.

The isolation of the work site, the severe competition for cases, and the acute divisions within the nursing ranks all worked to limit the forms of control over their labor nurses as a group could exercise. Individual nurses, sometimes with the help of others, found ways to make the work less difficult and to achieve a modicum of autonomy. But they could do little collectively to alleviate the competition and animosity between the trained and nontrained or to transform the economic structures which underlay and created the private duty system.

Sporadic efforts were made to find a more organized solution to the private duty dilemma. While waiting for a case or isolated in a patient's home, however, few nurses could attend regular meetings or put the time into the effort to sustain the few private duty leagues which briefly flourished. The professional nursing associations, dominated by nursing educators and focused on registration laws and educational reforms, often discounted the private duty nursing problems or hoped they would fade into oblivion. As women workers, providing services in the private employ of individual patients, nurses were similarly ignored by both male and female trade unionists.[55]

By the 1920s, private duty was increasingly becoming a nursing backwater. Home-based private duty was relinquished to the nontrained nurses or the older graduates as the younger nurses sought employment in private duty as hospital "specials." There was a brief attempt to revitalize, rationalize, and reorganize private duty nursing at the end of the 1920s. But economics and hospital and nursing politics undermined these efforts. At the same time, secluded and many times embittered by their experiences, private duty nurses became increasingly conservative and isolated.[56]

The history of private duty nursing suggests some insights into the broader historical concern to comprehend how different groups of workers sought to gain individual and collective control over their labor. Historians of male artisans and skilled workers, in particular, have often romanticized this labor and described the "degradation of work" under monopoly capitalism in the twentieth century. As historians of women's work, most notably Susan Porter Benson and Barbara Melosh have argued, however, this "lament" for the lost artisan past is a very different tune when sung on higher notes. The nature of the sex-segregated and overcrowded labor market for women workers, the skills their jobs required, and the conditions under which they labored give their history and struggles a very different character from that of male workers.[57]

The private duty nurse cannot be equated with the independent craftsman or the skilled worker whose consciousness and struggles are the focus of much historical writing today.[58] Nor, however, was the private duty nurse in a position similar to women who labored next to one another in factories or department stores. The work site of private duty, the ideology of altruism and caring which pervaded this form of service work, the competition and lack of cases, in sum both the labor process and the social relations of production of this form of health care made the private duty nurse a particular kind of worker. She was indeed "neither for the drawing room nor for the kitchen," but also neither for the trade unions and professional associations nor for the powerful informal work groups.

It may well be that private duty nurses were sui generis. However, while private duty nursing was unique in its isolation and ambiguities, other women workers, workers of color, and the unskilled similarly faced high unemployment, competition for jobs, and lack of collective control over their work. Until we understand the political economy and work cultures of these less autonomous groups, we will have as partial an understanding of the American working class as we have of the health work force when we focus only on physicians.

NOTES

This is a revised version of a paper presented at the Organization of American Historians Convention, April 15, 1978. Research for this article was supported by Grant Number 1 RO3 HS02879-01 from the National Center for Health Services Research, U.S. Department of Health and Human Services. As always, the trenchant comments of Diana Long Hall, David Rosner, Tim Sieber, and Lise Vogel were invaluable.

1 *Nursing Ethics: For Hospital and Private Use* (Cleveland: J. B. Savage, 1901), 32.

2 The exact date when Americans of all classes began to use the hospital cannot, of course, be set. For a careful analysis of this question, see both Morris Vogel, *The Invention of the Modern Hospital* (Chicago: University of Chicago Press, 1979), and David Rosner, *A Once Charitable Enterprise* (New York: Cambridge University Press, 1982). For a discussion of

the changes in private duty nursing in the 1920s, see Susan Reverby, *The Nursing Disorder: A Critical History of the Hospital-Nursing Relationship, 1860–1945* (New York: Cambridge University Press, forthcoming), and Barbara Melosh, *"The Physician's Hand": Work Culture and Conflict in American Nursing* (Philadelphia: Temple University Press, 1982), esp. chap. 3.

3 Boston Medical Library Directory for Nurses, vol. A–V (vol. J is missing), 1880–1914, Rare Books Room, Countway Medical Library, Harvard Medical School. The goodwill and support of Richard Wolfe, Carol Pine, and the entire Rare Books Room staff is gratefully acknowledged. The Directory claimed to represent the majority of Boston nurses (Boston Medical Library Association, *10th and 11th Annual Reports*, October 1887, 23). A comparison of the number of nurses in the Directory with the number in the census for Boston suggests that, in

any given year (with the exception of the closing years of the Directory), anywhere from one third to three quarters of Boston's nurses could be found in these volumes. The sample included 313 graduate nurses and 226 nongraduates. The sample included only the female nurses, although the Directory did provide work for a small number of male nurses.

4 See Reverby, *Nursing Disorder*, chap. 1.

5 Isabel Hampton Robb, "Educational standards for nurses," in *Nursing the Sick 1893*, by Isabel Hampton Robb et al. (New York: McGraw Hill, 1949), 5.

6 U.S. Bureau of the Census, *Historical Statistics of the United States* (Washington; U.S. Government Printing Office, 1959), ser. B. 192–194; Carroll D. Wright, *The Social, Commercial and Manufacturing Statistics of the City of Boston* (Boston: Rockwell and Churchill, 1892), 96–98; Secretary of the Commonwealth of Massachusetts, *Census of the Commonwealth of Massachusetts, 1905*, vol. 2: *Occupations and Defective Social and Physical Conditions* (Boston: Wright and Potter, 1909), 138.

7 See Nancy Tomes, "'Little world of our own': the Pennsylvania Hospital Training School for Nurses, 1895–1907," in *Women and Health in America*, ed. Judith Walzer Leavitt (Madison: University of Wisconsin Press, 1984), 467–481; Jo Ann Ashley, *Hospitals, Paternalism and the Role of the Nurse* (New York: Teachers College Press, 1976); Reverby, *Nursing Disorder*, chapters 4 and 6.

8 Gertrude Harding, *The Higher Aspects of Nursing* (Philadelphia: W. B. Saunders and Co., 1919), 109; Sara E. Parsons, *Nursing Problems and Obligations* (Boston: Barrows, 1916), 115.

9 Katherine DeWitt, *Private Duty Nursing* (Philadelphia: J. P. Lippincott, 1913); Emily A. M. Stoney, *Practical Points in Nursing for Nurses in Private Practice* (Philadelphia: W. B. Saunders, 1897); Elinor Lason, "Characteristic requisites for a private duty nurse," *Report of the 18th Convention of the American Nurses Association*, June 1915, 65; E. B. M., Letter to the Editor of the *Transcript*, October 24, 1881, clipping in vol. B, Boston Medical Library Directory for Nurses.

10 Annie E. Hutchinson, "Practical nursing in private practice," *Trained Nurse and Hospital Review* 37 (August 1906): 83. "Don't imagine that you can discipline a patient in his own home as you would in a hospital ward. It can't be done," the *Trained Nurse* cautioned. "Don'ts for the Private Duty Nurse," 47 (October 1911): 202.

11 DeWitt, *Private Duty Nursing*, 70.

12 Hutchison, "Practical nursing," 84.

13 A random sample was drawn of one hundred patients who hired nurses through the Boston Medical Library Directory between 1880 and 1914. The occupations of the male patients, or males in the family of the patient, were obtained by checking names and addresses in the *Boston City Directory* for the appropriate years. The class base of those who used the services of private duty nurses continued to be an issue in nursing and health care. For further discussion of this, see Melosh *The Physician's Hand*, chap. 3, and Susan Reverby, "'Something besides waiting': the politics of private duty nursing reform in the Depression," in *Nursing History: New Perspectives, New Possibilities*, ed. Ellen Condliffe Lagemann (New York: Teachers College Press, 1983).

For quantitative evidence which suggests the changing class background of nurses, and the differences between the training schools, see Reverby, *Nursing Disorder*, chap. 5; Jane Mottus, *New York Nightingales* (Ann Arbor: University Microfilms Books, 1981); and Janet Wilson James, "Isabel Hampton and the professionalization of nursing in the 1890s," in *The Therapeutic Revolution*, ed. Morris Vogel and Charles Rosenberg (Philadelphia: University of Pennsylvania Press, 1979): 201–44.

14 Brennan Gill, *The Trouble of One House* (Garden City: Doubleday, 1950), 142. This quotation was given to me by Barbara Melosh.

15 Dr. J. Madison Taylor, Letter to the Editor, *American Journal of Nursing* 4 (May 1904): 658. Stoney, *Practical Points in Nursing*, 24.

16 Philadelphia, Letter to the Editor, *Trained Nurse and Hospital Review* 8 (September 1892):278.

17 William L. Richardson, *Address on the Duties and Conduct of Nurses in Private Nursing, June 18, 1886* (Boston: Press of George H. Ellis, 1886), 10. Richardson's address must have been very popular because it was printed and widely circulated.

18 The analysis of the nurses' class origins and migration patterns is based upon samples drawn from both nursing school student records and the records of the Home for Aged Women in Boston which admitted nontrained women nurses. Other analysis is based upon the Boston Medical Library Directory sample compared with the census and city directory evidence. The Directory does give addresses for about half the nurses. In vols. K, L, and O of the Boston Medical Library records, the hometowns for 220 nurses were given. These data were tabulated to determine where the graduates had gone to school before they came to Boston and to differentiate the nongraduates and graduates. For further discussion of class origins and migration patterns of nurses, see Reverby, *Nursing Disorder*, chap. 1, 5, and 6.

The Boston Medical Library Directory does not give ethnicity data. The migration of women from the Maritimes into nursing may be a continuation of their earlier migration into domestic service in Boston; see Alan Brookes, "Migration from the Maritime Provinces of Canada to Boston, Mass., 1860–1900," master's thesis, University of Hull, 1974, 213–19. The role of Canadian nurses in the United States has not yet been examined, but these women were an important minority, especially in the leadership, from the 1870s through the 1930s.

Nurses were listed in the Directory either as "Miss," "Mrs." or "Mr." Marital status of the nurses given as "Mrs." was inferred by checking a subsample in the city directory and by using national census data.

19 This figure is based on all deaths *reported* to the Boston Medical Library Directory and is therefore an underestimate since not all deaths were reported. For all those who died *while registered* with the directory, the average age at death for the men was only forty-one, for the women, thirty-seven. For comparative death rates for Massachusetts, see *Historical Statistics*, Bicentennial edition, ser. B. 201–213, 63.

20 There is a statistically significant relationship between training and wages at the .00 level from 1880 till the close of the Directory in 1914, but no significant relationship between wages and years of nursing experience. On how the wages were set, see Dr. Charles Putnam to Miss Gertrude Hamilton, August 7, 1913, Directory for Nurses, *Letterbook*, vol. 1, February 11, 1891–January 7, 1914.

21 This is not to suggest that the nontrained nurses willingly did whatever the family wanted. One such nurse, for example, complained that families expected her to do the "spring housecleaning" along with the nursing. "A graduate of the training school of life," Letter to the Editor, *Trained Nurse and Hospital Review* 51 (March 1916): 369.

22 Richard Bradley, "Large part of hospital work performed in the home," *Modern Hospital* 1 (November 1913): 227–31; Charlotte Aikens, "The Committee on Grading of Nurses," *Trained Nurse and Hospital Review* 37 (March 1914): 168; Lavinia L. Dock, "Directories for nurses," *Report of the Second Annual Convention of the American Society of Superintendents of Training School for Nurses*, 1895, 59; Miss Hintze, Discussion on Dock paper, Ibid., 60.

23 There is a statistically significant relationship between training and total number of cases received throughout the Directory at the .01 level. Data on the number of nurses available to work, by type of training, were compared to the number, by type of training, who actually did the work, for the years 1889–93. The raw data for this latter calculation can be found in the *Letterbook*.

24 Miss E. L. Blanchard to Dr. Brigham, February 11, 1895, Boston Medical Library Directory, vol. E, 181; L. H., "What shall she do?" Letter to the Editor, *Trained Nurse and Hospital Review* 23 (January 1899): 42.

25 Massachusetts Bureau of Statistics, *The Decennial Census, 1915* (Boston: Wright and Potter, 1918), 510.

26 Dr. Putnam to Miss McBrien, January 20, 1894; Dr. Putnam to Dr. Brigham, September 10, 1894; Dr. Putnam to Miss Heintze, November 19, 1892, *Letterbook*.

27 Alice N. Lincoln, Secretary of the Board, Long Island Hospital, to Dr. E. W. Taylor, October 23, 1899, *Long Island Hospital Letterbook*, vol. 1, Long Island Hospital Collection, Rare Books Room, Countway Medical Library.

28 Dr. Brigham to Dr. Putnam, November 3, 1899, Boston Medical Library Directory for Nurses, *Letterbook*.

29 Case rates calculated from averages in the Boston Medical Library Directory sample. On the Cleveland nurse's experience, see James H. Rodabaugh and Mary Jane Rodabaugh, *Nursing in Ohio* (Columbus, Ohio: Ohio State Nurses' Association, 1951), 199–200. 1890 Massachusetts data were compiled from the 1890 Census and were given to me by Professor Alex Keyssar of Brandeis University from his ongoing study on unemployment in Massachusetts. Unemployment frequency was calculated by dividing the total unemployed by the total number in the occupation. Unfortunately, these numbers include both nontrained and trained nurses, as well as midwives.

30 In 1900, for example, the average weekly wage for domestic servants in Massachusetts was $3.61; in that year the nongraduates in the Directory were averaging $11.25; the graduates $20.93. For the domestic servant data, see Stanley Lebergott, *Manpower in Economic Growth* (New York: McGraw Hill, 1964), 542. For comparisons to other women workers in Boston, see Louise Marion Bosworth, *The Living Wage of Women Workers* (New York: Longmans, Green, 1911), 33–39.

31 "The social side of nursing," *Trained Nurse and Hospital Review* 2 (March 1888):95–97; Letter to the Editor, ibid. 39 (October 1916): 369.

32 Daniel Gormon to Dr. Putnam, June 14, 1894; Elizabeth Bowness to Miss McBrien, December 9, 1903, *Letterbook*. The raw data on the number of nurses requested each month for 1891 through 1914 can be found in ibid. This drop-off in demand was especially precipitous during the 1893 depression.

33 B. H. Giles to the Directory, March 10, 1895, vol. F, Boston Medical Library Directory, 176. For ex-

amples of letters which also reflect the nurses' bit-
terness at the lack of work, see Emilie Neale to Dr.
Brigham, January 8, 1894; Dr. Putnam to Miss
McBrien, February 24, 1900, *Letterbook*.

34 Stoney, *Practical Points*, 18; Henry Beates, Jr., M.D.,
    The Status of Nurses: A Sociologic Problem (Philadel-
    phia: National Board of Regents, 1909), 17.

35 Letter to the Editor, *Pacific Coast Nursing Journal* 3
    (March 1914): 130.

36 Gerald Markowitz and David Rosner, "Doctors in
    crisis: medical education and medical reform dur-
    ing the Progressive Ea, 1895–1915," in *Health Care
    in America*, ed. Susan Reverby and David Rosner
    (Philadelphia: Temple University Press, 1979), 185–
    205.

37 "Report of the monthly meeting of the New York
    City Training School Alumnae, June 8, 1897," *Trained
    Nurse and Hospital Review* 19 (July 1897): 34.

38 "The reasons and the remedy—a training school
    symposium," *National Hospital Record*, 1907, *11:* 14–
    20; "Boston trained nurse in Chicago," *Boston Tran-
    script*, October 24, 1881, clipping in vol. B, Boston
    Medical Library Directory.

39 Memo from Dr. George Rowe, superintendent of
    Boston City Hospital, to the Directory, no date; nu-
    merous other letters in 1892, *Letterbook*.

40 "The members of the family regard themselves as
    the nurse's employers. She very often does not."
    Editorial, "The patient's family," *Trained Nurse and
    Hospital Review* 46 (March 1911): 166.

41 See letters and clippings laid in the volumes of the
    Boston Medical Library Directory for Nurses.

42 *American Journal of Nursing*, 4 (October 1904): 2.

43 Dr. Charles Putnam to Dr. Edwin Brigham, Janu-
    ary 31, 1894, *Letterbook*. Putnam's position was "it is
    but fair that a nurse who offers her services at a
    lower price and intrusts her engagements to us
    should either reap some benefit from that low price
    or else be informed that she will reap no such ben-
    efit." Putnam and Brigham were the physicians from
    the Boston Medical Library in charge of the Direc-
    tory.

44 Quoted in A. S. Kavanogh, "The indispensable
    combination in hospital work," *Trained Nurse and
    Hospital Review* 37 (July 1914): 78.

45 Anna Maxwell to Miss C. C. McBrien, October 11,
    1899, *Letterbook*. In his reply to Maxwell's letter, Dr.
    Brigham was pessimistic about how much a lower-
    ing of the rates would help the overcrowding.
    Brigham to Maxwell, October 16, 1899, ibid.

46 A private duty nurse who secured her own cases
    did not of course have to keep to an agreed-upon
    wage rate (see Louise Darche, "The proper orga-
    nization of training schools in America," *Nursing the

Sick 1893, 106). But it was considered a "breach of
faith" with the Directory for a nurse to overcharge
the patients when she had agreed upon one price
(see Morton Prince to Miss C. C. McBrien, February 12, 1901, *Letterbook*). However, such overcharging was not uncommon (Florence LaFleur, Boston
Medical Library Directory, vol. R, 66; L. M. B. Russell, vol. I, 58; Annie Collimore to M. Adelaide
Nutting, June 15, 1925, M. Adelaide Nutting Papers, Teachers College, Columbia University, File
X, Folder 4, "Hours of work–domestic service."

47 Around 1904 the directory officials began to ask
    patients and physicians to inform them when they
    felt the nurse was "superior or first class." On their
    uncertainty about the legality of their position, see
    F. Morison to Dr. F. Shattuck, February 20, 1884,
    Letterbook.

48 "A graduate," Letter to the Editor, *Trained Nurse and
    Hospital Review* 15 (January 1895): 43.

49 On blacklists, see memo from Dr. George H. M.
    Rowe, Superintendent of Boston City Hospital, to
    the Directory, no date; numerous other letters in
    1892, *Letterbook*.

50 Letter to the Editor, *Trained Nurse and Hospital Re-
    view* 23 (February 1904): 103.

51 "Private duty problems," *Trained Nurse and Hospital
    Review* 73 (July 1924): 57–58.

52 Committee on the Grading of Nursing Schools,
    Nurses, Patients and Pocketbooks (New York: The
    Committee, 1928), 361.

53 "Overcrowding in the nursing profession," *Trained
    Nurse and Hospital Review* 21 (July 1898): 8–13.

54 Lucy Drown to Dr. Brigham, October 18, 1899, *Let-
    terbook*.

55 Recent historical scholarship has begun to docu-
    ment and analyze the divisions and conflict within
    American nursing. For examples, see, in addition
    to my previous work cited, Melosh, *The Physician's
    Hand;* Susan Armeny, "Resolute enthusiasts: the ef-
    fort to professionalize American nursing, 1893–
    1923," Ph.D. diss., University of Missouri, 1983;
    Lagemann, ed., *Nursing History.*

56 See Reverby, "'Something besides waiting.'"

57 The historiography on workers' control and capi-
    talist rationality is growing rapidly. For the key works
    and critiques, see David Montgomery, *Workers' Con-
    trol in America* (New York: Cambridge University
    Press, 1979); Harry Braverman, *Labor and Monopoly
    Capital: The Degradation of Work in the Twentieth Cen-
    tury* (New York: Monthly Review Press, 1974); Su-
    san Porter Benson, "The clerking sisterhood: ra-
    tionalization and the work culture of saleswomen in
    American department stores," *Radical America* 12
    (March–April 1978): 41-55; Melosh, *The Physician's

Hand; and James Green, "Culture, politics and workers' response to industrialization in the U.S.," *Radical America* 16 (January–April 1982): 101–30.

58 For an example of an analysis of private duty work in this mode, see David Wagner, "The proletarianization of nursing in the United States, 1932–1946," *International Journal of Health Services,* 1980, *10:* 271–90.

IMAGE AND INCOME

Despite some slippage in recent years, medicine remains one of the most trusted and envied of all professions. This was not always so. During the first 300 years of European settlement the practice of medicine ranked relatively low in professional status. The prestige of physicians reached its nadir in the middle third of the 19th century, with the proliferation of low-quality medical schools, the fragmentation of the profession into quarreling sects, and the repeal of laws regulating the practice of medicine. Even so, educated physicians, especially those in the larger cities, often did well for themselves. As the historian E. Brooks Holifield has shown, the average wealth of urban physicians in 1860 was over four times that of free white males.

The status of physicians improved during the later part of the century, as medicine grew more scientific, efficacious, and restrictive. By 1900 most states had enacted licensing laws to exclude incompetents from inflicting themselves on the public. Nevertheless, other occupations, such as law, business, and even the ministry, continued to attract the best and brightest students.

The relatively low income of physicians undoubtedly contributed to medicine's lack of appeal. Although a handful of 19th-century physicians acquired small fortunes, a large percentage of medical-school graduates eventually abandoned medicine to seek greener fields elsewhere. As late as 1913 the American Medical Association estimated that only about 10 percent of physicians in the United States were earning "a comfortable income." The tax records of Wisconsin for the next year lend credence to this statement. Fewer than 60 percent of the state's practitioners earned enough even to be eligible for taxes, and among those required to pay, the average earnings were $1,488 a year, compared to $3,581 for bankers and capitalists, $2,810 for manufacturers, and $2,568 for lawyers.

The financial advantages of medicine did not become readily apparent until about the time of World War II. As Ronald L. Numbers argues, health insurance—first voluntary, then government sponsored—played a crucial role in elevating the income of physicians. By 1990 the average American doctor earned more than $130,000 a year. With increased income came high status, especially in certain specialties, such as cardiovascular surgery, where practitioners averaged nearly $300,000 a year. General and family practitioners benefited the least, with incomes averaging below $100,000.

Like other professionals, American physicians in the late 20th century have suffered from a decline in public trust and respect and an erosion of their autonomy—characteristics associated with what John C. Burnham calls the end of the Golden Age of American medicine. Nevertheless, the majority of Americans profess satisfaction with their own doctors, and bright young people continue to vie for admission to medical school.

17

The Third Party: Health Insurance in America

RONALD L. NUMBERS

> No third party must be permitted to come between the patient and his physician in any medical matter.
>
> *American Medical Association, 1934*[1]

American medicine in the 19th century was essentially a two-party system: patients and physicians. Medical practice was relatively simple, and doctors —out of economic necessity more than to preserve an intimate physician-patient relationship— personally collected their bills. Most practitioners billed their patients annually or semi-annually, although those with office practices usually insisted on immediate payment.[2] They were not, however, always free to charge what they pleased. In many communities local medical societies established schedules of minimum fees and instructed members never to undercut their colleagues.[3] There was little objection to providing free care for the poor —or to overcharging the wealthy—but generally the American medical profession preferred fixed fees to the so-called "sliding scale."[4] When hospitals began to mushroom late in the century, they, too, charged patients directly according to fixed prices.

But even in the 19th century a small, but undetermined, number of Americans carried some insurance against sickness through an employer, fraternal order, trade union, or commercial insurance company. Most of these early plans, however, were designed primarily to provide income protection, with perhaps a fixed cash benefit for medi-

cal expenses; few provided medical care, and those that did, like the plans sponsored by remotely located lumber and mining companies, generally contracted with physicians at the lowest possible prices. This type of "contract" practice restricted the patient's choice of physician, allegedly commercialized the practice of medicine, sometimes resulted in shoddy medical care—and almost always elicited the opposition of organized medicine.[5] During the latter half of the century the American Medical Association (A.M.A.) repeatedly condemned arrangements that provided unlimited medical service for a fixed yearly sum and urged the profession to maintain "the old relations of perfect freedom between physicians and patients, with separate compensation for each separate service."[6]

Widespread interest in health insurance did not develop in the United States until the 1910s, and then the issue was compulsory, not voluntary, health insurance. During the late 19th and early 20th centuries rising costs and increased demands for medical care had prompted many European nations, beginning with Germany in 1883, to provide industrial workers with compulsory health insurance.[7] Americans, however, paid little attention to these foreign experiments before 1911, when the British parliament passed a National Insurance Act.

Inspired by developments abroad and the spirit of Progressive reform at home, the American Association for Labor Legislation in 1912 created a Committee on Social Insurance to prepare a model bill for introduction in state legislatures.[8] By the fall of 1915 this committee had completed a tentative draft and was laying plans for an extensive

RONALD L. NUMBERS is William Coleman Professor of the History of Science and Medicine, University of Wisconsin–Madison.

Reprinted with permission from *The Therapeutic Revolution: Essays in the Social History of American Medicine*, edited by Morris J. Vogel and Charles E. Rosenberg (Philadelphia: University of Pennsylvania Press, 1979), pp. 177–200.

legislative campaign. Its bill required the participation of virtually all manual laborers earning $100 a month or less, provided both income protection and complete medical care, and divided the payment of premiums among the state, the employer, and the employee.[9]

The medical profession's initial response to this proposal bordered on enthusiasm. Three progressive physicians—Alexander Lambert, Isaac M. Rubinow, and S. S. Goldwater—had served on the drafting committee, and for a brief period after the turn of the century organized medicine was in a reform-minded mood. Upon receiving a copy of the A.A.L.L.'s bill, Frederick R. Green, secretary of the A.M.A.'s Council on Health and Public Instruction, informed the bill's sponsors that their plan for compulsory health insurance was

> exactly in line with the views that I have held for a long time regarding the methods which should be followed in securing public health legislation. . . . Your plans are so entirely in line with our own that I want to be of every possible assistance.

Specifically, Green wanted to give the A.A.L.L. "the assistance and backing of the American Medical Association in some official way," and he proposed setting up an A.M.A. Committee on Social Insurance to cooperate with the A.A.L.L. in working out the medical provisions of the bill.[10] As a result of his efforts, the A.M.A. Board of Trustees early in 1916 appointed a three-man committee, with Lambert as chairman. He, in turn, hired Rubinow as executive secretary and set up committee headquarters in the same building with the A.A.L.L.

The *Journal of the American Medical Association* hailed the appearance of the model bill as "the inauguration of a great movement which ought to result in an improvement in the health of the industrial population and improve the conditions for medical service among the wage earners."[11] In the editor's opinion, "No other social movement in modern economic development is so pregnant with benefit to the public."[12] At the A.M.A.'s annual session in June 1916, President Rupert Blue called compulsory health insurance "the next step in social legislation,"[13] and Lambert, as chairman of the Committee on Social Insurance, presented a report that stopped just short of endorsing the measure.[14]

Physician support at the state level was similarly strong. In 1916 the state medical societies of both Pennsylvania and Wisconsin formally approved the principle of compulsory health insurance, and the Council of the Medical Society of the State of New York did likewise.[15] Reasons for favoring health insurance varied from physician to physician. According to the *Journal of the American Medical Association*, the most convincing argument was "the failure of many persons in this country at present to receive medical care";[16] but the average practitioner, who earned less than $2,000 a year, was probably more impressed by the prospect of a fixed income and no outstanding bills.[17] Besides, the coming of health insurance appeared inevitable, and most doctors preferred cooperating to fighting. "Whether one likes it or not," wrote the editor of the *Medical Record*,

> social health insurance is bound to come sooner or later, and it behooves the medical profession to meet this condition with dignity. . . . Blind condemnation will lead nowhere and may bring about a repetition of the humiliating experiences suffered by the medical profession in some of the European countries.[18]

By early 1917, however, medical opinion was beginning to shift, especially in New York, where the A.A.L.L. was concentrating its efforts. One after another of the county medical societies voted against compulsory health insurance, until finally the council of the state society rescinded its earlier endorsement.[19] Both friends and foes of the proposed legislation agreed on one point: the medical profession's chief objection was monetary in nature. As the exasperated secretary of the A.A.L.L. saw it, the "crux of the whole problem" was that physicians were constantly hearing the lie that the model bill would limit them to 25¢ a visit or about $1,200 a year.[20] "If you boil this health insurance matter down, it seems to be a question of the remuneration of the doctor," observed one New York physician, who believed that 99 out of 100 physicians had taken up the practice of medicine primarily "as a means of earning a livelihood."[21] Another New York practitioner, who opposed the A.A.L.L.'s bill, described all other objections besides payment as "merely camouflage for this one crucial thought." Medical opposition would melt away, he predicted, if adequate compensation were guaranteed.[22]

The medical profession was, of course, not alone in opposing compulsory health insurance. Commercial insurance companies, which would have been excluded from any participation, were especially critical; and some labor leaders, like Samuel Gompers, preferred higher wages to paternalistic social legislation.[23]

America's entry into World World I in April 1917 not only interrupted the campaign for compulsory health insurance, but touched off an epidemic of anti-German hysteria. Patriotic citizens lashed out at anything that smacked of Germany, including health insurance, which was reputed to have been "made in Germany." As the war progressed, Americans in increasing numbers began referring to compulsory health insurance as an "unAmerican" device that would lead to the "Prussianization of America."[24]

Shortly before the close of the war California voters, in the only referendum on compulsory health insurance, soundly defeated the measure by a vote of 358,324 to 133,858 and dampened the hopes of insurance advocates.[25] Their spirits revived briefly in the spring of 1919, when the New York State Senate passed a revised version of the model bill, but the bill subsequently died in the Assembly. By 1920 even the A.A.L.L. was rapidly losing interest in an obviously lost cause.

As the prospects for passage of the model bill declined, the stridency of anti-insurance doctors increased. "*Compulsory Health Insurance,*" declared one Brooklyn physician, "is an UnAmerican, Unsafe, Uneconomic, Unscientific, Unfair and Unscrupulous type of Legislation [supported by] Paid Professional Philanthropists, busybody Social Workers, Misguided Clergymen and Hysterical women."[26] In 1919 he and other critics launched a campaign to have the A.M.A.'s House of Delegates officially condemn compulsory health insurance. They failed on their first attempt, but the following year the delegates overwhelmingly approved a resolution stating

> That the American Medical Association declares its opposition to the institution of any plan embodying the system of compulsory contributory insurance against illness, or any other plan of compulsory insurance which provides for medical service to be rendered contributors or their dependents, provided, controlled, or regulated by any state or the Federal Government.[27]

This repudiation of compulsory health insurance was not, as one writer has suggested, the result of "an abdication of responsibility by the scientific and academic leaders of American medicine."[28] Nor was it primarily the product of a rank-and-file takeover by conservative physicians disgruntled with liberal leaders.[29] The doctors who rejected health insurance in 1920 were by and large the same ones who had welcomed — or at least accepted — it only four years earlier. Frederick Green, the person most responsible for the A.M.A.'s early support of compulsory health insurance, was by 1921 describing it as an "economically, socially and scientifically unsound" proposition favored only by "radicals."[30] And his experience was not atypical.

Many factors no doubt contributed to such changes of heart. Opportunism undoubtedly motivated some, and the political climate surely affected the attitudes of others. But more important, it seems, was the growing conviction that compulsory health insurance would lower the incomes of physicians rather than raise them, as many practitioners had earlier believed. With each legislative defeat of the model bill, the coming of compulsory health insurance seemed less and less inevitable, and the self-confidence of the profession grew correspondingly. "[T]his Health Insurance agitation has been good for us," concluded one prominent New York physician as the debate drew to a close. "If it goes no farther it will have brought us more firmly together than any other thing which has ever come to us."[31]

An additional factor affecting the medical profession's attitude toward compulsory health insurance was its recent experience with workmen's compensation, which was probably the most common form of health insurance in America from the 1910s to the 1940s. Beginning in 1911 many states passed laws making employers legally responsible for on-the-job injuries, but few of the early compensation acts provided comprehensive medical benefits. During the war, however, most states added such provisions or liberalized existing ones, giving American doctors their first taste of social insurance. For many, it was not pleasant. Employers often took out accident insurance with commercial companies, which either contracted with physicians to care for the injured or paid local practitioners according to an arbitrary fee schedule.[32] Neither arrangement pleased the medical profession, which complained that the abuses resulting from such practices "were akin to mayhem and murder."[33] It was evident from this experience, reported the A.M.A., that "pus and politics go together."[34]

In 1925 the New York State Medical Society reported that health insurance "is a dead issue in the United States. . . . It is not conceivable that any serious effort will again be made to subsidize medicine as the hand-maiden of the public."[35] The victorious New York physicians had every reason to be confident, but they failed to reckon with economic disaster. The Great Depression invalidated many assumptions about American society and threatened the financial security of both hospitals and physicians. Between 1929 and 1930 hospital receipts per patient declined from $236.12 to $59.26, and occupancy rates fell from 71.28 percent to 64.12 percent.[36] As the Depression continued, income from endowments and contributions decreased by nearly two-thirds, and the charity load almost quadrupled.[37] Particularly hard hit were the private voluntary hospitals, which had been expanding six times faster than the population.[38] The net income of physicians during the first year of the Depression dropped 17 percent, with general practitioners suffering the biggest losses. In some regions, particularly the cotton-growing states, collections from patients fell 50 percent, and the situation grew worse as the Depression continued.[39]

In response to this disaster, several hospitals began experimenting with insurance. Although not the first, the most influential of these experiments was the Baylor University Hospital plan, often described as the "father" of the Blue Cross movement. In December 1929 Baylor Vice-President Justin F. Kimball, former superintendent of the Dallas public schools, enrolled 1,250 public school teachers, who paid 50¢ a month for a maximum of 21 days of hospital care, an arrangement consciously modeled after the prepayment plans used in the lumber and railroad industries.[40] The success of single-hospital insurance at Baylor and other places soon led to the development of multiple-hospital plans that included all hospitals in a given area. The first of these appeared in Sacramento, California, in 1932, and by 1937, when the American Hospital Association began approving such programs, there were 26 in operation with 608,365 participating members.[41]

The motives behind these early endeavors are difficult to determine. In two recent studies of Blue Cross, for example, Odin W. Anderson stresses the altruistic spirit of the pioneers, while Sylvia Law emphasizes their economic interests.[42] There is, as one might expect, some evidence for both interpretations. Voluntary hospital insurance, said Michael M. Davis in 1931, has "the double aim of furnishing a new and broader base of support for hospitals and of helping small income people to meet their big sickness bills."[43] Economic concerns are, however, easier to document than altruism. It is significant that although financially disinterested civic organizations occasionally contributed funds to establish hospital insurance programs, "In most cases the initiative and main drive for the starting of the various plans came from the hospitals of the community—from hospital administrators and trustees."[44] In his 1932 survey of prepayment plans Pierce Williams concluded that hospitals had promoted insurance primarily "to put their finances on a sound basis."[45]

Physician reaction to these early experiments in hospital insurance was mixed. Those affected the most seemed pleased. A physician associated with a Grinnell, Iowa, plan described the attitude of local practitioners as "very cordial,"[46] and Kimball reported that Dallas doctors appreciated both the increased availability of hospital care for their patients and the fact that insurance got "the patient's hospital bill out of the way of the doctor's personal collections."[47] The A.M.A., however, was openly antagonistic, characterizing prepayment plans "as being economically unsound, unethical and inimical to the public interests."[48] According to the director of the Association's Bureau of Medical Economics, such schemes were largely "a result of 'tactics of desperation,' in which hard-pressed hospitals are seeking 'any port in a storm'"[49] The A.M.A.'s solution to the problem of financing health care was "to save for sickness."[50]

Despite these negative pronouncements, health insurance continued to grow—especially after the publication in 1932 of the final report of the Committee on the Costs of Medical Care. This group of 45 to 50 prominent Americans drawn from the fields of medicine, public health, and the social sciences set out in 1927 to ascertain the medical needs of the American people and the resources available to meet them. Ray Lyman Wilbur, a former president of the A.M.A., served as chairman, and over half of the members were physicians. At the end of five years of exhaustive study, funded by several philanthropic organizations, a majority of the committee, including the chairman, modestly recommended the adoption of group practice and voluntary health insurance as the best means of solving the nation's health care problems.[51]

But even this was too radical for eight of the physicians on the committee, who, with one other member, prepared a minority report denouncing "the thoroughly discredited method of voluntary insurance" as being more objectionable than compulsory health insurance. Health insurance, said the minority, would inevitably lead to the

> solicitation of patients, destructive competition among professional groups, inferior medical service, loss of personal relationship of patient and physician, and demoralization of the profession. It is clear that all such schemes are contrary to sound policy and that the shortest road to the commercialization of the practice of medicine is through the supposedly rosy path of insurance.

The dissenting doctors did, however, favor action to alleviate the financial plight of the medical profession, caused in part by having to provide free care to the poor. Thus they recommended that the government relieve physicians of this unfair "burden" by assuming financial responsibility for the care of the indigent. The results of such a plan, they said, "would be far reaching." In particular, the income of physicians would increase and young doctors would find it easier to begin the practice of medicine.[52]

Although some state medical societies (including those in Alabama and Massachusetts) endorsed the majority report,[53] and although more of the committee's physicians had voted with the majority than with the minority, the A.M.A.'s House of Delegates declared that the minority report represented "the collective opinion of the medical profession." Group practice and health insurance, said the delegates, "would be inimical to the best interests of all concerned."[54] Morris Fishbein, the outspoken editor of the association's *Journal*, characteristically reduced the issue to "Americanism versus sovietism for the American people."[55] "The alinement is clear," he wrote,

on the one side the forces representing the great foundations, public health officialdom, social theory—even socialism and communism—inciting to revolution; on the other side, the organized medical profession of this country urging an orderly evolution guided by controlled experimentation.[56]

The alignment may have seemed clear in 1932, but a revival of interest in compulsory health insurance soon blurred it. In 1934, President Franklin D. Roosevelt appointed a Committee on Economic Security to draft legislation for a social security program, which, everyone assumed, would include health insurance. Pressure from organized medicine, however, forced the President to drop health care from the bill he sent to Congress in 1935. Undaunted, progressive members of his administration continued to agitate for compulsory health insurance and in 1938 held a National Health Conference in Washington. This event aroused great popular interest in a government-sponsored health program, resulting the next year in Senator Robert F. Wagner's introduction of a bill to provide medical assistance for the poor, primarily through federal grants to the states.[57]

In view of these developments, the A.M.A. reversed its position on voluntary health insurance, hoping that such action would quiet demands for a compulsory system. In 1937 the House of Delegates approved group hospitalization plans that confined "their benefits strictly to the facilities ordinarily provided by hospitals; viz., hospital room, bed, board, nursing, routine drugs."[58] A short time later the association began taking credit for promoting the growth of hospitalization insurance, which it had so bitterly opposed only a few years before.[59]

At the same time it was giving its blessing to hospitalization insurance, the A.M.A. was working out a physician-controlled plan to provide medical care insurance. In 1934 the House of Delegates took a tentative step in that direction by agreeing on ten principles to govern "the conduct of any social experiments." These included complete physician control of medical services, free choice of physician, the inclusion of all qualified practitioners, and the exclusion of persons living above the "comfort level." The delegates stopped short of endorsing health insurance and made a point of emphasizing the traditional view that medical costs "should be borne by the patient if able to pay at the time the service is rendered."[60]

In February 1935, shortly after the Committee on Economic Security reported to the President, the House of Delegates met in special session — the first since World War I — to reaffirm its opposition to "all forms of compulsory sickness insurance." Recognizing the need to offer an alternative to government-sponsored insurance, the delegates encouraged "local medical organizations to establish plans for the provision of adequate medical service for all of the people . . . by voluntary budgeting to meet the costs of illness."[61] The language was vague, but the intention was clearly to foster the creation of society-controlled medical insurance plans.

In the aftermath of the National Health Conference of 1938 the A.M.A. called a second special session on insurance. This time the House of Delegates approved the development of "cash indemnity insurance plans" for low-income groups, controlled by local medical societies.[62] By offering cash benefits instead of service benefits, physicians hoped to retain their freedom to charge fees higher than the insurance benefits whenever it seemed appropriate.[63] In 1942, to meet competition from commercial insurance companies, the A.M.A. took the final step of approving medical service plans.[64]

By the late 1930s a number of local medical societies, particularly in the Northwest, had already organized "medical service bureaus" offering medical care for a fixed amount per year.[65] In 1939 the California Medical Association, in an effort to stave off compulsory health insurance, established the first statewide medical service plan.[66] Seven years later, when the A.M.A. created Associated Medical Care Plans, the precursor of Blue Shield, there were 43 medical society plans with a combined enrollment of three million members.[67] In most places coverage was limited to low-income families, who would otherwise have been among the least able to pay physicians' fees.

The threat of "socialized medicine" was no doubt the most compelling reason why organized medicine decided to embrace health insurance. As the demand for compulsory health insurance grew, more and more physicians came to see voluntary plans as their "only telling answer to federalization and regimentation."[68] "[I]t is better to inaugurate a voluntary payment plan," advised the secretary of the State Medical Society of Wisconsin, "rather than wait for a state controlled compulsory plan."[69]

But fear of compulsory health insurance was not the only reason why the medical profession changed its mind. By the late 1930s many physicians were also discerning potential benefits in health insurance.[70] A 1938 Gallup poll showed that nearly three-fourths of American doctors favored voluntary medical insurance, and over half were confident that it would increase their incomes.[71] Health insurance, predicted one Milwaukee physician, "would do away with the uncollectible accounts. . . . It would offer to the physician an opportunity of earning a living commensurate with the value of the service that he performs."[72] Furthermore, by paying for expensive services like x-rays and laboratory tests, it would enable doctors to practice a better quality of medicine.[73]

Once the profession recognized these possible benefits, it sought absolute control over medical service plans. In many states physicians won the right to monopolize medical care insurance through special enabling acts, which critics ironically regarded as "un-American."[74] In other places organized medicine tried to discourage physicians from participating in nonsociety plans by threatening expulsion and the denial of hospital privileges. In 1938 such heavy-handed tactics brought the A.M.A. an indictment (and eventual conviction) for violation of antitrust laws.[75]

Despite a genuine concern for the welfare of their patients, doctors did not embrace health insurance primarily to assist the public in obtaining better medical care. In fact, throughout the 1930s spokesmen for organized medicine repeatedly denied that health care in America was inadequate and attributed the good health of Americans to "the present system of medical practice," that is, to the traditional two-party system.[76] The physicians of Massachusetts may, as they claimed, have supported a medical service plan in recognition of "a problem in the distribution of the cost of decent medical care." But even in that progressive state competition from consumer cooperatives was just as important.[77]

Proudly displaying the medical profession's stamp of approval, health insurance entered a period of unprecedented growth (see Fig. 1). By 1952 over half of all Americans had purchased some

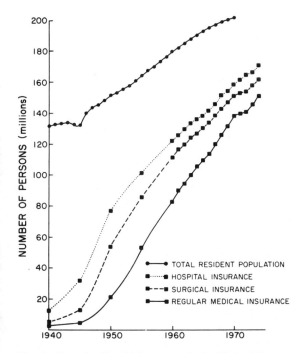

Fig. 1: Growth of health insurance in the United States. Sources: *Source Book of Health Insurance Data, 1975–76* (New York: Health Insurance Institute, 1976), p. 22; U.S. Bureau of the Census, *Historical Statistics of the United States: Colonial Times to 1970* (Washington, D.C.: Government Printing Office, 1975), Part 1, p. 8.

health insurance, and prepayment plans were being described as "the medical success story of the past 15 years."[78] Behind this growth was consumer demand, especially from labor unions, which after the war began bargaining for health insurance to meet rapidly rising medical costs that were making the prospect of sickness the "principal worry" of industrial workers. Following a 1948 Supreme Court ruling that health insurance benefits could be included in collective bargaining, "the engine of the voluntary health insurance movement," to use Raymond Munts's metaphor, moved out under a full head of steam. Within a period of three months the steel industry alone signed 236 contracts for group health insurance, and auto workers were not far behind.[79]

Growth statistics, however, do not tell the whole story. Although most Americans did have some health insurance by mid-century, coverage remained spotty. In 1952 insurance benefits paid only

15 percent of all private expenditures for health care (see Fig. 2). Besides, the persons most likely to be insured were employed workers living in urban, industrial areas, while the unemployed, the poor, the rural, the aged, and the chronically ill — those who needed it the most — went uninsured.[80]

With voluntary plans failing to protect so many Americans, the perennial debate over compulsory health insurance flared up again. Encouraged by organized labor, the Social Security Board in 1943 drafted a bill — named after its congressional sponsors, Senators Robert Wagner and James Murray and Representative John Dingell — providing health insurance to all persons paying social security taxes, as well as to their families. The time, however, was inauspicious. World War II was diverting the nation's attention to other issues, and without the President's active support the bill died quietly in committee.[81]

Two years later, with the war over and Harry S. Truman in the White House, prospects for passage appeared much brighter. Since his days as a county judge in Missouri, Truman had been concerned about the health needs of the poor, and within a few weeks of assuming the presidency he decided to lend his support to the health insurance campaign. Following a strategy session with the President, Wagner, Murray, and Dingell reintroduced their bill, this time adding dental and nursing care to the proposed benefits.[82]

These developments terrified the A.M.A., which viewed the Wagner-Murray-Dingell bill as the first step toward a totalitarian state, where American doctors would become "clock watchers and slaves of a system."[83] To head off passage of such legislation, the A.M.A. in 1946 began backing a substitute bill, sponsored by Senator Robert A. Taft, which authorized federal grants to the states to subsidize private health insurance for the indigent.[84]

The basic problem, as the Association's spokesman Morris Fishbein defined it, was one of "public relations." The medical profession had "to convince the American people that a voluntary sickness insurance system . . . is better for the American people than a federally controlled compulsory sickness insurance system."[85] Actually, most Americans needed little convincing. A 1946 Gallup poll showed that only 12 percent of the public favored extending social security to include health insurance, and more individuals thought the Wagner-

Murray-Dingell proposal would have a negative effect on health care than believed that it would be beneficial.[86]

Truman's surprise victory in 1948, at the close of a campaign that featured health insurance as a major issue, convinced the A.M.A. that it was time to declare all-out war. Shortly after the election returns were in, the House of Delegates voted to assess each member $25 to raise a war chest for combatting socialized medicine, which was defined as "a system of medical administration by which the government promises or attempts to provide for the medical needs of the entire population or a large part thereof."[87] Within a year $2,250,000 had been raised, and the public relations firm of Whitaker and Baxter was putting it to effective use in an effort to "educate" the American people.[88] The showdown came in 1950 when organized medicine won a stunning victory in the off-year elections, forcing many candidates to renounce their earlier support of compulsory health insurance and defeating "nearly 90 percent" of those who refused to back down.[89]

Throughout this controversy representatives of organized medicine insisted that the country did not need compulsory health insurance, just as they had insisted in the early 1930s that voluntary insurance was unnecessary. "There is no health emer-

gency in this country," said a complacent A.M.A. president in 1952. "The health of the American people has never been better."[90] If some individuals could not afford proper medical care, it was probably the result of self-indulgence rather than genuine need:

> Since one out of every four persons in the United States has a motor car, one out of two a radio, and since our people find funds available for such substances as liquors and tobacco in amounts almost as great as the total bill for medical care, one cannot but refer to the priorities and to the lack of suitable education which makes people choose to spend their money for such items rather than for the securing of medical care.[91]

What Americans needed, said the doctors, was more voluntary insurance, which had worked out so well that most physicians by the early 1950s no longer thought coverage should be restricted to low-income groups.[92] The financial and political benefits of health insurance were so great, the medical profession jealously protected it. When rumors began circulating that some surgeons were doubling their fees to insured patients, the A.M.A. called for an immediate crackdown. "Voluntary prepayment plans are the medical profession's greatest bulwark against the socialization of medi-

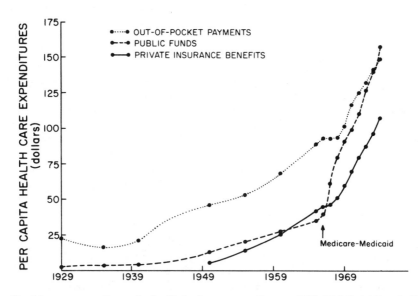

Fig. 2: Sources of health-care expenditures in the United States. Source: Nancy L. Worthington, "National health expenditures, 1929–74," *Social Security Bulletin*, February 1975, *38*: 16.

cine," said one official. "This program must not be jeopardized by avaricious physicians."[93]

The election of a Republican administration in 1952 effectively ended the debate over compulsory health insurance, and organized medicine breathed a sigh of relief. "As far as the medical profession is concerned," wrote the A.M.A. president, "there is general agreement that we are in less danger of socialization than for a number of years. . . . We have been given the opportunity to solve the problems of health in a truly American way."[94] The "American way," it went without saying, was the way of voluntary health insurance.

The Eisenhower years indeed proved to be tranquil ones for the medical profession. Encouraged by their physicians and by the constantly rising costs of medical care, an increasing number of Americans purchased health insurance, until by the early 1960s nearly three-fourths of all American families had some coverage (see Fig. 1). Still, this paid for only 27 percent of their medical bills, and many citizens, especially the poor and the elderly, had no protection at all.[95]

This problem led Representative Aime Forand in the late 1960s to reopen the debate over compulsory health insurance with a proposal limiting coverage to social security beneficiaries. In 1960 Senator John F. Kennedy introduced a similar measure in the Senate.[96] To organized medicine, even such restricted coverage amounted to "creeping socialism,"[97] and the A.M.A. would have none of it. The association's "strongest objection" continued to be that "it is unnecessary and would lower the quality of care rendered," the same argument it had been using since the 1910s. Its only concession was to approve a government plan providing assistance to "the indigent or near indigent," which would benefit physicians as much as the poor.[98] Thus in 1960 Congress, with A.M.A. approval, passed the Kerr-Mills amendment to the Social Security Act, granting federal assistance to the states to meet the health needs of the indigent and the elderly who qualified as "medically indigent."

If the medical profession hoped to forestall the coming of compulsory health insurance by this small compromise, Senator Kennedy's election to the presidency that fall soon convinced them otherwise. Upon occupying the White House, he immediately began laying plans to extend health insurance protection to all persons on social security, whether "medically indigent" or not. The A.M.A. denounced his plans as a "cruel hoax" that would disrupt the doctor-patient relationship, interfere with the free choice of physician, impose centralized control, and — worst of all — undermine the financial incentive to practice medicine. They would not only endanger the quality of medical care, but would discourage the best young people from entering the field.[99] Despite these ominous predictions, Congress in 1965 voted to include health insurance as a social security benefit (Medicare) and to provide for the indigent through grants to the states (Medicaid). Thus, after 50 years of debate, compulsory health insurance finally came to America.

In 1967, just two years after the passage of Medicare, third parties for the first time paid more than half of the nation's medical bills.[100] Many Americans continued to be without health insurance coverage, but seldom by choice.[101] Although critics frequently attacked the insurance business, no one advocated returning to a two-party system. In the opinion of one observer, the acceptance of health insurance was a phenomenon "without parallel in contemporary American life."[102] Prepayment plans benefitted both providers and consumers of medical care, but especially the providers.

Hospitals, the pioneers of voluntary health insurance, profited from the start. In 1947 Louis Reed reported that hospital administrators agreed unanimously that insurance plans had reduced their volume of free care and increased revenues.[103] In the years between 1939 and 1951 the amount of charity care provided by Philadelphia hospitals, for example, fell from 60 percent to 24 percent.[104] Later, in the 1960s and 1970s, the windfall from Medicare and Medicaid enabled many hospitals to improve — or at least to expand — their facilities.

Health insurance also proved advantageous to physicians, especially financially. In the period following the development of medical service plans, their incomes climbed dramatically (see Fig. 3); and, according to some analysts, the "most significant factor" contributing to this increase was third-party payments, which rose from 15.5 percent to nearly 50 percent of physicians' incomes in the two decades between 1950 and 1969.[105] Proving a cause-and-effect relationship is difficult, but the testimony of physicians themselves supports this

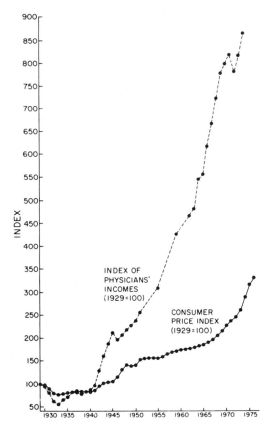

Fig. 3: The income of American physicians, 1929–1975.
Sources: U.S. Bureau of the Census, *Historical Statistics of the United States: Colonial Times to 1970* (Washington, D.C.: Government Printing Office, 1975), Part 1, pp. 175–176, 210–211; U.S. Bureau of the Census, *Statistical Abstract of the United States, 1975* (Washington, D.C.: Government Printing Office, 1975), p. 77; *Medical Economics,* Oct. 24, 1960, *37*: 40, and Nov. 10, 1975, *52*: 184; U.S. Bureau of Labor Statistics, *Consumer Price Index, 1971–76.* Graph prepared by Lawrence D. Lynch.

view. In 1957, for example, over half of the doctors in Michigan reported increased incomes as a result of prepaid medical care, with smalltown physicians and general practitioners registering the greatest gains. The most frequently cited explanations were better bill collecting and more patients.[106]

Certainly there can be little doubt that Medicare and Medicaid benefitted the medical profession handsomely. In fact, Robert and Rosemary Stevens concluded that "it seemed to be the phy-

sicians who gained the most."[107] After complaining for years that compulsory health insurance would beggar the profession and reduce the financial incentive to practice medicine, physicians discovered the results to be just the opposite. In the first year under Medicare the rate of increase in physician fees more than doubled (from 3.8 percent to 7.8 percent), while the rate of increase of the Consumer Price Index only rose from 2.0 percent to 3.3. percent.[108] A 1970 Senate Finance Committee investigation turned up at least 4,300 individual physicians who had received $25,000 or more from Medicare in 1968, and 68 of these had gotten over $100,000 each. Although most of this money was earned fairly, reports of questionable practices abounded. Some physicians allegedly saw patients more often than necessary, billed for care never given, and on occasion even resorted to the notorious "gang visit," charging $300 or $400 for one cursory sweep through a hospital ward or nursing home.[109] Such flagrant abuses prompted the president of one local medical society to warn his colleagues "to quit strangling the goose that can lay those golden eggs."[110]

Compared with the relatively tangible benefits of health insurance for hospitals and physicians, those to patients are more difficult to calculate. Prepayment plans undeniably gave Americans greater access to medical care than ever before, eased the financial strain of paying medical bills, and brought peace of mind to millions of policyholders. A grateful public showed its appreciation by buying increasingly comprehensive coverage. But it is not certain that they enjoyed better health for it. On the one hand, there are studies showing that "those who were eligible for Medicaid were likely to have better health than similar groups who were not."[111] But other studies indicate that although Medicare apparently encouraged more expensive types of treatment, like surgery rather than radiation for breast cancer, recovery rates remained roughly the same.[112]

Under health insurance from 1941 to 1970, life expectancy at birth in America did increase from 64.8 years to 70.9 years.[113] But again it is hard to determine how much — if any — of this should be credited to improved medical care, much less to the way in which it was financed. By the early 1970s even organized medicine was downplaying the ability of the medical profession to prolong life and preserve health. As Max H. Parrott of the A.M.A. testified in 1971, choice of life-style had become as important as medical care in determining the nation's health: "No matter how drastic a change is made in our medical care system, no matter how massive a program of national health insurance is undertaken, no matter what sort of system evolves, many of the really significant, underlying causes of ill health will remain largely unaffected."[114] In a society in which heart disease, cancer, accidents, and cirrhosis of the liver all ranked among the top ten killers,[115] it was indeed unrealistic to expect health insurance to cure the nation's ills.

NOTES

1 Minutes of the 85th annual session, June 11–15, 1934, *J.A.M.A.*, 1934, *102*: 2200.

2 D. W. Cathell, *The Physician Himself and What He Should Add to His Scientific Acquirements*, 3rd ed. (Baltimore: Cushings & Bailey, 1883), pp. 16, 175–176; Charles E. Rosenberg, "The practice of medicine in New York a century ago," *Bull. Hist. Med.*, 1967, *41*: 229–230.

3 George Rosen, *Fees and Fee Bills: Some Economic Aspects of Medical Practice in Nineteenth Century America* (Supplement No. 6, *Bull. Hist. Med.*; Baltimore: Johns Hopkins Press, 1946).

4 Jeffrey Lionel Berlant, *Profession and Monopoly: A Study of Medicine in the United States and Great Britain* (Berkeley: Univ. of California Press, 1975), pp. 101–102.

5 Pierce Williams, *The Purchase of Medical Care through Fixed Periodic Payment* (New York: National Bureau of Economic Research, 1932); Jerome L. Schwartz, "Early history of prepaid medical care plans," *Bull. Hist. Med.*, 1965, *39*: 450–475.

6 *1846–1958 Digest of Official Actions: American Medical Association* (Chicago: A.M.A., 1959), pp. 121–122.

7 For a summary of the European experience, see Richard Harrison Shryock, *The Development of Modern Medicine* (New York: Hafner Publishing Co., 1969), pp. 381–402.

8 This account of the first American debate over compulsory health insurance is based on Ronald L. Numbers, *Almost Persuaded: American Physicians and Compulsory Health Insurance, 1912–1920* (Baltimore: Johns Hopkins Univ. Press, 1978).

9 *Health Insurance: Standards and Tentative Draft of an Act* (New York: American Association for Labor Legislation, 1916).

10 F. R. Green to J. B. Andrews, Nov. 11, 1915, A.A.L.L. Papers, Cornell University.

11 "Industrial insurance," *J.A.M.A.*, 1916, *66*: 433.

12 "Cooperation in social insurance investigation," *ibid.*, pp. 1469–1470.

13 Rupert Blue, "Some of the larger problems of the medical profession," *ibid.*, p. 1901.

14 Report of the Committee on Social Insurance, *ibid.*, 1951–1985.

15 Proceedings of the Medical Society of the State of Pennsylvania, Sept. 18–21, 1916, *Penn. Med. J.*, 1916, *20*: 135, 143; Proceedings of the House of Delegates, State Medical Society of Wisconsin, Oct. 5, 1916, *Wis. Med. J.*, 1916, *15*: 288; Minutes of the Council, Medical Society of the State of New York, Dec. 9, 1916, *N.Y. St. J. Med.*, 1917, *17*: 47–48.

16 "Social insurance in California," *J.A.M.A.*, 1915, *65*: 1560.

17 Income statistics are scarce for this period, but a 1915 survey of physicians and surgeons in Richmond, Virginia, showed that "the very large proportion of physicians were earning less than $2,000," and income tax records for Wisconsin in 1914 indicate that the average income of taxed physicians was $1,488. Committee on Social Insurance, *Statistics Regarding the Medical Profession* (Social Insurance Series Pamphlet No. 7; Chicago: A.M.A., n.d.), pp. 81, 87.

18 "Opposition to the Health Insurance Bill," *Med. Rec.*, 1916, *89*: 424.

19 Report of the Committee on Legislation, *N.Y. St. J. Med.*, 1917, *17*: 234.

20 J. B. Andrews to New York Members of the A.A.L.L., Nov. 3, 1919, A.A.L.L. Papers.

21 "A Symposium on compulsory health insurance presented before the Medical Society of the County of Kings, Oct. 21, 1919," *Long Island Med. J.*, 1919, *13*: 434. George W. Kosmak made the statement.

22 M. Schulman to J. B. Andrews, Feb. 22, 1919, A.A.L.L. Papers.

23 See Gompers' testimony before the *Commission to Study Social Insurance and Unemployment: Hearings before the Committee on Labor, House of Representatives, 64th Congress, First Session, on H. J. Res. 159, April 6 and 11, 1916* (Washington, D.C.: Government Printing Office, 1918), p. 129.

24 See Roy Lubove, *The Struggle for Social Security, 1900–1935* (Cambridge, Mass.: Harvard Univ. Press, 1968), pp. 66–90.

25 On the California debate, see Arthur J. Viseltear, "Compulsory health insurance in California, 1915–18," *J. Hist. Med.*, 1969, *24*: 151–182.

26 "A Symposium on compulsory health insurance,"
    p. 445. John J. A. O'Reilly made the statement.
27 Minutes of the House of Delegates, *J.A.M.A.*, 1920,
    74: 1319.
28 John Gordon Freymann, "Leadership in American
    medicine: a matter of personal responsibility," *New
    Eng. J. Med.*, 1964, *270*: 710–715.
29 Elton Rayack, *Professional Power and American Medi-
    cine: The Economics of the American Medical Association*
    (Cleveland: World Publishing Co., 1967), pp. 143–
    146. For a similar view, see Carleton B. Chapman
    and John M. Talmadge, "The evolution of the right
    to health concept in the United States," *Pharos*, 1971,
    34: 39.
30 Frederick R. Green, "The social responsibilities of
    modern medicine," *Tr. Med. Soc. St. N.C.*, 1921, pp.
    401–403.
31 Henry Lyle Winter, "Social insurance," *N.Y. St. J.
    Med.*, 1920, *20*: 20.
32 On the early history of workmen's compensation
    in America, see Harry Weiss, "The development
    of workmen's compensation legislation in the United
    States" (Ph.D. dissertation, Univ. of Wisconsin,
    1933); and Lubove, *The Struggle for Social Security*,
    pp. 45–65.
33 Bureau of Medical Economics, *An Introduction to
    Medical Economics* (Chicago: A.M.A., 1935), p. 80.
34 Committee on Social Insurance, *Workmen's Compen-
    sation Laws* (Social Insurance Series Pamphlet No. 1;
    Chicago: A.M.A., [1915]), p. 60.
35 Report of the Committee on Medical Economics,
    N.Y. St. J. Med., 1925, *25*: 789.
36 Sylvia A. Law, *Blue Cross: What Went Wrong?*, 2nd
    ed. (New Haven: Yale Univ. Press, 1976), p. 6. Ac-
    cording to *J.A.M.A.*, the percentage of occupied
    beds in nongovernmental hospitals declined from
    64.6 percent to 63.2 percent between 1929 and 1930.
    "Hospital service in the United States," *J.A.M.A.*,
    1933, *100*: 892.
37 J. T. Richardson, *The Origin and Development of Group
    Hospitalization in the United States, 1890–1940* (Univ.
    of Missouri Studies, Vol. XX, No. 3; Columbia:
    Univ. of Missouri, 1945), p. 12.
38 "Hospital service in the United States," pp. 892–894.
39 Maurice Leven, *The Incomes of Physicians: An Eco-
    nomic and Statistical Analysis* (Committee on the Costs
    of Medical Care, Publication No. 24; Chicago:
    Univ. of Chicago Press, 1932), pp. 76–81. The frac-
    tion of California doctors earning less than $6,000
    a year rose from approximately one-half in 1929
    to three-fourths in 1933. Arthur J. Viseltear, "Com-
    pulsory health insurance in California, 1934–1935,"
    Am. J. Public Hlth., 1971, *61*: 2117.
40 J. F. Kimball, "Group hospitalization," *Tr. Am. Hosp.
    Assn.*, 1931, *33*: 667–668; J. F. Kimball, "Prepay-
    ment plan of hospital care,"*American Hospital Asso-
    ciation Bulletin*, 1934, *8*: 42–47; Odin W. Anderson,
    Blue Cross Since 1929: Accountability and the Public Trust
    (Cambridge, Mass.: Ballinger Publishing Co.,
    1975), pp. 18–19.
41 Louis S. Reed, *Blue Cross and Medical Service Plans*
    (Washington, D.C.: Government Printing Office,
    1947), pp. 10–12.
42 Anderson, *Blue Cross Since 1929*, pp. 29–44; Law,
    Blue Cross, pp. 6–8. Anderson quotes one pioneer,
    J. Douglas Colman, as saying that "All this notion
    that it was going to solve the financial problems of
    hospitals was farthest from their [the Blue Cross
    founders'] minds." But Colman himself became in-
    volved with prepayment plans because hospitals
    might have to close without them. "An interview
    with J. Douglas Colman," *Hospitals*, 1965, *39*: 45–46.
43 Michael M. Davis, "Effects of health insurance on
    hospitals abroad," *Tr. Am. Hosp. Assn.*, 1931, *33*: 585.
    At the same meeting where Davis read this paper,
    the president of the A.H.A. called for insurance as
    a partial answer to the problem of decreasing oc-
    cupancy rates; *ibid.*, pp. 195–197.
44 Reed, *Blue Cross and Medical Service Plans*, pp. 13–14.
45 Williams, *The Purchase of Medical Care*, p. 219.
46 Letter from E. E. Harris, Dec. 10, 1930, quoted
    ibid., p. 238.
47 Kimball, "Prepayment plan of hospital care," p. 45.
48 *1846–1958 Digest of Official Actions*, p. 313.
49 R. G. Leland. "Prepayment plans for hospital care,"
    J.A.M.A., 1933, *100*: 871. For similar expressions,
    see the address of President-Elect Dean Lewis,
    Minutes of the 84th annual session, June 12–16,
    1933, *ibid.*, p. 2021; and the editorial "Hospital in-
    surance and medical care," *ibid.*, p. 973.
50 *1846–1958 Digest of Official Actions*, p. 313.
51 *Medical Care for the American People: The Final Report
    of the Committee on the Costs of Medical Care* (Commit-
    tee on the Costs of Medical Care, Publication No.
    28; Chicago: Univ. of Chicago Press, 1932), pp.
    v–viii, 120.
52 *Ibid.*, pp. 164–65, 171–72. The committee's study
    of the incomes of physicians revealed that "the aver-
    age volume of free work furnished by physicians
    throughout the country is only 5 percent of the
    total." Leven, *The Incomes of Physicians*, p. 66.
53 Oliver Garceau, *The Political Life of the American
    Medical Association* (Hamden, Conn.: Archon Books,
    1961), p. 138.
54 Minutes of the 84th annual session, June 12–16,
    1933, p. 48.
55 "The Report of the Committee on the Costs of
    Medical Care," *J.A.M.A.*, 1932, *99*: 2035.
56 *Ibid.*, p. 1952.
57 The fullest account of this second debate over com-

pulsory health insurance is Daniel S. Hirshfield, *The Lost Reform: The Campaign for Compulsory Health Insurance in the United States from 1932 to 1943* (Cambridge, Mass.: Harvard Univ. Press, 1970). But see also Roy Lubove, "The New Deal and national health," *Current History*, August, 1963, *45*: 77–86, 117; Edwin E. Witte, *The Development of the Social Security Act* (Madison: Univ. of Wisconsin Press, 1962); Arthur J. Altmeyer, *The Formative Years of Social Security* (Madison: Univ. of Wisconsin Press, 1966); and James G. Burrow, *AMA: Voice of American Medicine* (Baltimore: Johns Hopkins Press, 1963), pp. 185–252.

58 Minutes of the 88th annual session, June 7–11, 1937, *J.A.M.A.*, 1937, *108*: 2219.

59 Minutes of the special session, Sept. 16–17, 1938, *ibid.*, 1938, *111*: 1193.

60 Minutes of the 85th annual session, June 11–15, 1934, *ibid.*, 1934, *102*: 2199–2201.

61 Minutes of the special session, Feb. 15–16, 1935, *ibid.*, 1935, *104*: 751.

62 Minutes of the special session, Sept. 16–17, 1938, *ibid.*, 1938, *111*: 1216; *1846–1958 Digest of Official Actions*, pp. 321–322. At this session black physicians representing the National Medical Association pledged to join the struggle against compulsory health insurance, even though it might not be in the best interest of their race. *J.A.M.A.*, 1938, *111*: 1211–1212.

63 Nathan Sinai, Odin W. Anderson, and Melvin L. Dollar, *Health Insurance in the United States* (New York: Commonwealth Fund, 1946), pp. 64–65.

64 Minutes of the 93rd annual session, June 8–12, 1942, *J.A.M.A.*, 1942, *119*: 728.

65 Reed, *Blue Cross and Medical Service Plans*, pp. 136–146.

66 Viseltear, "Compulsory health insurance in California, 1934–1935," pp. 2115–2126; Arthur J. Viseltear, "The California Medical-Economic Survey: Paul A. Dodd versus the California Medical Association," *Bull. Hist. Med.*, 1970, *44*: 151. Although the second debate over compulsory health insurance took place primarily on the national level, many compulsory health insurance bills were also introduced in state legislatures. See Carl W. Strow and Gerhard Hirschfeld, "Health insurance," *J.A.M.A.*, 1945, *128*: 871.

67 Anderson, *Blue Cross Since 1929*, p. 54.

68 R. L. Novy, "In retrospect: changing attitude of the medical profession," *J. Mich. St. Med. Soc.*, 1950, *49*: 708. See also George Farrell, "Development of voluntary nonprofit medical care insurance plans," *N.Y. St. J. Med.*, 1957, *57*: 560–564.

69 J. G. Crownhart, "The economic status of medicine," *Wis. Med. J.*, 1934, *33*: 230. I wish to thank

Jennifer Latham for her assistance in locating this and other documents relating to health insurance in Wisconsin.

70 E. Minihan and T. Levi develop this point in their unpublished paper, "The political economy of health care financing: the foundation for medical care in Wisconsin" (April 1975), pp. 22–23.

71 George H. Gallup, *The Gallup Poll: Public Opinion, 1935–1971* (New York: Random House, 1972), I, 107.

72 James C. Sargent, "Shall medicine be socialized?" *Wis. Med. J.*, 1933, *32*: 562. See also Donald K. Freedman and Elinor B. Harvey, "Development of voluntary health insurance in the United States," *N.Y. St. J. Med.*, 1940, *40*: 1704.

73 Reed, *Blue Cross and Medical Service Plans*, p. 230.

74 "Wisconsin Cooperative Association assails State Medical Society," *Wis. Med. J.*, 1946, *45*: 3.

75 *The United States of America, Appellants, vs. The American Medical Association . . . Appellees* (Chicago: A.M.A., 1941).

76 Minutes of the 90th annual session, May 15–19, 1939, *J.A.M.A.*, 1939, *112*: 2295–2296. See also the comments of President-Elect J. H. J. Upham, Minutes of the 88th annual session, June 7–11, 1937, *ibid.*, 1937, *108*: 2132.

77 James C. McCann, "Medical service plans," *ibid.*, 1942, *120*: 1318.

78 President's Commission on the Health Needs of the Nation, *Building America's Health* (Washington, D.C.: Government Printing Office, [1952]), I, 43; II, 257.

79 Raymond Munts, *Bargaining for Health: Labor Unions, Health Insurance, and Medical Care* (Madison: Univ. of Wisconsin Press, 1967), pp. 10–12, 49, 250. See also Frank G. Dickinson, "The trend toward labor health and welfare programs," *J.A.M.A.*, 1947, *133*: 1285–1286.

80 President's Commission on the Health Needs of the Nation, *Building America's Health*, I, 43; II, 253–254; Reed, *Blue Cross and Medical Service Plans*, pp. 28, 119; Sinai, Anderson, and Dollar, *Health Insurance in the United States*, pp. 57–58, 73; Odin W. Anderson and Jacob J. Feldman, *Family Medical Costs and Voluntary Health Insurance: A Nationwide Survey* (New York: McGraw-Hill Book Co., 1956), pp. 14–20.

81 Altmeyer, *The Formative Years of Social Security*, p. 146; Peter A. Corning, *The Evolution of Medicare: From Idea to Law* (Washington, D.C.: Government Printing Office, 1969), pp. 53–55.

82 Monte Mac Poen, "The Truman administration and national health insurance" (Ph.D. dissertation, Univ. of Missouri, 1967), pp. 54–63.

83 "The President's national health program and the

new Wagner bill," *J.A.M.A.*, 1945, *129*: 950–953. See also "Senator Wagner's comments," *ibid.*, 1945, *128*: 667–668.

84   Burrow, *AMA*, p. 347.

85   Morris Fishbein, "The public relations of American medicine," *J.A.M.A.*, 1946, *130*: 511.

86   Gallup, *The Gallup Poll*, I, 578. See also *ibid.*, II, 801–804, 862–863, 886.

87   *1846–1958 Digest of Official Actions*, p. 331; Minutes of the interim session, Nov. 30–Dec. 1, 1948, *J.A.M.A.*, 1948, *138*: 1241; "A call to action against nationalization of medicine," *ibid.*, pp. 1098–1099; "Reply by officers and trustees," *ibid.*, 1949, *139*: 532. A.M.A. officers later referred to this action as "American Medicine's Declaration of Independence"; Report of Co-ordinating Committee, *ibid.*, 1951, *147*: 1692.

88   Burrow, *AMA*, pp. 361–364.

89   R. Cragin Lewis, "New power at the polls," *Medical Economics*, January 1951, *28*: 76.

90   John W. Cline, "The president's page: a special message," *J.A.M.A.*, 1952, *148*: 208.

91   "A call to action against nationalization of medicine," *ibid.*, 1948, *138*: 1098. This comment was made in response to the Federal Security Administrator's statement that millions of Americans could not afford proper medical care. See Oscar R. Ewing, *The Nation's Health: A Report to the President* (September, 1948).

92   Odin W. Anderson, *The Uneasy Equilibrium: Private and Public Financing of Health Services in the United States, 1875–1965* (New Haven: College & Univ. Press, 1968), p. 140.

93   Cline, "The president's page: a special message," p. 1036.

94   Louis H. Bauer, "The president's page," *J.A.M.A.*, 1952, *150*: 1675.

95   Ronald Andersen and Odin W. Anderson, *A Decade of Health Services: Social Survey Trends in Use and Expenditure* (Chicago: Univ. of Chicago Press, 1967), pp. 75, 109, 153. See also Ethel Shanas, *The Health of Older People: A Social Survey* (Cambridge, Mass.: Harvard Univ. Press, 1962).

96   On the events leading up to Medicare, see Max J. Skidmore, *Medicine and the American Rhetoric of Reconciliation* (University: Univ. of Alabama Press, 1970), pp. 75–95.

97   J. H. Houghton, "President's message to the House of Delegates," *Wis. Med. J.*, 1965, *64*: 208.

98   "New drive for compulsory health insurance," *J.A.M.A.*, 1960, *172*: 344–345. See also Edward R. Annis, "House of Delegates report," *ibid.*, 1963, *185*: 202.

99   Donovan F. Ward, "Are 200,000 doctors wrong?" *ibid.*, 1965, *191*: 661–663; *The Case against the King-*

Anderson Bill (*H.R. 3820*) (Chicago: A.M.A., 1963), pp. 17, 118–119.

100 Nancy L. Worthington, "National health expenditures, 1929–74," *Social Security Bull.*, February 1975, *38*: 13–14.

101 Estimates of the number of uninsured in the early 1970s varied between 17 and 41 million; see Marjorie Smith Mueller, "Private health insurance in 1973: a review of coverage, enrollment, and financial experience," *ibid.*, p. 21. The likelihood of having health insurance corresponded directly with income. Over 90 percent of families earning above $10,000 in 1970 carried hospital insurance, for example, while less than 40 percent of families with incomes under $3,000 had it. Cambridge Research Institute, *Trends Affecting the U.S. Health Care System* (DHEW Publication No. HRA 76-14503; Washington, D.C.: Government Printing Office, 1976), p. 188.

102 President's Commission on the Health Needs of the Nation, *Building America's Health*, IV, 43.

103 Reed, *Blue Cross and Medical Service Plans*, p. 230.

104 President's Commission on the Health Needs of the Nation, *Building America's Health*, V, 390–391.

105 John Krizay and Andrew Wilson, *The Patient as Consumer: Health Care Financing in the United States* (Lexington, Mass.: Lexington Books, 1974), p. 111. During the 1960s physicians' incomes increased faster than those of other professionals, including chief accountants, attorneys, chemists, and engineers. *Ibid.*, p. 109.

106 *An Opinion Study of Prepaid Medical Care Coverage in Michigan* (Michigan State Med. Soc., 1957), p. 140.

107 Robert Stevens and Rosemary Stevens, *Welfare Medicine in America: A Case Study of Medicaid* (New York: The Free Press, 1974), p. 191. "One unforeseen result of Medicare and Medicaid," say the Stevenses, "was that in formalizing the system of doctors' charges by developing profiles of the 'usual and customary' fees prevailing in each area, some physicians became aware of what others were charging. Quite clearly, there was some 'standardizing-up'. . . ." *Ibid.*, p. 194.

108 Theodore R. Marmor, *The Politics of Medicare* (London: Routledge & Kegan Paul, 1970), p. 89.

109 *Medicare and Medicaid: Problems, Issues, and Alternatives*, Report of the Staff to the Committee on Finance, U.S. Senate (Washington, D.C.: Government Printing Office, 1970), pp. 9–10, 13.

110 Quoted in Stevens and Stevens, *Welfare Medicine in America*, p. 197.

111 *Ibid.*, p. 202.

112 Victor R. Fuchs, *Who Shall Live? Health, Economics, and Social Change* (New York: Basic Books, 1974), pp. 94–95.

113 U.S. Bureau of the Census, *Historical Statistics of the United States: Colonial Times to 1970* (Washington, D.C.: Government Printing Office, 1975), Part 1, p. 55. The great gains came before the 1960s; between 1961 and 1970 life expectancy only increased from 70.2 to 70.9, and actually decreased slightly for black males.

114 *National Health Insurance Proposals: Hearings before the Committee on Ways and Means, House of Representatives, Ninety-Second Congress, First Session on the Subject of National Health Insurance Proposals, Oct–Nov., 1971* (Washington, D.C.: Government Printing Office, 1972), p. 1950.

115 Monroe Lerner and Odin W. Anderson, *Health Progress in the United States, 1900–1960* (Chicago: Univ. of Chicago Press, 1963), p. 16.

18

American Medicine's Golden Age: What Happened to It?

JOHN C. BURNHAM

During the first half of the 20th century, up until the late 1950s, American physicians enjoyed social esteem and prestige along with an admiration for their work that was unprecedented in any age. Medicine was the model profession, and public opinion polls from the 1930s to the 1950s consistently confirmed that physicians were among the most highly admired individuals, comparable to or better than Supreme Court justices.[1] Highbrow and mass media commentators alike associated medical practice with the "miracles" of science and made few adverse comments on the profession.[2] By the 1970s, however, statesmen of medicine were writing unhappily about being "deprofessionalized" in the wake of attacks by articulate and knowledgeable critics, attacks that by 1981 were reflected specifically in substantial mistrust of the profession among the public at large.[3] One can conduct a historical postmortem of this unexpected turn of events by examining changes in direct public depreciations of the medical profession, using the different kinds and levels of criticism of M.D.'s as indicators of what happened.

The attitudes of leaders and shapers of opinion and of the public toward physicians did not translate directly into the behavior of patients. For economic and social reasons, amounts of money spent by Americans on medicine continued to increase dramatically even when attitudes changed. But, as was revealed both by polls and by a resurgence of alternatives to conventional medical practice, over time the critics not only affected doctors' sensibilities but also demonstrably damaged the social credi-

bility of the profession as a whole.[4] Since public acceptance is necessary for a profession to function, the criticism had tangible effects.

A long and honorable tradition of denigrating doctors was known to Aristophanes and Molière and continued to flourish in 19th-century America.[5] As late as 1908 a set of satirical "Medical Maxims" in this tradition included, for example:

> Diagnose for the rich neurasthenia, brainstorm, gout and appendicitis; for the poor insanity, delirium tremens, rheumatism and gall-stones . . . fatten the thin, thin the fat; stimulate the depressed, depress the stimulated; cure the sick, sicken the cured; but above all, keep them alive or you won't get your money.[6]

But in those same early years of the 20th century, the tradition of doctor baiting tended to die out as the golden age of medicine dawned. Whereas the post-1950s resurgence of criticism that culminated in Ivan Illich's *Medical Nemesis*[7] recalled traditional themes such as physician greed, pretension, and imposition, the later critics were also responding to new and untraditional characteristics of both medical practice and American society.[8] Moreover, the few particular criticisms that survived in the golden age helped shape and define the new deluge.

EVOLUTION OF THE MEDICAL IMAGE

During the 19th century, physicians seeking to professionalize their calling were fair game for hostile comment, with quacks and sectarians on one side and the practitioners' actual therapeutic impotence on the other. Some aristocrats of medicine and the medical ideal they represented did enjoy high prestige, but most (often deservedly) did not. Occa-

JOHN C. BURNHAM is Professor of History and Lecturer in Psychiatry at Ohio State University, Columbus, Ohio.

Reprinted with permission from *Science*, March 19, 1982, *215*: 1474–1479. Copyright © 1982 American Association for the Advancement of Science.

sionally, antimedical diatribes based on these earlier struggles persisted after the 1890s, along with other anachronisms like attacks on the germ theory of disease. But by and large, in the wake of medical, and particularly surgical, successes, publicity about the profession was favorable, and leaders of the American medical profession succeeded by the early 20th century in their campaign to persuade the public to want and expect uniformly well-trained, well-paid physicians who themselves set standards of practice.[9]

So effective was favorable publicity about both science and doctors that Americans in general began to view extensive medical care as a life necessity. Expansion of hospital care at the beginning of the century was an important indication of the change.

After some years, publications of the Committee on the Costs of Medical Care (1928–1933) and other surveys generated much criticism of the medical profession—not for members' inferior technical performance or misbehavior but for their failing to make physician services of any kind available to more people through economic and organizational means.[10] By the late 1930s the modern campaign for "socialized medicine" or compulsory health insurance had begun, and for many decades organized medical groups opposed any change in the structuring and financing of health care delivery.[11] All parties to the controversy, however, continued to agree that medical care was highly desirable.

While many public figures attacked the American Medical Association (A.M.A.) and state and local medical groups for their political activities, the public image of scientific medicine improved constantly.[12] By the 1940s virtually everyone had heard of miracle drugs and many people knew that they owed their lives to them. As writer Evelyn Barkins observed in 1952, "Most patients are as completely under the supposedly scientific yoke of modern medicine as any primitive savage is under the superstitious serfdom of the tribal witch doctor."[13]

Ultimately, however, the socialized medicine debates undermined public confidence in medicine as a profession. The heavily financed publicity campaigns undertaken in the name of the A.M.A. generated political statements that few people could take seriously and raised questions about the claims of members of the profession acting in scientific and clinical roles.[14] Even before World War II the evident social insensitivity of physician groups such as the A.M.A. tended to tarnish the doctor as a public figure, and many people began to associate the physician with another familiar stereotype, the small businessman, who was presumably not only grasping but slightly dishonest.[15] As one writer of the early 1940s observed of organized medicine, its "social outlook turns out to be . . . scarcely distinguishable from that of a plumber's union."[16] Indeed, the actions of physician groups caused the Supreme Court in 1943 officially to refuse to recognize doctors' professional claims and instead to find physician groups, including the A.M.A., guilty of restraint of "trade."[17]

Beginning in the 1940s, a number of reformers within the medical profession worked to expose inferior medical practice and upgrade medicine to a level appropriate for the age of penicillin and high technology. Some of the self-criticism revealed through these efforts was repeated by the general press. The combination of internal criticism and external distrust eventually had a negative effect, just as the social environment for all professions turned from favorable to unfavorable.

THE END OF THE GOLDEN AGE

The rare public doubters of the medical profession in the late 1940s and early 1950s gradually increased in number. By 1954, Herrymon Maurer, writing in *Fortune*, could cite a series of sensational articles in mass media magazines attacking not only money-making but incompetence in medical practice. Maurer's article was entitled "The M.D.'s Are Off Their Pedestal."[18] A few more years had to pass, however, before the number of recriminations reached the threshold that marked the end of an era.

Despite the ineptitude of the campaigns against socialized medicine, the public image of the physician *per se* was very favorable in the proscientific post-World War II period. This image was reflected, for example, in the activities of Dr. Kildare (a stereotype later known as Marcus Welby, M.D.) who moved from the novel and motion picture to the television screen. Physicians showed up in over half of 800 Hollywood films surveyed in 1949 and 1950. But in only 25 instances was the doctor portrayed as a bad person, and when he was bad there were often extenuating circumstances. He was almost never a humorous character, either.[19]

Around 1950 many physician organizations across the country began systematic campaigns to reduce the number of legitimate complaints of the public against physicians. Leaders in the profession had concluded that actual experiences of everyday Americans with medical care were the source of much of the antipathy directed toward the profession. An early and exemplary effort was that of the California Medical Association, which conducted a double program. First, California M.D.'s made medical care available (but on their own terms) to answer complaints about access to it. Second, and more important, they carried out a campaign to protect the public by hearing complaints against four types of abuses: (i) malpractice; (ii) "unnecessary or incompetent procedures"; (iii) excessive fees; and (iv) unethical acts of physicians. All over the United States grievance committees of local medical societies tried to adjust physician-patient disputes and effect some of the professional self-policing so notoriously absent theretofore.[20]

Grievance committees were, in fact, but one facet of a major attempt of reformer physicians to get each practitioner to emphasize and upgrade his or her personal relationships with patients. The doctors set out to fight bad public relations as one did syphilis, one case at a time but with a cumulative effect.[21] California M.D.'s in 1951 employed the psychologist Ernest Dichter to suggest how each practitioner should manage his or her patients. Every encounter between a physician and patient is, of course, an intensely and unabashedly narcissistic experience for the patient and therefore eminently suitable for psychological manipulation. A patient's gripes about high fees, for example, may mask a real grievance related to some personal slight inflicted by the doctor. Psychological studies and systematic research on patients, analogous to consumer surveys, both gave specificity to concerns about the individual doctor-patient relationship and helped inspire and shape programs to improve such relationships.[22] As an osteopath concluded in 1955, the trust of every patient had to be gained in order to overcome the belief that medicine was emphasizing business and quantity rather than service or quality.[23] The popular press also soon reflected the medical campaigns, elements of which were familiar from earlier A.M.A. publicity favoring the old family doctor as opposed to the cold, impersonal specialist. By 1959 an article in *Life* was popularizing this idea, portraying physicians favorably but still strongly emphasizing how much they needed to add sympathy to their science.[24]

At the same time that physicians were working on their public relations in the 1950s, overt popular indictments were pushing the profession off the "pedestal." Exactly where and when the final shove came is not certain. In the third quarter of the 20th century there were no fewer than 20 investigations of the New York City health system, and in 1966, after it was clear that the medical profession was in trouble, journalist Martin Gross[25] traced the new criticism to the first of these investigations in the cultural center of the country.[26] In 1965 an anonymous writer in *Consumer Reports*[27] dated modern criticism from the publication of a study conducted in 1956 in which investigators actually rated physician performance. Perhaps the most important date was 1958, when Richard Carter's *The Doctor Business*,[28] the first of a number of muckraking books, appeared. Carter's exposé and others that followed it drew heavily on both public investigations and exposés that members of the profession had written for internal professional purposes. Whatever the source, clearly adverse criticism had entered a novel phase by the end of the 1950s, reflecting and also creating new social circumstances within which physicians practiced.

Indignant lay writers and reformer M.D.'s shared an elevated opinion about what physicians ought to be. They were, wrote a journalist in 1954, supposed to be part of a double picture: "on the one hand, a group of dedicated and white-coated scientists, bending over test tubes and producing marvelous cures for various ailments, and, on the other, equally dedicated practitioners of medicine and surgery, devoting themselves to easing pain and prolonging human life, without thought of personal gain and at considerable self-sacrifice." Both the public and the profession, he noted, were beginning to notice substantial deviations from this widely held ideal and to become filled with "disillusionment . . . tinged with a bitterness which breeds public hostility."[29] Other observers traced the rising level of adverse comment to unrealistic hopes. As the 1950s ended, columnist Dorothy Thompson summarized for readers of the *Ladies' Home Journal* this growing public criticism of American physicians. There was bad hospital care, there were bad doctors, and there were excessive medical costs. But she went on to note the cause:

> In a rather profound sense the current attacks on the medical profession compliment it. People, it seems, *expect* more of physicians than they do of other professional men with the possible exception of the clergy. The medical profession has invited that expectation, and in the opinion of this writer, and with exceptions that only prove the rule, has deserved it.[30]

In later decades, as Americans came to expect the medical profession to furnish comfort, happiness, and well-behaved children as well as health, the disillusionment grew.

ADAPTING TO CHANGE

Since ancient times, critics—and the public at large —have usually discriminated sharply between their own personal physicians, who command professional trust, and the medical profession as a whole, which does not and which is susceptible to harsh judgments.[31] In the mid-20th century, however, doubts about medicine in general or "the doctor" intensified so much that even personal professional trust was often impaired, especially when a patient could not get the attention that he or she wanted. Critics at all levels who started by blaming the system, particularly the clinic and hospital, inadvertently raised questions about the M.D.'s who collaborated in the faulty operation of the institutions.

As professionals, physicians always functioned in part on the social level. When, in the 20th century, major changes occurred in the immediate social context within which medicine operated, the profession did not adapt quickly in either the formalities of practice or the self image it produced. One of the major new forces was the startling increase of chronic (as opposed to acute) diseases as the dominant concern in practice. A second new force was the growth of huge bureaucratic institutions, particularly hospitals, in the regular health care system. A third force was the greatly increased sophistication of consumers. And a fourth was the rise of psychological explanations for illness, leaving the physician dealing with the uncertainties of psychosomatics. All of these changes were well under way before the 1950s, and each helps to explain what happened to the golden age of medicine.

Critics and reformers outside the profession were also slow to respond to the changed situation. Carter's *The Doctor Business*, for instance, was targeted chiefly on the fee-for-service organization of medicine, and at most only a quarter of the volume was devoted to actual faults in health care. Even in 1960 in perhaps the most crucial of the new critical publications, *The Crisis in American Medicine*, the authors still tended to emphasize the economics of medicine even while recognizing that "millions of people are bitterly dissatisfied with the medical care they are getting."[32]

What eventually transformed the criticism was the addition of another ingredient from society as a whole: widespread anti-institutional sentiment along with a general disillusionment with many aspects of American life.[33] Among the target institutions were the professions, particularly professions based on expertise. In the mid-1950s writers in the highbrow and mass media began to paint negative or at least ambivalent images of many American institutions that in the 1940s had been beyond reproach: the city, the automobile, the large family—and the doctor. In making their unfavorable remarks about doctors, various kinds of public commentators drew from both past and then current concerns to focus on three aspects of the physician's function: the priestly, or sacerdotal, role; the technical role; and the role of the physician as a member of the health care system.

THE SACERDOTAL ROLE

In the first half of the 20th century, when medical intervention was becoming increasingly effective, such critics as there were tended to concentrate not on the technical role of physicians but on their priestly functioning as they went through medical ceremonies and acted as wise and trusted personages. In this preoccupation, commentators reflected basic popular attitudes. In novels, for example, despite the shift of physician characters from priestly and scholarly roles to scientific, their most important duties still centered on nonphysical problems and relationships.[34] Regardless of the passing of the old-fashioned family doctor, there was a well-understood public demand for a sympathetic personal relationship such as that furnished by the idealized country practitioner. "His successors have much to learn from him," observed an editorial writer in a typical comment as early as 1908. "At all events they must learn to be men, not merely scientists."[35] And even as the socialized medicine debate heated up, the impersonal system rather than individual M.D. performance was the subject of adverse comment.

In all of the criticism during the golden age, the emphasis on priestly personal functions of the physician, as opposed to effectiveness or even competence, is striking. As late as the 1950s, lists of common criticisms to which physicians were sensitive included most prominently "A failure to take a personal interest in the patient and his family," "Inability to get a doctor in cases of emergency," "Waiting time in doctors' offices," and other such items reflecting the continuing demand for personal attention.[36] The only other conspicuous categories of complaint had to do with fees and failure to communicate with the patient. Only in later decades did the demand for competence become very conspicuous.[37]

It is against this background of emphasis on the sacerdotal function of medical personnel that the great constant of criticism, greed, has to be viewed. Greed on the part of a physician violated a sacerdotal stereotype because most Americans expected that under ideal circumstances a physician was a dedicated professional who provided a service because the service was needed, not because it was profitable.[38] Greed showed up earlier as a concern in attacks on quackery, fee-splitting, and then, to a small extent, physician financial interest in laboratory and drug store enterprises.[39] But it was only after physicians had in general substantially increased their incomes that critics fastened on the evident wealth rather than specific fees of M.D.'s as evidence of unseemly grasping. This recent phase had to wait for the development of what David Horrobin has called "the politics of envy" in the late 20th century.[40]

That physician greed was a constant in criticism meant that even in the recent period, when technical as well as priestly performance in medicine was again subject to question, the motive that critics identified in errant physicians was avariciousness. Why else would a rational M.D. commit undesirable acts and reduce the quality of the medical care that he was delivering? And in the continuing socialized medicine controversy, when the physician as entrepreneur was an issue, greed was, again, imputed to medical advocates of *laissez-faire*.[41]

One area in which the public could and did react to physicians in their nontechnical roles was indifference to patients, epitomized in the contrast between house calls and clinic or hospital practice. Personal attention was the theme of the solo practice advocates both inside and outside the profession. It was the chief complaint of detractors of specialization, before and after the late 1950s. It was the object of the local grievance committees set up after World War II. And it was the subject of studies after mid-century by members of a new subspecialty, medical sociologists.

In an era of high technology, when the secrets of medicine became increasingly inaccessible and incomprehensible to the public, responsiveness to the patient remained the one aspect of practice by which most people could judge the M.D. By the 1960s, case histories of patient mistreatment on a social, not technical, level were standard in the growing literature of criticism. But the critics who wanted attention and care from the physician still did not usually specify what the care consisted of until well into the age of malpractice suits.[42]

THE TECHNICAL ROLE

Although the technical performance of the physician called forth little adverse comment before the 1950s, both the application of medical science and the individual competence of the M.D. in applying it had earlier been traditional and continuing subjects of recrimination. Kept alive for a time in the campaign against obviously incompetent nonphysician quacks, the theme of pretension and ineffective treatment continued to be an issue in occasional attacks on unnecessary surgery. Remarkable, however, was the fact that one type of criticism, that directed toward the laziness, negligence, and incompetence of M.D.'s, remained largely undeveloped for over half a century. There were a few stories about outright malpractice, and there were suggestions (usually made by M.D.'s trying to upgrade the profession) that many physicians were not keeping up with scientific literature.[43] But no rash of damaging exposés appeared until after the 1950s.

One dark side of the physician as technologist was the fear that practitioners would impose too much medicine, not only forcing inoculations and surgery on unwilling persons but, indeed, using patients for experimental purposes. In the 1920s,

Sinclair Lewis's *Arrowsmith* helped keep this traditional fear alive, but the physician as scientist who imposed on patients in the name of technique remained largely a literary figure. For decades, serious critics restricted themselves to the impersonality of the specialist, not his mania for medical intervention and innovation. Lay commentators, in fact, tended to write about fads in medicine in terms of progress and to ignore the discarded fashions. Publicists who did discuss faddism did so gently, like the 1928 humorist in *Collier's* who commented,

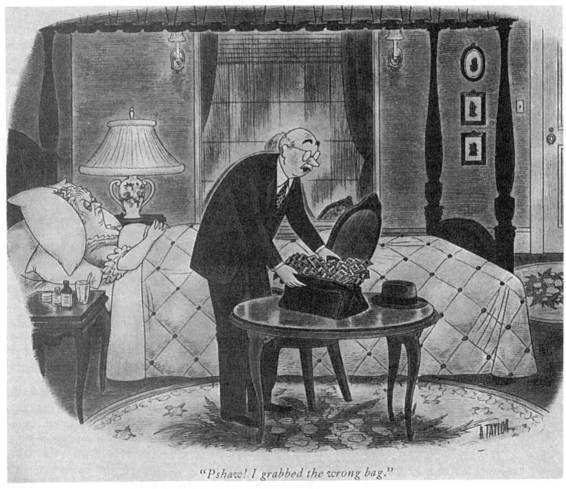

"Pshaw! I grabbed the wrong bag."

Drawing by R. Taylor; © 1959 The New Yorker Magazine, Inc.

An' now it's the gall bladder. Doctors are *mad* over it. The appendix, tonsils, teeth, auto-intoxication, acidosis—all are forgotten; an' the gall bladder is now the undisputed belle of the body. For a medical man it has all the lure an' emotional appeal of a Swinburne poem, a Ziegfeld chorus or a moonlight party in Hollywood.[44]

By the 1960s and 1970s critics were saying that, as one of them put it, medical faddism reflected "the underlying bias of the technological mindset and its activity orientation . . . that newer must be better and that doing more must be better than doing less; hence the possibility of harm is always a second thought. . . ."[45] By this time, then, deliberate risk had been added to lack of knowledge and skill. Moreover, the public ultimately developed a very high level of distrust of what critics had been characterizing as excessive use of drugs and surgery.[46]

THE SOCIAL ROLE

Beyond the priestly and technical requirements of medical practice, one of the well-understood demands society makes of any professionals in granting them special status has been that their activities be harmless to society (this is one reason that advertising, for example, cannot qualify as a profession). The traditional issue of whether the monopoly granted physicians was or was not antisocial became a crucial one in the 20th century. The reorganizers of American medicine at the turn of the century took pains to show that the newly licensed monopoly, "the medical trust," as early critics characterized it, that outlawed quacks and sectarians and vested licensure in the profession, *was* in the public interest.[47]

Medical leaders succeeded in winning the public's trust and approval.[48] Not even the failure of the self-policing that was a direct (though not essential) concomitant of the monopoly elicited much comment before the 1960s. Only insofar as physicians as a group failed to take positive action to provide medical care for all who wanted it, or as medical groups opposed institutional arrangements designed to improve and extend medical care, did criticism fall on the monopoly. Then, attribution of greed to physicians was one aspect of the accusation, but so also was conservatism, which was

a characteristic of other monopolies that consistently drew criticism in modern America. It was not until the 1960s and 1970s that new, well-educated groups tried to break the monopoly by developing new kinds of "health care deliverers" and by introducing lay control. Such developments grew out of distrust of the intentions and customs of the medical profession.

Attention to the social aspects of medicine was the qualitative characteristic that most clearly differentiated detractors of medicine before and after the 1950s. More recent critics not only decried the monopoly and maldistribution of medical care but also loaded physicians with responsibility for any number of social transgressions: exploiting menials, failing to provide incentives for improving health care delivery, encouraging unnecessary bureaucracies, increasingly setting arbitrary boundaries to illness, ignoring "positive" health, and in general, to use the term of the leading critic, Illich, "medicalizing" the whole society to the detriment of individual dignity and well being.[49]

THE EROSION OF PROFESSIONAL STATUS

Physicians have always been sensitive to criticism.[50] For half a century they were relatively free from public censure or actual interference in clinical and professional activities, and they enjoyed great public and personal admiration. Few people other than doctors knew about iatrogenic disease or the placebo effect. Criticism—and lack of it—reflected both the impression conveyed in public about the miracles of medicine and the persistence of the sacerdotal role of the physician, demanded by the public at all levels. But the physician as priest was already in some trouble by the 1930s. Attacks on impersonal specialism and on well-meaning social reformers' attempts to spread the technical benefits of medicine through prepayment (that is, insurance) and institutional reorganization laid a basis for doubts about the whole profession. Demand for a priest was still intense, as surveys even in the 1950s showed, but the profession in general was by then set in place to be the object of a more general social attack. This attack portended the end of generous funding for medical research and the end of such extremes of freedom of action as professionals might aspire to.[51]

Commentators with a sense of the tragic, or even just of the ironic, can find in the 20th-century physician ample justification for their views. As sociologist Eliot Freidson pointed out at the beginning of the 1960s, conflict between patient and physician was inevitable because the function of the physician was to apply general knowledge to a particular individual, the patient.[52] Applying knowledge involved trying to control the patient, and the patient in turn was interested in controlling his or her destiny.[53] In attempting to maximize the client-professional trust that would permit patients to yield control, physicians emphasized the validity of their science—and in so doing created a sophisticated public. That public in turn became increasingly competent to expose shortcomings of the profession and to react when physician reformers spoke out about their colleagues' failures.[54] "I wrote about . . . abuses and asked for changes," wrote District of Columbia internist Michael J. Halberstam in the mid-1970s. "And now changes are coming, but alas . . . they will probably be the wrong ones."[55]

One of the major results of the new criticism of the 1960s and 1970s, in which the technical as well as the sacerdotal function of the physician came into question, was therefore a series of demands for greater patient participation in the medical relationship, demands exacerbated by a resurgence of romantic individualism in the culture as a whole.[56] By 1972 one analyst[57] could add to the "engineering" and "priestly" models of health care and delivery two more, the "collegial" and the "contractual." Both of these last models involved patient participation and were flourishing in various settings.[58]

Insofar as the entire society was moving toward social leveling, the high status necessary for professional authority was being eroded throughout most of the century.[59] By the 1960s even the popular image of the physician as portrayed on television reflected a change from a charismatic figure, who used mysterious powers to resolve problems, to a new type of hero, one with only ordinary endowments and who potentially could behave unheroically.[60] But as early as the 1930s the sociologists who surveyed Muncie, Indiana, as "Middletown," had commented that physicians, and lawyers, too, were increasingly less visible as independent community leaders. Older physicians

continued to be aware of a change, but few could cite convincing detail as did J. A. Lundy of Worcester, Massachusetts, who in 1952 recalled the time when townspeople customarily tipped their hats to the physician.[61] Another perceived sign of erosion of the physician's place was the fact that patients felt increasingly free to shop around for an M.D. who suited them.[62] The loss was felt not by the technically oriented specialist whose bedside manner might be imperfect, but by the traditional family doctor. By the 1960s and 1970s physicians were complaining not only of lack of deference but of lay interference and assaults on professional privileges. The politics of envy were building in new ways upon traditions of criticism that had been muted in the first half of the 20th century but had not died.

CONCLUSION

The golden days of the medical profession can be defined by the amount and the content of criticism that the profession received—what little adverse comment there was, was often to the effect that highly desirable professional services were insufficiently available or that physicians had lapsed from their sacerdotal roles. In both cases the critics tended to fasten on the old theme of the doctor whose greed overcame his more professionally disinterested concern. The practice of medicine always involved M.D.'s in ambivalent relationships with both individual patients and society, and high-status professionals who could not or would not respond to patients' personal and selfish concerns of course generated complaints and could even become both personal and social scapegoats.[63] But it was the continuing politics of the socialized medicine debate that first planted the seeds of major and pervasive mistrust. When, after World War II, physicians themselves spoke out to increase the beneficent results of medicine and upgrade the profession in the direction of the professional ideal, they unwittingly opened the door for the latter-day critics who attacked not only priestly pretension but technical performance. The influence of these critics combined with other social forces in movements that in the 1960s and 1970s tended to impair the trust and freedom that had once marked medical practice.[64]

NOTES

A draft on this subject was originally prepared in connection with National Endowment for the Humanities grant FP-0013-79-54 (Seminar for the Professions); the writing was supported in part by a Special Research Assignment from The Ohio State University. For suggestions, thanks are due to members of the NEH seminar and to K. J. Andrien and J. R. Bartholomew.

1 G. H. Gallup, *The Gallup Poll: Public Opinion, 1935–1971* (New York: Random House, 1972), pp. 1152 and 1779–1780.

2 M. J. Halberstam, *Prism,* July–Aug. 1975: 15. Halberstam's is a casual observation: the literature contains no general surveys of the changing fortune of the physician in American organs of information and opinion in the last century, and not even the *Reader's Guide* has been systematically exploited for information about the place of physicians in American society. Sociologists' work on their medical contemporaries dates only from mid-century.

3 F. J. Ingelfinger, *New Eng. J. Med.,* 1976, *294*: 335; American Osteopathic Association, *A Survey of Public Attitudes Toward Medical Care and Medical Professionals* (Chicago: American Osteopathic Association, 1981).

4 American Osteopathic Association, *A Survey of Public Attitudes*; R. W. Hodge, P. M. Siegel, P. H. Rossi, *Am. J. Sociol.,* 1964, *70*: 286; A. Tyree and B. G. Smith, *Soc. Forces,* 1978, *56*: 881.

5 R. H. Shryock, *Ann. Med. Hist.,* 1930, n.s., *2*: 308.

6 Anonymous, *Life,* 1908, *52*: 196.

7 I. Illich, *Medical Nemesis: The Expropriation of Health* (New York: Random House, 1976).

8 The most delightful example is E. Berman, *The Solid Gold Stethoscope* (New York: Macmillan, 1976).

9 J. Duffy, *The Healers: The Rise of the Medical Establishment* (New York: McGraw-Hill, 1976); *J.A.M.A.,* 1967, *200*: 136; M. R. Kaufman, *Mt. Sinai J. Med. N.Y.,* 1976, *43*: 76; G. H. Brieger, *New Physician,* 1970, *19*: 845; B. Rosenkrantz, *Proc. Int. Cong. Hist. Sci.,* 1974, *14*: 113; J. S. Haller, *American Medicine in Transition, 1840–1910* (Urbana: Univ. of Illinois Press, 1981); J. G. Burrow, *Organized Medicine in the Progressive Era: The Move Toward Monopoly* (Baltimore: Johns Hopkins Univ. Press, 1977); *Med. Rev. Rev.,* 1917, *23*: 1; B. Sicherman, in *Nourishing the Humanistic in Medicine,* ed. W. R. Rogers and D. Barnard (Pittsburgh: Univ. of Pittsburgh Press, 1979), p. 95. The famous criticisms in the Flexner Report [A. Flexner, *Medical Education in the United States and Canada* (New York: Carnegie Foundation, 1910)] were aimed at diploma mills, not well-trained M.D.'s; the report spoke for, not against, the profession.

10 F. A. Walker, *Bull. Hist. Med.,* 1979, *53*: 489.

11 The group practice and socialized medicine controversy has been widely researched and is not covered in the present article. Standard sources include S. Kelley, Jr., *Professional Public Relations and Political Power* (Baltimore: Johns Hopkins Press, 1956), pp. 67–106; J. G. Burrow, *AMA: Voice of American Medicine* (Baltimore: Johns Hopkins Press, 1963); E. Rayack, *Professional Power and American Medicine: The Economics of the American Medical Association* (Cleveland: World, 1967); D. S. Hirshfield, *The Lost Reform: The Campaign for Compulsory Health Insurance in the United States from 1932 to 1943* (Cambridge, Mass.: Harvard Univ. Press, 1970); R. Harris, *A Sacred Trust* (New York: New American Library, 1966); R. Numbers, *Almost Persuaded: American Physicians and Compulsory Health Insurance, 1912–1920* (Baltimore: Johns Hopkins Univ. Press, 1978).

12 D. W. Blumhagen, *Ann. Intern. Med.* 1979, *91*: 111.

13 E. Barkins, *Are These Our Doctors?* (New York: Fell, 1952), pp. 171–172. See also a popular work, D. G. Cooley, *The Science Book of Wonder Drugs* (New York: Franklin Watts, 1954).

14 M. J. Gaughan (Thesis, Boston Univ., 1977). For a famous contemporary comment, see B. DeVoto, *Harper's Mag.,* 1951, *202*: 56.

15 A. M. Lee, *Psychiatry,* 1944, *7*: 371.

16 W. Kaempffert, *Am. Mercury,* 1943, *57*: 557.

17 American Medical Association v. United States, 317 *U.S. Reports* 519; P. S. Ward, unpublished paper read at American Association for the History of Medicine Meetings, Pittsburgh, May 1978.

18 H. Maurer, *Fortune,* Feb. 1954: 138.

19 R. R. Malmsheimer (Thesis, Univ. of Minnesota, 1978); "Doctors as Hollywood sees them," *Sci. Digest,* Oct. 1953: 60; J. Spears, *Films Rev.,* 1955, *6*: 437; E. H. Vincent, *Quart. Bull. Northwestern Univ. Med. Sch.,* 1950, *24*: 305.

20 J. Hunton, *GP,* 1951, *4*: 110; R. Carter, *The Doctor Business* (New York: Doubleday, 1958), pp. 235–238; G. B. Risse, in *Responsibility in Health Care,* ed. G. J. Agich (Dordrecht: Reidel, 1982).

21 R. W. Elwell, *Ohio State Med. J.,* 1950, *46*: 581.

22 Risse, *Responsibility in Health Care*; R. Waterson and W. Tibbits, *GP,* Oct. 1951, *4*: 93; M. Amrine, *Am. Psychol.,* 1958, *13*: 248. For a summary of the research, see S. Greenberg, *The Troubled Calling: Crisis in the Medical Establishment* (New York: Macmillan, 1965), Ch. 4.

23 *New York Times,* Oct. 16, 1955: 72.

24 *Life,* Oct. 12, 1959: 144; M. Austin, *Look,* Mar. 15, 1949: 34.

25 M. L. Gross, *The Doctors* (New York: Random House, 1966), p. 7.

26 R. R. Alford, *Health Care Politics: Ideological and Interest Group Barriers to Reform* (Chicago: Univ. of Chicago Press, 1975), p. 22. R. Bayer [*Homosexuality and American Psychiatry: The Politics of Diagnosis* (New York: Basic Books, 1981), p. 10] maintains that the public attack on psychiatry prefigured the attacks on medicine as an institution.

27 *Consumer Reports,* Mar. 1965: 146.

28 Previously cited in n. 20.

29 B. McKelway, *Med. Ann. District of Columbia,* 1954, *23*: 457.

30 D. Thompson, *Ladies' Home Journal,* Apr. 1959: 11. Later, television productions greatly intensified unrealistic expectations; G. Gerbner and others, *New Eng. J. Med.,* 1981, *305*: 901.

31 A striking modern survey, showing the generality of the phenomenon in all segments of the population, is Ben Gaffin & Associates, *What Americans Think of the Medical Profession . . . , Report on a Public Opinion Survey* (Chicago: American Medical Association, 1955).

32 M. K. Sanders, ed., *The Crisis in American Medicine* (New York: Harper, 1961), p. vii.

33 R. C. Maulitz, unpublished paper.

34 A. J. Cameron (Thesis, Univ. of Notre Dame, 1973), especially p. 157.

35 *New York Times,* Aug. 21, 1908: 6. An early sociological survey of patients (E. L. Koos, *Am. J. Public Hlth.,* 1955, *45*: 1551) showed that the young modern patients as well as the old who had, for example, actually seen house calls, responded negatively to impersonality in practice. There were probably also changes in the social expectations for the "sick" role; see, for example, E. Kendall, *Harper's Mag.,* 1959, *219*: 29.

36 J. T. T. Hundley, *Va. Med. Mon.,* 1952, *79*: 540; E. Stanton, *J. Maine Med. Assn.,* 1954, *45*: 56.

37 American Osteopathic Association, *A Survey of Public Attitudes.* By 1957 patients were almost evenly divided in wanting most kindly attention or technical skills and results (getting better); G. G. Reader, L. Pratt, M. C. Mudd, *Mod. Hosp.,* 1957, *89*: 88.

38 No attempt is made in this article to deal directly with the issue of professionalization and professional status; there is already a large special literature on the subject.

39 For a particularly good example of the restricted criticism, see an anonymous editorial, "Unprofessional conduct," *J. Med. Soc. N.J.,* 1929, *26*: 326. In the present discussion I treat the explicit content of the criticism and do not utilize the suggestion that complaints about fees were substituted for expressing other grievances.

40 D. F. Horrobin, *Medical Hubris: A Reply to Ivan Illich* (Montreal: Eden, 1977), p. 27. Ironically, high income was one of the factors that contributed to physicians' high prestige (Hodge, Siegel, and Rossi, "Occupational prestige in the United States, 1925–63;" Tyree and Smith, "Hierarchy in the United States: 1789–1969"). P. Starr (*Daedalus,* 1978, *107*: 175) suggests that this type of criticism did not appear conspicuously until after publicity about Medicaid abuses.

41 Stupidity was also an issue, being part of the argument that a reorganized, socialized physician would be better off economically.

42 For example, D. B. Smith and A. D. Kaluzny, *The White Labyrinth: Understanding the Organization of Health Care* (Berkeley: McCutchan, 1975).

43 R. S. Halle, in an article entitled, "Unfit doctors must go" (*Scribner's Mag.,* 1931, *90*: 514), dealt entirely with clear cases of malpractice; and H. M. Robinson (*Am. Mercury,* 1936, *38*: 321) blamed lawyers, patients, and unrealistic expectations for a rash of lawsuits.

44 *Collier's,* Aug. 4, 1928: 27.

45 L. Lander, *Defective Medicine: Risk, Anger, and the Malpractice Crisis* (New York: Farrar, Straus & Giroux, 1978), p. 41.

46 American Osteopathic Association, *A Survey of Public Attitudes.*

47 A. D. Bevan, *A.M.A. Bull.,* 1910, *5*: 243.

48 Burrow, *Organized Medicine in the Progressive Era.*

49 See n. 7 above.

50 O. W. Anderson, *Mich. Med.,* 1968, *67*: 455.

51 P. B. Hutt, *Daedalus,* 1978, *107*: 157.

52 E. Freidson, *Patients' Views of Medical Practice — A Study of Subscribers to a Prepaid Medical Plan in the Bronx* (New York: Russell Sage Foundation, 1961), p. 175.

53 Risse, *Responsibility in Health Care.*

54 T. H. Stubbs, *Emory Univ. Quart.,* 1947, *3*: 137.

55 See n. 2 above.

56 D. Nelkin, *Daedalus,* 1978, *107*: 191; E. Dichter (*N.Y. St. J. Med.,* 1954, *54*: 222) is an important example.

57 R. M. Veatch, *Hastings Center Rep.,* June 1972, *2*: 5.

58 *Ibid.*

59 Freidson, *Patients' Views of Medical Practice,* p. 187.

60 B. Myerhoff and W. R. Larson, *Hum. Organ. Clgh. Bull.,* 1964, *24*: 188.

61 J. A. Lundy, *New Eng. J. Med.,* 1952, *246*: 446.

62 R. S. Lynd and H. M. Lynd, *Middletown in Transition: A Study in Cultural Conflicts* (New York: Harcourt, Brace, 1937), p. 427 n; J. Kasteler, R. L. Kane, D. M. Olsen, C. Thetford, *J. Hlth. Soc. Behav.,* 1976, *17*: 328; F. W. Mann, *J. Maine Med. Assn.,* 1925, *16*: 137.

63 N. Y. Hoffman, *J.A.M.A.,* 1972, *220*: 58.

Another dimension—unchanging pop culture and the dangerous remoteness of the scientist—is not explored in this present article; see G. Basalla, in *Science and Its Public: The Changing Relationship,* ed. G. Holton and W. H. Blanpied (Dordrecht: Reidel, 1976), p. 261.

64 R. Branson, *Hastings Center Stud.,* 1973, *1*: (No. 2): 17.

Harper's Weekly, 1895, *39:* 586 (State Historical Society of Wisconsin)

These photographs, taken by the muckraking journalist Jacob Riis, illustrate the transformation brought about by New York City Street Commissioner George Waring after he took office in 1895. Both pictures show the same paved block in front of 212 Sullivan Street. The top photo, taken in March, 1893, depicts a scene typical of American cities in the late 19th century.

Harper's Weekly, 1895, *39:* 586 (State Historical Society of Wisconsin)

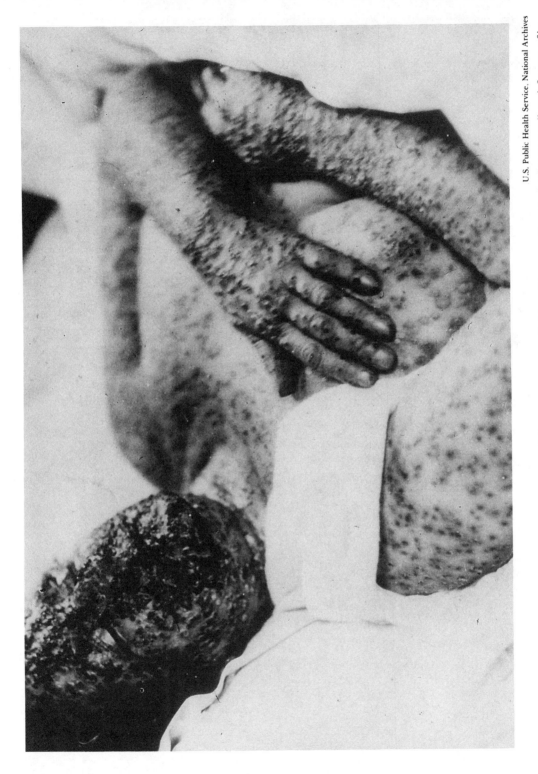

Of all 19th-century threats to health, epidemic diseases aroused the greatest concern. Besides taking many lives, the loathsome smallpox left many of its victims permanently disfigured.

Contaminated milk and overflowing privies contributed to the poor health of American city dwellers. The 1858 drawing (below) illustrates one dairy's attempt to obtain the last drop of milk from an obviously sick and dying cow. The 1913 photograph of a privy in Yonkers, New York, graphically portrays the unsanitary condition of one American community in the early 20th century.

"There Ain't No Law," pamphlet of the National Housing Association (New York, 1913). Courtesy of Clay McShane

Leslie's Illustrated Weekly Newspaper, 1858, 5: 369

Quarantines and school medical inspections were two of the measures urban health departments employed to reduce sickness among children. These photographs show a Milwaukee health inspector placarding the home of a little boy suffering from mumps and a New York City public health nurse examining the cleanliness of school children.

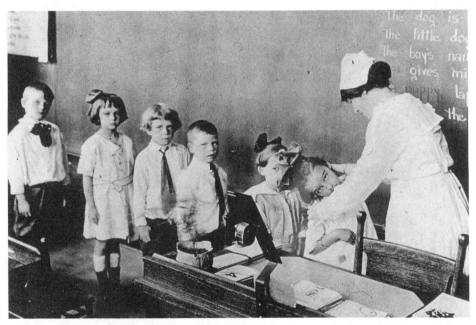

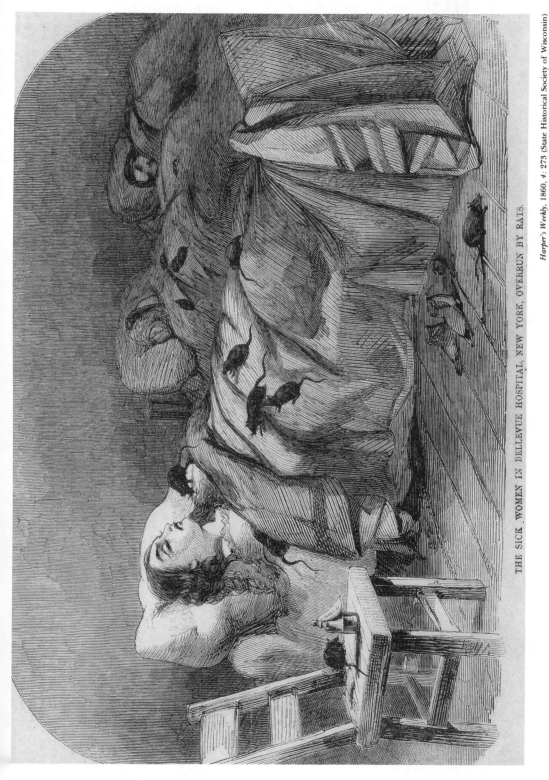

THE SICK WOMEN IN BELLEVUE HOSPITAL, NEW YORK, OVERRUN BY RATS.

Harper's Weekly, 1860, *4*: 273 (State Historical Society of Wisconsin)

Nineteenth-century hospitals sheltered the suffering poor, but their filthy and understaffed facilities sometimes did more harm than good. In 1860 a woman in New York City's Bellevue Hospital actually lost her newborn baby to the institution's rats.

J. P. Maygrier, *Midwifery Illustrated* (New York: Harper Bros., 1834), p. 90

Modesty frequently inhibited women from seeking assistance for sensitive ailments and sometimes affected the treatment they received, as this pelvic examination illustrates. Several health problems resulted from the fashionable 19th-century custom of wearing waist-restricting corsets.

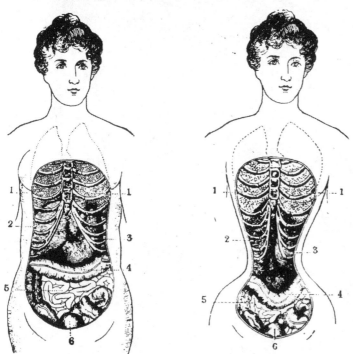

From *The Unfashionable Human Body* by Bernard Rudofsky. Copyright © 1971 by Bernard Rudofsky. Used by permission of Doubleday and Company, Inc.

FEE-BILL,

Adopted by the Western Medical Society of the State of Wisconsin, December, 1849.

Ordinary office prescription,	$ 0 50
Venesection, or extracting tooth at office,	50
Opening abscess,	50 *to* 5 00
Dresing wound,	50 " 5 00
Vaccination,	" 1 00
Dividing Fraenum,	50 " 2 00
Cupping,	1 00 " 2 00
Introducing Seton or Issue	1 00 " 2 00
Scarifying Eye,	1 00 " 5 00
Verbal advice,	1 00
Written advice,	2 00 " 5 00
Ordinary visit in Town,	1 00
Visit after 10 o'clock P. M.,	2 00
Additional patients, same family, (each)	50
Consultation visit,	3 00 " 5 00
Malignant Contagious diseases, (first)	3 00 " 5 00
Subsequent visits, each,	2 00
Natural Parturition, (ten hours.)	6 00 " 10 00
Extra Detention, (pr hour.)	50
Unnecessary Detention, (pr hour.)	1 00
Twin cases,	10 00 " 15 00
Instrumental Labor, or Turning,	15 00 " 25 00
Removing Placenta,	6 00 " 10 00
Visits after two days, charged as ordinary.	
Visit in country under two miles,	2 00
Do. over two miles, pr mile,	50 " 1 00
Visit, same neighborhood half price.	
Gonorrhœa, (in advance.)	5 00 " 20 00
Syphilis, (Do.)	10 00 " 25 00
Introducing Catheter, (first time,)	2 00
Do. Subsequently,	1 00
Paracentesis,	5 00 " 20 00
Excision Tonsils,	5 00 " 10 00
Operation for Hydrocele,	5 00 " 25 00
Do. Phimosis, Paraphimosis,	5 00 " 10 00
Do. Fistula Lachrymalis,	10 00 " 30 00
Do. " in Ano,	10 00 " 35 00

Operation for Imperforate Anus.	$ 5 00 *to* 25 00
Do. Vagina,	10 00 " 50 00
Do. Hare-Lip,	10 00 " 30 00
Do. Hernia,	25 *to* 100 00
Do. Cataract,	50 " 100 00
Do. Strabismus,	10 " 25 00
Do. Club-Foot,	25 " 100 00
Do. Stone,	100 " 200 00
Ligating Arteries,	10 " 200 00
Extirpating Eye,	50 " 100 00
Do. Testicle,	25 " 100 00
Do. Tumors,	5 " 100 00
Trephining,	25 " 100 00
Reducing Hernia,	5 " 15 00
Do. Prolapsus Ani,	2 " 10 00
Do. Fracture of Thigh,	25 " 50 00
Do. " Leg,	10 " 30 00
Do. " Clavicle,	10 " 25 00
Do. " Arm, Forearm,	10 " 25 00
Do, " Fingers, Toes,	3 " 10 00
Dislocation Hip-Joint,	25 " 100 00
Do. Shoulder-Joint,	15 " 35 00
Do. Elbow, Wrist, Ankle,	10 " 25 00
Do. Finger, Toe,	3 " 10 00
Amputation Thigh,	50 " 100 00
Do. Leg, Foot,	25 " 100 00
Do. Finger, Toe,	5 " 15 00
Do. Arm, Forearm, Wrist,	25 " 75 00
Do. Hip, or Shoulder Joint,	100 " 200 00
Do. Breast,	25 " 100 00
Do. Penis,	10 " 50 00
Inducing Premature Labor,	50 " 100 00

In all Surgical cases, the charge for subsequent attendance, to be according to time occupied and trouble incurred.

Visits in the country after dark to be considered as night visits, and charged double.

Resolved, That the moral, professional and pecuniary interests of this Society, require of its members a uniformity in charges.

Resolved, That we, the undersigned, members of the Western Medical Society of the State of Wisconsin, mutually pledge ourselves faithfully to adhere to the foregoing rates of charges.

(SIGNED,)

J. W. CLARK,	H. VAN DUSEN,	AZEL P. LADD,	A. SAMPSON,
J. S. RUSSELL,	GEO. D. WILBER,	WM. STODDART,	C. A. MILLS,
EDWARD CRONIN,	DAVID ROSS,	GEO. W. PHILLIPS,	T. R KIBBE.

ATTEST, GEO. D. WILBER, *Secretary*.   J. W. CLARK, *President*

History of Medicine Department, University of Wisconsin-Madison

Local medical societies attempted to regulate the costs of medical care and stabilize the income of physicians by publishing schedules of fees like this one for the Western Medical Society of the State of Wisconsin in 1849. Such documents can often tell us as much about the practice of medicine as about medical economics.

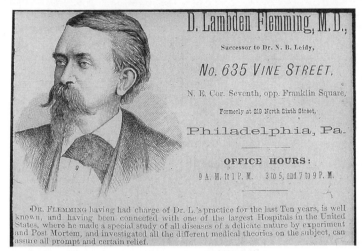

Picture Collection, New York Public Library

These illustrations suggest changes in the image of the American doctor: the businessman-physician of the 19th century, the family practitioner of the early 20th century, and the striking house staff of a major New York hospital in the 1970s.

State Historical Society of Colorado

Medical Dimensions, June, 1975, p. 14

Hydropathy was only one of several medical sects that flourished in 19th-century America. These photographs show the various treatments available at Dr. John Harvey Kellogg's Battle Creek Sanitarium, which prospered well into the 20th century.

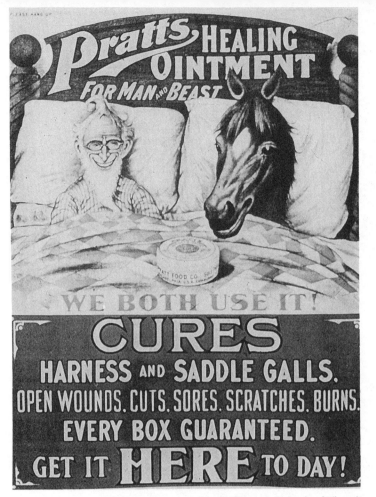

We BOTH USE IT!

School of Pharmacy, University of Wisconsin

In addition to its regular and sectarian physicians, America offered its sick an almost infinite variety of quacks and cures, from ubiquitous patent remedies like Pratt's Healing Ointment to more elaborate and costly devices like the worthless Electric Couch and Dry Bath.

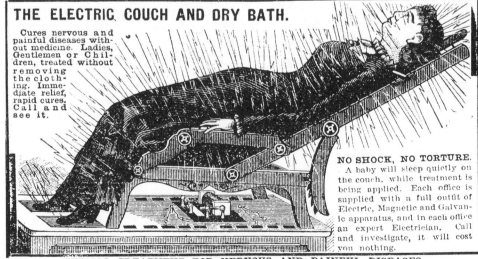

SPECIAL TREATMENT FOR NERVOUS AND PAINFUL DISEASES.

Smithsonian Institution

Bellevue Hospital Medical College.

CITY OF NEW YORK.

SESSION 1874-'75.

Admit

_____ is entitled

To all the Privileges of the Department of

PRACTICAL ANATOMY,

Until March 1, 1875.

A. Flint Jr— M. D., Sec'y of the Faculty.

☞ This Ticket is not an evidence that the holder of it has actually dissected, unless certified by the Professor of Practical Anatomy.

[OVER.]

Bellevue Hospital Medical College.

CITY OF NEW YORK.

SESSION 1874-'75.

LECTURES
ON

Physiology and Physiological Anatomy.

A. Flint Jr— M. D., Professor.

Admit

Bellevue Hospital Medical College.

SESSION 1874-'75.

LECTURES ON SURGERY.

Admit

Frank H. Hamilton M. D.,
Prof. of Practice of Surgery with Operations,

Lewis A. Sayre M. D.,
Prof. of Orthopedic Surgery,

Rev. W. B. Smith M. D.,
Prof. of Clinical and Operative Surgery.

Wm. Van Buren M. D.,
Prof. of the Principles of Surgery with Diseases of the Genito-Urinary System.

Bellevue Hospital Medical College.

CITY OF NEW YORK.

SESSION 1874-'75.

LECTURES
ON

THE PRINCIPLES AND PRACTICE OF MEDICINE.

Austin Flint, M. D., Professor.

Admit

Richard W. Schwarz Collection

Until late into the 19th century medical students paid for their tuition by purchasing tickets from each professor whose lectures they wanted to attend.

INSTITUTIONS

The hospital, so central to health care today, assumed little importance in American medical history until the late 19th century. Except for the almshouses and pesthouses found in large towns along the Atlantic seaboard, there were no hospitals in the British colonies of North America until 1751, when Benjamin Franklin and his Philadelphia friends founded the Pennsylvania Hospital. Modeled after the voluntary hospitals of England, this institution admitted both the mentally and physically ill and accepted those who could pay as well as those who could not.

For the urban poor in the 19th century, the most important medical institution was not the hospital but the dispensary, described by Charles E. Rosenberg. These institutions, which began to appear in the Northeast in the late 18th century, not only dispensed medicines and advice to needy patients but served as an important training ground for medical students and young physicians. With the rise of hospital outpatient departments in the 20th century and related changes in the values of the medical profession, however, independent dispensaries gave way to neighborhood health centers, a small part of the American medical scene.

Because of the nation's predominantly rural population and the social attitudes described by Morris J. Vogel, the idea of hospitals caught on slowly in America. Physicians throughout the 19th century continued to treat most of their patients at home and even to perform surgery there. As late as 1873 there were fewer than 200 hospitals in the United States, about a third of which were for the mentally ill. Yet only 50 years later the number of hospitals was approaching 7,000. This rapid growth resulted as much from the social changes associated with urbanization as from advances in medical technology, such as aseptic surgery and x-rays.

Mental hospitals in America date from 1773, when the colony of Virginia opened an institution in Williamsburg "for persons of insane and disordered minds." It was not until 50 years later, however, with the founding of the Worcester State Lunatic Hospital in Massachusetts, that public asylum building began in earnest. Thanks in large part to the efforts of Dorothea Dix and other reformers, most states by 1860 had established hospitals to care for the mentally ill. Although founded with the best intentions, these asylums soon found themselves playing a custodial rather than a curative role. As Gerald N. Grob suggests, this resulted less from neglect than from a changing patient population. In the 20th century, increasing use of medications and deinstitutionalization changed yet again how the mentally ill received care. Many of the elderly ended up being reinstitutionalized in for-profit nursing homes.

19

Social Class and Medical Care in 19th-Century America:
The Rise and Fall of the Dispensary

CHARLES E. ROSENBERG

To most mid-20th-century physicians, the term "dispensary" evokes the image of a hectic hospital pharmacy. To his mid-19th-century counterpart, it was both the primary means for providing the urban poor with medical care and a vital link in the prevailing system of medical education. These institutions had an effective life-span of roughly a hundred years. Founded in the closing decades of the 18th century, American dispensaries increased in scale and number throughout the 19th century and remained significant providers of health care well into the 20th century. By the 1920s, however, the dispensaries were on the road to extinction, increasingly submerged in the outpatient departments of urban hospitals. Historians have found the dispensary of little interest; even those contemporary medical activists seeking a usable past for experiments in the delivery of medical care, are hardly aware of their existence.[1] Yet a study of the dispensary illustrates not only an important aspect of medicine and philanthropy in the 19th-century city—but the social logic implicit in their rise and fall underlines permanently significant relationships between general social needs and values and the narrower world of medical men and ideas.

The dispensary was invented in late 18th-century England; it was an autonomous, free-standing institution, created in the hope of providing an alternative to the hospital in providing medical care for the urban poor. Like most such benevolent innovations, it was soon copied by socially conscious

CHARLES E. ROSENBERG is Janice and Julian Bers Professor of History and Sociology of Science at the University of Pennsylvania, Philadelphia, Pennsylvania.

Reprinted with permission from the *Journal of the History of Medicine and Allied Sciences*, 1974, 29: 32–54.

Americans; dispensaries were established in 1786 at Philadelphia, 1791 at New York, 1796 at Boston, and at Baltimore in 1800. Their growth was at first very slow. No additional dispensaries were established until 1816, when the managers of the Philadelphia Dispensary helped establish two new dispensaries, the Northern and Southern, to serve their city's rapidly developing fringes.[2] New Yorkers established the Northern Dispensary in 1827, the Eastern in 1832, the DeMilt in 1851, and North-Western in 1852. By 1874 there were 29 dispensaries in New York, by 1877, 33 in Philadelphia. Their growth was equally impressive in terms of number of patients treated; in New York, for example, the city's dispensaries treated 134,069 patients in 1860, roughly 180,000 in 1866, 213,000 in 1874 and 876,000 in 1900.[3]

The dispensaries shared certain organizational characteristics. Almost all had a central building —with the prominent exception of Boston which had none until the 1850s—and usually employed one full-time employee, an apothecary or house-physician who acted as steward, performed minor surgery, often vaccinated and pulled teeth—as well as prescribed for some patients. (Though most dispensaries limited their aid to prescriptions written by their own staff physicians, a few would fill prescriptions for the indigent patients of any regular physician.)[4] By mid-century the house-physicianship had in the larger dispensaries evolved into two separate positions, resident physician and druggist-apothecary. Most dispensaries also appointed younger physicians who visited patients too ill to attend the dispensary. Such "district visiting" was the principal task of the Philadelphia Dispensary when founded in 1786, remained the sole activity of the Boston Dispensary until 1856—and was continued by almost all urban dispensaries until the end of the 19th century, though the treatment of ambulatory patients grew proportionately more prominent in all. The dispensaries also appointed attending and consulting staffs from among their community's established practitioners, the attending staff treating patients well enough to visit the dispensary, the consulting staff serving a largely honorary role.

The dispensaries were shoe-string operations. Most, with the exception of those in New York which enjoyed state and city subventions, were supported by private contributions and the often-voluntary services of local physicians.[5] As late as the 1870s and 1880s—when a dispensary might treat over 25,000 patients a year—budgets of four or five thousand dollars were still common and annual reports vied in reporting how little had been spent for prescriptions—an average of under five cents per prescription was common. The Boston Dispensary and Philadelphia Dispensary gradually accumulated some endowment funds, though most others remained financially marginal. All, however, were sensitive to cyclical economic shifts, for contributions declined in periods of depression while patient pressure increased proportionately. As a result of the economy's downturn in 1857, for example, New York's Eastern Dispensary reported an increase of 22 percent in cases over 1858 and 42 percent over 1856.[6] A useful index to the shaky financial condition of many of the dispensaries was their frequent practice of renting a portion of their building to commercial tenants; such income often constituted a substantial portion of the institution's budget and could not be given up even when the dispensary needed room for expansion.[7]

Some of the dispensaries published detailed statistics of the numbers and kinds of ailments treated by their physicians; thus we can begin to reconstruct their everyday responsibilities. Most cases were, of course, relatively minor—for example, bronchitis, colds, or dyspepsia—and rarely were the numbers of deaths equal to more than 2 or 3 percent of the patients treated. Consistently enough, the number of female patients was always greater than that of males, in some instances as much as two to one; working men, that is, had necessarily to tolerate disease symptoms of far greater intensity before feeling able to consult a physician. In those cases serious enough to be treated at home by visiting physicians sex ratios tended to be more nearly equal. (It was not until the end of the century that dispensaries began to consider evening hours for workers.) Although the general level of mortality among all dispensary patients was low, mortality among patients treated in their homes approached the 10 or 11 percent normal for hospitals at the beginning of the century. Such death rates were particularly discouraging, for the district physician never treated many intractable cases. Chronic and degenerative ailments brought incapacity and eventual alms-house incarceration; these cases never found their way into the dispensary's mortality statistics. The dispensaries also performed minor surgery, treating fractures, contusions and lacerations—as well as casual if frequent dentistry, essentially the "indiscriminate extirpation" of offending teeth.[8]

The dispensaries also played an important public health role in providing vaccination for the poor and vaccine matter for the use of private practitioners. From a purely demographic point of view, indeed, vaccination was the most important function performed by the dispensaries. The dispensaries not only made vaccination available without cost, but some mounted door-to-door vaccination programs in their city's tenement districts. In periods of intense demand, most frequently at the outset or threat of a smallpox epidemic, the dispensaries were able to supply large amounts of vaccine matter at short notice. In the opening months of the Civil War for example, the New York Dispensary provided vaccine matter for all the state's recruits.[9]

Despite ventures into surgery, dentistry, and vaccination, dispensary therapeutics were generally synonymous with the writing of prescriptions; dispensaries dispensed. Throughout the first three-quarters of the 19th century, the phrase "prescribing for" was generally synonymous with seeing a patient; busy dispensary physicians could hardly be expected to do more than compose hasty and routine prescriptions. (Dispensary managers tended by mid-century to demand the use of formularies limited in both cost and variety; later in the century some dispensaries were charged with filling prescriptions by number, the dispensing physician being constrained by an abbreviated list of numbered and preformulated prescriptions.)[10] In this routine and exclusive dependence on drug therapy lay the principal difference between the care provided the urban poor and that paid for by the middle class. Physicians in private practice relied consistently in their therapeutics upon adjusting the regimen of their patients, especially in chronic ills; such injunctions were hardly appropriate in dispensary practice. The city poor could not very well vary their diet, take up horse-back riding, visit the seaside, or voyage to the West Indies.

Not surprisingly, the dispensaries tended to develop ties both formal and informal with other urban charities — in New York, for example, with the Commissioners of Emigration, Association for Improving the Condition of the Poor, and Children's Aid Society; in Philadelphia with the Board of Guardians for the Poor.[11] Dispensary physicians were in this sense *de facto* social workers. In New York, for example, a note from the dispensary physician was necessary if the commissioners were to issue a ration of coal; thus a mid-century whimsy referred to "coal fever" — an illness which struck suddenly during cold weather in the city's tenements.[12] In the post-Civil War decades, efforts to provide such physical amenities became somewhat more organized; dispensary physicians continued to work with existing philanthropic agencies and began as well to establish their own auxiliaries in hopes of providing food and nursing in deserving cases. In Philadelphia, the Lying-In and Nurse Charity and the Lying-In Department of the Northern Dispensary had provided some nursing service since the 1830s, while others had paid occasionally for nursing in selected cases since the opening years of the century. In a more contemporary idiom, the Instructive Visiting Nurse Service of the Boston Dispensary began in the 1880s to aid the dispensary's district physicians in their work, not only nursing, but educating the poor in hygiene and diet. In Boston and New York, diet kitchen associations provided nourishing food for patients bearing a dispensary physician's requisition. By 1883, the New York Diet Kitchen Association operated three kitchens in cooperation with the dispensaries and had fed 7,699 patients, filling 53,893 separate requisitions from dispensary physicians during the year.[13]

Another trend marking the 19th-century evolution of the dispensaries, reflecting and paralleling a more general development within the medical profession, was their internal reorganization along specialty lines. As early as 1826, the New York Dispensary reorganized itself, dividing patients treated at the dispensary into "classes" according to the nature of their ailment. Pioneering dispensaries for diseases of the eye and ear had come into being as early as the 1820s. By mid-century, the need for specialty differentiation was unquestioned. When the Brooklyn Dispensary opened in 1847, for example, it announced that patients would be distributed among the following classes: women and children, heart, lungs and throat, skin and vaccination, head and digestive organs, eye and ear, surgery and unclassified diseases. In the second half of the century, specialty designations became increasingly narrow and gradually closer to modern categories; nervous and genito-urinary diseases were, for example, among the most frequently created of such departments in the late 1870s and early 1880s. By 1905, the forward-looking Boston Dispensary boasted these impressively varied out-patient clinics: surgical, general medical, children, skin, nervous system, nose and throat, women, eye and ear, genito-urinary and x-ray.[14] An important related late-19th-century trend was the increasingly frequent establishment of specialized dispensaries, institutions that treated only particular ailments or ailments of particular organs.

These in brief outline were the chief characteristics which marked the growth of the dispensaries between the end of the 18th and last decades of the 19th centuries. Why did the founders and managers of our pioneer dispensaries find them so plausible a response to social need? What factors led to their initial adoption and subsequent growth?

In their appeals for public support, dispensary founders and supporters left abundant records of their conscious motives. Most prominent in the last years of the 18th and opening decades of the 19th centuries was a traditional sense of stewardship. "It is enough for us," as one physician-philanthropist put it, "to be assured that the poor are always with us, and that they are exposed to disease."

> Benevolence [he continued] is not that passive feeling which can be satisfied with doing no injury to our neighbor, or rest contented with mere good wishes for his well-being when he needs our assistance.

The poor, as a prominent New York clergyman explained the need for supporting the dispensary's work, "have feelings as well as we; they are bone of our bone and flesh of our flesh; men of like passions with ourselves."[15] Such sentiments remained deeply felt and were explicitly articulated throughout the first half of the century.

Other, more mundane, motives always coexisted with such humanitarian appeals. One was the familiar mercantilist contention that maintaining the health of the poor would not only save the tax dollars implied by the almshouse or hospital care of chronically ill workers, but would aid the economy more generally by helping maintain the labor force at optimum efficiency. (These appeals assumed, of course, the ability of the dispensary physicians to diagnose ills at a stage when they might still respond to available treatment.) A related argument urged the dispensaries' function as first-line of defense against epidemic disease; though such ills ordinarily began and reached epidemic proportions among the poor, once established they might attack even the comfortable and well-to-do. No household could feel immune when servants and artisans moved easily from the world of their betters to that of tenement-dwelling friends and families.[16] These arguments soon hardened into rhetorical formulae and were ritually intoned throughout the first two-thirds of the century. Thus, for example, a mid-century dispensary spokesman could, in appealing for support, argue that:[17]

> The political economist will find here cheapness and utility combined. The statesman will discover the greatest good of the greatest number combined promoted. The city official will find his sanitary police materially assisted. The heads of families will soon find how much the lives and health of their household are cared for and secured. The tax-payer will see his burdens diminished. The benevolent will have opened to his view in the Dispensary and its kindred and associated charities the widest field for the exercise of good will towards man; and the Christian will find a new proof of the truth that they do not love God less who love mankind more.

Finally, and matter-of-factly, their advocates always contended that dispensaries would serve as much-needed schools of clinical medicine.

But to catalogue the arguments of managers and fund-raisers is not precisely to explain the logic of their commitment. Why did the dispensaries grow so rapidly? Obviously because they worked, worked that is in terms of particular social realities and expectations. At least four such factors help explain the evolution of the dispensary in 19th-century America. First, they were entirely functional in terms of the internal organization of the medical profession. Second, they were entirely consistent with available therapeutic modalities. Third, they were effectively scaled to the needs of a small and comparatively homogeneous community; once established they became indispensable as urban growth dramatically increased their client constituency. Fourth, the dispensaries made sense in terms of their founders' expectations of the roles to be played both by government and private citizens.

Most fundamental was the relationship between the dispensary and the world of medical education and status. Without the initiative and voluntary support of the medical profession dispensaries would not have been created nor could they have survived. Physicians formed the core-group in the formation of almost every American dispensary from the end of the 18th to the beginning of the 20th centuries.[18]

In the first third of the 19th century, when formal clinical training could not be said to exist outside that presumed in the preceptorial relationship, the dispensary helped fill an important pedagogical void. Not only could visiting and attending physicians themselves accumulate experience and reputation while more firmly establishing their private practice—but they could use their dispensary appointment as a means of providing case materials for their apprentices. Thus Benjamin Rush could recommend Drs. Wistar and Griffits as preceptors since both held dispensary positions, "where a young man will see more pratice in a month than with most private physicians in a year." Almost from the first years of the dispensaries, indeed, critics often charged that students and apprentices were allowed to treat the poor. (In Philadelphia, for example, such complaints found their way into newspapers as early as 1791.)[19] In the second quarter of the 19th century, as the preceptorial system grew less significant, the role of the dispensaries in clinical training grew even more prominent; mid-century medical schools vied actively in establishing dispensaries for the benefit of their students.

Most significantly, dispensary physicianships served as a step in the career pattern of elite physicians. Despite the complaints of articulate mid-century critics as to the wretched state of medical education and practice, even a cursory analysis of the profession's structure indicates the existence of a well-defined elite, largely urban, often European-trained, and almost always enjoying the benefits of hospital and dispensary experience. It was just such ambitious young practitioners who served as dispensary visiting and attending physicians while they accumulated experience and gradually made the contacts so important to later success — contacts it should be emphasized, with older established physicians at least as much as with prospective patients.[20] (Prestigious and largely honorary consulting physicianships were normally reserved, in dispensaries as in hospitals, for a community's most influential and respected physicians.) Contemporaries never questioned the dispensary's teaching function. The trustees of the New York Dispensary admitted, for example, in 1854 that their institution served as "a practical school for physicians," but they contended, it was a perfectly defensible policy: "for, by this system, these Physicians must become accomplished practitioners, by the time the growth of their private practice shall oblige them to resign their posts at the Dispensary." With the growing importance of specialization as prerequisite to intellectual status and economic success after mid-century, the increasingly specialized dispensaries served as *de facto* residency programs, allowing ambitious — and often well-connected — young men to accumulate experience and reputation. Though formal statements by medical spokesmen uniformly disowned "exclusive" specialism until long after the Civil War, devotion to a pragmatic specialism was established much earlier in America's cities. In 1839, for example, the editor of the *Boston Medical & Surgical Journal* remarked, in commenting on the specialty organization of New York's Northern Dispensary, that "such is manifestly the tendency in our times, in the great cities, and it is the only way of becoming eminently qualified for rendering the best professional services — to learn to do one thing as well as it can be done."[21]

If the dispensary made excellent sense in terms of the institutional needs of American medicine, it was equally consistent with the technological means available — both at the end of the 18th century, and through the first half of the 19th. Beyond the stethoscope — not routinely applied before mid-century — no special aids to diagnosis were available to any physician, no therapeutics beyond bleeding, cupping, and administration of drugs. Surgery was ordinarily limited, for rich and poor alike, to the treatment of lacerations and fractures, the reduction of occasional dislocations, the lancing of boils and abscesses. Dispensaries seemed, for many decades into the 19th century, fully able to provide both adequate care for the poor and adequate training for their attendants.

The dispensaries seemed equally appropriate to the needs of a small and relatively homogeneous community. The world of the late 18th century assumed — even if it did not necessarily practice — face-to-face interaction between members of different social classes, interactions structured by customary relations of deference and stewardship. This social world-view is concretely illustrated in the acceptance by the dispensaries' founding generation of the contributor recommendation as basis for patient referrals. A certificate of recommendation was necessary, that is, before the dispensary would undertake treatment of a particular patient. This followed English hospital and dispensary practice. As the century progressed, however, the dispensaries which maintained the practice sometimes found it a cause of conflict between medical staff and lay managers. By-laws specified the privileges of recommendation accompanying each contributing membership; a typical arrangement was that which in exchange for a five dollar annual subscription offered the right to recommend two patients at any one time during the year. A 50 dollar subscription typically brought the same privilege for life. Similarly, early dispensary by-laws indicate that members of the boards of managers were expected to play an active and often personal role; the New York Dispensary, for example, created a trustees' committee to accompany visiting physicians on their rounds once a month.[22]

Equally revelatory of the world-view shared by the pious and benevolent Americans who founded the dispensaries was their assumption that a crucial difference separated the dispensary from the hospital patient; the dispensary patients would be drawn from among the worthy poor, hardworking and able to support themselves, except in periods of sickness or general unemployment. Such worthy poor might also include widows, orphans, and the handicapped. The lying-in department of Philadelphia's Northern Dispensary declared in 1835, for example, that it could aid only married women of respectable character, "such as require no aid when in health." Financial support for the dispensary would, the argument followed, keep such honest folk from alms-house residence and morally contaminating contact with those abandoned souls who were its natural inmates. Dispensary spokesmen tirelessly repeated these stylized categories by way of argument even as experience indicated that this neat and comforting ideological distinction failed to reflect reality. In 1830, a physician of the Boston Dispensary could complain indignantly that persons of the "most depraved and abandoned character frequently apply who think they have a right of choice between the Alms-House and the Dispensary." As late as 1869, the Philadelphia Dispensary could still explain that:[23]

> The principal object of this institution is to afford medical relief to the worthy (not the lowest class of) poor, in those cases where removal to a hospital would for any approved reason be ineligible. . . . In a thrifty population like our own, it is the exception . . . where removal to a hospital should be considered eligible.

The dispensaries were founded and grew, finally, because they were entirely consistent with the assumptions of most Americans in regard to the responsibilities of government and appropriate forms and functions of the public institutions which embodied such responsibilities. The prostitute, the drunkard, the lunatic and cripple were the city's responsibility—social subject matter for the alms-house or city physician. The dispensary, on the other hand, represented an appropriate response of humane and thoughtful Americans to the needs of hard-working fellow citizens, a response demanded both by Christian benevolence and community-oriented prudence; it was a form of social intervention limited, conservative and spiritually rewarding. In the second third of the 19th century, as demographic realities shifted inexorably, this traditional view still served to justify the now-expanded work of the dispensaries—

and at the same time to avoid systematic analysis of the changing nature and social condition of the constituency they served. It was only very slowly, and only in the minds of a minority of those associated with dispensary work, that it became clear that many of their city's honest and industrious laboring men were unable to pay for medical care even in times of prosperity.

The dispensary continued to change throughout the second half of the 19th century. We have already referred to their increase in numbers and degree of specialization. Equally significant was expansion of the dispensary form under new kinds of auspices. First, most urban—and even some small town—medical schools anxious to compete for students, established their own dispensaries so as to offer "clinical material" for their embryo physicians. Second, hospitals not only increased in number in the last third of the century, they also began to provide more outpatient care, in some localities duplicating services already offered by dispensaries. In Philadelphia with its flourishing medical schools the rivalry between hospitals and dispensaries emerged as early as 1845.[24] In certain areas outpatient facilities competed for patients, medical school clinics in particular advertising in newspapers and posting handbills. All these events were correlated, of course, with a growing demand within the medical profession for clinical training at every level, for the possession of attending and consulting physicanships, for the accumulation of specialty credentials. At the end of the century, finally, a growing public health movement used the by now familiar dispensary form to shape and deliver medical care and would-be prophylactic measures in slum areas—most conspicuously in the identification and treatment of tuberculosis.

Underlying these developments was a series of parallel changes, first in the scale of the human problems the dispensaries faced, second in the intellectual tools and social organization of the medical profession. First in time came an absolute increase in the numbers and shift in the social origins of those urban Americans calling upon the dispensary. Secondly, in terms of chronology if not significance, were shifts within the world of medicine which made the dispensary increasingly marginal in the priorities of medical men. One need hardly demonstrate the significance to medical practice of increasing specialization, the germ theory and antisepsis, the development of modern surgery, x-ray and clinical laboratory methods, the increasing centrality of the hospital; the way in which demographic and social changes reshaped the dispensaries is perhaps less familiar.

Whatever degree there had been in the original vision of a community bound by common ties of assumption and identity, this unifying vision corresponded less and less to reality as the 19th century progressed. The accustomed social distance between physician and charity patient seemed increasingly unbridgeable. A practical measure of this increasing social distance — and one which correlates with population and immigration statistics — was the growing disquietude of dispensary physicians in contemplating their patients. As early as 1828, New York's Northern Dispensary asked contributors to sympathize with their staff physicians'

> . . . great sacrifices of feeling and comfort, which they must necessarily make, by being forced into daily and hourly association with the miserable and degraded of our species, loathsome from disease, and often still more so by those disgusting habits which go to the utter extinction of decency in all its forms.

The traditional system in which dispensary patients or their messengers called first upon contributors seeking a recommendation and then upon visiting physicians at their regular homes or offices also showed signs of strain. In Boston, where the dispensary's lay managers had long opposed the establishment of a "central office," a major factor helping to overcome this reluctance in the 1850s was the unwillingness of district physicians to have their offices used by so "ignorant and degraded a class." "It is undesirable," as Henry J. Bigelow explained it, "for most physicians to receive at their own apartments the class of applicants who now form the mass of dispensary patients."[25]

The patients who seemed most familiar, closest to the physicians' own experience, were those most capable of evoking sympathy and understanding; thus the plight of those fallen in fortune, of the genteel widow, of the orphaned child of good parents were those which touched visiting physicians most deeply.

It is not infrequently that we witness much feeling manifested by those who have been able to employ their own physicians and purchase their own medicines, when through reverses of fortune they have for the first time applied for assistance from the Institution; such constitute the most interesting portion of our patients.

Other patients were far less interesting.[26]

There were, of course, the venereal and alcoholic; but these had always existed and their existence had always implied a certain conflict between morals and medical care. Far more unsettling by the 1840s were the new immigrants who streamed into America's cities and soon constituted a disproportionate part of the dispensary's clientele. By the early 1850s it was not uncommon for an absolute majority of a particular institution's patients to have been born in Ireland; in the districts of individual visiting physicians over 90 percent of those treated might be foreign born. Not surprisingly, the 1840s and 1850s saw dispensary administrators and trustees pointing again and again to the immigrant as they sought to explain the difficulties of their work and their ever increasing financial need.[27]

It was not only the numbers and the poverty, but the alienness of the immigrants which intensified the differences between them and their would-be medical attendants. It must be recalled that the desirability of dispensary appointments guaranteed their being filled by young physicians of at least middle class background — thus insuring as well a maximum social distance between physician and patient. As early as 1831, for example, Boston Dispensary visiting physicians, dismayed by the conditions they encountered, elected to survey the economic and moral status of their patients. In that age of temperance and pietism, it was only to have been expected that the district physicians found intemperance to be the most important single cause of disease in their patients — and intemperance to be most common among the foreign born. The Irish seemed particularly undesirable, filthy, drunken, generally inhospitable to middle-class standards of behavior. "Upon their habits — their mode of life," a dispensary physician explained in 1850, "depend the frequency and violence of disease. This I am fearful will continue to be the case, since no form of legislation can reach them, or force them to change their habits for those more conducive to cleanliness and health." "Deserving American poor," another Boston Dispensary physician complained, were "often deterred from seeking aid because they shrink from seeming to place themselves on a level with the degraded classes among the Irish."[28] The unfamiliar attitudes and habits of these patients often added to their troublesomeness; they ignored hygienic advice and often defied the physician's simplest requests. The Irish, for example, considered it dangerous to have lymph removed from the lesion of an individual vaccinated for smallpox; thus they refused to return to the dispensary after the required week to have the lesion checked (and to supply the lymph so useful in helping balance the dispensary's budget).[29] Later immigrant groups brought their peculiar beliefs and problems of communication; Jews and Italians replaced the Irish as objects of the dispensary physician's frustration and disdain.

A good many dispensary physicians were, of course, sympathetic to their patients, and in some cases not only sympathetic but convinced that environmental causes contributed to their clients' chronic ill-health. Yet even those individual physicians whose personal convictions made them most sensitive to the deprivation of their city's slum-dwellers, shared the ambivalence and even hostility of their peers. The same mid-century physicians, that is, who denounced basement dwellings, exploitative landlords, rotting meat and adulterated milk—shared a distaste for the intemperance, imprudence, filth, and apparent sexual immorality of those victimized by such conditions. One of the harsher dispensary critics of mid-century tenement conditions could, for example, contend that:

> . . . there is much squalor and other evidences of poverty which might be remedied had the patients more pride in cleanliness and more ambition to be doing well in the world.

As another mid-century physician explained, his patients' degradation and ignorance called "not for pity alone, but for the greatest exercise of patience and forbearance."[30]

A concern with social realities was, moreover, supported by and consistent with mid-19th-century etiological assumptions. Both acute and constitutional ills were seen as related closely to an individual's powers of resistance—itself a product of interaction between constitution and environment.

And the conditions encountered by dispensary physicians were exactly those which seemed to lower resistance and hence increase the incidence and virulence of disease. Thus a dispensary physician could note casually that scarlet fever was particularly virulent one year, since it proved as fatal to the rich as to the poor. Similarly, a pioneer ophthalmologist could urge the need for ophthalmological dispensaries because of the relationship between poverty and diseases of the eye:

> The sickly hue, and the toil worn features of these poor people are but the results of constitutional derangements . . . and as clearly reveal the inseparable union between the health of the body and the health of the eye, as between poverty and disease.

Throughout the century articulate dispensary spokesmen were aware of the need to provide food and clothing for their patients, convinced that medicines could be of only marginal help when patients had to return to work before their complete recovery, while their homes had no adequate heat, their tables only impure and decaying food. "No persons can more readily appreciate than we," as one put it, "the utter uselessness of drugs, if there is no possibility of nourishing and warming the patient."[31]

The attitudes of mid and late 19th-century physicians can best be described in terms not of hostility, but of ambivalence—and perhaps most importantly an ambivalence characterized by a world-view which related disease and morals alike to general social conditions. Both morality and morbidity were seen as resultants of the interactions between environmental circumstance and culpable moral decisions. This mixture of social concern, moralism, meliorism and deep seated antipathy was clearly apparent by mid-century and marked the writings of most dispensary spokesmen until the end of the century; it could not prove the basis of a long-lived commitment to the dispensary and the necessity of its peculiar social function.

Nevertheless, a handful of articulate spokesmen for the dispensary did elaborate a characteristic point of view by the century's end, in which disease was seen not only as related inextricably to environment, but which emphasized the dispensary's capacity to reach out into the homes of the sick poor, so as to deal with problems more fundamental than the symptoms which brought the patient to their attention. The ability of the dispensary to relate to the community surrounding it became in the arguments of such dispensary defenders an indispensable aspect of a socially adequate medical care system. Visiting physicians and nurses could simply not be replaced by a hospital outpatient department. Advocates of this higher dispensary calling argued again and again that one could not simply treat a patient's symptoms and do nothing about an environment which had much to do with causing those very symptoms. Such ideas were implemented perhaps most fully in the tuberculosis dispensaries created so widely in the first decade of the 20th century.[32]

Such would-be rationalizers of American medicine as Edward Corwin, S. S. Goldwater, Richard Cabot, and Michael Davis contended that the dispensary could, in addition to supplying primary treatment for the indigent, supplement the necessarily unfinished work of the general practitioner in those numerous cases where the patient could not afford a specialist's consultation or expensive x-rays and laboratory tests. The dispensary could, that is, serve a vast urban constituency able perhaps to afford the services of a general practitioner but unable to manage the cost of more extended or elaborate medical care. And such occasions increased steadily as the profession's ability to understand and even cure increased. Yet even as they urged such prudent considerations, these advocates of social medicine were well aware of the threat posed to the independence and ultimately to the existence of the dispensary by rapid changes in medical ideas, techniques, and institutional forms.

These arguments were consistent as well with the motivations and social assumptions of the contemporary settlement-house movement and other pioneer social welfare advocates. The settlement houses were often involved in dispensary-like programs themselves. But in a precisely timed irony, the dispensary as a viable independent institution was dying just as its most self-conscious advocates were formulating these brave contentions.

How did this come about? The dispensaries could hardly be said to have lost their social function; we have become quite conscious in recent years that their function is still not being adequately

fulfilled. In retrospect, however, their dissolution was inevitable. By the 1920s, most significantly, the dispensary had become as marginal to the needs of the medical profession as it had been central in the first two-thirds of the 19th century. A century of work in the city's slums, a growing — if always somewhat ambiguous — awareness of the relationship between health and environment, the conscious commitment of a small leadership group to the need for working in that human environment — all proved ultimately of little importance.

As hospital-centered interne and residency programs became a normal part of medical education — following inclusion of clinical training in the undergraduate years — it was inevitable that those elite physicians who would in earlier generations have been anxious to receive a dispensary appointment would now prefer hospital posts. Not only had hospitals increased greatly in number, but they contained beds, laboratory and x-ray facilities, and a cluster of appropriately trained specialists. The hospital's increasingly exclusive claims to practice the best, indeed the only adequate medicine seemed to grow more and more plausible. When, for example, in 1922 the Managers of the Philadelphia Dispensary decided to merge with the Pennsylvania Hospital, they explained that they had "found it practically impossible for an independent dispensary, unassociated with the facilities and specialists of a large modern hospital, to render the public adequate service."[33]

As the intellectual and institutional aspects of medicine changed, economic pressures also pointed toward the centralized and capital-intensive logic of the hospital. Expensive laboratory facilities, x-rays, modern operating rooms all demanded the investment of unprecedently large sums of money. The routine low-budget dosing which characterized the independent 19th-century dispensary seemed no longer a real option; dispensary boards had to face a growing and embarrassing asymmetry between their limited resources and the demands of high quality medical care. The hospital outpatient department seemed to many medical men a substantial and inevitable improvement over its predecessor institution. The growing tendency in the 20th century for medical schools to forge strong hospital ties only increased the centrality of the hospital.

Shorn of its relevance to the career needs of aspiring physicians, the dispensary was left with the clearly residual function of providing public health —charity—medical care, in itself a low-status occupation throughout the 19th century. Dispensary appointments had brought prestige and clinical opportunities in generations during which there were few other badges of status or roads to the acquisition of clinical skills; by the end of the century, there were other, more prestigious options for the ambitious young physician. Positions as municipal "out-door physicians" had a comparatively low status throughout the 19th century. The dominion of fee for service medicine remained essentially unchallenged by the liberal critics of the Progressive generation. Those ambitious young men incapable of remaining content with the mere accumulation of fees were—as the 20th century advanced —ordinarily attracted not by social medicine but increasingly by the "higher" and certainly less ambiguous demands of research; and even clinical investigation seemed in its most demanding forms to have little place in the dispensary.

If the dispensary had lost much of its appeal for the medical elite by the end of the 19th century, it had lost whatever goodwill it had had in the mind of the average practitioner decades earlier. There had always been occasional complaints in regard to the dispensaries intervening unfairly to compete with private physicians for a limited supply of paying patients. From the earliest years of their operation, American dispensaries had warned that their services were only for "such as are really necessitous." None however chose to investigate systematically the means of their patients until after the Civil War. Until the 1870s, criticism was comparatively muted; throughout the last third of the century, however, and into the 20th, the dispensaries were widely attacked as purveyors of ill-considered charity to the unworthy. The more constructive critics sought to find alternatives, the most popular —in addition to simply demanding a small fee— being the provident dispensary, a species of prepaid health plan which had proven workable in some areas in England. In city after city, local practitioners called meetings and commissioned reports predictably concluding that a goodly portion of those using dispensary services were quite capable of paying a private physician's fees.[34] Americans found it difficult to understand the social configuration of the society in which they lived; only abuse

by those in fact capable of paying medical bills could possibly explain the vast numbers who utilized dispensary services. To doubt this was to assume that large numbers of worthy and hard-working Americans were indeed too poor to pay for even minimally adequate medical care.[35]

Physicians were often unwilling to refer their paying patients—even if the payment were only 25 or 50 cents—to the more specialized facilities of neighboring dispensaries. As late as 1914 the director of Pennsylvania's tuberculosis program charged that local practitioners refused to refer patients in the early stages of the disease, unwilling to relinquish treatment until such working-people were too deteriorated to work—and pay. Attacks on the dispensary system were generally supported as well by the charity organization movement which, in city after city, attacked dispensary medicine as an excellent example of that undiscriminating alms-giving which served only to demoralize its recipients. (It should be noted that the majority of empirical studies of dispensary patients completed between the 1870s and World War I indicated that most dispensary patients were not in fact able to pay for medical care.)[36]

By the last quarter of the 19th century the dispensary patient no longer fit into that same vision of an ordered social universe which had guided and inspired the efforts of those benevolent Americans who had founded the first dispensaries a century earlier. Those older views of community and stewardship implied in the contributor-sponsorship system had faded by mid-century, paralleling changes in the environmental reality of America's cities. Similarly, it would have been hardly plausible to argue that New York or Boston tenement-dwellers should be visited in their homes so as to spare them the indignity of hospitalization. The constituency of both hospital and dispensary had changed. By the closing years of the 19th century, the dispensary had very clearly become the provider of charity medicine for a class who—if indeed worthy of such charity—were sharply differentiated from paying patients and who ordinarily lived in a section of the city removed physically from that of contributors, physicians, and private patients. Before mid-century and especially in the first quarter of the 19th century, dispensary managers still sought to enforce requirements that visiting physicians actually reside in the district they served—a natural enough sentiment in the 18th century but impracticable in post-bellum America.[37] The arguments employed by the end of the century to attract contributions had become almost exclusively prudential, appealing little either to explicitly religious convictions or to a feeling of identity with those at risk. Fund-raising circulars emphasized instead the need to avert crime, pauperism and prostitution.

Positive support for the dispensaries was, on the other hand, shaky indeed; aside from the support implicit in the inertia developed by all institutions, only a small group of socially active physicians and proto-social-welfare activists defended the dispensaries as a positive good. Many social workers, as we have indicated, evinced little affection for an institution which seemed to embody so casual and unscientific an approach to philanthropy. Even the oldest dispensaries did not survive as independent institutions past the early 1920s.

Historians have devoted little attention to the dispensary. Yet as our contemporaries begin to concern themselves with the delivery of medical care this neglect may end; for the dispensary provides such would-be reformers with a potentially usable past. The dispensary did at first provide a flexible, informal, and locally oriented framework for the delivery of public medicine. But the analogy to contemporary problems is limited; the flexibility and informality of the dispensary were a result of medicine's still primitive tools, its local orientation a consequence of the contributors being in some sense—or assuming themselves to be—part of the community served by the dispensary. Such conditions ceased to exist well before the end of the 19th century. And even within its own frame of reference, the 19th-century dispensary provided second-class, routine, episodic medicine, was a victim of shabby budgets, and even in its earliest decades marked by unquestioned distance between physician and patient. (A distance *perhaps* made tolerable by traditional attitudes of hierarchy and deference.)

Yet despite these imperfections, the death of the dispensary and the transfer of its functions and client constituency to general hospitals has not been an unqualified success. And though the history of the rise and fall of the dispensary provides no explicit program for contemporary medicine, it does underline a simple moral: any plan for the reordering of medical care must be based on the accommodation of at least three different factors. One is felt social need, felt, that is, by those with power to change social policy. A second factor is general social values and assumptions as they shape the world-view and thus help define the options available to such decision-makers. Third, there are the needs of the medical profession, needs expressed in the career decisions of particular physicians and needs defined by medicine's intellectual tools and institutional forms. Without a strong commitment to government intervention in health matters—a commitment impossible without an appropriate change in general social values—factors internal to the world of medicine have determined most forcefully the specific forms in which medical care has been provided for the American people. Thus the rise and fall of the dispensary; it was doomed neither by policy nor conspiracy but by a steadily shifting configuration of medical perceptions and priorities.

NOTES

1 The most valuable study of the dispensary is still that by Michael M. Davis, Jr., and Andrew R. Warner, *Dispensaries: Their Management and Development* (New York, 1918). The most useful account of the early years of any single dispensary is [William Lawrence], *A History of the Boston Dispensary* (Boston, 1859). For an example of contemporary interest, see George Rosen, "The first neighborhood health center movement—its rise and fall," *Am. J. Pub. Hlth.*, 1971, *61*: 1620–1637.

2 Philadelphia Dispensary, Minutebook 18, June 25, 1816, Archives of the Pennsylvania Hospital, Philadelphia (Hereafter A.P.H.). Cf. "Brief history of the Southern Dispensary," Southern Dispensary, *81st Annual Report* (Philadelphia, 1898), pp. 6–10.

3 Charles E. Rosenberg, "The practice of medicine in New York a century ago," *Bull. Hist. Med.*, 1967, *41*: 223–253, p. 236; F. B. Kirkbride, *The Dispensary Problem in Philadelphia: A Report made to the Hospital Association of Philadelphia, October 28, 1903* (Philadelphia, 1903). By 1900, Davis and Warner, *Dispensaries*, p. 10, estimated that there were roughly 100 dispensaries in the United States, 75 general and 25 special.

4 As late as 1899, the City of Baltimore still compensated the Baltimore Dispensary when it filled prescriptions for the indigent patients of any legal practitioner. Baltimore General Dispensary, *Character, By-Laws. &c. . . . Revised 1899* (Baltimore, 1899), p. 14.

5 New York's Eastern Dispensary reported in 1857 that the city's donation to the New York Dispensary had been set at $1000 in 1827. As other dispensaries were founded, these too received the same subvention. *23rd Annual Report, 1856* (New York, 1857), p. 19.

6 Eastern Dispensary (New York), *25th Annual Report, 1858* (New York, 1859), pp. 16–17. The panics of 1857, 1873, and 1893, as well as the Civil War years, all represented such periods of stress for the dispensaries. New York's North-Eastern Dispensary, for example, was so pressed by the Panic of 1873 that it could not even publish annual reports in 1874 and 1875. *15th Annual Report, 1876* (New York, 1877), p. 6.

7 As late as 1891, Philadelphia's Northern Dispensary bemoaned the fact that they could still not afford to stop renting their second floor, despite their establishment of five new specialty clinics and consequent need for space. Northern Dispensary, *74th Annual Report, 1891* (Philadelphia, 1892), p. 9. As early as 1803, the Philadelphia Dispensary was happy to rent its basement to a commercial tenant. Minutes, Dec. 12, 1803. The typical pattern was illustrated clearly by the New York Dispensary's decision in 1868 to build a four-story building, the basement, first, third, and fourth levels being rented, only the second used by the dispensary itself. *77th Annual Report, 1868* (New York, 1869), p. 12.

8 Eastern Dispensary (New York), *23rd Annual Report, 1856*, p. 22; S. L. Abbott to G. F. Thayer, April 6, 1844, Chronological File, Boston Dispensary Archives, New England Medical Center, Boston. (Hereinafter B.D.A.) The phrase describing the dispensary's casual dentistry is from New York Dispensary, *81st Annual Report, 1870* (New York, 1871), p. 17.

9 For a convenient summary of early vaccination work by the dispensaries, see DeMilt Dispensary (New York), *25th Annual Report, 1875* (1876), pp. 20–22. For the role of the dispensaries in the Civil War, see New York Dispensary, *72nd Annual Report, 1862* (New York, 1863), pp. 9–10; Eastern Dispensary (New York), *28th Annual Report, 1861* (New York, 1862), pp. 23–25. Though the poor were normally uninterested in vaccination, the threat of epidemics often created a sudden upsurge of interest; in one case, indeed, the New York Dispensary could refer to a "vaccination riot" on their premises. *76th Annual Report, 1865* (New York, 1865) p. 20. Many of the dispensaries were financially dependent on their sale of vaccine matter.

10 George Gould, "Abuse of a great charity," *Med. News, N.Y.,* 1890, *57*: 534–539; Medical College of the Pacific, Faculty Minutes, Jan. 29, 1878, July 22, 1881, Lane Medical Library, Stanford University. New York's Eastern Dispensary was so lacking in funds that its patients were given neither bottles nor printed instructions: "The patients universally bring a bottle or tea-cup to receive and hold the medicine." *28th Annual Report, 1861,* p. 14. There were only occasional conflicts between physicians and lay managers in regard to such cutting of corners. A revealing incident of this kind shook the Boston Dispensary in 1844 when the managers sought to compel their visiting physicians to employ scarification and bleeding instead of the far more expensive leeches. The physicians argued not only that the leeches had a different physiological effect—but that they had well-nigh banished more painful modes of bloodletting from private practice. G. T. Thayer to Visiting Physicians, Feb. 8, 1844; S. L. Abbott et al. to Thayer, April 6, 1844; S. L. Abbott et al. to President and Managers [February 1844], B.D.A.

11 For an example of such ties in a particular dispensary, see DeMilt Dispensary, *2nd Annual Report, 1852–53* (New York, 1853), p. 12; *4th Annual Report, 1855* (New York, 1856), pp. 10–11.

12 Eastern Dispensary (New York), *32nd Annual Report, 1865* (New York, 1866), p. 14.

13 New York Diet Kitchen Association, *11th Annual Report, 1883* (New York, 1884), p. 5. On nursing, see, for example, Philadelphia Dispensary, Minutes, Feb. 15, 1853, Oct. 17, 1854, A.P.H.

14 Brooklyn Dispensary, *Trustee's Report, April, 1847* (New York, 1847), p. 8; Boston Dispensary, *108th Annual Report* (Boston, 1905), pp. 10–12. For the crediting of the New York Dispensary with this particular first, see DeMilt Dispensary, *25th Annual Report, 1875,* p. 19n.

15 John G. Coffin, *An Address delivered before the Contributors of the Boston Dispensary, . . . October 21, 1813* (Boston, 1813), pp. 6, 15; John B. Romeyn, *The Good Samaritan: A Sermon, delivered in the Presbyterian Church, in Cedar-street, New York, . . . for the Benefit of the New York Dispensary* (New York, 1810), p. 16.

16 "Servants," one board of managers argued at midcentury, "who have relations and friends in the lower walks of life, and who are in the habit of visiting them, often in company with the children of their employers, would be subject to more danger than they are now exposed." DeMilt Dispensary, *3rd Annual Report, 1853–54* (New York, 1854), p. 10.

17 DeMilt Dispensary, *5th Annual Report, 1855–56* (New York, 1856), p. 11.

18 For typical examples later in the century, see Central Dispensary and Emergency Hospital of the District of Columbia, *24th Annual Report . . . Including an Historical Sketch of the Institution* (Washington, D.C., 1894), pp. 8–10; Camden City Dispensary, *26th Annual Report, 1892–93* (Camden, N.J., 1893), pp. 6–9.

19 Rush to John Dickinson, Oct. 4, 1791, *Letters of Benjamin Rush*, ed. L. H. Butterfield (Princeton, N.J., 1951), I, 610; Philadelphia *Dunlap's American Daily Advertiser*, Aug. 16, 18, 1791; Minutes, Philadelphia Dispensary, Aug. 26, 1791. Cf. [Lawrence], *History of Boston Dispensary*, pp. 90–91, 98–99. Another dispensary noted at the mid-century that they had "often been accused, as being rather the schools, where the young and inexperienced might find patients to their hands, than benevolent institutions where sufferings might be allayed and diseases cured." DeMilt Dispensary, *2nd Annual Report, 1852–53*, p. 8.

20 Surviving archives of the Boston Dispensary, for example, indicate in letters of recommendation for district physicians the pattern we have suggested: The Bigelows, James Jackson, and Oliver Wendell Holmes recommend and are recommended. Successful candidates had frequently studied in Europe and the Tremont Medical School or served as house physicians at the Massachusetts General Hospital. Cf. James Jackson to Board of Managers, Aug. 3, 1831; O. W. Holmes to William Gray, Sept. 3, 1845, or see letters in 1836 file from John Collins Warren, Jacob Bigelow, and George Hayward recommending O. W. Holmes as a visiting physician. B.D.A.

21 New York Dispensary, *Annual Report, 1854* (New York, 1855), p. 9. A year later, the same dispensary contended that their staff members "in a few years, hope to be the eminent physicians of New York, and it is their right to expect, and of the community to require, that the unequaled advantages, to be found here, should be freely offered them." *Annual Report, 1855* (New York, 1856), p. 10. *Boston Med. & Surg. J.*, 1839, 20: 351.

22 New York (City) Dispensary, *Charter and By-Laws . . .* (New York, 1814), p. 8. Another indication of the social assumptions of the generation which created the dispensaries was their concern over whether servants and apprentices were appropriate patients. John Bard argued in New York that servants should indeed be treated, but not at their place of work—which would have compelled "gentlemen to visit the servants of families in which they had no acquaintance with the Masters or Mis-

tresses." *A Letter from Dr. John Bard . . . to the Author of Thoughts on the Dispensary . . .* (New York, 1791), p. 20. See the entry for July 17, 1786, in the Minutes of the Philadelphia Dispensary for the question of treating apprentices.

23 Philadelphia Northern Dispensary, Philadelphia Lying-In Hospital, "Rules and Regulations, Adopted November 4, 1835," Historical Collections, College of Physicians of Philadelphia; [?] to Board of Managers, Oct. 1, 1830, B.D.A.; *Rules of the Philadelphia Dispensary with the Annual Report for 1869* (Philadelphia, 1870), p. 10. As late as 1879, the organizers of a specialized New York dispensary contended that they appealed to those patients able to pay a small fee and thus "saved the necessary associations of a public, free dispensary." *Report of the East Side Infirmary for Fistula and other Diseases of the Rectum* (New York, 1879), p. 5. Cf. Pittsburgh Free Dispensary, *3rd Annual Report, 1875* (Pittsburgh, 1876), p. 9.

24 Philadelphia Dispensary, Minutes, Dec. 26, 1845, A.P.H.

25 Northern Dispensary (New York), *1st Annual Report, 1828* (New York, 1828), cont. p. 10; DeMilt Dispensary, *2nd Annual Report, 1853*, p. 12; Bigelow to D. D. Slade, Aug. 22, 1855, B.D.A. Cf. D. D. Slade to My Dear Sir [William Lawrence], Sept. 3, 1855, B.D.A.; [Lawrence,] *History of the Boston Dispensary*, pp. 178–180.

26 Northern Dispensary (Philadelphia), *Annual Report, 1847* (Philadelphia, 1848), p. 10. Such sentiments were familiar ones. A "Contributor" to the Boston Dispensary explained in 1819 that its appropriate clients were those "many persons . . . who have been reduced from a state of competence to one little short of poverty, who while blessed with health, can, by industry, support themselves, but when attacked by sickness, and laid upon a bed of illness, find it impossible to pay the physician and apothecary." *New-England Palladium and Commercial Advertiser*, Jan. 12, 1819. The earliest rules of both Philadelphia and Boston dispensaries emphasized their wish to comfort "those who have seen better days . . . without being humiliated." Boston Dispensary, *Institution of . . . 1817* (Boston, 1817), p. 7.

27 In the New York Dispensary, for example, in 1853, of 7,188 patients treated, 1,582 were born in the United States and 4,886 in Ireland. At the Philadelphia Dispensary in 1857, 1,906 were born in the United States, 3,649 in Ireland. New York Dispensary, *64th Annual Report, 1853* (New York, 1854), p. 12; Philadelphia Dispensary, *Rules . . . with Annual Report for 1857* (Philadelphia, 1858), p. 14. Some dispensaries would not allow venereal cases to be treated, some imposed a special fee, while still others

allowed individual physicians to decide whether they would treat such errant souls.

28 Luther Parks, Jr., to Board of Managers, June 10, 1850, B.D.A. Referring to the Irish, another dispensary physician explained: "Upon their habits, —and mode of life, depend the frequency and violence of disease. This I am fearful will continue to be the case, since no form of legislation can reach them, or force them to change their habits for those more conducive to cleanliness and health." Charles W. Moore to Board of Managers, April 1, 1857, B.D.A. On the temperance question, see, for example, J. B. S. Jackson to Board of Managers, Oct. 8, 1853, B.D.A.

29 New York Dispensary, *64th Annual Report, 1853,* p. 10; New York Dispensary, *72nd Annual Report, 1862,* p. 20. When, in an effort to solve this problem New York's dispensaries initiated a small deposit to be refunded when the patient returned to have the vaccination checked, these intractable—and seemingly ungrateful—patients chose to regard it as a payment absolving them of any responsibility to the institution.

30 J. Trenor, Jr., physician to middle district, Eastern Dispensary (New York), *25th Annual Report, 1858,* p. 32; New York Dispensary, *Annual Report, 1837* (New York, 1838), p. 7.

31 Edward Reynolds, *An Address at the Dedication of the New Building of the Massachusetts Eye and Ear Infirmary, July 3, 1850* (Boston, 1850), p. 15; Mission Hospital and Dispensary for Women and Children, *2nd Annual Report, 1876* (Philadelphia, 1877), p. 10. The scarlet fever reference was by William Bibbins, DeMilt Dispensary, *6th Annual Report, 1856–57* (New York, 1857), p. 17.

32 For useful descriptions of the tuberculosis clinics, see, for example, F. Elisabeth Crowell, *The Work of New York's Tuberculosis Clinics . . .* (New York, 1910); Louis Hamman, "A brief report of the first two years' work in the Phipps Dispensary for tuberculosis of the Johns Hopkins Hospital," *Johns Hopkins Hosp. Bull.,* 1907, *18*: 293–297. For samples of the more positive defense of the dispensary and its appropriate role, see: S. S. Goldwater, "Dispensary ideals: with a plan for dispensary reform . . . ," *Am. J. Med. Sci.,* 1907, n.s. *134*: 313–335; Richard Cabot, "Why should hospitals neglect the care of chronic curable disease in out-patients?" *St. Paul Med. J.,*

1908, *10*: 110–120; Cabot, "Out-patient work: The most important and most neglected part of medical service," *J.A.M.A.,* 1912, *59*: 1688–1689; Good Samaritan Dispensary (New York), *29th Annual Report, 1919* (New York, 1920), p. 8. The most complete statement of a positive dispensary program is to be found in Davis and Warner, *Dispensaries.*

33 Philadelphia Dispensary, Minutes, Jan. 8, 1923, A.P.H. At the end of the 19th century, for example, the Boston Dispensary began a search for beds; it seemed a necessity if bright young men were to be kept on the staff. *Report of the Dinner Given to the Board of Managers of the Boston Dispensary by the Staff of Physicians . . . January 25th, 1909* [Boston, 1909], p. 13. Once allied with a hospital, the dispensary had invariably a lower status. Francis R. Packard charged in 1903 that hospitals would casually spend two or three hundred dollars for new surgical instruments yet balk at ten or fifteen for the dispensary. F. B. Kirkbride, *Dispensary Problem in Philadelphia,* p. 21.

34 Probably most significant is the tone of this debate. It was the ordinary practitioner who generally resented the way in which dispensaries with their elite house staffs attracted cases which might otherwise have remained in the hands of private practitioners. Discussions of "dispensary abuse" also served to express the resentment of many practitioners against the monopolization of hospital and dispensary posts by a minority of well-connected physicians. In its report on charity abuse, for example, the Medical Association of the District of Columbia also urged limited tenure in hospital staff appointments and access to hospital privileges for all "reputable members of the profession." *Report of the Special Committee . . . on the Hospital and Dispensary Abuse in the City of Washington* (Washington, D.C., 1896), pp. 15–16.

35 [William Lawrence], *Medical Relief to the Poor, September, 1877* (Boston, 1877), pp. 3–4; James Keiser, "The abuses in hospital and dispensary practice in Reading," *National Hospital Record,* 1899.

36 Albert P. Francine, "The state tuberculosis dispensaries," *Penn. Med. J.,* 1914, *17*: 940. See Davis and Warner, *Dispensaries,* pp. 42–58, for a brief discussion of patient eligibility.

37 G. F. Thayer to William Gray, April 12, 1838, B.D.A.

20

Patrons, Practitioners, and Patients:
The Voluntary Hospital in Mid-Victorian Boston

MORRIS J. VOGEL

The hospital of the immediate post-Civil War period differed little in some respects from its colonial and early 19th-century predecessor. It treated the same socially marginal constituency that American hospitals had always served. Its patients were the poor and those without roots in the community; dependence as much as disease still distinguished them from the public at large. Yet in some other respects the hospital of this era displayed concerns that were typically Victorian — concerns that shaped the transition of the institution into the hospital as we know it.

The general hospital of the 1870s was likely to be a charity, linking the voluntary efforts of doctors and donors in providing free medical care for those without any suitable alternative. For a hospital to exist, doctors had to be willing to provide gratuitous medical service for the sick poor while feeling sufficiently remunerated that they eagerly sought hospital positions. Donors had to be willing to support an institution that they themselves were never likely to use.

Traditionally, Boston's physicians had provided free care for the sick poor who had sought them out. Self-consciously advancing their claim to be professionals rather than businessmen, medical practitioners recognized a responsibility not to refuse advice or treatment to those who could not pay.[1] But in the hospitals and dispensaries organized up to the very end of the 19th century, many

MORRIS J. VOGEL is Professor of History at Temple University, Philadelphia, Pennsylvania.

physicians went well beyond their professional obligations and actively made themselves available to patients who could not, and in most instances were forbidden to, pay any fee. Not only did doctors seek duties in such institutions, but often actually founded them, as in the case of inexpensively operated dispensaries, providing only outpatient care. In the case of hospitals, doctors shared leading roles in organizing them with those who provided financial backing.[2]

Hospital and dispensary staff members were part of the city's medical and social elite. They were close in social origins to the donors who supported Boston's voluntary Protestant hospitals, if not directly related to them.[3] This background was part of the reason for their hospital work. The gratuitous treatment they rendered hospital patients was in the same tradition of stewardship as the charitable donations that supported voluntary hospitals.

But free medical treatment was much more than a charitable obligation. As a further consequence of social position, hospital practitioners had professional qualifications and interests that set them apart from their less fortunate medical brethren. In a period when locally available medical training was not advanced, men who later became associated with the city's hospitals were more likely than others to have enjoyed a European medical education after initial training in Boston. Once established in Boston, these upper-class doctors were more likely to assume positions in medical school faculties. And, though conservatives of their own class and background sometimes opposed specialization and even certain imported innovations, young physicians returning from Europe in the second half of the 19th century embraced specialization and the increasing scientific content of medicine more readily than Boston doctors less privileged by birth and social standing.[4]

Hospital positions furnished these upper-class doctors with the clinics that were becoming increasingly necessary for medical school teaching. Teaching brought financial benefits, as former students referred difficult cases to former professors for paying consultations. A hospital position also enabled a medical man to see and treat numbers of special cases, comparatively rare in private practice, and so develop a reputation that would itself be remunerative. Thus, though hospital patients did not pay fees to hospital practitioners, these men received what contemporaries referred to as "certain well-understood advantages."[5]

Hospital physicians earned their livelihoods in the care of well-to-do private patients who paid for the knowledge gained in hospital work. Private practice remained the norm. And because the 19th-century hospital was not the center of the doctor's work world, the few hours he put in there each day during his term of perhaps three months each year did not represent income lost.

Economic motives led nonelite doctors to complain about the abuse of charity they perceived in the medical care offered without fee in hospitals. They saw their natural clientele—the poor and working classes—drained off to the hospitals.[6] When the nature of the hospital patient population changed in the 1890s and in the first decade of the 20th century, complaints about abuse came from a new quarter—from doctors who treated the well-to-do patients beginning to enter hospitals at the turn of the century.[7] These complaints were a significant force in leading the hospital away from its purely charitable organization. But until late in the 19th century, Boston's hospitals, whether municipally or voluntarily supported, were charities.

In part, the wealthy supported these institutions because of their connections with the physicians who staffed them. Amos Lawrence, the mercantile prince, underwrote the entire cost of a children's hospital under the charge of his son, Dr. William R. Lawrence.[8] The staffing of South Boston's Roman Catholic Carney Hospital by Back Bay physicians brought in financial contributions from their friends and families.[9]

In part, too, the wealthy supported these institutions because enlightened selfishness led them to share certain of the physicians' goals. The knowledge and experience doctors gained in treating the poor "raised the standard of medical attainments"; hospital and dispensary practice thus "proved a blessing to rich and poor alike."[10] The Children's Hospital appealed "to all those who have children of their own," reminding them that they had a "double interest" in the institution; "not only on account of the great benefit it will confer on its little inmates, but also because of the advantages it offers for the study of special diseases by which their own offspring may be afflicted."[11] The *Boston Evening Transcript* warned the fortunate that their own well-being depended on the continued well-being of hospitals:

> The aids which society distributes to the hospitals are amply restored by the hospitals to society. . . . Mainly in these institutions the experience and insight, the methods of observation and treatment, the scientific research, are evolved which become employed for the general health of the country. . . . If we could imagine the hospitals abolished, the general death rate in all private practice would be increased.[12]

However, the "double interest" remained a divided interest, for the more fortunate classes did not expect to make direct use of the general hospital in the 1870s. Home care remained the norm. Accident victims, for example, though they might be injured outside the home, were likely to be brought home and cared for there. Speaking in 1864 at the dedication of the Boston City Hospital, its president, Thomas C. Amory, Jr., acknowledged that it was unlikely that hospitalization would replace the ideal of home care. Amory gave an example of what he regarded as a futile attempt to remove the prejudice against hospital care:

> One of our former governors . . . meeting with an accident in the street from which he narrowly escaped with his life, insisted, in order to remove this prejudice, upon being carried to the [Massachusetts General] Hospital. His example may have had its effect. But we doubt if many of our own people, born in Boston, when tolerably comfortable at home, will go, when ill, among strangers to be cured.[13]

When Amory himself was run down by a streetcar in 1886, "a doctor was called and the injured gentleman was removed to his home in carriage."[14]

The pattern of care obtaining at local railroad

accidents is revealing.[15] When a commuter train crashed near Roslindale in 1887, 24 passengers were killed and 14 hospitalized. But most of the nearly 100 victims were taken to their homes.

> The fact that the accident occurred in the midst of a settled suburban district, and that nobody upon the train was more than five miles away from home, made it possible to transport the dead and injured, so far as it was practicable under the circumstances, directly to their homes, and many were so taken.[16]

The severity of their injuries did not separate those hospitalized from those brought home. Only two of the six admitted to the Boston City Hospital were listed as seriously injured, and only one of eight at the Massachusetts General. Of the cases brought to their own homes, a doctor making 55 home visits the day after the wreck reported nine patients in dangerous condition.[17]

The hospital offered patients no medical advantages not available in the home; actually hospital treatment in the 1870s added the risks of sepsis or "hospitalism." The fact that the hospital offered no special medical benefits reinforced a resistance to hospitalization that stemmed from the role of the home as the traditional setting for those undergoing illness and from a negative image of the hospital. That image derived from the actual danger of hospitalism and the traditional identification of the hospital with the pesthole and almshouse. Thus even when home care was unavailable, hospital care was sometimes shunned.

The well-to-do might make use of a hospital in what were labeled peculiar circumstances. This category included individuals away from home because they were from out of town, and old people living alone. For these potential patients, limited separate facilities existed at both the Massachusetts General and Boston City hospitals.[18]

Even the sick poor would avoid the hospital if possible. One of the stated advantages of a dispensary was that outpatient care sidestepped the "dread" which the prospect of hospitalization evoked among many of the poor.[19] The city's two diet kitchens, founded in the 1870s, supplied home meals for dispensary patients too sick or poor to secure their own food.[20] The truly unfortunate shared with that minority of the prosperous classes who used the institution the problem of an inade-

quate or nonexistent home. The Boston Lying-in Hospital received some of its cases from dispensary physicians "who suddenly found themselves called upon to attend some poor woman in quarters utterly unfit for such purposes."[21]

The hospital offered shelter and attention to the sick poor. It replaced comfortless homes "in close courts, narrow alleys, damp cellars or filthy apartments, which the sunshine never enters, nor fresh air purifies." It made up for the absence of "natural protectors" for those without families. It provided relief for "helpless people, who would suffer tenfold more from neglect and ill treatment than they now suffer from disease, were it not for the shelter and care of the hospitals."[22] Indeed the role of hospital was defined in terms of the services it offered the sick and injured victims of a catalog of social ills.

The statistics of hospital use reflected these concerns. An analysis of nativity and occupation shows that patients treated at the Massachusetts General and Boston City Hospital in the 1870s were not a cross-section of the population.

Hospital annual reports listed the occupations admitted; these occupations may be organized according to the socioeconomic classification in Stephan Thernstrom's *The Other Bostonians* and then compared with Thernstrom's observations about the occupational structure of the city.[23] Male patients at the city's two major hospitals were divided into four categories: white-collar, skilled blue-collar, semi-skilled and service, and unskilled and menial. Such an analysis shows occupations with high socioeconomic status were underrepresented among hospital patients in 1870 and 1880, while those with low status were overrepresented.

At the Massachusetts General in 1870, 16.9 percent of the classifiable male patients were in white-collar occupations, while in 1880 that figure was 18.1 percent. In the city population, 32 percent of males were white-collar in 1880. Skilled blue-collar workers accounted for 41.9 percent of Massachusetts General patients in 1870 and 19.4 percent in 1880, while they provided 36 percent of the general male population in 1880. Among patients, 14.2 percent and 11.1 percent were in semi-skilled occupations in 1870 and 1880 respectively, while the city population contained 17 percent in that category in 1880. The unskilled accounted for 26.9 percent (1870) and 51.4 percent (1880) of the patient population and 15 percent of the city population in 1880.[24]

Much the same pattern prevailed at the Boston City Hospital. The largest single occupational category among patients was laborer, consisting of 524 of 1,419 men admitted in 1870/1871 and 792 of 2,696 in 1880/1881. Patients in white-collar occupations totaled 8.2 percent of the hospital's male admissions in 1870/1871 and 10.5 percent in 1880/1881. Skilled blue-collar workers accounted for 36.3 percent of Boston City patients in 1870/1871 and 31.6 percent in 1880/1881. Workers in semi-skilled and service occupations made up 11.8 percent of City Hospital patients in 1870/1871 and 21.7 percent in 1880/1881. As at the Massachusetts General, unskilled (including laborers) and menial workers — 43.5 percent and 36.1 percent — were disproportionately over-represented.[25]

Unfortunately, the listing of many women patients as simply wives or widows, and the absence of a satisfactory analysis of the female occupational structure, makes a comparison of female patients with the general female population more difficult. But the fact that nearly half the female patients admitted to both hospitals in the 1870s were identified as domestics reinforces the conclusion based on male employment patterns that hospital patients were drawn disproportionately from among the lower classes.[26]

Though the absolute and relative numbers undergoing hospitalization continued to increase in the 1870s as they had since the city's first hospital opened in 1821, the hospital's constituency remained largely the same, with the greater number of patients coming from an expanded lower class. The continued use of the institution by the stricken and helpless poor served to associate it with the almshouse and reinforced the negative image of the hospital held by society at large. Its image as a refuge for the unfortunate was further heightened by the fact that its patients were not just poor but, after the beginning of large-scale immigration at mid-century, largely foreign born. The Massachusetts General trustees had at first resisted allowing the Irish to enter the hospital as patients, claiming that "the admission of such patients creates in the minds of our citizens a prejudice against the Hospital, making them unwilling to enter it, — and thus tends directly to lower the general standing and character of its inmates." Feeling "the ex-

cess of foreigners among the patients" to be a bane, they had advised the admitting physician to use "the utmost vigilance," but found that "some such admissions must unavoidably take place." Hospital rules directed that all cases of sudden accident were to be admitted, thus bypassing the screening procedure; a very large proportion of accident cases was Irish. In time, the Massachusetts General trustees, "moved by a sense of duty and humanity," opened their wards to the foreign-born.[27]

In 1865, the hospital accepted 628 foreign as against 571 native-born patients. In 1870, the totals showed 718 foreign and 584 Americans, and in 1875 the figures were 1,042 and 799 respectively. The Irish made up the largest segment of the foreign-born population, maintaining at least a majority throughout the 1870s, with those born in the Canadian provinces second.[28] From the opening of the Boston City Hospital in 1864, a majority of its patients was foreign born. In 1865, 647 of its patients were born abroad and 459 in the United States, and in 1870/1871, 1,635 were foreign and 761 native-born. The foreign-born numbered 2,187 and native Americans 993 in 1875/1876. Throughout the period, Irish patients alone outnumbered the native-born.[29]

Just as Irish immigrants tarnished the image of the general hospital, so did the kind of women it cared for taint the image of the lying-in hospital. Maternity care would be among the last reasons causing the comfortable classes to enter hospitals. In the late 19th century women still considered childbirth a natural function, something that could, and should, be performed in the simplest and poorest home.[30] The hospital offered no specialized medical paraphernalia or contrivances, but instead threatened contagion, puerperal fever, and high maternal mortality. Generally, only the most desperate women entered hospitals to have their children. And perhaps the major cause of this desperation was illegitimacy. Small lying-ins, often no more than a few rooms in a tenement or boarding house, kept by midwives or the unscrupulous and untrained, served those seeking "to hide their shame" or having absolutely no alternative. These lying-ins, and the baby farms that sometimes accompanied them, were seen as accessories to vice and degradation, and as adjuncts to brothels. Lying-ins were the first hospitals needing licenses to operate in Massachusetts (1876), but the enforcement problems of the Boston Board of Health suggest that more lying-ins were operated without sanction of law than with it.[31]

Licensed and respectable lying-ins did exist, and did leave records, but their patients, too, were "unfortunate women." Cases included in the first volume of the maternity records of the New England Hospital for Women and Children, covering one-and-one-half years in the early 1870s, list 61 married and 57 unmarried mothers. Over 50 percent of the more than 1,300 mothers delivered in the 1870s at the Boston Lying-in were unmarried. And at St. Mary's Lying-in Hospital, only 20 of 550 patients cared for in the decade from 1874 to 1884 were married.[32] The lying-in hospitals of the period reaffirmed the notion that hospitals were institutions especially for the poor and desperate, and the illegitimacy intimately associated with them added the stigma of immorality.

The hospital was perceived as the kind of place all but the desperate would want to avoid. Yet, although it dealt primarily with the poor, its very nature — the onmipresence of death within its walls — imbued its concerns with a powerful attraction. The community at large was curious as to what went on inside it. This desire to know was heightened by the relative newness of the institution; though its history could be traced back to antiquity, hospitals began to emerge in numbers in Boston and the rest of the nation only after the Civil War. Finally, the curiosity as to what went on within hospitals derived from the fact that even the fortunate individual could not be certain that he would not someday be hospitalized.

Horror was a common response to such a prospect. Joseph Chamberlin, for many years the *Transcript's* "Listener,"[33] reacted strongly after visiting a hospitalized friend:

> If it should fall to the Listener's lot to be called upon to go in sickness to the very best of [hospitals], he would say, "Better a straw cot in an attic at home, with the clumsiest of unprofessional attendance, than the best private room in this place." . . . [T]here is something about the all-pervading presence of Sickness with a large S, this atmosphere of death, either just expected or just escaped, and all of this amiable perfunctoriness of nursing and medical attendance, that is simply horrible. The hospital . . . gives one sickness to think about morning, noon and night.[34]

A visitor might be acutely discomforted by the unnatural concentration of disease and death. But for the patient the environment was threatening:

> The doctors visit you incessantly, and, in spite of their courtesy, you feel as if you were not exactly an ailing human being, but merely a "case" that was being read as one reads a novel which is interesting enough, no doubt, but which is expected to develop a much more interesting phase, to wit, the catastrophe, at almost any moment. And then the grim disquieting presence of all these people like you in the ward around you![35]

The hospital reaffirmed the patient's mortality, but denied his humanity.

Chamberlin told the story of a patient hospitalized for an operation. After surgery, she was put to bed. She lived through a night punctuated by the "wailing and crying" of fellow patients, the death "in dreadful agony" of a neighboring patient, and the quiet but quick, and therefore ghostly, movement of attendants. It was terrifying: "Why it was like being dead and conscious of it!" The next day was quiet, but "spent in anticipating the coming of such another night, was almost as terrible." This particular hospital stay was cut short when a physician inquired "whether I had not any friend to whom I could go," found she had and "made immediate arrangements to have me taken away." Clearly, this patient and many of her contemporaries shared Chamberlin's conclusion: "What a matter for infinite sorrow it is that there should be homes in the world so dismal, so unhealthy, so ill attended, that their inmates are better off in the public wards of the hospital, when they are sick, than they are at home."[36]

Because the hospital was a strange and frightening place, the public welcomed reassurance from the informed. This might take the form of a newspaper article giving the generic history of the hospital and thus implying that it was not simply a modern aberration but an old institution that had proved its value in the past.[37] Or it might take the form of a correspondent's story of a hospital visit or a patient's description of his hospital stay.

Chamberlin's story was idiosyncratic: more common in the Boston press were counterphobic presentations that were almost uniformly formulaic. These addressed fears based on ignorance and substituted for them informed chronicles which denied the presence of death, disease, and pain in the hospital. Insanity, for example, was not mentioned in an extensive account of an insane asylum, though beautiful flowers and homelike accommodations in cottages were.[38] The smallpox hospital emerged from another narrative as a delightful place, serving wonderful food and providing comfortable beds, while smallpox itself, it was concluded, much improved the system.[39] For those hospitals which depended on the beneficence of the public to operate, reassurances that all was well within served a double function in that they encouraged contributions as well as disarmed anxieties.

Boston's City Hospital was supported as a municipal service, but the hospital tradition in Boston, as in the rest of the United States, had been set by the voluntary hospitals, with groups of private individuals undertaking the care of the sick poor as a public trust.[40] The Massachusetts General Hospital and Children's Hospital formed part of the complex of Boston's Protestant charities that owed their founding and existence, at least in part, to the religious doctrine of stewardship. Social and economic inequalities were legitimized by the notion that God meant for them to exist. But the elect, whose heavenly salvation was generally already demonstrated by their earthly riches, held their wealth only as God's trustees. With their wealth came the obligation to aid the less fortunate. The poor provided their economic betters the opportunity, the privilege actually, of spending God's wealth in a way that continually reemphasized their own chosen state.

The Children's Hospital was founded in 1869. It was intended for the poor, for "the little waifs who crowd our poorer streets." In its early years the institution stressed its spiritual role. Making their first annual report, the trustees stated that the institution would provide its patients "Christian nurture."[41] Sickness provided an opportunity for spiritual healing; the philosophy of the Children's Hospital reflected that of a local newspaper, which editorially downgraded the function of hospitals in furnishing medical treatment while commending them for giving patients the best gifts of all, "wrought through a ministry of sorrow."[42] The theological language in which all this was expressed was largely a carry-over from an earlier time; religious terminology provided a familiar and convenient vocabulary. Soon, society would no longer justify the hospital in traditional religious terms. One can already sense the beginning of a shift in the hospital's mission during Victorian times, from succoring the sick poor as its role in God's order, to denying that man had to accept God's diseases. Within a generation, this would give the hospital a drastically altered justification. But in the 1870s, these religious terms still symbolized real moral and social concerns.

The Children's Hospital defined its role in terms of the "moral benefit" it offered its patients. Socially, these benefits translated into a program of uplift and social control which it was hoped would help cope with the masses of threatening and increasingly alien poor crowding into the city. The trustees had expected that most of their patients would come from the poorest classes of the community. They found that many came "from the very lowest; from abodes of drunkeness, and vice in almost every form, where the most depressing and corrupting influences were acting both on the body and mind."[43] Hospitalization provided an opportunity to separate these children, at a most impressionable time in their lives, from corrupting influences that, if otherwise permitted to proceed unchecked, could perpetuate an impoverished and vicious class, permanently threatening society.

When a child entered, the hospital first decontaminated its new charge. "On their entrance they are immediately placed in a refreshing bath and clothed in the clean robes of the hospital." Uniform red flannel jackets replaced streetclothes.[44] The decontamination process went deeper; new influences were substituted for old in the hope that, in the few weeks it had, the institution could "help the child-soul to lift itself out of the mud in which it had been born, to assert its native purity in spite of unfortunate surroundings."[45]

Since treatment was not purely medical, the hospital did not restrict its practitioners to the medical profession. The entire Christian community was invited to participate in the healing process, to visit patients and encourage them "by word or counsel."[46] The hospital's first nurses were Episcopalian nuns. Their strength lay less in medical training than in the "Christian nurture" they provided patients. Sister Letitia was a model of this style of charity untainted by medical pretension. "Though enfeebled by disease of the lungs, which she knew must soon terminate her life, yet entirely forgetful of self," she continued nursing—all the while, of course, subjecting her charges to tuberculosis—until she died.[47]

The trustees wanted "to bring [their young patients] under the influence of order, purity and kindness." Among the means employed were tender nursing, books, pictures, "little works of art," and "the visits and attentions of the kind and cultivated."[48] Middle- and upper-class children outside the hospital were encouraged to undertake the painting, as wall decorations, of inspirational mottos that would "cultivate the devotional feelings" of the little sufferers inside. The fortunate who supported the hospital were encouraged to visit the children in its wards at any time of the day, to speak with them, provide role models, and in general to furnish that cultivated influence which the children of the poor had missed.[49] At the same time, parents having children in the hospital were severely restricted in the hours they could see their own children. The original parents' visiting hour allowed one relative at a time between eleven and twelve o'clock on weekdays only, raising difficulties for working fathers (or mothers) who wished to visit. Later, parent visiting was further restricted to the hour between eleven and twelve on Monday, Wednesday, and Friday only.[50] The trustees hoped that this regimen would change the children by "quickening their intellects, refining their manners, and encouraging and softening their hearts."[51]

Supporters of the institution hoped that a different child would leave the hospital than had entered it.[52] Children would leave having been "carefully taught cleanliness of habit, purity of thought and word" and with as much attention "paid to their moral training as can be found in any cultivated family," but the benefits of the hospital would not stop there:

> Think what a widespreading influence this becomes when the children return to their homes. . . . Even among the better class of poor people, the children soon notice the discomforts of careless, untidy habits, and are quick to compare such with the "so much better" at the hospital. In the joy of the child's homecoming, the parents are ready to gratify it by trying the new ways, and all unconsciously rise a little in the social scale by so doing.[53]

"In this wise," the hospital's founder wrote, the institution would "commence the education of the poorer classes."[54]

Even if the child did not go home and improve his family, he himself would be changed by the hospital in a way that would benefit society. The affluent and cultivated were told that they could not tell the difference between their own children and those within the hospital, even though the latter might be immigrant children from the North End. One visitor noted that "the faces of the children quickly lost the expression which we commonly meet in our little street Arabs, and become once more human and civilized."[55] Their hearts softened by kindness, mistrust and hostility evaporated from their faces and they no longer appeared as threatening as they had on the streets.

The hospital promised other far-reaching improvements. The health and strength gained during a hospital stay would not only aid the children, by enabling them to grow into "better men and women," but society as a whole—having escaped childhood invalidism, those healthier adults could support themselves. A promotional article mentioned the institution's success in educating its charges, even implying that it taught some how to read.[56] A hospital stay could help prepare a child for a socially desirable role in adult life.

These perceptions were colored, of course, by expectation. No doubt they express more than actually happened in the hospital in the way of having the children of the poor fulfill the fantasies of the rich. Further, these social expectations were less than the full rationale for the institution. The Children's Hospital was founded by physicians, in part for the sorts of professional reasons earlier suggested. At the same time, however, the founding physicians were responsible for much of this socially oriented promotional rhetoric. There is no reason to believe these doctors did not take their own language seriously. They were members of a social class as well as of a professional group, and shared the didactic concerns typical of Victorian culture.

The Massachusetts General Hospital, a secularly oriented Protestant hospital in the same sense as the Children's Hospital, also began with the mission of uplifting its patients. When founded, it had been intended for native American patients, and had offered to tide them over a bad time and send them on their way having meanwhile reinforced their view of a basically good society in which they could lead good lives.[57] But after the hospital had been overwhelmed by unappealing and apparently intractable and unimprovable adult immigrants, it gave up this aspect of its role. By the 1870s, its literature no longer expressed concern for the character of its patients, and the hospital continued to care for the sick poor with a diminished concern about what it was doing for its patients in a nonphysical sense. The hospital kept the support of its donors for a variety of reasons, the chief (probably) being an inertia in which benefactions served as a quiet reaffirmation of stewardship. An obligation to keep the hospital going because it served the needs of medical practitioners was recognized. Finally, the fact that the McLean Asylum for the Insane was a branch of Massachusetts General and served the upper classes in a very direct way maintained their interest in the corporation. Since the asylum generally met its operating expenses from patient revenues, the contributions it generated helped support the hospital.

Yet the loss of reforming zeal brought no relaxation of discipline within Massachusetts General; if anything, it reinforced it. The "influence of order" which pervaded the Children's Hospital furthered that institution's resocialization of its young patients. Order was a concern in Massachusetts General, too, but there it was a reflection of social reality, not part of a vision of social change. Many of the hospital's patients were not bedridden, but able to move around the wards and grounds, and expected to be able to leave the institution to walk about the city or enjoy the carriage rides into the countryside furnished by the Young Men's Christian Union. The hospital treated people new to urban life (through the 1870s the percentage of its patients born in Boston never approached 10 percent) and to the demands of institutional living. To help maintain discipline, its grounds were surrounded by a high wall and an always guarded gate through which patients and visitors had to pass.[58] Patients needed signed passes to leave and reenter the hospital, and visitors were carefully screened.[59] This discipline was maintained for internal reasons; rather than reform a patient who misbehaved, the hospital expelled him.[60]

In one sense, Massachusetts General helped keep order in the general community. Like other hospitals, it functioned as a guarantor of social

stability, or as one supporter of the Children's Hospital put it, "There is a practical side to this charity, which may commend it to thoughtful men." Hospitals provided the working classes with evidence that the wealthy were aware of their responsibilities: "the only sure way to reconcile labor to capital is to show the laborer by actual deeds that the rich man regards himself as the steward of the Master."[61] Until workmen's compensation went into effect, corporations, especially railroads and street railways, underwrote free beds at the Massachusetts General to which they sent employees injured on the job. These accident-prone enterprises provided a paternalistic form of insurance, absolved themselves of responsibility to their injured employees, and attempted to defuse issues that might otherwise build up workers' grievances.[62]

Concern for social order was apparent in the community's support of hospitals in general. One observer noted that "the hospitals act as a kind of insurance system for the laboring classes. They take the risks incidental to their position the more cheerfully, because they know that if injured they are assured of a special provision for their need in our hospitals."[63] Similar reasoning was used to elicit support for the Children's Hospital. Were it not there, or were it unable to admit a suffering child, there would be no telling what even the most respectable worker, distraught over his inability to secure aid for his child, might do. "It is under such circumstances the iron enters a man's soul, and he is ready for a 'strike' or any other desperate remedy that promises better times and money with which to provide good nursing and delicacies for his suffering children." A mother, turned down when applying for admission for her sick child, might go "fiercely on her way, ripe for any evil deed. . . ." But assured that the hospital would care for their sick children, the poor would respond with gratitude rather than violence.[64]

Beside acting as a guarantor of social stability, the hospital was perceived by some as contributing to community prosperity, both through the more tractable labor force it ensured and through the fact that the health of the population was directly translatable into material wealth. In encouraging support for the city's voluntary hospitals, the *Transcript* editorially assured "those who look into the matter [that they would] see that our hospitals are among the very bases of national health and prosperity, and the working of these institutions is, therefore, a matter of general interest and public importance."[65] The hospital thus served much the same function as the public school, to which it was sometimes likened by those arguing that the institution served the entire community and those needing it should use it as a guaranteed right with any cost borne by the community. But of course only a minority fully appreciated all the functions of a hospital; Boston's City Registrar complained that too many of those "by mere fortuitous circumstance different situated" had yet to learn that "the material condition of the whole community is involved in this subject."[66]

Though hospital treatment of the poor protected the established order and added to the wealth of the community, the people for whom it was intended were made to feel recipients of charity and reminded repeatedly that they were enjoying a privilege and their gratitude was expected in return. This attitude was embodied in law. In one case, a man treated gratuitously sued the Massachusetts General Hospital, claiming his broken leg was set improperly. The courts held that even if he had been treated incompetently and negligently, he was not entitled to recover because the institution was a charity.[67] In another case, a woman charity patient operated on at the Free Hospital for Women sued, claiming her operation was not successful. During the course of protracted litigation, "A Friend to our Charities" wrote the *Transcript* complaining that such hospital malpractice suits arose because "there are some patients so wholly devoid of ordinary gratitude for favors to which they had not a shadow of a claim, as to make their benefactors suffer by reason of their very kindness." When a verdict for the hospital was finally returned, the *Transcript's* headline, "A Victory for Charity," translated the jury's decision into a reaffirmation of the status of the hospital, though only meaning to imply that money given hospitals would not be drained by lawsuits.[68]

It is from this background that the modern hospital emerged. Changes in social attitudes and medical practice—products of social change and scientific progress—have reshaped the institution. But the hospital that was transformed by these forces was itself a shaping force, a product of its own past. It has influenced medical organization and the kind of medical care available. The hospital now cares for patients of all classes, but it is not a classless institution. The hospital allows physicians to practice the best medicine available, but its clinical setting sometimes discourages the human component of caring. And while government financing and third party payment have redistributed the burden of hospital support throughout society, the institution has often remained unresponsive to the mass of its patients. Finally, the hospital has not evolved toward any foreordained perfection. It is no more the ideal form of medical organization today than it was of social consideration for the poor in the second half of the 19th century.

NOTES

1   See, in this regard, the career of George Cheyne Shattuck (1813–1893), Shattuck Papers, Massachusetts Historical Society.

2   Nathaniel I. Bowditch, *A History of the Massachusetts General Hospital*, 2nd ed. (Boston: The Bowditch Fund, 1872), p. 3; "A statement made by four physicians . . ." (Boston: 1869), in "Papers and Clippings" (hereafter P.&C.), a scrapbook kept by Dr. F. H. Brown about the Children's Hospital, 1869–1879, in the Countway Library, Boston; For the Dispensary for Skin Diseases and the Dispensary for Diseases of the Nervous System, *Boston Med. & Surg. J.*, 1872, *86*: 81, 82; 1872, *87*: 58; F. H. Brown, *Medical Register for the Cities of Boston, Cambridge, Charlestown, and Chelsea* (Boston: J. Wilson & Son, 1873), n.p.

3   Morris J. Vogel, "Boston's hospitals, 1870–1930: a social history" (Ph.D. dissertation, Univ. of Chicago, 1974), pp. 147–150.

4   James Clarke White, *Sketches from My Life, 1833–1913* (Cambridge, Mass.: Riverside Press, 1914), pp. 267–271.

5   *Boston Med. & Surg. J.*, 1882, *107*: 455. See also Henry J. Bigelow, "Fees in hospitals," *Boston Med. & Surg. J.*, 1889, *120*: 378. Bigelow noted: "It has been said, with truth, that these hospital offices would command a considerable premium in money from the best class of practitioners were they annually put up at auction." An annual report of the Boston Dispensary discussed why doctors served: "It is not to be supposed that the motives of the attending physicians have been wholly foreign from

considerations of personal advantage. They have doubtless been actuated by the hope of professional improvement and the prospect of building up an honest fame, as well as by the desires of fulfilling the benevolent intentions of this charity." Quoted in *Boston Med. & Surg. J.*, 1882, *106*: 137.

6   *Boston J. Hlth.*, March 1886, *1*: 100.

7   *Boston Med. & Surg. J.*, 1905, *152*: 295–320.

8   Bowditch, *M.G.H. History*, p. 415.

9   Carney Hospital, *Annual Reports*, 1879–1889.

10 Boston Dispensary, *Annual Report*, quoted in *Boston Med. & Surg. J.*, 1882, *106*: 137.

11 Children's Hospital, *Appeal*, 1869, in P.&C.

12 *Boston Evening Transcript*, April 20, 1881.

13 Boston, *Proceedings at the Dedication of the City Hospital* (Boston: J. E. Farwell, 1865), p. 58.

14 *Boston Evening Transcript*, June 10, 1886.

15 Detailed sources exist for who used hospitals, but since there is no reliable census of accident and illness, it is difficult to deduce what proportion of any different category of sickness or injury was hospitalized and what was not. Catastrophes, like train wrecks, can provide a rough idea. The pattern derived, while inconclusive, is reaffirmed by an examination of actual hospital use in the period.

16 *Boston Evening Transcript*, March 14, 1887.

17 *Boston Evening Transcript*, March 14, 17, 1887; *Boston Med. & Surg. J.*, 1887, *116*: 268. Another "frightful disaster" with the same general pattern occurred with the wreck of an excursion train on the Old Colony Railroad, *Transcript*, Oct. 9, 10, 1878. There was a different pattern after a crash on the Eastern Railroad in Revere. A greater proportion of the injured were hospitalized; probably because an express Pullman was involved, many of the survivors were from out of town. *Transcript*, Aug. 28, 1871.

18 M.G.H., *60th Annual Report, 1873* p. 9; George H. M. Rowe (superintendent, B.C.H.) to John Pratt (superintendent, M.G.H.), March 7, 1894, in M.G.H. archives, Phillips House file.

19 *Boston Med. & Surg. J.*, 1872, *86*: 81, 82.

20 North End Diet Kitchen founded 1874; South End Diet Kitchen a short time later. *Boston Evening Transcript*, Oct. 20, 1874; Nov. 9, 1875; Dec. 2, 1878.

21 Boston Lying-in Hospital, *Annual Report for 1881*, quoted in *Boston Med. & Surg. J.*, 1882, *106*: 462.

22 Children's Hospital, *Appeal*, 1869, in P.&C.; *Boston Evening Transcript*, Oct. 28, 1876; April 1, 1881; Nov. 17, 1883.

23 Thernstrom does not classify census data for 1870 into these socioeconomic categories, so hospital figures for both 1870 and 1880 are compared to data for the general population derived from the 1880 census. Since Thernstrom finds very little change in the city's occupational structure from 1880 to 1920, this is not unreasonable. Stephan Thernstrom, *The Other Bostonians: Poverty and Progress in the American Metropolis, 1880–1970* (Cambridge, Mass.: Harvard Univ. Press, 1973), pp. 50, 51, 289–302.

24 The category of skilled workers in both 1870 and 1880 consisted entirely of Massachusetts General patients listed as "mechanics." This is an ambiguous term and appears to have been applied differently in 1870 and 1880: The number of patients in this category fell from 291 in 1870 to 219 in 1880. Over the same years there was a large jump in laborers (the whole of the unskilled category) from 187 to 580. This suggests that mechanics were not all skilled workers in 1870 and that their 41.9 percent of the male population over-represents skilled workers among the hospital's patients. At the same time, the figure of 26.9 percent may undercount the proportion of hospital patients who were unskilled in 1870. The M.G.H. admitted 780 males (85 unclassified, all minors) in 1870 and 1,363 (235 unclassified) in 1880. Computed from M.G.H., *Annual Reports*.

25 Computed from B.C.H., *Annual Reports*.

26 Computed from M.G.H. and B.C.H., *Annual Reports*.

27 M.G.H., "The Report of a [trustees'] Committee on the Financial Condition of the M.G.H., February 16, 1865." The trustees dated their financial difficulties from the change from "the industrious classes of our native population," many of whom had paid something toward their board, to the foreign-born, who dramatically increased the numbers treated free. The trustees' committee recommended carefully restricting the number of nonpaying patients (read foreign-born) and segregating them in a distinct section of the hospital, so that the institution could get back to serving "the classes for whose advantage it was established." The trustees rejected that suggestion, apparently because of the urging of the medical staff which feared that decreasing the numbers of the really poor would hurt medical education. M.G.H. Trustees [printed letter], April 1, 1865, both in Countway Library; Bowditch, *M.G.H. History*, p. 454.

28 Computed from M.G.H., *Annual Reports*. In 1870, the United States Census listed 35.1 percent of Boston's population as foreign born.

29 Computed from B.C.H., *Annual Reports*.

30 Frances E. Kobrin, "The American midwife controversy: a crisis of professionalization," *Bull. Hist. Med.*, 1966, *40*: 350–363.

31 Boston, Board of Health, *Annual Report for 1879*, p. 20.

32 New England Hospital for Women and Children, MSS Maternity Records, Vol. 1, in Countway Li-

brary; Computed from Boston Lying-in Hospital, *Annual Report for 1930*, p. 58; for St. Mary's, *Boston Med. & Surg. J.*, 1884, *110*: 363, 364.

33 Joseph Edgar Chamberlin, *The Boston Transcript: A History of its First Hundred Years* (Boston: Houghton Mifflin Co., 1930), p. 165.

34 *Boston Evening Transcript*, Feb. 24, 1888.

35 *Ibid.*

36 *Ibid.*

37 E.g., a long feature article, "The origin of hospitals," *ibid.*, July 13, 1886.

38 *Ibid.*, July 10, 1885. This account is of the McLean Asylum, a branch of the M.G.H.

39 *Ibid.*, Nov. 29, 1881.

40 Odin Anderson, *The Uneasy Equilibrium: Public and Private Financing of Health Services in the United States, 1875–1965* (New Haven: College and Univ. Press, 1968), p. 29.

41 Children's Hospital, *1st Annual Report, 1869*, p. 10.

42 *Boston Evening Transcript*, April 20, 1881.

43 Children's Hospital, *3rd Annual Report, 1871*, pp. 7, 8.

44 Charlestown *Chronicle*, Nov. 11, 1871; Boston *Post*, March 1, 1872; both in P.&C.

45 "The Children's Hospital: what 'Fireside' thinks about it," *Boston Evening Transcript*, Jan. 22, 1879.

46 Children's Hospital, *1st Annual Report, 1869*, p. 13; *3rd Annual Report, 1871*, pp. 8–10.

47 Children's Hospital, *8th Annual Report, 1876*, p. 8. It is perhaps unfair to draw so sharp a distinction between medicine and charity in this case. Koch's work was yet to come, and there was no hard medical knowledge of the transmission of tuberculosis.

48 Children's Hospital, *Appeal*, 1869, in P.&C.; Children's Hospital, *1st Annual Report, 1869*, p. 13.

49 Children's Hospital, *3rd Annual Report, 1871*, p. 10; *5th Annual Report, 1873*, p. 9. Young readers of the *Christian Register* were invited to come with their mothers to visit "the dear little occupants." *Christian Register*, May 8, 1869, June 5, 1869; in P.&C.

50 Children's Hospital, *1st Annual Report, 1869*, back cover; *15th Annual Report, 1883*, p. 9.

51 Children's Hospital, *1st Annual Report, 1869*, p. 13.

52 Boston *Post*, March 1, 1872, in P.&C.

53 "Fireside" *Boston Evening Transcript*, Jan. 22, 1879.

54 *Boston Med. & Surg. J.*, 1870, *83*: 140, 141, editorial; the hospital's founder doubled as the editor of the *Journal*.

55 Boston *Sunday Times*, Dec. 29, 1872, in P.&C.; letter to the editor, "My visit to the Children's Hospital," by "A Lady," *Boston Evening Transcript*, Feb. 22, 1875.

56 Boston *Post*, March 1, 1872, in P.&C.; Children's Hospital, *7th Annual Report, 1875*, p. 13.

57 Drs. James Jackson and John C. Warren, "Circular letter," [1810], in Bowditch, *M.G.H. History*, pp. 3–9.

58 Grace W. Myers, *History of the Massachusetts General Hospital: June, 1872 to December, 1900* (Boston: Griffith-Stillings Press, 1929), p. 12.

59 Dr. D. B. St. John Roosa described the visitors lining up at the gate for the twice weekly visiting hour at the New York Hospital, and how visitors were searched before entering. *The Old Hospital and Other Papers* 2nd ed. (New York: W. Wood & Co., 1889), p. 12.

60 E.g., M.G.H. Trustees MSS Records, Aug. 3, 1877, in M.G.H. archives.

61 "Fireside," *Boston Evening Transcript*, Jan. 22, 1879.

62 M.G.H., *Annual Reports, 1870–1910*.

63 *Boston Evening Transcript*, April 20, 1881.

64 Children's Hospital, *7th Annual Report, 1875*, p. 7; *Boston Evening Transcript*, Jan. 22, 1879; a letter from a grateful parent, *Transcript*, Feb. 29 [*sic*], 1874.

65 Registrar's Report of the City of Boston, quoted in *Boston Med. & Surg. J.*, 1871, *85*: 83; *Boston Evening Transcript*, April 20, 1881.

66 Letter to the editor, *Boston Evening Transcript*, Feb. 29, 1888; Registrar's Report, in *Boston Med. & Surg. J.*, 1871, *85*: 84.

67 McDonald vs. M.G.H., 120 Mass. 432, in E. B. Callander, "Torts of hospitals," *Am. Law Rev.*, 1881, *15*: 640; *Boston Evening Transcript*, July 12, 1875.

68 Stogdale vs. Baker, reported in *Boston Evening Transcript*, Nov. 21, 1885, Dec. 12, 1887, Jan. 5, 1888.

21

The Severely and Chronically Mentally Ill in America: Retrospect and Prospect

GERALD N. GROB

During the decade of the 1980s a large number of severely and chronically mentally ill persons were living in urban communities. Some were housed with parents or relatives; some found shelter in residential facilities supported by a combination of public and private funds; and some were confined in penal and correctional institutions. Many, however, were found amidst a large mass of other homeless persons who lived on the streets and survived under tragic circumstances. The co-existence of homeless and mentally ill persons, paradoxically, recalled Dorothea L. Dix's famous memorials to state legislatures from the 1840s to the 1870s deploring conditions among the nation's mentally ill population and urging institutional solutions. "I come to present the strong claims of suffering humanity," she informed the members of the Massachusetts legislature in her first memorial in 1843. "I come as the advocate of helpless, forgotten, insane and idiotic men and women; of beings, sunk to a condition from which the most unconcerned would start with real horror." In New Jersey, Dix found "in jails and poorhouses, and wandering at will over the country, large numbers of insane and idiotic persons." Such a state of affairs, she told the Tennessee legislature, was inexcusable, since the remedy was available in the form of *"rightly organized Hospitals,* adapted to the special care of the peculiar malady of the Insane."[1]

That the mental health system in contemporary America no longer assumes responsibility for all of

its severely and chronically mentally ill persons is obvious. Scarcely a month passes without accounts of the travail of former mental patients discharged into urban communities that are unable to meet even their minimal needs. Such persons, Paul S. Appelbaum recently observed,

> are an inescapable presence in urban America. In New York City they live in subway tunnels and on steam grates, and die in cardboard boxes on wind-swept street corners. The Los Angeles City Council has opened its chambers to them, allowing them to seek refuge from the Southern California winter on its hard marble floors. Pioneer Square in Seattle, Lafayette Park in Washington, the old downtown in Atlanta have all become places of refuge for these pitiable figures, so hard to tell apart: clothes tattered, skins stained by the streets, backs bent in a perpetual search for something edible, smokable, or tradeable that may have found its way to the pavement below.[2]

In seeking an explanation for the tragic plight of many severely and chronically mentally ill persons, we are at the outset confronted with a paradox. In the early 19th century the absence of institutional alternatives led social activists to agitate for the establishment of public mental hospitals that would provide both care and curative treatment. After World War II, by way of contrast, the emphasis was on the creation of a community-based system of services and, by implication, the eventual abandonment of a vast institutional complex that by the 1950s had an inpatient population of about 550,000. Why did the "solution" of one era become the "problem" of another? What were the sources and underlying ideology of shifting public policies? Most importantly, how well or poorly did specific

GERALD N. GROB is Henry E. Sigerist Professor of the History of Medicine, Rutgers University, New Brunswick, New Jersey.

Reprinted from *Transactions and Studies of the College of Physicians of Philadelphia,* 1991, series 5, *13* (No. 4): 337–362.

policies serve the needs of severely and chronically mentally ill persons?

Before 1800 the problems for communities posed by mental illnesses were relatively minor and generally of little public concern. This is not to suggest that mental illnesses were nonexistent. It is only to note that population was relatively sparse and scattered in somewhat isolated rural and agricultural communities. As late as 1790 no urban area had more than 50,000 residents, and only two had 25,000 or more. The number of mentally ill persons was correspondingly small, and such persons were generally cared for by their families or by local officials who assumed responsibility for their welfare. Under existing traditions they were treated in accordance with the English Poor Law system whose foundations antedated passage of the famous Elizabethan Poor Law legislation written between 1595 and 1601. This system was based on the principle that society had a corporate obligation for poor and dependent persons. Virtually every American colony enacted laws that replicated the English system. Under this arrangement local communities, rather than the colony or mother country, had to assume fiscal and supervisory responsibility for those persons incapable of surviving without some form of assistance. Indeed, the very concept of *social policy* involving the creation of systematic structures to deal with individual and group distress and dependency was largely absent during the colonial era.

Confinement of the mentally ill was a rare exception before 1800. There was no systematic effort to restrict "Lunaticks or distracted persons," to use contemporary terminology. Madness was tolerated and family care characteristic. Mentally ill persons without families or private resources received the same treatment as sane paupers; they were either boarded out with families or kept in public almshouses. Insanity became an issue of public concern only when afflicted individuals did not have access to the basic necessities of life, or when their violent behavior threatened others. The rural character of American society in the Revolutionary and post-Revolutionary era precluded serious consideration of structural changes to deal with the mentally ill or other dependent groups.[3]

After 1800, however, new circumstances ultimately led to reliance on some form of institutional care of the mentally ill. The transformation of insanity into a *social* problem requiring state intervention—as contrasted with familial and community responsibility—was by no means a unique phenomenon. The 19th century was noted for its widespread use of institutional solutions for social problems and for the transfer of functions from families to public or quasi-public structures. In 1820 only one state hospital for the mentally ill existed in the United States; by the Civil War virtually every state had established one or more public institutions for that purpose.[4]

Paradoxically, the creation of institutions reflected an extraordinarily optimistic view of the nature and prognosis of mental illnesses. Most mid-19th-century psychiatrists conceived of disease in individual rather than general terms. Health was a consequence of a symbiotic relationship between nature, society, and the individual. Disease, by way of contrast, represented an imbalance that followed the violation of certain natural laws that governed human nature. To be sure, mental illnesses were indistinguishable from other physical illnesses and occurred when false impressions were conveyed to the mind because the brain or other sensory organs had been impaired. Nor did psychiatrists differ from their colleagues in general practice when they insisted that mental illnesses were precipitated by a combination of psychological and environmental etiological factors that were mediated by the constitution or predisposition of the individual. Thus insanity often followed the violation of the natural laws that governed human behavior and was linked as well with immorality, improper living conditions, or other stresses that upset the natural balance.[5] Yet within the psychiatric perspective there remained a fundamental distinction between mental health and mental disease. The presence of the latter was indicated by dramatic behavioral and somatic signs that deviated sharply from the prior "normal" behavior of the individual. Unlike their 20th-century psychodynamic successors, 19th-century alienists rarely suggested that early interventions before the onset of the acute stage of insanity were possible or even effective.[6] Indeed, the very term *insanity* implied severity; hence there was no need to qualify *insanity* by using the designation of *severe*.

Since insanity followed improper behavioral patterns associated with a defective environment, therapy had to begin with the creation of a new and presumably more appropriate environment. Home treatment was useless, for the physician had no means of eliminating undesirable environmental influences. Institutionalization was a *sine qua non* because it shattered the link between an improper environment and the patient. In a hospital, patients could be exposed to a judicious amalgam of medical and moral treatment. Medical treatment was intended to rebuild the body to improve the mind, and to calm violent behavior by the administration of narcotics. Indeed, asylum therapeutics bore a remarkable similarity to the therapeutics of those engaged in general medical practice. "Moral treatment"—to use 19th-century terminology—implied kind, individualized care within a small hospital, the use of occupational therapy, and resort to religious exercises, amusements, and games. There were to be no threats of physical violence; and only rarely were mechanical means of restraint to be employed. Moral treatment, in effect, involved the re-education of the patient within a more suitable environment. The role of the psychiatrist was not fundamentally different from the role of the stern, occasionally authoritarian, yet loving and concerned father.[7]

From the prevailing model of disease, psychiatrists drew an obvious conclusion; insanity was as curable as, if not more curable than, most somatic illnesses. "Insanity," proclaimed Samuel B. Woodward (first president of what is today the American Psychiatric Association [APA]) in 1839, "of all diseases the most fearful, is found to be among the most curable."[8] Indeed, it was common for mental hospital superintendents in the 1830s and 1840s to claim that 90 percent or more of all recent cases—cases defined as being insane for one year or less—could be cured if treated promptly. William M. Awl, a contemporary of Woodward and superintendent of the Ohio Lunatic Asylum, reported that in the first four years following its opening in 1838 the Asylum had received 171 persons insane for one year or less; 69 between one and two years; 85 between two and five years; and 44 between 5 and 10 years. The recovery rates within each group, respectively, were 80, 35, 14, and 9 percent.[9]

Within such an intellectual framework the concept of chronicity (chronic long-term mental illness) had a somewhat different meaning. Chronicity was neither inherent nor inevitable, but followed the failure to provide acute cases with the benefit of therapy in mental hospitals. The incurable insane, wrote Edward Jarvis in 1855, "remain standing and abiding monuments of the neglect of the State to provide the means of health, and place them within the reach or the comprehension of the friends and guardians who had immediate charge of them, or of the neglect of those friends and guardians to avail themselves of these opportunities of restoration when they were offered to them."[10] Psychiatrists, moreover, denied that a recurrence of insanity was simply a continuation of the original illness. Just as individuals could suffer numerous respiratory or intestinal diseases, so too could they have subsequent unrelated episodes of insanity. Woodward, for example, argued that if a patient discharged as recovered had been free of symptoms for a year or more, any recurrence was attributable to a new cause that had no relation to the previous attack. If individuals who recovered would avoid "known causes of disease . . . they might safely pass on, and, in most cases, continue well." "It is just as possible," observed Thomas S. Kirkbride in his classic work on mental hospitals, "for any one to have an attack of insanity, to recover from it, and to have another attack at a subsequent period of life, as it is of any other disease, or as any one is liable to have a first attack."[11]

In general, mid-19th-century patients discharged as recovered or improved tended to be institutionalized for only brief periods from three to nine months. Hence, the prevailing belief was that a mental hospital with 200 beds could treat approximately 600 patients during a 12-month period (assuming that the average stay would be about four months). Surviving evidence suggests that the claims of therapeutic successes had some validity. Although affirmations about curability rates were undoubtedly exaggerated, there is little doubt that many individuals appeared to benefit from hospitalization. In the 1880s an enterprising superintendent undertook a follow-up study of over a thousand patients discharged as recovered on their only or last admission. The study took more than a decade to complete, and in the end data were accumulated on 984 individuals. Of these, 317 were alive and well at the time of their last reply, while an additional 251 who had died had never again been institutionalized. Thus nearly 58 percent of those discharged as recovered had functioned in a community setting without any relapse.[12]

The absence of a large chronic institutional population further reinforced the faith in curability. Between the 1830s and 1870s the proportion of long-term or chronic cases in hospitals was relatively low—that is, if the figures are compared with the extraordinarily high percentage between 1890 and 1950. Although national data are lacking, a sample of individual hospitals reveals that their functions as custodial institutions had not yet become dominant. The experiences of Worcester State Hospital—the oldest and most important public institution in Massachusetts—are instructive. In 1842 (a decade after it opened), 46.4 percent of its patients had been hospitalized for less than a year; only 13.2 percent had been in the hospital for five or more years. In 1870 the comparable figures were 49.6 and 13.9 percent. Nor was Worcester atypical. In 1850 41.1 percent of patients at the Virginia Western Lunatic Asylum had been hospitalized for less than a year and 29.6 percent for five years or more. The respective figures for the California Insane Asylum in 1860 were 40.2 percent and 0.1 percent. Unlike their 20th-century counterparts, hospitals before 1880 did not have large numbers of two classes of patients; the aged (over 65) and paretics (the tertiary stage of syphilis). Between 1830 and 1875 aged patients accounted for perhaps 5 to 10 percent of the total. At the Utica State Lunatic Asylum only 2.7 percent of all admissions between 1850 and 1868 were given a diagnosis of paresis.[13]

The low proportion of chronic patients in mental hospitals was due in part to the pattern of funding and, to a lesser extent, the exclusion of senility from psychiatric diagnostic categories. In general, state legislatures provided the capital funds necessary for acquiring new sites and constructing, expanding, and renovating existing physical plants. Local communities, on the other hand, were required to pay hospitals a sum equal to the actual cost of care and treatment of each patient admitted. The system, moreover, did not assume that every mentally ill person would be cared for in a state institution. Laws generally required that only dangerous mentally ill persons had to be sent to state hospitals. Others who could presumably benefit from therapeutic interventions (and thus ultimately be removed from welfare rolls) could, at the discretion of local officials, also be institutionalized. The system, in short, involved divided responsibility.

For much of the 19th century, therefore, a significant proportion of insane persons continued either to live in the community or else were kept in municipal almshouses. Families with sufficient resources could commit their relatives to state institutions, provided they were willing to assume financial liability for their upkeep. States, moreover, had to reimburse hospitals for those patients who had not established legal residency, such as immigrants. The result was a variegated pattern. Edward Jarvis's census of all persons identified by others as insane in Massachusetts in 1854 provides some insight into the distribution of the mentally ill. In his monumental survey Jarvis identified by name every insane person in the Bay State. According to his final tabulations, there were 2,632 insane persons. Of this number, 1,522 were paupers, and 1,110 were supported by their own resources or by friends. At the time 1,284 were either at home or in almshouses; 1,141 were in hospitals; and 207 were in local receptacles for the insane, houses of correction, jails, or state almshouses. Only 435 were identified as curable, as compared with 2,018 as incurable (the prognosis of 179 was unknown).[14]

Divided responsibility for the mentally ill had significant repercussions. The system tended to promote competition and rivalries between overlapping governmental jurisdictions. In many states the stipulation that individual communities were financially liable for their poor and indigent insane residents created incentives for local officials to keep them in almshouses where costs were lower. Hospital officials often faced unremitting pressure from communities to discharge patients—irrespective of their condition—in order to save money. Local officials on occasion even attempted to force hospitals to reimburse the community for work performed by patients, though such labor was part of a therapeutic regimen. Ironically, divided fiscal and government authority had the paradoxical effect of keeping the chronic population in mental hospitals at relatively low levels. Moreover, aged senile persons without resources or support networks were cared for in almshouses, which were the 19th-century equivalent of old age homes.

Mid-19th-century psychiatrists generally opposed the establishment of separate institutions for chronic or incurable cases and accepted the legitimacy of custodial care. They were not certain of their ability to distinguish between curable and incurable cases, and feared that the prevention of abuses of patients within strictly custodial institutions would be overly difficult. Moreover, their experiences with local officials led them to the conclusion that confinement of the chronic insane in almshouses or comparable institutions was detrimental. In the first edition of his classic work, Kirkbride vigorously condemned the maintenance of chronic patients in separate institutions:

> The first grand objection to such a separation is, that no one can say with entire certainty who is incurable; and to condemn any one to an institution for this particular class is like dooming him to utter hopelessness . . . . When patients cannot be cured, they should still be considered under treatment, as long as life lasts; if not with the hope of restoring them to health, to do what is next in importance, to promote their comfort and happiness, and to keep them from sinking still lower in the scale of humanity. Fortunately, almost precisely the same class of means are generally required for the best management and treatment of the curable and incurable, and almost as much skill may be shown in caring judiciously for the latter as for the former. When the incurable are in the same institution as the curable, there is little danger of their being neglected; but when once consigned to receptacles specially provided for them, all experience leads us to believe that but little time will elapse before they will be found gradually sinking, mentally and physically, their care entrusted to persons actuated only by selfish motives—the grand object being to ascertain at how little cost per week soul and body can be kept together—and, sooner or later, cruelty, neglect and suffering are pretty sure to be the results of every such experiment.

When Kirkbride published a second edition of his work a quarter of a century later, his views had not changed. "What is best for the recent," he insisted, "is best for the chronic."[15]

As the number of chronic patients slowly increased, however, states began to reconsider their policies. Massachusetts and New York, which main-tained the most extensive hospital systems, introduced some significant modifications. In the 1850s Massachusetts created a separate department for chronic insane immigrants at its state almshouse. In 1869 New York opened the Willard Asylum, which was intended to relieve local welfare institutions of any responsibility for chronic pauper insane patients. By the 1880s Wisconsin had established a controversial system that provided for state subsidies for the care of chronic cases at county-run facilities. These and other policy innovations had two distinct goals: to ensure that hospitals retained their therapeutic character; and to minimize the role of local communities whose officials were preoccupied mainly with maintaining low tax rates. The effort to segregate the chronic population met with an ambivalent reception from psychiatrists. Senior figures opposed a policy that stigmatized large numbers of individuals and consigned them to custodial institutions. Those in favor of the new departure, on the other hand, argued that the retention of chronic patients in local poorhouses institutionalized an inferior system.

Paradoxically, the debate over a policy that divided curable and chronic cases was rendered all but moot by the turn of the century. Disillusioned by a system that divided authority, states—led once again by New York and Massachusetts—adopted legislation that relieved local communities of any role whatsoever in caring for the mentally ill. The assumption of those who favored centralization was that local care, although less expensive, was substandard and also fostered chronicity and dependency. Conversely, care and treatment in hospitals, though more costly initially, would in the long run be cheaper because it would enhance the odds of recovery for some and provide more humane care for others.

Although the intent of state assumption of responsibility was to ensure that the mentally ill would receive a higher quality of care and treatment, the consequences in actual practice turned out to be quite different. In brief, local officials saw in the new laws a golden opportunity to shift some of their financial obligations onto the state. The purpose of the new laws mandating state responsibility for all mentally ill persons was self-evident: namely, to end local jurisdiction over the chronic mentally ill. But local officials went beyond the intent of the law. Traditionally, 19th-century almshouses (which were supported and administered by local governments) served in part as old-age homes for senile and aged persons without any financial resources. The passage of state care acts provided local officials with an unexpected opportunity. They proceeded to re-define senility in psychiatric terms and thus began to transfer aged persons from local almshouses to state mental hospitals. Humanitarian concerns played a relatively minor role in this development; economic considerations were of paramount significance.

Faced with rapidly escalating expenditures, communities were more than happy to transfer responsibility for their aged residents to state-supported facilities. Between 1880 and 1920, therefore, the almshouse populations (for this and other reasons) dropped precipitously. Admissions fell from 99.5 to 58.4 per 100,000 between 1904 and 1922. The decline in the number of mentally ill persons aged 60 and over was even sharper, dropping from 24.3 percent in 1880 to 5.6 percent in 1923.[16] What occurred, however, was not a deinstitutionalization movement, but rather a lateral transfer of individuals from one institution to another. "We are receiving every year a large number of old people, some of them very old, who are simply suffering from the mental decay incident to extreme old age," observed Charles C. Wagner (superintendent of the Binghamton State Hospital in New York) in moving terms. "A little mental confusion, forgetfulness and garrulity are sometimes the only symptoms exhibited, but the patient is duly certified to us as insane and has no one at home capable or possessed of means to care for him. We are unable to refuse these patients without creating ill-feeling in the community where they reside, nor are we able to assert that they are not insane within the meaning of the statute, for many of them, judged by the ordinary standards of sanity, cannot be regarded as entirely sane."[17]

As a result, the character of mental hospitals underwent a dramatic transformation during the first half of the 20th century. Prior to that time hospitals had substantial turnover rates even though they retained patients who failed to improve or recover. In the four decades following its opening in the 1840s, the proportion of patients who left the Utica State Lunatic Asylum hovered around 40 percent. In the 20th century, by way of contrast, the pattern changed markedly as the proportion of short-term cases fell and that of long-term cases increased. By 1904 only 27.8 percent of the total patient population for the United States as a whole had been institutionalized for 12 months or less. Within six years this percentage had fallen to 12.7, although rising to 17.4 in 1923. The greatest change, however, came among patients hospitalized for five years or more. In 1904, 39.2 percent of patients fell into this category; in 1910 and 1923 the respective percentages were 52.0 and 54.0. Although data for the United States as a whole were unavailable after 1923, the experiences of Massachusetts are illustrative. By the 1930s nearly 80 percent of its mental hospital beds were occupied by chronic patients.[18]

Chronicity, however, is a somewhat misleading term, for the group that it described was heterogeneous. The aged (over 60 or 65) constituted by far the single largest component. By 1920, for example, 18 percent of all first admissions to New York State mental hospitals were diagnosed as psychotic because of senility or arteriosclerosis; 20 years later the figure had risen to 31 percent. A decade later 40 percent of all first admissions were aged 60 and over, as compared with only 13.2 percent of the state's total population. The increase in the absolute number also reflected a change in age-specific admission rates. In their classic study of rates of institutionalization covering more than a century, Herbert Goldhamer and Andrew W. Marshall found that the greatest increase occurred in the older category. In 1855 the age-specific first-admission rate in Massachusetts aged 60 and over was 70.4 for males and 65.5 for females (per 100,000); by the beginning of World War II the corresponding figures were 279.5 and 223.0. As late as 1958 nearly a third of all resident state hospital patients were over 65 years of age.[19]

The rising age distribution mirrored a different but related characteristic of the institutionalized, namely, the presence of large numbers of patients whose abnormal behavior reflected an underlying somatic etiology. Even allowing for imprecise diagnoses and an imperfect statistical reporting system, it was quite evident that a significant proportion of the hospitalized population suffered from severe organic disorders for which there were no effective treatments. Of 49,116 first admissions in 1922 institutionalized because of various psychoses, 16,407 suffered from a variety of identifiable somatic conditions (senility, cerebral arteriosclerosis, paresis,

Huntington's chorea, brain tumors, etc.). Between 1922 and 1940 the proportion of such patients increased from 33.4 to 42.4 percent. In 1946 various forms of senility and paresis accounted for about half of all first admissions. A study of 683 aged persons admitted to Maryland hospitals in the early 1940s revealed that 47 percent died within the first year; overall only 10 percent were ever discharged while still alive.[20]

The increasing proportion of elderly patients in mental hospitals, however, was not due solely to rising admission rates. High death rates ensured that the length of stay of newly admitted elderly persons would be limited. At Warren State Hospital in Pennsylvania, for example, 72 percent of such patients died within five years of their admission in the period from 1936 to 1945. The proportion of elderly was instead augmented by the accumulation of schizophrenics admitted at a younger age who grew old at the hospital. The presence of this group accounted for the fact that in 1962 the median length of stay for schizophrenics in 23 selected states was 12.8 years, whereas the comparable figure for those admitted with mental diseases of the senium was 3.0 years.[21]

The large numbers of chronic and aged patients, however, tended to obscure the fact that release rates for nonelderly patients were improving. In a pioneering study of more than 15,000 patient cohorts admitted to Warren State Hospital in Pennsylvania between 1916 and 1950, Morton Kramer and his colleagues found that, for functional psychotics, the probability of release of first admissions within 12 months increased from 41 to 62 percent between 1919–1925 and 1946–1950. These and other data suggested that popular perceptions of mental hospitals were not entirely accurate. To put it another way, a high proportion of chronic and aged patients in public mental hospitals led many to overlook the reality that substantial numbers of patients were admitted, treated, and discharged in less than a year.[22]

During the early and mid-1940s, journalists and mental health professionals alike published numerous critical accounts of mental hospitals that detailed tragic conditions. To be sure, a decade and a half of financial neglect, due largely to the combined impact of the Great Depression of the 1930s and ensuing global conflict, simply exacerbated already existing severe problems. The depressing state of mental hospitals, however, was as much a function of the nature of their patients as it was the result of parsimonious or callous policies. The large number of chronically ill patients was undoubtedly the single most significant element in shaping a milieu seemingly antithetical to therapeutic goals.

During and after World War II the prevailing consensus on mental health policy in America slowly began to dissolve. Developments converged to reshape public policy during these years, and mental hospitals—institutions that had been the cornerstone of public policy for nearly a century and a half—slowly began to lose their social and medical legitimacy. First, there was a shift in psychiatric thinking toward a psychodynamic and psychoanalytic model emphasizing life experiences and the role of socioenvironmental factors. Second, the experiences of World War II appeared to demonstrate the efficacy of community and outpatient treatment of disturbed persons. Third, the belief that early intervention in the community would be effective in preventing subsequent hospitalization became popular. Fourth, a faith developed that psychiatry could promote prevention by contributing toward the amelioration of social problems that allegedly fostered mental diseases. Fifth, the introduction of psychological and somatic therapies (including, but not limited to, psychotropic drugs) held out the promise of a more normal existence for patients outside of mental institutions. Finally, an enhanced social welfare role of the federal government not only began to diminish the authority of state governments, but also hastened the transition from an institutionally based to a community-oriented policy.[23]

Winds of change were evident well before the widespread use of psychotropic drugs or the advent of deinstitutionalization. The specialty of psychiatry, long synonymous with institutional care, rapidly changed its character in the postwar era. Admittedly, psychiatrists had begun to find careers outside of public institutions in the interwar decades. But after 1945 there was a mass exodus of psychiatrists from mental hospitals into private and community practice. Within a decade more than 80 percent of the 10,000 members of the APA were employed outside of mental hospitals. Their positions were filled by foreign medical graduates with little or no training in psychiatry.[24] Although the APA staff continued to work with public hospitals (especially through the Central Inspection Board, annual Mental Hospital Institutes, and by conducting surveys in individual states), most of its members were neither knowledgeable about nor sympathetic toward their institutional brethren and often emphasized the desirability of noninstitutional alternatives. Moreover, most psychiatrists in the community treated large numbers of patients with psychological problems and thus their contacts with the severely and chronically mentally ill were sharply reduced. That hospitals had a large proportion of chronic patients hardly accorded with the self-image of the psychiatrist as an active and successful therapist. In his APA presidential address in 1958, as a matter of fact, Harry C. Solomon even described the large mental hospital as "antiquated, outmoded, and rapidly becoming obsolete." Robert C. Hunt, director of the Hudson River State Hospital in New York and an individual deeply concerned with institutional problems, responded publicly in critical terms. His "private reactions are still unprintable," he wrote to Solomon. Hunt subsequently informed the APA Commission on Long Term Planning that the organization had not played a constructive role in countering the detrimental effects associated with "the state hospital stereotype." The majority of APA members, he added in revealing terms, had neither contacts with nor knowledge about mental hospitals. Hence its members were prone to identify the prevailing stereotype with reality; the result was a virtual abandonment of the hospital by American psychiatrists.[25]

The weakening of the long-established links between hospitals and psychiatrists was also accompanied by a movement to strengthen outpatient and community clinics. Before 1940 such clinics had dealt predominantly with children rather than adults. The postwar enthusiasm for clinics received momentum with the passage of the National Mental Health Act of 1946, which provided grants to states to support existing outpatient facilities or to establish new ones. The ultimate goal, according to Robert H. Felix, first director of the National Institute for Mental Health, was one outpatient facility for each 100,000 persons. Although appropriations were modest, their impact was dramatic. Before 1948 more than half of all states had no clinics; by 1949 all but five had one or more. Six years later there were about 1,234 outpatient psychiatric clinics, of which about two-thirds were state supported or aided. Psychiatrists proved staunch proponents of a community-oriented policy, for they insisted

that early identification and treatment in outpatient facilities or private offices diminished the need for subsequent hospitalization and were also cost effective.[26]

During the 1950s support for a community-based policy increased steadily; the Governor's Conference and Council of State Governments, as well as private foundations such as the Milbank Memorial Fund, played important roles in marshalling support for innovation. In 1954 New York enacted its influential Community Mental Health Services Act, which provided state funding for outpatient clinics; California followed suit shortly thereafter with the passage of the Short-Doyle Act. By 1959 there were more than 1,400 clinics serving about 502,000 individuals, of whom 294,000 were over the age of 18.[27] The expansion of community facilities was accompanied also by new services to schools, courts, and social agencies by nonmedical mental health professionals. This development offered further proof of the degree to which the public sought, if not demanded, access to psychiatric and psychological services in noninstitutional settings. During these years Robert Felix and his colleagues at the National Institute of Mental Health used their links with key congressional figures to enhance the policymaking authority of the federal government as a vehicle to strengthen community policies.

Many of the claims about the efficacy of community care and treatment, however, rested on extraordinarily shaky foundations. The presumption was that outpatient psychiatric clinics could identify early cases of mental disorders and also serve as alternatives to mental hospitals. But empirical data to validate such assertions were lacking. Indeed, a study of about 500 patients in three California state hospitals during the 1950s found most of these patients unsuited to treatment in clinics. The authors of the study concluded by noting "the marked discontinuities in functions of the participating hospitals and clinics and the difficulties in initiating outpatient treatment with hospitalized patients shortly after their admission." They also called attention to "the value of services to bridge the gap between the traditional functions of hospitals and clinics for already hospitalized patients."[28]

Data collected by Morton Kramer and his associates at the Biometrics Branch of the National Institute of Mental Health raised equally serious problems. A community policy was based on the

Table 1

Number, Percent Distribution, and Rate Per 100,000 of Inpatient and Outpatient Care Episodes in U.S. Selected Mental Health Facilities, 1955, 1965, 1968, 1971

Year	Total All Facilities	Inpatient Service						Outpatient Service		
		All In-Patient Services	State & County Mental Hospitals	Private Mental Hospitals	General Hospital Units (Non-VA)	VA In-Patient Services	CMHCs*	All Out-Patient Services	CMHCs*	Other
Number of Patient Care Episodes										
1955	1,675,352	1,296,352	818,832	123,231	265,934	88,355	—	379,000	—	379,000
1965	2,636,525	1,565,525	804,926	125,428	519,328	115,843	—	1,071,000	—	1,071,000
1968	3,380,818	1,602,238	791,819	118,126	558,790	133,503	—	1,778,590	271,590	1,507,000
1971	4,038,143	1,721,389	745,259	126,600	542,642	176,800	130,088	2,316,754	622,906	1,693,848
Percent Distribution										
1955	100.0	77.4	48.9	7.3	15.9	5.3	—	22.6	—	22.6
1965	100.0	59.4	30.5	4.8	19.7	4.4	—	40.6	—	40.6
1968	100.0	47.3	23.4	3.5	16.5	3.9	—	52.7	8.0	44.7
1971	100.0	42.6	18.5	3.1	13.4	4.4	3.2	57.4	15.4	42.0
Rate Per 100,000 of Population										
1955	1,028	795	502	76	163	54	—	233	—	233
1965	1,376	817	420	65	271	60	—	559	—	559
1968	1,713	812	401	60	283	68	—	901	138	763
1971	1,977	843	365	62	266	87	64	1,134	305	829

*Federally assisted CMHCs only.

expectation that patients could be treated outside of institutions. Underlying this belief were several assumptions: (1) patients had homes; (2) patients had sympathetic families or other persons willing and able to assume responsibility for their care; (3) the organization of the household would not impede rehabilitation; and (4) the patient's presence would not cause undue hardships for other family members. In 1960, however, 48 percent of the mental hospital population was unmarried, 12 percent were widowed, and 13 percent were divorced or separated. A large proportion of patients, in other words, may have had no families to care for them. Hence the assumption that patients could reside in the community with their families while undergoing rehabilitation was hardly realistic.[29]

Such findings fell on deaf ears; the rhetoric of community care and treatment carried the day in the 1950s and 1960s. Too often, exaggerated claims were overlooked or ignored. Yet rhetoric cannot be dismissed so easily: it shaped agendas and debates; it created expectations that in turn molded policies; and it informed the socialization, training, and education of those in professional occupations. From the creation of the Joint Commission on Mental Illness and Health in 1955 and the publication of its influential *Action for Mental Health: Final Report of the Joint Commission on Mental Illness and Health, 1961* to the passage of the Community Mental Health Centers Act of 1963, the advocates of a community-oriented policy succeeded in forging a consensus regarding the desirability of diminishing the central role of mental hospitals and strengthening community facilities. They were joined by a variety of other individuals and groups. Psychiatric critics (e.g., Thomas Szasz) attacked the very legitimacy of the concept of mental illnesses; civil rights advocates identified the mentally ill as a group systematically deprived of constitutional liberties; and social activists emphasized that institutions such as mental hospitals could never be other than repressive and dehumanizing institutions. The result was a determined and partially successful effort to reshape public policy by diminishing the role of hospitals and enhancing the significance of outpatient and community services.

During the 1960s the attack on the legitimacy of institutional care began to bear fruit. Hospital populations declined rapidly after 1965. A shift in

thinking had made community care and treatment, at least in theory, an acceptable alternative to institutionalization. Administrative and structural changes within institutions, including open-door policies, informal admissions, and efforts to prepare patients for early release, as well as the introduction of psychotropic drugs, reinforced the faith in the efficacy of community treatment. The passage of Medicaid and Medicare, moreover, hastened the exodus of aged patients from hospitals to chronic nursing homes. The rapid expansion of third-party reimbursement plans stimulated the use of inpatient and outpatient psychiatric services in general hospitals. Ironically, the reduction of the patient population no doubt had the effect of improving the lives of those who remained in public mental hospitals.

Nowhere were the changes in the mental health system during the 1960s more visible than in the aggregate data dealing with patient care episodes.[30] In 1955 there were 1,675,352 patient care episodes; 379,000 (22.6 percent) were treated in outpatient facilities; 818,832 (48.9 percent) in state mental hospitals; and the remainder in other institutions. Of 3,380,818 episodes in 1968, 52.7 percent were treated in outpatient facilities (of which 8 percent were in community mental health centers [CMHC]), 23.4 percent in state hospitals, and 23.9 percent in other institutions. To put it another way, 77.4 percent of episodes were treated in inpatient facilities in 1955 and 22.6 percent in outpatient settings; 13 years later the respective figures were 47.3 percent and 52.7 percent. In sum, there was a profound shift in the location of services as well as an increase in the *rate* of episodes. In 1955 there were 1,028 episodes per 100,000; by 1968 this figure had risen substantially to 1,713 episodes (see Table 1).[31]

The change in the location of services, however, did not mean that public mental hospitals were on the road to extinction and that community outpatient centers and clinics were assuming their functions. On the contrary, outpatient facilities grew rapidly because they were used by new groups that in the past had no access to the mental health system and who were for the most part not in the severely mentally ill category. Thus while the rate of inpatient care episodes at public hospitals declined from 502 to 401 per 100,000, outpatient care episodes leaped from 233 to 901 between 1955 and

1968. In absolute terms, inpatient care episodes at public institutions fell from 818,832 to 791,819, whereas outpatient care episodes increased from 379,000 to 1,778,590 in the same period. These data demonstrate that the growth in outpatient services was not at the expense of inpatient ones. Many of the changes in the mental health system, in other words, occurred because of the expansion of services and recruitment of a new clientele rather than the substitution of one service for another.

The dramatic growth of outpatient facilities diminished the relative significance of public mental hospitals, which for more than a century had been central to the mental health system. The number of resident patients fell slowly in the period from 1955 to 1965, and more rapidly thereafter. Yet at the same time the number of admissions was increasing. In 1955, 178,003 persons were admitted to state and county mental hospitals. A decade later the figure was 316,664 persons. The rapid decline in the resident population after 1965 did not alter this pattern; in 1970 there were 384,511 admissions (see Table 2). These figures suggest that an important change in the function of state hospitals had taken place. During the first half of the 20th century these institutions cared for large numbers of chronic cases drawn from several categories, including schizophrenic patients admitted during youth and early maturity and who remained for the rest of their lives, paretics, and senile aged persons. By the late 1960s the number of aged and chronic patients began to fall, and mental hospitals then began to provide more short- and intermediate-term care and treatment for severely mentally ill persons.[32]

To be sure, the number of patient care episodes treated in general hospitals (with and without psychiatric units) and federally funded CMHCs increased, although there were sharp variations from place to place. The available (and imperfect) data, however, indicates that these facilities did not generally treat individuals previously admitted or likely to be admitted to mental hospitals. There were, for example, some striking differences in diagnostic categories. In 1969 state hospitals had a higher proportion of patients with schizophrenic reactions, a group that constituted the core of the severely mentally ill group. Nearly 30 percent of its admissions were in this category; 11 percent were

in the organic brain syndrome and 10.2 percent in the depressive categories. General hospital inpatient services, by way of contrast, treated different kinds of patients. More than a third of their admissions suffered from depressive disorders; schizophrenic reactions accounted for 17.2 percent and organic brain syndromes 6.5 percent.[33]

The differences between state mental and general hospitals with specialized units become even clearer from length-of-stay data. The mean and median stay in general hospitals in 1963 was 20 and 17 days respectively. These figures fell slightly during the 1960s, the former to 17 days in 1969 and the latter to 11 days in 1971. By 1975 the mean stay was only 11 days and the median 6.7 days.[34]

The pattern in state mental hospitals differed substantially. Unfortunately, length-of-stay data were not reported before 1970. Other data, however, shed light on the functions of these institutions. Before 1965 public mental hospitals had a large chronic population. Data from 23 states in 1962 revealed that the median stay for patients resident at the end of the year was 8.4 years. The distribution was even more striking: 18.4 percent of patients were institutionalized for less than a year; 22 percent from 1 to 4 years; 14.6 percent from 5 to 9 years; 20.4 percent from 10 to 19 years; and 24.6 percent 20 years or more.[35] After 1965 the number of long-term patients at public institutions fell precipitously, largely because changes in funding patterns led to a sharp decline in elderly and chronic patients. This is not in any way to imply that state hospitals no longer provided long-term care. On the contrary, state hospitals remained what three investigators termed "the place of last resort" for perhaps 100,000 individuals for whom no alternative facility was available. Thus in 1969 the mean stay of discharged patients at public hospitals was 421 days; six years later the corresponding figure was 270 days. Median length-of-stay data, however, reveal a quite different situation. In 1970 the median length of stay for admissions (and excluding deaths) was 41 days; five years later this figure had dropped to 25 days. These data suggest that public institutions continued to treat and care for more severely and chronically ill persons than any other kind of institution. Indeed, in 1969 and 1975 they accounted for 79.4 and 67.2 percent, respectively, of all days of inpatient psychiatric care.[36]

For the seriously and chronically mentally ill,

Table 2
Resident Patients, Admissions, Net Releases, and Deaths in U.S. State
and County Mental Hospitals, 1950–1971

Year	Resident Patients at End of Year	Admissions	Net Releases*	Deaths
1950	512,501	152,286	99,659	41,280
1951	520,326	152,079	101,802	42,107
1952	531,981	162,908	107,647	44,303
1953	545,045	170,621	113,959	45,087
1954	553,979	171,682	118,775	42,652
1955	558,922	178,003	126,498	44,384
1956	551,390	185,597	145,313	48,236
1957	548,626	194,497	150,413	46,848
1958	545,182	209,823	161,884	51,383
1959	541,883	222,791	176,411	49,647
1960	535,540	234,791	192,818	49,748
1961	527,456	252,742	215,595	46,880
1962	515,640	269,854	230,158	49,563
1963	504,604	283,591	245,745	49,052
1964	490,449	299,561	268,616	44,824
1965	475,202	316,664	288,397	43,964
1966	452,089	328,564	310,370	42,753
1967	426,309	345,673	332,549	39,608
1968	399,152	367,461	354,996	39,677
1969	369,969	374,771	367,992	35,962
1970	337,619	384,511	386,937	30,804
1971	308,983	402,472	405,681	26,835

*Net Releases equals the resident patients at the beginning of the year plus admissions minus deaths and resident patients at end of year.

therefore, the decade of the 1960s represented a watershed. In the preceding century mental health policy had rested on the presumption that the care and treatment of the mentally ill should take place in public hospitals financed and administered by the states. During the 1960s this consensus slowly but surely dissolved. Nevertheless, the expectation that the creation of community-oriented policies would lead to the decline and perhaps eventual demise of public mental hospitals proved premature. To be sure, the growing importance of general hospital inpatient units, community mental health centers, and chronic care nursing homes dimin-

ished the relative significance of public mental hospitals within the mental health system. Yet these new institutions (with the exception of chronic care nursing homes) often catered to new categories of previously untreated groups. Innovative policies and institutions, therefore, did not necessarily serve all mentally ill persons, many of whom remained dependent upon services provided at public mental hospitals.

The consequences of the innovations that transformed the mental health system, like those of all human activities, were at best mixed. When the emphasis on treatment in the community was combined with an expansion of services to new groups, the result was a policy that often overlooked the need to provide supportive services for the seriously and chronically mentally ill. In this sense mental health activists all but ignored what their 19th-century predecessors had perhaps instinctively grasped, namely, that care and treatment, although conceptually separate, were not mutually exclusive. Under certain conditions, for example, care was a form of treatment.

A policy that emphasized therapy, but left care unassigned, appealed to both the public and mental health professionals. Americans generally have had a favorable perception of medical therapies, which were linked with the objective findings of medical science. Their view of care (often equated with welfare) has been more ambivalent and reflected the pervasive belief that dependency in part was a function of character deficiencies, which in turn resulted in social and economic failure. It was not surprising, therefore, that mental health professionals identified themselves with medicine rather than welfare. Moreover, professional ser-

vices required relatively simple organizational structures. Welfare services (excluding the distribution of money), by way of contrast, depended on extraordinarily complex administrative systems, few of which came close to approximating the goals of their designers.

Thus one of the driving forces of these years—antipathy toward traditional mental hospitals and a faith in community-oriented policies and institutions—gave rise to a bifurcated system with weak institutional linkages or mechanisms to ensure continuity and coordination of services. Severely and chronically mentally ill persons were often released from public hospitals after relatively brief periods of time into communities without adequate support mechanisms. The implications of the absence of longitudinal responsibility for meeting some of their basic needs in the community—including housing, medical care, welfare, and social support services—would become painfully evident during the 1980s when the contraction of public welfare and housing programs exacerbated an already difficult situation.

NOTES

The research for this paper was supported by a grant from the National Institute for Mental Health (MH39030), Public Health Service, U.S. Department of Health and Human Services. Many of the data and generalizations are taken from my previous books, *Mental Institutions in America: Social Policy to 1875* (New York: Free Press, 1973), *Mental Illness and American Society, 1975–1940* (Princeton, N.J.: Princeton Univ. Press, 1983), and *From Asylum to Community: Mental Health Policy in Modern America* (Princeton, N.J.: Princeton Univ. Press, 1991).

1   Dorothea L. Dix, *Memorial to the Legislature of Massachusetts, 1843* (Boston: Munroe & Francis, 1843), p. 4; *Memorial Soliciting a State Hospital for the Insane Submitted to the Legislature of New Jersey, January 23, 1845* (Trenton, N.J.: n.p., 1845), p. 3; *Memorial Soliciting Enlarged and Improved Accommodation for the Insane of the State of Tennessee* (Nashville, Tenn.: B. R. M'Kennie, 1847), pp. 1–2.

2   Paul S. Appelbaum, "Crazy in the streets," *Commentary,* 1987, *83*: 34.

3   Grob, *Mental Institutions in America,* ch. 1; Mary Ann Jimenez, *Changing Faces of Madness: Early American Attitudes and Treatment of the Insane* (Hanover, N.H.: Univ. Press of New England, 1987).

4   To specify the factors that led to the creation of mental hospitals is beyond the scope of this essay; the literature dealing with the origins of institutional policies is both large and controversial.

5   For general discussions, see Charles E. Rosenberg, "The therapeutic revolution: medicine, meaning, and social change in nineteenth-century America," in *The Therapeutic Revolution: Essays in the Social History of American Medicine,* ed. Morris J. Vogel and Charles E. Rosenberg (Philadelphia: Univ. of Pennsylvania Press, 1979), pp. 3–25, and John H. Warner, *The Therapeutic Perspective: Medical Practice, Knowledge, and Identity in America, 1820–1885* (Cambridge, Mass.: Harvard Univ. Press, 1986).

6   Charles E. Rosenberg, "Body and mind in nineteenth-century medicine: some clinical origins of the neurosis construct," *Bull. Hist. Med.,* 1989, *63*: 185–197.

7   For discussions of early- and mid-19th-century psychiatric therapies, see Norman Dain, *Concepts of Insanity in the United States, 1789–1865* (New Brunswick, N.J.: Rutgers Univ. Press, 1964); Grob, *Mental Institutions in America;* Nancy Tomes, *A Generous Confidence: Thomas Story Kirkbride and the Art of Asylum-Keeping, 1840–1883* (London and New York: Cambridge Univ. Press, 1984); and Samuel B. Thielman, "Madness and medicine: trends in American medical therapeutics for insanity, 1820–1860," *Bull. Hist. Med.,* 1987, *61*: 25–46.

8 Worcester State Lunatic Hospital, *Annual Report,* 1839, 7: 65.

9 Ohio Lunatic Asylum, *Annual Report,* 1842, 4: 57.

10 [Edward Jarvis], *Report on Insanity and Idiocy in Massachusetts by the Commission on Lunacy Under Resolve of the Legislature of 1854* (Massachusetts House Document No. 144 [1855]: Boston: William White, 1855), p. 70.

11 Worcester State Lunatic Hospital, *Annual Report,* 1840, 8: 47, 1841, 9: 68, 1841, 10: 62; Thomas S. Kirkbride, *On the Construction, Organization, and General Arrangements of Hospitals for the Insane* (Philadelphia: Lindsay & Blakiston, 1854; 2nd ed., Philadelphia: J. B. Lippincott, 1880), p. 23.

12 Worcester State Lunatic Hospital, *Annual Report,* 1893, 61: 70.

13 *Ibid.,* 1842, 10: 17–27, 1857, 25: 55–56, 1866, 34: 67, 1867, 35: 34, 1870, 38: 38–60; Virginia Western Lunatic Asylum, *Annual Report,* 1850, 23: 14–23; California Insane Asylum, *Annual Report,* 1860, 8: 16–32; Insane Asylum of Louisiana, *Annual Report,* 1858, p. 13; New York City Lunatic Asylum, Blackwell's Island, *Annual Report,* 1861, p. 17; New Hampshire Asylum for the Insane, *Annual Report,* 1864, 23: 16; Utica State Lunatic Asylum, *Annual Report,* 1870, 38: 18–19; Western Pennsylvania Hospital, *Annual Report,* 1871, pp. 18–19. See also Ellen Dwyer, *Homes for the Mad: Life inside Two Nineteenth-Century Asylums* (New Brunswick, N.J.: Rutgers Univ. Press, 1987), p. 150.

14 *Report on Insanity and Idiocy,* pp. 18, 73. All of the manuscript returns listing every person by name can be found in "Report of the Physicians of Massachusetts. Superintendents of Hospitals . . . and Others Describing the Insane and Idiotic Persons in the State of Massachusetts in 1855, Made to the Commissioners on Lunacy," manuscript volume in the Countway Library of Medicine, Harvard Medical School, Boston, Massachusetts.

15 For the long block quotation, see Kirkbride, *On the Construction . . . of Hospitals for the Insane,* 1st ed., p. 59. The shorter quotation is found in the 2nd ed., p. 248.

16 Data on almshouse populations are drawn from the following U.S. Bureau of the Census publications: *Paupers in Almshouses, 1904* (Washington, D.C.: Government Printing Office, 1906), pp. 182, 184; *Paupers in Almshouses, 1910* (Washington, D.C.: Government Printing Office, 1915), pp. 42–43; *Paupers in Almshouses, 1923* (Washington, D.C.: Government Printing Office, 1925), pp. 5, 8, 33; *Insane and Feeble-Minded in Hospitals and Institutions, 1904* (Washington, D.C.: Government Printing Office, 1906), p. 29; *Patients in Hospitals for Mental Diseases, 1923* (Washington, D.C.: Government Printing Office, 1926), p. 7.

17 New York State Commission in Lunacy, *Annual Report,* 1900, 12: 29–30.

18 Dwyer, *Homes for the Mad,* pp. 150–151; *Insane and Feeble-Minded in Hospitals and Institutions, 1904,* p. 37; U.S. Bureau of the Census, *Insane and Feeble-Minded in Institutions, 1910* (Washington, D.C.: Government Printing Office, 1914), p. 59; *Patients in Hospitals for Mental Disease, 1923,* p. 36; Neil A. Dayton, *New Facts on Mental Disorders: Study of 89,190 Cases* (Springfield, Ill.: Charles C. Thomas, 1940), pp. 414–429.

19 Benjamin Malzberg, "A statistical analysis of the ages of first admissions to hospitals for mental disease in New York State," *Psychiatric Quarterly,* 1949, 20: 344–366; Benjamin Malzberg, "A comparison of first admissions to the New York civil state hospitals during 1919–1921 and 1949–1951," *Psychiatric Quarterly,* 1954, 28: 312–319; New York State Department of Mental Hygiene, *Annual Report,* 1939–1940, 42: 174–175; Herbert Goldhamer and Andrew W. Marshall, *Psychosis and Civilization: Two Studies in the Frequency of Mental Disease* (Glencoe, Ill.: Free Press, 1953), pp. 54, 91; American Psychiatric Association, *Report on Patients Over 65 in Public Mental Hospitals* (Washington, D.C.: American Psychiatric Association, 1960), p. 5.

20 Statistics compiled from U.S. Bureau of the Census, *Patients in Hospitals for Mental Disease, 1923, Mental Patients in State Hospitals, 1926 and 1927* (Washington, D.C.: Government Printing Office, 1930), *Patients in Mental Hospitals, 1940* (Washington, D.C.: Government Printing Office, 1943), and Morton Kramer, *Psychiatric Services and the Changing Institutional Scene, 1950–1985,* DHEW Publication no. [ADM] 76-374 (Washington, D.C.: Government Printing Office, 1976), p. 75; Oswaldo Camargo and George H. Preston, "What happens to patients who are hospitalized for the first time when over sixty-five years of age," *Am. J. Psychiatry,* 1945, 102: 168–173.

21 Earl S. Pollock, Ben Z. Locke, and Morton Kramer, "Trends in hospitalization and patterns of care of the aged mentally ill," in *Psychopathology of Aging,* ed. Paul H. Hoch and Joseph Zubin (New York: Grune & Stratton, 1961), p. 36; Morton Kramer et al., *A Historical Study of the Disposition of First Admissions to a State Mental Hospital: Experience of the Warren State Hospital during the Period 1916–1950,* U.S. Public Health Service Publication 445 (Washington, D.C.: Government Printing Office, 1955), p. 12; Morton Kramer et al., *Mental Disorders/Suicide* (Cambridge, Mass.: Harvard Univ. Press, 1972), pp. 27–28.

22 Kramer et al., *Historical Study of the Disposition of First Admissions,* p. 16.

23 Postwar developments are covered in Grob, *From Asylum to Community.*

24 Data on the location of psychiatrists is taken from the following: *Biographical Directory of Fellows & Members of the American Psychiatric Association as of October 1, 1957* (New York: R. R. Bowker, 1958); "Distribution of

members of the American Psychiatric Association, 1910–1960," Miscellaneous Papers, Box 1, Archives of the APA, Washington, D.C.; Joint Information Service, *Fact Sheet No. 2,* May 1957 and *No. 10,* August 1959; David A. Boyd, "Current and future trends in psychiatric residency training," *J. Med. Educ.,* 1958, *33:* 345–346.

25   Harry C. Solomon, "The American Psychiatric Association in relation to American psychiatry," *Am. J. Psychiatry,* 1958, *115:* 1–9; *New York Times,* 13 and 16 May 1958; Robert C. Hunt to Solomon, 17 June 1958, Solomon to Hunt, 19 June 1958, Solomon Papers, Archives of the APA; Robert C. Hunt, "The state hospital stereotype" (statement before the APA Commission on Long Term Planning, 30 Oct. 1959), Records of the Medical Director's Office, 200-11, Archives of the APA.

26   *New York Times,* 4 April 1947; NIMH, "Annual Report, Fiscal 1949," pp. 9–10, NIMH Records, Subject Files, 1940–1951, Box 82, Washington National Records Center, Suitland, Maryland; Anita K. Bahn and Vivian B. Norman, *Outpatient Psychiatric Clinics in the United States, 1954–1955,* U.S. Public Health Service *Public Health Monograph 49* (Washington, D.C.: Government Printing Office, 1957), p. 38.

27   "Gains in outpatient psychiatric services, 1959," *Public Health Reports,* 1960, *75:* 1092–1094; Vivian B. Norman, Beatrice M. Rosen, and Anita K. Bahn, "Psychiatric clinic outpatients in the United States, 1959," *Mental Hygiene,* 1962, *46:* 321–343.

28   Harold Sampson, David Ross, Bernice Engle, and Florine Livson, "Feasibility of community clinic treatment for state mental hospital patients," *Arch. Neurol. & Psychiat.,* 1958, *80:* 77. A larger version of this study appeared under the title *A Study of Suitability for Outpatient Clinic Treatment of State Mental Hospital Admissions, 1957,* California Department of Mental Hygiene, *Research Report No. 1* (1957).

29   See Morton Kramer, *Some Implications of Trends in the Usage of Psychiatric Facilities for Community Mental Health Programs and Related Research,* U.S. Public Health Service, *Publication 1434* (Washington, D.C.: Government Printing Office, 1967); Morton Kramer, "Epidemiology, biostatistics, and mental health planning," in APA *Psychiatric Research Reports,* 1967, *22:* 1–68; Morton Kramer, C. Taube, and S. Starr, "Patterns of use of psychiatric facilities by the aged: current status, trends, and implications," in APA *Psychiatric Research Reports,* 1968, *23:* 89–150.

30   The term "patient care episode" represents the sum of two numbers: residents at the beginning of the year or on the active roll of outpatient clinics, and admissions during the year. The first is an unduplicated count; the second includes duplications, since some individuals had multiple admissions.

31   Data taken from NIMH, *Statistical Note 23* (April 1970): 1–4, and *Statistical Note 154* (September 1980): 12. Slight differences in totals are due to the rounding out of fractions. It should be noted that NIMH data usually did not count patient care episodes in general hospitals without specialized psychiatric units.

32   NIMH data in Morton Kramer, *Psychiatric Services and the Changing Institutional Scene, 1950–1985,* DHEW Publication no. [ADM] 77-433 (Washington, D.C.: Government Printing Office, 1977), p. 78.

33   NIMH, *Psychiatric Services in General Hospitals, 1969–1970* (NIMH, *Mental Health Statistics,* Series A No. 11 [1972]), p. 21; Charles Kanno and P. L. Scheidemandel, *Psychiatric Treatment in the Community: A National Survey of General Hospital Psychiatry and Private Psychiatric Hospitals* (Washington, D.C.: American Psychiatric Association, 1974), p. 35; Kramer, *Psychiatric Services and the Changing Institutional Scene,* p. 17. See also James W. Thompson, R. D. Bass, and M. J. Witkin, "Fifty years of psychiatric services: 1940–1990," *Hospital & Community Psychiatry,* 1982, *33:* 714.

34   Raymond M. Glasscote and C. K. Kanno, *General Hospital Psychiatric Units: A National Survey* (Washington, D.C.: Joint Information Service, 1965), p. 16; Kanno and Scheidemandel, *Psychiatric Treatment in the Community,* p. 34; NIMH, *Statistical Note No. 70* (February 1973); Texas Division of Mental Health, *Psychiatric Inpatient Units in Texas General Hospitals: A Report . . . 1958* (n.p., 1958), p. 21.

35   Kramer et al., *Mental Disorders/Suicide,* p. 28. See also Pollack, P. H. Person, Jr., Morton Kramer, and Goldstein, *Patterns of Retention, Release, and Death of First Admissions to State Mental Hospitals,* U.S. Public Health Service *Publication 672* (Washington, D.C.: Government Printing Office, 1959), and NIMH, *Statistical Note 74* (1973).

36   Howard H. Goldman, N. H. Adams, and C. Taube, "Deinstitutionalization: the data demythologized," *Hospital & Community Psychiatry,* 1983, *34:* 133; Charles A. Kiesler and Amy E. Sibulkin, *Mental Hospitalization: Myths and Facts About a National Crisis* (Newbury Park: Sage Publications, 1987), pp. 86, 95; NIMH, *Mental Health, United States, 1985,* ed. Carl A. Taube and Sally A. Barrett, DHHS Pub. no. [ADM] 85-1378 (Washington, D.C.: Government Printing Office, 1985), pp. 33, 53. See also NIMH, *Statistical Note 74* (1973), and Fritz Redlich and S. R. Kellert, "Trends in American mental health," *Am. J. Psychiatry,* 1978, *135:* 24.

RACE AND MEDICINE

Apologists for slavery and racial segregation in the 19th century argued that black Americans were medically different from white Americans. They argued that the physical constitution of African Americans fitted them for a life of bondage and required the use of unique medical care. While some people today similarly posit innate racial differences, present-day analyses of the concept of race generally challenge the validity of such an argument based so strongly on biology. Many recent studies continue to suggest that, despite the preponderance of basic anatomical and physiological similarities between blacks and whites, some medical differences seem to exist. Various genetic diseases are more prominent in particular racial or ethnic groups; racial and ethnic differences in responses to medications also have been identified. Yet sickle-cell anemia, which has been labeled a "black" disease, is not confined to African Americans; it is also prevalent among people of Mediterranean, Middle Eastern, and Indian ancestry. Indeed, many alleged biological racial differences do not hold up to rigorous analysis and can only be explained in a specific historical and environmental context. Any racial differences we might observe in health status, scholars today agree, need to be understood in terms of social, as well as possible biological, factors.

The care sick slaves received before the Civil War, as Todd L. Savitt demonstrates, depended less on theories of racial differences than on the circumstances of their daily lives. Some planters, eager to protect their investment, contracted with local physicians to treat their slaves for a fixed yearly fee. More commonly, the master, mistress, or overseer—sometimes assisted by a black nurse—treated ailing slaves with home remedies, calling in physicians only for the most serious cases. Enslaved men and women often preferred or had to rely on their own folk remedies or seek help from their traditional healers.

Because of racial prejudice and their enslaved condition, African Americans were systematically denied the opportunity to become trained physicians. In the years immediately preceding the Civil War, a few blacks gained admittance to predominantly white medical schools, but opportunities remained scant until after the war, when black medical schools, beginning with Howard University School of Medicine in 1876, first appeared. Next, black physicians established their own hospitals and medical societies, including the National Medical Association, created in 1895 because of the discriminatory policies of the American Medical Association. As Vanessa Northington Gamble relates, black hospitals, originally founded to provide health care and education within a segregated society, evolved to become symbols of black pride and achievement. Such institutions have now become peripheral to the lives of most black people and are on the brink of extinction.

African Americans and other powerless groups in American society have been particularly vulnerable to exploitation by physicians seeking living bodies for experimentation or cadavers for dissection. Although human experimentation often produced little of value, at least three surgical breakthroughs in the first half of the 19th century depended in part on the use of southern black bodies: Ephraim McDowell's removal of ovaries, Crawford Long's co-discovery of surgical anesthesia, and James Marion Sims's repair of vesico-vaginal fistulae. As Allan M. Brandt shows, the use of blacks for experimental purposes did not end with emancipation. The notorious Tuskegee syphilis study, undertaken by the U.S. Public Health Service in 1932 and involving hundreds of black men, continued until 1972, when public revelation and indignation brought it to a halt.

22

Black Health on the Plantation: Masters, Slaves, and Physicians

TODD L. SAVITT

Sickness and death were constant worries of 19th-century Americans, especially in the disease-ridden antebellum South, where, by 1860, over 4 million blacks, most of them enslaved, composed much of the work force. Their presence created a special situation for two reasons: first, whites, responsible for slave medical care, in large part dictated the living and working conditions that promoted or destroyed blacks' health; second, some white southerners claimed (and many others believed) that blacks were medically different from whites and so in need of special treatment. This paper will look at the three groups of southerners most directly involved in issues of black health and care —masters, physicians, and the slaves themselves —and consider their relationships to the special problem of slavery.[1]

Section I discusses how southern physicians used their own and others' observations of black medical differences to develop both a partial rationale for enslaving blacks in the American South and an approach to medical care and treatment of Negroes. Section II describes the various ways in which a master's treatment of his plantation slaves affected black health. Finally, Section III deals with black medical care provided by masters, physicians, and slaves and with the interactions of the three groups.

TODD L. SAVITT is Professor of Medical Humanities and History at East Carolina University, Greenville, North Carolina.

Reprinted with permission from *Science and Medicine in the Old South,* edited by Ronald L. Numbers and Todd L. Savitt (Baton Rouge: Louisiana State University Press, 1989), pp. 327–355.

WERE BLACKS MEDICALLY DIFFERENT FROM WHITES?

"Scarcely any observant medical man, having charge of negro estates, fails to discover, by experience, important modifications in the diseases and appropriate treatment of the white and black race respectively."[2] Thus wrote the editor of a prominent Virginia medical journal in 1856, in an attempt to impress upon the state's physicians the importance of providing adequate health care to slaves. He could very well have been speaking to all southern doctors, many of whom believed that medical differences existed between blacks and whites. The issue was of both practical and political importance: it involved not only the health care of an entire racial group in the South, but also the partial justification for enslaving them.

Physicians who treated slave diseases had a pecuniary and professional concern with the subject. Their recorded opinions in medical, commercial, and agricultural journals, as well as in personal correspondence, attest to the seriousness with which they approached issues of black health. The politically minded physicians (all of whom practiced medicine) were also resolutely committed to explaining the southern position on slavery. They, too, published for the medical, commercial, and lay public. Men like Josiah Clark Nott of Mobile and Samuel A. Cartwright of New Orleans utilized their knowledge of black medicine to rationalize the necessity and usefulness of slavery. These apologists for the peculiar institution, in order to prove that slavery was humane and economically viable in the South, argued that blacks possessed immunity to certain diseases which devastated whites. Slaveowners, they said, did not sacrifice blacks every time they sent them into the rice fields or cane-brakes. Nor could physicians adequately treat blacks without knowledge of their anatomical and physiological peculiarities and disease proclivities. Blacks were medically different and mentally inferior to whites, they asserted.

In certain obvious physical ways Negroes did vary greatly from Caucasians. Old South writers particularly commented on facial features, hair, posture and gait, skin color, and odor. One school of American scientists spent much time and effort investigating cranium and brain size, as well as other characteristics, in part to discover whether blacks were inferior to whites. Physicians, slave-owners, and other interested persons also detected distinctions in the physical reactions of the two races both to diseases and to treatments. Their observations were often accurate, but at times they allowed racial prejudice to cloud their views. These observers remarked particularly about the relative immunity of blacks to southern fevers, especially intermittent fever (malaria) and yellow fever, susceptibility to intestinal and respiratory diseases, and tolerance of heat and intolerance of cold. Black children, they noted, died more frequently of marasmus (wasting away), convulsions, "teething," and suffocation or overlaying than did whites.[3]

One of the most fascinating subjects about which Old South medical authors wrote was black resistance to malaria, the focus of constant comment during the antebellum period because blacks' liability to it appeared to vary from region to region, plantation to plantation, and individual to individual. In dispute was the degree of susceptibility and the virulence of the disease in Negroes: Could slaves acquire some resistance to malaria by living in constant proximity to its supposed source? Were some slaves naturally immune? Did slaves have milder attacks of the disease than whites?

Modern science has answered many of the questions of immunity and prevention of malaria with which doctors and planters struggled in the antebellum period. Several factors contributed to the phenomenon of malarial immunity. As will be discussed below, many blacks did possess an inherited immunity to one or another form of malaria. But for Caucasians and those Negroes without natural resistance to a particular plasmodium (the

organism that causes malaria) type, it was possible to acquire malarial immunity or tolerance only under the conditions stated by the author of an article in the *New Orleans Medical News and Hospital Gazette* (1858–1859) — by suffering repeated infections of the disease over a period of several years. For this to occur, one of the four species of plasmodium — falciparum, vivax, malariae, and ovale — had to be present constantly in the endemic region, so that with each attack a person's supply of antibodies was strengthened against further parasitic invasions. Interruption of this process, such as removal to nonendemic areas for the summer (when exposure to the parasite was useful in building immunity) or for several years (during schooling or travel), prohibited the aggregation of sufficient antibodies to resist infection. In truly endemic areas, acquiring immunity this way was a risky affair: unprotected children struggled for their lives, adults suffered from relapses of infections contracted years before, and partially immune adults worked through mild cases. It is no wonder, then, that slaves sold from, say, Virginia, where one form of malaria was prevalent, to a Louisiana bayou or South Carolina rice plantation, where a different species or strain of plasmodium was endemic, had a high incidence of the disease. Even adult slaves from Africa had to go through a "seasoning" period, because the strains of malarial parasites in this country differed from those in their native lands.

Generally speaking, most malaria in the upper South and in inland piedmont areas was of the milder vivax type, while vivax and the more dangerous falciparum malaria coexisted in coastal and swampy inland portions of both the lower and upper South. Rare pockets of the malariae type (quartan fever) were scattered across the South. Ovale malaria appears not to have been present in the United States. Of course, epidemics of one type or another could strike any neighborhood, resulting in sickness and death even to those who had acquired resistance to the endemic variety. *Plasmodium falciparum* usually caused such epidemics, especially in the temperate regions of the South.[4]

The major reason for black immunity to vivax or falciparum malaria relates not to acquired resistance, but to selective genetic factors. At least three hereditary conditions prevalent among blacks in parts of modern Africa appear to confer immunity to malaria upon their bearers. Recent medical research indicates that the red blood cells of persons lacking a specific factor called Duffy antigen are resistant to invasion of *Plasmodium vivax.* Approximately 90 percent of West Africans lack Duffy antigens, as do about 70 percent of Afro-Americans. This inherited, symptomless, hematologic condition is extremely rare in other racial groups. All evidence points to the conclusion that infection by *Plasmodium vivax* requires the presence of Duffy-positive red blood cells. Since most members of the Negro race do not possess this factor, they are immune to vivax malaria. It can be safely assumed that the vast majority of American slaves and free blacks were likewise resistant to this form of the disease.

Some antebellum blacks had additional protection against malaria resulting from two abnormal genetic hemoglobin conditions, sickle cell disease (a form of anemia) and sickle cell trait (the symptomless carrier state of the sickling gene). People with either of these conditions had milder cases of, and decreased risk of mortality from, the most malignant form of malaria, falciparum. Many of those who had sickle cell *disease* died from its consequences before or during adolescence; however, blacks who had sickle cell *trait* lived entirely normal lives and could then transmit the gene for sickling to their offspring. Since people with the trait had one normal gene and one abnormal gene for sickling (in contrast to those with the disease, who had two of the abnormal genes), their offspring could inherit sickle cell anemia only when each parent contributed a gene for sickling. Because the sickle cell condition was not discovered until 1910, physicians in the antebellum South were unaware that this was one reason why some slaves on plantations appeared to be immune to malarial infections. One other genetic condition with a high incidence within the former slave-trading region probably affords some malarial resistance: deficiency of the enzyme glucose-6-phosphate dehydrogenase (G-6-PD deficiency).

It is impossible to provide an exact calculation of sickle cell and G-6-PD deficiency gene prevalence among antebellum southern blacks. However, an estimate might be ventured based on known gene frequencies among present-day Afro-Americans and those Africans residing in former slave-trading areas. One leading medical authority on abnormal hemoglobins has estimated that at least 22 percent of Africans first brought to this country possessed genes for sickling. Other medical scientists have determined that approximately 20 percent of West Africans have genes for G-6-PD deficiency. Overall, then, using conservative figures, approximately 30–40 percent of newly arrived slaves had one or both of these genes. Recent evidence points to a higher-than-expected frequency of the G-6-PD gene in patients with sickle cell disease, which might reduce this estimate by a few percent. Thus a large proportion of Negroes were immune to the severe effects of falciparum malaria and to the less virulent vivax malaria, facts which planters and physicians in the South could not help but notice and discuss openly.

As with malaria, planters and physicians speculated publicly on blacks' intolerance of cold climates but could never adequately prove their contentions. It was the confirmed opinion of many white southerners that blacks could not withstand cold weather to the same degree as whites because of their dark skin and equatorial origins. Their major concern was that blacks seemed to resist and tolerate respiratory infections less well than whites. Since the germ theory of disease was years in the future, these men and others explained their observations with a combination of then-current though not universally accepted medical, anthropological, and scientific logic—and occasionally with unfounded theories. Blacks, natives of a tropical climate, were physiologically ill suited for the cold winter weather and cool spring and fall nights of the temperate zone. They breathed less air, dissipated a greater amount of "animal heat" through the skin, and eliminated larger quantities of carbon via liver and skin than whites. In addition, blacks were exposed to the elements for much of the year, placing a strain on heat production within the body. One medical extremist, Samuel A. Cartwright of New Orleans, even claimed that Negroes' lungs functioned inefficiently, causing "defective atmospherization of the blood." Some noted that slaves often slept with their heads (rather than their feet) next to the fire and entirely covered by a blanket; this was seen as proof that they required warm, moist air to breathe and to survive in this climate. Blacks were, these men concluded, physiologically different from whites.

Even today there is some confusion among medical authorities regarding the susceptibility of Negroes to severe pulmonary infections. Some claim a racial or genetic predisposition, while others deny it. Historically, blacks have shown a higher incidence and more severe manifestations of respiratory illness than have whites. Explanations for this phenomenon are numerous. First, Negroes did not experience bacterial pneumonias until the coming of the Caucasian. The entire newly exposed population was thus exquisitely sensitive to these infections, and developed much more serious cases than whites, who had had frequent contact with the bacteria since childhood. Second, black African laborers who today move from moist tropical to temperate climates (e.g., to the gold mines of South Africa) contract pneumonia at a much higher rate than whites. Though the incidence of disease decreases with time, it always remains at a more elevated level than among Caucasians. The same phenomenon probably operated during slavery. At first the mortality rate from pneumonia among newly arrived slaves was probably inordinately high, but with time the figure decreased somewhat, though it remains higher than in Caucasians. Third, there appears to be a close relationship between resistance to pulmonary infection and exposure to cool, wet weather. Slaves, who worked outdoors in all seasons and often lived in drafty, damp cabins, were therefore more likely to suffer from respiratory diseases than their masters. Fourth, poor diet, a common slave problem, predisposes people to infections like pneumonia and other respiratory problems.[5] Finally, overcrowding and unsanitary living conditions caused an increased incidence of respiratory diseases. Slaves living in small cottages or grouped together in a large community at the quarters, where intimate and frequent visiting was common, stood a greater chance of contracting airborne infections than did the more isolated whites. Undoubtedly, all these factors combined to increase the occurrence of respiratory illness among southern blacks.

The most serious nonfatal manifestation of cold intolerance was frostbite. At least one proslavery apologist claimed that the Negro race was more susceptible than whites to this condition: "Almost every one has seen negroes in Northern cities, who have lost their legs by frost at sea—a thing rarely witnessed among whites, and yet where a single negro has been thus exposed doubtless a thousand of the former have."[6] The condition was a serious one, especially for slaveowners who stood to lose the labor of valuable workers.

Blacks are in fact more susceptible to cold injury than whites. Studies conducted during and after the Korean War indicate that blacks have a poorer adaptive response to cold exposure than do whites in the following ways: their metabolic rates increase more slowly and not as much as whites; their first shivers (one of the body's defensive responses to cold) occur at a lower skin temperature than for whites; and their incidence of frostbite is higher and their cases more severe than those of whites. Even after blacks have acclimated to cold (and they do so in a manner physiologically similar to whites), they are then only slightly less liable to sustain cold injury than they had been previously. Those antebellum observers who warned against overexposure of slaves to cold were essentially correct.

Racial differences in tolerance to heat also exist but may be modified under certain conditions. Again, antebellum observers agreed that blacks, having originated in an area known for its heat and humidity, were ideally suited for labor in the damp, warm South. One northern physician, John H. Van Evrie, explained blacks' resistance to heat in both religious and physiological terms:

> God has adapted him, both in his physical and mental structure, to the tropics. . . . His head is protected from the rays of a vertical sun by a dense mat of woolly hair, wholly impervious to its fiercest heats, while his entire surface, studded with innumerable sebaceous glands, forming a complete excretory system, relieves him from all those climatic influences so fatal, under the same circumstances, to the sensitive and highly organized white man. Instead of seeking to shelter himself from the burning sun of the tropics, he courts it, enjoys it, delights in its fiercest heats.[7]

Modern medical investigators would not agree with Van Evrie's reasoning. But they have discovered that under normal living conditions Negroes in Africa and the United States are better equipped to tolerate humid heat than whites. However, both races possess the same capacity to acclimatize to hot, humid conditions. The physiological mechanisms by which the human body acclimatizes to heat can be readily observed and measured. Increased external temperature causes the body to perspire more, resulting in a greater evaporative heat loss, a decline in skin and rectal temperatures, and a drop in the heart rate from its initially more rapid pace. When whites and blacks are equally active in the same environment over a period of time, there is little difference in heat tolerance.

From this information it can be assumed that in the Old South slaves and free blacks possessed a higher *natural* tolerance to humid heat stress than did whites. In addition, Negroes became quickly acclimatized to performance of their particular tasks under the prevailing climatic conditions of the region. This natural and acquired acclimatization enabled black laborers to withstand the damp heat of summer better than whites, who were unused to physical exertion under such severe conditions. White farm and general laborers, however, also must have adjusted to the heat and fared as well as blacks. One physiological difference between Caucasians and Negroes which might have affected work performance in the hot, humid South was the latter's inherent ability to discharge smaller amounts of sodium chloride and other vital body salts (electrolytes) into sweat and urine. Excessive loss of these salts leads to heat prostration and heatstroke. Thus conservation of needed electrolytes provided slaves with an advantage over laboring whites, whose requirements for replacement of the substances were greater. In the case of heat tolerance, then, white observers were correct in noting a racial difference, but they tended to ignore the fact that many whites did become acclimatized to the hot, humid environment.

Whites detected, or thought they detected, distinctive variances in black susceptibility to several other medical conditions common in the antebellum South. Many believed that slave women developed prolapsed uteri at a higher rate than white women, though modern anatomists have shown that Negroes are actually less prone to this affliction than Caucasians. Observers also noted that slaves were frequent sufferers of typhoid fever, worms, and dysentery, though we now know that the reason for this high prevalence was environmental rather than racial or genetic. Blacks did, however, have a greater resistance to the yellow fever virus than whites.

One disease that drew great attention because of its frequency and virulence was consumption (pulmonary tuberculosis), a leading cause of death in 19th-century America for members of both races. A particular form of the disease — characterized by extreme difficulty in breathing, unexplained abdominal pain around the navel, and rapidly progressing debility and emaciation, usually resulting in death — struck blacks so commonly that it came to be known as Negro Consumption or *Struma Africana*. In all likelihood most of the cases which white southerners described as Negro Consumption were miliary tuberculosis, the most serious and fatal form of the disease known, in which tubercles are found in many body organs simultaneously, overwhelming what natural defenses exist. The reason that rapidly fatal varieties of the disease (so-called galloping consumption) afflicted Negroes more frequently than Caucasians may be related to the fact that Caucasians (like Mongolians) had suffered from tuberculosis for many hundreds of years and had developed a strong immune response to the infection, whereas Africans (and American Indians and Eskimos) had been exposed to tuberculosis only since the coming of the white man and had not yet built up this same effective resistance. Others have discounted racial immunity as an explanation and have argued that, as a "virgin" population, blacks were highly susceptible to serious first attacks of tuberculosis. Additional factors such as malnourishment, preexisting illness, or general debility also contributed to the apparent black predisposition to tuberculosis.

Neonatal tetanus (also called *trismus nascentium*) was a common cause of death among newborn slaves throughout the South. Slaveowners and physicians, who recognized its origin in the improper handling of the umbilical stump, often discussed it in their writings. It still kills large numbers of children in undeveloped countries. *Clostridium tetani*, the same bacterium which caused tetanus in older children and adults, also infected newborns through the unwashed and frequently touched umbilical stump. In a typical antebellum case, related by Dr. Albert Snead of Richmond to his colleagues at a medical society meeting in 1853, an eight-day-old black child first refused her mother's breast and gave a few convulsive hand jerks. Soon the baby's entire muscular system was rigid, with her head bent back, fists and jaws clenched, and feet tightly flexed, as the bacterial toxin affected central nervous system tissue. In this case death from suffocation owing to respiratory muscle paralysis did not intervene until the 18th day (though it usually occurred within 7 to 10 days).

One cause of death which Virginians did not consider a disease and which seemed to occur almost exclusively among the slave population was "smothering," "overlaying," or "suffocation." Observers assumed that sleeping mothers simply rolled onto or pressed snugly against their infants, cutting off the air supply, or that angry, fearful parents intentionally destroyed their offspring rather than have them raised in slavery. Modern medical evidence strongly indicates that most of these deaths may be ascribed to a condition presently known as Sudden Infant Death Syndrome (SIDS) or "crib death" which, for reasons yet unexplained, affects blacks more frequently than whites.[8]

The diseases and conditions discussed above represent only some of the several which whites noted affected blacks and whites differently. Others are difficult to trace back to slavery times either through direct records or by implication through comparative West African medicine. Among those not mentioned, the most important is hypertension (high blood pressure). Others include polydactyly (six or more fingers per hand), umbilical hernia, cancers of the cervix, stomach, lungs, esophagus, and prostate, and toxemia of pregnancy. At the same time, Negroes are much less susceptible to hookworm disease, cystic fibrosis, and skin cancer than Caucasians.

Though blacks are not the only racial or ethnic group possessing increased immunity and susceptibility to specific diseases, they were the only group whose medical differences mattered to white residents of the Old South. Reports of black medical "peculiarities" appeared regularly in periodicals and pamphlets, presumably to alert southern physicians of problems they might encounter in practice. Agricultural and medical journal articles and medical student dissertations also discussed racial differences in responses to medical treatments. Most writers agreed that blacks withstood the heroic, depletive therapies of the day (bleeding, purging, vomiting, blistering) less well than whites.

Despite many writings on the subject of black diseases and treatments, no comprehensive discussion existed in any standard textbook for student doctors or practitioners. Some articles on the subject of treatment provided vague information, such as: "The Caucasian seems to yield more readily to remedies . . . than the African," or, "It is much more difficult to form a just diagnosis or prognosis with the latter [African] than the former [white], consequently the treatment is often more dubious."[9] In 1855 an editor of the *Virginia Medical and Surgical Journal* suggested that a Virginian write a book on "the modifications of disease in the Negro constitution." The subject, he proclaimed, stood "invitingly open"; no medical student, "fresh from Watson or Wood [textbooks of the day], with his new lancet and his armory of antiphlogistics," had been properly trained to treat many of the diseases to which the black man was subject: "Has he been taught that the African constitution sinks before the heavy blows of the 'heroic school' and runs down under the action of purgatives; that when the books say blood letting and calomel, the black man needs nourishment and opium?"[10] Many other southern medical writers put forth urgent pleas for medical school courses and books on black medicine.

John Stainbach Wilson, a Columbus, Georgia, physician who had spent years practicing medicine in southern Alabama, came closest to actually producing a textbook on black health. Its advertised title indicated the wide scope of the proposed contents: *The Plantation and Family Physician; A Work for Families Generally and for Southern Slaveowners Especially; Embracing the Peculiarities and Diseases, the Medical and Hygienic Management of Negroes, Together with the Causes, Symptoms, and Treatment of the Principal Diseases to Whites and Blacks.* But apparently the outbreak of hostilities in 1861 interrupted Wilson's plans. It was not until more than 100 years later, in 1975, that the first *Textbook of Black-Related Diseases* was finally published, written this time by black physicians.

For southern whites, black medical problems and health care had political as well as medical ramifications. Men like Cartwright, Nott, and Wilson were writing for an audience who wished to hear that blacks were distinct from whites. This was, after all, a part of the proslavery argument. Medical theory and practice were still in such a state of flux in the late 18th and early 19th centuries that there was little risk of any true scientific challenge to a medical system based on racial differences. Observers were correct in noting that blacks showed differing susceptibility and immunity to a few specific diseases and conditions. They capitalized on these conditions to illustrate the inferiority of blacks to whites, to rationalize the use of this "less fit" racial group as slaves, to justify subjecting Negro slaves to harsh working conditions in extreme dampness and heat in the malarious regions of the South, and to prove to their critics that they recognized the special medical weaknesses of blacks and took these failings into account when providing for their human chattel. But in terms of an overall theory of medical care predicated on racial inferiority, the issue was a false one. It is instructive to note here, for example, that no writer ventured beyond vague and cautious statements about bleeding or purging blacks less than whites. None presented an account of the amount of blood loss or the dose of medicine which was optimal for blacks. Remarks on the subject were always couched in terms which placed whites in a position of medical and physical superiority over Negroes, perfect for polemics and useless to the practitioner.

LIVING AND WORKING CONDITIONS

The state of slave health depended not only on disease immunities and susceptibilities but also on living and working conditions. How masters provided for sanitation, housing, food, clothing, and children's and women's special needs, and how they worked and disciplined their human chattel, had a great effect on the health of the black work force.[11] (See Table 1.)

Most bondsmen lived on plantations or farms in a well-defined area known as the quarters. Here was an ideal setting for the spread of disease, similar to the situation which existed in antebellum urban areas. What might have been considered a personal illness in the isolated, white, rural family dwelling became in a three- or ten- or thirty-home slave community a matter of public health and group concern.

At the slave quarters sneezing, coughing, or contact with improperly washed eating utensils and personal belongings promoted transmission of disease-causing microorganisms among family members. Poor ventilation, lack of sufficient windows for sunshine, and damp earthen floors merely added to the problem by aiding the growth of fungus and bacteria on food, clothing, floors, and

utensils, and the development of worm and insect larvae. Improper personal hygiene (infrequent baths, hairbrushings, and haircuts, unwashed clothes, unclean beds) led to such nuisances as bedbugs, body lice (which also carried typhus germs), ringworm of skin and scalp, and pinworms. In a household cramped for space, these diseases became family, not individual, problems. And when two or more families shared homes and facilities, the problem of contagion became further aggravated.

Contacts outside the home also facilitated the dissemination of disease. Children who played together all day under the supervision of a few older women, and then returned to their cabins in the evening, spread their day's accumulation of germs to other family members. Even mere Sunday and evening socializing in an ill neighbor's cabin was enough to "seed" the unsuspecting with disease. Contaminated water, unwashed or poorly cooked food, worm-larvae-infested soil, and disease-carrying farm animals and rodents also contributed their share to the unhealthfulness of the quarters.

The two major types of seasonal diseases which afflicted Old Dominion blacks—respiratory and intestinal—reflected living conditions within most slave communities. Respiratory illnesses prevailed during the cold months, when slaves were forced to spend much time indoors in intimate contact with their families and friends. Several important contagious diseases were spread through contact with respiratory system secretions: tuberculosis, diphtheria, colds and upper respiratory infections, influenza, pneumonia, and streptococcal infections (including sore throats and scarlet fever). The community life of the slave quarters also provided excellent surroundings for dissemination of several year-round diseases contracted through respiratory secretions. People today tend to regard these illnesses—whooping cough, measles, chicken pox, and mumps—as limited to the younger population, but adult slaves who had never experienced an outbreak of, say, measles, in Africa or the United States were quite susceptible to infection and even death. Measles and whooping cough are still important causes of fatality in developing countries, as they were in antebellum Virginia.

As warm weather arrived and workers spent more time outdoors, intestinal diseases caused by

Table 1
Leading Known Causes of Death in Virginia, 1850[a], by Race[b]

White (N = 8,014)[c]	Percent	Slave (N = 6,284)[c]	Percent	Free Black (N = 558)[c]	Percent
1. Tuberculosis[d]	13.8	1. Respiratory diseases[e]	16.1	1. Cholera[j]	17.2
2. Respiratory diseases[e]	9.2	2. Tuberculosis[d]	10.7	2. Tuberculosis[d]	13.8
3. Nervous system diseases[f]	8.6	3. Nervous system diseases[f]	7.9	3. Respiratory diseases[e]	10.0
4. Diarrhea[g]	8.5	4. Old age	7.3	4. Old age	7.9
5. Scarlet fever[h]	6.9	5. Dropsy[i]	6.0	5. Dropsy[i]	7.5
6. Dropsy[i]	5.5	6. Typhoid[k]	5.7	6. Nervous system diseases[f]	6.5
7. Cholera[j]	4.9	7. Diarrhea[g]	5.5	7. Diarrhea[g]	5.9
8. Typhoid[k]	4.7	8. Cholera[j]	5.5	8. Accidents[n]	5.8
9. Old age	4.5	9. Accidents[n]	4.5	9. Typhoid[k]	2.9
10. Digestive system diseases[l]	4.5	10. Digestive system diseases[l]	3.2	10. Digestive system diseases[l]	2.5
11. Diphtheria[m]	3.5	11. Diphtheria[m]	2.3	11. Whooping cough	2.5
12. Accidents[n]	3.4	12. Suffocation[q]	2.2	12. Maternity[o]	2.0
13. Maternity[o]	2.1	13. Scarlet fever[h]	2.0	13. Worms	2.0
14. Measles	1.5	14. Worms	1.8	14. Teething	1.3
15. Whooping cough	1.4	15. Maternity[o]	1.8	15. Diphtheria[m]	1.1
16. Neoplasms[p]	1.2	16. Measles	1.6	16. Intemperance	1.1
17. Heart disease	0.8	17. Teething	1.5	17. Malaria[r]	0.7
18. Teething	0.8	18. Whooping cough	1.5	18. Scarlet fever[h]	0.7
19. Worms	0.8	19. Homicide	0.9	19. Measles	0.5
20. Erysipelas	0.7	20. Heart disease	0.6	20. Homicide	0.5
Fevers (unclassifiable)[s]	6.5	Fevers (unclassifiable)[s]	5.5	Fevers (unclassifiable)[s]	4.6

[a]Includes figures from the entire state. Compiled from *Mortality Statistics of the Seventh Census of the United States, 1850* (Washington, D.C., 1855), pp. 291–295.

[b]All figures are percentages of the total number of deaths for that race during the designated time period. Causes listed as "unknown" were excluded from the computations.

[c]The numbers in parentheses reflect the total number of deaths from all known causes for that race. Only leading causes are listed for each race.

[d]Consumption, scrofula

[e]Asthma, bronchitis, catarrh, catarrhal fever, influenza, disease of lungs, pleurisy, pneumonia

[f]Apoplexy, disease of brain, chorea, congestion of brain, convulsions, epilepsy, brain fever, inflammation of brain, insanity, neuralgia, paralysis, disease of spine

[g]Cholera infantum, diarrhea, dysentery, summer complaint

[h]Scarlet fever, disease of throat, quinsy

[i]Dropsy, hydrothorax, ascites

[j]Cholera epidemics struck Virgina during the period covered by this census.

[k]Typhoid fever was referred to as typhus fever in the list of causes of death in the 1850 census.

[l]Disease of bowels, colic, cramp, dyspepsia, hernia, inflammation of bowels, inflammation of stomach, jaundice, disease of liver, piles, disease of stomach

[m]Croup

[n]Burns, drownings, scaldings, explosions, shootings, railroad, unspecified

[o]Childbirth, puerperal fever

[p]Cancer, tumor

[q]Many of these deaths are probably attributable to what is known now as crib death or Sudden Infant Death Syndrome (SIDS).

[r]Intermittent fever, remittent fever

[s]Bilious fever, congestive fever, inflammatory fever, fever not specified

poor outdoor sanitation and close contact with the earth became common. Respiratory diseases decreased in frequency and insects became important culprits in the spread of disease — particularly maladies of the digestive tract and various "fevers." What more could a mosquito or housefly wish than a large concentration of human beings, decaying leftover food scraps, scattered human feces, or a compost heap? Mosquitoes discharged yellow fever or malarial parasites while obtaining fresh blood from their victims. Flies transported bacteria such as *Vibrio* (cholera), *Salmonella* (food poisoning and typhoid), and *Shigella* (bacillary dysentery), the virus which causes infectious hepatitis, and the protozoan *Entameba histolytica* (amebic dysentery) from feces to food. Trichina worms, embedded in the muscles of hogs inhabiting yards often shared with bondsmen, were released into a slave's body when the meat was not completely cooked. Finally, there were the large parasitic worms, a concomitant of primitive sanitation.

Intestinal disorders were at least as common among antebellum Virginia's blacks as were respiratory diseases. The human alimentary tract is distinguished among all other body systems in that it receives daily large amounts of foreign material, usually in the form of food, which it must sort and assimilate into a usable form. In the Old South, where living conditions were generally unhygienic, seemingly good food and drink often concealed pathogenic organisms ranging from viruses to worm larvae. Hands entering mouths sometimes contained the germs that others had cast off in feces, urine, or contaminated food. It is not surprising to find that dysentery, typhoid fever, food poisoning, and worm diseases afflicted large numbers of southerners, especially those living in the poorest, most crowded circumstances, without sanitary facilities or time to prepare food properly. Slaves often fit into this category.

Syphilis and its bacteriologically unrelated associate, gonorrhea, are not transmitted through filth or unsanitary living conditions, but they were public health problems in the antebellum South nonetheless. Though morbidity figures for whites and blacks do not exist, the frequency with which these two diseases appeared in physicians' casebooks points to a rate which may be, as one historian has stated, "startling." Masters knew that venereal disease was disseminated by intimate sexual con-

tact, but they also knew how difficult it was to prevent contagion by recognition and isolation of infected individuals.

Old South physicians' records tell only part of the story of gonorrhea and syphilis. For each case that was recorded, many were not, because people treated the disease themselves or waited until it disappeared. They did not realize the potential hazards of concealment to personal and public health. Venereal diseases probably occurred with a frequency which could be considered epidemic. Each newly reported case arose from a person already harboring the microorganisms; that person, in turn, probably had spread the germs to several other unsuspecting victims.

Though the major medical problems at the quarters were communicable diseases, in the course of each day bondsmen faced other health problems unrelated to contagion or parasites. Inadequate clothing and food, poor working conditions, harsh physical punishment, pregnancy, and bodily disorders also made slaves sick or uncomfortable, often rendering them useless to their masters and burdensome to their friends and family.

Though adequate clothing was important for slaves, it did not play as crucial a role in the maintenance of health as did housing. Of course clothing covered and protected the body from exposure to wind, sun, rain, snow, cold, and insects. It also limited the severity of many minor falls to cuts, scrapes, and bruises, and of some industrial accidents to burns over small areas. But only a few disorders are spread by contact with infected clothing (smallpox, body lice, impetigo, typhus) or by contact of exposed skin with other objects (tetanus, yaws, hookworm, brucellosis).

Except in cases where slaves were truly underclothed in winter, possibly causing decreased resistance to respiratory ailments, the danger of contracting disease owing to inadequate or dirty wearing apparel was relatively small. Of articles of clothing which masters provided their bondsmen, shoes were probably the most important in terms of health and disease. Not only did they provide warmth in the winter to feet and toes highly susceptible to frostbite, but they also protected slaves against hookworm penetration, scrapes, scratches, burns, and some puncture wounds which would otherwise have caused tetanus.

Did slaves receive a diet adequate to keep them

healthy, laboring, and producing vigorous off-spring? Opinions vary on this question, for diets differed from individual to individual and our understanding of nutritional needs has changed often. Based on current dietary standards, the daily typical ration, one quart of whole ground, dry, bolted cornmeal, prepared from white corn (the South's favorite), and half a pound of cured, medium-fat ham with no bone or skin, could not have provided enough essential nutrients to sustain a moderately active, 22-year-old male or female, much less a hard-working laborer or a pregnant or lactating woman. Fieldhands fed this diet alone (with water) would soon have become emaciated and sickly and would have shown symptoms of several nutrient deficiencies. It is highly unlikely that any slave could have survived very long on a diet consisting solely of pork and cornmeal.

Most Old Dominion masters provided supplements to the basic hogmeat and cornmeal, a practice most urgently recommended by agricultural writers throughout the state. Vegetables topped the list of required additional foods. Planters could, if they planned ahead, have a ready supply of at least one or two varieties throughout the year. These writers also suggested adding, when available, fish, fresh meat, molasses, milk, and buttermilk to slave diets.

Many agricultural authors and slavemasters indicated in their writings that blacks often raised vegetables, poultry, and even pigs on their own plots of land near the quarters. The assumption was that extra food from this source would supplement rations supplied by the master. Surprisingly, however, some of these same writers also pointed out that slaves usually sold what they raised to the master or at the marketplace, thereby defeating the major purpose of the plan. In all likelihood the slaves did not dispose of all their produce, but saved some for future needs. The fact that there were bondsmen who sold rather than kept food indicates that, other than those saving every available penny to purchase their freedom, some slaves received sufficient nutrition from their regular rations, supplemented by homegrown food, to feel quite comfortable relying on their masters for proper nutrition.

Kenneth Kiple and Virginia H. King, in their recent book on the subject, *Another Dimension to the Black Diaspora*, assess the adequacy of a slave diet consisting primarily of pork and cornmeal supplemented occasionally with other foods.[12] Using recently discovered knowledge of human nutritional needs and nutrient actions and interactions in the body, they conclude that slaves received sufficient amounts of carbohydrates and calories but generally lacked some essential amino acids, vitamin C, riboflavin, niacin, thiamine, vitamin D, calcium, and iron. And slave children, who all too often began life with nutritional deficiencies owing to their mothers' poor pre- and postnatal diets, also lacked sufficient magnesium, calories, and protein in their diets. Not surprisingly, Kiple and King assert, slave adults and children suffered higher morbidity and mortality than whites because of lower resistance to infection and disrupted basic metabolic pathways. Among the most common resultant black health problems were respiratory and intestinal diseases, skin and eye afflictions, "teething," tetanus, "fits and seizures," and rickets. Diet, controlled to a large degree by the master, greatly affected slave health.

In addition to providing food, clothing, and housing, slaveowners also directed working conditions, punishments, and care of women and children. Though warm weather helped the crops grow, it did not always have the same effect on the black Virginians who tended them. Planters recording the effects of excessive heat on their fieldhands made it clear that even though Negroes originated in tropical Africa, they were not immune to sunstroke. Hill Carter of Shirley Plantation wrote, for instance, during the 1825 wheat harvest, "Hotest day ever felt—men gave out & some fainted."[13]

Slaveowners also recognized the potential hazards of overexertion and exposure. Some indulged their slaves by easing their tasks; others found this impossible, especially at certain times of the year. Hill Carter and no doubt many others worked their blacks in intense heat when necessary to harvest a crop. Charles Friend of Prince George County, on the other hand, had second thoughts when the ditching operation to which he had assigned many slaves evolved into a messy and unhealthy job: "We have the ditchers knee deep in water and mud. If I had known how bad it was I should not have put them to work at it but hired labor to do it."[14]

Farm accidents also took their toll on slaves. Falls, overturned carts, runaway wagons, drownings, limbs caught in farm machines, kicks from animals, and cuts from axes or scythe blades were the commonest types. Occasionally slaves suffered more remarkable mishaps, as when a 260-pound Culpepper County fieldhand jumped eight feet from a hay loft onto a pile of hay in which a wooden pitchfork lay concealed. The point punctured the man's scrotum and passed into his abdominal cavity, but miraculously pierced no internal organs. Thanks to prompt surgical attention he was doing "light work" around the plantation about three weeks later.

The whip was an integral part of slave life in the Old Dominion and the Old South. Those bondsmen who had not experienced its sting firsthand were acquainted with persons, usually friends or relatives, who had. Whites held out the threat of whipping as a means of maintaining order. When strong discipline was called for, so, very often, was the lash. Even the mildest and most God-fearing of masters permitted application of this painful instrument in extreme cases, though some insisted that the slave's skin not be cut or that there be a responsible witness present when punishment was administered.

From a medical point of view, whipping inflicted cruel and often permanent injuries upon its victims. Laying stripes across the bare back or buttocks caused indescribable pain, especially when each stroke dug deeper into previously opened wounds. During the interval between lashes, victims anticipated the next in anguish, wishing for postponement or for all due speed, though neither alternative brought relief. In addition to multiple lacerations of the skin, whipping caused loss of blood, injury to muscles (and internal organs, if the lash reached that deep), and shock. (Rubbing salt into these wounds, often complained of as a further mode of torture, actually cleansed the injured, exposed tissues and helped ward off infection). The paddle jarred every part of the body by the violence of the blow, and raised blisters from repeated strokes. In addition to the possibility of death (uncommon), there was the danger that muscle damage inflicted by these instruments might permanently incapacitate a slave or deform him for life. An Old Dominion slave who experienced the sting of the paddle recalled years later: "You be jes' as raw as a piece of beef an' hit eats you up. He loose you an' you go to house no work done

dat day."[15] No work done that day or, in many cases, for several days. Ellick, a rebellious member of Charles Friend's White Hill Plantation slave force, was slapped one day "for not being at the stable in time this morning," and "soundly whipped" the next day for running away and for not submitting to a flogging earlier that morning. He spent the next week recovering in bed, only to receive another whipping upon his return to work. This time he ran off for two days before settling back into the plantation routine.

The daily routines of slave women and children were often upset by health conditions peculiar to these groups. Female slaves probably lost more time from work for menstrual pain, discomfort, and disorders than for any other cause. Planters rarely named illnesses in their diaries or daybooks, but the frequency and regularity with which women of childbearing age appeared on sick lists indicates that menstrual conditions were a leading complaint. A Fauquier County physician considered the loss of four to eight weekdays per month not unusual for slave women. Among the menstrual maladies which afflicted bondswomen most often were amenorrhea (lack of menstrual flow), abnormal bleeding between cycles (sometimes caused by benign and malignant tumors), and abnormal discharges (resulting from such conditions as gonorrhea, tumors, and prolapsed uterus).

Some servants took advantage of their masters by complaining falsely of female indispositions. One unnamed Virginian who owned numerous slaves complained to Frederick Law Olmsted about such malingering women:

> The women on a plantation . . . will hardly earn their salt, after they come to the breeding age; they don't come to the field, and you go to the quarters and ask the old nurse what's the matter, and she says, "oh, she's not well, master; she's not fit to work, sir;" and what can you do? You have to take her word for it that something or other is the matter with her, and you dare not set her to work; and so she lays up till she feels like taking the air again, and plays the lady at your expense.[16]

Masters found it difficult to separate the sick from the falsely ill; as a result they often indulged their breeding-aged women rather than risk unknown complications. Thomas Jefferson, for instance, ordered his overseer not to coerce the female workers into exerting themselves, because "women . . . are destroyed by exposure to wet at certain periodical indispositions to which nature has subjected them."[17]

If white Virginians treated women's gynecological complaints with a certain delicacy, they regarded pregnancy as almost holy. In addition to receiving time off from work and avoiding whippings, expectant women were protected from execution in capital offenses until after parturition.[18] At least three cases arose between the Revolution and the Civil War in which slave women obtained execution postponements for this reason, though all were presumably put to death following delivery.

Children, like women, were exposed to certain unique disorders which caused illness or death. Though their labor did not usually account for much, young slaves' serious illnesses did mean time lost from work for mothers watching over them at home or distractedly worrying about them while performing daily tasks. Slave children suffered more frequently from most illnesses than their white counterparts, especially diarrhea, neonatal tetanus, convulsions, "teething" (not really a disease, but considered a cause of sickness and death prior to the 20th century), diphtheria, respiratory diseases, and whooping cough, owing to poorer living conditions and diet.

Worms occurred frequently in black children. The poor sanitary conditions at many antebellum Virginia slave quarters were conducive to the development of these parasites in the soil. Children playing in the dirt inevitably picked up worm larvae as they put fingers in mouths. Failure to use, or lack of, privy facilities only served to spread worm diseases to other residents of the quarters and to visiting slaves, who then carried these parasites to their own plantation quarters.

Some black children had overt sickle cell disease (noted above) with irregular hemolytic crises, severe joint pains, chronic leg ulcers, and abdominal pains. The medical records relating to antebellum Virginia do not provide any clear descriptions of the disease, probably because its symptoms resemble so many other conditions and because the sickness was not known until 1910. These children were often the "sickly" ones, useless for field work or heavy household duties, expensive to maintain because of frequent infections,

and often unable to bear children if they survived puberty. Their lot was a poor and painful one.

Slave children also developed diseases which no one could identify or treat. John Walker's young servants appeared one day with "head ach sweled faces & belly diseases";[19] Colonel John Ambler's evidenced swollen feet and faces, and bones cutting through the skin; and Landon Carter's had "swelling of the almonds of . . . [their] ears which burst inward and choaked . . . [them]."[20] The white tutor at Nomini Hall, Philip Vickers Fithian, noticed that one slave mother on this Westmoreland County estate had lost seven children successively, none of whom had even reached the age of ten: "The Negroes all seem much alarm'd. . . ."[21] Childhood was generally the least healthy period of a slave's life in antebellum Virginia.

SLAVE MEDICAL CARE

Bondage placed slaves in a difficult position with regard to health care. When taken ill they had a limited range of choices. Masters usually insisted that their slaves, legally an article of property, immediately inform the person in charge of any sickness so the malady might be arrested before it worsened. But some bondsmen, as people, felt reluctant to submit to the often harsh prescriptions and remedies of 18th- and 19th-century white medical practice. They preferred self-treatment or reliance on cures recommended by friends and older relatives. They depended on Negro herb and root doctors, or on influential conjurers among the local black population. This desire to treat oneself, or at least to have the freedom to choose one's mode of care, came into direct conflict with the demands and wishes of white masters, whose trust in black medicine was usually slight and whose main concern was keeping the slave force intact.

To further compound the problem, unannounced illnesses did not entitle bondsmen to time off from work. To treat their own illnesses slaves had to conceal them or pass them off to the master as less serious than they actually were. Masters who complained that blacks tended to report sickness only after the disease had progressed to a serious stage often discovered that slaves had treated illnesses at home first. The blacks' dilemma, then, was whether to delay reporting illnesses and treat those diseases at home, risking white reprisal; or to submit at once to the medicines of white America and, in a sense, surrender their bodies to their masters. The result was a dual system in which some slaves received treatment both from whites and blacks.

When illness afflicted a slave, white Virginians responded in several ways. They almost always applied treatments derived from European experience. Most often the master, mistress, or overseer first attempted to treat the ailment with home remedies. If the patient failed to respond to these home ministrations, the family physician was summoned. Some slaveowners distrusted "regular" doctors and called instead "irregular" practitioners: Thomsonians, homeopaths, hydropaths, empirics, eclectics, etc. Masters who hired out their bondsmen to others for a period of time arranged for medical care when signing the hiring bond. Whatever the situation, Virginians often displayed concern for the health of blacks in bondage. The reasons were threefold: slaves represented a financial investment which required protection; many masters felt a true humanitarian commitment toward their slaves; and whites realized that certain illnesses could easily spread to their own families if not properly treated and contained.

Those responsible for the care of sick slaves made home treatment the first step in the restorative process. White Virginians recognized that physicians, though possessed of great knowledge of the human body and the effects of certain medicines on it, were severely limited in the amount of good they could perform. Because no one understood the etiology of most diseases, no one could effectively cure them. Astute nonmedical observers could make diagnoses as well as doctors, and could even treat patients just as effectively. Physicians played their most crucial roles in executing certain surgical procedures, assisting mothers at childbirth, and instilling confidence in sick patients through an effective bedside manner. At other times their excessive use of drugs, overready cups and leeches, and ever-present lancets produced positive harm in depleting the body of blood and nourishment and exhausting the already weakened patient with frequent purges, vomits, sweats, and diuretics. Laymen often merely followed the same course of treatment that they had observed their physicians using or that they had read about in one of the ubiquitous domestic medical guides. Anyone could

practice blood-letting or dosing with a little experience. And a physician's services cost money, even when no treatment or cure resulted from the consultation.

Home care was not an innovation of 18th-century Virginians, but one that stemmed from man's natural instinct to relieve his own or his family's illness as quickly as possible. The unavailability of physicians, the inaccessibility of many farms to main highways, and the lack of good roads and speedy means of transportation reinforced such thinking among rural Virginians. Even when a doctor was summoned, hours or even a day passed before his arrival, during which time something had to be done to ease the patient's discomfort. People learned to tolerate pain and to cope with death, but the mitigation of suffering was still a primary goal. To that end most Virginians stocked their cabinets with favorite remedies (or the ingredients required for their preparation) in order to be well equipped when relief was demanded. On large plantations with many slaves this was a necessity, as Catherine C. Hopley, tutor at Forest Rill near Tappahannock, Essex County, noted: "A capacious medicine chest is an inseparable part of a Southern establishment; and I have seen medicines enough dispensed to furnish good occupation for an assistant, when colds or epidemics have prevailed."[22] Some physicians made a living selling medicine chests and domestic health guides designed specifically for use on southern plantations. Self-sufficiency in medical care was desirable on farms and even in urban households, especially when financial considerations were important.

An additional feature of home medical care for slaves was the plantation hospital or infirmary. Its form varied from farm to farm and existed primarily on the larger slaveholdings. It was quicker and more efficient to place ailing slaves in one building, where care could be tendered with a minimum amount of wasted movement and where all medicines, special equipment, and other necessary stores could be maintained. Of course, infectious diseases could spread quite rapidly through a hospital, subjecting those with noncontagious conditions to further sickness.

Armed with drugs from the plantation or home dispensary, one person, usually white, had the responsibility of dosing and treating ill slaves. The master, mistress, or overseer spent time each day with those claiming bodily disorders and soon developed a certain facility in handling both patients and drugs. The approach was empirical — if a particular medication or combination of drugs succeeded in arresting symptoms, it became the standard treatment for that malady in that household until a better one came along. Overseers and owners inscribed useful medical recipes into their diaries or journals and clipped suggestions from newspapers, almanacs, and books.

An overseer's or owner's incompetence or negligence was the slave's loss. New and inexperienced farm managers, unskilled in the treatment of illness, necessarily used bondsmen as guinea pigs for their "on-the-job" training. As a consequence of living on the wrong plantation at the wrong time, some slaves probably lost their lives or became invalids at the hands of new, poorly trained, or simply inhumane overseers or masters.

Despite many masters' policy of delaying a call to the physician until late in the course of a slave's disease, there were times when owners desperately wished for the doctor's presence. More practitioners should have retained in their files the numerous hastily scrawled notes from frantic slaveowners begging for medical assistance, or kept a record of each verbal summons to a sick slave at a distant farm or village household. For physicians did play important roles, both physiological and psychological, in the treatment of illness. Dr. Charles Brown of Charlottesville, for instance, had a thriving country practice during the early 19th century. He handled many types of problems: James Old wanted him to determine whether his slave woman, then "in a strange way," was pregnant or not; Bezaleel Brown needed his opinion "if I must bleed her [Jane, who had a pain in her side and suppression of urine] either large or small in quantity"; and Jemima Fretwell wished Brown to "cutt of[f] the arm" of a four-month-old slave which had been "so very badly burnt" that "the [elbow] joint appears like it will drap of[f]."[23] Sometimes physicians made daily visits to dress slaves' wounds or to keep track of household epidemics. In emergencies some owners panicked and fretted away many hours after learning of their physician's temporary absence.

Between the remedies of the household and the standard treatments of the physicians stood "ir-regular" medicine, often as important but only partially accepted in Virginia. The impact of alternative movements on the medical care of blacks in Virginia was greater than historians have recognized. Most slaveowners either treated with conventional medicines or called in regular doctors, rejecting the new cults as quackery; but a sizable minority, difficult to estimate, became enthusiastic proponents of at least one system — Thomsonianism. This movement with practitioners in areas with heavy slave concentrations (64 percent of the Tidewater counties and 66 percent of the Piedmont counties during the 1830s and 1840s) appealed to masters who were fed up with the ineffective and expensive treatments of their regular physicians. One Tidewater resident turned to Thomsonianism after experienced Norfolk physicians had unsuccessfully managed a household scarlet fever outbreak. All 20 cases, the happy slaveowner reported to the editors of a Thomsonian journal, had been cured. Another man, in Goochland County, stated that a local Thomsonian practitioner had cured his slave of a disease which one of the most respected regular physicians of the area had found intractable to the usual blister and salivation treatments. And a Prince Edward County Thomsonian doctor claimed to have cured a ten-year-old slave who had been suffering from rabies (a misdiagnosis, no doubt). With adherents to the sect so widely diffused throughout the state, the services or success stories of practitioners no doubt reached at least a portion of the slaveholding class and influenced its thinking.

Beyond the master's and overseer's eyes, back in the slaves' cabins, some Virginia blacks took medical matters into their own hands. When under the surveillance of whites, slaves usually (but not always) accepted their treatments. Some even administered them in the name of the master. But others developed or retained from an ancient African heritage their own brand of care, complete with special remedies, medical practitioners, and rituals. The result was a dual system of health care, the two parts of which often conflicted with each other.

Masters did not appreciate slaves overusing the plantation infirmary, medicines, or the family doctor, but they preferred this to black self-care for several reasons. Their quarrel with the bondsmen was the same as the physicians' with the masters:

slaves waited too long before seeking medical assistance and often misdiagnosed illnesses. Most owners permitted blacks a small amount of freedom in treating minor ailments at home, but lost their patience when sickness got out of hand. James L. Hubard, in charge of his father's lands during the latter's vacation at Alleghany Springs, reported that Daphny had treated her own son with vermifuges (worm medicines) for several days before realizing that the boy was suffering not from worms but from dysentery. Hubard quickly altered the medication and summoned a doctor, blaming the entire affair on "the stupidity of Daphny."[24] An enraged Landon Carter found a suckling child with measles at the slave quarters. "The mother," he wrote in his diary, "let nobody know of it until it was almost dead."[25]

Whites also accused slaves of negligence or incompetence in the care of their fellow bondsmen. Dr. G. Lane Corbin of Warwick County, for instance, promoted slaves' use of collodion, a syrupy dressing, because it required so little attention once applied: "This I consider of moment in regard to our slave population, whose negligence and inattention to such matters [as the proper dressing of wounds] must have attracted the notice of the most superficial observer."[26] Negroes frequently were charged with irresponsibility, ignorance, slovenliness, and indifference in the management of other blacks' illnesses. "They will never do right, left to themselves,"[27] declared one Franklin County planter.

Furthermore, some whites argued, slaves did not even care for their own personal health properly. Recovery was retarded and even reversed, Dr. W. S. Morton of Cumberland County remarked, "by their [slaves'] own stupid perversity in refusing confinement to bed, and to follow other important directions when in a very dangerous condition."[28] Masters and physicians often confirmed this but were powerless to combat it. It was difficult for whites, unless they were present at all times, to force ailing blacks to take medicines or to remain constantly in bed. A most spectacular instance of death following defiance of medical orders occurred in Portsmouth when a black male patient of Dr. John W. Trugien, confined to bed with a stab wound of the heart, sustained a massive effusion of blood from that organ upon exerting himself by rising from his pallet.[29]

To offset the failures and harshness of white remedies, or the negligence of masters, or, perhaps, to exert some control over their lives, some slaves treated their own diseases and disorders or turned to other trusted blacks for medical assistance, with or without the master's knowledge.[30] Black home remedies circulated secretly through the slave quarters and were passed down privately from generation to generation. Most of these cures were derived from local plants, though some medicines contained ingredients that had magical value only. Occasionally whites would learn of a particularly effective medicine and adopt it, as when Dr. Richard S. Cauthorn announced in the *Monthly Stethoscope* (1857) that an old folk remedy (milk weed or silk weed, *Asclepias syriaca* in the United States Dispensatory) which had been used for years by blacks in the counties north of Richmond worked almost as well as quinine for agues and fevers.[31] Otherwise most whites simply ignored or tolerated the black medical world until something occurred to bring their attention to it—either a great medical discovery or a slave death caused by abuse.

Because blacks practiced medicine in virtually every portion of the Old Dominion and because their methods were based partially on magic, problems occasionally arose. The main source of trouble was usually not the misuse of home remedies, but the "prescriptions" and activities of so-called conjure doctors. These men and women used trickery, violence, persuasion, and medical proficiency to gain their reputations among local black communities. They were viewed as healers of illness that white doctors could not touch with their medicines, and as perpetrators of sicknesses on any person they wished—all through "spells."

Superstition was a powerful force within the slave community, and a difficult one for white nonadherents to understand or overcome. For instance, the older brother of a slave patient of Dr. A. D. Galt of Williamsburg observed to the doctor that his medicines were useless because Gabriel "had been tricked" and "must have a Negro Doctor" to reverse the progress of the illness. Galt soon claimed to have cured the man, though he did admit that Gabriel suffered frequent relapses, "probably from intemperance in drink."[32] In another case, a slave woman took sick and eventually died on a plantation near Petersburg from what her fellow bondsmen believed were the effects of a conjurer. Some slaves speculated that the young man whom she had refused to marry "poisoned or tricked" her, though the overseer attributed her death to consumption.[33] Virginia Hayes Shepherd, a former slave interviewed at age 83 in 1939, described an incident to illustrate how superstitious her stepfather had been: "He believed he had a bunch something like boils. White doctor bathed it. After a few days it burst and live things came out of the boil and crawled on the floor. He thought he was conjured. He said an enemy of his put something on the horse's back and he rode it and got it on his buttocks and broke him out."[34]

Old Dominion whites did permit blacks to fulfill certain medical functions. Some planters assigned "trusted" slaves to the task of rendering medical assistance to all ailing bondsmen on the farm. In most cases, these blacks simply dispensed white remedies and performed venesection and cupping as learned from the master. Though not complete black self-care, this activity did represent a transitional stage in which slaves had the opportunity to apply some of their own knowledge of herbs, etc., gained from elders, in addition to white remedies. These nurses, predominantly women, usually won the respect of both blacks and whites for their curative skills. "Uncle" Bacchus White, an 89-year-old former slave interviewed in Fredericksburg in 1939, attested, "Aunt Judy uster to tend us when we uns were sic' and anything Aunt Judy couldn't do 'hit won't worth doin."[35] A white lady writing at about the same time provided a similarly romantic view of the black plantation nurse: "One of the house-servants, Amy Green—'Aunt Amy' we children called her—was a skilled nurse. My father kept a store of medicine, his scales, etc. so with Aunt Amy's poultices of horseradish and plattain-leaves and her various cuppings and plasters the ailments of the hundred negroes were well taken in hand."[36] Given such high testimony and devotion from plantation folk, one could hardly dispute the novelist Louise Clarke Pyrnelle's depiction of Aunt Nancy, a fictional antebellum household nurse who claimed, while dosing several young slaves, "Ef'n hit want fur dat furmifuge [vermifuge—worm medicine], den Marster wouldn't hab all dem niggers w'at yer see hyear."[37]

To Negro women often fell another task: prenatal and obstetrical care of whites and blacks, especially in rural areas. At least one slave on most large Virginia plantations learned and practiced the art of midwifery, not only at home but also throughout the neighborhood. Masters preferred to employ these skilled accouchers in uncomplicated cases rather than pay the relatively high fees of trained physicians. Doctors, remarked one member of the medical profession, attended at less than half of all births in the state. He estimated that 9/10 of all deliveries among the black population (another physician set it at 5/6) were conducted by midwives, most of whom were also black. He further asserted that midwives attended half the white women. Physicians often saw obstetrical cases only when problems arose. As a result of this demand for competent nonprofessional obstetrical services, Negro midwives flourished in the countryside.

Blacks did play a significant role in the health care system of the Old Dominion. They assisted whites and blacks in delivering children, letting blood, pulling teeth, administering medicines, and nursing the sick. The techniques and drugs they used were overtly derived from white medical practices. But unknown to masters, overseers, health officers, or physicians, blacks did also resort to their own treatments derived from their own heritage and experience. Occasionally the white and black medical worlds merged or openly clashed, but usually they remained silently separate.

NOTES

This paper is extracted in large part from Todd L. Savitt, *Medicine and Slavery: The Diseases and Health Care of Blacks in Antebellum Virginia* (Urbana: Univ. of Illinois Press, 1978), pp. 1–184. I acknowledge the helpful suggestions of Paul Escott, Kenneth Kipple, Ronald Numbers, and James Harvey Young.

1 Except in the first section, where a general overview of black-related diseases is presented, the focus is on Virginia from the Revolution to the Civil War. Health conditions in the Old Dominion at that time were, in many respects, typical of those prevailing throughout the antebellum South. Residents

suffered from malaria, parasitic worm diseases, and dysentery just as Mississippians and Georgians did. Yellow fever struck its major ports, though not as severely or as frequently as at Charleston, Mobile, and New Orleans. Virginia's position on the northern fringe of the slave South perhaps lessened the intensity and duration of warm-weather diseases, but not enough to render its diseases significantly different from those in the lower South.

During the time span under consideration the black population and the health picture in Virginia were relatively stable. The slave trade had ended, there was little black immigration into the state, and tropical diseases brought from Africa and unable to survive in the new environment had all but disappeared.

2   Editorial, *Monthly Steth. & Med. Reptr.,* 1856, *1:* 162–163.

3   These and other diseases and conditions are discussed in a recent, important book on black medical differences: Kenneth Kiple and Virginia H. King, *Another Dimension to the Black Diaspora: Diet, Disease and Racism* (Cambridge, England, 1981).

4   For a discussion of the southern disease environment from colonial times to the present, see Albert Cowdrey, *This Land, This South: An Environmental History* (Lexington, Ky., 1983).

5   Kiple and King (*Another Dimension*) emphasize dietary considerations to explain many slave health problems.

6   John H. Van Evrie, *Negroes and Negro "Slavery"* (New York, 1861), p. 25.

7   Van Evrie, pp. 251, 256.

8   Recent historical discussions of SIDS include Kiple and King, *Another Dimension,* pp. 107–110, and Michael P. Johnson, "Smothered slave infants: Were slave mothers at fault?" *J. Southern Hist.,* 1981, *47:* 493–520.

9   E. M. Pendleton, "On the susceptibility of the Caucasian and African races to the different classes of disease," *Southern Med. Rep.,* 1849, *1:* 336–337.

10 Editorial, "The Medical Society of Virginia," *Va. Med. & Surg. J.,* 1855, *4:* 256–258.

11 For examples of typical planters' writings on the management of slaves, see James O. Breeden, ed., *Advice Among Masters: The Ideal in Slave Management in the Old South* (Westport, Conn., 1980).

12 See notes 3 and 5 above.

13 Shirley on the James Farm Journals, June 23, 1825, Manuscript Room, Library of Congress.

14 Quoted in Wyndham B. Blanton, *Medicine in Virginia in the Eighteenth Century* (Richmond, 1931), p. 161.

15 Interview of William Lee, n.d., WPA Folklore File, Univ. of Virginia, Manuscript Room, Alderman Library.

16 Frederick Law Olmsted, *A Journey in the Seaboard Slave States, with Remarks on Their Economy* (New York, 1856), p. 190.

17 Thomas Jefferson to Joel Yancy, Jan. 17, 1819, reproduced in *Thomas Jefferson's Farm Book,* ed. Edwin M. Betts (Princeton, N.J., 1953), p. 43.

18 There were, of course, exceptions to these statements. See, for example, Johnson, "Smothered slave infants," pp. 511–520.

19 John Walker Diary, Apr. 23, 1853, Southern Historical Collection, Univ. of North Carolina at Chapel Hill.

20 Jack P. Greene, ed., *The Diary of Colonel Landon Carter of Sabine Hall, 1752–1778* (Charlottesville, Va., 1965), I, p. 377 (Mar. 31, 1770).

21 Hunter Dickinson Farish, ed., *Journal and Letters of Philip Vickers Fithian, 1773–1774: A Plantation Tutor of the Old Dominion* (Williamsburg, Va., 1957), p. 182.

22 [Catherine C. Hopley], *Life in the South: From the Commencement of the War* (London, England, 1863), I, p. 103.

23 For more examples of such notes, see Todd L. Savitt, "Patient letters to an early nineteenth century Virginia physician," *J. Florida Med. Assn.,* Aug. 1982, *69* (No. 8): 688–694.

24 James L. Hubard to Robert T. Hubard, Aug. 4, 1857, Robert T. Hubard Papers, Univ. of Virginia, Manuscript Room, Alderman Library.

25 Greene, ed., *Carter Diary,* II, p. 812 (May 20, 1774).

26 G. Lane Corbin, "Collodion on stumps of amputated limbs," *Steth. & Va. Med. Gaz.,* 1851, *1:* 489.

27 L. G. Cabell to Bowker Preston, Oct. 8, 1834, John Hook Collection, Univ. of Virginia, Manuscript Room, Alderman Library.

28 W. S. Morton, "Causes of mortality amongst Negroes," *Monthly Steth. & Med. Reptr.,* 1856, *1:* 290.

29 John W. H. Trugien, "A case of wound to the left ventricle of the heart.—Patient survived five days; —with remarks," *Am. J. Med. Sci.,* 1850, n.s., *20:* 99–102.

30 See, for more information and references, Lawrence W. Levine, *Black Culture and Black Consciousness: Afro-American Folk Thought from Slavery to Freedom* (New York, 1977), pp. 55–80.

31 Richard S. Cauthorn, "A new anti-periodic and a substitute for quinia," *Monthly Steth. & Med. Reptr.,* 1857, *2:* 7–14.

32 [A. D. Galt], *Practical Medicine: Illustrated by Cases of the Most Important Diseases,* ed. John M. Galt (Philadelphia, 1843), pp. 295–296.

33 [William McKean to James Dunlap], July 17, 1810, Roslin Plantation Records, Virginia State Library.

34   Interview of Virginia Hayes Shepherd, 1939, WPA Folklore File, Univ. of Virginia, Manuscript Room, Alderman Library.

35   Interview of Uncle Bacchus White, 1939, WPA Folklore File, Univ. of Virginia, Manuscript Room, Alderman Library.

36   White Hill Plantation Books, Vol. I, p. 8, Southern Historical Collection, Univ. of North Carolina at Chapel Hill.

37   Louise Carter Pyrnelle, *Diddie, Dumps, and Tot; or Plantation Child-Life* (New York, 1882), quoted in Blanton, *Medicine in Virginia,* p. 49.

23

Roots of the Black Hospital
Reform Movement

VANESSA NORTHINGTON GAMBLE

In a 1900 address, Dr. Daniel Hale Williams, the founder of Chicago's Provident Hospital and Nurse Training School, the nation's first black-controlled hospital, urged other African Americans to build their own hospitals. The existing racial discrimination against black physicians, nurses, and patients had prompted his call. In his 23 January 1900 speech before the Phillis Wheatley Club of Nashville, Tennessee, Williams contended,

> In view of this cruel ostracism, affecting so vitally the race, our duty seems plain. Institute Hospitals and Training Schools. Let us no longer sit idly and inanely deploring existing conditions. Let us not waste time trying to effect changes or modifications in the institutions unfriendly to us, but rather let us seek to promote the doctrine of helping and stimulating our race.

During the last decade of the 19th century and the first two decades of the 20th century, various segments of the black community did indeed take Williams's advice and create hospitals for themselves. Members of the white community also opened black hospitals. The number of these institutions grew rapidly. In 1912, 63 existed; seven years later, 118 black hospitals were in operation. Racial discrimination, white self-interest, black professional

VANESSA NORTHINGTON GAMBLE is Associate Professor of the History of Medicine and Family Medicine, and Director of the Center for Race and Ethnicity in Medicine, University of Wisconsin–Madison.

From *Making a Place for Ourselves: The Black Hospital Movement, 1920–1945* by Vanessa Northington Gamble, pp. 3–34. Copyright © 1995 by Vanessa Northington Gamble. Reprinted by permission of Oxford University Press, Inc.

concerns, divergent strategies for black social advancement, and changes in hospital care and medical practice all played major roles in the development of these institutions. Any understanding of the character and goals of the later black hospital reform movement must be rooted in an analysis of the evolution of black hospitals.[1]

The early 19th-century American hospital differed markedly from its mid-20th-century successor. The 19th-century institution was peripheral to the provision of medical care, medical education, and medical research. It operated primarily as a traditional welfare institution that cared for a variety of indigent and dependent persons, including (but by no means restricted to) the sick. Hospitals offered limited therapeutics and functioned, in part, to maintain the social order by isolating the socially marginal and by serving as loci for moral as well as medical care. Poverty and dependence were the main criteria for admission. Municipal hospitals, such as New York City's Bellevue and Philadelphia's General, began their institutional lives as medical departments in almshouses. The social elite in many cities also established voluntary or not-for-profit hospitals as charities to keep the "worthy poor" out of the almshouse and to protect them from the stigma and demoralizing influence of the almshouse. Principles of Christian stewardship regarding upper-class obligations to the poor encouraged the development of many voluntary hospitals and even governed their day-to-day operations. For example, the certification of one's worthiness by a lay trustee, not a physician's diagnosis, was the primary prerequisite to admission. The home served as the primary site of medical care for persons with resources and roots in the community. Hospital care offered no therapeutic advantages over domestic care; indeed, most people tried to avoid hospitals because of their identification as charitable institutions and because they feared contracting hospital-originated infections and fevers.[2]

In the early 19th century, hospitals also played an insignificant role in the professional development of the average practitioner. Medical education did not require clinical training in hospitals, and contemporary medical practice did not demand that physicians serve internships or residences or hold hospital appointments. The average physician could complete his or her medical education and have a successful practice without setting foot in a hospital. Hospital appointments were part of the career paths of a small group of elite, urban physicians, rather than of the general medical profession. Furthermore, it was the values and interests of lay trustees that dominated hospital decision-making, rather than those of the physician.

Early in the 19th century, communities, especially in the South, organized institutions to provide medical care to sick black people. The creation of slave hospitals represented one response. Those rudimentary facilities usually contained only a few beds and were located in buildings separate from the main slave quarters. They provided an efficient means to care for sick slaves because all medicines could be housed in one location. The facilities also isolated slaves with contagious diseases. A Tennessee physician urged in 1853,

> Every plantation should be provided with a hospital. . . . By bringing all the sick into the same house, the convenience of the physician is subserved, the time of the nurses economized, and better attendance secured. There should always be two rooms, so that the sexes may be kept separate, and there should be a water closet attached to each room. These rooms require to be close and warm. . . .
>
> The hospital should be located near the dwelling of the owner or overseer, surrounded by shade trees, kept neat and clean, and conducted in a manner to cheer and enliven the drooping spirits of the sick. Single beds should be used, with good mattresses, and made as comfortable as possible.

Such hospitals, however, only existed on the larger plantations. Slaveowners, their families, and their overseers usually staffed them. Physicians were only called in for the most serious cases. The hospitals were also frequently tended by slave women whose tasks included nursing, cleaning, and cooking.[3]

A few free-standing hospitals and infirmaries for African Americans, free and bond, also existed in the antebellum South. Those infirmaries, many of which had been established by physicians, operated away from the plantations and provided, for a fee, medical attention, board, and lodging. Georgia Infirmary, established in 1832 in Savannah, "for the relief and protection of aged and afflicted Africans," was the first hospital established by whites for the care of blacks. It had an endowment to provide care for free Africans, but owners were financially responsible for the care of their slaves. Twenty years later in the same city, three white physicians established an infirmary "for the reception of negroes requiring medical and surgical treatment." The infirmary accepted maternity cases, but not patients with contagious diseases. An advertisement for the infirmary described it as a "well-appointed establishment" with competent nurses, comfortable beds, well-ventilated wards, extensive pleasure grounds, and good food. Another institution, Mississippi State Hospital in Natchez, cared only for slaves and charged their owners a dollar a day for its services.[4]

These hospitals were not common. Few institutions in antebellum America existed specifically for the care of African Americans, and most medical facilities in the South excluded them. Contemporary racial customs and mores also restricted black access to hospital care in the North. Hospitals either denied African Americans admission or accommodated them, almost universally, in segregated wards often placed in undesirable locations such as unheated attics or damp basements. At Philadelphia's Pennsylvania Hospital, for example, attending physicians were unwilling to visit certain wards, particularly the black and venereal corridors. Syphilis and gonorrhea were considered to be contagious—at a time when most diseases were not thought to be—and were associated with immorality. Many whites believed that African Americans possessed intrinsic racial characteristics such as excessive sexual desire, immorality, and overindulgence which resulted in high rates of venereal disease. A similar contempt, it appears, existed in the care of patients with black skin and those, no matter their color, with venereal disease.[5]

Emancipation left open the question as to who would assume the responsibility for the medical care of the former slaves. One mechanism was the Freedmen's Bureau, established in 1865, which temporarily filled the void by setting up hospitals in the Southern and border states. However, its medical department suffered from a lack of funding and from organizational weaknesses. Because of low pay, the Bureau could not attract physicians with superior qualifications, and turnover was high. Financial constraints also forced it to use abandoned houses and old army hospitals, rather be able to build new facilities. Many of the facilities were not permanent, but frequently opened, closed, and relocated. The Freedmen's Bureau hospitals provided at least some medical care to the large number of dependent freedpeople who migrated from the plantations to urban areas. But the Freedmen's Bureau's efforts were grossly inadequate to meet the medical needs of five million former slaves. At its peak in September 1867, the Bureau operated 45 hospitals with a capacity of 5,292 beds. Furthermore, the Freedmen's Bureau had been created to aid the former slaves only until local governments could do so. By October 1868, only 11 of its hospitals remained, even though most local governments had not made provisions to care for the freedpeople. Public and private hospitals in many communities continued to exclude black patients. By 1872, all of the hospitals established by the Freedmen's Bureau had closed except the one in Washington, D.C., which was allowed to remain open because the Bureau thought that the disproportionately large number of freedpeople who had fled to the city would unduly burden local institutions. A special congressional mandate allowed the facility to continue as Freedmen's Hospital, first under the auspices of the War Department, and later, under the Department of the Interior.[6]

During and following Reconstruction, the white community gradually began to supply hospital care to African Americans, usually in separate, and not equal, facilities. Historian Howard N. Rabinowitz has argued that the establishment of these segregated facilities represented an improvement over previous conditions in which African Americans had been totally excluded. African Americans now had at least some access to hospitals. Several factors spurred the establishment of these white-sponsored hospitals. Some founders expressed a genuine, if paternalistic, interest in supplying health care to black people and in offering training opportunities to black health professionals. However, white self-interest was also at work. The germ theory of disease, popularized by the end of the 19th century, recognized that "germs have no color line." Medical and public health journals at the turn of the century portrayed African Americans as carriers of disease who posed a threat to white people, because "bacteria have a disconcerting fashion of ignoring segregation edicts." Despite Jim Crow, the argument went, white people remained at risk from diseases of black people. An editorial in the *Atlanta Constitution* warned that residential segregation would not protect white people

> because from that segregated district negro nurses would still emerge from diseased homes, to come into our homes and hold our children in their arms; negro cooks would still bring bacilli from the segregated district into the homes of the poor and rich white Atlantan; negro chauffeurs, negro butlers, negro laborers would come from within the pale and scatter disease.

The health of black people and that of white people were inextricably linked, so, if not for humanitarian reasons then at least for self-protection, whites needed to pay attention to the medical problems of African Americans. The white community also established black hospitals in order to escape the embarrassment of entirely neglecting the black sick, but without having to take care of them in white institutions. Regardless of motive, the goal behind the establishment of these hospitals was the same—to maintain and create a segregated hospital system.[7]

Segregated hospitals existed predominantly, but not invariably, in the South. They included Dixie Hospital and Nurse Training School, Hampton Institute, Virginia, established in 1891; MacVicar, Spelman Seminary, Atlanta, Georgia, established in 1896; St. Agnes, St. Augustine School, Raleigh, North Carolina, established in 1896; Kansas City (Missouri) General Hospital, No. 2 (Kansas City Colored), established in 1908; and St. Louis (Missouri) City Hospital, No. 2, established in 1919. Dixie Hospital and MacVicar Hospital were founded as adjuncts to two white-run educational institutions for African Americans, Hampton Institute and Spelman Seminary, respectively. The existence of nurse training schools at these institutions and the need to provide the nursing students with clinical experience provided the impetus for their establishment. As in other segregated black hospitals, Dixie Hospital and MacVicar Hospital did not allow black physicians to practice.[8]

In contrast, St. Agnes Hospital did provide access to black physicians. The hospital's founder was Sara Hunter, a white Episcopal churchwoman and wife of the principal of St. Augustine School, a black junior college. She saw providing black physicians a place in which to hospitalize their patients as part of her mission to improve the health of black people. However, black physicians had to work under the supervision of white ones; at least until 1925, all the chiefs of various services were white. Hunter had raised funds for the hospital from other members of the women's auxiliary of the Episcopal Church. On 18 October 1896, the hospital opened on the grounds of the school in a small building that had previously been a physician's residence. According to staff physician Mary Glenton, the hospital contained:

> No water in the house, except one faucet in the kitchen.
> No hot water, but what could be heated on the ward stoves. . . .
> Two small steamers for sterilizers . . . formed the operating room equipment. . . .
> No screens in windows or doors, and flying things innumerable, with wings small and great. . . .
> No plumbing anywhere—only earth closets.
> No Diet kitchen—the trays kept on a shelf in the kitchen.
> No gas for cooking nor for lighting, simply oil lamps.
> Not always enough food for patients; nor the proper kind for nurses and staff.

This makeshift facility was not atypical of small hospitals and infirmaries—black and white—in the late 19th century. The hospital, despite its rudimentary facilities, did provide a place for black physicians, nurses, and patients. It was the only hospital in the vicinity open to black patients, and many had to travel great distances to make use of its services. In 1898 two black women graduated from its nurse training school.[9]

Local governments also established separate hospitals to care for their black citizens. The push for such a facility in Kansas City, Missouri, came from the black community. In 1903, black physicians, led by Dr. Thomas C. Unthank, initiated efforts to establish a municipal hospital for African Americans because the existing one had very limited and inadequate accommodations for black patients, who often found themselves housed in the basement. The hospital also had no black staff. Unthank, an 1898 graduate of Howard University School of Medicine, had ample experience in the hospital field, having previously been involved in the founding of two small black private hospitals on both sides of the state line: in 1898 he established Douglass Hospital in Kansas City, Kansas, and in 1903 he opened Lange Hospital in Kansas City, Missouri. Unthank, in addition to his private practice, worked in Kansas City, Missouri, as an assistant city physician assigned to the city's African-American population. In this position he developed ties to City Hall that would later prove invaluable.[10]

In 1908, the city did establish a black hospital after it constructed a new municipal hospital and transferred most of the white patients to the new facility. Dr. J. Park Neal, first superintendent of the new hospital, reported that

> upon taking charge of the . . . Hospital, there were one hundred and twenty-six patients in the old hospital wards. On October 8, 1908, sixty-eight of these patients, all white, were transferred to eight wards of the new hospital. All the colored patients and those white patients suffering from tuberculosis and contagious diseases were left in the old hospital buildings for the care of all classes of colored patients and for tuberculosis and contagious diseases.

Again, we see the stigma attached to caring for black patients and those with contagious diseases. The old hospital was renamed Kansas City General Hospital, No. 2, or Kansas City Colored Hospital. The facilities at the two hospitals were by no means equal. The older hospital, built in 1873, lacked diagnostic facilities and clinical laboratories, and the funds spent on patients at the two facilities also demonstrated inequities. In 1908, costs per patient day were estimated to be $1.86 for whites and $.86 for blacks. From the city's point of view, the black hospital was not intended to provide clinical care equal to that of the white hospital, but to supply separate care.[11]

Moreover, Kansas City General Hospital, No. 2 did not initially meet the goals that the black physicians had set for professional development. Although it did provide care to indigent black patients, during its first three years of operation it had no black physicians and nurses on staff. However, in 1911 the hospital administration appointed four black physicians, including Unthank, to staff positions, selected its first black intern, and opened a training school for nurses. The hospital gradually came under black administration, but remained municipally financed and controlled. In 1914, the hospital became the first municipal hospital to be managed by African Americans when Dr. William J. Thompkins became superintendent and Mrs. Mary K. Hampton-Brown was named superintendent of nurses. By 1924, black people had assumed responsibility for all departments of the hospital even though the city was not completely convinced of the competence of African Americans to run the facility. White physicians from Kansas City General Hospital, No. 1 continued to serve as supervisors and consultants.[12]

During its early years, Thomas C. Unthank played a major role in the direction of the hospital. In a 1924 study of black nurse training schools commissioned by the Rockefeller Foundation, Ethel Johns, an Englishwoman, described the physician's remarkable political acumen. She wrote:

> He is reported to be a "political boss" of the more benevolent Tammany type, but honest and sincere upon the whole. The feeling at the white hospital is "that without him nothing could have been done at No. 2" . . . He undoubtedly possesses executive capacity and the ability "to get things out of the City Hall."

The gradual evolution of Kansas City General Hospital, No. 2 to an institution under black administration depended, in large measure, on Unthank's political connections.[13]

A segregated municipal hospital was also organized in St. Louis. As in Kansas City, black citizens protested the fact that the existing hospital barred black physicians and housed black patients in the rear sections of two floors. African Americans contended that as taxpayers they should have wider access to publicly financed facilities. In 1919, the city purchased and renovated the vacated hospital of Barnes Medical College to serve as St. Louis City

Hospital, No. 2. Several years later, Dr. W. Montague Cobb, editor of the *Journal of the National Medical Association* and a prominent medical civil rights activist, argued that the establishment of the two segregated municipal hospitals in Missouri represented the "old-clothes-to-Sam" pattern of hospital development. That is, the "transfer of . . . second-hand products of modern American culture to Negro hands as the brown population increases in an urban community." Both hospitals had been created out of vacated and outmoded facilities.[14]

During the last decade of the 19th century and the first two decades of the 20th century, black physicians, educational institutions, churches, and fraternal organizations also established hospitals. The creation of these facilities demonstrates the strength and resilience of the black community. African Americans were not passive victims in the face of oppression; rather, they developed mechanisms to take care of themselves. Confronted with the racism in American medicine, black people responded by establishing their own institutions. It should also be noted that black-created hospitals arose within the context of the solidification of Jim Crow laws in the South and increased racial tensions in the North. Historians have shown that an emphasis on black self-reliance and the development of black institutions are frequently found during periods of black discouragement and increased racist oppression. However, black-controlled hospitals should not be viewed solely as reactions to a segregated, exclusionary society, but also as growing out of the African-American community's longstanding tradition of providing for its members. The first black-controlled hospitals included Provident Hospital and Nurse Training School, Chicago, established in 1891; Tuskegee Institute and Nurse Training School, Tuskegee Institute, Alabama, established in 1892; Frederick Douglass Memorial Hospital and Training School, Philadelphia, established in 1895; and Home Infirmary, Clarksville, Tennessee, established in 1906.[15]

The motives behind the establishment of these black-controlled hospitals varied. Leaders of the National Medical Association (NMA) contended that physicians needed access to hospitals to keep abreast of professional developments in scientific medicine. The late 19th century saw the establishment of the germ theory of disease, breakthroughs in therapeutics and diagnostics, the development of technologies such as the X-ray, and the growth of surgery as a specialty because of the advent of asepsis, anesthesia, and new surgical techniques. These advances led to increases in the prestige, power, and reputations of physicians, and hospitals became essential components of this new scientific medicine, as sites of clinical practice and medical education for all physicians, not just for the urban elite physician. In hospitals, physicians could exchange professional knowledge and have access to expensive hospital-based technologies. In other words, as Charles E. Rosenberg has argued, "the hospital had been medicalized" and "the medical profession had been hospitalized."[16]

The transformation of the hospital paralleled a rapid growth in the number of black physicians from about 900 in 1890 to about 3,500 in 1920. Regardless of their credentials, most found it difficult to obtain hospital admitting privileges. Not all well-qualified white physicians found it easy to obtain hospital connections in this period either, and if white physicians had difficulty, it is clear that the situation for black physicians was all the more bleak.[17]

Black physicians also began hospitals to help themselves economically. They needed the facilities in order to survive professionally and to establish their legitimacy in the community. They also had to compete with white physicians, if not for white patients, then at least for the small pool of paying black patients. Physicians of both races cared for black patients, but white physicians had an advantage in that they could promise continuity of professional attendance upon hospitalization, which their black colleagues could not because they lacked admitting privileges at most facilities. Black physicians had to relinquish care of their hospitalized patients to white physicians, even in some white-run black hospitals, a practice that undermined confidence in black physicians and contributed to a perception that they were inferior to their white peers. In the case of paying patients, it also adversely affected the black physicians' pockets. W. E. B. DuBois, in his classic study *The Philadelphia Negro*, accurately described this professional dilemma:

> At first thought it would seem natural for Negroes to patronize Negro merchants, lawyers, and physicians, from a sense of pride and as a protest against race feeling among whites. When, however, we

come to think further, we can see many hindrances. If a child is sick, the father wants a good physician; he knows plenty of good white physicians; he knows nothing of the skill of the black doctor, for the black doctor has had no opportunity to exercise his skill. Consequently for many years the colored physicians had to sit idly by and see the 40,000 Negroes healed principally by white practitioners.

Thus, black physicians maintained that the creation of black-controlled hospitals was necessary for their financial and professional well-being.[18]

The black community also organized hospitals in order to train black women to be nurses. The late 19th century saw a transformation in the role of nursing, as well as that of hospitals. In the earlier part of the century, nursing was a menial occupation associated with lower-class, poorly educated women; in fact, many hospital nurses were actually convalescing patients. The professionalization of nursing began in 1873 with the establishment of the first training schools for nurses, founded originally to provide respectable work for middle-class women and to improve the moral climate of hospitals by the introduction of middle-class women to the wards. By 1900, there were 432 training schools, most controlled and run by hospitals. Nursing curricula emphasized efficiency and discipline. Pupil nurses provided hospitals with a source of cheap labor. As was true with medical schools, most programs shut their doors to black women, but consequently, black communities established hospitals and nurse training schools to provide their daughters with a career.[19]

The continued exclusion and segregation of black patients also forced the African-American community to act. In Newport News, Virginia, for example, black patients were housed in the city jail until the 1914 establishment of a black hospital, Whittaker Memorial Hospital. Other hospitals worked to maintain the color line and even consciously developed mechanisms to do it efficiently. In 1907, one hospital administrator advocated the following creative scheme:

> the negro department must always be as far from the white and executive as possible. . . . The equipments in all departments are practically the same; however, the linen, gowns and every individual article of the different departments must be kept as

separate as if in a different part of the city. Say, for instance, at a glance you can note where each article belongs, as all are marked in large letters— cream and white blankets for private rooms and white wards; slate-colored blankets for colored wards, and red blankets for ambulance service.

Some founders of black hospitals contended that their hospitals were necessary to provide black patients with much-needed services but also to assure that they were treated with respect and dignity. Black–controlled hospitals, they argued, could offer black patients "a high type of service along with the courtesy which was lacking in the majority of hospitals where colored patients were admitted." For many middle-class black people, these hospitals also represented a less onerous option since, despite their resources, these patients were usually placed in municipal hospitals or in the same inferior facilities as poor African Americans in voluntary hospitals. At the time, race more than class determined the nature and quality of hospital care.[20]

Organizers of black hospitals also claimed that only black physicians possessed the skills required to treat black patients optimally and that black hospitals could provide black patients with the best possible care. Mixed hospitals, with their segregated wards and mistreatment of black people, did not have the "heart" of black-controlled hospitals. Furthermore, they argued that the patients preferred to be treated in black hospitals. "Just as the German Hospital is the choice above others for the Germans," the 1898 annual report of Philadelphia's Douglass Hospital noted, "so is the Douglass Hospital especially preferred by a large class of our colored citizens." One reason, it was argued, that black patients favored black-operated facilities was that in majority institutions they were "apt to be subjected to experimental work." Many African Americans feared that if they entered a white hospital that they would be used as subjects for medical experimentation and demonstration. Such sentiments were not based on unwarranted paranoia. Slaves had been used as experimental subjects by antebellum physicians. Although slavery had ended, black people knew that, given the nation's racial and political climate, they still might not be able to refuse to participate in such practices.[21]

Discrimination and mistrust of the larger society had in fact prompted the formation of other ethnic and religious hospitals. Catholics and Jews founded denominational hospitals because they feared religious conversion and had encountered religious intolerance at Protestant institutions. The desire of Jewish patients to keep kosher and to speak Yiddish when hospitalized and the wishes of Catholic patients to receive the sacraments had been major factors in the establishment of sectarian hospitals. As Charles E. Rosenberg has argued, "Not the peculiar technology of medical diagnosis and therapeutics, but rather the need for good and familiar food, for warmth, cleanliness, and dignity justified the founding of such ethnic and religious hospitals." In creating their own hospitals, African Americans were following a precedent set by other ethnic groups to maintain institutions that answered their particular needs.[22]

The establishment of these black-controlled hospitals represented, in part, the institutionalization of Booker T. Washington's political ideology. His accommodationist philosophy, popular at the turn of the century, emphasized self-help, racial solidarity, and economic development as more productive strategies for racial advancement than politics, agitation, and the demand for immediate integration. Washington also stressed social uplift of the race as the key to racial equality. With the acquisition of wealth and morality, the philosophy went, African Americans would gain the respect of white people and consequently be accorded their rights as citizens. The creation of hospitals would contribute to racial uplift by improving the health status of black people, by demonstrating that black people could take care of themselves, and by contributing to the development of a black professional class.[23]

As was true of hospitals in general, most of the black medical institutions, established as for-profit facilities that would enable their owners to care for patients and to perform surgery, were small and did not offer training programs. Given their size, equipment, resources, and the training of their owners, it is doubtful whether such institutions provided patients with superior medical care. Examples of these private black hospitals included Home Infirmary in Clarksville, Tennessee, and Fair Haven Infirmary in Atlanta. Dr. Robert T. Burt opened the Home Infirmary in 1906, serving as both sole owner and surgeon-in-chief. Most patients entered the infirmary to have surgery: in 1911, surgical cases represented two-thirds of the case load. In 1909, a group of black physicians opened Atlanta's Fair Haven Infirmary, a 12-bed facility even featuring an operating room in which physicians performed major surgery, including abdominal cases. White physicians also utilized the infirmary to hospitalize their black patients.[24]

The establishment and development of the early black-controlled hospitals, however, were not without controversy within the black community, especially in the North. Tensions emerged over the role and mission of the institutions. An examination of the evolution of Provident Hospital in Chicago and Douglass Hospital in Philadelphia will demonstrate some of the obstacles faced by two of the first black-controlled hospitals. It also shows how some of the problems that these institutions encountered remained unsolved and would later plague the black hospital reformers.

The racially exclusionist policies of Chicago nursing schools had provided the primary impetus for the establishment of Provident Hospital and Training School in 1891. After a young black woman, Emma Reynolds, was refused admission to all of Chicago's nursing schools solely because of her race, her brother, the Reverend Louis H. Reynolds, turned to the prominent black surgeon Dr. Daniel Hale Williams (1856–1931) for help. A committee of physicians, ministers, and community leaders failed to find a place for Reynolds at any one of the city's nurse training schools. The experience spurred Williams to organize a biracial association of medical and civic leaders to establish an interracial hospital and nurse training school in Chicago.[25]

Williams successfully solicited funds and supplies from both black and white citizens. His venture was helped immensely by black clubwomen who embarked on fund-raising activities of their own. Fannie Barrier Williams (no relation to Dr. Williams), a noted clubwoman and journalist, became the official solicitor for the campaign raising more than $2,000 from black and white donors. Another clubwoman, Mrs. Cora Scott Pond, coordinated a fund-raising event that raised over $1,300 for the endeavor. Individual donors from the black community also contributed supplies to the hospital, including a parlor stove, a clothes wringer, chiffon lace for nurses' caps, books, and portraits. The donation of such gifts to hospitals was common practice at the turn of the century. Nanahyoke Sockum

Curtis, a nurse and the wife of Dr. Austin M. Curtis, who would later become Provident's first intern, successfully enlisted the support of meat-packing industrialist Philip D. Armour in the campaign to open the hospital. Other prominent white businessmen, including Cyrus H. McCormick, Herman Kohlsaat, and George M. Pullman, also gave funds. Self-interest undoubtedly played a role in the charitable actions of these men, since the maintenance of a healthy black work force required that black employees have access to medical facilities. In June 1891, only five months after the creation of the hospital association, a 14-bed hospital opened in a two-story frame house.[26]

Daniel Hale Williams's enterprise did not escape censure. Some members of the black community accused him of perpetuating segregation. One minister went so far as to curse the building and pray that it would burn to the ground. Williams was able to overcome such criticism because of the wide biracial support that the hospital received. Furthermore, the founder perceived the hospital not as an exclusively black enterprise, but as an interracial one that would not practice racial discrimination with regard to staff privileges, nurse training school applicants, and the admission of patients. Williams did not want black physicians to be separated from medicine's mainstream, nor did he want Provident Hospital to be used as an apology for the continued racism at other Chicago hospitals. At the time, the only other hospital to accept black patients was Cook County Hospital on the west side of the city, far from the black South Side.[27]

Williams's personal biography is crucial to understanding how he came to establish Provident Hospital. He had been born in Hollidaysburg, Pennsylvania, in 1856. His father, Daniel Jr., was a barber and well-to-do landowner. His mother, Sarah, was a housewife. After his father's death, the family moved to Rockford, Illinois. At the age of 17 Williams settled in nearby Janesville, Wisconsin, where, while he attended school, he worked as a barber just as his late father had done. In 1878 he graduated from Haire's Classical Academy. Shortly thereafter he decided that he wanted to become a physician and he apprenticed himself to a prominent white Janesville physician, Dr. Henry Palmer, who had previously served as Wisconsin Surgeon General for 10 years.

Williams's medical education continued at the

Chicago Medical College, a predominantly white school, from which he was graduated in 1883. At a time when postgraduate training was not the norm for the average practitioner and practically inaccessible for black physicians, Williams completed an internship at Mercy Hospital, a Catholic hospital associated with his alma mater. His qualifications thus equalled those of elite physicians of either race and exceeded those of many other doctors in late 19th-century America. Williams went on to enjoy wide contact and much prestige within the city's white medical community. His connections with this group surely made him aware of the increasing importance of hospitals to physicians' careers, and he later obtained appointments at the South Side Dispensary and the Protestant Orphan Asylum. In 1889 he became the first black physician named to the Illinois State Board of Health, while also maintaining a lucrative interracial private practice. Williams's career was anomalous. It was probably aided by the physician's fair skin—he was light enough to pass for white. Other black physicians found themselves barred from Chicago medical institutions, and therefore the establishment of Provident Hospital would certainly benefit their careers.

Although Williams had access to the white medical world and called for Provident Hospital to be interracial, he was not isolated from the black medical world. He was one of the founders of the National Medical Association, established in Atlanta in 1895. He was offered the presidency of that organization, but declined, serving instead as its first vice president. After the establishment of Provident Hospital, Williams conducted surgical clinics at Meharry Medical College and at various black hospitals throughout the South. He acknowledged that racism, especially in the South, seriously limited the opportunities afforded black nurses and physicians and adversely affected the health care received by black patients. He urged that the black community establish its own hospitals and nurse training schools because "a people who do not make provision for their sick and suffering are not worthy of civilization."[28]

Provident Hospital in its first years reflected Williams's interracial vision. He used his connections to enlist white medical leaders such as Frank Billings and Christian Fenger to serve as consultants and trustees at Provident. The medical staff, although biracial, had a racial division of labor: white physicians served primarily as consulting staff and black physicians as attending and resident staff. Dr. Austin Maurice Curtis, later the first black physician appointed to the staff of Cook County Hospital, was Provident's first intern. However, during the hospital's first 21 years most interns were white. The patient census also reflected biracial patronage. In its first year, the small hospital admitted 189 patients, 34 (18 percent) of them white.[29]

Provident Hospital flourished. In 1893, two years after its opening, Daniel Hale Williams performed one of the first successful open-heart surgeries there. Five years later, the hospital moved into a new 65-bed facility, the construction of which had been financed by white benefactors including Kohlsaat and Armour. By 1912, the hospital had no debts, an endowment of $50,000 (of which $45,000 had been donated by blacks), and a plant valued at $125,000.[30]

In spite of Williams's original intentions, Provident Hospital gradually evolved into a black institution. By 1915, some 93 percent of the patients were black. By 1916, all of the nurses (except the supervisor) and almost all of the staff physicians were black. This evolution reflected the deterioration of race relations in Chicago, as a rise in white hostility and institutional racism had greeted the growing number of black migrants from the South. In response, black leaders created separate institutions, organizations, and businesses to serve Chicago's burgeoning black population, which was predominantly confined, because of residential segregation, to the city's South Side. The 1920 census estimated that 109,458 African Americans lived in Chicago—an increase of almost 150 percent over 1910. The well-established Provident Hospital, emerging as an important center of this "black metropolis," was accordingly expected by the black community to meet its needs and aspirations.[31]

A change in the leadership at Provident Hospital had also contributed to its development as a black institution. Daniel Hale Williams resigned as medical director in 1894 when he became surgeon-in-chief at Freedmen's Hospital in Washington, D.C. He returned to the staff of Provident Hospital four years later; however, during his absence, a professional and personal rival, Dr. George Cleveland Hall, had become medical director. In addition, Hall's allies had gained control of the hospital's board of trustees. The longstanding feud between

Williams and Hall may well have stemmed initially from Williams's unsuccessful attempt to block Hall's staff appointment to Provident. Hall graduated in 1888 from the sectarian Bennett Eclectic Medical College, a proprietary school nominally affiliated with Chicago's Loyola University. Williams did not view Provident Hospital as an egalitarian institution open to all black physicians regardless of their credentials. He wanted the hospital to be a first-class, competitive institution, and insisted that all staff members meet the highest professional standards. He did not consider Hall, a graduate of an inferior sectarian school, to be a qualified candidate. Nonetheless, Hall's allies in the black community and on the Provident board secured him an appointment in spite of Williams's objections.[32]

The antagonism between Hall and Williams extended beyond the walls of Provident Hospital and reflected divisions within the black community at large. The historian Allan Spear views the feud between the two physicians as a manifestation of the power struggle between two factions within the black community—the old elite and the new leadership class. Williams represented the former, which had dominated black community life in Chicago until the first decade of the 20th century. This group espoused integration and had economic and professional ties in the white community. After 1900, a new leadership class, of which Hall was a prominent member, rose to power. Advocating black self-help and solidarity, this group relied on its economic and professional ties in the black community. Unlike Williams, Hall maintained close association with several black community organizations, including the National Negro Business League, the Wabash Avenue YMCA, and the NAACP. His conception of Provident Hospital differed significantly from that of its founder. Hall viewed the institution not as a model of interracial cooperation, but as "the Colored people's hospital . . . an establishment in which, more and more each year, the Colored folks are taking personal pride."[33]

Williams remained on staff until 1912, when he resigned bitterly. Accounts of the reasons behind his departure vary. One centers on the passage by the board of trustees of a Hall-sponsored resolution requiring all staff physicians to limit their practice to Provident. Such a measure would have undoubtedly been aimed at Williams, who had been ap-pointed associate attending surgeon at the predominantly white St. Luke's Hospital. In this version, Williams chose to resign from Provident rather than limit his practice and isolate himself from his white colleagues. In another account, he resigned after the board of trustees of the hospital refused to meet his demands that Hall be removed from the hospital staff and its board. Whatever the reason, after his resignation Provident's founder never again took an active role in its activities. The year after he broke ties with Provident Hospital, he gained additional status in the world of elite white physicians as the only African American out of the 1,059 physicians named charter members of the American College of Surgeons.[34]

Tensions over the role and direction of the black hospital also emerged in Philadelphia at the Frederick Douglass Memorial Hospital and Training School. Unlike the situation in Chicago, the resolution was not the departure of the hospital's founder, but the establishment of a new hospital. The history of Douglass Hospital was inextricably bound to the career of its founder, Dr. Nathan Francis Mossell (1856–1946). Mossell's medical and racial philosophies, personality, and drive dominated hospital policy during his 38-year tenure as medical director. Much of his racial philosophy, which prompted him to found Douglass Hospital, can be traced to his parents. Aaron and Eliza Bowers Mossell were free blacks who migrated to Canada from Baltimore because of racism and the paucity of educational opportunities in the South for their children. Their third child, Nathan, was born on 27 July 1856 in Hamilton, Ontario. Nine years later, when Aaron Mossell lost the brickyard he had established in Hamilton, the family moved to Lockport, New York. In this small town, not far from Rochester, the elder Mossell worked as a day laborer until he reestablished himself in business. As his business flourished, with even whites in his employ, Aaron Mossell became a respected and prosperous member of the Lockport community. This, however, did not shield his family from racism. Lockport maintained a separate school for black children. Mossell refused to allow his children to attend it, but pressured the school board to abolish its segregated school system. Nathan Mossell later remembered his childhood as being "conditioned by an environment which propelled me toward a manly approach to life. The only man for whom I ever worked was father. I had no opportunity to become a white man's 'boy.'"[35]

In 1871, 15-year-old Nathan Mossell left Lockport to join his older brother, Charles, at Lincoln University, a black institution in southeastern Pennsylvania. There he enrolled in the college preparatory department. Three years later he matriculated in the undergraduate program at Lincoln, from which he graduated with honors in 1879.

Mossell had decided to become a physician during his last year at Lincoln. In the fall of 1879, he called on Dr. James Tyson, Dean of the University of Pennsylvania School of Medicine, to request admission to the school. No African American had previously enrolled in the medical school, and Tyson, a Quaker, favored Mossell's admission. He noted that both Harvard and Yale had admitted black students, without major problems, and that Pennsylvania should be "equally liberal." Subsequently, Dean Tyson obtained the necessary faculty approval for Mossell's admission.[36]

Nathan Mossell began his medical education on 15 October 1879. His welcome to the University of Pennsylvania was not exactly warm. When he walked into a lecture hall students stomped their feet and hissed, "Put the damn nigger out." In the following weeks students sent letters protesting his admission. The opposition was not merely verbal. On one occasion, while Mossell walked along the Schuylkill River, just east of the campus, some students attempted to push him into it. Despite his cruel introduction to medical school, Mossell stayed and graduated with honors in 1882. At his graduation, he received a standing ovation from his classmates. Mossell wrote, much later, that the racism that he had encountered had not disturbed him because he considered himself "better prepared than most of the students. In those days one needed only a high school diploma to study medicine. Out of a class of 140-odd, about thirty of us had Bachelor of Arts degrees." Mossell's claim of indifference may be questioned; however, his courage and determination cannot be doubted.[37]

As was the case with Daniel Hale Williams, Mossell took a career path that was not the norm for the average practitioner, black or white. Following graduation, he actively pursued postgraduate education. Mossell worked for two years at the out-patient surgical clinic at the Hospital of the University of Pennsylvania. He took a postgraduate course at the Philadelphia Polyclinic. In 1889, he spent four months in London studying surgery at the prestigious Guy's Hospital and at St. Thomas Hospital. Mossell's career received a further boost in 1888 with his election as the first black member of the Philadelphia County Medical Society. Nominated by J. Britton Massey, a Southerner and a specialist in electrotherapy, his endorsers also included the distinguished surgeon D. Hayes Agnew and his former sponsor at the University of Pennsylvania, James Tyson. Initially there was resistance to Mossell's admission because some members questioned whether a black man could have the requisite credentials. Opposition subsided after Dean Tyson pointed out that "Dr. Mossell graduated with an average higher than three-fourths of his class." Despite his ties to the city's white medical establishment, there were limits to Mossell's ability to break racial barriers. He could not secure admitting privileges at any Philadelphia hospital, which surely must have disturbed a man of his training and professional aspirations.[38]

In addition to his medical activities, Mossell participated in many professional, political, and civic organizations devoted to racial equality and the social and economic advancement of African Americans. He was instrumental in the founding of the National Medical Association and in 1907 served as its eighth president. In 1900, Mossell helped establish the Philadelphia Academy of Medical and Allied Sciences, a constituent chapter of the National Medical Association. Outside of the medical world, he led a protest in Philadelphia against the dramatization of Thomas Dixon's racist novel *The Clansman*. The novel, the basis of the film *The Birth of a Nation*, praised the Ku Klux Klan and belittled African Americans. The following year, Mossell was one of the organizers of the meeting that launched the Niagara Movement, the forerunner of the National Association for the Advancement of Colored People.

On the evening of 25 June 1895, Dr. Mossell convened a meeting of black Philadelphians to discuss the feasibility of establishing a nurse training school. The group demonstrated the unusual strength and diversity of Philadelphia's black middle-class community and included ministers, business owners, and teachers. Significantly, Mossell was the only physician at the meeting. As was true of the organizers of Provident Hospital, this group of prominent African Americans considered it their obligation to establish a nurse training school because young black women who wished to study nursing faced bleak prospects, alleging that only one Philadelphia school, the Training School of Philadelphia Hospital, admitted them. The training school organizers maintained that nursing education offered black women an opportunity to develop a profession and that the development of such a class of women would greatly contribute to racial uplift. Nursing education would discourage "idleness and profligacy" in young black women and would provide them with a career that would be "useful and [a] benefit to themselves, to the sick, and to the community." The organizers also realized that a training school could not operate independently, but that a hospital was a necessary adjunct. A hospital would also provide much-needed facilities for the city's black physicians. They pledged $1,000 to build a hospital and a nurse training school. Mossell later reminisced:

> With all, we wish it had not been necessary to establish the Douglass Hospital. We deprecate the present trend in the dominant public mind to create in many sections of the country hospitals and medical schools, especially for colored people—it means extravagance, inefficiency, duplication of effort, and is undemocratic in that it establishes caste.

However, the needs of the black community and the racism at the other hospitals and training schools in Philadelphia forced them to act.[39]

Plans for the hospital progressed. A board of managers was elected and, not surprisingly, Dr. Mossell was chosen to be medical director of the yet unnamed institution. A relative of Mossell's wife Gertrude, Jacob C. White, Jr., was named to head the board. Most of the board's members were African Americans. As the historian Roger Lane has noted, the board's racial composition represented a break with the usual tradition of having rich white supporters chairing the city's black philanthropies. This was not the case at Douglass, which would not be white-dominated. Whites were a minority on the board and served in no position higher than treasurer. Philadelphia attorney and president of the board of education Samuel B. Huey took that office. At the 5 August 1895 meeting, the board of managers named the institution in memory of Frederick Douglass, the abolitionist and human rights champion who had died a few months earlier.[40]

Although the hospital was to be black-controlled, the organizers did not perceive the hospital as an exclusively black institution. Mossell and the board insisted that the hospital "not be an apology for the discriminatory practices of the other public agencies." It would not discriminate with regard to staff privileges, training school applicants, or the treatment of patients. Whites provided assistance in planning the hospital, served on the board of managers, and donated funds.[41]

The proposed hospital prompted some opposition from both the white and black communities. White opponents complained that the hospital represented an unnecessary addition to the already large number of charitable institutions in Philadelphia. At the forefront of black criticism was *The Weekly Tribune*, a black-operated newspaper whose editors viewed the establishment of Douglass Hospital as a concession to race prejudice. They wrote that it was "the quintessence of foolhardiness to continually prate about breaking down color barriers and then go on rearing them ourselves." Furthermore, they portrayed a rosier picture of the status of black patients and nurses than the supporters of the hospital had advanced. Critics claimed that none of the city's hospitals barred black patients and that black women, in the past, had graduated from the nurse training programs at Women's Medical College, the Hospital of the University of Pennsylvania, and Philadelphia Hospital. They also questioned whether the black community had the financial resources to support a hospital and pointed out that a number of black-run enterprises had failed because of a lack of funds. "We are not so enthused over the opening of the so-called Colored Hospital," a 7 September 1895 *Weekly Tribune* editorial noted,

> because it should be borne in mind that a hospital is a very expensive institution to manage successfully. As a people, are we in a position to support such an undertaking? Or do the projectors intend to depend more upon the philanthropy of our white friends.

As had been the case in Chicago, the strength and breadth of support, including that of the African Methodist Episcopal Preachers' Meeting and the Baptist Minister's Union, enabled the hospital organizers to overcome such opposition.[42]

The Frederick Douglass Memorial Hospital and Training School opened on 31 October 1895, only four months after the first organizational meeting. The 15-bed facility occupied a three-story leased building in South Philadelphia. The basement of the remodelled house served as the dispensary, the first floor as offices, and the second and third floors as patient wards. In addition, a small room on the second floor was outfitted to serve as an operating room—evidence that Mossell recognized the increasing importance of surgery to a hospital's function.

The nurse training school opened a few months after the hospital. During its first year of operation, the Douglass Hospital received 58 applications and admitted six young African-American women. The training school's impact stretched beyond the black population of Philadelphia. The first two women who graduated from the program (on 22 May 1897) were Hattie E. Mosely of Iowa and Mary E. Wilson of Virginia. By 1904, 15 women, all black, had graduated from the school. The graduates secured a variety of positions: Mary E. Wilson, class of 1897, became the head of a private sanitarium in Philadelphia; Carrie B. Earley, class of 1899, later became head nurse and matron at Douglass. She was succeeded by Marie J. Narcisse of the class of 1904. Graduates were also employed by such institutions as the Home for Aged and Infirmed Colored Persons in Philadelphia and the Home for Colored Persons in Atlantic City. Most of its graduates, as did the majority of those from white training schools, sought and obtained private employment. Both white and black families hired them.[43]

The hospital also succeeded in offering professional opportunities to black physicians. Of the 20 professionals, including the pharmacist and head nurse, comprising the original medical staff at Douglass, 12 were African Americans. The medical staff was organized into consulting and attending staffs. The roster of attending physicians was mostly black, while that of consulting physicians was exclusively white. The consulting physicians and surgeons, including prominent members of the Philadelphia medical establishment such as James

Tyson, Roland G. Curtin, Thomas S. K. Morton, and John B. Deaver, provided a reassuring legitimacy to the young institution. It is not clear whether their involvement with the hospital went much beyond this. For instance, it is not documented whether they ever admitted patients to the hospital. By 1900, the hospital created associate staff positions to include black physicians who had suffered from "caste disabilities" in the suburbs and neighboring cities such as Wilmington, Delaware; Chester, Pennsylvania; and Atlantic City, New Jersey.[44]

Douglass Hospital also offered postgraduate training to black medical school graduates. During the hospital's first 10 years, many of its residents had trained at Mossell's alma mater, the University of Pennsylvania. By 1904, 17 black physicians, all men, had completed residencies at Douglass. After residency, some stayed on at the hospital. J. Q. McDougald became assistant gynecologist; R. W. Bailey, assistant·pathologist, and Arthyr T. Boyer, assistant ophthalmologist. Some obtained positions at black hospitals in other cities with large black populations. Samuel P. Stafford became chief of staff at Provident Hospital, St. Louis, and J. E. Dibble became attending surgeon at Douglass Hospital, Kansas City, Kansas. Graduates also entered private practice, mostly in the Philadelphia area.

Despite these successes, Douglass Hospital had financial difficulties. The majority of its patients were poor African Americans who could not afford to pay the hospital's five-dollar-per-week fee. They either paid a reduced fee or were treated free of charge. In order to increase revenues, the hospital used a strategy employed by many contemporary hospitals; it attempted to increase the number of paying patients. But this tactic ultimately failed because Douglass could not greatly expand the number of paying patients because of space limitations.[45]

The hospital also tried to solicit funds and support from the white and black communities. However, the appeals to the two groups differed. The campaign to the white community stressed its relative wealth and its Christian obligation to those less fortunate. "We appeal to our white friends," one solicitation implored,

> because God, who has so arranged it in His economy that the wealth of the land is theirs, and the

meagre salaries of the employed will not permit a contribution large enough to help other than modestly an institution, even though it tends to their own vital interest.

Campaigns to African Americans stressed the hospital's importance to racial uplift. "It is your Institution, the only one in its kind in the State," they were reminded. "You cannot afford to sit idly by and wink only as opportunities pass; help this work, help educate your girls and your young men."[46]

Gertrude Mossell, Nathan Mossell's wife, headed the hospital's fund-raising activities, once again indicating the importance of black women to the development of the race's hospitals. During the hospital's first year, a sum of over $5,000 was raised, of which $4,300 came from members of the black community. Some contributions were as small as 25 cents. In 1896, Zion Wesley African Methodist Episcopal Church endowed the hospital's first bed for charity patients. Members of the black community responded not only with funds, but by donating needed items such as food, linens, and magazines, and by organizing charitable activities—balls, poetry readings, recitals at the Academy of Music, and classical music concerts. One especially publicized event was the February 1896 grand ball and concert featuring the famous soprano Sisseretta Jones ("Black Patti").[47]

As the hospital grew, the percentage of contributions coming from the black community decreased, and the need for supplemental funds became apparent. In 1896, the year in which the hospital was incorporated, representatives of the State Board of Charities visited Douglass Hospital and approved the institution to receive funds from the state legislature. Subsequently, the legislature appropriated $10,000 for the hospital's maintenance, but for unclear reasons, this appropriation initially was delayed. Mossell himself attributed the problem to "misguided" black politicians who resented his political independence. Mossell, however, was able to muster some political support of his own and defeat his critics. Between 1896 and 1904, Douglass Hospital received funds from the state ranging from $5,000 to $10,000 per year.[48]

Despite its fund-raising activities, by 1904 the hospital had accumulated an $18,000 deficit. However, its financial situation did not stop Douglass from making plans to enlarge. Expansion had be-

come a necessity because the hospital had quickly outgrown its original facilities. In 1896, the hospital treated 61 inpatients; in 1904 it treated 242. Similar growth had taken place in the outpatient department: from 987 patients treated in 1896 to 1,997 in 1902. Mossell also insisted that the new hospital be equipped with pathology and bacteriology laboratories in order to keep in step with contemporary developments in medicine. In 1904, Douglass purchased two houses for $8,000 to use as a new facility. Plans called for a 75-bed hospital with more accommodations for private patients and a larger operating room.

The hopeful expectations produced by the proposed expansion temporarily masked a growing controversy at the hospital that centered on its internal control. A headline in the 21 January 1905 *Weekly Tribune* announcing, "Dr. N. F. Mossell, Chief of Staff Asked to Resign His Place," gave some indication of the depths of the problem. Although the board of managers later denied this report, the headline did indeed herald the development of sharp and divisive tensions at Douglass Hospital.[49]

The conflicts ostensibly revolved around the hospital's operating room practices. Its policy, printed in each annual report, was that

> in order to facilitate Nurse Training and to extend the usefulness of the Institution, we permit all physicians to place their patients in the Hospital and attend them while there—except in operative cases—the Hospital reserves the right to use discretion as to the operator.

At the time, hospitals, especially those affiliated with university medical schools, commonly used such closed-staff arrangements as a way of controlling surgery. Mossell, it seems, was trying to duplicate procedures employed at elite hospitals. However, the criteria used for staff privileges were often unclear and appointments were frequently made on the basis of "favoritism over skill." The use of a closed-staff policy at Douglass Hospital led several members of the medical staff to question whether the goal of the institution was to provide professional opportunities for black physicians in general or for Nathan Mossell exclusively. During the hospital's first year, 31 inpatient operations were performed, two-thirds of them by Mossell. It is not clear whether Mossell continued to perform such a large percentage of the surgery, but it *is* evident that some black members of the medical staff perceived that he did. Mossell's critics alleged that his domination of the operating room limited their clinical experience and threatened their professional advancement. They became dissatisfied with the professional opportunities offered by Douglass Hospital, and Nathan Mossell became the target of their discontent. This controversy needs to be understood on two levels. On the one hand, some physicians clearly chafed under Mossell's undisputed medical control of the hospital. But a deeper tension existed as well. As we shall see later, Mossell saw himself and his hospital as representatives of new professional standards.[50]

Factions developed within the hospital and soon extended beyond its walls. The black press closely followed the staff problems and took sides. The *Weekly Tribune* had originally opposed the establishment of Douglass Hospital, but conceded by January 1903 that it was an essential community institution. The newspaper, however, supported Mossell's critics by portraying the medical director as an autocrat who ran a private hospital—not a hospital that benefited the general black community. Furthermore, it alleged that his continued presence threatened the hospital's existence. A 21 January 1905 article in the newspaper argued that

> the public at large are fast losing confidence in the Douglass Hospital because they plainly see the methods of its head are faulty. Hence public condemnation follows in its wake, making unpopular an institution which should be the joy of every charitably inclined and patriotic citizen.

In contrast, *The Courant,* another black publication, blamed the problems at Douglass on the managers, who it contended had "failed to perform their duty, and are a set of hedgers, unworthy to conduct either a hospital or any other public charity."[51]

Nathan Mossell attributed the hospital's problems to professional jealousies. In a 1908 paper, "The Modern Hospital: Its Construction, Organization, and Management," which he presented at the annual meeting of the National Medical Association, Mossell reflected on the tensions at the hospital. Mossell asserted:

> the ideal arrangement . . . would seem to be a combination between all the reputable physicians located in the town and all the leading citizens in the

management of the enterprise, but experience has shown that local jealousies first between the physicians and finally extending to the private patients and sympathizers of those who are disgruntled are sure to greatly hamper the work of the institution, if not disrupt it.[52]

As the months went by, conditions at Douglass continued to deteriorate. In October 1905, 15 of the 24 practicing black physicians in Philadelphia petitioned the board of managers for a reorganization of the medical staff. Shortly afterwards, the board of managers voted to remove Mossell as medical director and replace him with Dr. Edwin C. Howard, a graduate of Harvard Medical School. The coup had been led by board member Dr. Eugene Theodore Hinson, an 1898 graduate of the University of Pennsylvania School of Medicine. Howard, however, held the position only briefly: later in the month, the board met again and nullified his election. Dr. Mossell had contended that the election was invalid because the newly elected members of the board had not been informed of the pivotal meeting. He was able to elicit support for his position from the majority of the board and was reinstated as medical director. This action so enraged Howard's supporters that they threatened court action, but the threat proved to be empty and a suit did not follow.[53]

The power struggle at Douglass Hospital resulted in the establishment of a second black hospital in Philadelphia. On 5 December 1905, a group of black physicians and lay people, believing that Douglass Hospital had become a "privately managed, narrow, unprogressive institution," met at Dr. Howard's office to make plans for a new hospital. The group proceeded with their plans, and on 12 February 1906 (Lincoln's Birthday), Mercy Hospital and Nurse Training School opened only four blocks from Douglass Hospital.[54]

Although Douglass had been founded to provide clinical opportunities to black medical professionals, Nathan Mossell did not visualize the hospital as a democratic institution. In a 1906 article, "An Institution That's Doing a Great Job," he wrote that the hospital had "never for a moment considered it wise or desirable to have every colored physician's name associated with the hospital simply because he is colored." Mossell's operating room policy was designed to limit access. His actions must therefore be viewed from the perspective of his professional aspirations and goals. Wanting to be accepted as a skillful surgeon and Douglass Hospital to be seen as a first-class hospital, he had to limit access to those he deemed to be qualified, no matter how ill-defined and arbitrary his criteria. This undemocratic process was commonplace in American hospitals and also led to tensions among physicians at other institutions. Ironically, the procedure that Mossell supported had been used to exclude black physicians from majority hospitals. Perhaps he wanted to demonstrate that black physicians could meet the growing demands of scientific medical practice and satisfy the standards established by elite, technologically sophisticated white hospitals. Mossell's critics did not control the hospital board as had George Cleveland Hall's supporters in Chicago, and therefore they could not force the hospital's founder from his sinecure. Frustrated at Douglass and barred from other hospitals, Mossell's opponents had no other choice but to organize another black hospital in Philadelphia.[55]

Between 1912 and 1919, the number of black hospitals and nurse training schools, white and black controlled, increased from approximately 63 to 118. About 80 percent of these facilities operated in the South. This regional concentration can be attributed to two factors. First, the black population lived primarily in the South. Most black physicians went to school and practiced in the region. Second, the establishment of black hospitals in the North often met with stiff opposition from segments of the African-American community who argued that the less restrictive racial climate of the North did not require it. Opponents conceded that the stringent racial situation in the South necessitated the establishment of black-controlled institutions in that region.[56]

Supporters of black hospitals responded that racism did indeed exist in the North and that the region also needed black-controlled institutions. The National Medical Association made its position clear in a 1909 editorial in the *Journal of the National Medical Association,* lamenting that

> it is needless for our brethren in the North to longer hide behind the mantle of inactivity and continue to say that these things are needless—they inject the race question—they draw the color line. Not so. The race question is already drawn, and with such a heavy, dark stroke as to not be recognized.

Leaders of the NMA pointed out that even in the North hospitals barred patients or admitted them to segregated and inferior wards and excluded black physicians from their staffs. Black hospitals, they maintained, should not be limited to the South.[57]

The inability of many black physicians to obtain membership in the American Medical Association (AMA) worked to bar them from hospital appointments. Many hospitals' criteria for staff positions included membership in the AMA or its local affiliate. In order to join, a physician had just to be accepted by a local affiliate. However, many local medical societies, especially in the South, excluded black physicians until the 1950s. Those black physicians fortunate enough to gain admitting privileges often had to work under the supervision of white physicians, causing one Georgia physician to complain angrily, "These white surgeons don't want Negroes to learn. And even when he is operating on my own patient he works in an incision the size of a rathole and I can't even see what he is doing." Black medical leaders feared that their lack of hospital opportunities would relegate them to second-class citizenship within medicine.[58]

Some historians have interpreted the plight of black physicians as due to their race. On one level this certainly is true: it would be absurd to attempt to write a "color-blind" history of black physicians. But to say that "race did get in the way of success," obscures a crucial point. It was not the color of a physician's skin that blocked his or her path. It was white reaction to skin color: the *racism* of the society in which they lived, including the racism that permeated the medical profession. This racism was not limited to the prejudiced attitudes of isolated individuals. It was institutionalized in the structure of American medical practice. And, sadly, this institutionalized racism actually intensified as medicine became more scientific and as new standards for medical practice developed in the first decades of the 20th century.[59]

The transformation of the hospital and the rise of scientific medicine led to efforts to standardize and improve hospitals. Previously, hospitals had reflected community needs and values more than national medical standards. Indeed, there had been no formal criteria governing the provision of hospital care and the management of hospitals. The American College of Surgeons, founded in 1912, attempted to remedy this problem by establishing national yardsticks for the evaluation of hospitals. The College hoped that adherence to its guidelines would upgrade or eliminate inferior hospitals. In 1919, it issued its first set of rules that specified minimum standards for staff, equipment, facilities, and basic operating procedures. These guidelines had to be met before a hospital could be placed on the American College of Surgeons' list of approved hospitals.[60]

In 1910, the Carnegie Foundation for the Advancement of Teaching had released Abraham Flexner's highly critical study of American medical education. Several historians have explored the many significant reforms in American medical education during the years surrounding the publication of the Flexner Report. One key change that they have identified was a new emphasis on clinical training in hospitals as an integral component of medical education. The American Medical Association's Council on Medical Education had recommended in 1905 that medical school graduates complete an internship as part of their basic medical education. By 1914, an estimated 80 percent of medical school graduates served an internship; five medical colleges required an internship as a prerequisite for a medical degree; and one state board, the Pennsylvania board, required it as a prerequisite for licensure. The reforms in medical education also led to efforts to improve the quality of postgraduate training. In 1919, the Council on Medical Education started issuing "The Essentials of an Approved Internship," a list of minimal criteria that approved internships had to meet in order to warrant approval. In addition to the internship, specialty training and other forms of hospital-based medical training began to expand.[61]

Black physicians, however, continued to find it difficult to obtain postgraduate training. In a 1917 *Woman's Medical Journal* article, Dr. Isabella Vanderwall candidly discussed her inability to obtain an internship. This black physician, a 1915 graduate of New York Medical College for Women, was rejected by four hospitals, including the one affiliated with her medical school, not because she was unqualified—she had graduated first in her class—but because she was black. Vanderwall's story reveals much about racial discrimination in medicine and the pain and frustration it brought. It also makes plain the young physician's determination and perseverance in the face of racism. She wrote:

I had almost given up hope of securing an internship when one day, I saw a notice on the college bulletin board saying the Hospital for Women and Children in Syracuse, New York, wanted an interne. Here I thought was another chance. So I wrote, sent in my application, and was accepted without parley. . . . So to Syracuse I went with bag and baggage enough to last me for a year. I found the hospital; I found the superintendent. She asked me what I wanted. I told her I was Dr. Vanderwall, the new interne. She simply stared and said not a word. Finally when she came to her senses, she said to me: "You can't come here; we can't have you here! You are colored! You will have to go back." Go back! . . . No, indeed, not I! So for three days I stayed in Syracuse, trotting hither and thither, seeing this authority and the other official. Everywhere I met the same answer: "No, we cannot have you. We do not doubt your ability, but you are colored, therefore we will not have you." So they sent me home without any internship.

Vanderwall contended that the inability of black physicians to obtain internships threatened their future in medicine. She, herself, was able to practice, having obtained her licenses in New York and New Jersey before the laws on compulsory internship went into effect.[62]

The expected track for black physicians wanting to obtain an internship was to pursue it at one of the black hospitals. However, not all black hospitals admitted them. Dr. Roscoe C. Giles, the first black graduate of Cornell University School of Medicine, recalled that "excuses, alibis, and subterfuges" were used to deny him a position at several New York City hospitals, including Lincoln Hospital. Lincoln Hospital, a municipally operated hospital, had begun its institutional life in 1842 as the Home for Worthy, Aged, Indigent Colored People. The work of the institution increasingly included medical care, and in 1882 it changed its name to the Colored Home and Hospital. In 1898, the hospital moved from Manhattan's East Side to the then semi-rural and predominantly white South Bronx, and in 1912 it changed its name again to Lincoln Hospital and Home. The hospital's relocation resulted in a change in its patient population. The chronic care and nursing home wards remained exclusively white. The hospital wards became progressively white. The hospital, however, did not fully abandon its commitment to African Americans. Black patients were not barred from admission to the facility. In addition, it continued to operate the training school for black nurses it had begun in 1898. Although the hospital trained black nurses and offered them employment, it denied such opportunities to black physicians. Commenting on their status at the black division of Atlanta's Grady Hospital, another institution that admitted black patients and nurses, one black physician lamented, "We envy the nurses, they at least get a chance—we have none." The situations at Lincoln and Grady hospitals make clear, once again, that not all institutions considered to be black hospitals, usually those operated by whites, supported the advancement of black physicians. Dr. Giles, who in 1938 became the first black physician certified by the American Board of Surgery, finally did obtain an internship—at Provident Hospital.[63]

The changes in medical education that took place during the first two decades of the 20th century mandated that medical schools provide clinical training to all students. The University of Pennsylvania School of Medicine, Dr. Nathan Mossell's alma mater, attempted in 1916 to forge a relationship with Douglass Hospital. The university wanted to use the hospital's wards to train black medical students in obstetrics. The affiliation with Douglass Hospital was to be for black students only. It is clear that racial and sexual taboos about black men touching white women were at work here, since the arrangement was not suggested for any other clinical clerkship. Mossell adamantly refused the proposition, stating that it would be acceptable only if the medical school would send both black and white students. Mossell's confrontation with the University of Pennsylvania was dramatized later in the 1947 play "Within Our Gates," written by Sylvia James to honor the physician and his work. In her account, James used the character Dr. Browning to represent Dr. William Pepper, III, who was the actual dean of the medical school. The play read in part:

BROWNING   (Dr. Browning of the U. of Pa.):
            It's an unpleasant duty that brings me here. . . . You know that we have a few colored students at Penn . . .
MOSSELL   Yes, I know I helped to get some of them in.
BROWNING   That's right . . . you did. Well, they seem to be doing all right. . . . I'm sure they'll make fine doctors. These boys have to serve their internship some place, Dr. Mossell . . . there's been increasing pressure on us to

bar them from interning at our hospital.
MOSSELL   I'm sorry . . . I don't get that. Why?
BROWNING   Well . . . some of the white patients object to being treated by colored doctors . . . our white boys are resentful at having to work so closely with them. It's bad . . . I don't like it . . . but there it is.
MOSSELL   I see. And I suppose you're going to ask me to permit them to come here.
BROWNING   That's right. Douglass is the only adequately equipped colored hospital in the state.
MOSSELL   Dr. Browning . . . we don't claim here at Douglass that we are a *colored* hospital. True, most of our staff are colored, most of our patients too. But this hospital was organized to protest against racial segregation, not to encourage it. We have no intention of aiding any Jim Crow tendencies. And we won't become a dumping ground for other schools, who are as well equipped as we to train their own students.

The discriminatory practices of hospitals in Philadelphia had forced the creation of Douglass Hospital. Mossell argued that it had been established to protest segregation, not to foster it. If Mossell had accepted the offer from the University of Pennsylvania, it would have undermined his intention that the hospital be interracial. Furthermore, he would have been an accomplice in the creation of a segregated program for black medical students. Such an act would have been contrary to his views on racial equality.[64]

The hospital's stance was not without consequences. The Chamber of Commerce and the State Board of Charities withdrew their endorsements because the hospital had not cooperated with the university. These actions decreased the hospital's chances of raising money. In 1919, the state legislature attached a rider to Douglass Hospital's appropriation: the institution would not receive its $22,000 appropriation unless Mossell were removed as medical director. The hospital remained firm and, for two years, it went without state funding. Mossell alleged that its rival institution, Mercy Hospital, did not have its financial stability jeopardized because it had gone along with the university's plan. He believed that the physicians at Mercy had "sold their birthright of freedom for a mess of pottage, i.e. white recognition and support." State funding to Douglass was finally reinstated after a battle waged by Edwin Vare, a Philadelphia political leader, resulted in the removal of the rider. Although Douglass had been financially damaged, Mossell had proved his point: "Douglass Hospital [would] not accept subordinate management [nor be] forced, therefore, to help the larger and more powerful institution carry out schemes of race segregation." It was one thing for the black community to establish institutions as a response to racism. It was another for the white community to attempt to use those institutions to serve and perpetuate segregation.[65]

By 1920, ambiguities and conflicts remained over the role and mission of black hospitals, as illustrated in the controversy at Douglass Hospital and the subsequent establishment of Mercy Hospital. Were these hospitals to be democratic institutions that offered opportunities to all black physicians? Or were they to be elitist institutions that excluded physicians who did not meet particular, and sometimes arbitrary, standards? Certainly, the medical profession as a whole grappled with this tension between breadth of opportunity and the desire to raise standards. But the stakes were even higher for black physicians. Unless they secured appointment to a black facility, they usually had no hospital affiliation, and hence no firm anchor to the institutional foundation of modern medicine.

Other questions remained. Were black hospitals necessary or were they concessions to segregation? In the case of white-controlled institutions, were they to be open to black physicians at all? Whose expectations and interests should they meet? The conflict between Douglass Hospital and the University of Pennsylvania illustrates some of the dilemmas faced by black hospitals. Douglass Hospital had been established to provide black physicians and nurses with educational and professional opportunities. However, when a white institution attempted to use the hospital for that purpose, charges of Jim Crowism surfaced.

By 1920, black hospitals faced additional challenges. Their agendas would be greatly influenced by the changes that had taken place in medical practice and hospital care. Regardless of sponsorship or individual history, most black hospitals were ill-equipped small facilities lacking clinical training programs. Consequently, they were inadequately prepared to survive contemporary changes in scientific medicine, hospital technology, hospital standardization, and hospital accreditation. Of course, many other American hospitals faced a similar predicament, but the stakes were extraordinarily high for black institutions. In many communities they provided the only place in which black patients could be treated and black health care professionals could receive training and opportunities to practice. Compelled by community and professional needs, their options limited by racism, and haunted by unsolved questions about the role of black hospitals, black medical leaders went to work to improve their hospitals.

NOTES

1   Daniel Hale Williams, "The need of hospitals and training schools for colored people in the South," *National Hospital and Sanitarium Record*, April 1900, *3*: 4; John A. Kenney, *The Negro in Medicine* (Tuskegee, Ala., 1912), pp. 42–44; "Hospitals and nurse training schools," in *The Negro Year Book and Annual Encyclopedia of the Negro*, ed. Monroe N. Work (Tuskegee Institute, Ala.: Negro Yearbook Co., 1918/19), pp. 424–426.

2   Charles E. Rosenberg, *The Care of Strangers: The Rise of America's Health Care System* (New York: Basic Books, 1987); David Rosner, *A Once Charitable Enterprise: Hospitals and Health Care in Brooklyn and New York, 1885–1915* (Cambridge: Cambridge Univ. Press, 1982); Morris Vogel, *The Invention of the Modern Hospital: Boston, 1870–1930* (Chicago: Univ. of Chicago Press, 1980); Harry Dowling, *City Hospitals: The Undercare of*

the Underprivileged (Cambridge, Mass.: Harvard Univ. Press, 1982); and Joan E. Lynaugh, *The Community Hospitals of Kansas City, Missouri, 1870–1915* (New York: Garland Publishing, 1989).

3   James O. Breeden, ed., *Advice among Masters: The Ideal in Slave Management in the Old South* (Westport, Conn.: Greenwood Press, 1980), p. 183. Studies of plantation medicine include Weymouth T. Jordan, "Plantation medicine in the Old South," *Alabama Review*, 1950, *3*: 83–107; Elizabeth Barnaby Keeney, "Unless powerful sick: domestic medicine in the Old South," in *Science and Medicine in the Old South*, ed. Ronald L. Numbers and Todd L. Savitt (Baton Rouge: Louisiana State Univ. Press, 1989), pp. 276–294; William Dosite Postell, *The Health of Slaves on Southern Plantations* (1951; reprint, Gloucester, Mass.: Peter Smith, 1970); Todd L. Savitt, *Medicine and Slav-*

ery: *The Diseases and Health Care of Blacks in Antebellum Virginia* (Urbana: Univ. of Illinois Press, 1978); Richard Harrison Shryock, "Medical practice in the Old South," *South Atlantic Quart.*, April 1930, *29*: 160–178 [reprinted in Shryock, *Medicine in America: Historical Essays* (Baltimore: Johns Hopkins Univ. Press, 1966), pp. 49–70]; and Felice Swados, "Negro health on the antebellum plantation," *Bull. Hist. Med.*, 1941, *10*: 460–472.

4 Georgia infirmary," pamphlet celebrating its 100th anniversary, pp. 8–11, copy available at Tuskegee University Archives; "Infirmary for Negroes at Savannah, Geo.," *Charleston Med. J. & Rev.*, 1852, 7: 724; and Postell, *The Health of Slaves*, p. 140.

5 Charles E. Rosenberg, "And heal the sick: hospital and patient in nineteenth century America," *J. Soc. Hist.*, 1977, *10*: 432.

6 Paul Skeels Peirce, *The Freedmen's Bureau: A Chapter in the History of Reconstruction* (Iowa City: The University of Iowa, 1904), pp. 87–94; Todd L. Savitt, "Politics in medicine: the Georgia Freedmen's Bureau and the organization of health care, 1865–66," *Civil War History*, 1982, *28*: 45–64; Gail S. Hasson, "Health and welfare of freedmen in reconstruction Alabama," *Alabama Review*, 1982, *35*: 94–110; J. Thomas May, "The Louisiana Negro in transition: an appraisal of the medical activities of the Freedmen's Bureau," *Bulletin of the Tulane University Medical Faculty*, 1967, *26*: 29–36; Marshall Scott Legan, "Disease and the freedmen in Mississippi during Reconstruction," *J. Hist. Med.*, 1973, *28*: 257–267; Gaines M. Foster, "The limitations of federal health care for freedmen, 1862–1868," *J. Soc. Hist.*, 1982, *48*: 350–372; and Alan Raphael, "Health and social welfare of Kentucky black people," *Societas*, 1972, *2*: 143–157. For information on Freedmen's Hospital see W. Montague Cobb, "A short history of Freedmen's Hospital," *J. Nat. Med. Assn.*, 1962, *54*: 271–287; Thomas Holt, Cassandra Smith-Parker, and Rosalyn Terborg-Penn, *A Special Mission: The Story of Freedmen's Hospital, 1862–1962* (Washington, D.C.: Academic Affairs Division, Howard University, 1975); and William A. Warfield, "A brief history of Freedmen's Hospital," *Freedmen's Hospital Bulletin*, 1934, *1*: 1–2.

7 Howard N. Rabinowitz, *Race Relations in the Urban South 1865–1890* (Urbana: Univ. of Illinois Press, 1980), pp. 128–151. "Germs Have No Color Line" served as the slogan for a fund-raising campaign in the 1920s to create a center for black medical education at Chicago's Provident Hospital. Stuart Galishoff, "Germs know no color line: black health and public policy in Atlanta, 1900–1918," *J. Hist. Med.*, 1985, *40*: 29. For an extensive discussion of the "Negro health problem" see Vanessa Northington Gamble, ed., *Germs Have No Color Line: Blacks and American Medicine, 1900–1940* (New York: Garland Publishing, 1989). Julius Rosenwald Fund, compiled by Harrison L. Harris and Margaret L. Plumley, *Negro Hospitals: A Compilation of Available Statistics* (Chicago: Julius Rosenwald Fund, 1931), p. 15.

8 Ethel Johns, "A study of the present status of the Negro woman in nursing, 1925," Exhibit K (Atlanta), pp. K-7–8; Box 122, Record Group 1.1, Series 200C, Rockefeller Foundation Archives; Alice Mabel Bacon, "The Dixie and its work," *Southern Workman*, Nov. 1891, *20*: 244; Cora M. Folsom, "The Dixie in the beginning," *Southern Workman*, Mar. 1926, *55*: 121–126; Patricia A. Sloan, "Commitment to equality: a view of early black nursing schools," in *Historical Studies in Nursing*, ed. Louise Fitzpatrick (New York: Teachers College Press, 1978), p. 76; and Johns, "Negro woman in nursing," Exhibit H (Hampton, Virginia), pp. H-1–3.

9 Johns, "Negro woman in nursing," Exhibit J (Raleigh, North Carolina), pp. H-1–3; W. Montague Cobb, "St. Agnes Hospital, Raleigh, North Carolina, 1896–1961," *J. Nat. Med. Assn.*, 1961, *53*: 441–442.

10 Accounts of the founding of the hospital and biographical information on Unthank can be found in Clyde Reed Bradford, "History of Kansas City General Hospital, Colored Division," *Jackson County Medical Journal*, Oct. 8, 1932, *26*: 6–15; and in Samuel U. Rodgers, "Kansas City General Hospital, No. 2: a historical summary," *J. Nat. Med. Assn.*, 1962, *54*: 523–544.

11 *Annual Report of the Board of Hospital and Health for the Year Ending April 14, 1909* (Kansas City, Mo.), p. 83.

12 Bradford, "History of Kansas City General Hospital," p. 9.

13 Johns, "Negro woman in nursing," Exhibit C (Kansas City, Mo.), p. C-2.

14 W. Montague Cobb, *Medical Care and the Plight of the Negro* (New York: The National Association for the Advancement of Colored People, 1947), pp. 20–29. For histories on the development of the hospital see Homer G. Phillips Hospital, *The History and Development of Homer G. Phillips Hospital* (St. Louis: 1945); H. Phillip Venable, "The history of Homer G. Phillips Hospital," *J. Nat. Med. Assn.*, 1961, *53*: 541–551; and Frank O. Richards, "The St. Louis story: the training of black surgeons in St. Louis, Missouri," in *A Century of Black Surgeons: The U.S.A. Experience*, ed. Claude H. Organ, Jr., and Margaret Kosiba (Norman, Okla.: Transcript Press, 1987), pp. 197–247.

15 August Meier, *Negro Thought in America 1880–1915: Racial Ideologies in the Age of Booker T. Washington* (Ann Arbor: Univ. of Michigan Press, 1966); Herbert Shapiro, *White Violence and Black Response: From Reconstruction to Montgomery* (Amherst: Univ. of Massachusetts Press, 1988); Allan H. Spear, *Black Chicago: The*

Making of a Negro Ghetto, 1890–1920 (Chicago: Univ. of Chicago Press, 1967); and St. Clair Drake and Horace R. Cayton, *Black Metropolis*, rev. ed. (New York: Harper and Row, 1962). For discussions of the self-help tradition in the African-American community see Richard W. Thomas, "The historical roots of contemporary urban black self-help in the United States," in *Contemporary Urban America: Problems, Issues, and Alternatives*, ed. Marvel Lang (Lanham, Md.: Univ. Press of America, 1991), pp. 253–291; and Lenwood G. Davis, "The politics of black self-help in the United States: an historical overview," in *Black Organizations: Issues on Survival Techniques* (Washington, D.C.: Univ. Press of America, 1980), pp. 37–50.

16 Rosenberg, *The Care of Strangers*, p. 346. For analyses of the transformation of the American hospital and the rise of scientific medicine see George Rosen, "The impact of the hospital on the physician, the patient, and the community," *Hosp. Admin.*, Fall 1964, *9*: 15–33; Rosemary Stevens, *In Sickness and in Wealth* (New York: Basic Books, 1989); Paul Starr, *The Social Transformation of American Medicine* (New York: Basic Books, 1982); Rosenberg, *The Care of Strangers;* Rosner, *A Once Charitable Enterprise;* and Vogel, *The Invention of the Modern Hospital.*

17 For tables providing statistics on the number of black physicians see Todd L. Savitt, "Entering a white profession: black physicians in the new south, 1880–1920," *Bull. Hist. Med.*, 1987, *61*: 510–511.

18 W. E. B. DuBois, *The Philadelphia Negro: A Social Study* (1899; reprint, New York: Schocken, 1967), p. 113.

19 Darlene Clark Hine, *Black Women in White: Racial Conflict and Cooperation in the Nursing Profession, 1890–1950* (Bloomington: Indiana Univ. Press, 1989). For extensive discussions of the professionalization of nursing see Barbara Melosh, *The Physician's Hand: Work Culture and Conflict in American Nursing* (Philadelphia: Temple Univ. Press, 1982); and Susan M. Reverby, *Ordered to Care: The Dilemma of American Nursing, 1850–1945* (Cambridge: Cambridge Univ. Press, 1987).

20 "Proceedings of the Imhotep National Conference on Hospital Integration," *J. Nat. Med. Assn.*, 1957, *49*: 197; Eugene B. Elder, "The management of the race question in hospitals," *Tr. Am. Hosp. Assn.*, 1907, *9*: 128; and "Carson's private hospital, Washington, D.C.," *J. Nat. Med. Assn.*, 1930, *22*: 148.

21 H. M. Green, *A More or Less Critical Review of the Hospital Situation Among Negroes in the United States*, n.d. (circa 1930), pp. 4–5; and *Third Annual Report of the Frederick Douglass Memorial Hospital and Training School* (Philadelphia: The Hospital, 1898), p. 9. The annual reports of Douglass Hospital are available at The Library of the College of Physicians of Philadelphia. Thomas Wallace Swan, "Pennsylvania's memorial to

Frederick Douglass," *Howard's Magazine*, Oct. 1899, *4*: 6. For more information on medical experimentation on slaves see Savitt, *Medicine and Slavery*, pp. 281–307; Savitt, "The use of blacks for medical experimentation and demonstration in the Old South," *J. Southern Hist.*, 1982, *48*: 819–827; Diana E. Axelsen, "Women as victims of medical experimentation: J. Marion Sims' surgery on slave women, 1845–1850," *Sage*, Fall 1985, *2*: 10–13; David C. Humphrey, "Dissection and discrimination: the social origins of cadavers in America, 1760–1915," *Bull. N.Y. Acad. Med.*, 1973, *49*: 819–827; and F. N. Boney, "Slaves as guinea pigs: Georgia and Alabama episodes," *Alabama Review*, 1984, *37*: 45–51.

22 Rosenberg, *The Care of Strangers*, p. 112.

23 Meier, *Negro Thought;* and Louis R. Harlan, *Booker T. Washington: The Wizard of Tuskegee, 1901–1915* (New York: Oxford Univ. Press, 1983).

24 John A. Kenney, "Home infirmary, Clarksville, Tennessee" and "Fair Haven infirmary," in *The Negro in Medicine* (Tuskegee, Ala., 1912), pp. 47–48.

25 Accounts of the founding of Provident Hospital and biographical information on its founder can be found in Cassius Ellis, "Daniel Hale Williams, M.D., F.A.C.S.," in *A Century of Black Surgeons*, pp. 311–332; Helen Buckler, *Daniel Hale Williams: Negro Surgeon*, 2nd ed. (New York: Pitman, 1968); W. Montague Cobb, "Daniel Hale Williams, M.D., 1858–1931," *J. Nat. Med. Assn.*, 1953, *45*: 379–385; Ulysses Grant Dailey, "Daniel Hale Williams: pioneer surgeon and father of Negro hospitals" (Paper presented at the meeting of the National Hospital Association, Chicago, Ill., 18 Aug. 1941), Provident Medical Center, Chicago, Ill.; Henry B. Matthews, "Provident Hospital then and now," *J. Nat. Med. Assn.*, 1961, *53*: 209–224; Theresita Norris, "An historical account of Provident Hospital," Chicago Medical Society, unpublished, n.d. (circa 1940), State Historical Society of Wisconsin, Madison, Wis.; and Rayford W. Logan and Michael R. Winston, eds., *Dictionary of American Negro Biography* (New York: W. W. Norton, 1982), pp. 654–656. For a general history of Chicago medicine see Thomas Neville Bonner, *Medicine in Chicago, 1850–1950*, 2nd ed. (Urbana: Univ. of Illinois Press, 1991).

26 Susan Lynn Smith, "The Black Women's Club Movement: self-improvement and sisterhood, 1890–1915" (M.A. thesis, Univ. of Wisconsin–Madison, 1986), pp. 79–100; and Buckler, *Daniel Hale Williams*, pp. 70–73.

27 Williams, "Need of hospitals," p. 7.

28 *Ibid.*, p. 4.

29 Norris, "Provident Hospital," p. 3; Robert McMurdy to William C. Graves, 30 Dec. 1912, Rosenwald Papers, cited in Spear, *Black Chicago*, p. 98.

30 Williams did not perform the first successful open-

heart surgery, as has often been stated. His 1893 operation on the pericardium followed, by two years, that of the St. Louis physician H. C. Dalton. See C. Walton Lillehei, "Invited commentary," in *A Century of Black Surgeons*, pp. 332–334. McMurdy to Graves, 30 Dec. 1912, cited in Spear, *Black Chicago*, p. 98.

31 Norris, "An historical account," p. 5; and Spear, *Black Chicago*, pp. 99, 141. The classic study of black social development in Chicago is Drake and Cayton, *Black Metropolis*. See also Spear, *Black Chicago*.

32 Buckler, *Daniel Hale Williams*, pp. 175–176; and Spear, *Black Chicago*, pp. 72–73, 99–100. For biographical information on Hall see Spear, *Black Chicago*, pp. 72–73; Buckler, *Daniel Hale Williams*, p. 77; "Some Chicagoans of note," *Crisis*, Sept. 1915, *10*: 241; "Doctor George C. Hall," *J. Nat. Med. Assn.*, 1922, *14*: 216; and John W. Lawlah, "George Cleveland Hall, 1864–1930: a profile," *ibid.*, 1954, *46*: 207–210.

33 Spear, *Black Chicago*, pp. 51–89; and Charles Bentley to Julius Rosenwald, 16 Oct. 1917, Rosenwald Papers, cited in Spear, *Black Chicago*, p. 174.

34 Dailey, "Daniel Hale Williams," p. 2; and Lawlah, "George Cleveland Hall," p. 208.

35 Nathan Francis Mossell, "Biographical sketch," unpublished manuscript, n.d. (circa 1946), p. 7, Mossell Papers. When the author used these papers, the manuscripts were in the possession of Dr. Mossell's granddaughter, Mrs. Gertrude Cunningham. However, the papers have now been deposited in the University of Pennsylvania Archives. For additional information about Mossell and the founding of Douglass Hospital see W. Montague Cobb, "Nathan Francis Mossell, M.D., 1856–1946," *J. Nat. Med. Assn.*, 1954, *46*: 118–130; Edward S. Cooper, "Mercy-Douglass Hospital: historical perspective," *ibid.*, 1961, *53*: 1–7; Elliot M. Rudwick, "A brief history of Mercy-Douglass Hospital in Philadelphia," *Journal of Negro Education*, Winter 1951, *20*: 50–66; and Rayford W. Logan and Michael R. Winston, eds., *Dictionary of American Negro Biography* (New York: W. W. Norton, 1972), pp. 457–458.

36 Mossell, "Biographical sketch," p. 15.

37 *Ibid.*, p. 16; and Cobb, "Mossell," p. 122.

38 Cobb, "Mossell," p. 118.

39 *Seventh Annual Report of the Frederick Douglass Memorial Hospital and Training School* (Philadelphia: The Hospital, 1902), p. 12. The hospital's annual reports are available at the Library of the College of Physicians of Philadelphia. Swan, "Pennsylvania's memorial," p. 4; and Alfred Gordon, "Frederick Douglass Memorial Hospital and Training School," in *Philadelphia—World's Greatest Medical Centre* (Philadelphia, n.d.), p. 59. The group that met to organize the hospital included Rev. R. Heywood Stitt, pastor, Zion Wes-

ley AME Church; Jacob C. White, Jr., retired principal, Robert Vaux Grammar School; Rev. Matthew Anderson, pastor, Berean Presbyterian Church and founder of Berean Building and Loan Association; Rev. L. G. Jordan, pastor, Union Baptist Church; S. J. M. Brock, a leading black businessman; Dr. William A. Jackson, a dental surgeon; P. A. Dutreuille, caterer; Aaron A. Mossell, attorney and youngest brother of the physician; Mrs. Bishop B. T. Tanner, the widow of Bishop Tanner of the AME Church; Henry M. Minton, a pharmacist; and Alma G. Sommerville, a volunteer in many charitable causes.

40 Roger Lane, *William Dorsey's Philadelphia & Ours: On the Past and Future of the Black City in America* (New York: Oxford Univ. Press, 1991), p. 181.

41 Nathan Francis Mossell, "Frederick Douglass Memorial Hospital," unpublished manuscript, n.d. (circa 1946), p. 3, Mossell Papers.

42 DuBois, *The Philadelphia Negro*, p. 230; "A fair question," *The Weekly Tribune*, 7 Sept. 1895, n.p.; "That hospital," *The Weekly Tribune*, 21 Sept. 1895; and clipping from *Weekly Tribune*, n.d. All the newspaper articles are contained in scrapbooks owned by Dr. Mossell, and are among the Mossell Papers.

43 *Second Annual Report of the Frederick Douglass Memorial Hospital and Training School* (Philadelphia: The Hospital, 1897). The information on the early history of the hospital, except where noted, was obtained from the hospital's annual reports from 1896 to 1904, not including 1899, which were not in the collection at the College of Physicians of Philadelphia.

44 "A hospital for colored people," *Colored American* (Washington, D.C.), 7 Sept. 1895, n.p., Mossell Scrapbooks; and *Fifth Annual Report of the Frederick Douglass Memorial Hospital and Training School* (Philadelphia: The Hospital, 1900), p. 13.

45 *Fifth Annual Report*, p. 13.

46 *Eighth and Ninth Annual Report of the Frederick Douglass Memorial Hospital and Training School* (Philadelphia: The Hospital, 1903–1904), p. 16; and *Seventh Annual Report of the Frederick Douglass Memorial Hospital and Training School* (Philadelphia: The Hospital, 1902), p. 12.

47 *Eighth and Ninth Annual Report*, p. 16; *Seventh Annual Report*, p. 12; and *First Annual Report*, pp. 10–12.

48 Mossell, "Douglass Hospital," p. 15.

49 "Dr. N. F. Mossell, chief of staff asked to resign his place," *The Weekly Tribune*, 21 Jan. 1905, n.p., Mossell Scrapbooks, Mossell Papers.

50 *First Annual Report*, p. 14; Stevens, *In Sickness and in Wealth*, p. 53; and Cobb, "Mossell," p. 21.

51 "Friends of Douglass Hospital rallying to aid it," *Weekly Tribune*, Jan. 1903, n.p.; "Dr. N. F. Mossell, chief of staff asked to resign his place," *Weekly Tribune*, 21 Jan. 1905, n.p.; and "Dr. Mossell not responsible,"

Courant, 4 Feb. 1905, n.p., Mossell Scrapbooks, Mossell Papers.

52 Nathan Francis Mossell, "The modern hospital: its construction, organization, and management," *J. Nat. Med. Assn.,* 1909, *1*: 98.

53 Cobb, "Mossell," pp. 124–125; and "Colored physicians start a new hospital," *Weekly Tribune,* 31 Mar. 1906, n.p., Mossell Scrapbooks, Mossell Papers.

54 Harold E. Farmer, "An account of the earliest colored gentlemen in medical science in the United States," *Bull. Hist. Med.,* 1940, *8*: 615. After years of struggling independently, Mercy Hospital and Douglass Hospital merged in 1949 to form Mercy-Douglass Hospital.

55 Nathan Francis Mossell, "An institution that's doing a great job," *Christian Banner,* 13 Apr. 1906, n.p., Mossell Scrapbooks, Mossell Papers.

56 John A. Kenney, "Hospitals and nurse training schools, etc.," in his *The Negro in Medicine* (Tuskegee, Ala., 1912), pp. 42–44; Monroe N. Work, ed., "Hospitals and nurse training schools," in *The Negro Year Book and Annual Encyclopedia of the Negro* (Tuskegee Institute, Ala.: Negro Year Book Co., 1912), pp. 155–157, and 1921–1922, pp. 370–372. The exact number of black hospitals is unclear. Kenney listed 144 for 1922. Both authors listed hospitals and training schools without distinguishing between the institutions. Therefore, the estimate for the number of hospitals may be high.

57 "Editorial," *J. Nat. Med. Assn.,* Apr.–June 1909, *1*: 105.

58 Peter Marshall Murray, "Memoirs—clinic" (Mar. 1954), p. 1, Box 6, Murray Papers.

59 Todd L. Savitt, "Entering a white profession," p. 532.

60 For extensive discussion of hospital standardization see Edward T. Morman, ed., *Efficiency, Scientific Management, and Hospital Standardization* (New York: Garland Publishing, 1989); Stevens, *In Sickness and in Wealth,* pp. 52–79.

61 Rosemary Stevens, *American Medicine and the Public Interest* (New Haven: Yale Univ. Press, 1971), p. 118. For additional information on the Flexner Report and its impact on medical education see Kenneth M. Ludmerer, *Learning to Heal: The Development of American Medical Education* (New York: Basic Books, 1985); Barbara Barzansky and Norman Gevitz, eds., *Beyond Flexner: Medical Education in the Twentieth Century* (New York: Greenwood Press, 1992); Robert P. Hudson, "Abraham Flexner in perspective: American medical education, 1865–1910," *Bull. Hist. Med.,* 1972, *56*: 545–561 (ch. 12, this book); and Howard Berliner, "A larger perspective on the Flexner Report," *International Journal of Health Services,* 1975: 573–592.

62 Isabella Vanderwall, "Some problems of the colored woman physician," *Woman's Med. J.,* 1917, *27*: 156–158.

63 Roscoe C. Giles to Peter Marshall Murray, 1 Mar. 1931, Box 5, Murray Papers; W. Montague Cobb, "Roscoe Conkling Giles, M.D., F.A.C.S., F.I.C.S., 1890–1970," *J. Nat. Med. Assn.,* 1970, *62*: 254–256; Fitzhugh Mullan, *White Coat, Clenched Fist: The Political Education of an American Physician* (New York: Macmillan, 1976), pp. 117–121; and Johns, "The Negro woman in nursing," Exhibit K (Atlanta), pp. K-2–3.

64 Cobb, "Mossell," pp. 125–127; Gordon, "Frederick Douglass Memorial Hospital," p. 60; Mossell, "Douglass Hospital," pp. 17–26; and "Racial 'Jim Crow' inspired Douglass," *Philadelphia Tribune,* 8 Aug. 1936, p. 1.

65 Johns, "Negro woman in nursing," Exhibit E (Philadelphia), pp. E-8–9; and Mossell, "Douglass Hospital," p. 26. Mossell's comment on selling one's birthright for a mess of pottage is a biblical allusion. It refers to Esau's selling his birthright to his brother Jacob (Genesis, 25: 29–34).

24

Racism and Research: The Case of the Tuskegee Syphilis Study

ALLAN M. BRANDT

In 1932 the U.S. Public Health Service (USPHS) initiated an experiment in Macon County, Alabama, to determine the natural course of untreated, latent syphilis in black males. The test comprised 400 syphilitic men, as well as 200 uninfected men who served as controls. The first published report of the study appeared in 1936 with subsequent papers issued every four to six years, through the 1960s. When penicillin became widely available by the early 1950s as the preferred treatment for syphilis, the men did not receive therapy. In fact on several occasions, the USPHS actually sought to prevent treatment. Moreover, a committee at the federally operated Center for Disease Control decided in 1969 that the study should be continued. Only in 1972, when accounts of the study first appeared in the national press, did the Department of Health, Education and Welfare halt the experiment. At that time 74 of the test subjects were still alive; at least 28, but perhaps more than 100, had died directly from advanced syphilitic lesions.[1] In August 1972, HEW appointed an investigatory panel which issued a report the following year. The panel found the study to have been "ethically unjustified," and argued that penicillin should have been provided to the men.[2]

This article attempts to place the Tuskegee Study in a historical context and to assess its ethical implications. Despite the media attention which the study received, the HEW *Final Report*, and the criticism expressed by several professional organizations, the experiment has been largely misunderstood. The most basic questions of *how* the study

was undertaken in the first place and *why* it continued for 40 years were never addressed by the HEW investigation. Moreover, the panel misconstrued the nature of the experiment, failing to consult important documents available at the National Archives which bear significantly on its ethical assessment. Only by examining the specific ways in which values are engaged in scientific research can the study be understood.

RACISM AND MEDICAL OPINION

A brief review of the prevailing scientific thought regarding race and heredity in the early 20th century is fundamental for an understanding of the Tuskegee Study. By the turn of the century, Darwinism had provided a new rationale for American racism.[3] Essentially primitive peoples, it was argued, could not be assimilated into a complex, white civilization. Scientists speculated that in the struggle for survival the Negro in America was doomed. Particularly prone to disease, vice, and crime, black Americans could not be helped by education or philanthropy. Social Darwinists analyzed census data to predict the virtual extinction of the Negro in the 20th century, for they believed the Negro race in America was in the throes of a degenerative evolutionary process.[4]

The medical profession supported these findings of late 19th- and early 20th-century anthropologists, ethnologists, and biologists. Physicians studying the effects of emancipation on health concluded almost universally that freedom had caused the mental, moral, and physical deterioration of the black population.[5] They substantiated this argument by citing examples in the comparative anatomy of the black and white races. As Dr. W. T. English wrote: "A careful inspection reveals the

ALLAN M. BRANDT is Amalie Moses Kass Professor of the History of Medicine and Professor of the History of Science at Harvard University, Cambridge, Massachusetts.

Reprinted with permission from *Hastings Center Report*, 1978, *8* (No. 6): 21–29.

body of the negro a mass of minor defects and imperfections from the crown of the head to the soles
of the feet. . . ."[6] Cranial structures, wide nasal
apertures, receding chins, projecting jaws, all typed
the Negro as the lowest species in the Darwinian
hierarchy.[7]

Interest in racial differences centered on the sexual nature of blacks. The Negro, doctors explained,
possessed an excessive sexual desire, which threatened the very foundations of white society. As one
physician noted in the *Journal of the American Medical Association*, "The negro springs from a southern race, and as such his sexual appetite is strong;
all of his environments stimulate this appetite, and
as a general rule his emotional type of religion certainly does not decrease it."[8] Doctors reported a
complete lack of morality on the part of blacks:

> Virtue in the negro race is like angels' visits—
> few and far between. In a practice of sixteen years
> I have never examined a virgin negro over four-
> teen years of age.[9]

A particularly ominous feature of this overzealous
sexuality, doctors argued, was the black males' desire for white women. "A perversion from which
most races are exempt," wrote Dr. English,
"prompts the negro's inclination towards white
women, whereas other races incline towards females of their own."[10] Though English estimated
the "gray matter of the negro brain" to be at least
1,000 years behind that of the white races, his genital organs were overdeveloped. As Dr. William Lee
Howard noted:

> The attacks on defenseless white women are evi-
> dences of racial instincts that are about as amen-
> able to ethical culture as is the inherent odor of
> the race. . . . When education will reduce the size
> of the negro's penis as well as bring about the sen-
> sitiveness of the terminal fibers which exist in the
> Caucasian, then will it also be able to prevent the
> African's birthright to sexual madness and
> excess.[11]

One southern medical journal proposed "Castration Instead of Lynching," as retribution for black
sexual crimes. "An impressive trial by a ghost-like
kuklux klan [sic] and a 'ghost' physician or surgeon to perform the operation would make it an
event the 'patient' would never forget," noted the
editorial.[12]

According to these physicians, lust and immorality, unstable families, and reversion to barbaric
tendencies made blacks especially prone to venereal diseases. One doctor estimated that over 50
percent of all Negroes over the age of 25 were syphilitic.[13] Virtually free of disease as slaves, they were
now overwhelmed by it, according to informed
medical opinion. Moreover, doctors believed that
treatment for venereal disease among blacks was
impossible, particularly because in its latent stage
the symptoms of syphilis become quiescent. As Dr.
Thomas W. Murrell wrote:

> They come for treatment at the beginning and
> at the end. When there are visible manifestations
> or when harried by pain, they readily come, for
> as a race they are not averse to physic; but tell
> them not, though they look well and feel well, that
> they are still diseased. Here ignorance rates sci-
> ence a fool. . . .[14]

Even the best-educated black, according to Murrell, could not be convinced to seek treatment for
syphilis.[15] Venereal disease, according to some doctors, threatened the future of the race. The medical profession attributed the low birth rate among
blacks to the high prevalence of venereal disease
which caused stillbirths and miscarriages. Moreover, the high rates of syphilis were thought to lead
to increased insanity and crime. One doctor writing at the turn of the century estimated that the
number of insane Negroes had increased 13-fold
since the end of the Civil War.[16] Dr. Murrell's conclusion echoed the most informed anthropological and ethnological data:

> So the scourge sweeps among them. Those that
> are treated are only half cured, and the effort to
> assimilate a complex civilization driving their dis-
> eased minds until the results are criminal records.
> Perhaps here, in conjunction with tuberculosis,
> will be the end of the negro problem. Disease will
> accomplish what man cannot do.[17]

This particular configuration of ideas formed the
core of medical opinion concerning blacks, sex, and
disease in the early 20th century. Doctors generally discounted socioeconomic explanations of the
state of black health, arguing that better medical
care could not alter the evolutionary scheme.[18]
These assumptions provide the backdrop for examining the Tuskegee Syphilis Study.

THE ORIGINS OF THE EXPERIMENT

In 1929, under a grant from the Julius Rosenwald Fund, the USPHS conducted studies in the rural South to determine the prevalence of syphilis among blacks and explore the possibilities for mass treatment. The USPHS found Macon County, Alabama, in which the town of Tuskegee is located, to have the highest syphilis rate of the six counties surveyed. The Rosenwald Study concluded that mass treatment could be successfully implemented among rural blacks.[19] Although it is doubtful that the necessary funds would have been allocated even in the best economic conditions, after the economy collapsed in 1929, the findings were ignored. It is, however, ironic that the Tuskegee Study came to be based on findings of the Rosenwald Study that demonstrated the possibilities of mass treatment.

Three years later, in 1932, Dr. Taliaferro Clark, chief of the USPHS Venereal Disease Division and author of the Rosenwald Study report, decided that conditions in Macon County merited renewed attention. Clark believed the high prevalence of syphilis offered an "unusual opportunity" for observation. From its inception, the USPHS regarded the Tuskegee Study as a classic "study in nature,"* rather than an experiment.[20] As long as syphilis was so prevalent in Macon and most of the blacks went untreated throughout life, it seemed only natural to Clark that it would be valuable to observe the consequences. He described it as a "ready-made situation."[21] Surgeon General H. S. Cumming wrote to R. R. Moton, director of the Tuskegee Institute:

> The recent syphilis control demonstration carried out in Macon County, with the financial assistance of the Julius Rosenwald Fund, revealed the

*In 1865, Claude Bernard, the famous French physiologist, outlined the distinction between a "study in nature" and experimentation. A study in nature required simple observation, an essentially passive act, while experimentation demanded intervention which altered the original condition. The Tuskegee Study was thus clearly not a study in nature. The very act of diagnosis altered the original conditions. "It is on this very possibility of acting or not acting on a body," wrote Bernard, "that the distinction will exclusively rest between sciences called sciences of observation and sciences called experimental."

presence of an unusually high rate in this county and, what is more remarkable, the fact that 99 percent of this group was entirely without previous treatment. This combination, together with the expected cooperation of your hospital, offers an unparalleled opportunity for carrying on this piece of scientific research which probably cannot be duplicated anywhere else in the world.[22]

Although no formal protocol appears to have been written, several letters of Clark and Cumming suggest what the USPHS hoped to find. Clark indicated that it would be important to see how disease affected the daily lives of the men:

> The results of these studies of case records suggest the desirability of making a further study of the effect of untreated syphilis on the human economy among people now living and engaged in their daily pursuits.[23]

It also seems that the USPHS believed the experiment might demonstrate that antisyphilitic treatment was unnecessary. As Cumming noted: "It is expected the results of this study may have a marked bearing on the treatment, or conversely the non-necessity of treatment, of cases of latent syphilis."[24]

The immediate source of Cumming's hypothesis appears to have been the famous Oslo Study of untreated syphilis. Between 1890 and 1910, Professor C. Boeck, the chief of the Oslo Venereal Clinic, withheld treatment from almost 2,000 patients infected with syphilis. He was convinced that therapies then available, primarily mercurial ointment, were of no value. When arsenic therapy became widely available by 1910, after Paul Ehrlich's historic discovery of "606," the study was abandoned. E. Bruusgaard, Boeck's successor, conducted a follow-up study of 473 of the untreated patients from 1925 to 1927. He found that 27.9 percent of these patients had undergone a "spontaneous cure," and now manifested no symptoms of the disease. Moreover, he estimated that as many as 70 percent of all syphilitics went through life without inconvenience from the disease.[25] His study, however, clearly acknowledged the dangers of untreated syphilis for the remaining 30 percent.

Thus every major textbook of syphilis at the time of the Tuskegee Study's inception strongly advocated treating syphilis even in its latent stages, which follow the initial inflammatory reaction. In discussing the Oslo Study, Dr. J. E. Moore, one of the nation's leading venereologists wrote, "This summary of Bruusgaard's study is by no means intended to suggest that syphilis be allowed to pass untreated."[26] If a complete cure could not be effected, at least the most devastating effects of the disease could be avoided. Although the standard therapies of the time, arsenical compounds and bismuth injection, involved certain dangers because of their toxicity, the alternatives were much worse. As the Oslo Study had shown, untreated syphilis could lead to cardiovascular disease, insanity, and premature death.[27] Moore wrote in his 1933 textbook:

> Though it imposes a slight though measurable risk of its own, treatment markedly diminishes the risk from syphilis. In latent syphilis, as I shall show, the probability of progression, relapse, or death is reduced from a probable 25–30 percent without treatment to about 5 percent with it; and the gravity of the relapse if it occurs, is markedly diminished.[28]

"Another compelling reason for treatment," noted Moore, "exists in the fact that every patient with latent syphilis may be, and perhaps is, infectious for others."[29] In 1932, the year in which the Tuskegee Study began, the USPHS sponsored and published a paper by Moore and six other syphilis experts that strongly argued for treating latent syphilis.[30]

The Oslo Study, therefore, could not have provided justification for the USPHS to undertake a study that did not entail treatment. Rather, the suppositions that conditions in Tuskegee existed "naturally" and that the men would not be treated anyway provided the experiment's rationale. In turn, these two assumptions rested on the prevailing medical attitudes concerning blacks, sex, and disease. For example, Clark explained the prevalence of venereal disease in Macon County by emphasizing promiscuity among blacks:

> This state of affairs is due to the paucity of doctors, rather low intelligence of the Negro population in this section, depressed economic conditions, and the very common promiscuous sex relations of this population group which not only contribute to the spread of syphilis but also contribute to the prevailing indifference with regard to treatment.[31]

In fact, Moore, who had written so persuasively in favor of treating latent syphilis, suggested that existing knowledge did not apply to Negroes. Although he had called the Oslo Study "a never-to-be-repeated human experiment,"[32] he served as an expert consultant to the Tuskegee Study:

> I think that such a study as you have contemplated would be of immense value. It will be necessary of course in the consideration of the results to evaluate the special factors introduced by a selection of the material from negro males. Syphilis in the negro is in many respects almost a different disease from syphilis in the white.[33]

Dr. O. C. Wenger, chief of the federally operated venereal disease clinic at Hot Springs, Arkansas, praised Moore's judgment, adding, "This study will emphasize those differences."[34] On another occasion he advised Clark, "We must remember we are dealing with a group of people who are illiterate, have no conception of time, and whose personal history is always indefinite."[35]

The doctors who devised and directed the Tuskegee Study accepted the mainstream assumptions regarding blacks and venereal disease. The premise that blacks, promiscuous and lustful, would not seek or continue treatment, shaped the study. A test of untreated syphilis seemed "natural" because the USPHS presumed the men would never be treated; the Tuskegee Study made that a self-fulfilling prophecy.

SELECTING THE SUBJECTS

Clark sent Dr. Raymond Vonderlehr to Tuskegee in September 1932 to assemble a sample of men with latent syphilis for the experiment. The basic design of the study called for the selection of syphilitic black males between the ages of 25 and 60, a thorough physical examination including x-rays, and finally, a spinal tap to determine the incidence of neuro-syphilis.[36] They had no intention of providing any treatment for the infected men.[37] The USPHS originally scheduled the whole experiment to last six months; it seemed to be both a simple and inexpensive project.

The task of collecting the sample, however, proved to be more difficult than the USPHS had supposed. Vonderlehr canvassed the largely illiterate, poverty-stricken population of sharecroppers and tenant farmers in search of test subjects. If his circulars requested only men over 25 to attend his clinics, none would appear, suspecting he was conducting draft physicals. Therefore, he was forced to test large numbers of women and men who did not fit the experiment's specifications. This involved considerable expense, since the USPHS had promised the Macon County Board of Health that it would treat those who were infected, but not included in the study.[38] Clark wrote to Vonderlehr about the situation: "It never once occured to me that we would be called upon to treat a large part of the county as return for the privilege of making this study. . . . I am anxious to keep the expenditures for treatment down to the lowest possible point because it is the one item of expenditure in connection with the study most difficult to defend despite our knowledge of the need therefor."[39] Vonderlehr responded: "If we could find from 100 to 200 cases . . . we would not have to do another Wassermann on useless individuals. . . ."[40]

Significantly, the attempt to develop the sample contradicted the prediction the USPHS had made initially regarding the prevalence of the disease in Macon County. Overall rates of syphilis fell well below expectations; as opposed to the USPHS projection of 35 percent, 20 percent of those tested were actually diseased.[41] Moreover, those who had sought and received previous treatment far exceeded the expectations of the USPHS. Clark noted in a letter to Vonderlehr:

> I find your report of March 6th quite interesting but regret the necessity for Wassermanning [sic] . . . such a large number of individuals in order to uncover this relatively limited number of untreated cases.[42]

Further difficulties arose in enlisting the subjects to participate in the experiment, to be "Wassermanned," and to return for a subsequent series of examinations. Vonderlehr found that only the offer of treatment elicited the cooperation of the men. They were told they were ill and were promised free care. Offered therapy, they became willing subjects.[43] The USPHS did not tell the men that they were participants in an experiment; on the contrary, the subjects believed they were being treated for "bad blood"—the rural South's colloquialism for syphilis. They thought they were participating in a public health demonstration similar to the one

that had been conducted by the Julius Rosenwald Fund in Tuskegee several years earlier. In the end, the men were so eager for medical care that the number of defaulters in the experiment proved to be insignificant.[44]

To preserve the subjects' interest, Vonderlehr gave most of the men mercurial ointment, a noneffective drug, while some of the younger men apparently received inadequate dosages of neoarsphenamine.[45] This required Vonderlehr to write frequently to Clark requesting supplies. He feared the experiment would fail if the men were not offered treatment.

> It is desirable and essential if the study is to be a success to maintain the interest of each of the cases examined by me through to the time when the spinal puncture can be completed. Expenditure of several hundred dollars for drugs for these men would be well worth while if their interest and cooperation would be maintained in so doing. . . . It is my desire to keep the main purpose of the work from the negroes in the county and continue their interest in treatment. That is what the vast majority wants and the examination seems relatively unimportant to them in comparison. It would probably cause the entire experiment to collapse if the clinics were stopped before the work is completed.[46]

On another occasion he explained:

> Dozens of patients have been sent away without treatment during the past two weeks and it would have been impossible to continue without the free distribution of drugs because of the unfavorable impression made on the negro.[47]

The readiness of the test subjects to participate, of course, contradicted the notion that blacks would not seek or continue therapy.

The final procedure of the experiment was to be a spinal tap to test for evidence of neurosyphilis. The USPHS presented this purely diagnostic exam, which often entails considerable pain and complications, to the men as a "special treatment." Clark explained to Moore:

> We have not yet commenced the spinal punctures. This operation will be deferred to the last in order not to unduly disturb our field work by any adverse reports by the patients subjected to spinal puncture because of some disagreeable sensations following this procedure. These negroes are very

ignorant and easily influenced by things that would be of minor significance in a more intelligent group.[48]

The letter to the subjects announcing the spinal tap read:

> Some time ago you were given a thorough examination and since that time we hope you have gotten a great deal of treatment for bad blood. You will now be given your last chance to get a second examination. This examination is a very special one and after it is finished you will be given a special treatment if it is believed you are in a condition to stand it. . . .
>
> REMEMBER THIS IS YOUR LAST CHANCE FOR SPECIAL FREE TREATMENT. BE SURE TO MEET THE NURSE.[49]

The HEW investigation did not uncover this crucial fact: the men participated in the study under the guise of treatment.

Despite the fact that their assumption regarding prevalence and black attitudes toward treatment had proved wrong, the USPHS decided in the summer of 1933 to continue the study. Once again, it seemed only "natural" to pursue the research since the sample already existed, and with a depressed economy, the cost of treatment appeared prohibitive—although there is no indication it was ever considered. Vonderlehr first suggested extending the study in letters to Clark and Wenger:

> At the end of this project we shall have a considerable number of cases presenting various complications of syphilis, who have received only mercury and may still be considered untreated in the modern sense of therapy. Should these cases be followed over a period of from five to ten years many interesting facts could be learned regarding the course and complications of untreated syphilis.[50]

"As I see it," responded Wenger, "we have no further interest in these patients *until they die.*"[51] Apparently, the physicians engaged in the experiment believed that only autopsies could scientifically confirm the findings of the study. Surgeon General Cumming explained this in a letter to R. R. Moton, requesting the continued cooperation of the Tuskegee Institute Hospital:

> This study which was predominantly clinical in character points to the frequent occurrence of severe complications involving the various vital organs of the body and indicates that syphilis as a disease does a great deal of damage. Since clinical observations are not considered final in the medical world, it is our desire to continue observation on the cases selected for the recent study and if possible to bring a percentage of these cases to autopsy so that pathological confirmation may be made of the disease processes.[52]

Bringing the men to autopsy required the USPHS to devise a further series of deceptions and inducements. Wenger warned Vonderlehr that the men must not realize that they would be autopsied:

> There is one danger in the latter plan and that is if the colored population become aware that accepting free hospital care means a post-mortem, every darkey will leave Macon County and it will hurt [Dr. Eugene] Dibble's hospital.[53]

"Naturally," responded Vonderlehr, "It is not my intention to let it be generally known that the main object of the present activities is the bringing of the men to necropsy."[54] The subjects' trust in the USPHS made the plan viable. The USPHS gave Dr. Dibble, the director of the Tuskegee Institute Hospital, an interim appointment to the Public Health Service. As Wenger noted:

> One thing is certain. The only way we are going to get post-mortems is to have the demise take place in Dibble's hospital and when these colored folks are told that Doctor Dibble is now a Government doctor too they will have more confidence.[55]*

---

*The degree of black cooperation in conducting the study remains unclear and would be impossible to properly assess in an article of this length. It seems certain that some members of the Tuskegee Institute staff such as R. R. Moton and Eugene Dibble understood the nature of the experiment and gave their support to it. There is, however, evidence that some blacks who assisted the USPHS physicians were not aware of the deceptive nature of the experiment. Dr. Joshua Williams, an intern at the John A. Andrew Memorial Hospital (Tuskegee Institute) in 1932, assisted Vonderlehr in taking blood samples of the test subjects. In 1973 he told the HEW panel: "I know we thought it was merely a service group organized to help the people in the area. We didn't know it was a research project at all at the time." (See "Transcript of proceedings," Tuskegee Syphilis Study Ad Hoc Advisory Panel, Feb. 23, 1973, unpublished typescript,

After the USPHS approved the continuation of the experiment in 1933, Vonderlehr decided that it would be necessary to select a group of healthy, uninfected men to serve as controls. Vonderlehr, who had succeeded Clark as chief of the Venereal Disease Division, sent Dr. J. R. Heller to Tuskegee to gather the control group. Heller distributed drugs (noneffective) to these men, which suggests that they also believed they were undergoing treatment.[56] Control subjects who became syphilitic were simply transferred to the test group — a strikingly inept violation of standard research procedure.[57]

The USPHS offered several inducements to maintain contact and to procure the continued cooperation of the men. Eunice Rivers, a black nurse, was hired to follow their health and to secure approval for autopsies. She gave the men noneffective medicines — "spring tonic" and aspirin — as well as transportation and hot meals on the days of their examinations.[58] More important, Nurse Rivers provided continuity to the project over the entire 40-year period. By supplying "medicinals," the USPHS was able to continue to deceive the participants, who believed that they were receiving therapy from the government doctors. Deceit was integral to the study. When the test subjects complained about spinal taps one doctor wrote:

> They simply do not like spinal punctures. A few of those who were tapped are enthusiastic over the results but to most, the suggestion causes violent shaking of the head; others claim they were robbed of their procreative powers (regardless of the fact that I claim it stimulates them).[59]

Letters to the subjects announcing an impending USPHS visit to Tuskegee explained: "[The doctor] wants to make a special examination to find out how you have been feeling and whether the treatment has improved your health."[60] In fact, after the first six months of the study, the USPHS had furnished no treatment whatsoever.

National Library of Medicine, Bethesda, Maryland.) It is also apparent that Eunice Rivers, the black nurse who had primary responsibility for maintaining contact with the men over the 40 years, did not fully understand the dangers of the experiment. In any event, black involvement in the study in no way mitigates the racial assumptions of the experiment, but rather, demonstrates their power.

Finally, because it proved difficult to persuade the men to come to the hospital when they became severely ill, the USPHS promised to cover their burial expenses. The Milbank Memorial Fund provided approximately $50 per man for this purpose beginning in 1935. This was a particularly strong inducement as funeral rites constituted an important component of the cultural life of rural blacks.[61] One report of the study concluded, "Without this suasion it would, we believe, have been impossible to secure the cooperation of the group and their families."[62]

Reports of the study's findings, which appeared regularly in the medical press beginning in 1936, consistently cited the ravages of untreated syphilis. The first paper, read at the 1936 American Medical Association annual meeting, found "that syphilis in this period [latency] tends to greatly increase the frequency of manifestations of cardiovascular disease."[63] Only 16 percent of the subjects gave no sign of morbidity as opposed to 61 percent of the controls. Ten years later, a report noted coldly, "The fact that nearly twice as large a proportion of the syphilitic individuals as of the control group has died is a very striking one." Life expectancy, concluded the doctors, is reduced by about 20 percent.[64]

A 1955 article found that slightly more than 30 percent of the test group autopsied had died *directly* from advanced syphilitic lesions of either the cardiovascular or the central nervous system.[65] Another published account stated, "Review of those still living reveals that an appreciable number have late complications of syphilis which probably will result, for some at least, in contributing materially to the ultimate cause of death."[66] In 1950, Dr. Wenger had concluded, "We now know, where we could only surmise before, that we have contributed to their ailments and shortened their lives."[67] As black physician Vernal Cave, a member of the HEW panel, later wrote, "They proved a point, then proved a point, then proved a point."[68]

During the 40 years of the experiment the USPHS had sought on several occasions to ensure that the subjects did not receive treatment from other sources. To this end, Vonderlehr met with groups of local black doctors in 1934, to ask their cooperation in not treating the men. Lists of subjects were distributed to Macon County physicians along with letters requesting them to refer these men back to the USPHS if they sought care.[69] The USPHS warned the Alabama Health Department not to treat the test subjects when they took a mobile VD unit into Tuskegee in the early 1940s.[70] In 1941, the army drafted several subjects and told them to begin antisyphilitic treatment immediately. The USPHS supplied the draft board with a list of 256 names they desired to have excluded from treatment, and the board complied.[71]

In spite of these efforts, by the early 1950s many of the men had secured some treatment on their own. By 1952, almost 30 percent of the test subjects had received some penicillin, although only 7.5 percent had received what could be considered adequate doses.[72] Vonderlehr wrote to one of the participating physicians, "I hope that the availability of antibiotics has not interfered too much with this project."[73] A report published in 1955 considered whether the treatment that some of the men had obtained had "defeated" the study. The article attempted to explain the relatively low exposure to penicillin in an age of antibiotics, suggesting as a reason: "the stoicism of these men as a group; they still regard hospitals and medicines with suspicion and prefer an occasional dose of time-honored herbs or tonics to modern drugs."[74] The authors failed to note that the men believed they already were under the care of the government doctors and thus saw no need to seek treatment elsewhere. Any treatment which the men might have received, concluded the report, had been insufficient to compromise the experiment.

When the USPHS evaluated the status of the study in the 1960s they continued to rationalize the racial aspects of the experiment. For example, the minutes of a 1965 meeting at the Center for Disease Control recorded:

> Racial issue was mentioned briefly. Will not affect the study. Any questions can be handled by saying these people were at the point that therapy would no longer help them. They are getting better medical care than they would under any other circumstances.[75]

A group of physicians met again at the CDC in 1969 to decide whether or not to terminate the study. Although one doctor argued that the study should be stopped and the men treated, the consensus was to continue. Dr. J. Lawton Smith remarked, "You will never have another study like

this; take advantage of it."[76] A memo prepared by Dr. James B. Lucas, assistant chief of the Venereal Disease Branch stated: "Nothing learned will prevent, find, or cure a single case of infectious syphilis or bring us closer to our basic mission of controlling venereal disease in the United States."[77] He concluded, however, that the study should be continued "along its present lines." When the first accounts of the experiment appeared in the national press in July 1972, data were still being collected and autopsies performed.[78]

THE HEW FINAL REPORT

HEW finally formed the Tuskegee Syphilis Study Ad Hoc Advisory Panel on August 28, 1972, in response to criticism that the press descriptions of the experiment had triggered. The panel, composed of nine members, five of them black, concentrated on two issues. First, was the study justified in 1932 and had the men given their informed consent? Second, should penicillin have been provided when it became available in the early 1950s? The panel was also charged with determining if the study should be terminated and assessing current policies regarding experimentation with human subjects.[79] The group issued their report in June 1973.

By focusing on the issues of penicillin therapy and informed consent, the *Final Report* and the investigation betrayed a basic misunderstanding of the experiment's purposes and design. The HEW report implied that the failure to provide penicillin constituted the study's major ethical misjudgment; implicit was the assumption that no adequate therapy existed prior to penicillin. Nonetheless medical authorities firmly believed in the efficacy of arsenotherapy for treating syphilis at the time of the experiment's inception in 1932. The panel further failed to recognize that the entire study had been predicated on nontreatment. Provision of effective medication would have violated the rationale of the experiment—to study the natural course of the disease until death. On several occasions, in fact, the USPHS had prevented the men from receiving proper treatment. Indeed, there is no evidence that the USPHS ever considered providing penicillin.

The other focus of the *Final Report*—informed consent—also served to obscure the historical facts of the experiment. In light of the deceptions and exploitations which the experiment perpetrated, it is an understatement to declare, as the *Report* did, that the experiment was "ethically unjustified," because it failed to obtain informed consent from the subjects. The *Final Report*'s statement, "Submitting voluntarily is not informed consent," indicated that the panel believed that the men had volunteered *for the experiment*.[80] The records in the National Archives make clear that the men did not submit voluntarily to an experiment; they were told and they believed that they were getting free treatment from expert government doctors for a serious disease. The failure of the HEW *Final Report* to expose this critical fact — that the USPHS lied to the subjects — calls into question the thoroughness and credibility of their investigation.

Failure to place the study in a historical context also made it impossible for the investigation to deal with the essentially racist nature of the experiment. The panel treated the study as an aberration, well intentioned but misguided.[81] Moreover, concern that the *Final Report* might be viewed as a critique of human experimentation in general seems to have severely limited the scope of the inquiry. The *Final Report* is quick to remind the reader on two occasions: "The position of the Panel must not be construed to be a general repudiation of scientific research with human subjects."[82] The *Report* assures

us that a better designed experiment could have been justified:

> It is possible that a scientific study in 1932 of untreated syphilis, properly conceived with a clear protocol and conducted with suitable subjects who fully understood the implications of their involvement, might have been justified in the prepenicillin era. This is especially true when one considers the uncertain nature of the results of treatment of late latent syphilis and the highly toxic nature of therapeutic agents then available.[83]

This statement is questionable in view of the proven dangers of untreated syphilis known in 1932.

Since the publication of the HEW *Final Report*, a defense of the Tuskegee Study has emerged. These arguments, most clearly articulated by Dr. R. H. Kampmeier in the *Southern Medical Journal*, center on the limited knowledge of effective therapy for latent syphilis when the experiment began. Kampmeier argues that by 1950, penicillin would have been of no value for these men.[84] Others have suggested that the men were fortunate to have been spared the highly toxic treatments of the earlier period.[85] Moreover, even these contemporary defenses assume that the men never would have been treated anyway. As Dr. Charles Barnett of Stanford University wrote in 1974, "The lack of treat-

---

Claude Bernard on Human Experimentation (1865)
Experiments, then, may be performed on man, but within what limits? It is our duty and our right to perform an experiment on man whenever it can save his life, cure him or gain him some personal benefit. The principle of medical and surgical morality, therefore, consists in never performing on man an experiment which might be harmful to him to any extent, even though the result might be highly advantageous to science, i.e., to the health of others. But performing experiments and operations exclusively from the point of view of the patient's own advantage does not prevent their turning out profitably to science. . . . For we must not deceive ourselves, morals do not forbid making experiments on one's neighbor or on one's self. Christian morals forbid only one thing, doing ill to one's neighbor. So, among the experiments that may be tried on man, those that can only harm are forbidden, those that are innocent are permissible, and those that may do good are obligatory. Claude Bernard, *An Introduction to the Study of Experimental Medicine* (1865). Trans. Henry C. Green (New York: Dover Publications, 1957).

From the HEW Final Report (1973)
1. In retrospect, the Public Health Service Study of Untreated Syphilis in the Male Negro in Macon County, Alabama, was ethically unjustified in 1932. This judgment made in 1973 about the conduct of the study in 1932 is made with the advantage of hindsight acutely sharpened over some forty years, concerning an activity in a different age with different social standards. Nevertheless, one fundamental ethical rule is that a person should not be subjected to avoidable risk of death or physical harm unless he freely and intelligently consents. There is no evidence that such consent was obtained from the participants in this study.

2. Because of the paucity of information available today on the manner in which the study was conceived, designed and sustained, a scientific justification for a short term demonstration study cannot be ruled out. However, the conduct of the longitudinal study as initially reported in 1936 and through the years is judged to be scientifically unsound and its results are disproportionately meager compared with known risks to human subjects involved. . . .

ment was not contrived by the USPHS but was an established fact of which they proposed to take advantage."[86] Several doctors who participated in the study continued to justify the experiment. Dr. J. R. Heller, who on one occasion had referred to the test subjects as the "Ethiopian population," told reporters in 1972:

> I don't see why they should be shocked or horrified. There was no racial side to this. It just happened to be in a black community. I feel this was a perfectly straightforward study, perfectly ethical, with controls. Part of our mission as physicians is to find out what happens to individuals with disease and without disease.[87]

These apologies, as well as the HEW *Final Report*, ignore many of the essential ethical issues which the study poses. The Tuskegee Study reveals the persistence of beliefs within the medical profession about the nature of blacks, sex, and disease — beliefs that had tragic repercussions long after their alleged "scientific" bases were known to be incorrect. Most strikingly, the entire health of a community was jeopardized by leaving a communicable disease untreated.[88] There can be little doubt that the Tuskegee researchers regarded their subjects as less than human.[89] As a result, the ethical canons of experimenting on human subjects were completely disregarded.

The study also raises significant questions about professional self-regulation and scientific bureaucracy. Once the USPHS decided to extend the experiment in the summer of 1933, it was unlikely that the test would be halted short of the men's deaths. The experiment was widely reported for 40 years without evoking any significant protest within the medical community. Nor did any bureaucratic mechanism exist within the government for the periodic reassessment of the Tuskegee experiment's ethics and scientific value. The USPHS sent physicians to Tuskegee every several years to check on the study's progress, but never subjected the morality or usefulness of the experiment to serious scrutiny. Only the press accounts of 1972 finally punctured the continued rationalizations of the USPHS and brought the study to an end. Even the HEW investigation was compromised by fear that it would be considered a threat to future human experimentation.

In retrospect the Tuskegee Study revealed more about the pathology of racism than it did about the pathology of syphilis; more about the nature of scientific inquiry than the nature of the disease process. The injustice committed by the experiment went well beyond the facts outlined in the press and the HEW *Final Report*. The degree of deception and damages have been seriously underestimated. As this history of the study suggests, the notion that science is a value-free discipline must be rejected. The need for greater vigilance in assessing the specific ways in which social values and attitudes affect professional behavior is clearly indicated.[90]

NOTES

1 The best general accounts of the study are "The 40-year death watch," *Medical World News*, Aug. 18, 1972: 15–17; and Dolores Katz, "Why 430 blacks with syphilis went uncured for 40 years," Detroit *Free Press*, Nov. 5, 1972. The mortality figure is based on a published report of the study which appeared in 1955. See Jesse J. Peters, James H. Peers, Sidney Olansky, John C. Cutler, and Geraldine Gleeson, "Untreated syphilis in the male Negro: pathologic findings in syphilitic and nonsyphilitic patients," *J. Chron. Dis.*, Feb. 1955, *1*: 127–148. The article estimated that 30.4 percent of the untreated men would die from syphilitic lesions.

2 *Final Report* of the Tuskegee Syphilis Study Ad Hoc Advisory Panel, Department of Health, Education,

and Welfare (Washington, D.C.: Government Printing Office, 1973). (Hereafter, HEW *Final Report*).

3 See George M. Fredrickson, *The Black Image in the White Mind* (New York: Harper and Row, 1971), pp. 228–255. Also, John H. Haller, *Outcasts From Evolution* (Urbana: Univ. of Illinois Press, 1971), pp. 40–68.

4 Frederickson, *The Black Image*, pp. 247–249.

5 "Deterioration of the American Negro," *Atlanta J.-Rec. Med.*, July 1903, *5*: 287–288. See also J. A. Rodgers, "The effect of freedom upon the psychological development of the Negro," *Proc. Am. Medico-Psychological Assn.*, 1900, *7*: 88–99. "From the most healthy race in the country forty years ago," concluded Dr. Henry McHatton, "he is today the most

diseased." "The sexual status of the Negro—past and present," *Am. J. Dermatology & Genito-Urinary Dis.,* Jan. 1906, *10:* 7–9.

6 W. T. English, "The Negro problem from the physician's point of view," *Atlanta J.-Rec. Med.,* Oct. 1903, *5:* 461. See also, "Racial anatomical peculiarities," *N.Y. Med. J.,* Apr. 1896, *63:* 500–501.

7 "Racial anatomical peculiarities," p. 501. Also, Charles S. Bacon, "The race problem," *Medicine* (Detroit), May 1903, *9:* 338–343.

8 H. H. Hazen, "Syphilis in the American Negro," *J.A.M.A.,* Aug. 8, 1914, *63:* 463. For deeper background into the historical relationship of racism and sexuality, see Winthrop D. Jordan, *White Over Black* (Chapel Hill: Univ. of North Carolina Press, 1968; Pelican Books, 1969), pp. 32–40.

9 Daniel David Quillian, "Racial peculiarities: a cause of the prevalence of syphilis in Negroes," *Am. J. Dermatology & Genito-Urinary Dis.,* July 1906, *10:* 277.

10 English, "The Negro problem . . . ," p. 463.

11 William Lee Howard, "The Negro as a distinct ethnic factor in civilization," *Medicine* (Detroit), June 1903, *9:* 424. See also, Thomas W. Murrell, "Syphilis in the American Negro," *J.A.M.A.,* Mar. 12, 1910, *54:* 848.

12 "Castration instead of lynching," *Atlanta J.-Rec. Med.,* Oct. 1906, *8:* 457. The editorial added: "The badge of disgrace and emasculation might be branded upon the face or forehead, as a warning, in the form of an 'R,' emblematic of the crime for which this punishment was and will be inflicted."

13 Searle Harris, "The future of the Negro from the standpoint of the southern physician," *Alabama Med. J.,* Jan. 1902, *14:* 62. Other articles on the prevalence of venereal disease among blacks are: H. L. McNeil, "Syphilis in the southern Negro," *J.A.M.A.,* Sept. 30, 1916, *67:* 1001–1004; Ernest Philip Boas, "The relative prevalence of syphilis among Negroes and whites," *Social Hygiene,* Sept. 1915, *1:* 610–616. Doctors went to considerable trouble to distinguish the morbidity and mortality of various diseases among blacks and whites. See, for example, Marion M. Torchia, "Tuberculosis among American Negroes: medical research on a racial disease, 1830–1950," *J. Hist. Med.,* July 1977, *32:* 252–279.

14 Thomas W. Murrell, "Syphilis in the Negro: its bearing on the race problem," *Am. J. Dermatology & Genito-Urinary Dis.,* Aug. 1906, *10:* 307.

15 "Even among the educated, only a very few will carry out the most elementary instructions as to personal hygiene. One thing you cannot do, and that is to convince the negro that he has a disease that he cannot see or feel. This is due to lack of concentration rather than lack of faith; even if he

does believe, he does not care; a child of fancy, the sensations of the passing hour are his only guides to the future." Murrell, "Syphilis in the American Negro," p. 847.

16 "Deterioration of the American Negro," *Atlanta J.-Rec. Med.,* July 1903, *5:* 288.

17 Murrell, "Syphilis in the Negro; its bearing on the race problem," p. 307.

18 "The anatomical and physiological conditions of the African must be understood, his place in the anthropological scale realized, and his biological basis accepted as being unchangeable by man, before we shall be able to govern his natural uncontrollable sexual passions." See, "As ye sow that shall ye also reap," *Atlanta J.-Rec. Med.,* June 1899, *1:* 266.

19 Taliaferro Clark, *The Control of Syphilis in Southern Rural Areas* (Chicago: Julius Rosenwald Fund, 1932), pp. 53–58. Approximately 35 percent of the inhabitants of Macon County who were examined were found to be syphilitic.

20 See Claude Bernard, *An Introduction to the Study of Experimental Medicine* (New York: Dover, 1865, 1957), pp. 5–26.

21 Taliaferro Clark to M. M. Davis, Oct. 29, 1932. Records of the USPHS Venereal Disease Division, Record Group 90, Box 239, National Archives, Washington National Record Center, Suitland, Maryland. (Hereafter, NA-WNRC). Materials in this collection which relate to the early history of the study were apparently never consulted by the HEW investigation. Included are letters, reports, and memoranda written by the physicians engaged in the study.

22 H. S. Cumming to R. R. Moton, Sept. 20, 1932, NA-WNRC.

23 Clark to Davis, Oct. 29, 1932, NA-WNRC.

24 Cumming to Moton, Sept. 20, 1932, NA-WNRC.

25 Bruusgaard was able to locate 309 living patients, as well as records from 164 who were deceased. His findings were published as "Ueber das Schicksal der nicht specifizch behandelten Luetiken," *Arch. Dermatology & Syphilis,* 1929, *157:* 309–332. The best discussion of the Boeck-Bruusgaard data is E. Gurney Clark and Niels Danbolt, "The Oslo Study of the natural history of untreated syphilis," *J. Chron. Dis.,* Sept. 1955, *2:* 311–344.

26 Joseph Earle Moore, *The Modern Treatment of Syphilis* (Baltimore: Charles C. Thomas, 1933), p. 24.

27 *Ibid.,* pp. 231–247; see also John H. Stokes, *Modern Clinical Syphilology* (Philadelphia: W. B. Saunders, 1928), pp. 231–239.

28 Moore, *Modern Treatment of Syphilis,* p. 237.

29 *Ibid.,* p. 236.

30 J. E. Moore, H. N. Cole, P. A. O'Leary, J. H.

Stokes, U. J. Wile, T. Clark, T. Parran, J. H. Usilton, "Cooperative clinical studies in the treatment of syphilis: latent syphilis," *Venereal Disease Information,* (Sept. 20, 1932), *13*: 351. The authors also concluded that the latently syphilitic were potential carriers of the disease, thus meriting treatment.

31 Clark to Paul A. O'Leary, Sept. 27, 1932, NA-WNRC. O'Leary, of the Mayo Clinic, misunderstood the design of the study, replying: "The investigation which you are planning in Alabama is indeed an intriguing one, particularly because of the opportunity it affords of observing treatment in a previously untreated group. I assure you such a study is of interest to me, and I shall look forward to its report in the future." O'Leary to Clark, Oct. 3, 1932, NA-WNRC.

32 Joseph Earle Moore, "Latent syphilis," unpublished typescript, n.d., p. 7. American Social Hygiene Association Papers, Social Welfare History Archives Center, University of Minnesota, Minneapolis.

33 Moore to Clark, Sept. 28, 1932, NA-WNRC. Moore had written in his textbook, "In late syphilis the negro is particularly prone to the development of bone or cardiovascular lesions." See Moore, *The Modern Treatment of Syphilis,* p. 35.

34 O. C. Wenger to Clark, Oct. 3, 1932, NA-WNRC.

35 *Ibid.,* Sept. 29, 1932.

36 Clark memorandum, Sept. 26, 1932, NA-WNRC. See also, Clark to Davis, Oct. 29, 1932, NA-WNRC.

37 As Clark wrote: "You will observe that our plan has nothing to do with treatment. It is purely a diagnostic procedure carried out to determine what has happened to the syphilitic Negro who has had no treatment." Clark to Paul A. O'Leary, Sept. 27, 1932, NA-WNRC.

38 D. G. Gill to O. C. Wenger, Oct. 10, 1932, NA-WNRC.

39 Clark to Vonderlehr, Jan. 25, 1933, NA-WNRC.

40 Vonderlehr to Clark, Feb. 28, 1933, NA-WNRC.

41 *Ibid.,* Nov. 2, 1932. Also, *ibid.,* Feb. 6, 1933.

42 Clark to Vonderlehr, Mar. 9, 1933, NA-WNRC.

43 Vonderlehr later explained: "The reason treatment was given to many of these men was twofold: First, when the study was started in the fall of 1932, no plans had been made for its continuation and a few of the patients were treated before we fully realized the need for continuing the project on a permanent basis. Second it was difficult to hold the interest of the group of Negroes in Macon County unless some treatment was given." Vonderlehr to Austin V. Diebert, Dec. 5, 1938, Tuskegee Syphilis Study Ad Hoc Advisory Panel Papers, Box 1, National Library of Medicine, Bethesda, Maryland. (Hereafter, TSS-NLM.) This collection contains the materials assembled by the HEW investigation in 1972.

44 Vonderlehr to Clark, Feb. 6, 1933, NA-WNRC.

45 H. S. Cumming to J. N. Baker, Aug. 5, 1933, NA-WNRC.

46 Vonderlehr to Clark, Jan. 22, 1933; Jan. 12, 1933, NA-WNRC.

47 Vonderlehr to Clark, Jan. 28, 1933, NA-WNRC.

48 Clark to Moore, Mar. 25, 1933, NA-WNRC.

49 Macon County Health Department, "Letter to subjects," n.d., NA-WNRC.

50 Vonderlehr to Clark, Apr. 8, 1933, NA-WNRC. See also, Vonderlehr to Wenger, July 18, 1933, NA-WNRC.

51 Wenger to Vonderlehr, July 21, 1933, NA-WNRC. The italics are Wenger's.

52 Cumming to Moton, July 27, 1933, NA-WNRC.

53 Wenger to Vonderlehr, July 21, 1933, NA-WNRC.

54 Vonderlehr to Murray Smith, July 27, 1933, NA-WNRC.

55 Wenger to Vonderlehr, Aug. 5, 1933, NA-WNRC.

56 Vonderlehr to Wenger, Oct. 24, 1933, NA-WNRC. Controls were given salicylates.

57 Austin V. Diebert and Martha C. Bruyere, "Untreated syphilis in the male Negro, III," *Venereal Disease Information,* Dec. 1946, *27*: 301–314.

58 Eunice Rivers, Stanley Schuman, Lloyd Simpson, Sidney Olansky, "Twenty-years of followup experience in a long-range medical study," *Public Hlth. Rep.,* Apr. 1953, *68*: 391–395. In this article Nurse Rivers explains her role in the experiment. She wrote: "Because of the low educational status of the majority of the patients, it was impossible to appeal to them from a purely scientific approach. Therefore, various methods were used to maintain their interest. Free medicines, burial assistance or insurance (the project being referred to as 'Miss Rivers' Lodge'), free hot meals on the days of examination, transportation to and from the hospital, and an opportunity to stop in town on the return trip to shop or visit with their friends on the streets all helped. In spite of these attractions, there were some who refused their examinations because they were not sick and did not see that they were being benefitted" (p. 393).

59 Austin V. Diebert to Raymond Vonderlehr, Mar. 29, 1939, TSS-NLM, Box 1.

60 Murray Smith to subjects, 1938, TSS-NLM, Box 1. See also, Sidney Olansky to John C. Cutler, Nov. 6, 1951, TSS-NLM, Box 2.

61 The USPHS originally requested that the Julius Rosenwald Fund meet this expense. See Cumming to Davis, Oct. 4, 1934, NA-WNRC. This money was usually divided between the undertaker, pathologist, and hospital. Lloyd Isaacs to Raymond Vonderlehr, Apr. 23, 1940, TSS-NLM, Box 1.

62 Stanley H. Schuman, Sidney Olansky, Eunice

Rivers, C. A. Smith, Dorothy S. Rambo, "Untreated syphilis in the male Negro: background and current status of patients in the Tuskegee study," *J. Chron. Dis.,* Nov. 1955, *2:* 555.

63   R. A. Vonderlehr and Taliaferro Clark, "Untreated syphilis in the male Negro," *Venereal Disease Information,* Sept. 1936, *17:* 262.

64   J. R. Heller and P. T. Bruyere, "Untreated syphilis in the male Negro: II. Mortality during 12 years of observation," *Venereal Disease Information,* Feb. 1946, *27:* 34–38.

65   Jesse J. Peters, James H. Peers, Sidney Olansky, John C. Cutler, and Geraldine Gleeson, "Untreated syphilis in the male Negro: pathologic findings in syphilitic and non-syphilitic patients," *J. Chron. Dis.,* Feb. 1955, *1:* 127–148.

66   Sidney Olansky, Stanley H. Schuman, Jesse J. Peters, C. A. Smith, and Dorothy S. Rambo, "Untreated syphilis in the male Negro, X. Twenty years of clinical observation of untreated syphilitic and presumably nonsyphilitic groups," *J. Chron. Dis.,* Aug. 1956, *4:* 184.

67   O. C. Wenger, "Untreated syphilis in male Negro," unpublished typescript, 1950, p. 3. Tuskegee Files, Center for Disease Control, Atlanta, Georgia. (Hereafter TF-CDC).

68   Vernal G. Cave, "Proper uses and abuses of the health care delivery system for minorities with special reference to the Tuskegee syphilis study," *J. Nat. Med. Assn.,* Jan. 1975, *67:* 83.

69   See for example, Vonderlehr to B. W. Booth, Apr. 18, 1934; and Vonderlehr to E. R. Lett, Nov. 20, 1933, NA-WNRC.

70   "Transcript of proceedings—Tuskegee Syphilis Ad Hoc Advisory Panel," Feb. 23, 1973, unpublished typescript, TSS-NLM, Box 1.

71   Raymond Vonderlehr to Murray Smith, Apr. 30, 1942; and Smith to Vonderlehr, June 8, 1942, TSS-NLM, Box 1.

72   Stanley H. Schuman, Sidney Olansky, Eunice Rivers, C. A. Smith, and Dorothy S. Rambo, "Untreated syphilis in the male Negro: background and current status of patients in the Tuskegee study," *J. Chron. Dis.,* Nov. 1955, *2:* 550–553.

73   Raymond Vonderlehr to Stanley H. Schuman, Feb. 5, 1952, TSS-NLM, Box 2.

74   Schuman and others, "Untreated syphilis . . .," p. 550.

75   "Minutes, April 5, 1965" unpublished typescript, TSS-NLM, Box 1.

76   "Tuskegee Ad Hoc Committee meeting—minutes, February 6, 1969," TF-CDC.

77   James B. Lucas to William J. Brown, Sept. 10, 1970, TF-CDC.

78   Elizabeth M. Kennebrew to Arnold C. Schroeter, Feb. 24, 1971, TSS-NLM, Box 1.

79   See *Medical Tribune,* Sept. 13, 1972: 1, 20; and Report on HEW's Tuskegee Report," *Medical World News,* Sept. 14, 1973: 57–58.

80   HEW *Final Report,* p. 7.

81   The notable exception is Jay Katz's eloquent "Reservations about the panel report on Charge 1," HEW *Final Report,* pp. 14–15.

82   HEW *Final Report,* pp. 8, 12.

83   *Ibid.*

84   See R. H. Kampmeier, "The Tuskegee Study of untreated syphilis," *Southern Med. J.,* Oct. 1972, *65:* 1247–1251; and "Final report on the 'Tuskegee Syphilis Study,'" *Southern Med. J.,* Nov. 1974, *67:* 1349–1353.

85   Leonard J. Goldwater, "The Tuskegee Study in historical perspective," unpublished typescript, TSS-NLM; see also "Treponemes and Tuskegee," *Lancet,* June 23, 1973: 1438; and Louis Lasagna, *The VD Epidemic* (Philadelphia: Temple Univ. Press, 1975), pp. 64–66.

86   Quoted in "Debate revives on the PHS study," *Medical World News,* Apr. 19, 1974: 37.

87   Heller to Vonderlehr, Nov. 28, 1933, NA-WNRC; quoted in *Medical Tribune,* Aug. 23, 1972: 14.

88   Although it is now known that syphilis is rarely infectious after its early phase, at the time of the study's inception latent syphilis was thought to be communicable. The fact that members of the control group were placed in the test group when they became syphilitic proves that at least some infectious men were denied treatment.

89   When the subjects are drawn from minority groups, especially those with which the researcher cannot identify, basic human rights may be compromised. Hans Jonas has clearly explicated the problem in his "Philosophical reflections on experimenting with human subjects," *Daedalus,* Spring 1969, *98:* 234–237. As Jonas writes: "If the properties we adduced as the particular qualifications of the members of the scientific fraternity itself are taken as general criteria of selection, then one should look for additional subjects where a maximum of identification, understanding, and spontaneity can be expected— that is, among the most highly motivated, the most highly educated, and the least 'captive' members of the community."

90   Since the original publication of this article, a full-length study of the Tuskegee Experiment has appeared. See James H. Jones, *Bad Blood: The Tuskegee Syphilis Experiment* (New York: Free Press, 1981).

EPIDEMICS

Although epidemics have seldom taken more lives than endemic diseases, they have always attracted more attention. Sweeping inexorably through town after town, these unpredictable plagues aroused more fear and anxiety than more deadly but common everyday killers such as pneumonia and tuberculosis. Cholera visited the United States only five times—in 1832, 1849, 1866, 1873, and 1892—but the mere thought of this viciously dehydrating disease struck fear in the heart of virtually every American throughout the 19th century. Among the most feared of epidemics, in addition to cholera, was yellow fever. John Duffy illustrates how the two diseases brought terror and death to American cities in the 18th and 19th centuries and acted as significant spurs to public health reform and the development of ameliorative institutions.

Smallpox, another epidemic disease, similarly spawned significant public health responses. A physically repulsive and often disfiguring disease, it arrived in the New World with the European settlers, slaying many American Indians and taking the lives of numerous immigrants. When smallpox epidemics struck during the colonial period, officials typically quarantined the affected area and isolated the sick, either at home or in a pesthouse. In 1721 Cotton Mather introduced another means of stopping this loathsome disease, inoculation (or variolation). Controversial at first, this procedure, which introduced a mild case of the disease in order to produce long-term immunity, became standard practice in 18th-century America. Then, at the beginning of the 19th century, Benjamin Waterhouse brought William Jenner's method of vaccination with cowpox virus to America. Although safer than variolation, this procedure, too, evoked considerable debate. Judith Walzer Leavitt relates New York City's experiences with smallpox and the health officials' attempts to control the disease, concluding the story with the 1947 outbreak and the successful vaccination of almost six and one-half million people. In 1979, the World Health Organization certified global eradication of this disease, marking perhaps one of the world's most significant public health achievements.

With the exception of the 1918–1919 influenza epidemic, which killed more Americans than World War I, the United States has been remarkably free of major epidemics for much of the 20th century. As the understanding of germ transmission grew and especially after antibiotics became available at mid-century, people became convinced that many infectious diseases could be contained and cured. Americans relaxed their fears about epidemics. Unfortunately, any optimism that we had permanently conquered epidemic diseases was short-lived. The swine flu scare of 1976 demonstrated that we still can live in fear of an attack by a mysterious new killer. Although that epidemic never materialized, its threat raised the country's consciousness and fears anew.

The end of the century is proving that the threat of major killer epidemics is far from over. The recent examples of HIV infection and AIDS—which by the mid-1990s had attacked well over one million Americans—and of drug-resistant tuberculosis, the incidence of which is climbing, as well as new emerging viruses like Hanta, Ebola, and Sabia, serve as contemporary examples of the helplessness and fear that can still be engendered by epidemic disasters. Allan M. Brandt uses historical experiences with sexually transmitted diseases to examine America's early experiences with AIDS and to conclude that present-day efforts to control the epidemic must take into account a full range of social, cultural, and biological aspects of the disease.

25

"Be Safe. Be Sure."
New York City's Experience with
Epidemic Smallpox

JUDITH WALZER LEAVITT

The lines of people stretched as far as the eye could see. Men, women, and children, numbering each day in the thousands, stood for hours, sometimes getting drenched in the rain, as they waited their turn. The mood of the crowds was cheerful for the most part, except when word of delay was passed along the lines; then the police trying to control the masses of people had some difficulty. The people were not in line for tickets to a concert or a World Series game: they queued up to receive free small-pox vaccinations. It was April 1947, and the New Yorkers were responding to a threat of an epidemic of smallpox, a loathsome and often fatal disease that most of them had never seen.

The characteristic rash of the disease had become so rare in New York that doctors misdiagnosed the early cases. Smallpox was so unusual, in fact, that citizens had lost their worry about contracting it, and most New Yorkers in that spring of 1947 had not protected themselves through the sure-preventive vaccination. Only an estimated two million of the city's seven and one half million inhabitants had any degree of immunity. The population was thus at risk to contract the disease if exposed, and possibly to die from it. The danger was real, and health officials recognized it when Dr.

JUDITH WALZER LEAVITT is Professor of the History of Medicine, History of Science, and Women's Studies, and Associate Dean for Faculty, University of Wisconsin–Madison.

From *Hives of Sickness: Public Health and Epidemics in New York City,* edited by David Rosner (New Brunswick, N.J.: Rutgers University Press, 1995), pp. 95–114. Copyright © 1995 by the Museum of the City of New York. Reprinted by permission of Rutgers University Press.

Dorothea M. Tolle, the medical superintendent of Willard Parker Hospital, the city's reception point for people with contagious diseases, telephoned the City Health Department on March 28, 1947, and reported two suspected cases of smallpox.

THE 1947 SMALLPOX OUTBREAK

Exposure had arrived, innocently, on March 1 on a cross-country bus that had stopped in Texas, Missouri, Ohio, and Pennsylvania, on its way from Mexico City to New York City. Two passengers on that bus, Mr. and Mrs. Eugene Le Bar, had traveled the whole distance, planning to continue on to Maine. By the time the bus reached New York, however, Eugene Le Bar was not feeling well, and the two delayed their journey and checked in at a midtown Manhattan hotel for a rest. During the hours when Eugene felt well enough, the couple explored the city, walking along Fifth Avenue and stopping in a few shops. But by March 5, Le Bar felt worse, and he went for help to Bellevue Hospital, where he was admitted to the dermatology service. His condition continued to deteriorate, however, and baffled Bellevue's skin doctors. Le Bar was transferred to Willard Parker Hospital with an unknown but suspected contagious condition. On March 10, Eugene Le Bar died. His doctor's diagnosis was bronchitis with hemorrhages. His wife continued on her way to Maine.

The public knew nothing of these events. Even when Dr. Tolle tentatively diagnosed smallpox in two other patients and reported them to the Health Department on March 28, the news media were not apprised of the situation. At that time, the health officials vaccinated the staff and patients at Willard Parker, isolated the suspected cases, and forwarded specimens from them to the Army Medical School Laboratory in Washington, D.C., for a confirming diagnosis. While they awaited the results of laboratory tests, officials meanwhile tried to establish a link between the two sick people, one a 22-month-old black child from the Bronx and the other a 25-year-old Puerto Rican man from Harlem. They soon established that both had been patients at Willard Parker previously, the baby with tonsillitis and the man with mumps, overlapping in time with Eugene Le Bar's admission. Both were unvaccinated. On April 4, the Health Department received back from Washington the laboratory report that matter taken from these patients, and from Le Bar, unmistakably revealed smallpox. On April 5, New Yorkers opened their morning newspapers to learn of their risk.[1]

Health Commissioner Israel Weinstein weighed his words to the press carefully. On the one hand, Dr. Weinstein assured the public that the chances of a full-scale virulent epidemic were "slight." He did not want to promote panic. On the other hand, Weinstein and Hospital Commissioner Edward M. Bernecker called for all who "have never been vaccinated or who have not been vaccinated since childhood to go at once to their doctors to receive this protection."[2] The health officials wanted to give a strong message so that people would actually take the time to gain protection, but not so strong that social order would break down. The motto, repeated over and over again, was "Be safe. Be sure. Get vaccinated." More than four weeks had passed since Le Bar had frequented the downtown hotel, walked the city streets, and lain sick and dying (but not isolated) in two city hospitals. His contacts had had ample time to spread the virus unwittingly but liberally throughout and beyond the five boroughs.

In the days following the initial revelation, the newspapers reported other cases of smallpox, suspected and confirmed. On April 7 a man suffering from smallpox was admitted to Willard Parker Hospital. The next day two more people suspected of having smallpox were hospitalized. On April 13, Mrs. Carmen Acosta, the pregnant wife of one of the earliest sufferers, died at Willard Parker. On April 14 New Yorkers learned that the disease had spread to Dutchess County, by way of a young boy who had brought smallpox to the village of Millbrook from his stay at Willard Parker as a scarlet fever patient. And a case of smallpox traced to New York appeared in Bremerhaven, Germany. Within two weeks of becoming established in New York City, smallpox had infected 12 persons; with each new identification, the health commissioner stepped up his activities.

Efforts to prevent an epidemic from sweeping the city moved in two directions simultaneously. First was the mass vaccination campaign; second was the case-tracing epidemiological work. Health officials saw both as essential to ultimate success.

To encourage everyone to seek a vaccination, the Health Department provided free vaccinations at each of its 21 neighborhood health centers and 60 child health stations. The 13 municipal hospitals provided the service in their out-patient clinics. Once the campaign was in full swing, Weinstein organized free vaccinations at the 84 police precincts throughout the city. All public and parochial schools offered vaccinations to their pupils. Many people received their vaccinations from private physicians or at clinics established through labor unions or industry. Mayor William O'Dwyer, amid great publicity, submitted to the procedure while urging his fellow citizens to follow his example. President Harry S. Truman, planning a visit to New York, underwent vaccination as well. The press and radio provided vaccination center schedules and helpful hints for easing the pain of sore arms. Each day the newspapers trumpeted the numbers vaccinated: at the height of the program half a million New Yorkers got their arms scratched with the virus in a single day. In the two weeks following the discovery of smallpox in the city, 5,000,000 New Yorkers had been vaccinated; by the end of one month, with the danger over and the campaign completed, 6,350,000 people had had their arms pricked in the name of disease prevention.[3] More people had been vaccinated in a shorter period of time than ever before in history.

The vaccination campaign owed its success not just to a well-planned and well-executed Health Department response, but also to the federal and local government. Midway through the month of April, the city and private laboratories that had been producing vaccine reached their limit. People waiting in line to be vaccinated were told to come back another time. An emergency call to the Navy, Army, and the U.S. Public Health Service provided new supplies of the vital fluid. Simultaneously, Mayor O'Dwyer put pressure on the drug companies to produce more and to sell the larger quantities to the city at a cheaper price. Dr. Tom Rivers, a Board of Health member and head of the Rockefeller Institute for Medical Research, advised the mayor during the epidemic. He attended a meeting during which O'Dwyer virtually locked the pharmaceutical representatives into City Hall until they agreed to comply with his demands.[4]

Once the vaccine itself became available in sufficient quantity, another deficit in people power emerged. It took a lot of people, more than the city government had on ready call, to organize and execute a massive vaccination campaign. The Health Department sought volunteer help. At least 3,000 individuals came forward to offer aid: many came from the American Red Cross, the Civilian Defense Volunteer Organization, and the American Women's Voluntary Services. Organizations active during the recent war mobilization quickly reassembled their volunteers to help the nurses and physicians who were galvanized to help puncture citizens' arms.

Widespread public compliance with these efforts can be attributed, in large part, to the postwar period in which the epidemic occurred. Not only were wartime volunteers still available, but the public also retained something of an emergency mentality and easily followed government advice. Furthermore, the city health officials consciously and carefully kept the vaccination program voluntary, appealing to citizens' judgment instead of trying to force compliance, a stance that heightened reasoned responses and kept public panic at a minimum.

In addition to the mass vaccination campaign in 1947, the New York City Health Department initiated a second stage of its smallpox prevention efforts: tracing possible contacts of the people who had confirmed cases of the disease. Because Eugene Le Bar had traveled throughout the country during the period when he was infectious, and because bus passengers, hotel patrons, hospital patients, and others with whom he might have had contact were spread throughout the country as well as beyond its borders by the time the danger became known, a coordinated effort among government agencies at the local, state, and federal level became necessary. In order to make sure that small-

pox did not spread far from this target case, all people who were exposed to Le Bar or the other people already identified and hospitalized had to be vaccinated and observed for a full 21 days. Hundreds of people had stayed in the (unnamed) midtown hotel during the five days the Le Bars were there. They were scattered around the country and had to be traced and watched. Mrs. Le Bar herself was followed to Maine by the U.S. Public Health Service. All the people who had gotten on and off the bus in Texas, Missouri, Ohio, and Pennsylvania needed to be found and warned. Anyone who had been a patient or visited a patient at Bellevue or Willard Parker during the critical dates had to be located and protected. What about the people the Le Bars and their contacts had passed on the street or jostled in the subway? The task was enormous. That it proceeded systematically and methodically is a testament to how far public health in general and smallpox control in particular had progressed in New York by the middle of the 20th century.

HISTORY OF SMALLPOX IN NEW YORK CITY

Smallpox had a long history on Manhattan Island. It first appeared during the Dutch occupation, probably in 1649, and recurred throughout the colonial period.[5] In the first half of the 19th century, smallpox seemed endemic, although it was not one of the city's major killers. During the second half of the 19th century, smallpox appeared more sporadically, but when it flared up its impact could be severe. In the 1870s it sometimes accounted for more than 1,000 deaths yearly; it resumed its devastation for a few years during the 1880s and again briefly in the 1890s. At the very beginning of the 20th century—in 1901 and 1902—New York suffered another epidemic, during which over 2,500 people contracted the disease and more than 700 of them died. But then it seemed to be finished. Two cases were discovered in 1922, six in 1925, and one more in 1939: they seemed anomalous to a population already feeling beyond reach of smallpox's dangers. It was the mirage of safety, possibly, that led many New Yorkers not to get vaccinated before a threat reappeared in April 1947, and the weight of previous experience that allowed them to accept the medical protection readily when offered it.

From the colonial period through the creation of

the Metropolitan Board of Health in 1866 and into the 20th century, New York confronted issues in disease control that had severely tested both medical and civic authority. In the process the city developed modes of disease control that were acceptable to most New Yorkers and effective in coping with epidemic emergencies. It is worth examining two of the issues that arose historically relative to smallpox in order to comprehend the meaning of the triumphant show of private and public cooperation in 1947. The first concerns the safety and efficacy of variolation and later vaccination as preventives to this disease; the second concerns governmental authority.

Smallpox always elicited a public response when it appeared: it was a horrible disease that could not be ignored. Succumbing to the infection, a sufferer felt flulike symptoms of headache, fever, chills, and nausea; later, in the fair-skinned, a scarlet color crept over the body. Then the victim developed red spots, usually more densely on the face and extremities, which evolved from flat to raised pimples, finally blistering into oozing pustules. In addition to the terrifying outward manifestation, the disease ravaged inside the body, attacking the throat, lungs, heart, liver, and intestines, sometimes causing convulsions, hemorrhages, or delirium. On the average, about one-quarter of the people attacked by smallpox died; lucky victims survived the horror and watched the pustules crust over and the scabs drop off, leaving scars, lifelong reminders of their ordeal. People with smallpox shed literally millions of the disease's viruses into the air; they remained infectious from just before the rash appeared until the scabs disappeared, a period of about three weeks.[6] It is no wonder that when smallpox appeared in a community, inhabitants rushed to try to protect themselves.

Limiting the Disease

Before the 18th century, New Yorkers could do little more than quarantine in the harbor ships containing smallpox sufferers and isolate those people who caught the disease on land. In 1721, however, another alternative, inoculation, was tried in Boston, and its use spread southward in time for the epidemic that struck New York City in 1731.[7] Long used in Africa and Asia, inoculation with matter

from a smallpox pustule, when introduced under the skin of a person who had never had the disease, seemed to protect effectively against a virulent attack. The procedure, known technically as variolation, produced a mild case of smallpox but only rarely led to death. It produced, as did smallpox acquired naturally, lifetime immunity. In 1732, the procedure appeared to be quite popular in the New York area: 160 people submitted to inoculation in Jamaica in that year, and only one of them died, a significantly smaller number than the expected one-quarter of naturally acquired cases.[8]

Variolation had its drawbacks, unfortunately. Not only was there a small chance of death in people who might never have contracted the natural disease, but inoculation with a case of smallpox, no matter how mild, produced a person just as infectious as someone who had contracted the disease through the ordinary routes. Thus, communities that adopted inoculation policies risked spreading cases of smallpox much more widely than might have occurred had they not started the practice. In so doing, officials put people who did not choose to protect themselves through variolation (from religious beliefs, for example, that prohibited tampering with God's will) at greater risk of catching the infection.

Although hotly contested in its early years, variolation rapidly gained advocates, and its use spread throughout the English colonies. Benjamin Franklin, having lost a child to naturally acquired smallpox in 1736, became one of its most ardent supporters. In his autobiography he wrote down his feelings about his son's death: "A fine boy of four years old, by the smallpox, taken in the common way. I long regretted him, and still regret that I had not given it to him by inoculation. This I mention for the sake of parents who omit that operation, on the supposition that they should never forgive themselves, if a child died under it, my example showing that the regret may be the same either way, and therefore the safer should be chosen."[9] Other parents used arguments similar to Franklin's to convince themselves to inoculate their children, and soon more people caught smallpox from the knife than from exposure. The more people were inoculated with the disease, the more rapidly it spread; the more it spread, the more people would try to protect themselves through inoculation. The infection was thus kept alive by inoculation; it also was prevented from wreaking its worst ravages.

The cycle of infection and variolation was broken in 1801, when Drs. Valentine Seaman and David Hosack made available in New York a new procedure, imported from Great Britain.[10] Vaccination—inoculation with cowpox matter—was far safer than introducing smallpox directly; it effectively provided immunity from smallpox but did not lead to its dissemination. Vaccination quickly replaced variolation as the procedure of choice for those seeking protection from smallpox.

One of the ironies of American medical history is that despite the availability of effective and safe vaccination, smallpox continued for almost 150 years periodically to scourge communities around the country. This is in great part because of enemies of vaccination who worked hard to convince people that the procedure was dangerous and ineffective: dangerous because vicious infections or other diseases such as syphilis could be transferred from one person to another through the procedure and ineffective because cases of smallpox could be traced to people who thought themselves protected by vaccination.[11] New York harbored its share of the so-called antivaccinationists, who were busy in the 19th century trying to convince people not to submit to the procedure. They were often successful in turning people against vaccination. During the New York City epidemic of 1853 and 1854, for example, one mother deliberately exposed her four children to smallpox instead of vaccinating them, "in order that they might have the disease at once, and thus relieve her from the anxiety." Three of the children died.[12]

The antivaccinationists spoke directly to a concern that all citizens, and especially immigrant groups, harbored about violations of personal liberty. Many people had come to America to escape governmental tyranny: was governmental repression reemerging on these shores in the guise of "free" vaccinations? Seeing the issue in these terms, and fearing vaccination as a first step toward greater governmental intervention in private lives and decision-making, many people joined the ranks of the antivaccinationists and resisted public programs aimed to protect them from smallpox.

New York's antivaccinationists stepped up their activity after the Metropolitan Board of Health was formed in 1866.[13] An Anti-Vaccination League was begun in 1885, and it challenged the Board of Health to debate the "evils of vaccination."[14] The board refused, but groups around the city did use their meetings to discuss the issues raised in the vaccination controversy. The Society of Medical Jurisprudence and State Medicine, for example, discussed smallpox at one of its meetings, agreeing among themselves that vaccination was "an excellent thing."[15] During the virulent epidemic of 1893–94, when health officials brandished their power to detain forcibly any who refused vaccination, the Anti-Compulsory Vaccination League, a Brooklyn-based group, successfully sued the Brooklyn Department of Health. On appeal the suit was overthrown, but the incident put before the public the issue of infringement of personal liberty, which the antivaccinationists emphasized. Thereafter New York health officials were wont to highlight educational persuasion instead of their police powers.[16]

One of the reasons the antivaccinationists were successful in their campaign to convince some people not to undergo vaccination was the truth of their arguments about the risks of the procedure as it was practiced in the 19th century. Indeed, many doctors, so-called regulars and irregulars alike, agreed that vaccination as it was practiced could transmit other diseases and lead to a false sense of security. Some practitioners still used "humanized" virus during the second half of the 19th century, matter taken from the pustules of inoculated humans, which could carry with it the added danger of infecting recipients with other diseases, such as syphilis. If transmitted through too many people, the virus lost its powers of inducing immunity. Most inoculators by the last third of the century used bovine matter, which they considered safer as well as more effective than the humanized material, but which, when kept too long, also gave minimal protection against smallpox. Furthermore, the inoculation procedure itself could be a cause for worry. Scarification, puncture, or abrasion prepared the skin for the introduction of the viral matter from the "point" on which it was stored. If inoculators were in a hurry, or if they were not sufficiently trained, they might fail to clean the puncture site adequately or use a single point to lance many people. It was not surprising that many laypeople found it difficult to interpret their degree of safety in the face of varied practices and conflicting information.[17]

The Politics of Prevention

The questions raised by the uncertainty of smallpox inoculation practices in the second half of the 19th century significantly involved the issue of state power. Should the government step in to regulate the practice of vaccination itself, thus insuring its safety and quelling citizen doubts? Should it at the same time insist that citizens must protect themselves, thus making vaccination compulsory? What level of government had legal authority to regulate these practices, and how far could it impinge on personal liberty? Turning now to this issue of the authority of the government, we can examine the evolution of public authority to protect the public's health, specifically through smallpox prevention. Through this example, we can begin to understand how citizens, working through their worries about possible violation of individual liberties, were convinced to allow the government to take control to the extent witnessed during the 1947 smallpox epidemic in New York City.

The year before the formation of the New York Metropolitan Board of Health in 1866, Dr. J. M. Toner, a noted physician and writer, published a paper on the "propriety and necessity of compulsory vaccination," in which he challenged physicians and health officials to use their powers to protect the public health through vaccination.[18] He maintained that it was the duty of government to protect the health of its citizens if they would not do it themselves, and that officials should "secure effective vaccine protection to every person, without distinction, residing within the sphere of their authority." Parents, too, had duties in Toner's plan. He insisted that vaccination was as important for parents to provide for children as were food and clothing: "failure to provide protection against the smallpox seems to be more maliciously wicked than to neglect either food or clothing, as the former may not only cause the death of the child, but be the means of spreading disease and death among many others." Thus, holding parents responsible but putting on government the burden of providing the services and enforcing vaccination, Toner's

proposal set the stage for debate for the rest of the century.[19]

The issues Toner raised were discussed widely in the medical and lay literature of the second half of the 19th century. It became essential for health departments to devise strategies to try to prevent smallpox and to define the role of the state unambiguously. In New York, before the 1866 Metropolitan Health Act was passed, health officers repeatedly voiced their frustration about their lack of authority to handle smallpox emergencies. In 1862, for example, officials sent 320 people to the Smallpox Hospital on Blackwell Island, but another 83 refused to go and could not be pressured. In 1865, the press noted that "there is no adequate legal power to prevent [smallpox] spreading. . . . [P]ersons taken with this disease are allowed to travel in the cars and stages which are constantly running in the principal thoroughfares."[20] The *New York Times* concluded, "The prophylactic against smallpox is vaccination; but this end cannot be attained except through the agency of strong laws energetically enforced."[21] The seemingly intractable problem was one of the factors moving New Yorkers to support the imposition of stronger health authority.

The first response of the more powerful Board of Health was to regularize the production and accessibility of smallpox vaccine. In its first *Annual Report,* the new board noted that the recent dispensing of 70,000 vaccinations free of charge had resulted in a mild epidemic (35 people had died of smallpox in the board's first eight months of operation) and that city physicians hoped that their success would "lead to the adoption of systematic methods for the vaccination of all children and unprotected persons."[22] Indeed, the health ordinances adopted in the early months of the board's operation provided "that every person, being the parent or guardian, or having the care, custody or control of any minor or other individual, shall . . . cause and procure such minor or individual to be so promptly, frequently and effectively vaccinated, that such minor or individual shall not take, or be liable to take, the smallpox."[23]

Despite these firm words and intentions, New York officials did not portray their vaccination policy then or later in the century as compulsory. Insisting that "compulsory" measures could be enforced only under "imperial" governments, and that New York City's policy was not compulsory because such rules "would result in antagonism to the work, which would defeat the object," health officials seemed to be obfuscating reality. Not only did the board actively encourage vaccination and provide it free to all who wanted it, it also cooperated with the Board of Education in enforcing an ordinance that made vaccination a prerequisite to school attendance at all schools in the city. By the time of the 1947 smallpox scare, New York was one of only 12 states to require such school-related vaccination even in the absence of a threat of smallpox. During the years when smallpox actually threatened the city, the board went much further, insisting that all inhabitants be vaccinated or else submit to isolation measures to protect those around them.[24]

From 1866 to the beginning of the new century, New York tested and refined its smallpox strategies. In addition to establishing school-based programs and offering free vaccination to all, the board instituted house-to-house vaccination visits in 1871. The same year, the board systematically began to remove to the Smallpox Hospital all those suffering from smallpox who could not be "properly isolated" at home.[25] Yet smallpox recurred. In the words of one observer, its continued presence was due to "the positive refusal of large numbers of our foreign population to submit to vaccination."[26] Limits in budget and personnel left the health department incapable of full-scale actions.

The legal status of vaccination rested on prevailing, and by the 20th century increasingly unanimous, medical opinion. In the words of sanitarian M. J. Rosenau, "Vaccination affords a high degree of immunity to the individual, and well-nigh perfect protection to the community. To remain unvaccinated is selfish in that by so doing a person steals a certain measure of protection from the community on account of the barrier of vaccinated persons around him."[27] Rosenau advocated compulsory vaccination. Yet he and probably most other Americans by the early 20th century did not equate "compulsory" with forcible policies. The question of the invasion of personal liberties of citizens—the degree to which such liberties could be abridged by government in the name of protecting the public health—remained of paramount concern. A policy

mandating vaccination was not the same as one forcing citizens to undergo such a procedure.

Vaccination requirements as promulgated by state legislatures, municipal ordinances, and school board regulations emanated from the police power of the states. They were not delegated to the federal government in the Constitution. State legislatures could require vaccination of all citizens under conditions that they stipulated, or they could delegate such power to municipalities or other governmental bodies such as boards of health. Throughout the second half of the 19th century, the extent of police powers was becoming understood largely through court rulings. Public health officials and private citizens continued to debate the constitutionality of compulsory vaccination, a debate especially spurred on by antivaccination groups. The issues were ultimately settled in law if not in everyone's mind by a United States Supreme Court decision, *Jacobson* v. *Massachusetts*, handed down in 1905.[28]

The differences between compulsory and forcible measures in vaccination regulations became clear in Justice Harlan's opinion. The case involved a citizen of the Commonwealth of Massachusetts, Henning Jacobson, who did not want to submit to vaccination. The state legislature had empowered boards of health to require vaccination if an epidemic threatened. When smallpox appeared in the city, the Cambridge Board of Health adopted such a regulation and ordered all inhabitants to be vaccinated. Jacobson argued that the ruling was inconsistent with the spirit of the Constitution and violated his individual rights. Justice Harlan and six of his colleagues on the Supreme Court thought otherwise: "The liberty secured by the Constitution of the United States to every person within its jurisdiction does not import an absolute right in each person to be, at all times and in all circumstances, wholly freed from restraint. There are manifold restraints to which every person is necessarily subject for the common good. . . . Real liberty for all could not exist under the operation of the principle which recognizes the right of each individual person to use his own, whether in respect of his person or his property, regardless of the injury that may be done to others." Harlan based his judgment, in part, on a prior New York case in which the exclusion of children from school unless vaccinated had been upheld in those instances in which smallpox threatened the community.[29]

The most important issue for Justice Harlan was the danger to other people posed by individuals or groups who would not cooperate with health officials. But there were limits to how far the state could go in protecting the public health. Nowhere in this decision, or in any other, did the court rule that people could be forcibly vaccinated against their will. As legal expert James Tobey concluded from his analysis of vaccination laws, "Compulsory vaccination means that all persons may be required to submit to vaccination for the common good, and if they refuse . . . they may be arrested, fined, imprisoned, quarantined, isolated, or excluded from school . . . but they cannot be forcibly vaccinated, desirable as such a procedure might be from the standpoint of public health protection."[30] Individual liberty and public health could be protected simultaneously under compulsory vaccination provisions, but there were limits on each.

Resolution of the seeming incompatibility between individual liberty and public health at the beginning of the 20th century made it much easier for subsequent boards of health to develop and effectuate policies to control smallpox. Although New York City was not threatened by smallpox again in a major way after 1901, it did encounter a few threats. In June 1911, for example, a woman arrived in New York from Washington sick with smallpox. Her physician reported the case to the Health Department, which promptly removed the woman to an isolation hospital and then launched its vaccination efforts. Seventeen police officers and 15 doctors arrived in the West Side neighborhood armed with vaccination tools. The newspaper reported their efforts: "The work began on the ground floor of each building that none of those on the upper floors should escape. All the apartments were then visited in turn and the tenants informed of the danger of infection. . . . Many of the grown-ups objected to being vaccinated. . . . In families where there were young children, however, there was little opposition on the part of parents."[31]

Three years later a man appeared at the Health Department office suffering from smallpox. The elevator and the rooms he had visited were fumigated and all employees were revaccinated.[32] The existence of a few cases of smallpox in 1925 elicited a citywide vaccination campaign, headed by Mayor

Hylan, who was vaccinated in front of the press. The warning Health Commissioner Monaghan issued in that year could have been repeated in 1947: "It is a curious consequence of the efficacy of vaccination that numberless communities have almost forgotten that there is such a thing as smallpox, and many physicians never have seen a case of it. Therein lies the danger at present, for the malady is far from extinct, though it ought to be and might have been if universal vaccination had been practiced for a sufficient period all over the country."[33]

In many respects, the experience in 1925 predicted what would follow 22 years later. The city contributed extra funds for a wide-scale vaccination campaign, vaccination remained voluntary, businesses offered the procedure to employees, and publicity was widespread. The *New York Times* editorialized that any inhabitants who had not been vaccinated "ought to feel a sense of duty unperformed." The newspaper revealed the dangers when it told of one undiagnosed smallpox victim who had traveled to his physician in a crowded subway train; the only way to protect against such unwitting transmission of the disease, the paper concluded, was to submit to "the most trivial of 'operations.'"[34] Volunteers from organizations such as the Association for Improving the Condition of the Poor, the New York Diet Kitchen Association, the Maternity Center Association, and the Henry Street Settlement helped. The Health Department put especial effort into vaccinating schoolchildren, giving each pupil a blue-and-white button to wear: on it the seal of the city was surrounded with the words, "I am protected by vaccination. Are you?"[35] When the feared epidemic did not appear, his political foes accused the health commissioner of creating the emergency to produce jobs for his friends. But threatened epidemics and outbreaks elsewhere in the country in 1925 provided evidence of New York's prudent behavior."[36]

A Balance of Power

Both the efficacy of the vaccination procedure and an understanding of the limits of governmental authority were necessary prerequisites to successful epidemic smallpox prevention . and control. New Yorkers, and Americans in general, learned through the ups and downs of the 19th- and 20th-century experiences that vaccination, if the practice could be carefully controlled (which became significantly easier after the federal government regulated vaccine production in 1902), was safe. Epidemics could be forestalled. Local government and health departments had important roles to play in such epidemic control; in fact, eradication of the threat of epidemics could not occur without effective governmental actions.

The specific actions planned by Health Commissioner Weinstein and his advisors in 1947 were telling. Despite broad, well-defined police powers that could have been invoked in the instance of a threat from smallpox, these health officials decided to emphasize voluntary vaccination. Furthermore, they relied heavily on public education through the press and the radio to convince the public of its duty, and they effectively utilized contacts with health officials at various levels of government throughout the country to tackle the tough job of case tracing. Tom Rivers's account of how Mayor O'Dwyer, behind the scenes, strong-armed the companies producing the vaccine to get them to agree to produce the necessary quantities and charge reasonable rates indicates that interagency cooperation was not always immediately forthcoming. But it did materialize under strategically placed pressure.

No doubt contributing to the successes of 1947 was the fact that the scare occurred in the aftermath of the Second World War. New Yorkers had already become accustomed to regulations that affected the behavior of the whole population, from food rationing to blackout drills. Throughout the war period, health officials had urged vaccination (although most people did not seem to have complied with this) because of the possibility of mass evacuation and the potential dangers from new environments.[37] Wartime voluntary agencies had been active and could be quickly reassembled.

Perhaps most important in the 1947 effort, health officials encouraged a network of citizen conduct and volunteer activity to support their efforts, thus reaching into the community to sustain the prevention work, gaining cooperation without requiring it. With the power of the law behind him, Weinstein did not need to use it. New Yorkers had learned through a long history with smallpox how devastating the disease could be and how vaccination could provide protection against its worst ravages. People had become accustomed to the exercise of governmental authority. On its part, the Board of Health had learned to value citizen cooperation and to seek it in broadly acceptable ways, largely through publicity and convincing argument.

In assessing the successes of April 1947, Weinstein made note of what might have happened. If the experience had mirrored the 1901–1902 epidemic, with the same case and mortality rate, New York would have reported 4,310 cases of smallpox and 902 deaths. By comparison, the statistics of 12 cases and 2 deaths looked fine. The health commissioner spread the credit widely, thanking the press and radio, his staff, the nurses, other city employees, and the mayor. Most of all, he credited the "intelligent cooperation of the public and the generosity of private physicians and volunteer workers," without whom, he thought, "it would have been impossible to have achieved this remarkable record."[38] He was right.

NOTES

I am grateful for the timely research assistance of Sarah A. Leavitt and Nancy Ralph.

1 The events are reconstructed from the *New York Times* and the *Daily Mirror* from April 1947. For a particularly vivid account of the epidemic, see Berton Rouché, "A man from Mexico," which first appeared in the pages of *The New Yorker*, reprinted in *Eleven Blue Men* (New York: Berkley Medallion Books, 1965), pp. 91–108.

2 Quotation from the *Daily Mirror*, Apr. 6, 1947, p. 5. The relationship between the Health Department and the press is described in Karl Pretshold and Car-

olyn C. Sulzer, "Speed, action and candor: the public relations story of New York's smallpox emergency," *Channels*, Sept. 1947, *25*: 3–6.

3 Israel Weinstein submitted his report to the mayor on May 11, giving the statistics: 1.6 million had received their vaccinations at public and private hospitals; 1.9 million at Health Department clinics and police stations; 1.2 million received vaccinations at clinics at their place of work or labor union; one million from private physicians; and 650,000 in the city's schools. See the *New York Times*, May 12, 1947. Typical during the campaign was vaccination of 200,000

people a day. This number represented the usual number of vaccinations in a whole year of Health Department work. See also Israel Weinstein, "An outbreak of smallpox in New York City," *Am. J. Public Hlth.*, 1947, *37*: 1376–1384.

4   *Tom Rivers: Reflections on a Life in Medicine and Science,* An oral history memoir prepared by Saul Benison (Cambridge, Mass.: MIT Press, 1967), pp. 385–388. Rivers did not think highly of Israel Weinstein's ability to administer New York's Health Department.

5   John Duffy, *A History of Public Health in New York City, 1625–1866* (New York: Russell Sage Foundation, 1968). This book and Duffy's second volume, *A History of Public Health in New York City, 1866–1966* (New York: Russell Sage Foundation, 1974), provided the background for this historical discussion of New York's experience with smallpox.

6   An excellent book describing modern smallpox, with a lengthy historical section, is C. W. Dixon, *Smallpox* (London: J. and A. Churchill Ltd., 1962). See also Donald R. Hopkins, *Princes and Peasants: Smallpox in History* (Chicago: Univ. of Chicago Press, 1983).

7   For an excellent account of the controversy over inoculation in Boston, see John B. Blake, "The inoculation controversy in Boston, 1721–1722," *New England Quarterly,* 1952, *25*: 489–506; reprinted in Judith Walzer Leavitt and Ronald L. Numbers, eds., *Sickness and Health in America: Historical Readings,* 2nd ed. rev. (Madison: Univ. of Wisconsin Press, 1985), pp. 347–355.

8   Duffy, *Public Health in New York City, 1625–1866,* p. 55.

9   Quoted in Francis R. Packard, *History of Medicine in the United States* (New York: Paul B. Hoeber, 1931), vol. 1, p. 79.

10 The story of Jenner's discovery and the introduction of vaccination into the United States can be read in Hopkins, *Princes and Peasants;* Genevieve Miller, *The Adoption of Inoculation for Smallpox in England and France* (Philadelphia: Univ. of Pennsylvania Press, 1957); John Blake, *Benjamin Waterhouse and the Introduction of Vaccination* (Philadelphia: Univ. of Pennsylvania Press, 1957); and O. E. Winslow, *A Destroying Angel* (Boston: Houghton Mifflin, 1974).

11 Antivaccinationists can be followed in Martin Kaufman, "The American antivaccinationists and their arguments," *Bull. Hist. Med.,* 1967, *41*: 463–478, and in Judith Walzer Leavitt, *The Healthiest City: Milwaukee and the Politics of Health Reform* (Princeton: Princeton Univ. Press, 1982; Madison: Univ. of Wisconsin Press, 1996).

12 Quoted from J. C. Hutchinson, "On vaccination and the causes of the prevalence of small-pox in New York in 1853–4," *N.Y. J. Med.,* 1854, *12*: 349–368, in Hopkins, *Princes and Peasants,* pp. 269–270.

13 The story of the creation of the Metropolitan Board of Health is well told in Gert H. Brieger, "Sanitary reform in New York City: Stephen Smith and the passage of the metropolitan health bill," *Bull. Hist. Med.,* 1966, *40*: 407–429 (ch. 28, this book); and in Duffy, *History of Public Health in New York City, 1625–1866,* pp. 540–571.

14 *New York Times,* Oct. 27, 1885.

15 *New York Times,* Nov. 13, 1885.

16 The incident is described in Duffy, *A History of Public Health in New York City, 1625–1866,* pp. 153–154. At the same time as this legal action, New Yorkers learned of the smallpox troubles in Milwaukee, during which citizens rioted in the streets to object to health department policies about vaccination, and the health commissioner was impeached. See Leavitt, *The Healthiest City: Milwaukee;* and *Leslie's Weekly Illustrated Newspaper,* Sept. 27, 1894, *79*: 207. The *New York Times* frequently referred to the antivaccinationists as "public enemies"; see, for example, May 6, 1914.

17 For a contemporary account of the vaccination procedure, see W. A. Hardaway, *Essentials of Vaccination; a Compilation of Facts Relating to Vaccine Inoculation and Its Influence in the Prevention of Smallpox* (Chicago: Jansen, McClury and Company, 1882). See also a lengthy discussion of 19th-century practices in Dixon, *Smallpox,* pp. 249–295.

18 J. M. Toner, "A paper on the propriety and necessity of compulsory vaccination," *Tr. A.M.A.,* 1865, *16*: 307–330.

19 Quotations from *ibid.,* pp. 314, 317. See also William H. Richardson, "Smallpox in New York City, with some statistics and remarks on vaccination," *Transactions of the New York State Medical Society* (1865), pp. 143–156.

20 *New York Times,* Apr. 29, 1865.

21 *Ibid.*

22 [First] *Annual Report of the Metropolitan Board of Health, 1866* (Albany: Van Benthuysen & Sons' Steam Printing House, 1867), p. 309.

23 *Ibid.,* p. 706.

24 See, for example, the *New York Times,* Jan. 24, 1863. In 1862 the Board of Education passed a bylaw regulating vaccination for schoolchildren, which was endorsed by the health department. To trace vaccination laws, consult William Fowler, "Principal provisions of smallpox vaccination laws and regulations in the United States," *Public Hlth. Rep.,* Jan. 1941, *56*: 167–189; Clark Bell, "Compulsory vaccination; should it be enforced in law?" *J.A.M.A.,* Jan. 9, 1897, *28*: 49–53; James A. Tobey, *Public Health Law* (New York: The Commonwealth Fund, 1947), pp. 233–250; and the various Annual Reports of the Board of

Health. The quotations are from the *Annual Report* (1866), p. 310, and Bell, "Compulsory vaccinations," p. 49.

25 *New York Times,* Mar. 16, 23, 1871. Throughout the 1871–1873 epidemic, newspapers were filled with stories of the terrible conditions within the Smallpox Hospital. Exposés and accusations filled pages of newsprint; the Board of Health tried to dissociate itself from the institution, which was run by the Commissioner of Charities and Corrections, and the issue emerged as highly political. Control was transferred to the Board of Health in 1874.

26 *New York Times,* May 30, 1875.

27 Milton J. Rosenau, *Preventive Medicine and Hygiene,* 6th ed. (New York: D. Appleton-Century, 1935), p. 21.

28 197 U.S. 11. On this issue, consult Hampton L. Carson, "The legal aspects of vaccination," *N.Y. Med. J.,* 1909, *89:* 106–111. See also Tobey, *Public Health Law,* pp. 238–240; the decision is reproduced in its en-

tirety on pages 359–375 and the quotations here are from that printing.

29 *Viemeister* v. *White* (1904), 179 N.Y. 235; cited in Tobey, *Public Health Law,* p. 239.

30 Tobey, *Public Health Law,* p. 240.

31 *New York Times,* June 25, 1911.

32 *New York Times,* June 13, 1914.

33 Quoted in the *New York Times,* May 16, 1925.

34 *New York Times,* June 4, 1925.

35 *New York Times,* Aug. 9, 1925.

36 See, for example, John P. Koehler, "How Milwaukee aborted its smallpox epidemic," *Wis. Med. J.,* 1925, *24:* 324.

37 See, for example, *New York Times,* Apr. 27, 1942, referring to directives issued by the health department.

38 Weinstein, "An outbreak of smallpox," p. 1381. See also the *Report of the Department of Health, City of New York for the Years 1941–1948* (New York: Department of Health, 1949).

26

Social Impact of Disease in the Late 19th Century

JOHN DUFFY

The late 19th century witnessed the bacteriological revolution, without doubt one of the most significant events in the history of medicine. Prior to this, epidemic and endemic diseases were as inexplicable and mysterious to man as they had been to his most primitive forebears. A few empirical discoveries, such as vaccination for smallpox, had led to some improvement in conditions of health, but the origin and transmission of diseases were as obscure as ever. Acrimonious debates characterized medical meetings as late as the 1880s as theory vied with theory, and theorist with theorist. The greatest advance in knowledge of infectious diseases until then had come from the general recognition that such diseases flourished in filthy, overcrowded conditions. This development, for which the medical profession deserved only partial credit, resulted in the movement for sanitation, which began reducing the urban death rate well before bacteriology provided health officials with a sound rationale.

Although the movement for public and personal hygiene was firmly established in the second half of the 19th century, and Pasteur, Koch, and their colleagues were unraveling the tangled skein of bacteriology, communicable diseases still remained the leading health problem. The health records of every city show that tuberculosis, diphtheria, scarlet fever, whooping cough, enteric disorders, measles, smallpox, and even malaria were endemic. Infant mortality—largely attributed to such vague causes as summer fever and diarrhea, teething, colic, and convulsions—was a major component of the high total death rate. The loss of so many children,

however, was accepted as the inexorable working of fate.

Smallpox, the one disease for which a fairly effective preventive measure was available, should have created no difficulty, yet it continued to flare up in every American city. A series of outbreaks in New York City during the 1870s caused 805 deaths in 1871, 929 in 1872, 484 in 1874, and 1,280 in 1875.[1] During three of these same years the annual death toll from smallpox in New Orleans was more than 500, and Dr. Joseph Jones, president of the Louisiana State Board of Health, later declared that 6,432 residents of New Orleans had died of smallpox in the years from 1863 to 1883. As late as the winter of 1899–1900, 3 of 12 medical students at Tulane University infected during a widespread outbreak died of the disease.[2]

Compared with other communicable infections such as diphtheria, for which little could be done, smallpox was only a minor cause of death. Diphtheria, a fearful disorder with an equally high fatality rate, was a major epidemic disease throughout most of this period. Earlier, during the 1850s and 1860s, it had been merely one of many children's complaints, but its incidence took a startling upturn in the 1870s. From 1866 to 1872 diphtheria deaths in New York averaged about 325 per year. In 1873 the figure jumped to 1,151, increased to more than 1,600 in 1874, and then reached a new high of 2,329 in 1875. From 1880 to 1896 the annual deaths from diphtheria never fell below 1,000; on three occasions the total was well in excess of 2,000. The peak period for diphtheria in New York City came during the 1890s, the years when throat cultures and antitoxin therapy were introduced. New York's problems with diphtheria were in no sense unique.[3] In New Orleans a health official informed a joint meeting of the city's two

The late JOHN DUFFY was Priscilla Alden Burke Professor Emeritus of History, University of Maryland, College Park, Maryland, and Professor Emeritus of the History of Medicine at the Tulane University School of Medicine.

Reprinted with permission from *Bulletin of the New York Academy of Medicine*, 1971, 47: 797–811.

medical societies in 1887 that diphtheria had long existed there, but never before had it been "so widespread and abundant as now."[4] By this date diphtheria had spread throughout America, ravaging town and country alike. Since many deaths from diphtheria went unrecorded, and the hundreds of infant deaths attributed to croup and other vague causes undoubtedly included some cases of diphtheria, the actual toll was probably larger than the statistics of mortality show.

The most surprising aspect of diphtheria was that it aroused so little concern. One of the few newspaper editorials about it came after an 1873–1874 epidemic which killed 1,344 people in New York City. On this occasion the editor of the *New York Times* declared: "Had a tithe of the number died from anything resembling cholera or yellow fever we should have had a public scare which would have compelled such a cleaning out of tenements, flushing of sewers, and clearing away of street filth as had not been witnessed for many years."[5] Occasional discussions can be found in medical journals and transactions of societies but these centered chiefly around methods of treatment. The casual public reaction to diphtheria contrasts sharply with the attitude of colonists a century or so earlier. When a virulent form of the disease suddenly burst upon western Europe and the American colonies in the 1730s, it aroused widespread apprehension. By the 1870s, however, diphtheria was a familiar disorder to which the population had become accustomed, and its annual toll among the young had come to be taken as a matter of course. The doctors could do little about it, and the public attitude was one of resignation.

This same fatalistic attitude also characterized the public reaction to scarlet fever, tuberculosis, typhoid, and the other perennial disorders. Dr. Abraham Jacobi, reporting for the Committee on Hygiene of the New York County Medical Society, pointed out that between 1866 and 1890 about 43,000 residents of New York had died of diphtheria and croup and that more than 18,000 had succumbed to scarlet fever. Despite this enormous mortality, the city had made virtually no public provision for the sick. Nine years before, in 1882, he continued, the municipal hospital facilities were so crowded with cases of smallpox, typhus, and typhoid that there had been no room for patients with diphtheria or scarlet fever. Since that time

nothing had been done except to open one hospital with 70 beds. Almost in despair, Dr. Jacobi exclaimed: "Seventy beds, and twenty-five hundred cases are permitted to die annually."[6] Dr. Jacobi's statement takes on added significance when one considers that New York City had one of the best health departments in the United States.

In terms of mortality, two diseases, phthisis, or consumption (tuberculosis of the lungs), and pneumonia, should have caused the greatest outcry. Both, however, were considered "constitutional" diseases, and their very frequence dispelled the fears one might expect to be associated with them. In 1870 tuberculosis of the lungs was responsible for about 4,000 deaths in New York City; this figure rose steadily in the ensuing years until about 1890, when almost 5,500 deaths were reported. Deaths from pneumonia rose even more sharply — from 1,836 in 1870 to 6,487 in 1893. Despite their enormous death toll, these familiar and chronic complaints lacked the drama of the great pestilences, and they went largely unnoticed by the general public.[7]

Although most of these statistics have been drawn from New York and New Orleans, the conditions that they reflect prevailed in all major American cities. New Orleans and other southern urban areas differed from the North only with respect to malaria and yellow fever. As in the North, tuberculosis and the respiratory diseases were the number one killers, while diphtheria, scarlet fever, smallpox, measles, and other disorders contributed to the general mortality.

Although gradually receding southward, malaria was a major problem in the United States throughout the 19th century. In New York City 457 deaths were attributed to malaria during 1881, and it was 1895 before the city's annual number of deaths from the disease fell below 100.[8] In terms of total mortality, malaria was of little significance to New York and most northern cities, but it was a major factor in the South. In 1888 Dr. Stanford Chaillé surveyed the causes of death in New Orleans and concluded that tuberculosis, malaria, and dysentery were the chief culprits. Bearing out Dr. Chaillé's statement, the records of the New Orleans Charity Hospital for 1883 show that 45 percent of the 8,000 patients admitted were treated for malaria. But malaria, too, was an old and familiar complaint, and in those areas where it was endemic its recurrence each spring and fall was accepted almost as inevitably as the seasonal cycle itself.[9]

In sharp contrast to this casual acceptance of the diseases mentioned thus far was the public reaction to Asiatic cholera and yellow fever. Although both disorders had reached their peak in the 1850s and henceforth were only a minor cause of morbidity and mortality, they dominated newspaper stories relating to health, preoccupied a good share of the time of the medical profession, and were important factors in promoting public health measures. Had either disease gained a permanent foothold in the United States, it might well have been among the ranking causes of mortality and morbidity, but at the same time it would have become familiar and in the process would have lost its capacity to inspire terror. As it was, outbreaks of cholera in any part of the world or the appearance of a case of cholera or yellow fever in quarantine was enough to arouse the newspapers, medical societies, and civic authorities in every American port.

Of the two diseases, yellow fever had a much longer history in the United States. It first appeared in the late 17th century in Boston and then plagued every American port from Boston southward until the beginning of the 19th century. After a series of major epidemics from 1793 to 1805, the northeastern section of the United States was virtually free of the disease. Attacks on the South Atlantic and Gulf Coast areas, however, intensified in the first half of the 19th century and reached their peak in the 1850s. The number and intensity of the outbreaks, with one or two exceptions, tapered off sharply after the Civil War, although the disease continued to be a real threat to every southern port.[10]

Yellow fever is a fatal and frightening disease; its attacks on the cities of the eastern seaboard from 1793 to 1805 left a vivid imprint upon the public mind. Throughout the remainder of the century, memories of this pestilence were constantly revived by grim accounts of the recurrent outbreaks in southern ports. Moreover, the disease was endemic in the West Indies, and it was a rare summer when one or more cases were not discovered by northern quarantine officials. In 1856 lax enforcement of quarantine laws resulted in more than 500 cases of yellow fever on Staten Island and the western

end of Long Island. The New York City quarantine station was located on Staten Island at this time, and outraged local residents barricaded all entrances to it. When the New York authorities responded in 1857 by buying a new site several miles away, an armed mob vandalized the buildings. The following summer, when additional yellow fever patients were landed, another mob burned the quarantine hospital to the ground. Determined opposition by local citizens at all proposed new sites forced the quarantine officials to buy an old steamer to use as a floating hospital for yellow fever.[11] Although the fever never gained a foothold in Manhattan, every summer New York newspapers carried stories of its ravages in the South, and they rarely failed to editorialize upon its danger whenever cases were reported on incoming vessels.

In southern ports it was not necessary to revive old memories, since most residents had experienced close contact with the disease. In 1866–1867 the fever struck coastal towns from Wilmington and New Bern in North Carolina all the way to Brownsville, Texas. Desultory attacks continued until 1878, when the disease was once again widespread. On this occasion it traveled up the Mississippi Valley as far as St. Louis, Chattanooga, and Louisville. Aside from a major outbreak in Florida during 1888, only scattered cases were reported until 1897–1899 and 1905, when minor epidemics occurred in New Orleans and the surrounding areas. The 1878 outbreak, by far the most severe in the postwar years, resulted in 27,000 cases and over 4,000 deaths in New Orleans and wiped out almost 10 percent of the populations of Memphis and Vicksburg.[12]

Considering these statistics, it is not to be wondered that rumors of yellow jack or the "saffron scourge," as it was sometimes called in New Orleans, was enough to cause panic. When a reported outbreak of yellow fever in Ocean Springs, Mississippi, in 1897 led the New Orleans Board of Health to proclaim a quarantine against all Gulf Coast towns, a panic-stricken mob of New Orleans residents vacationing in one of the resorts seized control of a train and brought it to the Louisiana state line. Here the train was held up until the health officials, recognizing the hungry and desperate condition of the passengers, reluctantly permitted them to enter New Orleans. This act of mercy by the Board of Health was assailed bitterly and was a factor in the subsequent resignation of the entire board.[13]

When the disease appeared in New Orleans, the mayor arranged for one of the schools to be used as a temporary yellow fever hospital. The following night an armed mob, objecting to the presence of a hospital in their neighborhood, set fire to the building. When firemen arrived, onlookers cut the hoses, precipitating a fight between the mob and the firemen and policemen. Even as late as 1905 the reaction to the presence of yellow fever was one of profound shock. The president of the local medical society in New Orleans wrote: "When the first knowledge reached our city of the presence of this dread disease in our midst, there was almost a panic—stocks and bonds went begging, a pall seemed to be thrown on all things, a general exodus of those who could afford it took place, and the commercial interests seem paralyzed."[14]

Asiatic cholera, the most feared of all diseases in the 19th century, arrived in the western world as a by-product of the industrial revolution. Because of its short incubation period and rapid course, the disease was restricted to the Far East almost until the advent of steam power and rapid transportation. At the same time, industrialism brought massive urbanization with all its concomitant problems: crowded slums, limited and contaminated water supplies, hopelessly ineffectual methods for eliminating sewage and garbage, and city governments ill-equipped to deal with the explosive growth of population. Thus the industrial revolution provided both the rapid transportation necessary for spreading the disease and seed beds where it could flourish in the crowded cities.

Improvements in communication contributed further to enhancing the role played by cholera, for no disease in American history was so widely heralded at its first appearance (1832). The introduction of cheap newspapers and journals had made it possible for the American public to follow the disastrous course of this pestilence as it advanced through Russia, eastern Europe, and pushed northwestward to the Atlantic. The accounts of its destructive progress built up growing apprehensions which were intensified by urgent warnings from health authorities and medical societies that the filthy state of American communities had already set the stage for explosive outbursts of disease. Cholera struck the United States first

in 1832 and returned in 1848–1849. On both occasions it swept through cities and towns within a few weeks, killing thousands. In 1866 and 1873 the disease again threatened, but prompt sanitary measures limited its effect. Without knowing precisely why, health authorities recognized that the infection was spread through the feces of infected persons, and they resorted successfully to disinfecting procedures.[15]

Unlike yellow fever, which periodically demonstrated the reality of its threat, Asiatic cholera was never more than a potential danger in the years which followed the Civil War, yet it received an inordinate amount of attention from newspapers and journals in all sections of the United States. Most of the civic cleanups and sanitary campaigns were sparked by what was considered to be the imminent danger from this disease. It shared with yellow fever the capacity for creating panic and brutalizing decent citizens. Victims of Asiatic cholera were often dumped ashore by crews and passengers of river boats, much to the dismay of local residents, who occasionally left them there to die. When the disease appeared in Pittsburgh in 1849 and the Sisters of Mercy opened their hospital to its victims, meetings were held by indignant neighborhood residents and local newspaper correspondents attacked the sisters bitterly. In nearby Allegheny the same situation held true for the Reverend Passavant when he, too, offered help to cholera patients.[16]

The reaction of Americans to a threatened cholera outbreak in 1873 shows how the apprehensions aroused by earlier epidemics carried over into the postwar years. As the disease began spreading into Europe, the newspapers were filled with cholera stories, and the *New York Times* editorialized on "cholera panics." The editor of a medical journal declared that in the United States cholera was the "all-absorbing topic." Responding to demands from newspapers and medical societies, the New York City Health Department promptly began a major effort to alleviate the worst sanitary conditions within the city.[17]

A few years later, when cholera broke out in Toulon and Marseilles, American newspapers once again carried daily front-page reports of the disease. In July 1884 President Chester Arthur reflected national concern by issuing a proclamation warning state officials to be on guard. Throughout the following winter cholera continued to pre-occupy public attention. In January a group of New York businessmen organized the Sanitary Protective Society to mobilize all existing health agencies within the city. As the public clamor for action increased, the city board of health secured a special appropriation of $50,000. When the expected epidemic did not materialize, the board was given permission to retain the fund for future use. The following year Asiatic cholera was reported in Italy, and President S. Grover Cleveland was requested to prohibit all Italian immigration until the danger was over.[18]

The last major cholera scare came in 1892. Once again a state of alarm characterized the entire American seaboard. Daily front-page stories reported enormous casualties in Russia and hinted of comparable figures in western European cities. Municipal authorities, collaborating with health officials, initiated massive sanitary campaigns, checked on food and water supplies, and made preparations for the expected assault. In New York the city health department retained its summer corps of 50 physicians on an emergency basis; the St. John's Guild lent its "floating hospital" for the use of cholera cases; J. P. Morgan offered the use of a steamship to house cabin passengers from immigrant vessels during the quarantine period; and the directors of St. Mark's Hospital organized a volunteer medical and nursing corps. On the national scene President Benjamin Harrison responded to the crisis by ordering all immigrant vessels to perform a minimum 20-day quarantine. To facilitate the procedure of quarantine, the state of New York leased buildings on Fire Island for the use of healthy cabin passengers during the quarantine period. On hearing this news, the local board of health promptly deputized all citizens and prepared to resist. An armed mob lined the pier, and it was not until the governor mobilized the National Guard that the mob dispersed and passengers were able to land without being molested.[19]

Since most societies tend to operate on a crisis basis, the diseases which were most effective in precipitating social change were those with the greatest shock value. In this category it is clear that Asiatic cholera and yellow fever stood by themselves, with smallpox a poor third, and the other disorders ranking well behind. The outbreaks of yellow fever which struck the eastern seaboard from 1793 to 1795 had the immediate effect of bringing into existence temporary boards of health, which had surprisingly wide powers. In New York City, for example, the Board of Health was given the authority and funds to evacuate large sections of the city and to provide food, housing, and medical care for the poor. A permanent result of these outbreaks was the creation of the office of City Inspector, a forerunner of New York's health department. Throughout the century yellow fever scares continued to give impetus to health reform. The outbreaks in the 1850s in New Orleans and the southern states had repercussions in every eastern port and greatly strengthened the position of reformers fighting for permanent boards of health.

In the southern states, which bore the brunt of the attacks in the 19th century, yellow fever provided the chief stimulus to health reform. Two major epidemics in Louisiana in 1853 and 1854, the first of which killed almost 9,000 residents of New Orleans and the second another 2,500, were directly responsible for the creation of the Louisiana State Board of Health, the first such agency in the United States.[20] Successive epidemics strengthened this board until 1897, when the consternation aroused by the reappearance of yellow fever after an absence of several years forced the members of the board to resign and led to a reorganization of the state board and the establishment of a separate board of health for New Orleans. In 1878 the disastrous outbreak, which affected almost every major town on the South Atlantic and Gulf coasts and spread far up the Mississippi Valley, aroused the entire nation. In Memphis, a city which had not recovered from the Civil War, the loss of 3,500 residents to yellow fever brought a major social and political upheaval.[21] On the national scene, Congress reacted by passing the first national quarantine act. As the full impact of the 1878 epidemic was felt, health reformers were able to secure from Congress a second measure creating the National Board of Health. Neither of these laws proved effective; the quarantine law was weak, and the National Board of Health, after a stormy existence, virtually disappeared in 1883 when Congress eliminated its appropriation. Nonetheless, during its brief lifetime the National Board of Health did help to arouse a public health consciousness, and it paved the way for the creation of the United States Public Health Service a few years later.

Asiatic cholera, because it constituted a threat to all areas, was possibly even more significant than yellow fever. The first two waves of this disorder, 1832–1835 and 1848–1855, struck at the coastal cities and then followed the unexcelled waterways of North America. In their wake they left not only a trail of death and suffering but also a host of temporary health boards. During the first attack on Pittsburgh, for example, a ten-man sanitary board was appointed and given an appropriation of $10,000. The following year the funds were reduced to $6,000 and, as the threat of cholera receded, the board disappeared and the funds for sanitation were virtually eliminated from the municipal budget.[22] The second wave of Asiatic cholera at the mid-century coincided with the emerging sanitary movement and the peak years of yellow fever. The two diseases were largely responsible for the organization of the National Sanitary Conventions which met from 1856 to 1860. These gatherings of state and municipal health officials and representatives of medical societies were the first attempts to devise national quarantine and public health programs, and they helped lay the basis for the subsequent establishment of the American Public Health Association.

The second and third waves of Asiatic cholera played a significant role in the establishment of the Health Department of New York City. More than 5,000 New Yorkers died of cholera during 1849 and several hundred more died of it in 1854. Since sanitationists argued that cholera was the product of crowding, and the filth-and-quarantine faction believed that it was a specific communicable disease which could be kept out of the city, cholera supplied both factions in the health movement with ammunition in their effort to obtain a permanent health agency for the city. In the years following the cholera outbreaks of 1849–1854, campaigns to educate the public gradually gained momentum. Several health bills for New York City were introduced into the state legislature during the early 1860s but they all failed. At this stage the third epidemic wave of Asiatic cholera appeared, and its threat in the winter of 1865–66 led to the passage of a Metropolitan Board of Health Act for New York City. The first problem confronting the Metropolitan Board was to deal with the imminent danger from cholera. An energetic sanitary campaign combined with rigid isolation, quarantine, and disinfection measures kept the number of cases to a minimum. This 1866 attack on the United States was relatively mild and probably would have had a minor effect on New York City. New Yorkers, remembering the 5,000 deaths a few years earlier, gave full credit to the Metropolitan Board of Health. This auspicious start left a residue of good will which resulted in strong public support for the health department for many years.[23]

Repeated cholera scares continued to remind New York officials and the general public of the need for a strong health department, but it was not until 1892 that the disorder again made a permanent impact on the city. The widespread alarm touched off by cholera in that year has already been mentioned. For several years prior to it Drs. Hermann M. Biggs and T. Mitchell Prudden had been advocating the establishment of a bacteriological laboratory. Capitalizing on the general apprehension, Dr. Biggs won his point with the city Board of Estimate, and, in September 1892, New York City established the first laboratory to be used for the routine diagnosis of disease.

Possibly more important than the direct effect of epidemic diseases upon social and political reform was their indirect impact. The middle and upper classes sought to insulate themselves from the deplorable condition of the working class, but for those members who encountered the appalling infant mortality and the ravages of disease among the lower economic groups the experience was often traumatic. Moreover, as conditions in the urban slums worsened, the diseases of the poor could not be contained, and public health became a matter of concern for all the people.

Members of the medical profession were among the first to encounter the disease and misery of the poor. It was recognized that clinics and dispensaries catering to the poor were essential to medical training and research, and young physicians and surgeons were thrown into direct contact with the realities of poverty. Not surprisingly, in America physicians were among the leading advocates of public health. More significantly, since the integral relation between poverty and disease was all too obvious, they were also among the leaders of social reform.

During the terrible epidemics of Asiatic cholera and yellow fever, volunteer groups of all sorts came in contact with dire poverty, and many individuals seeking to help the deserving poor gradually came to realize that even the undeserving poor were the product of their brutalizing environment. In the South a notable example of the volunteer groups was the Howard Association, named after John Howard, the famous English reformer. Originating in New Orleans during a yellow fever epidemic in 1837, its program gradually spread to other southern cities and towns. The members were young businessmen who volunteered their services during major epidemics. The Howards, as they were called, organized massive relief programs to provide medical care for the sick poor and housing and food for their families. The willingness of these men to volunteer for work with the Howard Association evidences some degree of social conscience, but their intimate contact with poverty created a new awareness of social needs.

As far back as the 16th century it had been argued that a country's population was a major form of wealth. By the mid-19th century demography was emerging as a science, and improvements in the collection of vital statistics began to reveal the high morbidity and mortality rates in urban areas. One of the major arguments used by health and social reformers was the economic cost of sickness and death. Estimating the productivity per adult worker, they calculated the loss of productivity caused by the many deaths and added to it the cost of medical care for the sick. The validity of this argument was demonstrated clearly by the repeated epidemics of yellow fever which effectively closed down southern cities, and brought all economic activities to a halt. Throughout the 19th century most physicians and laymen believed that epidemic diseases were either propagated or nurtured in conditions of dirt and overcrowding. This environmental concept led to an assault on the atrocious tenement conditions, nuisance trades, deplorable working conditions, and other abuses.

Late in the century the bacteriological revolution turned the medical profession away from environmentalism and focused its attention upon pathogenic organisms. The germ theory had the beneficent effect of awakening the upper classes to the realization that bacteria were no respecters of economic or social position and that a man's health was dependent to some extent on the health of his fellowmen. The knowledge that the diseases of the workers who sewed clothes in their filthy tenement homes or who processed food could be spread to decent, clean, and respectable citizens served as a powerful incentive to the reform of public health. Since public health could not be separated from social conditions, the net result was an attack on poverty.

The best evidence that a concern for public health underlay much of the effort for social reform is to be found in the multiplicity of volunteer sanitary associations which sprang up in the late 19th century. In every city private groups worked to establish or improve water and sewerage systems, to clean streets, to provide pure milk for the infant poor, to remedy abuses in municipal hospitals and other institutions, and to establish dispensaries, clinics, and hospitals. Examples of these groups in New York City were the Association for Improving the Condition of the Poor, the New York Sanitary Reform Society, the Ladies Health Protective Association, the St. John's Guild, the Sanitary Protective League, the Sanitary Aid Society, and the New York Society for the Prevention of Contagious Diseases. Of the many voluntary organizations operating in New York during this period, some sought only one immediate objective and disbanded after a brief existence, others created organizations that survived for many years. What they all shared in common was the belief that a healthy population was basic to a sound society.

In glancing back over the 19th century one can safely conclude that the rapid expansion of urban areas provided fertile grounds for communicable diseases, and that these diseases were both a cause and effect of the desperate poverty which characterized so many of the cities. At the same time the frightening sickness and death rates drew attention to the deplorable condition of the poor. Dramatic outbreaks of yellow fever and cholera profoundly stirred public opinion and directly and indirectly contributed to the growth of public health institutions. Meanwhile statistical evidence was developing which showed an even heavier toll from chronic and endemic disorders. The net effect, as shown by even the most cursory reading of late-19th-century newspapers, was that public health and sanitary reform became major public issues. And for nearly all social reformers, whether their concern was with infant welfare, tenement conditions, or even political reform, the elimination of sickness and disease became a major aim.

NOTES

1   See the *Annual Report of the New York City Health Department, 1871–75* (the title varies, sometimes designated as the *Annual Report of the Board of Health . . .*).

2   *The Rudolph Matas History of Medicine in Louisiana,* ed. J. Duffy, 2 vols. (Baton Rouge, 1958), II, 438, 442–443.

3   *Annual Report N.Y.C. Health Dept., 1866–1896.*

4   *New Orleans Med. & Surg. J.,* 1887–1888, *15*: 470–474.

5   *New York Times,* July 14, 1874.

6   A. Jacobi, "The unsanitary condition of the primary schools of the City of New York," *Sanitarian,* 1892, *28*: 331–324.

7   *Annual Report N.Y.C. Health Dept. 1870–1893.*

8   *Ibid.*

9   S. E. Chaillé, "Life and death rates; New Orleans and other cities compared," *New Orleans Med. & Surg. J.,* 1888–1889, *16*: 85–100. *Ibid.,* 1884–1885, *12*: 716–717.

10 J. Duffy, "Yellow fever in the continental United States during the nineteenth century," *Bull. N.Y. Acad. Med.,* 1968, *54*: 687–701.

11 J. Duffy, *A History of Public Health in New York City, 1625–1866* (New York, 1968), pp. 101–123, 440–460.

12 Duffy, "Yellow fever in the continental United States," pp. 639–696.

13 *The Rudolph Matas History of Medicine in Louisiana,* ed. Duffy, II, 430.

14 G. Augustin, *History of Yellow Fever* (New Orleans, 1909), pp. 1061–1062.

15 For an excellent account, see C. E. Rosenberg, *The Cholera Years: The United States in 1832, 1849 and 1866* (Chicago, 1962).

16 J. Duffy, "The impact of Asiatic cholera on Pittsburgh, Wheeling, and Charleston," *Western Penn. Hist. Mag.,* 1964, *58*: 199–211.

17 *Sanitarian,* 1873, *1*: 228–229.

18 This material was taken from chapter 7 of the author's second volume on the public health history of New York City: *A History of Public Health in New York City, 1866–1966* (New York, 1974).

19 *Ibid.*

20 J. Duffy, *Sword of Pestilence: The New Orleans Yellow Fever Epidemic of 1853* (Baton Rouge, 1966), pp. 139, 167.

21 J. H. Ellis, "Memphis' sanitary revolution, 1880–1890," *Tenn. Hist. Quart.,* 1964, *23*: 59–72.

22 Duffy, "Impact of Asiatic cholera on Pittsburgh, Wheeling, and Charleston," pp. 202–203.

23 Duffy, *History of Public Health in New York City, 1625–1866,* pp. 441–446.

27

AIDS in Historical Perspective: Four Lessons from the History of Sexually Transmitted Diseases

INTRODUCTION

It has become abundantly clear in the first six years of the AIDS (acquired immunodeficiency syndrome) epidemic that there will be no simple answer to this health crisis. The obstacles to establishing effective public health policies are considerable. AIDS is a new disease with a unique set of public health problems. The medical, social, and political aspects of the disease present American society and the world community with an awesome task.

The United States has relatively little recent experience dealing with health crises. Since the introduction of antibiotics during World War II, health priorities shifted to chronic, systemic diseases. We had come to believe that the problem of infectious, epidemic disease had passed—a topic of concern only to the developing world and historians.

In this respect, it is not surprising that in these first years of the epidemic there has been a desire to look for historical models as a means of dealing with the AIDS epidemic. Many have pointed to past and contemporary public health approaches to sexually transmitted diseases (STDs) as important precedents for the fight against AIDS.[1] And indeed, there are significant similarities between AIDS and other sexually transmitted infections which go beyond the mere fact of sexual transmission. Syphilis, for example, also may have severe pathological effects. In the first half of the 20th century, it was both greatly feared and highly stigmatized. In light

of these analogues, the social history of efforts to control syphilis and other STDs may serve to inform our assessments of the current epidemic.

But history holds no simple truths. AIDS is not syphilis; our responses to the current epidemic will be shaped by contemporary science, politics, and culture. Yet the history of disease does offer an important set of perspectives on current proposals and strategies. Moreover, history points to the range of variables that will need to be addressed if we are to create effective and just policies.

In these early years of the AIDS epidemic, there has been a tendency to use analogy as a means of devising policy. It makes sense to draw upon past policies and institutional arrangements to address the problems posed by the current crisis. But we need to be sophisticated in drawing analogues, to recognize not only how AIDS is like past epidemics, but the precise ways in which it is different. This article draws four "lessons" from the social history of sexually transmitted disease in the United States and assesses their relevance for the current epidemic.

LESSON #1—FEAR OF DISEASE WILL POWERFULLY INFLUENCE MEDICAL APPROACHES AND PUBLIC HEALTH POLICY

The last years of the 19th century and first of the 20th witnessed considerable fear of sexually transmitted infection, not unlike that which we are experiencing today. A series of important discoveries about the pathology of syphilis and gonorrhea had revealed a range of alarming pathological consequences from debility, insanity, and paralysis, to sterility and blindness. In this age of antibiotics, it is

ALLAN M. BRANDT is Amalie Moses Kass Professor of the History of Medicine and Professor of the History of Science at Harvard University, Cambridge, Massachusetts.

Reprinted with permission from *American Journal of Public Health*, 1988, 78: 367–371. Copyright © 1988 by the American Public Health Association.

easy to forget the fear and dread that syphilis invoked in the past.

Among the reasons that syphilis was so greatly feared was the assumption that it could be casually transmitted. Doctors at the turn of the 20th century catalogued the various modes of transmission: pens, pencils, toothbrushes, towels and bedding, and medical procedures were all identified as potential means of communication.[2] As one woman explained in an anonymous essay in 1912:

> At first it was unbelievable. I knew of the disease only through newspaper advertisements [for patent medicines]. I had understood that it was the result of sin and that it originated and was contracted only in the underworld of the city. I felt sure that my friend was mistaken in diagnosis when he exclaimed, "Another tragedy of the public drinking cup!" I eagerly met his remark with the assurance that I did not use public drinking cups, that I had used my own cup for years. He led me to review my summer. After recalling a number of times when my thirst had forced me to go to the public fountain, I came at least to realize that what he had told me was true.[3]

The doctor, of course, had diagnosed syphilis. One indication of how seriously these casual modes of transmission were taken is the fact that the U.S. Navy removed doorknobs from its battleships during World War I, claiming that they had become a source of infection for many of its sailors. We now know, of course, that syphilis cannot be contracted in these ways. This poses a difficult historical problem: Why did physicians believe that they could be?

Theories of casual transmission reflected deep cultural fears about disease and sexuality in the early 20th century. In these approaches to venereal disease, concerns about hygiene, contamination, and contagion were expressed, anxieties that reflected a great deal about the contemporary society and culture. Venereal disease was viewed as a threat to the entire late Victorian social and sexual system, which placed great value on discipline, restraint, and homogeneity. The sexual code of that era held that sex would receive social sanction only in marriage. But the concerns about venereal disease and casual transmission also reflected a pervasive fear of the urban masses, the growth of the cities, and the changing nature of familial relationships.[4]

Today, persistent fears about casual transmission of AIDS reflect a somewhat different, yet no less significant, social configuration. First, AIDS is strongly associated with behaviors which have been traditionally considered deviant. This is true for both homosexuality and intravenous drug use. After a generation of growing social tolerance for homosexuality, the epidemic has generated new fears and heightened old hostilities. Just as syphilis created a disease-oriented xenophobia in the early 20th century, AIDS has today generated a new homophobia. AIDS has recast anxiety about contamination in a new light. Among certain social critics, AIDS is seen as "proof" of a certain moral order.

Second, fears are fanned because we live in an era in which the authority of scientific expertise has eroded. This may well be an aspect of a broader decline in the legitimacy of social institutions, but it is clearly seen in the areas of science and medicine. Despite significant evidence that HIV (human immunodeficiency virus) is not casually transmitted, medical and public health experts have been unable to provide the categorical reassurances that the public would like. But without such guarantees, public fear has remained high. In part, this reflects a misunderstanding of the nature of science and its inherent uncertainty. While physicians and public health officials have experience tolerating such uncertainty, the public requires better education in order to effectively evaluate risks.[5]

Third, as a culture, we Americans are relatively unsophisticated in our assessments of relative risk. How are we to evaluate the risks of AIDS? How shall social policy be constructed around what are small or unknown risks? The ostracism of HIV-infected children from their schools in certain locales, the refusal of some physicians to treat AIDS patients, and job and housing discrimination against those infected (and those suspected of being infected) all reveal the pervasive fears surrounding the epidemic. Clearly, then, one public health goal must be to address these fears. Addressing such fears means understanding their etiology. They originate in the particular social meaning of AIDS—its "social construction." We will not be able to mitigate these concerns effectively until we understand their deeper meaning. The response to AIDS will be fundamentally shaped by these fears; therefore, we need to develop techniques to assist individuals to distinguish irrational fears of AIDS from realistic and legitimate concerns. In this respect, many have focused on the need for more education.

LESSON #2—EDUCATION WILL NOT CONTROL THE AIDS EPIDEMIC

Early in the 20th century, physicians, public health officials, and social reformers concerned about the problem of syphilis and gonorrhea called for a major educational campaign.[6] They cogently argued that the tide of infection could not be stemmed until the public had adequate knowledge about these diseases, their modes of transmission, and the means of prevention. They called for an end to "the conspiracy of silence"—the Victorian code of sexual ethics—that considered all discussion of sexuality and disease in respectable society inappropriate. Physicians had contributed to this state of affairs by hiding diagnoses from their patients and families, and upholding what came to be known as the "medical secret." One physician described the nature of the conventions surrounding sexually transmitted diseases:

> Medical men are walking with eyes wide open along the edge of despair so treacherous and so pitiless that the wonder can only be that they have failed to warn the world away. Not a signboard! Not a caution spoken above a whisper! All mystery and seclusion. . . . As a result of this studied propriety, a world more full of venereal infection than any other pestilence.[7]

Prince Morrow, the leader of the social hygiene movement, the antivenereal disease campaign, concluded, "Social sentiment holds that it is a greater violation of the properties of life publicly to mention venereal disease than privately to contract it."[8]

During this period, the press remained reticent on the subject of sexually transmitted infections, refusing to print accounts of their effects. Reporters employed euphemisms such as "rare blood disorder," when forced to include a reference to a venereal infection. Nevertheless, magazines and newspapers did accept advertisements for venereal nostrums and quacks. In 1912, the U.S. Post Office confiscated copies of birth control advocate Margaret Sanger's *What Every Girl Should Know,* because it considered the references to syphilis and gonor-

rhea "obscene" under the provisions of the Comstock law.[9]

Enlightened physicians vigorously called for an end to this hypocrisy. "We are dealing with the solution of a problem," explained Dr. Egbert Grandin, "where ignorance is *not* bliss but is misfortune, and where, therefore, it is folly not to be wise."[10] Social reformers viewed education and publicity as a panacea; forthright education would end the problem of sexually transmitted infection. If parents failed to perform their social responsibilities and inform their children, then the schools should include sex education. By 1919, the U.S. Public Health Service endorsed sex education in the schools, noting, "As in many instances the school must take up the burden neglected by others."[11] By 1922, almost half of all secondary schools offered some instruction in sex hygiene.

Educational programs devised by the social hygienists emphasized fear of infection. Prince Morrow, for example, called fear "the protective genius of the human body." Another physician explained, "The sexual instinct is imperative and will only listen to fear." Margaret Cleaves, a leading social hygienist, argued, "There should be taught such disgust and dread of these conditions that naught would induce the seeking of a polluted source for the sake of gratifying a controllable desire."[12]

In this sense, educational efforts may have actually contributed to the pervasive fears of infection, to the stigma associated with the diseases, and to the discrimination against its victims. Indeed, educational materials produced throughout the first decades of the 20th century emphasized the inherent dangers of all sexual activity, especially disease and unwanted pregnancy. In this respect, such educational programs, rather than being termed sex education were actually anti-sex education. Pamphlets and films repeatedly emphasized the "loathesome" and disfiguring aspects of sexually transmitted disease; the most drastic pathological consequences (insanity, paralysis, blindness, and death); as well as the disastrous impact on personal relations.

This orientation toward sex education reached its apogee during World War I, when American soldiers were told, "A German bullet is cleaner than a whore." Despite their threatening quality, these educational programs did not have the desired effect of reducing the rates of infection. And indeed, sexual mores in the 20th century have responded to a number of social and cultural forces more powerful than the fear of disease.

There are, nonetheless, some precedents for successful educational campaigns. During World War II, the military initiated a massive educational campaign against sexually transmitted disease. But unlike prior efforts, it reminded soldiers that disease could be prevented through the use of condoms, which were widely distributed. The military program recognized that sexual behaviors could be modified, but that calls for outright abstinence were likely to fail. Given the need for an efficient and healthy army, officials maintained a pragmatic posture that separated morals from the essential task of prevention. As one medical officer explained, "It is difficult to make the sex act unpopular."[13]

Today, calls for better education are frequently offered as the best hope for controlling the AIDS epidemic. But this will only be true if some resolution is reached concerning the specific content and nature of such educational efforts. The limited effectiveness of education which merely encourages fear is well documented. Moreover, AIDS education requires a forthright confrontation of aspects of human sexuality that are typically avoided. To be effective, AIDS education must be explicit, focused, and appropriately targeted to a range of at-risk social groups. As the history of sexually transmitted diseases makes clear, we need to study the nature of behavior and disease. If education is to have a positive impact, we need to be far more sophisticated, creative, and bold in devising and implementing programs.

Education is not a panacea for the AIDS epidemic, just as education did not solve the problem of other sexually transmitted diseases earlier in the 20th century. It is one critical aspect of a fully articulated program. As this historical vignette makes clear, we need to be far more explicit about what we mean when we say "education." Certainly education about AIDS is an important element of any public health approach to the crisis, but we need to substantively evaluate a range of educational programs and their impact on behavior for populations with a variety of needs.

Because the impact of education is unclear and the dangers of the epidemic are perceived as great (see lesson #1), there has been considerable interest in compulsory public health measures as a primary means of controlling AIDS.

LESSON #3—COMPULSORY PUBLIC HEALTH MEASURES WILL *NOT* CONTROL THE EPIDEMIC

Given the considerable fear that the epidemic has generated and its obvious dangers, demands have been voiced for the implementation of compulsory public health interventions. The history of efforts to control syphilis during the 20th century indicates the limits of compulsory measures which range from required premarital testing to quarantine of infected individuals.

Next to programs for compulsory vaccination, compulsory programs for premarital syphilis serologies are probably the most widely known of all compulsory public health measures in the 20th-century United States. The development of effective laboratory diagnostic measures stands as a signal contribution in the history of the control of sexually transmitted diseases. With the development of the Wassermann test in 1906, there was a generally reliable way of detecting the presence of syphilis. The achievement of such a test offered a new series of public health potentials. No longer would diagnosis depend on strictly clinical criteria. Diagnosis among the asymptomatic was now possible, as was the ability to test the effectiveness of treatments. The availability of the test led to the development of programs for compulsory testing.

Significantly, calls for compulsory screening for syphilis predated the Wassermann exam. Beginning in the last years of the 19th century, several states began to mandate premarital medical examinations to assure that sexually transmitted infections were not communicated in marriage. But without a definitive test, such examinations were of limited use. With a laboratory test, however, calls were voiced for requiring premarital blood tests. In 1935, Connecticut became the first state to mandate premarital serologies of all prospective brides and grooms. The rationale for premarital screening was clear. If every individual about to be married were tested, and, if found to be infected, treated, the transmission of infection to marital partners and offspring would be halted. The legislation was vigorously supported by the public health establishment, organized women's groups, magazines, and the news media. Many clinicians, however, argued against the legislation, suggesting that diagnosis should not rely exclusively on laboratory findings which were, in some instances, incorrect. N. A. Nelson of the Massachusetts Department of Public Health explained, "Today, it is becoming the fashion to support, by law, the too common notion that the laboratory is infallible."[14] Despite such objections, by the end of World War II, virtually all the states had enacted provisions mandating premarital serologies.

Legislation is currently pending in 35 state legislatures that would require premarital HIV serologies. The rationale for such programs is often the historical precedent of syphilis screening. The logic seems intuitively correct: We screen for syphilis; AIDS is a far more serious disease; we should therefore screen for AIDS. In this respect it is worth reviewing the effectiveness of premarital syphilis screening as well as those factors that distinguish syphilis from AIDS.

Mandatory premarital serologies never proved to be a particularly effective mechanism for finding new cases of syphilis. First, physicians and public health officials recognized that there was a significant rate of false positive tests which occurred because of technical inadequacies of the tests themselves or as a result of biological phenomena (such as other infections). As the concepts of sensitivity (the test's performance among those with the disease) and specificity (the test's performance among those free of infections) came to be more fully understood in the 1930s, the oversensitivity of tests like the Wassermann was revealed. As many as 25 percent of individuals determined to be infected with syphilis by the Wassermann test were actually free of infection; nevertheless, these individuals often underwent toxic treatment with arsenical drugs, assuming the tests were correct. Beyond this, individuals with false positive tests often suffered the social repercussions of being infected: deep stigma and disrupted relationships. As many physicians pointed out, a positive serology did not always mean that an individual could transmit the disease. Because the tests tended to be mandated for a population at relatively low risk of infection, their accuracy was further compromised. Some individuals reportedly avoided the test altogether.[15]

Many of the difficulties associated with the high numbers of false positives were alleviated as new,

more specific tests were developed in the 1940s and 1950s, but the central problem remained. Premarital syphilis serologies failed to identify a significant percentage of the infected population. In 1978, for example, premarital screening accounted for only 1.27 percent of all national tests found to be positive for syphilis. The costs of these programs were estimated at $80 million annually.[16] Another study in California projected the costs per case found through premarital screening to be $240,000.[17] Moreover, premarital screening for syphilis continued to find a significant number of false positives. As these studies indicated, the benefits of screening programs are dependent on the prevalence of the disease in the population being screened. In this respect, it seems unlikely that premarital screening effectively served the function of preventing infections within marriage that its advocates assumed it would. These data led a number of states to repeal mandatory premarital serologies in the early 1980s.

Compulsory premarital syphilis serologies thus offer a dubious precedent for required HIV screening. The point, of course, is *not* that the test is inaccurate. ELISA (enzyme-linked immunosorbent assay) testing coupled with the Western Blot *can* be quite reliable, but only when applied to populations which are likely to have been infected. Screening of low-prevalence populations, like premarital couples, is unlikely to have significant impact on the course of the epidemic. Not only will such programs find relatively few new cases, they will also reveal large numbers of false positives. A recent study concluded that a national mandatory premarital screening program would find approximately 1,200 new cases of HIV infection, one-tenth of 1 percent of those currently infected. But it would also incorrectly identify as many as 380 individuals—actually free of infection—as infected, even with supplementary Western blot tests. Such a program would also falsely reassure as many as 120 individuals with false negative results.[18] Moreover, the inability to treat and render noninfectious those individuals who are found to be infected severely limits the potential benefits of such mandatory measures. With syphilis serologies, the rationale of the program was to treat infected individuals.

This, of course, is *not* to argue that testing has no role in an effective AIDS public health campaign. During the late 1930s, a massive voluntary testing campaign heightened consciousness of syphilis in Chicago, bringing thousands of new cases into treatment. AIDS testing, conducted voluntarily and confidentially, targeted to individuals who have specific risk factors for infection, may have significant public health benefits. Compulsory screening, however, could merely discourage infected individuals from being tested. This makes clear the need to enact legislation guaranteeing the confidentiality of those who volunteer to be tested and prohibiting discrimination against HIV-infected individuals.

As a mandatory measure, premarital screening is a relatively modest proposal. During the course of the 20th century, more radical and intrusive compulsory measures to control STDs, such as quarantine, have also been attempted. These, too, have failed. During World War I, as hysteria about the impact of STDs rose, Congress passed legislation to support the quarantine of prostitutes suspected of spreading disease. The Act held that anyone suspected of harboring a venereal infection could be detained and incarcerated until determined to be noninfectious. During the course of the War, more than 20,000 women were held in camps because they were suspected of being "spreaders" of venereal disease.

The program had no apparent impact on rates of infection, which actually climbed substantially during the War. In sexually transmitted infections, the reservoir of infection is relatively high, modes of transmission are specific, and infected individuals may be healthy. In the case of AIDS, where there is no medical intervention to render individuals noninfectious, quarantine is totally impractical because it would require life-long incarceration of the infected.

Compulsory measures often generate critics because such policies may infringe on basic civil liberties. From an ethical and legal viewpoint, the first question that must be asked about any potential policy intervention is: Is it likely to work? Only if there is clear evidence to suggest the program would be effective does it make sense to evaluate the civil liberties implications. Then it is possible to evaluate the constitutional question: Is the public health benefit to be derived worthy of the possible costs in civil liberties? Is the proposed compulsory program the least restrictive of the range of potential measures available to achieve the public good?[19]

In this respect, it is worth noting that compulsory measures may actually be counterproductive. First, they require substantial resources that could be more effectively allocated. Second, they have often had the effect of driving the very individuals that the program hopes to reach farther away from public health institutions. Ineffective draconian measures would serve only to augment the AIDS crisis. Nevertheless, despite the fact that such programs offer no benefits, they may have substantial political and cultural appeal (see lesson #1).

Because compulsory measures are controversial and unlikely to control the epidemic, there is considerable hope that we will soon have a "magic bullet"—a biomedical "fix" to free us of the hazards of AIDS.

LESSON #4—THE DEVELOPMENT OF EFFECTIVE TREATMENTS AND VACCINES WILL *NOT* IMMEDIATELY OR EASILY END THE AIDS EPIDEMIC

As the history of efforts to control other sexually transmitted diseases makes clear, effective treatment has not always led to control. In 1909, German Nobel laureate Paul Ehrlich discovered Salvarsan (arsphenamine), an arsenic compound which killed the spirochete, the organism which causes syphilis. Salvarsan was the first effective chemotherapeutic agent for a specific disease. Ehrlich called Salvarsan a "magic bullet," a drug which would seek out and destroy its mark.[20] He claimed that modern medicine would seek the discovery of a series of such drugs to eliminate the microorganisms which cause disease. Although Salvarsan was an effective treatment, it was toxic and difficult to administer. Patients required a painful regimen of injections, sometimes for as long as two years.

Unlike the arsphenamines, penicillin was truly a wonder drug. In early 1943, Dr. John S. Mahoney of the U.S. Public Health Service found that penicillin was effective in treating rabbits infected with syphilis. After repeating his experiments with human subjects, his findings were announced and the massive production of penicillin began.[21]

With a single shot, the scourge of syphilis could be avoided. Incidence fell from a high of 72 cases per 100,000 in 1943 to about 4 per 100,000 in 1956.[22] In 1949, Mahoney wrote, "as a result of antibiotic therapy, gonorrhea has almost passed from the scene as an important clinical and public entity."[23] An article in the *American Journal of Syphilis* in 1951 asked, "Are Venereal Diseases Disappearing?" Although the article concluded that it was too soon to know, by 1955 the *Journal* itself had disappeared. *The Journal of Social Hygiene,* for half a century the leading publication on social dimensions of the problem, also ceased publication. As rates reached all time lows, it appeared that venereal diseases would join the ranks of other infectious diseases that had come under the control of modern medicine.

Although there is no question that the nature and meaning of syphilis and gonorrhea underwent a fundamental change with the introduction of antibiotic therapy, the decline of venereal diseases proved short-lived. Rates of infection began to climb in the early 1960s. By the late 1950s much of the machinery, especially procedures for public education, case-finding, tracing and diagnosis, had been severely reduced.[24]

In 1987, the Centers for Disease Control (CDC) reported an increase in cases of primary and secondary syphilis. The estimated annual rate per 100,000 population rose from 10.9 to 13.3 cases, the largest increase in 10 years. These figures are particularly striking in that they come in the midst of the AIDS epidemic which many have assumed has led to a substantial decline in sexual encounters. Moreover, after an eight-year decline, rates of congenital syphilis have also reportedly risen since 1983. The CDC concluded that individuals with a history of sexually transmitted infection are at increased risk for infection with the AIDS virus.[25]

Despite the effectiveness of penicillin as a cure for syphilis, the disease has persisted. The issue, therefore, is not merely the development of effective treatments but the *process* by which they are deployed; the means by which they move from laboratory to full allocation to those affected. Effective treatments without adequate education, counseling, and funding may not reach those who most need them. Even "magic bullets" need to be effectively delivered. Obviously effective treatments should be a priority in a multifaceted approach to AIDS and will ultimately be an important component in its control; but even a magic bullet will not quickly or completely solve the problem.

No doubt, new and more effective treatments for AIDS will be developed in the years ahead, but their deployment will raise a series of complex issues ranging from human subject research to actual allocation. And while effective treatments may help to control further infection, as they do for syphilis and gonorrhea, treatments which prolong the lives of AIDS patients may have little or no impact on the rates of transmission of the virus, which occurs principally among individuals who have no symptoms of disease.

This suggests certain fundamental flaws in the biomedical model of disease. Diseases are complex bio-ecological problems that may be mitigated only by addressing a range of scientific, social, and political considerations. No single intervention—even an effective vaccine—will adequately address the complexities of the AIDS epidemic.

CONCLUSIONS

As these historical lessons make clear, in the context of fear surrounding the epidemic (lesson #1), the principal proposals for eradicating AIDS (lessons #2–4) are unlikely to be effective, at least in the immediate future. These lessons should not imply, however, that nothing will work; they make evident that no single avenue is likely to lead to success. Moreover, they suggest that in considering any intervention we will require sophisticated research to understand its potential impact on the epidemic. While education, testing, and biomedical research all offer some hope, in each instance we will need to fully consider their particular effectiveness as measures to control disease.

Simple answers based upon historical precedents are unlikely to alleviate the AIDS crisis. History does, however, point to a range of variables which influence disease, and [to] those factors which require attention if it is to be effectively addressed. Any successful approach to the epidemic will require a full recognition of the important social, cultural, and biological aspects of AIDS. A public health priority will be to lead in the process of discerning those programs likely to have a beneficial impact from those with considerable political and cultural appeal, but unlikely to positively affect the course of the epidemic. Only in this way will we be able to devise effective and humane public policies.

NOTES

1 J. C. Cutler and R. C. Arnold, "Venereal disease control by health departments in the past: lessons for the present," *Am. J. Public Hlth.*, 1988, *78*: 372–376.

2 L. D. Bulkey, *Syphilis of the Innocent* (New York: Bailey and Fairchild, 1894).

3 Anon., "What one woman has had to bear," *Forum*, 1912, *68*: 451–454.

4 A. M. Brandt, *No Magic Bullet: A Social History of Venereal Disease in the United States since 1880* (New York: Oxford Univ. Press, 1985; rev. ed. 1987).

5 L. E. Eisenberg, "The genesis of fear," *Law, Medicine and Health Care*, 1986, *14*: 243–249; and M. H. Becker and J. G. Joseph, "AIDS and behavioral change to reduce risk: a review," *Am. J. Public Hlth.*, 1988, *78*: 394–410.

6 A. Yankauer, "AIDS and public health," *Am. J. Public Hlth.*, 1988, *78*: 364–366.

7 R. N. Wilson, "The relation of the medical profession to the social evil," *J.A.M.A.*, 1906, *47*: 32.

8 P. A. Morrow, "Publicity as a factor in venereal prophylaxis," *J.A.M.A.*, 1906, *47*: 1246.

9 Brandt, *No Magic Bullet*, p. 24.

10 E. Grandin, "Should the great body of the general public be enlightened?" *Charities and Commons*, Feb. 24, 1906.

11 Public Health Service, *The Problem of Sex Education in the Schools* (Washington, D.C.: GPO, 1919), p. 9.

12 M. Cleaves, *Transactions of the American Society for Social and Moral Prophylaxis*, 1910, *3*: 31.

13 J. P. Pappas, "The venereal problem in the US Army," *Military Surgeon*, Aug. 1943, *93*: 182.

14 N. A. Nelson, "Marriage and the laboratory," *American Journal of Syphilis*, 1939, *23*: 289.

15 J. A. Kolmer, "The problem of falsely doubtful and positive reactions in the serology of syphilis," *Am. J. Public Hlth.*, 1944, *34*: 510–526.

16 Y. Felman, "Repeal of mandated premarital tests for syphilis: a survey of state health officers," *Am. J. Public Hlth.*, 1981, *71*: 155–159.

17 R. J. Haskell, "A cost-benefit analysis of California's mandatory premarital screening program for syphilis," *Western J. Med.*, 1984, *141*: 538–541.

18 P. D. Cleary et al., "Compulsory premarital screening for the Human Immunodeficiency Virus," *J.A.M.A.*, 1987, *258*: 1757–1762.

19 L. Gostin and W. Curran, "The limits of compulsion in controlling AIDS," *Hastings Cent. Rep.*, 1986, *16*, suppl., pp. 24–29.

20 M. Marquardt, *Paul Ehrlich* (New York: Henry Schuman, 1951).

21 H. Dowling, *Fighting Infection* (Cambridge, Mass.: Harvard Univ. Press, 1977).

22 W. J. Brown et al., *Syphilis and Other Venereal Diseases* (Cambridge, Mass.: Harvard Univ. Press, 1970).

23 J. F. Mahoney, "The effect of antibiotics on the concepts and practices of public health," in *The Impact of the Antibiotics on Medicine and Society*, ed. I. Galdston (New York: International Universities Press, 1958).

24 Cutler and Arnold, "Venereal disease control."

25 Public Health Service, Centers for Disease Control, *Morbidity and Mortality Weekly Report* (Washington, D.C.: GPO, 1987), 36: 393.

PUBLIC HEALTH REFORM

America's urban health problems reached frightening proportions during the middle years of the 19th century. Mortality rates in cities such as New York, Boston, and Philadelphia rose alarmingly, as infectious diseases raged out of control. Hordes of immigrants crowded into ill-ventilated tenements and basement hovels, while workers of all ages toiled long hours in stifling sweatshops. Although Americans living below the subsistence level were the most vulnerable, the public health movement arose primarily from concerns of the middle and upper classes and their attempts to protect their own lives and businesses.

The American movement for public health reform benefited from the English experience with sanitary reform, although our country's political institutions created unique problems. Gert H. Brieger describes the political struggle in New York to establish a permanent, effective board of health, which culminated in 1866 with the creation of the Metropolitan Board of Health, a model of its kind in America. Within a year of its birth this organization successfully met its first challenge in reducing the death rate during a cholera epidemic.

Following New York's lead, other cities and states established their own boards, and for a brief period, from 1879 to 1883, there was even a National Board of Health. But the public health movement generally remained a municipal or state affair in the 19th century, led by public-spirited physicians, sanitarians, engineers, nurses, and philanthropic groups. By the 1870s their efforts seemed to be paying off, as urban death rates declined and allocations for public health stabilized.

As we have seen, epidemics stirred cities to spend money for needed, yet expensive, projects such as sewerage, water systems, and garbage-disposal works. Many factors determined whether an epidemic would have positive or negative impact and whether support for public health would result. Existing medical knowledge defined the limits of health activity, but that alone did not determine action. Because appropriations had to pass through city councils and thus be debated in the public arena, politicians had to be convinced that health measures were both expedient and popular. Economic considerations, class, race, ethnicity, timing with regard to elections, and personalities all played a part in their decision and determined how a particular city would respond to a given threat to the public's health. Gretchen A. Condran, Henry Williams, and Rose A. Cheney analyze the effects of public health measures on lowering urban death rates.

Little progress occurred in the area of occupational health until the 20th century. David Rosner and Gerald Markowitz describe the early movement for occupational safety and health and analyze the composition of 20th-century reforms. The occupational health movement represented a coalition of interest groups: workers, reformers, journalists, politicians, business people, and professionals—crossing political boundaries—joined together to bring about changes that were perceived to be in the public interest. The authors agree that "many aspects of this movement were paternalistic," but they conclude that, nonetheless, the health of industrial workers was finally addressed because of the success in building a broad coalition.

28

Sanitary Reform in New York City: Stephen Smith and the Passage of the Metropolitan Health Bill

GERT H. BRIEGER

I

"Frenzy in the South," proclaimed the bold headline of the *New York Times* on March 10, 1865. These were the last days of the long and bitter struggle, and New Yorkers welcomed the news. Each day they read about Grant and Sheridan in Virginia and of Sherman's impressive march up through the Carolinas. On March 16, however, a different subject dominated the first two pages of the paper. Instead of the usual fare of war news, the *Times* provided its readers with the testimony of Dr. Stephen Smith presented before a joint committee of the New York State Legislature. He had given evidence about the sanitary conditions of New York City and on the urgent need for new health laws. Smith gave a stinging indictment of the municipal authorities responsible for the public health and described the miserable conditions of the city's streets and tenements. This testimony, based on a massive effort by a large group of reform-dedicated physicians, was the culmination of years of effort by numerous citizens of America's major metropolis.[1]

The horrible conditions which existed at the close of the Civil War were not peculiar to New York or to that period. As Ford has pointed out, the relationship between cellar-dwellings and disease production in New York had been discussed as early as the 1790s.[2] In 1820 David Hosack told the medical students of the College of Physicians and Surgeons that the filth in various parts of the city

was marked and that amelioration could probably not be achieved without the aid of the state legislature.[3] Benjamin W. McCready, in 1837, pointed to poor housing and insufficient space as a major source of ill-health among the workers.[4] In 1842, John H. Griscom, one of the truly important figures in the story of sanitary reform, appended to his annual report of the City Inspector's office a pamphlet entitled "A Brief View of the Sanitary Condition of the City," in which he described living conditions and the destitution and misery of cellar-dwellers. He urged upon the Common Council the necessity of better housing laws.[5]

In December 1844, Griscom delivered a discourse at the American Institute which he published during the next year as *The Sanitary Condition of the Laboring Population of New York*. He reported the health problems which faced many tenement-dwelling New Yorkers, and he urged "SANITARY REFORM." He was anxious to profit from the experience of English and French sanitarians, and their influence is evident in the text as well as the title.[6]

Griscom was one of the first to show that the system of subtenancy and the rental extortions of the sublandlord were among the principal causes of misery of so many of the city's poor. This system enabled the owner of one or several houses to rent them to a sublandlord, who in turn divided them into as many apartments as possible. He then extracted as much rent as he could, often from helpless immigrants. By making few repairs and providing little maintenance he realized a great profit. Often he owned the local store as well and fixed prices at relatively high levels. The working classes were virtually restricted to lower Manhattan because there was no means of inexpensive,

GERT H. BRIEGER is William H. Welch Professor and Chair of the Department of History of Science, Medicine, and Technology at the Johns Hopkins University, Baltimore, Maryland.

From the *Bulletin of the History of Medicine*, 1966, *40*: 407–429. Copyright © 1966 by the Johns Hopkins University Press. Reprinted by permission of the Johns Hopkins University Press.

rapid transportation which would have freed them from the packed tenement conditions.[7]

Griscom also clearly described the evils of cellar-dwellings, often soggy and lacking ventilation. Many of the cellar-apartments were below sea level. When high tides came, the rooms were submerged! During heavy rains, the streets drained into the cellars. The cellar-dwellers, or Troglodytes as they were often called, lived and slept on planks suspended well above the floors.

In contrast to McCready, who seems to have practiced and taught medicine in comparative quiet, Griscom continued active agitation for improved public health laws. He was one of the first to stress that the physicians in the public dispensaries of the city would be ideally suited for the jobs of health wardens or sanitary inspectors.[8] He was the first witness before the Select Committee of the New York Senate appointed in 1858 to investigate the health department of New York City. He then sat with the committee during its interrogation of over 20 witnesses, often interjecting lively questions and barbed remarks.[9]

This committee was appointed to investigate the "assertion that great defects exist, and great improvements are practicable in the health department and sanitary laws of the city of New York."[10] Three major questions were asked: (1) Whether the allegations were true that New York had a higher ratio of mortality than other large cities. (2) If true, what were the causes of this excess mortality. (3) What were the possible remedies.[11]

Almost unanimously the witnesses gave an affirmative answer to the first question. As to the causes, this report established what was to become a constantly recurring refrain. Almost all the witnesses ascribed the excessive mortality to overcrowded tenement houses, improper light, ventilation, and food, filthy streets, insufficient sewerage, and an almost total lack of a regularly constituted and effective department of health.

Contrary to the arguments put forward by the City Inspector, Mr. Morton, the committee stated that a properly constituted health department would require the talents of the best educated men, well versed in the recent advances of medical science. Not since 1844 had a physician been City Inspector. "A man when he wants his watch repaired," argued Dr. John McNulty, "does not take it to a shoemaker. . . ."[12]

The City Inspector and his men disagreed. Mr. Richard Downing, Superintendent of Sanitary Inspection, claimed that knowledge of the law, not medicine, was necessary. It did not require medical knowledge, he said, to smell an odor or to recognize a filthy street.[13] Mr. Morton added his feeling that physicians would not do the job; they would think it undignified to "go running through tenement houses and sticking their noses down privies, to see if they were healthy or not. . . ."[14]

For two years prior to the Senate committee's investigation, the leading medical society, the New York Academy of Medicine, petitioned the legislature for modifications of the health laws. In 1856 the academy sent a memorial to Albany which stated that "a large portion of the annual mortality of this city results from diseases, whose causes are more or less within our control, but which are totally unchecked by any public administration of proper sanitary precautions, and that from this neglect, in addition to a very great and unnecessary loss of life, the city and State endure an incalculable detriment to their commercial and moral interests."[15] No bill was passed at the 1857 session.[16] Nor was a renewal of the petition successful the following year.

In the meantime, however, the legislature did pass the Metropolitan Police Bill, in 1857, an important model and precedent for future health legislation. By transferring to state control the city's police force, the Republican state legislature created a new police district comprising the counties of New York, Kings, Richmond, and Westchester. The Board of Police was to consist of five commissioners, plus the mayors of New York and Brooklyn. New York's Mayor Fernando Wood resisted the new law and kept control of his original municipal force—most of whom had voted for him on the Democratic ticket. Only after rioting in the streets with the municipal faction, the arrest of Mayor Wood, and the use of the state militia, did the Metropolitan Police finally win their right to act as the city's legally constituted guardians of the peace.[17]

A recent historian of this episode pointed out how the situation was complicated by a growing political and social cleavage between New York City and the rest of the state.[18] Each mayor, in his annual messages of the succeeding years, used the argument that local problems, such as sanitation, should be kept under local control. Thus the sanitary reform measures, suggested repeatedly between 1856 and 1866 by the leading physicians of New York, took on increasingly complex political overtones. Health bill advocates found themselves fighting, not only for good health and efficient sanitary administration, but against political corruption and the control of Tammany Hall over the city.[19]

In the meantime, the New York Sanitary Association had been founded in January 1859. The members immediately took up the fight for the health bill before the legislature, then in session. John Griscom and Elisha Harris were officers of the group, and Stephen Smith, Joseph M. Smith, and Peter Cooper were among the members of the council.[20]

That winter the association impressed upon the legislators the urgent need for a health bill. They used established arguments: New York's ratio of mortality was greater than that of most cities in the United States and western Europe; those diseases which most contributed to this mortality were due to the absence of proper sanitary administrations and were just those diseases thought to be preventable; New York had three separate health authorities, none of which functioned properly. These were the Board of Health, composed of the Mayor, Aldermen, and Councilmen and rarely in session; the Commissioners of Health, including the Mayor, the Presidents of the Boards of Aldermen and Councilmen, the Resident Physician, and Health Officer of the Port; and the City Inspector's Department.[21] The Sanitary Association pointed out that there were about 112 individuals directly and indirectly supposedly concerned for the health of the people, but "that there is not one who feels it to be required of him to take note of, or to use any effort whatever to check the immense amount of disease. . . ."[22]

Once again the story was the same. Their bill failed to pass.

By the end of 1860, the Sanitary Association could say that its meetings had been well attended and that it considered itself a permanent organization. It continued to act as a lobby group in Albany. The membership increased to over 250, representing the professions of law, medicine, education, and divinity.[23]

As late as 1862 the group continued to hear interesting papers, but the pressures of the war seem to have caused a cessation of its activities sometime in that year.[24]

While I have concerned myself primarily with matters of the public health, problems of personal health were not ignored. The sanitarians had the twin aims of improving the health laws and of educating the people, especially the poor, in the ways of proper hygiene. This, indeed, was also the aim of the numerous philanthropic organizations that flourished in New York at mid-century. The popular magazines and the newspapers often contained articles on health or advice on matters pertaining to diet or epidemics. One author, in 1856, thought that Americans should be the healthiest people in the world. If they were not, it was due to the hectic pace of life, with too much work and too little play.[25]

Medical teachers did not neglect the subject of hygiene, although often admitting that, in this area, ". . . the profession of medicine has hitherto grievously failed"[26] — this, even though the medical profession had expended an incredible amount of time and talent upon the subject of public and private hygiene in the quarter century before the Civil War.[27] And so, too, many doctors believed that hygiene and not quarantine was the true law of health.[28]

II

When the *New York Journal of Medicine* ceased publication in 1860, Stephen Smith shifted his editorial chair to the newly established *American Medical Times*, and he began four years of vigorous crusading in a large variety of areas. The older journal had contained few editorial comments in its bimonthly numbers, while its successor, a weekly, published lengthy editorials in every issue. In his writings he concerned himself with the role of medicine and the medical profession in society, including frequent expositions on wartime problems.[29]

Since Smith was a New Yorker, and because his journal was published in that city and presumably found most of its readers there, he usually devoted himself to local sanitary problems — primarily the need for legislative reform and the necessity for vigorous action on the part of the medical profession. In his first editorial in the new journal he set forth many of the precepts he planned to follow. He singled out the subject of hygiene to receive vigilant and faithful attention.[30] In the third number he elaborated on these ideas and stated his intention periodically to examine the more important questions relating to sanitary and quarantine systems of American cities and particularly the role and duty of the medical profession.[31] Despite the potential usefulness of his public health editorials nationally, he consistently focused on New York City.

During the summer of 1860 things looked bleak indeed. The country was threatened by division, political feelings ran high, and local sanitary problems continued to increase. More and more immigrants had to be fed and housed as each month passed. Smith shared the pessimistic mood of late summer when he noted that amidst legislative corruption it was not at all certain that improved health laws could be obtained: "And such are the necessities of the people, such the jeopardy of life and health as well as commercial interests, that our population cannot safely await the good time coming, when good laws and municipal reform shall effect the sanitary improvements now demanded. From various quarters the question comes up — What shall be done?"[32]

His answer was clear: he believed that more pressing than questions of quarantine were those relating to civic hygiene. More attention had to be devoted to municipal sanitary arrangements, especially to those of New York.[33]

The cause of sanitary reform moved forward slowly during the Civil War. Although official population figures showed a slight decrease in the 1865 census as compared with 1860, the city's municipal problems became worse.[34] Stokes noted that the city's growth during the war had been checked, but not stopped entirely. Fewer buildings were erected and the misery of crowded tenements grew. The increase in production and the resultant higher wages probably did elevate the general standard of living, but often prices increased too so the poorest classes were left with a net loss.[35] The poor did not share in the profits of rising real estate values and the general business expansion — in fact these things operated to their disadvantage. The increase in luxury which "struck every observer" falls short of describing the condition of more than half of the population — the tenement dwellers.[36]

Although wages rose, prices rose faster. Eggs, 15 cents in 1861, rose to 25 cents by the end of 1863; potatoes went from $1.50 to $2.25 per bushel during the same period. The increase in wages was generally about 25 percent, or less than half the increase of prices.[37]

Citizens interested in sanitary reform worked on through the war years. They helped introduce a bill into the state legislature each winter, but it failed to pass with each succeeding session. The daily papers and popular journals continued to clamor for both municipal and sanitary reform, building up to quite a pitch in the year prior to the success of 1866 — the Metropolitan Health Bill.

The medical press was also active, especially the influential *American Medical Times*. Others, besides Stephen Smith, participated in the effort; but it is on Smith's role I wish to focus. I should note that his work in the early 1860s was of much broader scope than my emphasis on sanitation would indicate.

Prior to 1864, Smith's efforts in behalf of a Metropolitan Health Bill were confined chiefly to the numerous editorials praising each bill, exhorting his fellow physicians to exert influence upon the legislature, and bemoaning the lack of a properly constituted health department in the city.

In December 1860, he asked: Will the next legislature provide a sanitary code for the city?[38] He pointed out that more than a fourth of the state's population resided in New York and Brooklyn. He noted that it was widely acknowledged that these million and a quarter people were "living under one of the most corrupt and corrupting municipal governments in the civilized world, and that reform without the interposition of State legislation is impractical." To call the existing Health Department by that name was a misnomer. "It does little for health, but much for disease and death."[39]

Early in 1861, Smith aimed his editorial guns in a violent attack on the City Inspector, who was to become a favorite target for the next five years.[40]

The City Inspector was really the only active health official in New York. It is true that there was a Physician of the Port and a Resident Physician, who, with the Mayor and Presidents of the Boards of Councilmen and Aldermen, constituted a Commission of Health. In fact, however, in matters other than quarantine, it was mainly the City Inspector and his 44 health wardens who looked after the sanitation of the city.[41]

In his report for 1860, the inspector, Mr. Daniel E. Delavan, a two-year incumbent in the post, claimed a healthy condition for New York when compared to European cities.[42] What problems there were he attributed mainly to immigrants. His report admitted many of the sanitary problems of New York and he gave some reasons for them. In the first place, he was very critical of the medical profession for their failure to cooperate in the proper registration of vital statistics. Thus he noted the paradox of "opposition among the very class whose leading spirits have been most active for some years past in this city in urging the cause of sanitary reform."[43] He admitted to the filthy condition of the streets, stating that the Common Council had removed $50,000 from his budget and that in November 1860 money for street cleaning ran out; 300 miles of paved streets was a lot to sweep.[44] Mr. Delavan also was opposed to having the state legislature do for the citizens what they could best do for themselves. With a final thrust aimed at those working for reform, he said, "Nor is it necessary for the further efficiency of this department that it should become the nursery of students of medicine. . . ."[45]

Smith dealt harshly with this report. He noted that it contained ". . . its usual variety of loose and often absurd statements in regard to the public health, and deductions, the result of the most profound ignorance of sanitary science."[46] He objected mostly to the assertion that New York was a healthy city. The ratio of 1 death in 36 of its population made the mortality rate the highest of civilized cities.

Early in 1862, Smith bemoaned the singular indifference, which he felt was evident in the city, toward the fearful living conditions of most of its people.[47] The *Medical and Surgical Reporter* of Philadelphia agreed and said that the Augean stable had to be cleansed and a Hercules of sanitary science was needed to do the job.[48]

In February 1862 Smith became more optimistic about the possibility of a health bill. There was good evidence at last that a reorganization of the health department was to take place. Perhaps he saw a turning point in the road to victory when he noted that, "the question which is presented this winter is not, Shall there be a reform, but, What shall be its character?"[49]

Optimism was short-lived. In early May he be-

gan an editorial plaintively announcing the adjournment of the legislature without the enactment of a health bill. The metropolitan concept introduced into the proposed bill was distasteful to the mayor. Smith decided that Mayor Opdyke had really joined the "Ring" and thereby had aided the defeat of "this most righteous measure."[50]

Others, too, "confessed to a very great disappointment" at the failure of the bill. The *Medical and Surgical Reporter* entitled its editorial, "Health Bills and a Diseased Body Politic." "Politicians, those curses of our country . . ." doubtless were to blame, said the *Reporter*.[51] And so again New York was left to its own sanitary devices, devices that were due to receive shocking public description and denunciation in the succeeding few years.

In what had become a cycle of editorial moods, Smith seemed most discouraged toward the end of 1862. It was an extremely busy year for him, personally. Besides the weekly editorial writing and editing of the *Medical Times*, he wrote a manual for military surgeons, used in the Civil War; he continued active teaching and practice at the Bellevue Hospital and its new Medical College, where he was the first Professor of the Principles of Surgery; and he served a stint in the military hospitals of Virginia as an acting assistant surgeon. But the sanitary reform of New York was still a pressing concern for him.

In November he wrote that prospects for a bill seemed discouraging, that many had been led to believe that subsequent efforts would lead nowhere and hence should be abandoned indefinitely. Smith disagreed. While thousands of New Yorkers were dying annually of what he firmly believed were preventable diseases, and while half the city's population lived in the cheerless, sunless, and airless tenements, it was unthinkable to yield in the struggle. Ceaseless agitation would be required. His feelings were perhaps neatly summed up when he noted in November of 1862, "we should not, however, lose sight of the fact that we are striving to accomplish a reform which in importance and in magnitude rises superior to all civil, social, religious, or political questions of the time."[52] Allowing for the exaggeration and zeal of the reformer, this statement still, I believe, illustrates the extent of his commitment to sanitary reform, especially in view of the state of the nation's political and economic health.

He described the efforts he envisioned from legislative enactments: First, they should protect the citizen, especially the impoverished one, from disease and thereby lengthen life; second, they would develop a strong and healthy generation of citizens; and third, all health reforms would add greatly to the sum of human happiness. So fully impressed was he with the importance of sanitary reform that he felt, even though the prospect of success was not as great as it had been the year before, "we ought to put forth increased energy instead of relaxing our efforts."[53]

Public apathy was a major obstacle in the path of sanitary reform. This unconcern was as prevalent in much of the medical profession as it seems to have been in the general public. Numerous writers appealed to the educated, the rich, the influential, to raise their voices in protest. *Harper's Weekly* noted, for instance, "There is certainly no city in the world where intelligent and decent people surrender themselves to a band of knaves with such good humor as in New York."[54] The *Medical Times* noted that: "The country is horrified when a thousand fall victims in an ill-fought battle, but in this city 10,000 die annually of diseases which the city authorities have the power to remove, and no one is shocked."[55]

December 12, 1863, marked a turning point. On that day a group of the leading lawyers and merchants of the city, including Peter Cooper, John Jacob Astor, Jr., August Belmont, and Hamilton Fish, formed the Citizens' Association. They were organized "for purposes of public usefulness."[56]

At an association meeting two months later, a committee was formed to solicit from the medical profession the "fullest and most reliable information relative to the public health." Shortly, 24 of the city's leading medical men received a letter from the association—among them: Valentine Mott, Willard Parker, Stephen Smith, John H. Griscom, Elisha Harris, Austin Flint, Frank H. Hamilton, and Gurdon Buck.

Only a week later, on March 9, the physicians answered the committee's request for information. The doctors pointed out that although New York with its many natural advantages ought to be one of the healthiest cities, the exact reverse was the case. They believed the high mortality rate was a reliable index of the city's miserable health conditions. They provided comparative statistics: mortality in New York was 1 in every 35.7 of the population, while in Philadelphia it was 1 in 43.6, in Boston 1 in 41.2, and in Hartford as low as 1 in 54.8. The city fared poorly in comparison to London and Liverpool as well. They pointed to Lyon Playfair's figures in Great Britain, which showed that for every death there were at least 28 cases of illness.[57]

In the meantime, the Citizens' Association busied itself with the task of lobbying for the health bill then being considered in Albany. Representatives appeared before the legislature on March 15 and 16, only to meet resistance from both Democratic and Republican members. The former regarded the sweeping measures proposed under a metropolitan, nonpolitical health department as being aimed at their friends, which indeed it was. The Republicans were reluctant to interfere in city affairs, thereby hurting their chances in the upcoming presidential election that fall. The members of the Citizens' Association found, according to Stephen Smith, that to many members of the legislature, "the death of five thousand citizens was not so serious as the possibility of a presidential defeat."[58]

It must have been clear, at this point, to the Citizens' Association and the friends of sanitary reform, that what they really needed was a set of clear and extensive facts about the health conditions of the city—facts that would overcome the inertia of some legislators and facts which would once and for all clearly disprove the data which the City Inspector's Office had for so long been bringing to Albany to controvert any proposal for better laws.

A Council of Hygiene and Public Health was formed in April within the Citizens' Association. Its president was Joseph M. Smith, a prominent physician and writer who had made a monumental study of the epidemics of New York State.[59] Elisha Harris, a close friend and neighbor of Stephen Smith, was made secretary. The council consisted of 16 physicians, many of whom had been recipients and signers of the letters noted above.[60]

To gather the necessary facts about the true health and sanitary conditions of New York and its three-quarters of a million residents, an extensive survey was planned. This survey was organized and directed primarily by Stephen Smith.[61] It was to become, according to numerous commenters, the most complete sanitary survey ever made, and certainly an important landmark in the history of public health in America.

The survey began early in May, got under full swing by July, and was completed by mid-November.[62] The city was divided into 31 districts, each inspected thoroughly by a physician. It was intended, by means of the survey, to arrive at "positive knowledge of the amount of preventible disease existing in New York, the location of insalubrious quarters, the peculiar habitats of typhus, smallpox, . . . and the conditions on which the alarming prevalence of these diseases depend."[63] A month later, in late July, Smith noted, "It is nothing less than a full and accurate inquiry into the causes of disease in this city by competent medical men."[64] And thus it should also be credited as a landmark in the history of epidemiology.

Although there were numerous etiological theories current in 1864, in the sanitarians' view the environment played the most important part, especially the so-called localizing causes, or those which promoted the prevalence of disease in particular localities. Each of the 31 inspectors reported on his district and described his findings mainly in terms of cleanliness and filth.[65]

The inspectors were, for the most part, young physicians who were employed by one of the public (charity supported) dispensaries of the city. They were ideally qualified, for their patients mostly came from the poorer districts; moreover, they had had experience in visiting the tenements. They were given only token compensation ($40 per month) for their labor. Their reports, charts, maps, and diagrams filled 17 folio volumes. On his return from his work on behalf of the U.S. Sanitary Commission, Dr. Harris edited the 360-page report and added a 143-page introduction. The report was published in April 1865. The reviews of the book were uniformly laudatory, and many pointed out the great importance of the survey and the *Report* for the future of sanitary government in New York.[66]

The *Report* has been frequently cited in discussions of American public health and of housing problems. Indeed, it deserves to be ranked very high among primary documents, not only in the history of public health reform but in epidemiology as well. Furthermore, it affords an extremely detailed look at some aspects of the way of life in each of the wards of New York in 1864, and it should be of great interest in the study of urban history.

Epidemiology has been defined as "the study of the distribution and determinants of disease prevalence in man."[67] According to this concept, then, the *Report* belongs among the most important of 19th century epidemiological studies. It contains graphic, statistical, and descriptive information on population, number and size of tenement houses, prevailing diseases, schools, churches, stores, slaughter-houses, factories, brothels, drinking establishments, sewerage, streets, and topography.

Its impact was manifold. Certainly it did not "drop stillborn from the printer's hand," as had been the case with the Shattuck *Report* 15 years earlier. Instead, as Kramer has pointed out for the Chadwick *Report*, it was a document that was alive; it aroused indignation and wonderment; it had emotional appeal beyond its intellectual content. And it led to effective legislation.[68]

Besides the descriptions of overflowing privies with their nauseous odors, garbage, offal, ashes, and generally dirty streets, the inspectors also wrote about slaughterhouses, fat- and bone-boiling establishments, and stables, all dispersed among the tenements. The streets were the main focus of complaint from the magazines and newspapers of the day as well. Youngsters, it is said, could easily earn nickels by standing along Broadway and sweeping a path through the muck for those who wanted to cross.[69] On Thirty-ninth Street, the inspector reported that blood and liquid animal remains flowed for two blocks from a slaughterhouse to the river.[70] Another great source of nuisance was the wooden garbage box. These usually rotted, allowing liquid contents to flow out. It seems they also provided a ready source of wood for political bonfires.[71]

Occasionally the survey itself was responsible for immediate improvements. In parts of the third district the inspector noted progress with each succeeding visit he made.[72]

It was armed with the data from this comprehensive epidemiological study of New York City that Stephen Smith appeared before the legislature in Albany on February 13, 1865. The report had not yet been published, but Smith in his testimony, to which I alluded at the opening of this paper, quoted widely from it. He spoke before the joint committee of the Senate and Assembly, presided over by Andrew D. White. The committee had already heard testimony from Mr. Dorman B. Eaton on the legal aspects of the proposed bill.[73] Smith told them that he and the Citizens' Association had been inspired and aided by the work of similar organizations in Great Britain, notably the Health of Towns Association.

He further told the legislators that the best method of arriving at a complete understanding of the existing causes of disease was by a house-to-house inspection. Since it was disease that was the object of study, it could only have been carried out properly by sufficiently trained men, viz., physicians.

He described conditions in general terms of cleanliness, stating that the degree of public health of a town was to be measured by its cleanliness and that in no way was the sanitary government of New York to be commended. He called the City Inspector's department a "gigantic imposture." The 22 health wardens and an equal number of assistants were grossly ignorant of sanitary matters, but that was to be expected in view of their backgrounds, since they were liquor store owners, local politicians, stonemasons, and carpenters. Not only were they ignorant of medical matters, but Smith also accused them of unwillingness to visit houses where known cases of disease existed. He told of one health warden who sent for an attendant of a smallpox patient in an upper room. Ordering the attendant not to approach too closely, he then advised: "Burn camphor on the stove, and hang bags of camphor about the necks of the children." Smith then asked the members of the Senate and Assembly:

> To what depth of humiliation must that community have descended, which tolerates as its sanitary officers men who are not only utterly disqualified by education, business, and moral character, but who have not even the poor qualification of courage to perform their duties?

He ended his long testimony with the recommendation that New York heed the experience of other large cities in establishing a well-organized health board. That board, he opined, should be independent of politics and above partisan control. Furthermore the board must combine administrative ability with a knowledge of disease and its prevention. For this reason he felt that the composition

of the board should include medical and nonmedical members. His testimony, although well organized and at times forceful as well as eloquent, was not original. It represented the consensus of most of his coworkers. It is to his credit, however, that the problems of sanitary reform were continually held up before the medical profession through his editorials in the *American Medical Times* and now were brought before the general public, with the aid of the *New York Times*. Equally to his credit was it that he had helped to write the medical portion of the proposed bill and, together with Dorman B. Eaton, a lawyer and a keen student of sanitary laws, had drafted the final version.

Why the *Times* hesitated for over a month before publishing his speech I cannot explain with certainty. It is entirely possible that when the testimony was given on February 13, the prospect of legislation was bright; but that by March 16, when the *Times* printed the speech, the bill was already in dire straits.[74] Henry Raymond and the *Times* were deeply committed to a health reform measure, as is attested by frequent editorials during the winter of 1865 and again during the next session in 1866. It may well be that the *Times* felt that publication of the facts would serve to bring pressure to bear upon the reluctant legislators.[75] That it did not achieve this result is a matter of record. But now the issues and the facts were clearly before the public. Because part of the Citizens' Association survey was published in the *Times* it achieved a wider public circulation for writing on the subject of sanitation than had ever been the case before. This was a milestone in New York history.

Unfortunately, Smith's testimony in February, its publication in March, and the subsequent appearance of the printed *Report* were not enough to sway the lawmakers in Albany. Frequent editorials in the *Times*, the *Tribune*, and the *Citizen*, a weekly paper founded by the Citizens' Association in 1864, did not seem to help either.

On January 16, 1865, the *Times* noted that typhus and smallpox were running rife. The "ignorant men called 'Health Wardens'" were receiving annually an aggregate of nearly $50,000, but in the opinion of the newspaper, "The only persons who are doing anything for the public health are the agents of the Citizens' Association."

On March 3, the *Tribune* reported the continuing discussion in Albany and the testimony that had been given for the bill by the members of the Citizens' Association and against it by Francis I. A. Boole, the City Inspector; Lewis A. Sayre, the resident physician; and Cyrus Ramsey, the Registrar in the City Inspector's department. According to the *Tribune*, Ramsey attempted ineffectually to controvert the statements made by Smith, Eaton, and the friends of reform. The *Tribune*, somewhat incredulously, noted that Ramsey was driven to extremes in support of the existing corrupt system when he even went so far as to ridicule the idea that cleanliness was an important source of health.[76]

As March progressed, the *Times* pointed out that the Democratic members in the legislature hung together on every question, but not the Republicans. The health bill had, despite pleas from many sides, once again become a political and partisan issue.[77] The *Tribune* said, "To lose the rich *placer* of the City Inspector's department is to cut the winds of scores of active workers, whose only duty is to sign the payrolls and work for the party that gives them fat sinecures."[78]

On April 12, three days after Appomattox, the bill finally came out of committee into the House. Two days later, amid "perfect bedlam," it was defeated. Most New Yorkers, however, were probably much too dazed and saddened to read the account from Albany on the morning of April 15. The headline that day proclaimed "Awful Event."[79] It is not likely that in the days following the tragic and shocking death of Lincoln, those who had labored so hard that year for a health bill had any time for grief on its account.

In 1865 cholera threatened New York again, and the press of the city became increasingly alarmed over the prospects of another epidemic. The summer passed, however, and with a sigh of relief those concerned felt that with the approaching cold season the city would be safe, at least temporarily. But the need for legislative action became more acute.[80]

On November 9, the *Nation* claimed that New York was nearly as filthy as the Asiatic towns from which the cholera came.

> It is awful yet comical to learn that the Board of Health, which at such a crisis ought to reign supreme, is such a disreputable body that, nobody having the power to adjourn it if once organized, the Mayor is afraid to call it together.[81]

The *Times*, a day later, noted that the Mayor considered the cholera the lesser of the two evils, and so left the Board alone.[82]

As 1866 opened, Mayor John T. Hoffman, a Tammany leader, stated in his first message to the Common Council that the Board of Commissioners of Health, would be "able to accomplish all that may be required of it." He was against a metropolitan bill, and he used the old argument that such a bill would be an interference with the municipal rights of New York.[83] The *Times* retorted that the only branches of the city government managed with honesty were Central Park, the Police Department, and the charities on Blackwell's Island — all created by the state legislature. As for the *Times*, it felt that, "We would prefer to live under a 'Legislative Commission' to dying prematurely and painfully under the pure Democratic rule of a city constituency."[84]

All through January and February the *Times*, and occasionally the *Tribune*, continued to press for a bill. After a complicated fight between the Senate, which passed a bill including the names of four physicians who were to be commissioners, and the Assembly, which wanted to allow the governor to name the members of the board, the whole thing was nearly scuttled once again.[85] Success was finally achieved on February 15. The Assembly passed an amended bill allowing the governor to name the commissioners. With the aid of an impassioned speech by Senator Andrew D. White, using some of the testimony Stephen Smith had presented the previous year, the Conference Committee of the two houses settled their differences and received assurances from Governor Fenton that none of his appointments would be on a political basis.[86] On February 21, 1866, the *Times* felt that the final victory for a health bill was not to be passed over without public notice. The dedicated physicians and philanthropists who, for ten years, had come each winter to Albany received due praise.

The official date of passage was February 26, and the title of the law was "An Act to Create A Metropolitan Sanitary District and Board of Health therein for the Preservation of Life and Health and to Prevent the Spread of Disease."[87]

The law, which in essence had been drafted by Stephen Smith and Dorman B. Eaton the year before, created a health department for the metropolitan area of New York. The new Metropolitan Board of Health, as it was called, was given extremely broad powers to make laws, to carry them out, and to sit in judgment of them, all at the same time. Questions of constitutionality would soon arise.[88]

The general implications of this bill of 1866 are several. It was undoubtedly a major triumph in the history of public health, as noted by Rosen.[89] Locally it served to give a great city the beginning of really effective sanitary government, carried out by professionals. Furthermore, as a major piece of reform legislation it may have been one of the beginning moves against a thread of corruption so strong that it was not broken until the Tweed "Ring" was finally deposed in the early 1870s. The Reverend Samuel Osgood may have had the health bill in mind when he said:

> Careful legislation, with intelligent suffrage and a city government more on the plan of the national, and taking from the Common Council its temptations to base jobs, will set us right, and free us from being subject to the dynasty of dirt and sovereignty of sots.[90]

On a broader scale, the sanitary reform of New York City had several more specific results. In the first place it was the first comprehensive health legislation of its kind in the United States, and later it was to serve as a model for numerous local and state bills. In this respect too the reform movement seems to have united the sanitary interests of numerous physicians and laymen alike. Sanitary science was becoming a specialty in this country, as in Europe. The work in New York also led to the formation of the most important national health group, the American Public Health Association. Stephen Smith, who was one of the prime movers in its founding and its first president, gave credit to his work for the Metropolitan Health Bill and later as a commissioner on the board for providing their inception of the A.P.H.A. in 1872.[91]

Finally, the sanitary reform work in New York also played a role in the changing status or image of the physician and of medicine as a whole. Kramer has noted that before public health could be undertaken, medicine had to put its house in order.[92] But the reverse may have been even more the case: Effective sanitary legislation and the organization of competent health departments played a great part in helping medicine to reestablish the much needed order in the house.

NOTES

A portion of this paper was presented to The Johns Hopkins Medical History Club, March 14, 1966, and at the Hixon Hour, University of Kansas Medical Center, April 18, 1966. It is part of a larger project, a biographical study of Stephen Smith, in which I am presently engaged. See *Bull. Hist. Med.,* 1965, *39:* 85.

This investigation was supported by U.S. Public Health Service Training Grant number 9T1-LM-105-06.

1   Smith had presented the evidence to the legislative committee on Feb. 13, 1865. It was published in the *New York Times* a month later and was reprinted in Stephen Smith, *The City That Was* (New York: Allaben, 1911), ch. 4. The latter version contained only very minor changes.

2   James Ford, *Slums and Housing, with Special Reference to New York City, History, Conditions, Policy,* 2 vols. (Cambridge, Mass.: Harvard Univ. Press, 1936), I, 17–204. Ford describes many of the early health ordinances. The amount of space he has devoted to sanitary matters is an indication of the close relationship of health and housing. There are a number of other works that deal with health conditions and sanitary laws prior to 1866. A few examples are Susan Wade Peabody, "Historical study of legislation regarding public health in the states of New York and Massachusetts," *J. Infect. Dis.,* 1909, Suppl. No. 4, 156 pp.; Charles F. Bolduan, "Over a century of health administration in New York City," *Department of Health Monograph Series,* No. 13, 1916; John Blake, "Historical study of the development of the New York City Department of Health," typescript, *c.* 1952, 128 pp.; Charles E. Rosenberg, *The Cholera Years: The United States in 1832, 1849, and 1866* (Chicago: Univ. of Chicago Press, 1962); George Rosen, "Public health problems in New York City during the nineteenth century," *N.Y. St. J. Med.,* 1950, *50:* 73–78; and Howard D. Kramer, "Early municipal and state boards of health," *Bull. Hist. Med.,* 1950, *24:* 503–509.

I should also say at the outset that the city did have a health organization during the years prior to the Metropolitan Health Bill of 1866. A history of the department is in the process of being compiled by Professor John Duffy, who will give in great detail what I have perhaps too much simplified in this paper. [Editors' note: Since the original publication of this article, Duffy's book has been published: John Duffy, *A History of Public Health in New York City,* 2 vols. (New York: Russell Sage Foundation, 1968–1974).] It was the "felt reality" of the time, however, according to many physicians, that New York did not indeed have a health department worthy of that name. *Reports, Resolutions, and Proceedings of the Commissioners of Health of the City of New York for the Years 1856–1859* (New York: Clark, 1860), reveals that meetings were frequent, sometimes even daily, but that mostly they dealt with quarantine matters and occasionally with removal of nuisances.

3 David Hosack, "Observations on the means of improving the medical police of the city of New York," in *Essays on Various Subjects of Medical Science,* 2 vols. (New York: Seymour, 1824), II, 9–86.

4 Benjamin W. McCready, *On the Influence of Trades, Professions, and Occupations in the United States, in the Production of Disease,* ed. Genevieve Miller (Baltimore: Johns Hopkins Press, 1943), pp. 41–45.

5 John H. Griscom, *Annual Report of the Interments in the City and County of New York for the Year 1842, with Remarks Therein, And a Brief View of the Sanitary Condition of the City* (New York, 1843). See also Lawrence Veiller, "Tenement house reform in New York City, 1834–1900," in *The Tenement House Problem,* ed. Robert W. DeForest and Lawrence Veiller, 2 vols. (New York: Macmillan, 1903), I, 71–75, in which Veiller has included long quotes from Griscom.

6 Not only were New Yorkers influenced and inspired by the work of the Parisian and London sanitarians, but there were frequent allusions to the mortality rates in these cities, as compared to New York. New York usually fared second best. See also "Health: New York versus London," *Hunt's Merchant's Magazine,* 1863, *48*: 120–124; and "Health of New York, Philadelphia, and Baltimore, for 1860," *Am. Med. Monthly,* 1861, *15*: 312–316. It was especially galling to New Yorkers that the over-all mortality rate of the United States was far lower than that of England (15 per 1000 *v.* 23 per 1000) yet in New York City it was much higher (36 per 1000). See *Report of the Committee on the Incorporation of Cities and Villages, on the bill entitled "An Act concerning the Public Health of the counties of New York, Kings, and Richmond,"* New York State Legislature, Assembly Doc. No. 129, 1860.

7 The problem of tenements and housing reform has been fully dealt with by others. The role of housing in sanitary reform was a central one. See for instance, Ford, *Slums and Housing*; DeForest and Veiller, *The Tenement House Problem*; Gordon Atkins, *Health, Housing, and Poverty in New York City 1865–*

1898 (Ann Arbor: Edwards, 1947), which includes a good discussion of the sanitary reform of 1866. Roy Lubove, *The Progressives and the Slums* (Pittsburgh: Univ. of Pittsburgh Press, 1962), also deals with the formation of the Metropolitan Board of Health.

8 John H. Griscom, "Improvements of the public health, and the establishment of a sanitary police in the city of New York," *Tr. Med. Soc. St. N.Y.,* 1857, pp. 107–123.

9 *Report of the Select Committee Appointed to Investigate the Health Department of the City of New York,* New York State Legislature, Senate Doc. No. 49, 1859.

10 *Ibid.,* p. 1.

11 *Ibid.,* p. 3.

12 *Ibid.,* p. 52.

13 *Ibid.,* pp. 156–157.

14 *Ibid.,* p. 174.

15 New York State Legislature, Assembly Doc. No. 129, p. 1.

16 Not only was the bill refused, but, to add to the problems of sanitation, the City Inspector was at that time given supervision of street cleaning. As the New York Sanitary Association pointed out later, this was added to his already grossly neglected sanitary duties. *Reports of the Sanitary Association of the City of New York* (New York, 1859), p. 7.

17 See Denis T. Lynch, *"Boss" Tweed, the Story of a Grim Generation* (New York: Boni and Liverwright, 1927), pp. 187–199; Samuel A. Pleasants, *Fernando Wood of New York* (New York: Columbia Univ. Studies in History, Economics and Public Law, No. 536, 1948), ch. 5; James F. Richardson, "Mayor Fernando Wood and the New York Police force, 1855–1857," *N.-Y. Hist. Soc. Quart.,* 1966, *50*: 5–40.

18 Richardson, "Mayor Fernando Wood," p. 6.

19 Those interested in sanitary reform seem to have been well aware of their enemies. Stephen Smith frequently described corrupt practices such as the bribery to which the Tammany-controlled City Inspector's Office allegedly resorted. The 1857 bill had been "effectually defeated by the paid agents of corrupt officials who succeeded, at a late period of the session, in sequestering or destroying all traces of the bill, both manuscript and printed." *Am. Med. Times,* 1860, *1*: 423. It must also be noted that the friends of sanitary reform early realized that a health department with jurisdiction merely over New York, and not including Brooklyn or the other surrounding communities, would have been of little avail. The large interchange of people each day made the metropolitan concept a necessity. See "New York Health Bills," *Am. Med. Times,* 1862, *4*: 70–71. For a brief, general description of New York City government, see Seth Low, *New York in 1850*

and in 1890 (New York: New-York Historical Society, 1892).

20 N.Y. Sanitary Association, *Report*. Elisha Harris must have been everybody's favorite secretary. He held that job in the Sanitary Association, and later in the Council of Hygiene, the U.S. Sanitary Commission, and the American Public Health Association. See also, Wilson G. Smillie, *Public Health, Its Promise for the Future* (New York: Macmillan, 1955), pp. 289–290.

21 N.Y. Sanitary Association, *Report*, pp. 11–13.

22 *Ibid.*, p. 13.

23 *Second Annual Report of the N.Y. Sanitary Association for the year ending December 1860* (New York, 1860), pp. 1–23.

24 Louis Elsberg's "The domain of medical police," *Am. Med. Monthly*, 1862, *17*: 321–337, was delivered before the Association. This concept of medical police was a prominent idea in the writings of John Griscom too. See George Rosen, "The fate of the concept of medical police 1780–1890," *Centaurus*, 1957, *5*: 97–113. Actually the term "sanitary police" might be more applicable to the goals of the New York sanitarians. According to Shattuck, in the term "medical police," cure of disease is implied; while in the idea of "sanitary police" prevention is stressed. Lemuel Shattuck, *Report of the Sanitary Commission of Massachusetts 1850* (repr.; Cambridge, Mass.: Harvard Univ. Press, 1948).

25 [Robert Tomes], "Why we get sick," *Harper's Monthly*, 1856, *13*: 642–647. The author noted that, "A host of diseases of the heart, the brain, nerves, and stomach, which exhaust the doctor's skill and fill his pockets, came in with modern civilization. To these diseases the Americans are far more subject than any other people . . . ," p. 642.

26 Frank H. Hamilton, "Hygiene," *N.Y. J. Med.*, 1859, *7*: 60–74, p. 60. Hamilton, a renowned surgeon, had been Stephen Smith's teacher and preceptor and remained a close friend in later years.

27 *Ibid.*, p. 63.

28 "Quarantine and Hygiene," *North Am. Rev.*, 1860, *91*: 438–491, p. 491.

29 Although Elisha Harris and George Shrady were assistant editors, I have ascribed the editorials to Smith, throughout. Harris actually resigned in 1861 because of increasing work with the U.S. Sanitary Commission. Shrady apparently did most of the reports of medical society meetings. Furthermore, Smith published many of the editorials in his book *Doctor in Medicine and Other Papers on Professional Subjects* (New York: Wood, 1872), thereby claiming authorship for those included. In various letters to his wife, to be dealt with in a future study, Smith complained of the wearying task of his weekly editorials. In this paper I am concerned only with those editorials that dealt with sanitary reform.

30 *Am. Med. Times*, 1860, *1*: 15. Howard D. Kramer, "The beginnings of the public health movement in the United States," *Bull. Hist. Med.*, 1947, *21*: 352–376, gives Smith and the *American Medical Times* a great deal of credit for bringing about reform, p. 375. Also Kramer, "Early municipal and state boards."

31 "Our sanitary defences," *Am. Med. Times*, 1860, *1*: 46–47.

32 *Ibid.*, p. 100.

33 *Ibid.*, 1861, *2*: 47–48.

34 For a general discussion see Emerson D. Fite, *Social and Industrial Conditions in the North During the Civil War* (New York: Macmillan, 1910), p. 229.

35 I. N. Phelps Stokes, *The Iconography of Manhattan Island*, 6 vols. (New York, 1895–1928), III, 736–756; Milledge L. Bonham, Jr., "New York and the Civil War," in *History of the State of New York*, 10 vols., ed. Alexander C. Flick (New York: Columbia Univ. Press, 1933–1937), VII, 99–135.

36 Allan Nevins, *The Evening Post, A Century of Journalism* (New York: Boni and Liverwright, 1922), p. 364.

37 Fite, *Social and Industrial Conditions*, p. 184. Also Edgar W. Martin, *The Standard of Living in 1860* (Chicago: Univ. of Chicago Press, 1942), contains many useful data.

38 "Health Laws," *Am. Med. Times*, 1860, *1*: 423–424.

39 *Ibid.*

40 "Health of New York in 1860," *Am. Med. Times*, 1861, *2*: 63–64.

41 See *Proceedings of a Select Committee of The Senate . . . Appointed to Investigate Various Departments of the City of New York*, New York State Legislature, Senate Doc. No. 38, Feb. 9, 1865. In this 612-page report, which dealt only with the City Inspector's department, there is a wealth of information. Testimony revealed the buying and selling of jobs and the incompetence of the health wardens. The duties of the 22 wardens and an equal number of assistants were to report nuisances, inspect buildings, privies, and cesspools, report all diseases, and to prevent accumulation of garbage and offal on the streets and sidewalks (p. 345). Several health wardens admitted they did not go personally to see cases of disease. They generally admitted that some smallpox existed and "a few fevers," when in fact smallpox, typhus, typhoid, cholera infantum, and scarlatina were widespread (pp. 455–456). Several of the wardens also admitted that they "devoted" a month's pay—but usually claimed ignorance of the fact it was used to aid defeat of health bills in Albany (pp. 462, 467).

42 *Annual Report of the City Inspector . . . for the Year End-*

ing December 31, 1860 (New York: Board of Alder-men, Doc. No. 5, 1861), p. 10.

43 *Ibid.*, pp. 16–17.

44 *Ibid.*, pp. 19–22.

45 *Ibid.*, p. 60.

46 *Am. Med. Times*, 1861, *2*: 63.

47 "Sanitary legislation," *Am. Med. Times*, 1862, *4*: 28–29.

48 *Med. & Surg. Reptr.*, 1862, *7*: 349–351.

49 "New York health bills," *Am. Med. Times*, 1862, *4*: 70.

50 "Failure of the health bill," *ibid.*, pp. 250–251.

51 *Med. & Surg. Reptr.*, 1862, *8*: 124–125.

52 "The prospect of health-reform in New York," *Am. Med. Times*, 1862, *5*: 276.

53 *Ibid.*, p. 277.

54 *Harper's Weekly*, 1863, *7*: 786.

55 "Sanitary interests in New York," *Am. Med. Times*, 1863, *6*: 21–22.

56 For a description of the founding of the Citizens' Association, see Edward C. Mack, *Peter Cooper, Citizen of New York* (New York: Duell, Sloan, Pearce, 1949), ch. 19. See also the *New York Citizen*, Aug. 13, 1864.

57 These two letters may be found in *Report of the Council of Hygiene and Public Health of the Citizens' Association of New York, Upon the Sanitary Condition of the City* (New York: Appleton, 1865), pp. ix–xiii.

58 "Citizens' Association and health reform," *Am. Med. Times*, 1864, *8*: 200. Also at the beginning of 1864 there was an investigation into the affairs of the City Inspector's department. Smith published excerpts from a letter written by Thomas N. Carr, who had been superintendent of street cleaning. Carr said that New York simply had no sanitary department worthy of the name. He felt that the only concern the City Inspector had was for the streets. Carr continued: "On an examination of the annual sanitary reports of England or France, the mind is astonished by the vastness of research, investigation, and scientific elaboration which these reports contain, and yet, strange to say, street cleaning, instead of being the all absorbing feature of these documents, is not even mentioned." *Am. Med. Times*, 1864, *8*: 57.

59 Joseph M. Smith, "Report on the medical topography and epidemics of New York," *Tr. A.M.A.*, 1860, *13*: 81–269.

60 In 1865 Smith became the secretary. For the names of the members of the council, see the first part of the introduction to the *Report*.

61 The evidence concerning Smith's role lies mostly within his own writings, especially *The City That Was*. Charles F. Chandler, however, also gave Smith credit for organizing the survey. See *Stephen Smith: Addresses in Recognition of His Public Services on the Occasion of His Eighty-Eighth Birthday* (New York: New York Academy of Medicine, 1911), p. 19, in which Chandler, then an old man himself, noted, "This work was organized and supervised to its completion by Dr. Stephen Smith." In 1864 Joseph M. Smith, the president of the council, was 75 years old. Elisha Harris, the secretary, was mainly occupied by the U.S. Sanitary Commission.

62 *New York Times*, March 16, 1865, or *The City That Was*, p. 57.

63 *Am. Med. Times*, 1864, *8*: 307. The *New York Tribune*, June 3, 1864, had great praise for the efforts of the Citizens' Association.

64 *Am. Med. Times*, 1866, *9*: 47.

65 There is a long discussion of etiological factors in disease in the introductory portion of the *Report*, pp. xlvii–lxviii. See also Richard H. Shryock, "The origins and significance of the public health movement in the United States," *Ann. Med. Hist.*, 1929, *1*: 645–665, especially pp. 650–652; Charles E. Rosenberg, "The cause of cholera: aspects of etiological thought in nineteenth century America," *Bull. Hist. Med.*, 1960, *34*: 331–354, and his *The Cholera Years*; and John Simon, *Filth Diseases and Their Prevention*, 1st Am. ed. (Boston: Campbell, 1876).

66 The *New York Times* on July 7, 1865, said: "No volume of intenser interest has ever seen the light in this city. . . ." See also *Nation*, 1865, *1*: 250; *Am. J. Med. Sci.*, 1865, *100*: 419–428.

67 Brian MacMahon, Thomas F. Pugh, and Johannes Ipsen, *Epidemiologic Methods* (Boston: Little, Brown, 1960), p. 3.

68 Kramer, "Beginnings," pp. 361–362. There were occasional descriptions of living conditions among the poor in the general press. See, for instance, Samuel B. Halliday, *The Lost and Found; or Life Among the Poor* (New York: Blakeman & Mason, 1859). Halliday was a member of the N.Y. Sanitary Association. Soon after the survey of the Citizens' Association was published there was a vivid article in the *Nation* by Bayard Taylor, entitled "A descent into the depths," 1866, *2*: 302–304. Although I have focused attention on the two societies in which Smith played a role, this is not to say that they were the only ones active in sanitary reform at this time. The A.I.C.P. and the various missions and tract societies were also active. The A.I.C.P *Report* for 1853 contains a long discussion of sanitary needs. Its founder and leading spirit, Robert M. Hartley, was well known for his crusade against swill milk. See Roy Lubove, "The New York Association for Improving the Condition of the Poor: the formative years," *N.-Y. Hist. Soc. Quart.*, 1959, *43*: 307–327; and Atkins, *Health, Housing, and Poverty*. Carroll S. Rosenberg's "Protestants and Five Pointers: The Five Points House of Industry, 1850–1870," *N.-Y.*

Hist. Soc. Quart., 1964, *48*: 327–347, describes New York's most notorious slum and efforts toward amelioration. Also Allan Nevins has drawn attention to the reformers in his "The golden thread in the history of New York," *N.-Y. Hist. Soc. Quart.,* 1955, *39*: 5–22.

69   Israel Weinstein, "Eighty years of public health in New York City," *Bull. N.Y. Acad. Med.,* 1947, *23*: 221–237.

70   *Citizens' Association Report,* pp. 261–262.

71   *Ibid.,* p. 285; and *Annual Report of the City Inspector . . . for the Year Ending December 31, 1861* (New York: Board of Aldermen, Doc. no. 4, 1862), pp. 21–23.

72   *Citizens' Association Report,* p. 42.

73   Smith, *City That Was,* p. 46; Andrew Dickson White, *Autobiography,* 2 vols. (New York: Century, 1905), I, 107–110. White reported the oft-quoted testimony of one of the city's health wardens, who, when asked the meaning of the word "hygiene," answered that it referred to bad smells arising from standing water. White also sat on a Senate committee during the investigation of the City Inspector's department early in 1865. Despite the pleas of the Citizens' Association, City Inspector Boole was not dismissed. See Stokes, *Iconography,* V, 1912; and Senate Doc. No. 38, 1865, p. 467. Eaton's testimony was given Feb. 2, 1865, and was published by "Friends of the Bill" later that year. Together with an appendix, his remarks take up 56 printed pages. *Remarks of D. B. Eaton, Esq., at a Joint Meeting of the Committees of the Senate and Assembly* (New York: Nesbitt, 1865).

74   On March 10 and 11, 1865, the *Times* noted that opposition was brewing from quarters formerly friendly to the bill. The paper warned that delay in passage of the bill was dangerous, so late in the session. Smith gave a great deal of credit to Henry J. Raymond, editor of the *Times,* calling him an ardent reformer. *City That Was,* p. 173.

75   That some of the legislators were impressed by the testimony was attested to by at least one member of Smith's audience. After hearing the description of sweatshop conditions in the tenements and of clothing, in the process of manufacture, draped over cribs of children with active smallpox, one of the committee supposedly said to Smith: "Why I bought underwear at one of those stores a few days ago, and I believe I have got smallpox, for I begin to itch all over." *City That Was,* p. 156. It should be stressed that this episode was 46 years in the past when the book was published.

76   Ramsey was a physician but seems to have been completely under Tammany sway. Lewis A. Sayre was the Resident Physician of New York, as well as a leading teacher of orthopedic surgery. His difference of opinion with the members of the Citizens' Association seems, on the surface, to have been on intellectual grounds. Smith, who was a fellow faculty member of Sayre's at Bellevue, had once called the latter's job (as Resident Physician) a sinecure. *Am. Med. Times,* 1862, *4*: 252. It is, of course, quite possible that Sayre's own vested interests prompted his belief in the status quo. His salary was about $5,000 per year.

77   *New York Times,* March 20, 1865.

78   *New York Tribune,* March 20, 1865.

79   *New York Times,* April 13, 15, 17; *New York Tribune,* April 15. The *Times* on April 17 noted three reasons for the defeat of the bill: Several Union (Republican) members were absent owing to illness; several others were unwilling to create another commission for the Governor; and City Inspector F.I.A. Boole had spent nearly the whole winter in Albany, armed with sufficient funds to kill the bill. This occurred in an Assembly in which the Republicans had a majority of 24. For a discussion of Boole, see Gustavus Myers, *The History of Tammany Hall* (New York: Boni and Liveright, 1917), pp. 205–208. An informative discussion of the state political situation at this time may be found in Homer A. Stebbins, *A Political History of the State of New York 1865–1869* (New York: Columbia Univ. Studies in History, Economics and Public Law, No. 55, 1913).

80   *New York Times,* Oct. 31, Nov. 9, 1865; *Nation,* 1865, *1*: 609.

81   *Nation,* 1865, *1*: 577.

82   Quoted by Rosenberg, *Cholera Years,* p. 186. See also Smith, *City That Was,* p. 166.

83   The speech is printed in the *New York Times,* Jan. 3, 1866.

84   *Ibid.,* Jan. 7, 1866.

85   *Ibid.,* Feb. 7, 1866.

86   *Ibid.,* Feb. 16, 20, 1866.

87   *Laws of New York, 1866,* Ch. 74 (reprinted by Bergen & Tripp, Printers, 1866).

88   Smith described Eaton's role and his activities in other spheres in chapter 6 of *The City That Was.* This book, incidentally, was dedicated to the memory of Dorman B. Eaton. Eaton was perhaps best known for his work in civil service reform. See Ari Hoogenboom, *Outlawing The Spoils: A History of the Civil Service Reform Movement, 1865–1883* (Urbana: Univ. of Illinois Press, 1961). Eaton was credited by Smith with having written the legal aspects of that bill. Eaton's views can be seen in his "The essential conditions of good sanitary administration," *Reports and Papers, American Public Health Association,* 1874–1875, *2*: 498–514. He was a student of English sanitary law and patterned the New York bill on what he

had learned in England. See Dorman B. Eaton, *Sanitary Regulations in England and New York* (New York: Amerman, 1872). See also Stephen Smith, "Development of American public health endeavor," *Am. J. Public Hlth.*, 1915, *5*: 1115–1119; Stephen Smith, "The origin and organization of the Department of Health of the City of New York," *Med. Rec.*, 1918, *93*: 1115–1117; and his "The history of public health, 1871–1921," in *A Half Century of Public Health*, ed. M. P. Ravenel (New York: A.P.H.A., 1921), pp. 1–12; especially pp. 4–10 deal with the Metropolitan Health Bill.

89 George Rosen, *A History of Public Health* (New York: MD, 1958), p. 247.

90 Samuel Osgood, *New York in the Nineteenth Century* (New York: New-York Historical Society, 1866), pp. 40–41. See also a review of numerous documents of the Citizens' Association, including the *Report of the Council of Hygiene,* by James Parton, "The government of the City of New York," *North Am. Rev.,* 1866, *103*: 413–465. Deserving more work is an analysis of those who were involved in the health reform movement. What were their backgrounds, their motives, and how large a part did they actually play? Also, how was the health reform movement, if indeed one can call it a movement at all, related to other reforms and reformers of the time? Health legislation played an important role in the general amelioration of the urban environment and in the development of cities. This too is an aspect of 19th-century public health that deserves much more study. See particularly Charles N. Glaab, *The American City, A Documentary Study* (Homewood: Dorsey, 1963). Also Arthur M. Schlesinger, "The city in American history," *Miss. Valley Hist. Rev.,* 1940, *27*: 43–66, and Blake McKelvey, *The Urbanization of America 1860–1915* (New Brunswick: Rutgers Univ. Press, 1963).

91 Stephen Smith, "American public health endeavor," p. 1117. Eaton and Elisha Harris were also active in the early work of the A.P.H.A.

92 Kramer, "Beginnings," p. 370.

29

The Decline in Mortality in Philadelphia from 1870 to 1930: The Role of Municipal Services

GRETCHEN A. CONDRAN, HENRY WILLIAMS, AND ROSE A. CHENEY

The Christmas season of 1880 should have been a happy occasion for Charles and Caroline Kautz. Charles, a German immigrant in his mid-40s, had operated a three-person bakery in the Moyamensing section of Philadelphia over a decade. Caroline, several years his junior and his wife of over 15 years, had married him during her teens and soon assumed the responsibilities of parenthood. Though Pennsylvania born, Caroline also came from German stock and, with her husband, worshiped at a local German Lutheran church. Caroline and Charles Kautz and their three sons and three daughters must have eagerly anticipated a joyous holiday celebration.

But it was not their fate to enjoy this Christmas. In 3½ weeks from late November to mid-December, smallpox took the lives of five of the six Kautz children. Six-year-old Clara died on November 20, followed by Albert (aged 2) two weeks later, then Edward (aged 4) on December 10, Charles (aged 8) on December 11, and Bertha (aged 10) on December 14. Only 12-year-old Sophia survived. Burial in the Lutheran church cemetery followed each of the deaths in numbing succession.

Gretchen A. Condran is Associate Professor of Sociology at Temple University, Philadelphia, Pennsylvania, and Research Associate in the Population Studies Center at the University of Pennsylvania, Philadelphia.

Henry Williams is Director of Special Projects and Secretary of the University of Pennsylvania Health System Trustee Board and Office of Chief Executive Officer and Dean, University of Pennsylvania Health System, Philadelphia.

Rose A. Cheney is Senior Research Associate at Philadelphia Health Management Corporation.

From *The Pennsylvania Magazine of History and Biography,* Medical Philadelphia Issue, April 1984, *108*: 153–177. Courtesy of *The Pennsylvania Magazine of History and Biography.*

The tragedies that befell the Kautzes over a century ago illustrate in a concrete way part of the enormous gap separating the 20th and the 19th centuries in America. Epidemic diseases, childhood death, and low adult life expectancies dominated life. Indeed, in the space of a decade, the Kautzes had witnessed smallpox epidemics in Philadelphia in 1871–1872 and 1876 prior to the 1880–1881 epidemic that devastated their family. The difference between the two centuries is often measured in terms of the rapid diffusion of communications, transportation, and machine technologies throughout our society. These forces, without doubt, make our lives much different. But they are no more important to us than the lengthening of adult life and the elimination both of frequent early death and the constant anticipation of epidemic disease. Recurrent smallpox epidemics, for example, which in Philadelphia alone took 4,464 lives in 1871 and 1872 and 1,760 lives in 1880 and 1881, seem to be remnants of a distant past.[1] Illnesses that rarely confront us today were frequently present only a century ago.[2]

Yet, as significant as the reduction in mortality has been, scholars still search for the reasons for the "mortality transition." Rising personal income, improved diet, better sanitation and personal hygiene, public health activities, advancing medical technology are offered as competing, complementary, or interacting explanations.[3] How any or all of these factors affect the propensity for a disease to be present in the environment (exposure), or for a given population to avoid infection from a disease which is present (resistance), or for survival when individuals contract a disease (case survival) is also open to considerable debate.

Although the origins and sources of the mortality transition are not fully understood, scholars have developed a commonly accepted portrait of the change. It occurred during the 19th and early 20th centuries in most of Europe, the United States, Canada, Australia, and New Zealand. In American cities, for example, mortality levels were high at the beginning of the 19th century and remained high for most of the century. Although mortality rates for some age groups and from some diseases (e.g., cholera and typhus) began declining earlier, a substantial and sustained decline in overall mortality levels did not begin until late in the 19th century.[4]

The changes in mortality rates in Philadelphia closely paralleled in both timing and magnitude those in other large cities of the United States and in urban areas in other industrializing nations.[5] These trends can be traced fairly well after 1860 when the official registration of deaths with the issuance of a death certificate began. The deaths were aggregated and published in annual reports of the Board of Health and included in the mayor's yearly report. The published data provide a classification of the deaths by age and cause for the total population of the city. In addition, deaths by cause are published for the wards within the city.

Using an average of these registration data for three years surrounding each federal decennial census date from 1870 to 1930 and base population figures from the decennial censuses, one can observe the pattern of mortality change.[6] The description of changing mortality is expressed in two ways: as the probability that an individual in a particular age group will die during a given five-year age period and as the expectation of life at selected ages. For each year, the expectation of life for a given age is the average number of years that would be lived by a hypothetical group of persons who reached that age in the year and who were subjected to that year's probabilities of dying at each subsequent age.

The magnitude of the overall mortality decline in Philadelphia is apparent for all age groups (see Table 1). In 1870, for example, about 175 of every 1,000 children born never survived to their first birthday. By 1930, fewer than 75 of every 1,000 infants perished during the first year of life — a decline of about 175 percent in the probability of dying.

Table 1

Life Table Probabilities of Dying (q_x x 1000) and Expectations of Life at Selected Ages, Philadelphia, 1870 to 1930

Age	1870	1880	1890	1900	1910	1920	1930
	Probability of Dying (q_x x 1000)						
0–1*	174.0	159.7	152.9	136.5	119.7	95.5	74.9
1–5	141.9	120.6	111.1	89.3	68.5	40.8	23.1
5–10	41.9	37.8	36.8	28.3	19.8	17.6	10.9
10–15	19.8	19.2	17.0	16.9	11.9	11.2	7.7
15–20	32.9	30.0	28.7	31.2	21.6	17.9	12.5
20–30	114.2	97.8	82.3	71.7	55.7	53.8	19.2
30–40	117.9	109.7	96.3	95.6	84.5	68.8	28.8
40–50	133.7	131.9	127.3	127.2	117.1	98.4	50.4
50–60	194.5	183.5	190.9	198.2	194.4	169.7	99.6
60–70	319.7	303.0	326.5	350.5	354.7	314.9	182.9
	Expectation of Life						
e_0	39.6	42.3	43.7	45.8	49.6	54.4	57.9
e_1	46.8	49.3	50.5	52.0	55.4	59.1	61.5
e_5	50.4	51.8	52.6	52.9	55.3	57.6	58.9
e_{20}	39.8	41.0	41.6	41.6	43.0	45.0	45.6

*Estimated by fitting Coale and Demeny South Model Life tables to $_4q_1$.
Source: Gretchen A. Condran and Rose A. Cheney, "Mortality Trends in Philadelphia: Age- and Cause-Specific Death Rates 1870–1930," *Demography*, 1982, *9* (No. 1): 100.

Indeed, the late 19th century was still a difficult time for all children. Over a third of all infants born in 1870 died before they were ten. Sixty years later, about a tenth did not survive to that age, a substantial improvement in life chances. Young adults also suffered from a harsh mortality regime in 1870 — 114 of every 1,000 individuals reaching age 20 never lived to age 30 — but by 1930 fewer than 20 of 1,000 20-year-olds died within ten years. The elderly also enjoyed a moderation in their mortality rates. In 1870, about 320 of every 1,000 individuals reaching 60 years of age died within the next decade; by 1930, the corresponding figure was 183.

The declines in infant mortality were particularly important for raising life expectancies between 1870 and 1930. At birth in 1870 an individual could expect on average to reach age 40; by 1930, an infant could expect to reach age 58 — an increase in life expectancy of about 46 percent. The impact of the mortality transition, however, was not as substantial for those who could survive their childhood and adolescent years. In 1870, for example, individuals who made it to age 20 could on average expect to see age 60; by 1930, the group of young adults might expect to celebrate their 66th birthday — an increase of life expectancy of but 15 percent. Clearly, the major consequence of the transition was to increase the numbers who survived infancy and childhood. Those who had already reached adulthood gained fewer additional years of life.

Although mortality rates declined for each age group between 1870 and 1930, the timing of the reduction differed by group (see Table 1). Childhood and infant mortality levels declined throughout each of the six decades, but the declines were uneven. The expectation of life at birth and at age one in Philadelphia rose almost three years between 1870 and 1880, but increased at a slower rate between 1880 and 1900. After 1900, the rate of improvement in mortality levels increased again.

On the other hand, substantial declines in adult mortality did not occur in Philadelphia until after the turn of the century, except for 20-to-29-year-olds. This group exhibited the highest rates of death from tuberculosis, and declines in its overall death rates occurred steadily from 1870 to 1930, largely because the deaths from tuberculosis were declining. In contrast to the other groups, the el-derly actually exhibited rising mortality rates until the 1910–1920 decade, but then a precipitous drop to 1930.

Cause-of-death data provide the first clues to the explanation of these trends and age-group differentials, but these data must be used with caution. A correct assessment of the levels of a disease in a population is dependent upon the consistency of classification schemes and reporting procedures over time. In Philadelphia, data are available from 1870 to 1930 but are not of uniform quality for the entire period. Between 1870 and 1900, the number of causes of death reported by the Bureau of Health steadily increased. Much of the increase resulted from changes in diagnostic ability and the replacement of general cause-of-death categories with more specific descriptions of the sources of mortality. In 1904, the Bureau of Health began to group individual causes into broader nosological headings.[7]

Thirteen causes of death can be followed with reasonable certainty from 1870 to 1930. Although these infectious diseases represent but a minority of all causes of death reported in Philadelphia, they account for 45 percent of all deaths occurring in 1870, about 40 percent of those in 1900, and about 24 percent of those in 1930. As we will see, these declining proportions indicate that increasing control of infectious disease played a significant role in the overall mortality decline. Even for these selected causes of death, however, causes for which diagnosis was probably very accurate throughout the period must be separated from those in which trends may be obscured by changes in diagnosis over time (see Table 2). The mortality rates for each disease are standardized on the age distribution of Philadelphia's population in 1870 and, therefore, the rates for each cause are those which would have occurred if the age distribution of Philadelphia's population had not changed in the time period.

The group of infectious diseases for which diagnosis was very good showed a spectacular decline over the period. In 1870, about 600 of every 100,000 Philadelphians died from these causes; 60 years later fewer than 80 did. Scarlet fever, smallpox, typhoid, and diphtheria, all major killers in 1870, were virtually eliminated as causes of death by 1930. The timing in declines differed by specific disease, however. Most of the individual disease categories declined between 1870 and 1900 and

Table 2
Percent Contribution of Causes of Death to Overall Decline in Age-Standardized Death Rates Philadelphia, 1870, 1900, and 1930

Causes of Death	Percent Decline Explained by Cause*		
	1870–1930	1870–1900	1900–1930
Infectious— specific diagnosis	52.1	64.7	44.5
Diphtheria and croup	5.4	−5.1	11.8
Erysipelas	0.5	0.8	0.3
Malaria	0.4	0.8	0.3
Measles	0.7	0.8	0.6
Scarlet fever	8.8	18.5	2.8
Smallpox	8.5	21.5	0.6
Tuberculosis	22.5	26.8	19.8
Typhoid	4.6	1.1	6.8
Whooping cough	0.7	−0.6	1.5
Primarily Infectious— diagnosis less specific	14.1	−1.3	23.5
Diarrheal diseases	16.4	19.8	14.3
Epidemic meningitis	0.0	0.1	0.0
Influenza	−1.5	−2.8	−0.7
Pneumonia	−0.8	−18.4	10.0
Residual	33.7	36.6	32.0

*The percent decline was calculated using 1900 and 1930 death rates standardized on the age distribution of the 1870 Philadelphia population.

Source: Gretchen A. Condran and Rose A. Cheney, "Mortality Trends in Philadelphia: Age- and Cause-Specific Death Rates 1870–1930," *Demography*, 1982, 9 (No. 1): 103.

continued to decline between 1900 and 1930, but diphtheria and whooping cough exhibited declines only after 1900.

The second group of diseases, those whose diagnoses were more problematic in the 19th and early 20th centuries, showed less clear downward trends than the more easily diagnosed causes of death. Death rates from diarrheal diseases declined throughout the period, although the decline was much greater after 1900 than before. Epidemic meningitis changed little between 1870 and 1930, while influenza death rates rose for both 30-year

periods. Pneumonia death rates increased substantially between 1870 and 1900 but demonstrated a large decline in the next 30 years. The fairly steady decline which occurred in the residual category of cause of death is difficult to interpret. This category contains many different causes of death, including all those that are ill-defined such as marasmus and debility. Therefore, actual changes in the death rates cannot be distinguished from improvements in the diagnosis over time.

Examining the contribution of the 13 cause-of-death categories to the overall decline in mortality establishes that the control of these infectious diseases was the most important component of the mortality transition (see Table 2). The percentage of the total decline in mortality due to each cause of death was calculated for the whole period from 1870 to 1930 and for each of the two 30-year periods separately. Over the 60-year period, two-thirds of the mortality transition is accounted for by the decline in the combined 13 disease categories alone. Over half of the total reduction is explained solely by decreases in diseases subject to specific diagnoses. These diseases were particularly responsible for the decline from 1870 to 1900, but also contributed heavily to mortality decline after the turn of the century. In contrast, mortality rates from the group of infectious diseases which were less easily diagnosed actually rose until 1900, but became a significant part of the overall decline afterward.

The fluctuations in the trends of particular diseases bring us closer to the mechanisms governing the transition. For both time periods, tuberculosis explained the largest percentage of the decline in the age-standardized death rates. It accounted for 26.8 percent of the decline in mortality from 1870 to 1900, 19.8 percent of the decline from 1900 to 1930, and over 22 percent for the entire six decades. Decreasing death rates from diarrheal diseases explain about 16 percent of the mortality decline from 1870 to 1930, contributing slightly more to the overall decline during the initial 30 years. The severity of scarlet fever and smallpox epidemics declined so much from 1870 to 1900 that they represent the third and fourth most significant contributions. After 1900, however, they were no longer important causes of death, and they no longer contributed to the decline in mortality. Other diseases actually increased in destructiveness before becoming sources of the mortality decline. Deaths due to diphtheria and, particularly, to pneumonia rose between 1870 and 1900, but these diseases became important sources of mortality decline between 1900 and 1930. Quite clearly, the mortality transition was largely a consequence of the reduction in fatalities from infectious diseases. Why did this reduction occur?

Much of the demographic literature seeking to explain the mortality transition asserts that in the past mortality declines in more developed areas resulted from economic growth and changes in living standards and that the importation of modern technology has been responsible for recent mortality reductions in less-developed areas.[8] This contention is generally true, but it remains a general hypothesis implying specific relationships which have yet to be empirically verified. The specific determinants of declines in mortality remain a central question in studying the demographic changes of industrialized nations, historically, and of developing nations, presently.

Thomas McKeown and his colleagues[9] have emphasized the importance of economic development and accompanying increases in food supplies and per capita income as sources of the historical decline in mortality in the West. From an examination of the diseases which contributed to the decline in mortality in England and Wales during the 19th century, the McKeown group concludes that the major influence on declining death rates was the rising standard of living, the most significant feature of which was an improvement in diet. They argue that improvements in hygiene and public sanitation had some impact on mortality rates after 1850 and that minor changes in overall mortality levels resulted from a favorable shift in the relationship between micro-organisms and the human hosts. Advancing medical technology had but a limited impact on mortality decline. According to McKeown, therapies made no contribution and immunization accounted for only a small part of the reduction in the death rates before 1935. A comparison of other studies with that of McKeown is hampered by variations in the research methods and classification schemes used to elaborate the determinants of mortality decline. Regardless of the broad categories used, however, the literature suggests that the factors responsible for the spectacu-

lar decline in mortality which occurred in late 19th- and early 20th-century Europe and the United States were numerous and complex. The importance of each factor has, by no means, been firmly established.[10] In many previous discussions, economic development has been considered mainly in terms of increases in per capita income and has been related to mortality through advances in nutrition.[11] However, economic development and industrialization encompass not only additions to per capita income, but also technological changes, increases in knowledge (including medical knowledge), mass education, urbanization, and changes in the social organization of both governmental and other groups. All these aspects of economic development affect mortality levels by influencing a number of other factors which more immediately determine mortality levels: density, crowding, immigration, water supplies, food purity, per capita food supplies, expenditures on public health, and personal hygiene practices.

Many of these phenomena are related to each other, often in complicated ways. Explanations of the decline of mortality usually juxtapose the variables relating to changes in income, medical technology, public health activity, or the environment. Such explanations should be replaced by a model which sees income changes operating through a number of these variables. Increased per capita income leads to increases in the purchase of health-enhancing goods—especially better food and better housing. It also potentially improves an individual's health by increasing his access to medical care and services and, on a more aggregate level, by increasing the funds available for public expenditures on health activities. At the same time, many of the intermediate variables which influenced mortality were probably also affected by aspects of economic development other than the change in per capita income. Improvements in nutrition resulted from changes in per capita income but also from technological changes, particularly those in transportation and food processing.

Because the decline in mortality was largely a result of a decline in infectious diseases, we can begin an explanation of mortality decline with the relatively simple notion that the chances of dying from a particular disease depend on three things: the exposure of a population to the disease; the population's resistance to the disease; and, once the disease is contracted, the ability of the people to survive its effects. Although these three are not easily distinguished empirically within populations, they can help us to organize some of the more easily identified variables which have affected mortality levels.[12]

In general, the decline in mortality prior to 1930 resulted largely from a decrease in exposure, and/ or an increase in resistance, so that individuals were less likely to die from a given disease because they were less likely to contract it in the first place. This is not to say that there were no improvements in survival rates once a disease was contracted. Some researchers, for example, argue that improvements in case survival were a major part of the decline in the death rate from diphtheria.[13] Our analysis will concentrate on the prevention of illness, largely because it appears to have been a more significant factor in the early mortality decline than efforts to treat illness once it occurred.

Disease prevention resulted from a series of explicit public and private efforts in conjunction with economic forces that were beyond the public's immediate control. It is unwise to single out any particular factor as central to disease control, though we can conclude that a rising standard of living, promoted by rapid economic development, was a significant cause of this phenomenon.

The history of tuberculosis, the disease most important to mortality decline in Philadelphia, illustrates well the difficulties in explaining why the exposure of a given population to a particular disease declined. Overt efforts to contain this disease were not especially effective, largely because medical and public officials debated its contagious nature until nearly the turn of the century and took virtually no effective actions to limit it before 1930.[14] Nevertheless, its decline as a source of mortality is evident, both during the late 19th century, when little was known about this debilitating affliction, and during the first 30 years of the 20th century, when tuberculosis was at last recognized as a contagious disease. Explanations citing improvement in diet resulting from economic advance and the growing resistance of populations to the disease as a consequence of exposure both suggest possible answers, but it is clear that explicit attempts at containment do not account for its decline.[15]

Other diseases were effectively limited by direct and knowledgeable action, particularly by public health authorities. The emergence of the public health bureaucracy is an important aspect of the process of economic development and population growth. In Philadelphia from 1870 to 1930, the government apparatus concerned with the health and welfare of the population considerably expanded its scope. The growth of its power and influence helped produce lower mortality rates from a number of specific causes of death. Initially, the public health sector had an impact on medical science through the institution of data collection systems which informed the activities and pronouncements of public health officials. Later, as in other cities, public health efforts were concentrated primarily on reducing the population's exposure to disease by decreasing the number of pathogens in the environment. Public health activities encompassing quarantines, disinfection, purification of water supplies, and advocacy of particular child care practices all focused on prevention of illness, a focus that has to some extent been lost in the more recent emphasis on antibiotics and high technology medical treatment. Even in the instances where medical treatment played a central role in disease control, as with smallpox, public campaigns were largely responsible for promoting such medical remedies as vaccination.[16]

The history of public health activities in the 19th and early 20th centuries has been written for a number of cities and our aim is not to add Philadelphia to these histories. Rather, we seek to illuminate the connections between those activities and the shifts in mortality levels outlined above. To do that, we will focus on two causes of death on which public health activities were focused. By concentrating on one city and, at times, on small sections of Philadelphia, we will attempt to establish patterns and explanations that might apply to other urban areas during their mortality transitions.

The most dramatic impact of the city's activity can be observed in the efforts to combat high death rates from typhoid fever by cleaning up city water supplies. The pollution of water supplies quite often accompanied city growth. Impure water was not a problem in every large city, but Philadelphia shared with many cities the pollution of a central water system. The centralization of water systems allowed typhoid fever, transmitted through human excreta, to spread throughout the city.

As early as 1860, the Bureau of Health found two problems with the water supply system in Philadelphia. First, especially in summer, the quantity was inadequate to meet the needs of the population. The Bureau of Health argued that the shortage of water had adverse consequences, because more water was needed both for personal hygiene and to clean streets and sewers. In addition, the water supply was being polluted.[17] In 1861, the annual report of the Bureau of Health contained a scathing account of the pollutants of the water supply.[18] The city received water from several water stations, but all water came ultimately from the Schuylkill and Delaware rivers. According to the Bureau of Health, the Schuylkill was being polluted by sulphate of iron from coal mines, by the infusion of refuse dyes and wastes from the various factories on the eastern bank, and, most objectionable of all, by the emptying of city sewers into the river. The sewers received drainage from the streets and were connected to privies and cesspools.[19] The Delaware River was even more foul. In 1874, the Bureau of Health suggested that the Delaware be terminated as a source of water in favor of the Schuylkill, which, by that time, was protected from some pollutants.[20] The city had acquired land along both sides of the river for park land to prevent further growth of pollution-producing industries along the river banks.[21]

The pollution problems remained, however, and city officials linked impure water with diseases, particularly typhoid fever. Filtration of the water supply was first broached as early as 1853 but not implemented until after the turn of the century.[22] At the time filtration was begun, the city was divided into six water districts that received water from different waterworks. The timing of the introduction of water filters differed across the six districts. The death rates from typhoid fever from 1890 to 1920 have been calculated for each of the water districts and are shown in Fig. 1. The graph points out the importance of water filtration in combating the disease. The Roxborough (1902) and Belmont (1904) districts were the first to receive filtered water. In these two districts, the 1905–1906 typhoid epidemic that showed up in the other four water districts did not occur. The Tor-

resdale district had high rates of typhoid in 1905, 1906, and 1907 but showed a precipitous decline by 1908. Its water filtration system was put into operation in 1907. By the end of 1908, 56 percent of the city's water was being filtered. Subsequently, filtered water was supplied to the East Park, Queen Lane, and Fairmount water districts so that on March 1, 1909, "the entire city was supplied with filtered water." No epidemics of typhoid fever occurred after the filtration began.[23]

For the city as a whole the picture is clear. The year 1906 is the peak typhoid year for the city—over 1,000 deaths occurred in just one year. Epidemics had occurred with regularity before that. The number of deaths grew in each epidemic as the city's population grew and as an increasing proportion of the city's population was exposed to the polluted water supply. Epidemics ceased in 1906, for by the time the next epidemic would have likely occurred, the water supply was filtered.

While it was agreed that the purification of the polluted water supply affected the death rate from typhoid fever, contemporaries were intrigued by the notion, supported by data in several cities, that the death rates from diseases other than typhoid fever would decline following the purification of public water supplies. The idea arose from the work of two individuals, Hiram F. Mills of Lawrence, Massachusetts, and J. J. Reincke of Hamburg, Germany. In 1893–1894, they observed that the purification of polluted water supplies in Lawrence and Hamburg produced a significant decline in the general death rate. The relationship was named the "Mills-Reincke phenomenon." It was tested on a number of other cities by Allen Hazen, who in 1904 reached the conclusion that "where one death from typhoid fever has been avoided by the use of better water, a certain number of deaths, probably two or three, from other causes have been avoided." He hypothesized that the decline in the overall death rate following the water purification would be two or three times greater than that from the decline in typhoid fever alone.[24]

Although some of the excess decline in the overall death rates observed by Mills, Reincke, and Hazen was attributable to changes in other waterborne diseases, it was not totally accounted for by these. Changes in typhoid rates were related to pneumonia, tuberculosis, and other diseases which were not water-borne. In Philadelphia, during the period when water filtration was introduced, there was no constant multiplicative relationship between the decline in typhoid death rates and the decline in death rates from other causes. The existence of six water districts with differences in the timing of filtration affords us an opportunity to examine more carefully the relationship between the conditions of the water supplied to the population and the deaths from diseases other than typhoid. The water district data for Philadelphia show very little evidence of the Mills-Reincke phenomenon.

None of the causes of death which we examined showed the decline coincidental to water filtration which is seen in the typhoid fever death rates. There was little or no association between deaths from typhoid fever and deaths from tuberculosis or pneumonia, the leading causes of adult mortality. More surprisingly, however, even diseases like infant and childhood diarrhea, dysentery, and enteritis, which might be water-borne, did not decline in water districts immediately following filtration. Water filtration did put an end to epidemic

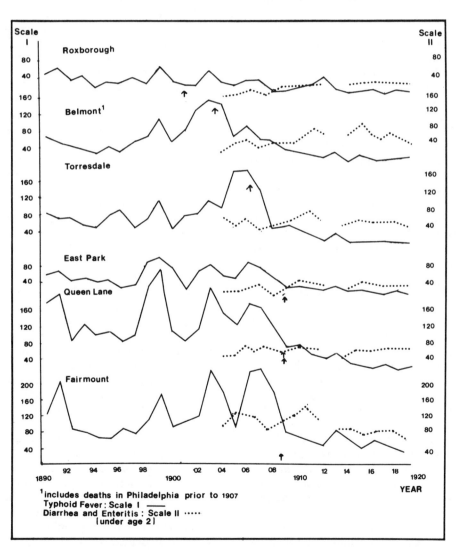

Fig. 1: Deaths from typhoid fever in Philadelphia water districts, 1890–1920.

outbreaks of typhoid fever, but it did not go beyond that.

The major cause of death among infants and young children was diarrheal disease. In these age groups, diarrhea is generally precipitated by contaminated food or water supplies and/or malnutrition. The immediate cause of death is dehydration, which occurs most quickly in very young infants with low body weights.[25] In 19th-century Philadelphia, there was a definite seasonal pattern to infant and childhood diarrheal deaths, the frequency having been much greater in the summer months.[26]

The decline in diarrheal death rates for the city as a whole is shown in Fig. 2. Mortality rates from infant and childhood diarrhea declined from the mid-1870s to 1900. An upsurge appeared shortly after 1900 and was followed by a relatively steep decline. The rise after 1900 corresponds to a shift in the classification scheme of causes of death, and therefore is likely to be an artifact of the changing nosology. The best estimate of the trend in infant and childhood diarrheal death rates is that they continued downward throughout the first decade of the 20th century and that the decline became steeper sometime about 1910. The sharp downward trend beginning about 1910 would be consistent with the positive influence of water filtration on the occurrence of the disease. However, as we have stated in the previous section, it is clear that the source of the decline was not the shifts in the water supply. Fig. 1, which contains the typhoid fever deaths, includes the numbers of deaths from infant and childhood diarrhea. There is no evidence of a decline in these deaths resulting from filtration of the city's water supply. The factors influencing the levels of mortality from diarrhea in the early years of life must be sought elsewhere.

The Bureau of Health in Philadelphia undertook three activities directed particularly at reducing infant and early childhood mortality rates. The first of these was the issuance of pamphlets on the care and feeding of children. In the mid-to-late 1870s, 40,000 copies of a circular urging mothers to breastfeed their babies and to postpone weaning until after the summer months were distributed. For mothers who did not breastfeed, the pamphlets suggested that clean bottles be used in feeding infants.[27] The advice contained in the pamphlet was sound. From the scanty data available, it appears that breast-fed babies had lower death rates from diarrheal diseases than did babies fed animal milk. The question which cannot be answered is whether the advice was heeded by enough mothers to have produced a decline in mortality rates. Changes in the seasonal pattern of diarrheal diseases suggests that the advice was heeded and may have resulted in an attenuation of the summer peak of mortality among one-year-olds prior to 1880, but a seasonal pattern, particularly for infants, continued until 1920.[28]

The impact created by the establishment of a Child Hygiene Bureau can be somewhat more easily assessed. The bureau was organized in 1910 and located child hygiene clinics in 8 wards between 1910 and 1914 and in 19 wards between 1915 and 1918.[29] Trends in death rates based on births for child hygiene districts are shown in Fig. 3. Wards were selected as sites of clinics because of the high infant and childhood death rates. Therefore, the wards with clinics had higher infant diarrheal death rates before the clinics opened than wards without clinics. The wards with clinics, however, showed declines in diarrheal death rates which were steeper than those without clinics, so that by 1930, the infant diarrheal rates in the latter were higher than those in the former. Also noticeable in Fig. 3 is that while infant diarrheal death rates were declining for all wards for the time period from 1904 to 1920, clinics appear to have accelerated the decline in the wards that had them.

Finally, the Bureau of Health worked long to improve the quality of the milk being fed to children in the city. In 1889, City Council created the post of inspector of milk to insure the wholesomeness of milk in the city. This appointment, however, by no means solved the problems.[30] Between 1890 and 1914, difficult legislative struggles were waged for the city councilmen's votes on various regulations affecting the milk supply. Milk producers favored little or no regulation, while the Bureau of Health argued that the poor quality of the milk supply had adverse health consequences and that regulation of the producers was needed.[31]

There were a number of problems with the milk supplied to the city. First, it was often skimmed and/or had added water. As a consequence, babies and young children fed the milk were likely to be inadequately nourished. Second, the milk itself was impure. Sloppy handling resulted in pollutants that

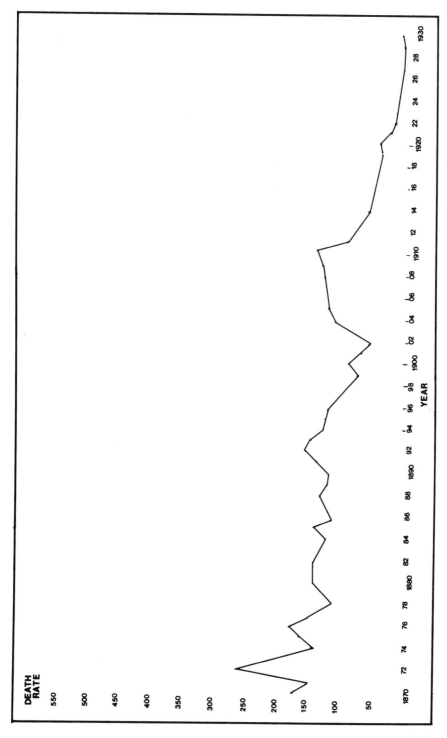

Fig. 2: Death rates from diarrheal diseases under age two (2), Philadelphia, 1870–1930 (death rates per 1,000 total population).

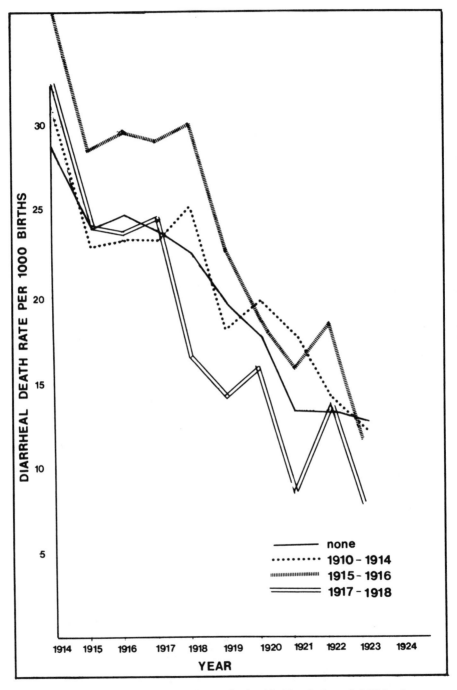

Fig. 3: Infant diarrheal death rates in wards classified by timing of child hygiene programs, Philadelphia, 1914–1923.

entered the milk, and lack of refrigeration, especially in the summer, resulted in the growth of bacteria. Third, some of the milk was obtained from tubercular cows and was a source of nonrespiratory tuberculosis in the population. The last problem seems not to have been severe; a system of inspecting herds for tuberculosis was instituted quite early. The first two problems with the milk supply were likely to have directly affected the incidence of diarrheal diseases in infants and young children. Bureau of Health efforts to prevent the adulteration of milk were only moderately successful between 1890 and 1895. In 1895, the inspector was given authority to check the bacteriological content of the milk.[32] Both home and commercial pasteurization of milk was being done by 1905.[33] As the result of legislation going into effect in 1911, milk was condemned and destroyed if its temperature was below standard. In that year alone almost 29,000 quarts of milk failed to meet the standard.[34] On July 1, 1914, an ordinance requiring all milk in the city to be pasteurized was passed.[35]

Limited statistics available on the quality of the milk supply indicate that improvements occurred over time. Between 1889 and 1893, an average of 8.5 percent of inspected milk was found to be skimmed or watered. During the next five years that percentage had dropped to 3.3. From 1899 through 1903, an average of about 1.5 percent was skimmed or watered, and for two five-year periods after that, average yearly percentages dropped to .5 and .2 percent. Statistics from 1899 to 1911 on the bacteriological content of milk show a much less consistent downward trend. A substantial drop occurred between 1908 and 1909 and in the years following.[36] These improvements happened before universal pasteurization became law. In summary, improvements in the milk supply came largely after the turn of the century, but were substantial enough to have accelerated the decline in infant and childhood diarrhea.

The decline in mortality experienced by Western industrialized nations in the late 19th and early 20th centuries creates a major historical question. As illustrated by our case study of Philadelphia, life expectancies at birth significantly increased and propensities to die from infectious disease, particularly tuberculosis and infant diarrhea, decreased during the period from 1870 to 1930. Certain epidemic diseases, notably typhoid fever, smallpox, and scarlet fever, virtually disappeared as causes of death. Of particular importance was the sharp drop in the frequency of infant and childhood death, which propelled the marked increase in life expectancies. The circumstances surrounding the deaths of the Kautz children 100 years ago have long been gone from the Western world.

The reasons for this remarkable change remain elusive. We can say with some assurance that mortality declines occurred because the population became less exposed and more resistant to infectious disease, rather than because medical technology increased chances of survival once a disease was contracted. But explaining this reduced exposure and greater resistance is difficult.

Certainly, no single factor can account for the mortality transition. Our brief discussion of the diseases central to the mortality decline indicates a multitude of possibilities. Tuberculosis, for example, declined in impact for no clearly discernible reason, although improved diet and growing resistance among the population to the disease were probable factors. Smallpox was eradicated through a combination of advancing medical technology and governmental crusades promoting vaccination. Typhoid fever was eliminated by water filtration, and infant and childhood diarrheal deaths were sharply decreased by several factors, including public dissemination of information about the diseases, the establishment of locally based child hygiene clinics, and improvement in the quality of milk.

Underlying the overall transition was a general improvement in living standards, concurrent with rapid economic development. Economic advance in itself, however, would not have produced the evident declines in death rates of certain diseases without the efforts of public officials who proposed, debated, and carried out many actions to further the health of their community. Typhoid fever, smallpox, and infant and childhood diarrhea, in particular, were severely curtailed by their activities.

Governmental intervention in Philadelphia occurred in spite of opposition from various special interest groups and constant bickering between the Board of Health and City Council. Regulation often came, too, with officials possessing only limited knowledge about the health problems against which their efforts were directed. But its impact was not small. Considering only typhoid fever, diarrheal diseases, and smallpox, direct public health activities probably account for almost one-third of the mortality decline in Philadelphia between 1870 and 1930. And this conservative estimate ignores possible indirect effects on other diseases.

Scepticism voiced today about the role of government too often neglects the very beneficial impact public agencies have made in the lives of their citizenry, particularly in the field of public health.

Many commentators also mistakenly think of public intervention and regulation as phenomena associated with the expansion of government in the 1930s. Yet by 1930 infectious epidemic disease and exceedingly high mortality rates had been brought largely under control with the direct assistance of a well-established municipal services bureaucracy. The health issues we face today are not the same as those a century ago, but we would be prudent to recognize past successes and how they were achieved.

NOTES

This research was done at the Philadelphia Social History Project, Theodore Hershberg, director, with support from the Center for Population Research, National Institute of Child Health and Human Development (RO 1HD 12413). The PSHP gratefully acknowledges the funding it has received from the following federal agencies: the National Institute of Mental Health; the National Science Foundation; and the National Endowment for the Humanities.

We also wish to thank William Kreider, Jeffrey Seaman, Wayne Dunlap, and Mara Weitzman for computer support, and Alicia Gilham for graphics and typing.

1 City of Philadelphia, *First Annual Message of Edwin S. Stuart, Mayor of the City of Philadelphia, with Annual Report of Abraham M. Beitler and the Annual Report of the Board of Health for the year ending December 31, 1891* (Philadelphia, 1892), p. 508.

2 In a recent review of Robert Gottfried's book on the Black Death, William H. McNeill takes a much longer view of how "the disappearance of epidemic disease as a serious factor in human life . . . separate[s] us from our ancestors." See "The plague of plagues," *N.Y. Rev. Books,* July 21, 1983, *30*: 28–29.

3 See for example: Thomas McKeown, *The Modern Rise of Population* (New York, 1976), pp. 128–142; Thomas McKeown and R. G. Record, "Reasons for the decline of mortality in England and Wales during the nineteenth century," *Population Stud.,* July 1962, *16*: (No. 2): 94–122; Samuel H. Preston and Etienne van de Walle, "Urban French mortality in the nineteenth century," *Population Stud.,* July 1978, *32* (No. 2): 275–297; Gretchen A. Condran and Rose A. Cheney, "Mortality trends in Philadelphia: age- and cause-specific death rates 1870–1930," *Demography,* Feb. 1982, *9* (No. 1): 97–123.

4 Gretchen A. Condran and Eileen Crimmins, "Mortality differentials between rural and urban areas of states in the northeastern United States, 1890–

1900," *J. Hist. Geography,* Apr. 1980, *6* (No. 2): 179–202; Robert Higgs, "Mortality in rural america, 1870–1920: estimates and conjectures," *Explorations Econ. Hist.,* 1973, *10* (No. 2): 177–193.

5 Condran and Cheney, "Mortality trends in Philadelphia," pp. 98–99.

6 The values of $_1q_0$ in Table 1 were estimated using the values of $_4q_1$ fitted to South Model Life tables. For a full explanation of the estimating procedure, see Condran and Cheney, "Mortality trends in Philadelphia," pp. 91–102.

7 City of Philadelphia, *Annual Report of the Bureau of Health of the City of Philadelphia for the year ending December 31, 1904* (Philadelphia, 1905), pp. 236–267.

8 Abdel R. Omran, "The epidemiological transition: a theory of the epidemiology of population change," *Milbank Mem. Fund Quart.,* 1971, *49* (No. 4, Part 1): 509–538; United Nations, *Population Bulletin of the United Nations, no. 6 (with special reference to the situation and recent trends of mortality in the world)* (New York, 1963), pp. 9–10; Kingsley Davis, "The amazing decline of mortality in underdeveloped areas," *Am. Econ. Rev.,* 1956, *46*: 305–318; George Stolnitz, "A century of international mortality trends," *Population Stud.,* July 1955, *9* (No. 1): 24–55.

9 McKeown, *The Modern Rise of Population,* p. 153; McKeown and Record, "Reasons for the decline of mortality," pp. 94–122; Thomas McKeown and R. G. Brown, "Medical evidence related to English population changes in the eighteenth century," *Population Stud.,* 1955, *9*: 119–141.

10 For a listing of factors influencing mortality levels, see United Nations, *The Determinants and Consequences of Population Trends* (New York, 1973), pp. 86–91.

11 McKeown, *The Modern Rise of Population,* pp. 153–154.

12 Eileen M. Crimmins and Gretchen A. Condran, "Mortality variation in U.S. cities in 1900: a two-level explanation by cause-of-death and underlying factors," *Soc. Sci. Hist.,* Winter, 1983, *7* (No. 1): 31–59.

13 See, for example, John Duffy, *A History of Public Health in New York City 1866–1966* (New York, 1974), p. 157.

14 City of Philadelphia, *First Annual Message of Charles F. Warwick, Mayor of the City of Philadelphia, with Annual Reports of Abraham M. Beitler and of the Board of Health for the year ending December 31, 1895* (Philadelphia, 1896), p. 80; City of Philadelphia, *Fourth Annual Message of Edwin S. Stuart, Mayor of the City of Philadelphia, with Annual Reports of Abraham M. Beitler and of the Board of Health for the year ending December 31, 1894* (Philadelphia, 1895), pp. 72–78.

15 See, for example, Edgar Sydenstriker, *Health and Environment* (New York, 1933), pp. 120–121; E. R. N. Grigg, "The arcana of tuberculosis" Part I, *Am. Rev. Tuberculosis & Pulmonary Dis.*, 1958, *78* (No. 2): 151–172.

16 McKeown, *The Modern Rise of Population*, p. 99; City of Philadelphia, *First Annual Message of Charles Warwick, 1895*, pp. 83–84; Preston and van de Walle, "Urban French mortality," p. 282.

17 City of Philadelphia, *Report of the Department of Health for 1860* (Philadelphia, 1861), pp. 39–41.

18 City of Philadelphia, *Report of the Board of Health of the City and Port of Philadelphia to the Mayor for 1861* (Philadelphia, 1862), pp. 16–17.

19 *Ibid.*

20 City of Philadelphia, *Report of the Board of Health of the City and Port of Philadelphia to the Mayor for the year 1874* (Philadelphia, 1875), pp. 45–47.

21 City of Philadelphia, *Report of the Board of Health of the City and Port of Philadelphia to the Mayor for the year 1872* (Philadelphia, 1873), pp. 99–101.

22 James C. Booth and T. H. Garrett, *Report of the Philadelphia Watering Committee . . . on Filtration . . . on Schuylkill Water* (Philadelphia, 1854).

23 City of Philadelphia, *Annual Report of the Bureau of Water for the year 1902* (Philadelphia, 1903), p. 94, found in *Fourth Annual Message of Samuel H. Ashbridge, Mayor of the City of Philadelphia, Vol. II;* City of Philadelphia, *Annual Report of the Bureau of Water for the year 1904* (Philadelphia, 1905), p. 69, found in *Second Annual Message of John Weaver, Mayor of the City of Philadelphia, Vol. II;* City of Philadelphia, *Annual Report of the Bureau of Water for the year 1907* (Philadelphia, 1908), p. 47, found in *First Annual Message of John E. Reyburn, Mayor of the City of Philadelphia, Vol. II;* City of Philadelphia, *Annual Report of the Bureau of Water for the year 1908* (Philadelphia, 1909), p. 50, found in *Second Annual Message of John E. Reyburn, Mayor of the City of Philadelphia, Vol. II*; City of Philadelphia, *Annual Report of the Bureau of Water for the year 1909* (Philadelphia, 1910), p. 47 (quotation), found in *Third Annual Message of John E. Reyburn, Mayor of the City of Philadelphia, Vol. II*. Although

the entire city received filtered water in 1909, problems with supplying the Queen Lane district meant that a dependable supply was not finally insured until 1911. See City of Philadelphia, *Annual Reports of the Director of the Department of Public Health and Charities and of the Chief of the Bureau of Health for the year ending December 31, 1911* (Philadelphia, 1912), p. 17.

24 W. T. Sedgwick and J. Scott MacNutt, "On the Mills-Reincke phenomenon and Hazen's theorem concerning the decrease in mortality from diseases other than typhoid fever following the purification of public water-supplies," *J. Infect. Dis.*, Aug. 1910, *7* (No. 4): 489–563.

25 Samuel H. Preston, *Mortality Patterns in National Populations: With Special Reference to Recorded Causes of Death* (New York, 1976), p. 38.

26 Rose A. Cheney, "Seasonal infant and childhood mortality in Philadelphia, 1865–1920," *J. Interdisc. Hist.*, Winter 1984, *14* (No. 3): 561–562.

27 City of Philadelphia, *Report of the Board of Health for the year 1874*, pp. 397–402.

28 Cheney, "Seasonal infant and childhood mortality," pp. 578–579.

29 City of Philadelphia, *Third Annual Message of Harry A. Markey, Mayor of Philadelphia, containing the reports of the various Departments of the City of Philadelphia for the year ending December 31, 1930* (Philadelphia, 1931), p. 391; City of Philadelphia, *Second Annual Message of Harry A. Markey, Mayor of Philadelphia, containing the reports of the various Departments of the City of Philadelphia, for the year ending December 31, 1929* (Philadelphia, 1930), pp. 233–238, 397; City of Philadelphia, *Annual Report of the Bureau of Health for the year ending December 31, 1923* (Philadelphia, 1924), p. 11; City of Philadelphia, *Annual Report of the Bureau of Health for the year ending December 31, 1922* (Philadelphia, 1923), p. 83.

30 City of Philadelphia, *Second Annual Message of Edwin H. Fitler, Mayor of the City of Philadelphia, with Annual Report of William S. Stokeley and Annual Report of the Bureau of Health for the year ending December 31, 1888* (Philadelphia, 1889), pp. 19–20.

31 A record of the battles waged can be found in the Annual Reports of the Bureau of Health for the years from 1890 to 1914.

32 City of Philadelphia, *First Annual Message of Charles F. Warwick, Mayor of the City of Philadelphia, with Annual Reports of Abraham M. Beitler and the Board of Health for the year ending December 31, 1895* (Philadelphia, 1896), p. 140.

33 City of Philadelphia, *Third Annual Message of John Weaver, Mayor of the City of Philadelphia, with the Annual Reports of the Director of the Department of Public Health and Charities and the Chief of the Bureau of Health*

for the year ending December 31, 1905 (Philadelphia, 1906), pp. 214–215.

34 City of Philadelphia, *Annual Reports of the Director of the Department of Public Health and Charities and of the Chief of the Bureau of Health for the year ending December 31, 1911* (Philadelphia, 1912), pp. 21–23.

35 City of Philadelphia, *Fourth Annual Message of Rudolph Blankenburg, Mayor of Philadelphia, Vol. III* (Philadelphia, 1915), p. 4.

36 The statistics presented are from the Annual Reports of the Bureau of Health for the years from 1889 to 1913.

30

The Early Movement for Occupational Safety and Health, 1900–1917

DAVID ROSNER AND GERALD MARKOWITZ

One of the most important aspects of the history of Progressivism and American labor was the struggle from the turn of the century to World War I to achieve a safe and healthful work environment. During these years a broad-based movement to reform working conditions arose; it succeeded in achieving some major legislative victories at the state and federal levels and heightened consciousness about workplace hazards among the population at large.

Unfortunately, the significance of this movement has been obscured for historians and those working in safety and health today, for we have concentrated on only one aspect of the effort to control work conditions: the political battles surrounding the passage of workers' compensation laws in the years 1908–1914. In this article we will analyze the broader movement for safety and health at the turn of the century. This early movement was successful in developing a multiclass and multi-interest constituency that succeeded in bringing workers' safety and health issues to the consciousness of millions of Americans. The movement built on existing state factory-inspection programs and efforts in the 1880s and 1890s to enact protective legislation for women and children. It reached its peak during the first two decades of the 20th century and garnered support among reformers working in public health, conservation, housing, and labor legislation, especially legislation aimed at protecting children and women. Contemporary discussion of occupational safety and health was wedded to broader social concerns regarding workers' hous-

ing, sanitation, and general living conditions. "In one aspect," reflected one observer, the early 20th-century movement was "a chapter in the general health movement . . . next of kin to the crusade against tuberculosis, infant mortality, hookworm, and typhoid." But, "in another aspect it is part and parcel of the movement for labor legislation, being intermingled with the program of child labor reform, factory legislation, and factory inspection, and standing as a counterpart of measures for the compensation of industrial accidents."[1] Workers, social workers, housing reformers, journalists, politicians, big business representatives, social scientists, professionals, socialists, wobblies, and charity workers, who rarely agreed on much of anything, found themselves in alliance to stop the slaughter of workers in American industry.

It is important to recognize, however, the serious limitations of this movement. Because it was a broad-based coalition, it brought together radical and conservative groups. There was always a tension between the goals of socialist leaders such as Crystal Eastman and business groups such as United States Steel. By 1914, the movement lost its radical edge and became a vehicle for voluntaristic and management-sponsored approaches to reform. The movement's energy became focused on state workers' compensation laws, thus channeling the broader environmental and political concerns into narrow legalistic and administrative efforts. Management itself began to set the agenda for improving working conditions through its sponsorship and domination of the National Safety Council. It was able to redefine the issue to be one of safety rather than health and to take the issue out of the political sphere and place it in the technical and engineering arena.

DAVID ROSNER is Distinguished Professor of History at Baruch College and the City University of New York Graduate Center.
GERALD MARKOWITZ is Professor of History at John Jay College of Criminal Justice, City University of New York.

I

The growing concern over safety and health issues in the first decade of the 20th century arose in the wake of the revolutionary social and economic changes that America had just undergone. In little more than three decades Americans had witnessed the virtual explosion of urban and manufacturing centers. This was shocking to Americans reared in rural settings. Before the Civil War, most Americans lived on farms or in small towns; the few factories that existed were scatttered in mill towns and villages in the Northeast. With the growth of the transcontinental railroads, however, the development of national markets, increased exploitation of natural resources such as coal and iron, and the massive immigration of labor from rural Europe to the cities of the East and Midwest, conditions of work changed dramatically. America moved from being a fourth-rate industrial power to the leading industrial producer in the world. But work for the vast majority of laborers deteriorated. Speed-ups, monotonous tasks, exposure to chemical toxins, metallic and organic dusts, and unprotected machinery made the American workplace among the most dangerous in the world. In mining, for instance, England, Germany, and France experienced death rates of fewer than 1.5 per 1,000 workers during the first years of this century. In the United States more than 3 miners in every 1,000 could expect to die while working in a mine during any given year.[2]

The enormous wealth produced by the new industrial plants was achieved at an inordinate social cost. "To unprecedented prosperity . . . there is a seemy side of which little is said," reported one observer in 1907. "Thousands of wage earners, men, women and children, [are] caught in the machinery of our record breaking production and turned out cripples. Other thousands [are] killed outright," he reported. "How many there [are] none can say exactly, for we [are] too busy making our record breaking production to count the dead."[3] In a theme that would repeatedly appear, reformers compared the toll of industrial accidents to an undeclared war. As early as 1904 *The Outlook*, a mass-circulation magazine, commented on the horrendous social effects of industrialization. "The frightful increase in the number of casualties of all kinds in this country during the last two or three

years is becoming a matter of the first importance. A greater number of people are killed every year by so-called accidents than are killed in many wars of considerable magnitude," it pointed out. "It is becoming as perilous to live in the United States as to participate in actual warfare." The editorial demanded that the state document the extent of industrial accidents "in order that the people of the United States may face the situation and understand how cheap human life has become under American conditions."[4]

The power of the early 20th-century movement depended on the widespread publicity provided by a group of journalists and writers. These "muckrakers" exposed the horrible conditions of work to millions of Americans through magazine articles, pamphlets and books. Their primary aim was to arouse the public through a widespread propaganda campaign aimed at forcing reform legislation through Congress and especially state legislatures. They also sought to force particularly dangerous industries to clean up their workplaces. William Hard was one such muckraker. His 1904 article, "Making Steel and Killing Men," in *Everybody's Magazine* detailed the horrible work conditions in the south Chicago plant of the United States Steel Corporation. Hard described the dangerous conditions that led to the deaths of 46 workers and the permanent disablement of 386 others in just one year. He accused the company of endangering workers' lives by failing to provide a variety of safeguards near the blast furnaces and cauldrons into which the molten metal was cast for steel rails and girders. In vivid detail he described how men dropped into vats of molten metal or were showered with steel by sudden explosions in the furnace. The article created a tremendous stir among the 3 million readers of *Everybody's* and forced the company to provide elementary safeguards for its workers. Looking back just three years after its publication, John Fitch, another popular writer, noted that Hard's article spurred the company to begin a safety campaign. Subsequently, U.S. Steel would boast of its impressive safety record without acknowledging the role that popular pressure played in forcing improvements.[5]

As bad as conditions are today for American workers, we often forget how much worse they were just a few decades ago. During the first part of the century, stories of the plight of workers were reported daily in popular magazines and newspapers. These stories attest to the pervasiveness of industrial hazards and to the heightened consciousness that then existed. Reports of severe accidents on the railroads, in the mines, and in the factories reminded Americans that industrialism was taking an enormous toll.

The Survey, Everybody's Magazine, and *The Outlook* served as the outlets through which muckrakers and others exposed conditions in the dangerous trades. In their pages authors detailed "the death roll of industry" and documented how workers died in the mines, on the railroads, and in the factories. In an article on the dangers of various trades, one author charged that industrialists sent "to the hospital or the graveyard one worker every minute of the year."[6] Others noted that "the price we pay in human lives for our industrial progress is . . . appalling. For nearly every floor of every skyscraper that goes to make up Manhattan's picturesque skyline a man gives up his life."[7] One of the most graphic popular descriptions of the dangers of the meat packing industry was Upton Sinclair's *The Jungle.*

The publishers of these magazines and books were not printing this material as a public service. Rather, they recognized that it could sell magazines, and they were therefore willing to devote space, time, and money to a variety of occupational safety and health issues. "We spent a great deal of money in getting the material for *Everybody's Magazine*," noted the owner, "but we want to go further. We want it to count in the industrial life of this nation. Our hope is that after reading it you will enroll yourself with those who feel that nothing which concerns man should be a matter of indifference to any man."[8]

Publicity for labor's plight was also provided by an extraordinary group of social activists, academics, and professionals such as Alice Hamilton, Paul Kellogg, and John R. Commons, who, as doctors, writers, lawyers, and social workers, were instrumental in popularizing this issue. They lectured, wrote articles, exposés, and books, sponsored meetings, and conducted independent investigations of conditions of Pennsylvania coal miners, Birmingham and Pittsburgh steel workers, New York City garment workers, Massachusetts textile workers, Minnesota iron miners, and numerous other industries. Their work was of the highest academic

standards, but they were not embarrassed to express their concern for workers and their commitment to reform. For instance, Dr. Alice Hamilton, a pioneer in the occupational safety and health movement, is remembered today for her important pathbreaking scientific investigations into conditions of work in a variety of industries, most notably lead. But she also saw herself as an advocate of reform. Presaging her later work around lead poisoning and other industrial toxins, she noted the terrible mortality from accidents and diseases that slowly ate away at the body and soul of the workforce.[9] Along with others such as Frances Perkins and Eleanor Roosevelt, who would later be instrumental in shaping labor legislation during the New Deal, Hamilton developed political skill and experience during the early movement for occupational safety and health.

What is especially exciting about the early 20th-century movement is that it gained strength from its heterogeneous constituency. "The most important economic movement underway in America today is the movement for the conservation of the human resources of the country," said Earl Mayo in an article entitled "The Work That Kills." This movement "has its humanitarian as well as its commercial side . . . it is a many-sided movement and its ultimate possibilities have not yet begun to be appreciated."[10] Thus, the movement was a multiclass effort that united groups of widely different interests. In the following pages, we will discuss four different themes that dominated the thinking and concerns of the various groups. First, we will outline the different ideological strands that made up this broad coalition. Second, we will look at their very different notion of occupational safety and health that helped give unity to the movement. All groups recognized that workers' health could be maintained only by improvements in both the home and work environments. Third, we will examine how different actors interpreted the obvious suffering that affected industrial workers. All parties were acutely aware that the extraordinary number of workers maimed and killed exacerbated the conflict between labor and capital. Fourth, we will outline the growing consensus that it was necessary to provide the working class some minimal protection. All parties came to recognize that workers' health could be safeguarded only through the intervention of the state.

II

The movement for occupational safety and health was made up of two major ideological strands. Most of the reformers used concepts and language borrowed from the lexicon of business when analyzing the problems of death, disability, and disease created at the workplace. Others, however, including socialists and moralists rejected a worldview that saw workers as machines or the raw materials for industry.

Those reformers that worked within a capitalist ideological framework used words such as *costs*, *efficiency*, *resources*, and *breakdowns* to describe the problem that accidents and disease posed for the society. Injured workers were compared to broken-down machines in need of repair, and workers killed on the job were seen as "wasted resources."

The most blatant statement of this ideology was provided by big business itself. Big business saw human conservation purely in terms of the long-term financial benefits of company stockholders. "A careful businessman sees that his property is maintained in excellent condition," said one propagandist for the International Harvester Company in 1912. "His buildings are kept in good repair and fully insured against loss from fire. His machinery is always maintained at a high point of efficiency, no rust is allowed to accumulate. . . . In short, every dollar he invests in his business is guarded and nursed so that it brings forth its full and legitimate earning power." Conserving the workforce "is simply applying the same business principles to his workers that he applies to the rest of his business."[11]

More important than the big-business view itself was the acceptance by many reformers of the underlying economic rationale for the protection of workers. In an article subtitled "Human Life as a National Asset," one physician noted that "all economists today agree that the greatest asset of any country is the vital efficiency of its citizens." Others asked, "What, in dollars and cents, may be roughly figured as a man's worth to the community." Still others noted that industrial accidents meant "incalculable loss of time and energy in business."[12] Clearly, reformers thought that it was the almighty dollar that would capture the public's imagination in a dynamic capitalist economy. One author suggested: "Starting on the lowest level let us formulate our initial axiom in terms of dollars. A sound man can do more than a sick man. Therefore, he can make more money." Similarly, "a sound city can do more work than a sick city. Therefore, in the long run, it can accumulate more wealth."[13] Propagandists used the rhetoric of business to justify their positions, because they knew that few reforms could be achieved without a strong commercial rationale. In a period dominated by the new corporate giants even the most basic human values had to be justified in terms of their price tags.

Others completely rejected the view of a worker's life as a commodity and distrusted the motives of corporate welfare programs. "Businessmen would rather risk their neighbor's lives than their own money," was the underlying philosophy of many. Ultimately, these critics believed that business would abandon its safety efforts and forsake the workforce. "Greed feels nothing, knows nothing, cares for nothing but profit. It fears nothing but the loss of dollars. . . . There is no . . . reason to suppose [employers] are distressed by the sight of human beings crushed under falling walls or leaping all aflame from tenth story windows."[14] Left-wing unions raised even more fundamental objections to the values of business civilization. "The average person is never tired of boasting about the wonderful achievements of modern civilization," noted a writer in *The Glass Worker* in 1908. "The huge factories dot the earth everywhere, polluting the landscape with their unsightly vomitings, but the question arises, are these mighty achievements worth the price that humanity is paying for them. To everyone who is inclined to draw his conclusions from the facts," the writer continued, "it is well known that the vast majority of the people in all civilized countries lead lives of arduous and ill-paid toil that cannot fail to result in physical and moral degeneration. . . . Driven as they are to the limit of human endurance in the factory, mill and sweat shop, . . . forced to crowd into narrow quarters, subsist upon a scanty supply of food of the cheapest and most inferior quality, the healthy physical and moral development of these urchins of modern industry is out of the question." *The Glass Worker* condemned modern capitalist industry and turned business rhetoric against itself: "The facts stand glaringly forth that social health and welfare are not being conserved. It is being forced into physical and moral bankruptcy because of the fearful price it has to pay for those achievements of which the average man is so boastful."[15]

It was not only the labor radicals who objected to the treatment of life as a commodity. Some middle-class reformers also defined such a view as innately immoral. "Products that are made . . . by hours longer than health endurance are anti-social and immoral products and express a ruinous social cost no matter what the selling cost may be," said Margaret Dreier Robbins in her presidential address to the Women's Trade Union League. "In shop and factory and mill all over our country women are working under conditions that weaken vitality and sap moral fiber — conditions that are destructive alike to physical health and moral development."[16]

Others agreed that "people are a natural resource," but rejected a purely commercial definition of that term. William Ludlow Chenery of Chicago's Hull House wrote: "Society . . . which has begun to feel with the great force of an elemental emotion the necessity for conserving the trees, the rivers, the lands and the mineral deposits of the nation hitherto has given but small thought to the protection of the people. . . . It is a kind of commentary on the American conception of things that no one thought of investigating these things until about three years ago. We were too busy to think about the human factor of industry, the thing, the product was the end and object." Chenery concluded that he hoped "the national emphasis [would] be shifted from the integrity of property to the welfare of people."[17]

Another group, frightened by industrialization itself, had little faith that industry's excesses could be contained by government or labor. Using moral and religious rhetoric, they longed for a return to a more bucolic past. "Surely our modern industrial civilization resembles a Frankenstein," noted Josiah Strong, a leader in the Social Gospel Movement. "And unless something is done to check the monster it is creating he will grow ever more murderous."[18]

For most religious and moral leaders the impetus for their concern was a fundamentalist fear that science and technology posed a threat to a spiritual life. This resistance to technological change led them to support efforts that reasserted human and spiritual values over materialistic ones. For in-

stance, they were ardent supporters of a six-day week. The Reverend George W. Grannis, general secretary of the Lord's Day Alliance of the United States, appeared before a New York State commission to advocate a shortened work week. Similarly, William P. Swartz of the New York Sabbath Committee testified that "thousands and thousands are robbed of the privilege and opportunity for that higher development of character[,] . . . one of the staple elements of our country and necessary for its existence, by the fact that they don't have . . . time . . . for the home or family ties, for the conditions that surround and make sound and elevate the manhood underlying and making the nation." Swartz pressed the commission to enact reforms and to "do everything we can to take off . . . the seven days yoke."[19] While these leaders were romantics and often religious fundamentalists in their beliefs, the logic of their views drove them to radical stances on a variety of labor issues. Their desire to limit the effects of industrialization, by initiating shorter hours and work weeks, put them in an alliance with radical elements of the trade union movement who were pushing for the eight-hour day.

One of the first and most important ways that muckrakers and other reformers sought to awaken the public to the horrible costs of industrialism was through popular discussion of the plight of women and children in mines, mills, and factories. Beginning in the late 19th century, magazines carried lurid stories of children awakening at dawn to toil at the looms, and women leaving their loved ones to work in the factory. By the early years of this century, there was hardly a journal that did not have articles such as the "Slavery of Childhood" or "Women's Factory Work." By 1900 over 1.7 million children worked for wages, many of them for 10 to 14 hours a day. Edwin Markham, a popular writer of the period, told of the little boys and girls who "sicken and faint" from "the heat and the odors . . . in the fancy box and candy factories." He also described how children in glass factories toiled in front of "red hot furnaces becoming numb and blistered."[20] Others described sickly women toiling in the bakeries and cracker factories. "Girl workers in the excessive heat, with the smell of dough, staring at the trays which pass slowly, continuously beneath their eyes, become light-headed and ill."[21]

The depiction of the plight of women and children served both progressive as well as reactionary ends. For some, the publicity was meant to force industries to improve working conditions by spurring unionization of women workers. For others, the publicity was seen as a means of securing for women protective legislation that could serve as an opening wedge to improve conditions for all workers. But others sought to use the issue of women's safety and health to force women to abandon work and return to the home.

The Women's Trade Union League, made up of union women and middle-class reformers, sought to organize women workers. In one pamphlet, it asked its readers, "Perhaps you have complained about the chemicals used in the washing of your clothes, which cause them to wear out quickly, but what do you suppose is the effect of these chemicals upon the workers?" The pamphlet detailed the long hours, intense heat, and bad air that these women workers faced every day and suggested that it was inevitable "that these workers became physical wrecks in a very short time, just because you and I never count the cost." What was the "way out" according to the league?—"The Laundry Workers Union" because "the great school of the Working People [is] the Trade Union." Margaret Dreier Robbins, the president, maintained that other unions, such as the United Textile Workers, had "never ceased in its demands for better sanitary conditions . . . ."[22]

Few reformers were willing to place the hope for women's salvation in the hands of a trade union, and instead put their faith in protective legislation which sought to limit the number of hours and night work, and to prohibit women entirely from certain trades. While we recognize today that this effort resulted in the displacement of women from industry, and especially from the higher paying jobs, some reformers had hoped for a different result. They believed that if shorter hours were required for women, if special requirements were enacted for women's employment, it would redound to the benefit of both men and women. William Hard and Rheta Childe Dorr argued that "legislative concern for women leads unavoidably to legislative concern for men." They continued, "We cannot debar women from Industry. We must make Industry fit for women. And when it is fit for them not many years will pass before it will become fit for men. Industry, the bachelor, needs in his bachelor quarters the touch, the effect, of feminine occupancy before he can really cease to be but masculine, before he can really become completely human."[23]

But Hard and Dorr still accepted that there were innate differences which made women the weaker sex. They suggested that women couldn't stand as long as men, they were not as strong, and a "woman's nervous system is more unstable than man's and more easily shaken from its equilibrium." They believed that women should be excluded from a few jobs involving "hideously severe, de-sexing physical toil."[24] Others seized upon the alleged inferiority of women to seek to exclude women from many fields of endeavor and to return them to the home or at least to the lowest paying jobs. *The Glass Worker*, a union publication radical on most issues, was outraged that women were employed in traditional male roles in some foundries. They used as an excuse for their stance the physical nature of the work. They claimed the work was unfeminine: "Their faces and hands are begrimed with black dust and grease, and were they not required to wash themselves before leaving the plant they would present an unseemly appearance on the streets."[25]

The major complaint against women working was that it caused irreparable harm to their ability to bear healthy children. The Woman's Label League, an organization of women formed to support the mostly male trade union movement, featured a report by the New York Department of Charities that said the increased number of "feeble-minded children is a direct result" of women's work in factories. The *Woman's Label League Journal* said, "Ten years of the work . . . unfits women for the duties and responsibilities of motherhood."[26] The National Child Labor Committee issued a flyer (see Fig. 1) which suggested that mothers who worked outside the home became a burden on both husbands and children, and were a direct cause of child labor. This, in turn, resulted in children who suffered from "impaired health, illiteracy, inefficiency, delinquency, (and) dependence." The reformers sought to "set it right" by taking working mothers "who neglect home and children" and making them into homemakers who could "enhance the efficiency of father and children."[27]

The focus on women and children was a double-edged sword. It could be used to arouse the country to the dangers that all workers faced; it was also used to exclude female workers by asserting that it was wrong to subject the weaker sex to horrible working conditions.

III

Underlying much of the agitation for reform was the belief that the high human costs of industrialism were producing severe class antagonisms. Social workers felt that the worker, especially the immigrant worker, might never become fully integrated into the fabric of American life if subjected to undue physical strain or intolerable conditions on the job. Settlement workers were fearful that industrialism and its attendant hardships were creating a generation of disaffected and dangerous classes. Graham Taylor, a leading social worker, warned, "Nothing is more dangerous in a democracy than to allow a sense of detachment to divide a class from the mass, a craft or individual from the community of interest, . . . personal and private instincts and ideals from public welfare."[28]

To many at the turn of the century, worker safety and health was a primary source of conflict between workers and owners. Some labor and socialist groups maintained that the extraordinary number of deaths and accidents that workers suffered was proof that capitalism was unable to serve workers' interests. But others, primarily from the middle and upper classes, argued that reform was possible, indeed necessary to restore the harmony of interests between workers and capitalists. All agreed with the findings of the United States Commission on Industrial Relations: dangerous and unhealthy working conditions were a crucial source of conflict and dissatisfaction in American industry. The commission found that labor was especially upset that "the rapid pace of modern industry . . . results in accidents and premature old age," and that there was a "lack of attention to sickness and accidents" as well as "difficulty and delay [in] securing compensation for accidents." The commission also reported a widespread "fear . . . of being driven to poverty by sickness, accident or involuntary loss of employment."[29]

The American Association for Labor Legisla-

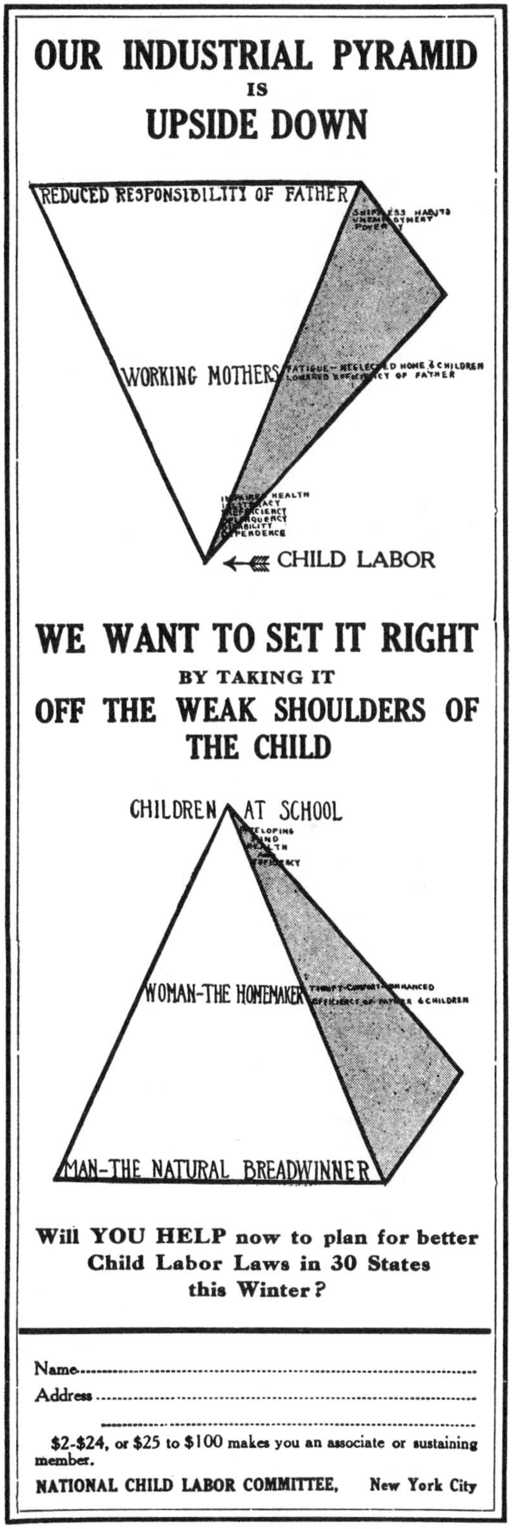

Fig. 1: From *The Survey*, Jan. 18, 1913, *29*: 534.

tion (A.A.L.L.) was one of the foremost organizations dedicated to reform. The members of the association were prominent labor leaders, corporate liberals, social reformers, and academics of the day, including Jane Addams, Samuel Gompers, John Mitchell, Robert Hunter, Florence Kelley, and Everit Macy. The A.A.L.L., funded by some of the biggest capitalists of the era, sought to find a model of change that would reconcile labor and capital.[30] In 1909 Henry W. Farnum, the president of the A.A.L.L., described the problem that his organization sought to address. "Some, exaggerating the incidental evils of progress, decry all efforts of betterment, and long for the good old times when there were no reformers," he noted in his presidential address. "Others, realizing strongly the evils which grow up without regulation, think that reform has not been carried far enough and advocate some extreme remedies such as Socialism." The A.A.L.L. sought to avoid both of these courses. Farnum wondered "if there is no principle based upon experience which will enable us to steer the Ship of State so as to avoid the Scylla of conservatism and the Charybdis of radicalism."[31] Its fears of the two extremes made it adopt as its slogan, "Social Justice Is the Best Insurance Against Social Unrest."[32]

The A.A.L.L. was active in organizing conferences on industrial work conditions, workers' compensation, various forms of tenement and social legislation, unemployment, and compulsory health insurance. It organized the first annual conferences on the growing problems of industrial diseases. True to its desire to build a consensus, the A.A.L.L. sought to "bring physicians, employees and employers together in a united effort to lessen the ravages of occupational diseases," and to instill them with the "zeal and enthusiasm of social workers." In addition to labor and capital the conferences were attended by insurance experts, efficiency engineers, state and federal public health officials, doctors, and settlement house workers.[33]

Occupational safety and health was the major focus of the association. In an article entitled "An Immediate Labor Program," the A.A.L.L. called for eight legislative initiatives, seven of which addressed workers' safety and health: protection from lead poisoning, a uniform system of reporting accidents and diseases, systematic investigations of industrial hygiene and safety, factory inspection

and labor law enforcement, revision of the Federal Employees Compensation Act, better state workers' compensation legislation, and one day's rest in seven. John B. Andrews, secretary of the association (and its major spokesperson), summarized the importance of occupational safety and health to the A.A.L.L.: "Since its organization in February 1906, the Association has regarded the prevention of industrial accidents, and the enactment of a just plan of compensation for industrial injuries, as the most pressing immediate problem in labor legislation."[34]

The association organized the National Commission on Industrial Hygiene in 1908, and a year later an A.A.L.L. representative gave a major address on industrial diseases and occupational standards at the National Conference on Charities and Correction. The association also published an influential investigation of phosphorus poisoning in the match industry which led to federal legislation eliminating the use of deadly white phosphorus in matches. This was the first federal action that proved effective in controlling an occupationally caused disease.

Middle-class Progressives and enlightened capitalists believed that it was possible to reform and improve dangerous working conditions. But others, most notably socialists, saw such conditions as an inevitable product of capitalist production. For them, the only way the workplace could be made safe was if the workers took over the management of the factory. At a mass meeting of the cloak makers union called by the New York Socialist party, Morris Hillquit proclaimed, "We protest against the wanton and ceaseless murder of our sons and daughters in factories, mines, and mills and we declare that if the capitalist class cannot conduct the industries of the country without murder the workers will assume management of the industries and conduct them with safety to the employees and for the good of the community."[35] Others suggested that workers could protect themselves only through political organization. In order to make their plants safe, they "should join the Socialist Party and help to abolish capitalism and introduce socialism."[36]

The horrible conditions that many workers faced every day led some socialists to call for revolution. "When I read . . . such records as this: 'Helper—flooring factory—age 19—clothing caught by setscrews in shafting; both arms and legs torn off;

death ensued in five hours,' my spirit revolts," declared Crystal Eastman, the famous socialist spokeswoman in 1911. "And when the dead bodies of girls are found piled up against locked doors . . . after a factory fire[,] . . . who wants to hear about a great relief fund? What we want is to start a revolution."[37] Eastman and other socialists saw industrial accidents as one sign of the war that management waged against the working class in industrializing America.

For many socialists the revolution was not necessarily one fought out with guns. Rather, what was needed was a radical redirection in its uses of science, technology, and spirit. Among some impassioned advocates of reform were many who had a fervent belief in the power of knowledge and information in the struggle to attain safe workplace conditions. "We must pause and consider what are the essential weapons in our campaign," Eastman said. Rather than guns, the "first thing we need is information, complete and accurate information about the accidents that are happening. It seems a tame thing to drop so suddenly from talk of revolutions to talk of statistics," she acknowledged, "but I believe in statistics just as firmly as I believe in revolutions. . . . And what is more, I believe statistics are good stuff to start a revolution with."[38] The passion with which Eastman called for accurate statistics reflected the fervent belief among this generation that information itself was an integral tool in the battle for social justice.

Labor's anger at such conditions was frequently expressed in strikes at unhealthy and dangerous shops. One long and dramatic example of the discord created by dangerous health conditions was the nine-week general strike of cloak makers in New York City. The strike was called because of the "unsanitary condition in a large number of shops." The settlement, reached after the bitter and acrimonious picketing and public pressure, included the creation of a Joint Board of Sanitary Control. This body, composed of employers and employees, sought to establish sanitary standards for the industry.[39] In one year the cloak makers union called 28 successful "sanitary" strikes in New York, and set the stage for the public outcry that followed the tragic deaths of 140 young women in New York's infamous Triangle Shirt Waist Fire in 1911.[40]

The shock of the Triangle fire prompted many to indict industrial capitalists for their callousness and

their greed (see Fig. 2). After a march and rally of 50,000 trade unionists, Rose Schneiderman, vice president of the Women's Trade Union League, argued, "I would be a traitor to these poor burned bodies if I came here to talk of good fellowship. The old Inquisition had its rack and thumb screws and its instruments of torture with iron teeth. We know what these things are today," she continued. "The iron teeth are the necessities, the thumb screws the high powered and swift machinery close to which we must work, and the rack is here in the fire-trap structures that will destroy us the minute they will catch fire."[41]

IV

The need to ameliorate the threat of class conflict and the nature of the problems that workers faced led to a broad conception of occupational safety and health. In the early years of the century labor joined with middle-class reformers to argue for a new definition of the intimate relationship between the health of workers and the health of the general community.

Social workers were some of the earliest and most articulate advocates of this broader vision. Their first-hand experience with the immigrant poor made them sensitive to the relationship between poverty, illness, and work. They knew that most workers lived in neighborhoods that bordered or surrounded the factories. The crowded conditions in the home, the lack of light and ventilation in the slum streets, and the horrible working conditions combined to produce unhealthy environments for workers and their families. Thus, social workers recognized that to improve the health of immigrant workers it was necessary to go beyond industrial hygiene and include housing conditions, diet, and individual morality.

The most important example of this view of disease causation was contemporary discussion of tuberculosis, the devastating lung disease which ravaged working-class families, incapacitating children and grown-ups alike. In 1904, *Charities*, the journal of the Charity Organization Society, noted that tuberculosis, one of the worst scourges of the working classes, went hand-in-hand with industrialism. It maintained that any effective program aimed at controlling the disease would have to acknowledge the relationship between homelife, neighborhood, and workplace. Social workers and other professionals believed that tuberculosis struck those weakened by overwork, poor nutrition, or crowded living conditions. When these conditions were combined with a dusty work environment, a worker's health was almost sure to be impaired. An effective program, therefore, to control consumption required close attention to all aspects of a worker's life. Consumption "must be checked by public sanitary preventive measures in street and shop and home." It could "be eradicated only by a movement democratic enough to reach intimately all three."[42]

For charity workers the issue of consumption illustrated the environmental, rather than the individual, roots of illness and dependence. "I would like to speak of the importance to the Charity Organization Society of health and safety conditions for labor," said Frederick Almy, secretary of Buf-

falo's chapter of the organization. "I think one-third or perhaps one-half of our work would disappear if labor conditions were all they might be. I mean if employers took as good care of their men as they do of their machinery." Almy continued by rejecting explanations that blamed the workers themselves for their poor health. It was obvious "that sickness [was] due to the housing and working conditions. . . ."[43] In contrast to the 19th-century charity notion that illness was a reflection of the personal worthiness of the sufferer, social workers in the Progressive Era recognized that social conditions were far more important determinants of health and illness.

Reflecting this broader ecological notion, Graham Taylor argued that there was no clear demarcation between occupational safety and health and other social and environmental problems. He said that the activities of social workers were "those

Sloan in New York Call.

"HERE IS THE REAL TRIANGLE."

Fig.2: Cartoon in the New York socialist daily newspaper, *The Call,* was designed to invoke the Triangle Waist Company. The mutilated employers' liability bill in the foreground was a comment on a New York State Court

of Appeals decision that workers were not constitutionally protected from the economic losses incurred in the 1911 fire. From *The Survey,* April 8, 1911, *26*: 81.

increasingly discerned, but illy defined connections between industrial conditions and relationships and the social, moral and civic interests of our times." He went on to note that "industrial casualties [were] among the deepest taproots of dependency."[44]

Social workers, forced by their experiences in the nation's growing slums, formed a *de facto* alliance with labor and turned to it for information and support. At the 25th Annual Convention of the American Federation of Labor in 1906, for instance, the delegates identified tuberculosis as one of its most pressing problems. In a dramatic chart showing the death rate from consumption in 53 occupations the A.F. of L. pointed out that marble and stone cutters, cigar makers and plasterers, printers and servants all had death rates well above 4 per 1,000, while bankers, brokers, and officials had the lowest death rates, below 1 per 1,000. (It is unclear whether the A.F. of L. accurately identified tuberculosis, or if they grouped other lung disorders such as silicosis, byssinosis, or other pneumoconioses together.) In an address before the convention one speaker spelled out the connection between work, wages, living conditions, and tuberculosis: "All this means, really, the regulation of factory conditions, the regulation of housing, and the passage of child labor laws" was essential for battling the "Great White Plague." Like the social workers, the convention demanded improved public accommodations and play areas as well as more traditional demands for shorter hours and better pay. They called for the "development of playgrounds adjacent to all public schools" and "large open 'breathing spaces' interspersed in all cities." This was in addition to their advocacy of the elimination of sweatshops, "rigid inspection of mines, mills, factories and workshops," and the "incorporation in trade agreements in collective bargaining, covering working conditions, for suitable sanitation and ventilation." Significantly, the A.F. of L. claimed that health was a direct reflection of the success of the trade union movement. "In the same degree that the trade union movement becomes powerful will it establish such improved conditions that will check and eliminate the ravages of consumption."[45] In the following years unions became more active in fighting tuberculosis. They recognized that it was particularly prevalent among the working class, and

as *The Glass Worker*'s editor noted, "At least one-third of the deaths during the chief working period of life are caused by pulmonary tuberculosis. . . . In some trades . . . from 35 to 50% of all deaths are caused by tuberculosis." The editorial noted that "no movement" that ignores the industrial workplace can mount an effective "campaign against tuberculosis."[46]

Union campaigns to make shop conditions more sanitary were linked to broader public health issues, most important, the battle against infectious disease. Cleaning up the workplace and keeping the workforce healthy were seen as benefits to both the worker and the public. In 1910 the greater New York local of the International Union of Bakers and Confectionary Workers conducted a successful strike to demand more sanitary working conditions. "Perhaps no phase of the trade union movement has ever affected the public so directly as the agitation for sanitary conditions in the bake shops," commented one leading periodical. In May of 1909, 3,000 Jewish bakers struck, and less than a year later 4,000 German workers followed suit.[47] The union identified unsanitary workshops and the spread of infectious disease with nonunion bakeries. They insisted "on proper ventilation and sanitary conditions" and demanded "the great bread eating public of this country should see to it that the bread they eat bears the bakers' union label."[48] Their ability to link unsanitary working conditions to the health of the public gave their strike a tremendous appeal and power.

The very prevalence of contagious diseases, such as diphtheria, influenza, tuberculosis, and typhoid, within the working class spurred consumer groups to take up the issue of health conditions on the job and in the home. In part the appeal was based on fear, to make sure the middle class and wealthy would not be infected by goods tainted by sick workers. In the growing garment industry of New York, many dresses, shirts, and trousers were sewn on a piece-work basis in tenement slums, raising the spectre that the same diseases infecting those in tenements would be transmitted to the men, women, and children of the middle class. It was this terror that led the National Consumers League to become active in tenement reform and anti-tuberculosis campaigns. Its label on goods came to represent the mark of clothing manufactured under hygienic conditions. The league became involved in a wide variety of labor-related issues, including fire safety, workers' compensation, and occupational disease legislation.[49]

The concern with public health also led the National Consumers League to confront the issue of piece work in tenements. It was felt that work at home led to "long hours of uncontrolled labor for women and little children — often under far worse conditions than those of regulated factories." This in turn undermined the "health and strength" of the workforce and, therefore, spurred the spread of contagious diseases.[50] Frances Perkins, who had served for two years as secretary of the New York City Consumers League as well as at Greenwich House, also condemned the system of home work as "the most practical example which we have of the ability of industry to enslave its workers." But she also noted that this work also had a deleterious effect on public health: "One of the things which I noticed in regard to the system of this home work is that it is a menace to the health of the community and the health of the worker."[51]

The special social conditions surrounding work at the turn of the century led to the broad conception of the meaning of occupational safety and health. The rapid unregulated growth of industry, the enormous immigration of foreign workers, the growing strength of the Socialist party, the fear of social unrest, and the terror engendered by infectious disease gave the movement a special appeal. Occupational safety and health was part and parcel of a larger movement to reform American society.

V

The movement to control workplace hazards was widespread, encompassing a variety of different groups. Among these groups were radicals such as Crystal Eastman and Morris Hillquit. At the other end of the political spectrum were "enlightened" owners of corporations such as International Harvester and United States Steel that sought to undermine the movement through voluntary welfare and safety programs. Between these two extremes were a wide variety of middle-class reformers and conservative labor representatives who recognized that uncontrolled capitalism was killing and maiming so many workers that it was undermining the legitimacy of capitalism itself. The

conflicts this generated were so severe that a program of reform was essential.

The movement was most successful in making occupational safety and health a national issue. For almost a decade, exposés of inhumane working conditions and demands for reform were regular features in newspapers and magazines across the country. The outcry against the hazards that workers faced on the job raised the public's consciousness to a level that was not to be reached again until the 1970s. In its early years, socialists and labor representatives enunciated a new program for reforming the workplace. They insisted that occupational health issues, especially the control of lead, phosphorus, and other industrial poisons, be an integral part of the movement's agenda. They recognized that a broad environmental approach had to be adopted if workers' health was really to be protected. Finally, they demanded that government play the decisive regulatory role in protecting the well-being of labor. After 1914, however, business groups came to dominate this movement through organizations such as the National Safety Council. Their approach was entirely different. They narrowed the concern to safety rather than health and turned to engineers and technicians rather than to labor and government. They relied on voluntary action by industry itself to improve safety conditions and emphasized the responsibility of the worker for many accidents and the need for workers to avoid carelessness on the job.

On a concrete level the movement was successful in achieving some legislative reform. Several states had already acted in the late 19th century to initiate state factory-inspection bureaus to control workplace hazards. But it must be remembered that there was hardly any federal involvement in this area in the 19th century.

Before the Progressive Era, the dominant *laissez-faire* ideology defined government's role extremely narrowly: most officials saw themselves merely as passive facilitators of trade and industrial development. By 1900 the boundaries of government had expanded to include such activities as the regulation of interstate commerce. But it was only exceptional public officials who openly supported workers in their dealings with management.

During the first two decades of this century, however, a number of public officials began to challenge this older 19th-century notion of government's passive role. Especially in the larger industrial states of the East and the Midwest, legislators and administrators such as Senators Robert Wagner and Robert LaFollette and President Teddy Roosevelt pressed for a more activist stance for government. Labor relations was one of their major foci, and they sought to reconcile the divergent views of labor and capital.

Underlying the efforts of progressive leaders was a new conception of the state, a state that regulated for the common good. Robert Wagner summarized this view: "No government is properly performing when it permits the working people within its bounds to be employed under unsanitary conditions, when it fails to protect them from preventable diseases and accidents, when it permits the premature employment of its young children, and the excessive toil of its women." For reformers the new industrial system demanded a different conception of the state. The new industrial system, dependent as it was on the harmonious relationship between labor and capital, demanded a healthy workforce. The role of the state was to protect this "most precious asset, the workers within its bounds."[52]

Industrial accidents and diseases proved to be an attractive issue for progressives interested in expanding the role of the state. The rapid increase in the industrial labor force and the social disorder that accompanied urban growth, immigration, and industrialization made workers' health and living conditions a natural concern for government. One official writing in the Department of Labor *Bulletin* in 1903 pointed to the link between industrial conditions and America's health: "The miserable hygienic conditions existing in the working places . . . are unjust to the working classes, and sometimes react with frightful results upon the public. . . . With the multiplication of factories the improvement in the lot of the laboring man has become a vital question of the day. . . . The health of society in general is both directly and indirectly menaced by insanitary conditions in any industry."[53]

From the modern perspective, the efforts by reformers to establish a governmental role in workplace regulation may seem trivial or even conservative. But at the time these efforts were supported by many radicals as significant reforms. In little more than two decades, we saw the establishment

of a meaningful federal Department of Labor, active women's and children's bureaus, the reinforcement and subsidization of state factory-inspection systems through state departments of labor, and the beginnings of a role in occupational safety and health within local health departments. We also saw the passage of the first significant child and women's labor legislation and a host of specific state acts regulating working conditions in tanneries, bakeries, foundries, and numerous other industries. Also, for the first time, there was a serious attempt to organize a more reliable method for collecting statistics on occupational injuries and deaths. Finally, it must be pointed out that in 1900 no state in the Union had a workers' compensation law. By 1915 every highly industrialized state had passed an act for some form of compensation.

On the federal level, the reform movement was successful in passing a number of significant pieces of legislation. The railroad workers' compensation act in 1907 and a broader federal employees' compensation act were two of the landmark legislative efforts. But other attempts at outlawing certain special hazards also were important milestones in this early period of governmental activity. The passage of the Esch Phosphorus bill, an act placing a prohibitive tax on kitchen matches that used deadly white phosphorus, was a dramatic victory of the early reform movement, even if it was later than European actions.[54]

The movement forced politicians to pay attention to certain elemental safety and health issues. In 1912, the yearly Conference of Charities and Correction issued a "Platform of Industrial Minimums" in anticipation of the upcoming presidential elections. Included among them were the traditional demands for "a living wage," an eight-hour day, and a six-day work week. But added to the list was a set of demands that had never appeared before. The conference called for mandatory governmental investigations of all industries "with a view to establishing standards of sanitation and safety and a basis for establishing compensation for injury." Also among these standards was a call to prohibit "the manufacture or sale of poisonous articles, dangerous to the life of the worker and his family" and a law that required the mandatory reporting of all "deaths, injuries and diseases due to industrial operations." The conference also called for workers to have a right to a "safe and sanitary home."[55]

The platform developed by the national conference received wide publicity and was used as the model for the industrial platform of Teddy Roosevelt's Progressive party in 1912: "The supreme duty of the national government is the conservation of human resources through an enlarged measure of social and industrial justice. . . ." They called for "effective legislation looking to the prevention of industrial accidents, occupational disease, over work, involuntary unemployment, and other injurious effects incident to modern industry."[56]

Despite the very real differences in emphasis and approach, the various elements of this movement were able to unite around a few legislative and programmatic goals. Workers' compensation was one especially important reform goal that has received a great deal of attention. But a host of other efforts also characterized the movement and, taken together, these efforts force us to reevaluate the meaning of the movement. A look at its impact on the consciousness of the American population gives it an exceedingly radical appearance.

On one level, the importance of the movement was reflected in the profound effect that propagandistic efforts had on the thinking of the majority of Americans. During the 19th century job-related injury and death were often seen as necessary by-products of industrialization. Personal health and safety were believed to be the responsibility of the workers themselves. Legal doctrines such as the "fellow servant" rule and the notion of "assumed risk" spoke to the prevailing 19th-century belief that the burden of responsibility for occupational injury and disease resided with the workforce rather than with management. The efforts of reformers altered the prevailing assumptions of many Americans. The notion that workers assumed the risks of work by taking the job in the first place was largely discredited. But workers continue to be maimed, poisoned, and killed at their workplaces. However limited the success in reforming the workplace, the true legacy of the early movement is the principle that workers have a right to a safe workplace, and it is the responsibility of society to guarantee that right.

NOTES

1   Paul S. Pierce, "Industrial diseases," *North Am. Rev.,* Oct. 1911, *194*: 530.

2   Arthur B. Reeve, "The death roll of industry," *Charities and the Commons,* 1907, *17*: 791.

3   *Ibid.*

4   Editorial, "Slaughter by accident," *The Outlook,* Oct. 8, 1904, *78*: 359.

5   William Hard, "Making steel and killing men," *Everybody's Magazine,* Nov. 1907, *17*: 579–592; John Fitch, "The human side of large outputs," *The Survey,* Nov. 4, 1911, *27*: 1149.

6   Reeve, "Death roll of industry," p. 807.

7   "The price, in blood, for industrial progress," *Current Literature,* Dec. 1911, *51*: 629.

8   William Hard and others, *Injured in the Course of Duty,* pamphlet (n.p., n.d.).

9   Alice Hamilton, *Exploring the Dangerous Trades: the Autobiography of Alice Hamilton* (Boston, 1943).

10 Earl Mayo, "The work that kills," *The Outlook,* Sept. 23, 1911, *99*: 203.

11 *International Harvester Company and Its Employees,* pamphlet (Chicago, 1912), pp. 3, 4.

12 Thomas F. Harrington, "Prevention of disease versus cost of living," *Scient. Am. Suppl.,* June 28, 1913, *75*: 402–403; Arthur B. Reeve, "Our industrial jug-

gernaut," *Everybody's Magazine,* Feb. 1907, *16*: 149; William Hard, *Injured in the Course of Business* (n.p., n.d.).

13 Paul S. Pierce, "Industrial diseases," *North Am. Rev.,* Oct. 1911, *194*: 530.

14 Editorial, "Business and manslaughter," *The Independent,* May 2, 1919, *72*: 964–965.

15 "Is it worth the price?" *The Glass Worker,* June 1908, 5.

16 Margaret Dreier Robbins, The President's Report, *Third Biennial Convention of the National Women's Trade Union League of America* (Boston, 1911).

17 William Ludlow Chenery, "Occupational diseases," *The Independent,* Feb. 9, 1911, *70*: 306–309.

18 Josiah Strong, "Our industrial juggernaut," *North Am. Rev.,* Nov. 16, 1906, *183*: 1034. See also Strong's commitment to changing work conditions in his founding and assuming the presidency of the American Museum of Safety Devices and Industrial Hygiene in 1907/1908: "Apparently little is done in this country for the protection of working men and women." Josiah Strong to Miss E. L. M. Tate, Feb. 25, 1908, American Association for Labor Legislation (A.A.L.L.) manuscript, Labor-Management Documentation Center, New York State School of

Industrial and Labor Relations, Cornell University, Microfilm R 2390.

19 New York, *Second Report of the Factory Investigating Commission,* 1913, Vol. IV, pp. 1361, 1364.

20 Quoted in Nora Gause, "The slavery of childhood," *The Woman's Label League Journal,* May 1907: 15.

21 Elizabeth Beardsley Butler, "The cracker industry in Pittsburgh," *Charities and the Commons,* Sept. 5, 1908, *20*: 648–655.

22 Margaret Dreier Robbins, "Foreward for the Women's Trade Union League," *The National Women's Trade Union League* (n.p., n.d.).

23 William Hard with Rheta Childe Dorr as collaborator, "The woman's invasion," *Everybody's Magazine,* 1910, *21*: 372–385.

24 *Ibid.*

25 *The Glass Worker,* Apr. 1912, 9.

26 *The Woman's Label League Journal,* Oct. 1913, *11*: 1.

27 *The Survey,* Jan. 18, 1913, *29*: 534. For Hull House's commitment to occupational safety and health, see Jane Addams to John Commons, Mar. 5, 1908, and Irene Osgood to Charles Henderson, Oct. 8, 1908, A.A.L.L. manuscript, Microfilm R 2390.

28 Graham Taylor, "Industrial basis for social interpretation," *The Survey,* Apr. 3, 1909, *22*: 10.

29 U.S. Commission on Industrial Relations, Preliminary Report, *The Survey,* Dec. 12, 1914, *33*: 288.

30 See Domhoff, *The Higher Circles* (New York, 1965) and Irwin Yellowitz, *New York and the Progressive Movement* (New York, 1968).

31 Henry W. Farnum, "Labor legislation and economic progress," address by the president at the annual meeting, Dec. 28, 1909, Columbia Univ., Butler Library.

32 See letterhead of A.A.L.L. correspondence.

33 Shelby M. Harrison, "2nd National Conference on Industrial Diseases," *The Survey,* June 15, 1912, *28*: 448–450. See also, the A.A.L.L.'s pamphlet, *The Menace of Lead* [c. 1911]: "What the absence of state sanitary control means to workers" is that "thousands of workers are daily exposed to [lead's] influence and thereby run the risk, not only of disability, but even of death." A.A.L.L. manuscript, Microfilm R 2453.

34 John B. Andrews, "Report of work, 1910," *Am. Labor Legis. Rev.,* 1911: 95.

35 "The responsibility is on all citizens," *The Survey,* Apr. 8, 1911, *26*: 85.

36 John M. Work, "Give us justice not charity," *The Glass Worker,* Aug. 1911, *9*: 6.

37 Crystal Eastman, "The three essentials for accident prevention," *Annals,* June 1911, *38*: 98–99.

38 *Ibid.*

39 "Joint sanitary standards in the cloak, suit and skirt industry," *The Survey,* Aug. 19, 1911, *26*: 734.

40 "Strikes for good health," *The Survey,* Jan. 20, 1912, *27*: 1592.

41 "Responsibility is on all citizens," p. 84.

42 "The tuberculosis fight in an industrial city," *Charities,* Dec. 17, 1904: 279.

43 Frederick Almy, "Transcript of Proceedings, New York," *Second Report of the New York Factory Investigating Commission,* 1913, Vol. IV, pp. 1831–1832.

44 Graham Taylor, "Industrial basis for social interpretation," p. 9.

45 "How to prevent consumption," *The International Woodworker,* May 1906, *16*: 137–139; "Paul Kennaday on tuberculosis," *ibid.,* pp. 139–140.

46 "Labor is against tuberculosis," *The Glass Worker,* Apr. 1909, *6*: 6.

47 "A strike for clean bread," *The Survey,* June 18, 1910, *24*: 483–488.

48 "Investigations have disclosed the fact that unhealthy and poisonous bread is made in non-union bake shops," *The Woman's Label League Journal,* June 1913: 13.

49 Charles Swan, "Enterprise liability for industrial injuries," *Ann. Am. Acad. Polit. & Soc. Sci.,* July 1911, *38*: 262–263; "The Consumers League label and its offspring," *The Survey,* Aug. 8, 1914, *32*: 478.

50 Mary H. Loines, chairman of the Brooklyn Auxiliary of the Consumers League of the City of New York, to Hon. Robert F. Wagner, Dec. 12, 1912, New York, *Second Report of the New York Factory Investigating Commission,* 1913, Vol. II, 1330–1331.

51 *Ibid.,* Vol. IV, 1576–1577.

52 Quoted in Paul Kennaday, "The Reorganization Bill," *The Survey,* Feb. 22, 1913, *29*: 727–728.

53 C. F. W. Doehring, "Factory sanitation and labor protection," U.S. Department of Labor *Bulletin,* Jan. 1903, *44*: 12.

54 Gordon Thayer, "Matches or men?," *Everybody's Magazine,* Apr. 1912, *26*: 490–498.

55 "A platform of industrial minimums," *The Survey,* July 6, 1912, *28*: 517–518.

56 Paul U. Kellogg, "The industrial platform of a new party," *The Survey,* Aug. 24, 1912, *28*: 668.

PUBLIC HEALTH AND PERSONAL HYGIENE

The importance of personal hygiene to maintaining good health is widely accepted today. The great majority of early Americans, however, rarely bathed and showed little appreciation for fresh air, sunlight, or exercise. Their eating habits, including the consumption of gargantuan amounts of meat, kept many stomachs continually upset. Fruits and most vegetables seldom appeared on the table, and butter or lard often saturated the food that did appear. It is no wonder that one writer called dyspepsia "the great endemic of the northern states."

In the 1830s Sylvester Graham launched his popular health crusade. To preserve health, Graham advocated subsisting "entirely on the products of the vegetable kingdom and pure water," while abstaining from all harmful activities such as drinking, smoking, and masturbating. Since stimulating substances were thought to arouse the sexual passions, the adoption of a bland, meatless diet seemed the best way to control these unwholesome urges. Graham lectured throughout the country, winning a large following among those Americans who had lost faith in the more traditional methods of preserving health.

For health reformers, cleanliness ranked next to meatlessness. Graham himself had recommended a sponge bath every morning upon rising or, better still, the "exceedingly great luxury" of standing in a tub and pouring a tumbler of water over the whole body. But despite Graham's efforts and the availability of bathtubs with attached plumbing after the 1820s, the notion of regular bathing caught on slowly. A survey conducted in 1877 for the Michigan State Board of Health revealed that, although the "better" classes customarily took a weekly bath, among mechanics and farming families "the regular and systematic use of the bath has not come into general appreciation." As David Glassberg shows, it was not until the turn of the 20th century that public baths, identified as beneficial for the public's health, became available for the urban poor who did not have plumbing facilities at home.

Domestic plumbing and the home environment raised their own problems at the end of the 19th century. Citing dangers from sewer gas to germs, public health advocates tried to enlist the public in family-based efforts to improve the nation's health. As Nancy Tomes relates, the sanitarian message found a ready audience among the house-proud middle and upper classes as Americans, women particularly, developed a heightened awareness of domestic sources of infection. This "private side of public health" illustrates the close connection between individual effort and public health.

Cigarette smoking offers another example of the intersection of personal and public health issues. As Allan M. Brandt discusses, the individual decision to smoke came to be seen, in the 20th century, as a collective concern about the nation's health. Although almost a century earlier Graham had identified smoking as dangerous to health, it was not until the 1920s that researchers focused on the specific health problems associated with smoking and not until the 1964 Surgeon General's report that the government took an active role in the effort to prevent diseases associated with the habit. Programs to stop people from smoking required changing individual choices and behaviors, and many Americans felt the federal government had no legitimate role in regulating such private concerns. Commercial interests of the tobacco companies as well as the social meaning of the cigarette in American culture also affected this particular public health activity.

Twentieth-century science has accumulated a large body of evidence supporting the intimate relationship between health and lifestyle. Although concern for health sometimes relates more to fad than science, people who do not abuse their bodies live demonstrably longer and healthier lives than those who do.

31

The Design of Reform: The Public Bath Movement in America

DAVID GLASSBERG

In 1904, the United States Bureau of Labor prepared an exhibit for the Louisiana Purchase Exposition in St. Louis on the nation's public baths. The Bureau reported that nearly 80 percent of the 99 indoor and outdoor public bathing facilities in America had been established between 1895 and 1904. An explosion of public bathing had occurred in schools, on beaches, and in industry. Indoor public bathhouses counted for over half of the boom, increasing from six to 49 in those 10 years.[1]

The decade which saw the establishment of public bathhouses also experienced the flowering of many other public institutions. By the turn of the century, many cities provided schools, libraries, museums, zoos, parks, playgrounds, and summer concerts, as well as police, fire protection, liquor licensing, sanitary inspection, garbage collection, paved streets and sidewalks, hospitals, insane asylums, and some direct poor relief. Made possible by the economies of scale that densely populated city offered, the institutions embodied a new civic ethos which sought to gather the disparate urban groups into one great community. "The city-dweller has become a citizen," proclaimed Frederic C. Howe; "His social sense is being organized and his demands upon the government have been rapidly increasing."[2] Reformers such as Howe hoped that all would use these new institutions and participate in a common civic life.

The reformers believed that in return for insuring the citizens' physical and moral well-being, city life required adherence to certain standards and

principles. Reformers created the new institutions not only to provide services but also to uphold the standards and teach the principles necessary for civic civilization to the entire society. The growth of public institutions at the turn of the century was as much the result of what Daniel Walker Howe describes as a characteristically "Victorian" defense of "threatened" values and beliefs as an optimistic embrace of the city's possibilities for new levels of cooperation.[3]

Cleanliness was one such standard upon which all decent citizens would agree. To be clean was to be a respectable member of the community; to remain unwashed was to be a physical and moral menace. Building public bathhouses for the poor institutionalized the reformers' faith in cleanliness and their desire to extend their baptismal rites of common citizenship to all residents of their city. Yet contrary to the promises of their founders, the bathhouses tended to emphasize, rather than to diminish, the distance between the "great unwashed" and the rest of society.

Poor Americans at the turn of the century faced a shortage of private bathing facilities. Although most middle- and upper-class single homes in the 1890s had bathtubs, few tenements came so equipped, and few poor could afford the 15 dollars for a tub even if bathwater were made free. The New York City Mayor's Committee on Public Baths and Comfort Stations reported in 1897 that well over 90 percent of the families in the tenement districts of the four largest American cities had no baths.[4]

Without private baths, tenement dwellers washed in courtyard hydrants, hall sinks, or in tubs shared with several other families. Investigators for the New York State Tenement House Commission

DAVID GLASSBERG is Associate Professor of History at the University of Massachusetts, Amherst, Massachusetts.

Reprinted with permission from *American Studies*, 1979, *20*: 5–21.

heard slum residents complain that they rarely used the common tubs for fear that their neighbors suffered from skin disease. Considering the lack of privacy, clean facilities, and water pressure to upper floors of tenements, Henry Moscowitz doubted that his fellow residents bathed more than six times a year.[5]

Some cities provided free "floating baths" each summer. Originating in Boston in the 1860s, the floating baths were wooden frames extended over a river, inside which people bathed. By 1889, New York City had 15 of these structures, administered by its Department of Health. Despite their popularity with the thousands who flocked to cool off on hot summer days, the floating baths had numerous problems. They occupied valuable river-front space. The polluted rivers gave the baths the reputation of being "floating sewers." The flimsy seasonal wooden structures needed almost constant repair. Most important, the floating baths were not available to the public year-round. The only remaining inexpensive solution to the bath shortage was to construct permanent indoor bathhouses.

Earlier attempts to establish indoor public bathhouses had generally failed. Despite a shortage of private baths, cities in mid-19th-century America did not feel bathhouses so necessary that they be erected at public expense. In England, an 1846 law enabled a town to tax its citizens to build public bathing facilities, and London opened 13 bathhouses by 1854. American cities, however, left the building of bathhouses entirely to private charities. Philadelphia's City Council in 1848 even refused to grant the Philadelphia Society for the Employment and Instruction of the Poor's new bathhouse a special rate with the municipal water company, forcing the agency to abandon its hope of providing free public baths.[6] Similar attempts by ill-funded private charities to establish baths were also short lived.

The lack of fervor for providing indoor public bathhouses in mid-19th-century America reflected the belief that bathing, although beneficial to health, was not essential. Wealthy Americans enjoyed resorts built around hot springs, such as those at Saratoga, New York, and Yellowstone, Wyoming. At home, they chose from among a variety of therapeutic hot and cold tub baths, shower baths, russian (hot vapor) baths, turkish (hot air) baths, "needle" baths, and "electric" baths. The needle bath surrounded the bather with a coil of perforated pipe that pinpointed jets of water over his body; the electric bath bathed one's body in light rays.[7] Bathing was to be encouraged—but few felt the provision of baths important enough to be a public responsibility, on the order of police and fire protection.

Moreover, some felt that even if public baths were provided, the poor would not use them. The New York Association for Improving the Condition of the Poor's bathhouse, established in 1849, soon closed, reportedly from insufficient patronage to meet expenses. Its directors complained that they were "too far in advance of the habits of the people."[8] Superintendent L. N. Case of the Detroit Water Works observed, "The class most desirable to reach . . . is not particularly fond of water as a lavation. The old saying that you can lead a horse to the water but cannot make him drink is very applicable in this connection."[9] A rumor circulated in New York City that the poor so little wanted to bathe that those few slum dwellers who owned tubs only used them for storing coal. Indoor public bathhouses seemed in mid-19th-century America to be an expensive, unnecessary service which the poor probably would not use anyway.

By the end of the century, however, many viewed providing public bathhouses as a matter of utmost urgency. The big city's contagious diseases, ugly slums, relaxed moral codes, and strange immigrant customs offered reformers new and compelling reasons for insisting that all should bathe. Moreover, the progressive civic ethos demanded that the solution to these problems was a public responsibility. Reformers believed that the poor must have baths, and that it was the duty of "those who [were] already washed," in the words of Boston Mayor Josiah Quincy, to provide them.[10]

One impetus to the bath movement came from the desire to upgrade public health. City health departments in the 1890s began to feel the full impact of the germ theory of disease. Where previously doctors blamed disease upon a poor general environment, proof emerged in the 1880s that specific microorganisms caused such illnesses as typhoid, tuberculosis, cholera, diphtheria, plague, and dysentery.[11] The possibility that these microorganisms hid in layers of dirt transformed every unclean individual into a potential disease-bearer. Dr. Moreau Morris of New York warned that "the body exhalations of an unwashed sample of humanity sitting next to us in our crowded cars may communicate a deadly typhus germ without our consciousness."[12] Seeing the lack of baths among the slum population as an invitation to city-wide epidemic, the *Philadelphia Ledger* proclaimed "Every dirty man or woman is a menace to the health of the community."[13] The healthy city so depended upon the clean, healthy individual that the *Philadelphia North American* looked forward to the day when "public baths will be as common as public schools, and bathing, like education, will be made compulsory."[14] Science mandated baths as essential for public health.

The second impetus to the bath movement came from the desire to upgrade public morality. Whereas earlier attempts to improve morality had concentrated upon reforming each individual's habits, usually through religion, and later attempts to upgrade morality concentrated upon reforming the degrading environment, usually through replacing "bad" institutions such as the saloon, the bath movement at the turn of the century sought to reform both the individual's habits and environment at once. The reformers believed that a reciprocal relationship existed between moral character and a clean environment. Cleanliness was the result of moral habits—New York State official Goodwin Brown observed that it was a sure way to distinguish the "honest" from the "idle" poor.[15] Yet at the same time, reformers believed that cleanliness encouraged morality, while a dirty environment bred moral decline. The poor's dirt was both a badge of immorality and one of its causes. Public baths, by promoting both the habit of cleanliness and a clean environment, could reverse the spiral of moral decline.

Reformers thus saw the baths not just as cleansing facilities, but as missions to the slums to spread the "gospel of cleanliness." The New York Tenement Commission insisted that "the cultivation of the habit of personal cleanliness [has] a favorable effect . . . upon character, tending toward self-respect and decency of life."[16] W. L. Ross, manager of Philadelphia's Gaskill Street Baths, explained that "the object is not only to promote bathing facilities, but to elevate taste and morals."[17] Reformers hoped that the bathhouse would soon replace the saloon as a community center. After all, reasoned New Yorker William Tolman, "It is morally better to give a man an opportunity to wash the outside

of his body with water, rather than the inside of his body with whiskey."[18] The Boston Bath Commission cited a decrease in the number of juvenile arrests as evidence of its baths' success.[19] As Boston Mayor Josiah Quincy proclaimed in 1898, "When physical dirt has been banished, a long step has been taken in the elimination of moral dirt."[20]

By promoting health and morality, baths had the power to transform "urban barbarism" into "civic civilization."[21] "The advance of civilization is largely measured by the victories of mankind over its greatest enemy—dirt," Mayor Quincy noted. "One of the chief and most fundamental differences in conditions between the savage and the civilized man is that the former is dirty while the latter is relatively clean."[22] William P. Gerhard observed in his building guide for *Modern Baths and Bathhouses* that "all cultured nations have practiced bathing, chiefly at a period in their history when they flourished most, and that with the decay of civilization and culture, baths also disappeared."[23] The New York Tenement Commission described the lack of public bathhouses as "a disgrace to the city and to the civilization of the 19th century."[24] One bath advocate in Chicago insisted, "The greatest civilizing power that can be brought to bear on these uncivilized Europeans crowding into our cities lies in the public bath."[25]

The gospel of cleanliness at the turn of the century thus represented an attempt to bolster "Victorian" moral standards which reformers felt were becoming polluted by the diversity of behavior in their city. Cleanliness was necessary for participation in a common city life—Goodwin Brown declared that "without a sense of cleanliness, a high degree of civic pride is impossible."[26] Moreover, cleanliness had a physical rationale, provided by germ theory. It was proof that society indeed possessed enduring values upon which all decent citizens would agree. Providing for the poor's cleanliness with a system of public bathhouses was essential for maintaining a healthy, moral, "civilized" society.

The first successful indoor public bathhouse in the United States opened in New York City in 1891. The "People's Baths" resulted from a coalition of private charities which included the New York Association for Improving the Condition of the Poor (AICP), the New York Mission Tract Society, the Protestant Episcopal City Mission, the Society for the Prevention of Cruelty to Children, the New York Academy of Medicine, the Charity Organization Society, and the St. John's Guild. Pooling resources, the agencies under AICP leadership erected a bathhouse at a cost of $27,000.[27] The building's cream-colored facade sharply contrasted with its dingy tenement house surroundings. An inscription above the large arch over the doorway proclaimed "Cleanliness is Next to Godliness." The bathhouse's opening moved one amateur bard to pen an ode of praise, which concluded:

> The man who is clean from his scalp to his toes,
> Should always be jolly, wherever he goes.
> To be clean without leads to pureness within,
> Where lurks germs, the vilest of terrible sin.
>
> So hurra! Yes, hurra! that this bathhouse is built,
> At sin and at filth to make a brave tilt.
> May the AICP by this right royal gift,
> Save many a soul now wrecked and adrift.[28]

Two innovations helped the People's Baths to succeed. Unlike earlier bathhouses, it had showers in place of tubs. Dr. Simon Baruch, whose tour of European bathhouses provided the basis for the AICP's bathhouse design, explained that the public shower-bath was more thoroughly cleansing and less likely to communicate disease than tubs and was much less expensive to operate. Showers used less water, less space, and took less time for each bather than tubs, allowing the AICP to bathe more patrons. Unlike tubs, Baruch argued, showers required neither scrubbing nor changing of water between bathers, nor did they wear out and require replacement. The second innovation introduced inexpensive, easily cleaned cement and iron building materials in place of wood. Architects designed everything to be hosed down frequently. Each shower-bath cost the AICP between three and four cents. The 70–80,000 patrons each year, either paying a nickel to use one of the 18 first-floor showers or else using one of the nine free basement showers, enabled the AICP nearly to break even.[29]

With the success of the People's Baths, other New York City charities, such as the Baron de Hirsch Fund, DeMilt Dispensary, and the Riverside Association, also built bathhouses. Reformers soon clamored for the New York City government to build municipal bathhouses. Mayor William L. Strong appointed a Mayor's Committee on Public Baths

and Comfort Stations in 1894. Secretary William Tolman declared, "Now that these philanthropies have demonstrated the need and the demand for cleansing baths, they have done their duty and the city should undertake that work which is clearly a municipal function."[30] Doctor Baruch added, "It is the duty of a municipality to prevent disease. It is the duty of a municipality to prevent immorality. I believe that money spent on baths raises the standard of health and morality."[31] But the New York City Council had yet to see the need for municipal baths and defeated a bill that would have established a New York City Bath Department with six bathhouses.

The state government of New York therefore acted before the city did. State Commissioner of Lunacy Goodwin Brown was so impressed with the sanitary results of replacing tubs with showers in all state asylums that he drafted a bill, which became law in 1892, authorizing local governments to use public money for municipal baths. When few towns took advantage of the provision, the state in 1895 enacted a new law, also drafted by Brown, compelling each municipality of over 50,000 people to establish a system of free public bathhouses. The law required each bathhouse to have hot and cold water and to be open not less than 14 hours each day. Generally cities were slow to comply with the law—the first municipal bathhouse in the state opened in Buffalo in 1897, and by 1904 only 13 had been built statewide.

Philadelphia's experience with bathhouses paralleled that of New York. Municipal bathhouses did not follow the initiative taken by private charity. Although Philadelphia's poor mobbed the three indoor bathhouses built by the Public Baths Association of Philadelphia between 1898 and 1903, the city of Philadelphia built no year-round indoor baths. Attempts to rally municipal support on behalf of a system of public bathhouses for the poor fell short of reformers' goals.[32]

Attempts to rally philanthropic support for bathhouses also usually disappointed the reformers. William Tolman had hoped that wealthy "merchant princes" would donate bathhouses to cities, just as they endowed schools, theatres, and museums. Some philanthropists did endow municipal baths. Baltimore's Henry Walters, best known for endowing the Walters Art Gallery, also funded four public bathhouses. Many other philanthropists, however, would not endow institutions exclusively for the poor. They echoed Pittsburgh industrialist Henry Phipps' complaint in 1902 that he was "tired of trying to wash the great unwashed" (though the next year Phipps did help to endow a municipal bathhouse).[33]

Chicago's experience differed from that of New York and Philadelphia. Instead of building bathhouses itself, the Municipal Order League (later the Chicago Free Bath and Sanitary League) pressured the Chicago City Council to provide municipal baths from the very first. In February 1893, Dr. Gertrude G. Wellington, anticipating the crush of people in Chicago for the World's Columbian Exposition, and noticing the vast army of workingmen already in the city to build the fair, asked Mayor Washburn to provide both temporary baths for the fair months and three permanent indoor bathhouses, one each on the west, north and south sides of the city. Her letter mentioned five reasons for building baths: (1) That the poor of the city, especially on the West Side, were without bathing facilities, (2) "That men are vicious when dirty as well as when hungry," (3) That the act would make him (the mayor) very popular, (4) That it will help prevent typhoid, cholera, and crime, and (5) "That it will inspire sweeter manners and a better observance of law."[34] With the help of Jane Addams, the reformers mobilized the press, immigrant organizations such as the *Turnverein* and the residents of the district, and successfully presented their case before the City Council. In January 1894, Chicago opened the first municipally run indoor bathhouse in America. The city named the bathhouse after Mayor Carter H. Harrison, following Wellington's shrewd suggestion that naming baths after political figures gave politicians added incentive to support future baths. Indeed, Chicago would soon not only have bathhouses named after politicians, but a politician, "Bathhouse John" Coughlin, nicknamed after his baths. Discovering that more municipal bathhouses would be built through political pressure than disinterested appeals, Chicago's reformers succeeded in establishing municipal bathhouses before their eastern counterparts. By 1910, the city operated 50 bathhouses.

The pattern of bathhouse reform across America was similar. Like many other reform movements of the period, it was a national network, sharing expertise through letters and social work journals such as *Charities Review* long before it had a national organization (the American Association for Hygiene and Public Baths was not established until 1912). Reformers either tried to build their own bathhouses, as in New York City and Philadelphia, or they joined with immigrant groups to pressure city governments for municipal baths. Chicago's experience demonstrates how successful the latter groups could be.

The bath advocates worked to establish not just bathhouses, but also to place shower-baths in schools, mines, and factories. Baltimore and Chicago led the way in establishing baths in public schools. Reformers sought to make bathing part of the school routine, hoping that a clean child would attempt to change his dirty home environment.[35] One advocate of industrial bathhouses, Brooklyn drop-forge owner J. H. Williams, installed 12 showers on the premises for his men. "As it is acknowledged that habitual bathing prevents disease and promotes health and morality, baths for working people affect all classes of society," he explained. "Employers are therefore under moral obligations to provide such facilities."[36]

The reform leaders were generally upper-middle-class social welfare professionals, many of them women. The Chicago Free Bath and Sanitary League's Dr. Gertrude Wellington proclaimed women "the natural housekeepers of a great city."[37] The New York Mayor's Committee on Public Baths and Comfort Stations was chaired by social work specialists William G. Hamilton of the AICP and Dr. Moreau Morris of the New York Tenement House Association. The reformers were generally the same class of people that pushed for public museums, libraries, and theatres. Chicago's Municipal Order League advocated not only bathhouses but also the paving of all streets in the Loop and the building of kiosks in city parks for summer concerts. Pittsburgh's baths were maintained under the auspices of the Allegheny County Civic Club. The bath advocates seemed optimistic about society's potential to realize traditional goals using the new scientific techniques. Their movement demonstrated the desire to maintain a standard of morality and respectability—symbolized by cleanliness—through the most modern, efficient means which they could find—a system of public shower-baths. The reformers believed that public baths could help bring society closer to their vision of the ideal city, free of corruption, immorality, dirt, and disease.

Yet despite their popularity, neither the municipally nor privately run indoor bathhouses became the civic institutions which the reformers hoped. Usually built exclusively for the poor, the bathhouse designs had little in common with the stately public schools, parks, libraries, and museums built for the rest of society. Although some reformers hoped that the bathhouse would replace the saloon as a community center, its actual design discouraged lingering and offered patrons little reason to visit except to get clean.

The desire to wash large numbers of patrons as cheaply as possible gave urban indoor public bathhouses throughout America a similar look. In general, the bathhouses were small, scattered throughout the city, each catering to its own neighborhood. This not only brought baths to the people's door, but also insured that the different races, classes, and ethnic groups would not mix.

The buildings generally were of two stories, with light brick and stone trim. The light-colored facades were to contrast the bathhouse with its dingy tenement neighborhood, to act as a model for the neighborhood. Reformers directed each bathhouse attendant to wash the sidewalk in front of his or her bathhouse in the hope that its neighbors would cleanse their walks. Still, the exteriors were deliberately quite plain, compared with other public buildings. A bathhouse "exterior must be modest," Dr. Baruch recommended, "so as not to repel the poor and lowly by [its] architectural pretensions."[38] The buildings seldom employed classical orders or attempted associations with the Roman baths.

The plainness of the bathhouses resulted not just from wanting to avoid scaring off the poor with grand facades but also from the reformers' desire to build at minimum cost. Despite the New York People's Baths' success, no other bathhouse built in a tenement district in America came close to breaking even financially. Most municipally run baths were free; the privately run ones generally charged no more than a nickel. Less money spent on buildings helped to minimize losses. Some also felt that the poor deserved no better than the least expensive facilities possible. When New York City's relatively ornate Seward Park Baths flooded in 1904 just four days after it had opened, the attendant blamed the mishap on vandalism, grumbling that

the place was "too good for the class of people who used the baths."[39] This attitude, coupled with the wish to save money, gave the bathhouses a businesslike, though not unpleasant, appearance.

Architects designed the bathhouses solely for speed and sanitation to process efficiently a steady flow of patrons. Nearly all bathhouses followed the example of New York City's successful People's Baths, using concrete interior materials and showers to the exclusion of tubs. New York's Dr. Baruch insisted that as well as being less expensive to maintain, a shower-bath was refreshing and invigorating, unlike the "dangerously relaxing" tub bath.[40] Dr. Moreau Morris explained that "these baths are for cleansing purposes only, and by the use of the individual spray-bath a degree of personal cleanness is attained that cannot be secured by any other means so thoroughly, so efficiently, so quickly, and so economically."[41] Harold Werner, architect of New York City's West Sixtieth Street Baths, noted that "ease of communication and rapidity of handling the bathers were the prime considerations" in his design. No tubs were used, Werner added, because they were "a source of jealousy and confusion."[42]

Bathhouse design also encouraged efficient bathing through the use of partitions and screens. Generally, each bather had his own two-chamber compartment—the first for the shower and the second for his or her clothes. (In New York's People's Baths, however, the bather used a single compartment with a rubber sheet to cover his or her clothes.) Men were separated from women not only in the baths but also in the waiting rooms and entrances. Philadelphia's Gaskill Street Baths even barred men from using the same laundry facilities as women. Builders placed wire mesh over the top of each bathing compartment, reportedly to prevent patrons from thieving.

The typical bathing experience involved taking a ticket, waiting for one's number to be called, then within the next 20 minutes, picking up a towel and a two-inch bar of soap, undressing, receiving a specified amount of water, often under the control of the attendant, for a specified amount of time, then dressing and leaving. The temperature of the Dover Street, Boston baths reached no higher than 73°F., "discouraging the tendency to indulge in the ennervating soak that a hot shower provides."[43] The New York Mayor's Committee assured its readers that to orchestrate the flow of patrons in the People's Baths, police maintained "perfect order" while a "competent matron" looked after the women and a "man of experience" looked after the men.[44] Smoking, swearing, and intoxicated people were prohibited from bathhouses; so was "loitering and loafing." Ironically, at a time when New York built Pennsylvania Station modeled on the luxurious Roman baths of Caracalla, the designs and experiences in American public bathhouses most nearly resembled those of a railway depot.

Despite the apparent constraints, the poor flocked to the baths. Between 1898 and 1908 the Public Baths Association of Philadelphia reported a seven-fold increase in patrons.[45] Posters, billboards, newspaper ads, and coupons for free baths brought thousands to the baths each year. The baths especially provided relief on hot summer days. In Chicago, indoor public bath attendance in the four months from June through September often equalled that for the other eight months of the year.[46] Although Irish immigrants and native white Americans were said to harbor a "repugnance" to public bathing, Jews, Russians, Slavs, Germans, and Italians crowded the baths, especially before the sabbath and holidays. Baltimore's Argyle Street Baths, built for blacks, reported heavy patronage.[47]

Many more men bathed than women. In New York City's municipal baths the ratio of men to women was two to one; in Chicago it was four to one. The imbalance reflected the bathhouses' design, which usually included more showers on the men's side of the house. Bathhouses which opened different days of the week for each sex, instead of having separate men's and women's sides, generally gave men more time than women. Chicago's Harrison Baths opened to women but two days a week. Also, new bathhouses built exclusively for men far outnumbered those built just for women. This suggests that some reformers saw public bathhouses as primarily for workingmen.

Once inside the bathhouse, however, the men and women were treated the same, except that the women's experience may have been slightly less vigorous. The women's side was more likely than the men's side to have a few tubs. Baltimore's Walters Bath Number One, built in 1900, had 18 showers for men and five showers, two tubs for women. Also, Chicago's baths maintained their hot water at

105°F. on women's days, 100°F. at other times. Still, the bathhouses seemed designed to give women as well as men a brief but thorough cleaning.

Perhaps the more rigorous aspects of the baths contributed to their appeal. Patrons could be reasonably sure that the public baths were clean, unlike the common tenement house tubs about which they complained. Each bather—if only for 10 minutes—had privacy, something which he or she would seldom find at home. The replacement, at Jane Addams' insistence, of 19 showers with a 20′ × 30′ plunge bath at Chicago's Harrison Baths proved unpopular, reportedly because the patrons did not want to bathe together in so small a space.[48] Public bathhouses offered the poor an inexpensive means to get clean, allowing them privacy, hot water, and clean facilities in good repair not available at home.

The baths also offered the poor a source of recreation. Although the vast majority of indoor bathhouses were administered by city health departments and designed solely to clean large numbers of people as quickly as possible, usage patterns and accounts of bathhouse behavior suggest that the poor used bathhouses for recreation as well. The superintendent of one Philadelphia bathhouse lamented that his patrons used the baths as "playhouses" rather than setting about the "serious matter" of getting clean.[49] That indoor bath attendance increased three-fold in summer over winter, even when outdoor bathing facilities were also available, suggests that some patrons were at least as concerned with cooling off at a nearby place as they were with getting clean.

Indoor public bathhouses remained popular as long as they provided cleansing and recreational opportunities not available elsewhere. They quickly declined as other recreation and cleansing possibilities emerged in the 20th century. In the same decade in which cities erected bathhouses, they also established public bathing beaches. Chicago, Boston, New York City, Cleveland, Detroit, and St. Paul all opened new beaches to bathers in the 1890s. City-dwellers preferred hanging out on the beach with friends to the bathhouses' ten-minute shower as a source of recreation.

Meanwhile, tenement house reforms diminished the necessity for public cleansing facilities. New York's Model Tenement House Reform Law of 1901, though not requiring bathing facilities, mandated that builders provide water for each floor (later amended to each apartment) in a tenement. Chicago, Philadelphia, Pittsburgh, Boston, Baltimore, Cleveland, and San Francisco soon followed with similar laws requiring a water-closet for each family or each three rooms.[50] These laws encouraged builders, since they were required to install plumbing anyway, to install bathtubs. Eighty-six percent of the new tenements built in New York City between 1901 and 1910 had bathtubs.[51] The spread of individual family tubs was further encouraged by the development in 1920 of a technique to mass-produce the one-piece, double-shell enamel tub (the type used today), reducing its cost approximately 20 percent. By 1934, only 11 percent of the dwelling units in New York City were without baths or showers.[52]

The rapid decline of public-bath building suggests that both patrons and reformers saw the bathhouses as no more than a stop-gap measure until each family would have its own bath. Patrons preferred the convenience of their own bathing facilities. Reformers also preferred that each family practice cleanliness at home, where it would feel individually responsible for keeping its facilities in good repair. Lawrence Veiller, instigator of New York's 1901 Tenement House Law and a driving force nationally for housing reform, observed that privately owned facilities deteriorated much less quickly than public ones.[53] Both patrons and reformers believed that each family was better off with its own bathtub; soon society could afford to provide it.

The movement for public bathhouses stemmed from the amalgam of "Victorian" didacticism and "progressive" civic faith that also produced a number of other public institutions at the turn of the century. Like the museums, libraries, and parks, the baths were to embody and teach "Victorian" values to the poor, exerting a moral influence that would help them to rise out of poverty. The reformers' "gospel of cleanliness" demonstrated the desire to promote a standard for respectability and "civilization"—a benchmark for participation in civic life. The reformers claimed that the poor could conform to those standards and share in that life.

The few public bathhouses built in middle-class areas most nearly substantiated the reformers' civic rhetoric. Brookline, Massachusetts, built its municipal bathhouse in 1895 for $40,000, about four times the cost of the Harrison Baths in Chicago. The large red-brick and stone-trim building was located near the public high school, at the center of town. The bathhouse included a gymnasium, and had marble stairs leading into its 26' × 80' swimming pool lined with white-glazed brick and adamantine mosaic. The well-appointed, centrally located structure, like libraries and museums, was designed to instill in its patrons a feeling of public order and achievement.

The majority of bathhouses, however, located in tenement areas, offered their patrons a bare-concrete shower-bath, designed only to "sanitize" them as quickly and efficiently as possible. The baths resulted not just from the reformers' compassion but also from their desire for control; they built baths not just to ameliorate the condition of the poor but also to help contain it. Reformers hoped that the baths would ensure a relatively sanitary urban population until society enacted extensive housing reform. Instead of extending a soapy hand out to the poor, inviting them to share in a common civic life, the bathhouses' design and experience suggest that the reformers' foremost consideration was to prevent the contamination of the rest of society by crime, immorality, and disease.

NOTES

Thanks to John Higham and Neil Harris for comments on earlier versions of this essay.

1   G. W. W. Hanger, "Public baths in the United States," *U.S. Bureau of Labor Bulletin*, Sept. 1904, 9: 1254–1261. All statistics and descriptions of bathhouses, unless otherwise noted, are from this report.

2   Frederick C. Howe, *The City: The Hope for Democracy* (1905; Seattle, 1967), p. 47.

3   Daniel Walker Howe, "Victorian culture in America,"
in *Victorian America*, ed. Daniel Walker Howe (Philadelphia, 1976), pp. 2–28.

4   *New York Mayor's Committee on Public Baths and Comfort Stations Report* (New York, 1897), p. 11.

5   Robert DeForest and Lawrence Veiller, eds., *The Tenement House Problem* (1903; New York, 1970), p. 415.

6   Philadelphia Society for the Employment and Instruction of the Poor, *Annual Report* (Philadelphia, 1849), p. 7.

7 See Siegfried Giedion, *Mechanization Takes Command,* Part VII: "The mechanization of the bath" (New York, 1948), pp. 628–712.

8 Harvey Fiske, "The introduction of public rain-baths in America: part two," *Sanitarian,* July 1896, *37*: 6.

9 "Free water for private baths," *Engineering Record,* June 18, 1898, *38*: 51.

10 Josiah Quincy, "Gymnasiums and playgrounds," *Sanitarian,* Oct. 1898, *41*: 309.

11 Howard D. Kramer, "The germ theory and the early public health program in the U.S.," *Bull. Hist. Med.,* 1948, *22*: 235–241.

12 Moreau Morris, "More about the public rain-baths," *Sanitarian,* July 1896, *37*: 11.

13 *Philadelphia Ledger,* June 11, 1898.

14 *Philadelphia North American,* July 5, 1902, p. 8.

15 Goodwin Brown, "Public baths," *Charities Review,* Jan. 1893, *2*: 144.

16 *N.Y. (State) Tenement House Committee Report* (Albany, 1895), p. 49.

17 *Philadelphia North American,* Dec. 30, 1901.

18 William Tolman, "Public baths, or the gospel of cleanliness," *Yale Review,* May 1897, *6*: 51.

19 Bertha H. Smith, "The public bath," *Outlook,* 1905, p. 576.

20 Quincy, "Gymnasiums and playgrounds," p. 306.

21 Tolman, "Public baths," p. 51.

22 Quincy, "Gymnasiums and playgrounds," p. 309.

23 William Gerhard, *Modern Baths and Bathhouses* (New York, 1908), p. 5.

24 New York Tenement Committee, p. 49.

25 Chicago Free Bath and Sanitary League, *Round-up for 1897* (Chicago, 1897), p. 16.

26 Brown, "Public baths," p. 144.

27 F. L. Longworth, "The People's Baths," *Charities Review,* Jan. 1893, *2*: 180–183.

28 "Bathhouse opens," *New York Tribune,* Aug. 18, 1891, p. 5.

29 Hanger, "Public baths in the U.S.," p. 1267.

30 Tolman, "Public baths," p. 54.

31 Smith, "The public bath," p. 575.

32 Public Baths Association of Philadelphia, *Scrapbook,* Historical Society of Pennsylvania, Philadelphia.

33 *Philadelphia North American,* July 5, 1902.

34 Chicago Free Bath and Sanitary League, pp. 15–16.

35 *Ibid.,* p. 39. Also see "Public school baths," *Charities Review,* May 1899, *9*: 98.

36 *Charities Review,* Nov. 1893, *3*: 53.

37 Chicago Free Bath and Sanitary League, pp. 12–15.

38 Harvey Fiske, "The introduction of public rain-baths in America: part one," *Sanitarian,* 1896, *36*: 485.

39 "Closed public baths a week after opening," *New York Times,* July 19, 1904, p. 5.

40 Fiske, "Introduction of public rain-baths, part one," p. 486.

41 Morris, "More about the public rain-baths," p. 12.

42 Hanger, "Public baths in the U.S.," p. 1334.

43 "A new public bathhouse," *Charities Review,* Nov. 1898, *8*: 391.

44 New York Mayor's Committee, p. 38.

45 Public Baths Association of Philadelphia, *Scrapbook.*

46 Chicago Department of Health, *Monthly Report* (Chicago, 1895–1901).

47 *Philadelphia Record,* Jan. 12, 1913; *Charities,* May 21, 1904, *12*: 523; Hanger, "Public baths in the U.S.," p. 1349.

48 Chicago Free Bath and Sanitary League, p. 22.

49 *Philadelphia Record,* July 9, 1898; also see Public Baths Association of Philadelphia, *Scrapbook.*

50 Ford H. MacGregor, *Tenement House Legislation: State and Local* (Madison, Wis., 1909). Also Roy Lubove, *The Progressives and the Slums: Tenement House Reform in New York City 1890–1917* (Pittsburgh, 1962).

51 Lawrence Veiller, *Housing Reform: A Handbook for Practical Use in American Cities* (New York, 1910), pp. 111–112.

52 James Ford, *Slums and Housing* (Cambridge, Mass., 1934), p. 308.

53 Veiller, *Housing Reform,* pp. 110–113.

32

The Cigarette, Risk, and American Culture

ALLAN M. BRANDT

On Saturday, 11 January 1964, Surgeon General Luther Terry stepped to the podium of the State Department auditorium to begin a nationally televised press conference. Seated directly behind him were 10 eminent physicians and scientists, the members of his Advisory Committee on Smoking and Health. This group of individuals had met regularly over the last 18 months to evaluate the evidence about the risks of cigarette smoking. Although the results of this investigation had been held top secret, signs prohibiting smoking hung in the auditorium, a harbinger of the coming announcement. In the outside corridors, members of the press puffed away. Reporters were offered copies of the report in the closed auditorium for an hour before the press conference. At the conclusion of the session, they rushed to phones to call in the story. The next day, the report received front-page coverage throughout the country.

For the seventy million regular smokers in the United States, the report constituted bad news. It found that among men who smoke cigarettes the death rate from cancer of the lung was 1,000 percent higher than among nonsmokers. The report also cited chronic bronchitis and emphysema to be of far greater incidence among smokers. Additionally, the committee found that the incidence of coronary artery disease, the leading cause of death in the United States, was 70 percent higher among smokers. In short, cigarette smokers placed themselves at much higher risk of serious disease than did nonsmokers.[1]

These findings, contained in the massive 387-page document, which cited thousands of research studies, held few surprises. In fact, the committee had conducted no new research. It had merely reviewed existing data. And indeed, since the early 20th century and beyond, physicians had pointed out hazards of cigarette smoking. As long as there have been cigarettes there has been concern about their impact on health. By the time of the release of the report, polls showed that most Americans already associated cigarettes with cancer. If such information was widely known, what is the meaning of the surgeon general's committee report, and what was its significance? What were the social and scientific forces which led to the report and what was its impact?

The report marked the beginning of a revolution in attitudes and behaviors relating to cigarettes. In the last quarter century, half of all living Americans who ever smoked have now quit. At the time of the 1964 report, 42 percent of all U.S. adults smoked; in 1989 only 26 percent were smokers. According to the most recent *Surgeon General's Report* (1989), approximately 750,000 smoking-related deaths have been avoided since 1964 because people have quit or not started smoking.[2] Terry's *Surgeon General's Report* signaled the beginning of a profound change in the meaning of the cigarette and spurred new interest more generally in the relationship of behavior, risk, and health.

This essay briefly traces the history of the debate about the risk of smoking and places the *Surgeon General's Report* in a broader context by examining the process by which the cigarette came to be defined as a major health risk. The report raised fundamental questions about the nature of biomedicine, public health, and especially causal inference; it profoundly altered the way we think about issues

ALLAN M. BRANDT is Amalie Moses Kass Professor of the History of Medicine and Professor of the History of Science at Harvard University, Cambridge, Massachusetts.

Reprinted by permission of *Daedalus*, Journal of the American Academy of Arts and Sciences, from the issue entitled "Risk," Fall 1990, *119* (No. 4): 155–176.

of health and disease. This was part of a broader debate in 20th-century science about the nature of evidence, proof, and causality, a debate about the epistemological foundations of biomedicine. How do we know what we know? What constitutes "proof" in modern science? What is the nature of causal relationships? And finally, how will risks be defined, measured, and regulated?

The cigarette provides a means of tracing an important watershed in medical "ways of knowing." But the issues raised go beyond the realm of biomedicine; the debate about smoking was shaped by the meaning of the cigarette in American culture, the nature of the tobacco industry, public health, and government. In short, the process by which risk is assessed and perceived reveals deep social, cultural, and political values.[3]

THE RISE OF THE CIGARETTE

In many ways the cigarette seems such a ubiquitous part of American culture that it is difficult to imagine that it is really a 20th-century phenomenon. Between 1900 and 1965, per capita consumption rose from 49 to 4,318.

Developments in agricultural technique, production technology, and industrial organization, as well as such factors as the introduction of the portable match, all contributed to the growth of the tobacco industry.[4] The cigarette marks the convergence of corporate capitalism, technology, mass marketing, and, in particular, the impact of advertising.[5] These forces induced new modes of individual and group behavior. With the rise of consumerism, a new behavioral ethic was defined. From a culture that promoted self-denial and self-discipline in the late 19th-century—one condemning indulgence in all forms—Americans were now encouraged to indulge.

As individuals came to fear the loss of autonomy in an individual world, cigarette smoking promised individual redemption. The Marlboro man was the first urban-industrial cowboy, a symbol of modernity, autonomy, power, and sexuality. Such advertising pointed away from the product toward the moral and psychological value of the patron.[6] Advertising promised consumers well-being and power.[7] Creating demand for relatively undifferentiated, nonessential items was the core of the new consumer culture, which the cigarette epitomizes.

The tobacco industry boomed, as did state revenues associated with the manufacture and sale of cigarettes. When demand for cigarettes rose, so too did concern about their impact on health.

As cigarette smoking became increasingly popular in the early 20th century, claims for its virtues and vices ran strong. Though many touted the positive and pleasureful aspects of the cigarette, the dramatic rise in smoking was accompanied by a powerful anti-cigarette movement which sought to identify both the moral and the health risks of tobacco. By the first decade of the 20th-century, concerns about the demoralizing impact of the cigarette were widely cited.

Boys were often caught sneaking off behind school buildings to smoke in groups, and "cigarette fiends" were identified as a major social problem of the growing cities. Among the most prominent anti-cigarette crusaders was Henry Ford. "If you will study the history of almost any criminal you will find that he is an inveterate cigarette smoker," advised Ford.[8] He donated the funds for the publication of a national journal which appeared under the title "The Case Against the Little White Slaver." On another occasion Ford explained, "With every breath of cigarette smoke they inhale imbecility and exhale manhood. . . . The yellow finger stain is an emblem of deeper degradation and enslavement than the ball and chain." Ford enlisted Thomas Edison to investigate scientifically the harms of smoking.

In addition to the concern expressed about young boys smoking, anti-cigarette activists centered attention on the detrimental consequences of smoking for women—now vigorously solicited as smokers—and its impact on their health and social mores.[9] As the movement for prohibition gathered momentum, cigarettes were frequently tied to the use of alcohol. By the First World War, some 13 states had enacted legislation prohibiting or regulating the sale of cigarettes; anti-cigarette activists often cited medical and scientific experts in support of such controls.[10]

By the 1920s, as consumption continued to rise, research into the consequences of smoking intensified. Researchers focused attention on the impact of tobacco on what they called "mental efficiency." Usually in such studies, smokers fared poorly.[11] But the problem with such research was clear; as one scientist explained, "It might be either that the smoking habit induces lethargy or that lazy men are the kind that find smoking agreeable."[12] This problem of inference would continue to plague the debate about smoking into our time. Other studies concentrated on the physical growth and development of smokers; did cigarette smoking stunt growth?[13] The moralistic nature of such experiments lay just below the surface.

It could be argued that moral reformers protesting the rising use of cigarettes in America hid behind the cloak of scientific authority in offering their arguments. But this would misrepresent their ideas and tactics. They simply saw no tension in seeing the cigarette as ungodly *and* unhealthy; they equated moral dangers and health risks. Moral reformers had absolutely no compunction about employing arguments based on weak data about the physically debilitating impact of smoking. Medical doctors and researchers moved easily between citing the moral and citing the health consequences of smoking; there was no attempt to differentiate such arguments.

MODERN EPIDEMIOLOGY AND STATISTICAL INFERENCE

By the late 1920s, researchers began to focus more precisely on the *specific* health consequences of smoking. As early as 1928, in a somewhat primitive epidemiological study, researchers associated heavy smoking with cancer.[14] In addition, surgeons published clinical reports associating cancer in their patients with their smoking habits.[15] In 1931, Frederick L. Hoffman, a well-known statistician for the Prudential Insurance Company, tied smoking to cancer. Hoffman noted the difficulties of conducting epidemiological studies in this area. The basic methodological questions of statistical research— issues of representativeness, sample size, and the construction of control groups—all presented researchers with a series of complex problems. Hoffman called for the exercise of moderation in all behavior, a truism of progressive hygiene, suggesting that "extreme moderation in smoking habits would certainly be advisable."[16]

In 1938, Raymond Pearl, the Johns Hopkins statistician and biometrician, published the first significant statistical analysis of the health impact of smoking. Pearl came to the conclusion that in

individuals it was difficult to assess the risks of such behaviors, especially when their impact was not immediate and when many intervening variables also affected health. Therefore, he concluded, the only precise way to evaluate their effect on health was to employ statistical methods after collecting data on large groups. Comparing the mortality curves of smokers and nonsmokers, Pearl found that individuals who smoked could expect shorter lives. He offered no explanation for why this might be so.[17]

During the 1920s and 1930s, as the first studies attempting to link cigarettes to cancer were conducted, the field of epidemiology stood at a crossroads. The bacteriological revolution of the late 19th and early 20th century had directed attention away from the traditional environmental questions which had brought epidemiology to the fore. Research came to center on mechanism: identifying causative agents, universally assumed to be microorganisms. Indeed, the notion that disease was actually "caused" by hazards in the environment fell into disrepute. Public health officers were compelled to demonstrate Robert Koch's postulates, the fundamental truths of the new germ theory.[18] There were, of course, exceptions to this trend, especially in the study of industrial and occupational health. But these fields for the most part were distant from the central concerns of medicine and public health. In fact, the major statistical work of the period came from population genetics and the actuarial studies of the insurance industry, rather than from the disciplines of public health. Neither Hoffman nor Pearl would have considered himself an epidemiologist.

The municipal laboratory had become the new focus of public health. Even when researchers identified environmental or behavioral risks, they generally focused on the mechanism of disease. The whole notion of statistical inference was questioned, as research centered on the cellular level. In this respect, exposure to a carcinogen was equated with exposure to an infectious organism. Identifying the health risks of a particular behavior like smoking fitted this model poorly. The length of time before the disease developed was protracted (and equated with an "incubation period"); in addition, the large number of intervening variables confounded notions of specific causality. Everyone "exposed" did not get the disease; indeed, most did not; and some who were not exposed did. Also, there was broad cultural discomfort with notions of comparative risk assessment. How dangerous was the cigarette? How did this danger rate vis-à-vis other risks? Finally, medical theory offered few persuasive models for understanding systemic and chronic diseases; the anomalies of cigarette smoking did not fit the biomedical model's ideal of specific causality.

Changing patterns of disease, however, forced researchers to search for other models of causality. By the end of the Second World War, concern about lung cancer had intensified. It seemed to statisticians and physicians to be a striking exception to many other disease patterns of the 20th century; deaths from lung cancer had risen from 4,000 in 1935 to 11,000 in 1945. By 1960 the number of annual lung cancer deaths would rise to 36,000.[19] By the mid-1980s carcinoma of the lung would become the most prevalent of all cancers, accounting for more than 140,000 deaths each year. Yet, at the turn of the 20th century, the disease was a relative rarity, with less than 400 cases recorded in 1900.

There were, of course, many theories to account for this shift. Some observers attributed the rise in cases to better reporting, more sophisticated diagnostic abilities, the widespread use of X-rays, and the ability to make precise pathological analyses. Others suggested that increasing life expectancy permitted the development of disease that in an earlier era would not have had the chance to wreak its havoc on victims who would die earlier deaths from other causes.[20]

But some physicians and public health officers pointed to one of the most dramatic behavioral changes in the history of American culture, the rise of cigarette smoking. By the late 1940s it was already known that prolonged exposure to certain industrial chemicals and vapors—chromate, nickel carbonyl, and radioactive dusts—could produce lung cancer. Some scientists now suggested that the inhalation of cigarette smoke might have similar effects. This hypothesis led to a series of epidemiological studies of the risk of smoking. These studies, in turn, would lead to a redefinition of risk, epidemiology, and public health.

First published in the 1950s, these investigations were based upon retrospective findings: in other words, individuals with lung cancer were identified in hospitals and interviewed regarding their smoking practices; they were then compared with a similar group who did not smoke. The findings revealed that cigarette smokers were at far higher risk for the development of lung cancer than were nonsmokers. But critics raised a series of objections to such studies. In particular, it was clear that there were a number of opportunities for bias in the construction of sample and control groups. For example, it was suggested that lung cancer patients were likely, because of the nature of their disease, to exaggerate their smoking habits.[21]

Given the methodological problems with retrospective studies, two major prospective studies on smoking and cancer were begun in 1951. Under the auspices of the British Medical Research Council, Richard Doll and Bradford Hill sent questionnaires on smoking practices to all British physicians.[22] When members of the profession died, Doll and Hill obtained data concerning the cause of their deaths. The results were consistent with the earlier findings from the retrospective studies.[23]

A second major prospective study, conducted by E. Cuyler Hammond under the auspices of the American Cancer Society, led to similar conclusions. Total death rates were far higher among smokers than among nonsmokers. Lung cancer deaths were 3 to 9 times as high among smokers as among nonsmokers, 5 to 16 times as high among heavy smokers. Among those who smoked two or more packs a day, the death rates were 2.25 times as high as for men who had never smoked, a strong indication of a dose effect. Excess mortality was even higher for coronary artery disease than for lung cancer; rates for smokers exceeded those for nonsmokers by 70 percent. Quitting, Hammond found, reduced risk; formerly a heavy smoker, he himself now quit.[24] By 1960, a range of epidemiological studies had all arrived at consistent findings: cigarette smoking significantly contributed to lung cancer and coronary artery disease.[25]

These epidemiological studies introduced the concept of large, population-based surveys. They focused attention on the definition of comparative risk and excess mortality. Implicit in such studies was a critique of the whole notion of specific causality; these researchers recognized that there were literally hundreds of variables affecting the incidence of disease. Therefore they sought to design studies which, by including many individuals, would be controlled except for a single variable—in this case, cigarette smoking.

This mode of research touched off an important debate within the scientific community about the nature of causality, proof, and risk. At stake were the very epistemological foundations of scientific knowledge: How do we know what we know? What is the reliability of causal inference from statistical data? Those committed to hereditarian, genetic views of cancer, for example, found fault with an epidemiologic approach which centered attention on behavioral effects.[26]

At the basis of the epidemiological argument was the clear limitation of laboratory experimentation for making determinations about probability and risk. The debate about smoking and health revealed an intraprofessional battle between epidemiology and lab science, a clash of values, assumptions, and expectations. Moreover, the debate revealed a deeper discomfort with statistical logic and quantitative methods in biomedicine, a trend which persists today.[27] Before any successful anticigarette campaign could be waged, the legitimacy of epidemiological data concerning risk for generating health policy would have to be established.

FROM EPIDEMIOLOGY TO PUBLIC POLICY

Knowledge of the risks of smoking—which continued to accrue throughout the 1950s—did not immediately lead to the formulation of public policy. Indeed, there was considerable debate about the implications of these findings for public health authorities. What was the appropriate role of the state vis-à-vis the risks of cigarette smoking? Should the government play a role in educating its citizens about the hazards of smoking? Recognizing the gravity of the hazard, should the government take steps to regulate the sale of cigarettes more aggressively, or restrict their use? These questions, of course, were complicated by the nature of the behavior itself: no one need be exposed to the hazards of smoking unless he or she so chose; the "voluntary" nature of the risks, it was argued, militated against any governmental intervention.

The first step which the federal government took—haltingly, in 1962—was to sponsor a com-

mission to study the evidence that cigarettes were harmful. In some respects, this was a curious way to proceed, given the quality of the evidence which already existed. But the creation of the Surgeon General's Advisory Commission on Smoking and Health revealed the political aspects of the debate.[28] First, powerful economic interests repeatedly called the epidemiological findings into question, suggesting that the relationship of cigarettes to disease was "merely statistical" and that no clear and objective findings confirmed these risks "in the laboratory."[29] The industry responded to the epidemiological data with advertising campaigns which assured the public their brands were safe, at the same time that it introduced filter cigarettes with expansive claims for health and safety.[30] The industry worked diligently to undermine, if not obscure, public perceptions of the risk of the cigarette. Second, there was no single, authoritative "reading" of the mounting evidence. Forces in the public health establishment, especially the voluntary health agencies, realized that the findings linking cigarettes to disease had to be legitimated in the medical and scientific communities, as well as among the public.[31] Identifying "risk factors" for disease would become an increasingly important aspect of the work of the "voluntaries," eager to assure the public—and especially contributors—of progress in finding "the cause" of serious chronic disease.[32]

The advisory committee, appointed in July 1962, explicitly avoided all questions of social policy; its charge was to determine whether or not smoking *caused* disease. But it conducted no new research. The committee reviewed some 7,000 publications, including 3,000 research reports published since 1950.[33] It sought to arrive at a clinical judgment on smoking. As one public health official explained, "What do we (that is, the surgeon general of the United States Public Health Service) advise our patient, the American public, about smoking?" Implicit in this question was a particular model of public health and the role of the state.

The report, despite the fact that it offered no new data, nevertheless made a fundamental contribution to the study of causal inference in epidemiological studies. What did it mean to say, for example, that cigarettes *caused* lung cancer? How should "cause" be distinguished from "associated with," "a factor," or "determinant"? Members of the committee realized the complexity of saying simply that smoking causes cancer. Many individuals could smoke heavily throughout their lives and apparently suffer no adverse consequences; "cause" implied a single process in which A, by necessity, would lead to B. Therefore, they acknowledged the complexity: "It should be said at once," the report explained, "that no member of this Committee used the word 'cause' in an absolute sense in the area of this study. Although various disciplines and fields of scientific knowledge were represented among the membership, all members shared a common conception of the multiple etiology of biological processes. No member was so naive as to insist upon mono-etiology in pathological processes or in vital phenomena."[34] Yet their conviction was clear: smoking presented a tremendous risk to health. The committee developed a set of criteria for evaluating causal relationships which has been widely applied since that time. Causal evidence had to be (1) consistent, (2) strong, (3) specific, (4) supportive of appropriate temporal relationship, and (5) coherent.[35] At the press conference announcing the committee's findings, Terry was asked whether he would now recommend to a patient to stop smoking. His answer was an unequivocal "yes."

The report served the political functions on which it was predicated. It provided power and legitimacy to the epidemiologic findings; indeed, the report was of fundamental importance in raising the stature of epidemiology as a discipline. It made clear that the government would accept broader responsibility for the determination of risks and for public education to prevent disease. The ability of self-interested parties such as the tobacco industry to disparage such findings was now delimited. With the first *Surgeon General's Report,* the battle against the cigarette was joined; less obvious was how the government would utilize this document in setting a public health agenda.

THE TOBACCO WARS

In retrospect, the immediate public and political response to the *Surgeon General's Report* appears strikingly naive. Newspapers reporting the findings speculated that the tobacco industry would wither away. The presumption was widely held that smokers—now apprised of the risks—would quickly quit. In Congress, such ideas influenced legislators,

who in 1965 passed the Federal Cigarette Labeling and Advertising Act. The legislation established the National Clearinghouse on Smoking and Health to encourage health education about the dangers of smoking. In addition it required that all packs of cigarettes carrying a warning: "Caution: Cigarette Smoking May Be Hazardous to Your Health." Given that the surgeon general had found that smoking *causes* lung cancer, the warning was remarkably weak, indicating the effectiveness of the tobacco lobby on Capitol Hill. It further reflected the relative lack of experience most legislators had had with scientific findings. At the hearings concerning this legislation, tobacco spokesmen challenged the findings of the surgeon general. By treating all perspectives as those of "interested" parties to be brokered in the political process, members of Congress sought compromise. Moreover, the powerful economic interests, especially of tobacco-growing states, acted forcefully to moderate any regulatory initiatives.[36] Nevertheless, as scientific studies collected in subsequent surgeon general's reports continued to indict the cigarette as a major cause of serious disease, Congress took additional action. In 1971, the label was changed to "Warning: the Surgeon General Has Determined that Cigarette Smoking Is Dangerous to your Health." And, in 1985, four rotating labels were mandated.[37]

Increasingly, the battle over the nature of the risks of smoking would be waged in the media. Luther Terry's effective control of the media, for example, greatly contributed to the success of his committee. First, Terry appointed a commission of elite scientists and clinicians to study the issue of smoking and health; he successfully obviated any easy dismissal of the report by requiring that none of its members had previously expressed positions on the dangers of the cigarette.[38] Second, he invited the tobacco industry to review a list of prospective committee members and reject anyone they desired to. This made it impossible for the industry to easily discredit the report. The "secret" meetings of the committee generated widespread speculation in the press during the 18 months of its deliberations.[39] This interest culminated in the nationally televised press conference of January 1964. Sunday newspapers throughout the country reported the story on front pages.

By 1964, in the aftermath of televised presidential debates, a presidential assassination, and a growing war in East Asia, all powerfully portrayed through the electronic media, the expanding role of the media and possibilities of exploiting them for a range of purposes, including public health education, were increasingly recognized. Terry's report and the nationally televised press conference made the surgeon general, for the first time, into a *public* figure with access to the media. It gave the office a new meaning and authority which subsequent surgeon generals would augment. Indeed, the surgeon general's principal role—given that the office has little funding or authority to initiate programs—is to speak effectively through the media.

In the struggle concerning the "meaning" of the cigarette, control of the media was bitterly contested. The tobacco industry had considerable resources to expend in this fight, attempting to allay the growing concerns about the impact of smoking on health. Advertisements, for example, continued to suggest that smokers were youthful, healthy, attractive, and sexually seductive. Although the Federal Trade Commission took action to demand a higher level of accountability from the industry, regulations were weak and difficult to enforce. The anti-tobacco forces thus pursued other strategies. A young consumer lawyer, John Banzhaf III, decided to attempt to get the FCC to apply the fairness doctrine (for equal air time) to cigarette advertising. He formed the group ASH (Action on Smoking and Health). After a court struggle, he forced the national networks to air anti-smoking spots in prime time. Anti-cigarette ads got approximately $40 million of free air time. These public-service announcements apparently did have an impact; per capita consumption fell from 4,197 in 1966 to 3,969 in 1970.[40] Given the success of this anti-cigarette media blitz, the industry then acquiesced to a legislative ban on broadcast advertising, thus averting the fairness doctrine.[41]

Congressional anti-smoking policy proved to be decidedly limited. Modest funding for public education, requiring warning labels on packages, and banning broadcast advertising constituted the entire federal program to reduce smoking. At the same time, tobacco subsidies were maintained, placing the government in the ambiguous position of working to limit cigarette smoking while simultaneously contributing to the growth of tobacco. The limits of the federal program revealed the ongoing power of the tobacco lobby and the economic interests that it represents.[42]

TRANSFORMING THE MEANING OF CIGARETTE SMOKING

Additional research findings about the nature of the risks of cigarette smoking served to tip the balance in favor of anti-smoking forces during the last decade. Despite considerable gains in stigmatizing the cigarette, the anti-smoking forces had, by the late 1970s, foundered on a traditional American libertarian ethic: "It's my body and I'll do with it as I please." In keeping with this powerful cultural ideal, further governmental interference relating to smoking was seen as constituting unjustifiable intrusion into individual decisions. The Tobacco Institute viewed such intervention as "health and safety fascism." It was one thing for the government to inform the public about the dangers of smoking, quite another to restrict or ban the behavior.

For this reason, scientific studies of the impact of "sidestream" smoke took on special significance. With the publication of studies which demonstrated the risks of exposure to other people's cigarette smoke—in particular, a higher risk of lung cancer—the anti-smoking movement was reinvigorated on the basis of a powerful communitarian ethic: "Do with your own body whatever you like, but you may not expose others to risks which they do not agree to take on themselves." As epidemiologist Michael J. Martin explained, "Many people are willing to take on risk, even an enormous risk, themselves. But few are willing to tolerate even a small risk imposed on them."[43] With the imprimatur of a new *Surgeon General's Report* (1986), the data on "involuntary" smoking led to remarkable changes in the effectiveness of efforts to restrict smoking in public places.[44] By mid-1988, 320 local communities had adopted laws restricting smoking in public places, up from 90 in 1985.[45] Cigarette smoking was banned from virtually all domestic airline flights beginning in early 1990.

Another *Surgeon General's Report* (1988) also called into question the voluntary nature of cigarette smoking, now for the smoker.[46] By documenting the addictive qualities of cigarette smoking, the report further undermined the notion of an individual voluntarily "deciding" to smoke. Not surprisingly, the tobacco industry challenged these findings. Walker Merryman, a spokesman for the industry's Tobacco Institute, offered a socially elastic definition of addiction: "I've not heard of anyone holding up a liquor store or mugging an old lady to get the money to buy cigarettes."[47] Nevertheless, the recognition that cigarette smoking subjects individuals to well-recognized biological processes of transient mood alterations, tolerance, and withdrawal symptoms led increasingly to the inclusion of nicotine addiction as one more aspect of substance abuse, a deviant behavior.[48] Moreover, the growing recognition of the difficulty of quitting undercut the notion that smoking was simply a matter of choice.

Studies of the risks of sidestream smoke and the addictive nature of cigarettes were promoted by a growing anti-smoking coalition which included physicians, public health experts, and aggressive consumer activists. This, of course, is not to question the scientific validity of such studies, but rather to emphasize the relationship between authoritative science and its social and political context. The new research agenda facilitated the ongoing process of delegitimizing cigarette smoking in American culture. The cigarette—the icon of our consumer culture, the symbol of pleasure and power, sexuality and individuality—had become suspect. The smoker would subsequently be redefined, in a process which we continue to see played out—from the independent Marlboro man or liberated Virginia Slim to a new vision of a weak, irrational, and now, addicted, individual. The innocuous habit had become the noxious addiction. The stigmatization of the cigarette became a critical aspect of a revolution in American values about personal health and behavior.

The stigma of the cigarette is now tainting its producers. Increasingly, the production and sale of such a clearly dangerous and damaging product is being viewed as a moral issue; the cigarette companies are losing their standing as "reputable" industries.[49] Major social institutions have moved in recent months to sever their ties with an industry increasingly associated in the public mind with "marketing death." The decisions of Harvard University and City University of New York, for example, to divest their endowment holdings in cigarette companies explicitly expands the moral valence of the cigarette issue, further isolating the

industry.[50] Despite the continued profitability of the cigarette, the industry is losing the tobacco wars, the battle to maintain a legitimate place for the cigarette in American culture.

It would nevertheless be premature to celebrate the decline in cigarette consumption. Cigarettes continue to exact an enormous toll on health in the United States, and, increasingly, throughout the world. According to recently revised figures, 390,000 deaths each year are attributed to cigarette smoking in the United States alone.[51] Smoking is estimated to cause 30 percent of all cancer deaths, 21 percent of all deaths from coronary artery disease, and 82 percent of all deaths from chronic obstructive pulmonary disease. Since 1986, lung cancer has become the leading cause of cancer deaths among American women, surpassing breast cancer, the epidemiological result of the rise in women smoking since the 1940s. Smoking remains the "single most important preventable cause of death" in the United States.[52] A recent federal study estimated that cigarettes cost the nation some $52 billion each year in health expenses and time lost from work.[53] Despite the decline in smoking, the tobacco industry remains highly profitable,[54] and the industry continues to spend more than $2.5 billion each year promoting the sale of cigarettes.[55]

As cigarette consumption in the United States has declined, multi-national tobacco companies have worked to market their product more vigorously in the developing world. Recent worldwide surveys of cigarette consumption show steep increases in Africa and Asia. The World Health Organization recently characterized the commercial marketing of cigarettes in developing nations as "intense and ruthless."[56] According to WHO, 600,000 new cases of lung cancer now occur worldwide every year; most are the result of cigarette smoking. By the year 2000, the annual number of lung cancer cases may be as high as 2 million, with 900,000 in China alone.[57] In this sense, the risks of tobacco consumption truly are global; changes in Western consumption have been a catalyst for accelerating sales in the developing world.[58]

WHOSE RISK IS IT, ANYWAY?

With cigarette smoking there remains a complex political and cultural conflict about risk and responsibility. Consensus about the risks of smoking touched off an important debate about the question of responsibility for risk: Who is responsible for the serious burden of disease imposed by cigarette smoking? In this respect, the first *Surgeon General's Report* marks an important watershed in the history of public health. The government accepted new responsibility for the elucidation of health risks through epidemiological studies. The report articulated an expanded vision of public health, suggesting that the government had an important regulatory function in protecting its citizens from harmful products by identifying risks and educating the public.

Nevertheless, the government has been caught in an ambiguous position in its efforts to control cigarette smoking. Given the history of the prohibition of alcohol, there is little support for an outright ban on cigarettes, even from the most aggressive anti-cigarette activists. Advertising remains a contested area of public policy, but even opponents of smoking have expressed concern that a total ban may conflict with First Amendment rights. Although education has been effective among the educated, the relationship between risk and behavior modification remains obscure.[59] More significant than any particular federal intervention have been the local bans on smoking in public areas and workplaces, which have created a powerful anti-smoking social environment. In a relatively short time, public space has been subdivided; cigarette smoking has become the most rigorously defined of all public behaviors.

Recognizing that the federal government's policy options regarding cigarettes are limited, some have called for a higher standard of corporate responsibility. Smokers who have incurred serious disease, acting as plaintiffs, have attempted to fix the burden of responsibility squarely on the tobacco industry itself. In the last decade, hundreds of civil suits have been filed making the claim that the tobacco companies persisted in selling a lethal product all the while knowing (and obscuring) the risks. Given the highly addictive nature of the cigarette and the slick promotion campaigns of the industry, plaintiffs' lawyers have contended that the companies should accept—in compensatory damages—responsibility for the debility and death their product has wreaked.

These liability suits have generally been unsuccessful. Although they appeal to a populist, anti-business strain of thought in American society, typically such suits have failed in spite of the ability of lawyers to portray the industry as cynical and profit driven. Within American society there is a powerful expectation regarding individual responsibility for risk taking. The labeling and educational activities of the government have served to reinforce these expectations. As consensus regarding the risks of the cigarette has grown, the industry has, ironically, been freed of responsibility for the risks of its product.

Increasingly, Americans have come to accept notions of individual responsibility for the systemic and chronic diseases. Because heart disease, cancer, and other diseases are powerfully influenced by a range of individual "lifestyle" behaviors, including diet, alcohol consumption, and smoking, many health care analysts have come to emphasize the significance of modifying behaviors to affect health status, and more generally, patterns of disease.[60]

Such views have particular appeal in the context of the American health culture, which has historically emphasized the significance of an individual's responsibility for disease. Americans, in this respect, have largely come to reject fatalistic explanatory models of disease and its causes. Social values have underscored norms which suggest that individuals *can* and *should* exert fundamental control over their own health through careful and rational avoidance of risks. The popularity, for example, of the "Just Say No" campaign against illicit drugs reflects an essentially "voluntaristic" notion of risk.

As effective as such values may be in serving to define healthful behaviors, they present an important political and cultural irony. According to this behavioral ethic, those who continue to take risks must be held accountable for the results; but this emphasis on individual responsibility may deny broader social responsibilities for health and disease. This view, which has developed increasingly powerful adherents in the last decade, actually misrepresents the history of cigarette smoking in the 20th century. Smoking is a complex behavior which has reflected deep social, cultural, and economic forces, as well as a powerful biological process of addiction. Simply identifying individual behavior as the primary vehicle of risk negates the fact that behavior itself is, at times, beyond the scope of individual agency. Behavior is shaped by powerful currents—cultural, psychological, as well as biological processes—not all immediately within the control of the individual. Behaviors such as cigarette smoking are sociocultural phenomena, not merely individual, or necessarily rational.

The emphasis on personal responsibility for risk taking and disease has come at the very moment when cigarette smoking is increasingly stratified by education, social class, and race. In 1985, 35 percent of blacks smoked compared with 29 percent of whites.[61] For college graduates the proportion of smokers fell from 28 percent in 1974 to 18 percent in 1985; for those without a college degree the decrease during the same period was from 36 to 34 percent.[62] Thus, to emphasize individual accountability is to deny that some groups may be more susceptible to certain behavioral risks, that the behavior itself is *not* simply a matter of choice.

In assessments of environmental risks, we have often considered who lives nearest the hazard; the recognition that such risks are externally imposed generates social concern for their victims. Such has not been the case with risks associated with individual behaviors; individuals who "take"—note the

voluntaristic bias—such risks are considered ignorant, stupid, or self-destructive. But perhaps we might begin to rethink behavioral risk; rather than simply hold individuals accountable for the risks they incur, we might ask who is at risk to become, or remain, a cigarette smoker and why.

To adequately understand the answers to such questions will, no doubt, require a new, multidisciplinary research agenda in which the relationship of social and cultural contexts (including powerful economic forces) will be related in a sophisticated way to individual psychological motivations to engage (or disengage) certain risk behaviors. We need to better understand the *meanings* of particular behaviors and risks to particular groups of populations. What are the biological, psychological, and social forces that make it possible for some individuals to quit smoking, for example, while others, eager to free themselves of addiction, nonetheless fail? And finally, how may we promote cultural shifts that enhance both personal efficacy and autonomy in the name of health? Until we can adequately answer these questions, the cigarette will continue to be a powerful, if not pervasive, risk.

NOTES

1 U.S. Department of Health, Education, and Welfare, *Surgeon General's Report, Smoking and Health: Report of the Advisory Committee to the Surgeon General of the Public Health Service*, PHS publication no. 1103 (Washington, D.C.: GPO, 1964); *New York Times*, Jan. 12, 1964.

2 U.S. Department of Health and Human Services, *Surgeon General's Report, Reducing the Health Consequences of Smoking: Twenty-Five Years of Progress*, DHHS publication no. (CDC) 90–8411 (Washington, D.C.: Public Health Service, 1989).

3 On the social and cultural nature of risk and risk assessment, see M. Douglass and A. Wildavsky, *Culture and Risk: An Essay on the Selection of Technical and Environmental Dangers* (Berkeley: Univ. of California Press, 1982).

4 W. Bennett, "The cigarette century," *Science*, Sept./Oct. 1980, *80*: 37–43.

5 M. Schudson, *Advertising the Uneasy Persuasion* (New York: Basic Books, 1985).

6 R. W. Fox and T. J. Lears, *The Culture of Consumption* (New York: Pantheon, 1983).

7 R. Marchand, *Advertising the American Dream* (Berkeley: Univ. of California Press, 1985).

8 H. Ford, introduction to *The Case against the Little White Slaver* (Detroit, 1914), p. 5.

9 See G. L. Dillow, "The hundred year war against the cigarette," *American Heritage*, Feb./Mar. 1981, *32*: 94–107. On the history of recruiting women smokers, see, for example, Schudson.

10 R. Sobel, *They Satisfy* (New York: Doubleday, 1978).

11 See, for example, M. V. O'Shea, *Tobacco and Mental Efficiency* (New York: Macmillan, 1923), a monograph sponsored by the Committee to Study the Tobacco Problem, which was organized in 1918.

12 J. R. Earp, "Tobacco and scholarship," *Scientific Monthly*, Apr. 1928, *26*: 335–336.

13 See E. L. Clarke, "Effect of smoking on Clark College students," *Clark College Record*, July 1909, *4*: 3191–3198; G. L. Meylan "The effect of smoking on college students," *Popular Science Monthly*, Aug. 1910, *77*: 169–178.

14 H. L. Lombard and C. R. Doering, "Cancer studies in Massachusetts. 2. Habits, characteristics, and environment of individuals with and without cancer," *New Eng. J. Med.*, 1928, *198*: 481–487.

15 A. Ochsner, "My first recognition of the relationship of smoking and lung cancer," *Preventive Medicine*, 1973, *2*: 611–614.

16 F. L. Hoffman, "Cancer and smoking habits," *Annals of Surgery*, Jan. 1931: 50–67.

17 R. Pearl, "Tobacco smoking and longevity," *Science,* Mar. 4, 1938, *87*: 216–217.

18 M. Susser, "Epidemiology in the United States after World War II: the evolution of technique," *Epidemiologic Reviews,* 1985, 7: 147–177.

19 E. C. Hammond, "The effects of smoking," *Scient. Am.,* July 1962, *207*: 39–51.

20 C. C. Little, "Some phases of the problem of smoking and lung cancer," *New Eng. J. Med.,* 1961, *264*: 1241–1245.

21 R. Doll and A. Hill, "A study of the aetiology of carcinoma of the lung," *Brit. Med. J.,* Dec. 13, 1952, *2*: 1271–1286.

22 C. Webster, *The Health Services since the War* (London: Her Majesty's Stationery Office, 1988), 1: 233–237.

23 R. Doll and A. Hill, "The mortality of doctors in relation to their smoking habits: a preliminary report," *Brit. Med. J.,* 1954, *1*: 1451–1455; Doll and Hill, "Lung cancer and other causes of death in relation to smoking: a second report on the mortality of British doctors," *Brit. Med. J.,* Nov. 1, 1956, *2*: 1071–1081.

24 E. C. Hammond and D. Horn, "Smoking and death rates—Report on forty-four months of follow-up on 187,783 men. 1. Total mortality," *J.A.M.A.,* 1958, *166*: 1159–1172.

25 J. Cornfield et al., "Smoking and lung cancer: recent evidence and discussion of some questions," *Journal of the National Cancer Institute,* 1959, *22*: 123–203.

26 R. A. Fisher, "Dangers of cigarette-smoking," *Brit. Med. J.,* 1957, part 2, no. 43: 297–298.

27 A. R. Feinstein, *Clinical Judgment* (Baltimore: Williams and Wilkins, 1967).

28 A. L. Fritschler, *Smoking and Politics: Policymaking and the Federal Bureaucracy* (New York: Appleton-Century-Crofts, 1969).

29 L. Terry, "The first surgeon general's report on smoking," *N.Y. St. J. Med.,* 1983, *13*: 1254–1255.

30 T. Whiteside, *Selling Death: Cigarette Advertising and Public Health* (New York: Liveright, 1971).

31 S. Wagner, *Cigarette Country* (New York: Praeger, 1981).

32 J. Patterson, *The Dread Disease: Cancer and American Culture* (Cambridge, Mass.: Harvard Univ. Press, 1987).

33 *Surgeon General's Report,* National Archives Manuscripts, record group 90, Washington, D.C.

34 U.S. Department of Health, Education and Welfare, *Surgeon General's Report, Smoking and Health.*

35 A. M. Lilienfeld, "The surgeon general's 'epidemiological criteria for causality,'" *J. Chron. Dis.,* 1983, *36*: 837–845.

36 E. M. Whelan, *A Smoking Gun* (Philadelphia: George F. Stickley, 1984).

37 J. K. Iglehart, "Smoking and public policy," *New Eng.*

38 L. Terry, "The first surgeon general's report on smoking," *N.Y. St. J. Med.,* 1983, *13*: 1254–1255.

39 L. M. Schuman, "The origins of the report of the advisory committee on smoking and health to the surgeon general," *Journal of Public Health Policy,* Mar. 1981, 2: 19–27.

40 K. E. Warner, "Cigarette smoking in the 1970s: the impact of the antismoking campaign on consumption," *Science,* 1981, *211*: 729–731.

41 Whelan, *Smoking Gun.*

42 H. M. Sapolsky, "The political obstacles to the control of cigarette smoking in the United States," *Journal of Health Politics, Policy and Law,* 1980, *5*(2): 277–290.

43 "Tobacco company profits just won't quit," *Business Week,* Dec. 22, 1986, pp. 66–67. For a full exposition of the ethical issues associated with cigarette smoking, see R. E. Goodin, *No Smoking: The Ethical Issues* (Chicago: Univ. of Chicago Press, 1989).

44 J. E. Fielding, "Smoking: health effects and control," *New Eng. J. Med.,* 1985, *313*: 491–498; 555–561.

45 U.S. Department of Health and Human Services, *Surgeon General's Report, Reducing the Health Consequences of Smoking.*

46 U.S. Department of Health and Human Services, *Surgeon General's Report, The Health Consequences of Smoking: Nicotine Addiction.* DHHS publication no. (CDC) 88–8406 (Washington, D.C.: Public Health Service, 1988).

47 "Surgeon general's stature is likely to add force to report on smoking as addiction," *Wall Street Journal,* May 13, 1988, p. 1.

48 J. E. Henningfield, "Pharmacologic basis and treatment of cigarette smoking," *Clinical Psychiatry,* 1984, *45*: 24–34.

49 M. Kinsley, "Smokenders," *New Republic,* May 14, 1990, pp. 4, 53.

50 *New York Times,* 24 May 1990.

51 "U.S. report raises estimate of smoking toll," *Washington Post,* Jan. 11, 1989.

52 U.S. Department of Health and Human Services, *Surgeon General's Report, Reducing the Health Consequences of Smoking.*

53 *New York Times,* Feb. 21, 1990.

54 "Tobacco company profits just won't quit," *Business Week,* Dec. 22, 1986, pp. 66–67.

55 R. M. Davis, "Current trends in cigarette advertising and marketing," *New Eng. J. Med.,* 1987, *316*: 725–732.

56 R. Herman, "Diseases of affluence," *Washington Post Health,* Jan. 3, 1989, pp. 12–15.

57 W. U. Chandler, "Banishing tobacco," *World Watch Paper* 68 (Washington, D.C.: World Watch Institute, 1986).

58 P. Schmeisser, "Pushing cigarettes overseas," *New York Times Magazine,* July 10, 1988, pp. 16–25.

59 H. Leventhal and P. D. Cleary, "The smoking problem: a review of the research and theory in behavioral risk modification," *Psychological Bulletin,* 1980, *88*(2): 370–405; and N. D. Weinstein, ed., *Taking Care: Understanding and Encouraging Self Protective Behavior* (New York: Cambridge Univ. Press, 1987).

60 J. Knowles, "The responsibility of the individual," *Daedalus,* Winter 1977, *106*: 57–80.

61 M. C. Fiore et al., "Trends in cigarette smoking in the United States: the changing influence of gender and race," *J.A.M.A.,* 1989, *261*: 49–55.

62 *Ibid.,* 56–60.

33

The Private Side of Public Health: Sanitary Science, Domestic Hygiene, and the Germ Theory, 1870–1900

NANCY TOMES

> Disease and grim death stalk through our fine dwelling houses, disease and grim death of our own making. . . . When we are soundly asleep, cradled in fancied security, the impalpable subtle enemy, malaria, arises from the outlets of the very utensils [water closets and sinks] introduced for the preservation of our health, for the prolongation of our lives, fanning us into deeper slumbers, like the wings of the vampire.
>
> *Leopold Brandeis, 1873–74*[1]

Warnings about the health dangers posed by domestic plumbing were commonplace in the popular American press of the 1870s and the 1880s. So pervasive were such dire cautions about "sewer gas," as malaria or impure air was more commonly known, that the "good people of New England feared it perhaps more than they did the Evil One," according to Charles V. Chapin, a prominent public health authority. Hardly had the alarm over sewer gas subsided before the earnest readers of advice literature were bombarded by information about another "insidious foe that stealthily enters our homes and destroys our health," namely the germ. With the rapid popularization of the germ theory in the late 1870s and the 1880s, literate Americans learned that "the higher life is . . . everywhere inter-penetrated . . . by the lower life," in the words of one microscopist, and that their damp cellars, dusty carpets, and dank water closets were domestic breeding grounds for these invisible agents of deadly disease.[2]

From sewer gas to germs, late 19th-century Americans became alert to a host of new dangers

lurking in the home. Their anxieties reflected the hard work of several generations of health reformers, loosely aligned under the banner of sanitary science, who claimed to have scientific proof that damp cellars, poor ventilation, dirty carpets, and untrapped soil pipes caused the spread of typhoid, diphtheria, scarlet fever, and other infectious diseases.[3] Initially focused on the unsanitary conditions of tenements and other dwellings of the poor, the campaign to improve domestic hygiene was broadened to include the abodes of the wealthy as well. For late 19th-century domestic sanitarians, as they might be called, the home became an important vector of disease among all classes of the citizenry.[4]

The new concern with preventing what were often collectively referred to as "house diseases" inspired a spate of popular literature, including advice books, magazines, newspaper articles, and health department circulars, designed to instruct homeowners and housewives in the principles of domestic hygiene. At a time when municipal public health services were still very primitive, the public was urged to take individual measures, such as improving household ventilation and plumbing, boiling and filtering drinking water, and isolating the sick within the household, to protect their loved ones from debility and death. Even as learned doctors continued to debate the dangers posed by

NANCY TOMES is Professor of History, State University of New York, Stony Brook, New York.

From the *Bulletin of the History of Medicine*, 1990, *64*: 509–539. Copyright © 1990 by the Johns Hopkins University Press. Reprinted by permission of the Johns Hopkins University Press.

sewer gas and germs, health reformers relentlessly promoted a simpler, less scrupulous version of sanitary science that emphasized the efficacy of individual action against disease.

For a variety of reasons, the sanitarian message gained an early and wide hearing among the urban middle and upper classes. Affluent Americans were a peculiarly house-proud people, for whom owning a home, furnishing it tastefully, and running it efficiently were badges of respectability. From the 1820s onward, they consumed reams of advice literature aimed at making their dwellings more beautiful, more functional, and more morally uplifting. An intense attachment to the home was part of a "cult of domesticity" that targeted middle-class women and invested family life, partricularly child rearing, with enormous moral and social significance. Compounding this significance was the peculiar health consciousness of 19th-century Americans, who avidly pursued new fads in diet, dress, and exercise. In an era of high geographic and social mobility, habits of personal hygiene became less a matter of following tradition and more a reflection of individual "enlightenment" and self-discipline.[5]

Given the middle class's commitment to improving their homes and their health, the steadily rising rates of both epidemic and endemic infectious diseases in antebellum American cities naturally caused great concern. In a culture where declining family size and heightened individualism made the emotional aspects of family life increasingly intense, the impact of disease on the city-born young, one-third to one-half of whom died before their 10th birthday, was particularly ghastly. To protect their families, affluent Americans, especially mothers, had good reason to pay heed to sanitarian reformers who linked rising rates of disease to individual sanitary failings.[6]

Under the sanitarians' tutelage, health-conscious Americans developed a heightened awareness of domestic sources of infection, and adopted various protective rituals—ranging from leaving windows open at night to boiling water and using chemical disinfectants—to ward off disease. As we now make efforts to evade carcinogens such as asbestos and radon, so they performed home tests of air and water, drank bottled water, purchased patent devices to filter out disease-causing agents, and sought a healthy lifestyle that would build their own, and

their children's, resistance to infection. While both sexes had duties to perform in protecting the sanitary state of the home, the burden of daily watchfulness fell more heavily on wives and mothers.[7] The brisk sale of "sanitary goods," from flush toilets to patent water filters and chemical disinfectants, attests to the growing anxieties of the Victorian *mater-* and *paterfamilias,* anxieties that were sometimes preyed upon by unscrupulous commercial interests.

This fascinating episode in the history of changing popular attitudes and beliefs about infectious disease has been curiously neglected by historians. While 19th-century scholars are very familiar with other aspects of the sanitarians' work, such as their persistent campaigns to found city and state health departments, pass stringent health legislation, and build modern sewer and water purification facilities, their equally strenuous efforts to revolutionize what I term the "private side" of public health have gone virtually unacknowledged and unexplored by both social and medical historians.

To be sure, social historians have freely invoked the sanitarians' formulation "Dirt equals disease" to explain the increasing rigor of both bathing and housecleaning. However, social historians have tended to treat the expressed concern about disease prevention as a rationalization for some other, more genuine, objective such as reinforcing gender roles, class differences, or ethnic prejudices. As Norbert Elias wrote in 1939 in his classic work on civilization and manners, "the primary impulse" for changes in personal hygiene "does not come from rational understanding of the causes of illness, but . . . from changes in the way people live together, in the structure of society." In other words, this school of analysis assumes that the decision to install water closets and use disinfectants in the home had more to do with upholding the conventions of social class and of gender roles than it did with the desire to evade disease.[8]

Unfortunately, historians of medicine and public health, who might be expected to accord more importance to changing scientific concepts of disease, have shied away from the topic as well, for reasons that are deeply rooted in their own historiographical traditions. The conventional portrait of the sanitarians emphasizes their devotion to municipal works on a grand scale rather than their passionate commitment to popular education and voluntary reform. The 19th-century public health movement is usually seen in terms of the drive to expand the state's power to regulate the larger urban environment; supposedly it was not until the rise of the bacteriologically based "new public health" of the early 20th century that popular education and personal hygiene became imperatives of the movement.[9] The crusade against "house diseases" (that is, diseases spread by improper domestic practices) does not fit conventional intellectual divisions into historical periods either; since Lloyd G. Stevenson's widely read article "Science down the Drain" (1955) first portrayed the sanitarians as hostile to experimental medicine and the germ theory, medical historians have tended to assume that the 1880s marked a deep intellectual divide between scientific world views. Before that decade, the miasma theory, which equated infection with atmospheric impurity, supposedly prevailed, whereas afterwards all right-minded physicians accepted the germ theory of specific contagion. Thus the persistence from 1870 to 1920 of sanitarian beliefs and practices about the importance of domestic hygiene appears as a curious anomaly, rather than a vital, integral aspect of the public health program.[10]

Historians also have a natural tendency to focus on the evolution of scientific measures that actually "worked," as defined by current standards. From the standpoint of modern knowledge, 19th-century concerns about deadly sewer gas and pathogenic carpets seem hardly to warrant the same attention as large-scale water purification or smallpox vaccination programs. Thomas McKeown's contention that improvements in nutrition and general living standards contributed most to the decline of mortality from infectious disease perhaps reinforced the inclination to dismiss domestic hygiene practices as negligible factors in the overall decline of mortality. No doubt the fact that much of the popular hygiene message was directed at women contributed to its perception as a quaint but unimportant topic.[11]

In this article, I will argue that, to the contrary, the late 19th-century campaign to reform households, as well as the behavior of individuals within them, is deserving of much more attention from both social and medical historians than it has customarily received. The drive to prevent infectious disease certainly became entangled with 19th-century social prejudices, yet it possessed an impetus of its own, born of both high death rates and

changing scientific knowledge of disease, that has not been sufficiently acknowledged or explored by social historians. Without denying the complexity of the cultural processes involved, I believe that changing scientific concepts of disease had more to do with transforming personal and domestic hygiene than is usually assumed. In fact, the sanitarian crusade against "house diseases" represents one of the earliest instances in which new scientific information—in this case, information about the origins and prevention of disease—led to widespread changes in popular behavior through the medium of mass education.

Far from being a tangential concern of the public health movement, I will argue that the educational work and voluntary reform of the 1870s and 1880s laid the groundwork for the better-known public works of succeeding decades. Adding the "private side" to the conventional narrative about the late 19th-century public health movement will make for a more balanced record of its emphases and achievements. It also highlights the growing influence of medical science on popular culture: an influence that preceded and helped to make possible the increasing prestige of medicine in Progressive Era America.[12]

By analyzing the campaign against house diseases, I hope to correct the facile opposition of "miasma versus germ theory" by showing that, at least at the level of popular science, the assimilation of sanitarian theories of infection and contagion paved the way for the rapid acceptance of the germ theory. By pathologizing the home in the decades from 1860 to 1880, the domestic sanitarians created a framework, in terms of both ideas and methods of popular education, for the relatively rapid dissemination of the belief that microorganisms were the agents of human disease. Domestic hygiene thus provides a fascinating glimpse of a more general process of intellectual change whereby new information about microorganisms was understood and acted upon in the framework of older ideas and behaviors. The study of popular preventive measures underscores the growing tension between elite scientific debates about disease, which often ridiculed simplistic concepts such as sewer gas and germs, and the version of scientific "truth" offered to the public by public health authorities, who downplayed dissension for the sake of justifying a clear and presumably comforting course of action.[13]

Finally, I will speculate on some interesting demographic implications of the history of changes in domestic hygiene. Demographers have recently become interested in the role of personal hygiene as a factor in lowering mortality rates from infectious diseases. Studies in Third World countries have shown that the mother's level of education and mastery of hygienic food preparation contribute significantly to the success of programs to reduce infant mortality. This modern finding has led to a reconsideration of the demographic puzzle posed by the late 19th-century "mortality transition": that is, why did a significant decline in mortality from infectious diseases take place *before* there was a large-scale purification of the supplies of water and milk? Historical demographers are now investigating whether changes in personal hygiene may have played a more important role in the 19th-century mortality transition than has been previously assumed. By studying the changing theory and practice of domestic hygiene, historians can make a useful contribution to current demographical debates.[14]

To that end, I offer here a preliminary exploration of the "private side" of the 19th-century public health movement. The first section of this article examines the origins of domestic sanitary science; the second section examines methods of popular health education; the third section examines the chief principles of domestic hygiene as popularized in the decades roughly from 1860 to 1900; the fourth section looks at the assimilation of the germ theory into the domestic sanitarians' vision of the house as a vector of disease; the fifth section assesses how deeply the new domestic hygiene penetrated middle-class thought and behavior in the late 19th century; and the sixth section deals with the transition from voluntary to compulsory initiatives in public health. In assaying this rich material, I have used printed primary sources representing the national, indeed international, culture of late 19th-century sanitary science; but for the sake of narrative focus, the manuscript material is drawn chiefly from Philadelphia.[15]

THE ORIGINS OF DOMESTIC SANITARY SCIENCE

Attempting to ward off infectious disease by a careful domestic regimen was by no means a revolutionary concept to 19th-century Americans.

Their Western European cultural heritage made them heirs to a centuries-old tradition of preserving health by keeping both body and home clean and infused with fresh air. The more specific ancestry of the concept of the "house disease" can be traced back to the practice of household cleanliness, disinfection, and quarantine common in premodern epidemics. In mid-18th-century France, physicians reworked these traditional concerns with new physiological and chemical knowledge to create the Enlightenment science of hygiene. The hygienists' emphasis on the atmosphere's role as a carrier of disease made the provision of pure air and the control of foul odors the foundation of the early 19th-century public health movements in France, England, and the United States.[16]

Still, before the middle of the 19th century, the public health movement had relatively little to offer the individual citizen concerned with avoiding infectious diseases. Hygiene manuals devoted only a few pages to measures for keeping clear of contagion, devoting the bulk of their text to the role of personal regimen and of constitutional tendencies in the production of illness. In other words, when advice givers considered the issue of individual responsibility for illness, infectious diseases seemed among the least preventable, given the pervasive nature of atmospheric contagion: the only way to avoid the dangers of a corrupt atmosphere was to maintain one's general health, live in a dry, well-ventilated house, and avoid people who were obviously ill. Benjamin Rush summed up the thrust of conventional wisdom when he wrote to his wife during the 1793 yellow fever epidemic, "There is but one preventative that is certain, and that is 'to fly from it.'"[17]

The Home Book of Health and Medicine, published by an anonymous "physician of Philadelphia" in 1835, suggests to what a small extent domestic conduct was seen to be linked with specific diseases. In a treatise of 619 pages, the author spent only a few paragraphs, under the headings "Air" and "Cleanliness," enjoining his readers to keep their homes, particularly their cellars, clean, dry, and well-ventilated. The only specific diseases mentioned in connection with poor domestic hygiene were typhus and cholera, which the author unequivocally categorized as diseases of "poverty and low-life." When "persons in easy circumstances" fell ill of cholera, he concluded, it was due not to poor home conditions but to some irregularity in personal regimen such as improper diet, intemperance, or fatigue. Before the 1860s, then, what public health authorities later began to call house diseases were chiefly seen as being associated with the poor, and the routines of preventive domestic hygiene remained simple and relatively insignificant.[18]

Between 1860 and 1880, changes in both the scientific understanding of disease and the material circumstances of middle-class home life elevated domestic prevention of disease to a new importance. New insights from physiology, pathological anatomy, and epidemiology allowed the disaggregation of the vague categories of fevers and fluxes inherited from the 18th century into discrete pathological entities. Appreciation of the specificity of infectious diseases was accompanied by a growing conviction of their preventability. All the evidence compiled by the clinical, pathological, and epidemiological investigations of the day seemed to verify the same focal points of infection: corrupted air and impure water. "Sanitary science," as the preventive formulations came to be termed, did not reject the older atmospheric theory of infection, but rather expanded and elaborated upon it. While air remained the first cause usually invoked to explain the spread of disease, water figured as an increasingly important factor after the famous studies of William Budd and John Snow in the late 1840s showed how cholera spread through tainted water supplies. Budd provided additional evidence of the same mode of transmission for typhoid fever. Public health authorities devised a new category of diseases—"zymotic diseases," which included cholera, typhoid, diphtheria, smallpox, measles, and scarlet fever—to denote illnesses caused by impure air and water.[19]

While Anglo-American physicians generally agreed on certain causal associations—that decaying organic matter led to putrefaction, which produced fermentation and its characteristic disease symptoms in the body—they still differed sharply over how zymotic diseases originated and spread. Some disorders, such as smallpox and other eruptive fevers, seemed clearly to spread from person to person through the direct transfer of contagious matter. Others, such as typhoid, seemed capable of originating *de novo* under certain atmospheric conditions and infecting individuals through the mechanism of corrupted air; once established, the disease then spread through

fecal contamination of the water supply. However, though doctors fiercely debated such matters in their medical societies and journals, when it came time to advise the public, most of them warned indiscriminately against the dangers of infection and contagion and advocated the same safeguards for both. The various doctrines of contingent contagionism, which sought to reconcile the critical role of the atmosphere with the growing evidence for individual transmission of contagion, became the foundation of domestic sanitary science in the 1870s and the 1880s.[20]

As public health authorities traced out the complex routes of atmospheric and water pollution, they became increasingly aware of the role that certain domestic practices played in the spread of disease. New scientific evidence implicated changes in the middle-class household, especially its plumbing system, as a key factor in the rising rates of zymotic diseases. Ironically, "progress," namely the greater availability of running water and the growing popularity of the water closet, had created a sewage crisis of frightening proportions. The civil engineer and authority on house drainage William Paul Gerhard succinctly explained the problem:

> In city dwellings the ample supply of water, which in turn serves as a vehicle for transporting refuse matters, and the more general introduction of the convenient plumbing fixtures, led, owing to the leaky condition of brick or earthenware drains under houses, to a sewage-sodden condition of the soil under basements. This is true not only of the vast number of buildings erected by shrewd speculators, but it applies alike to the palatial mansions of the rich. Indeed, the death-rate from zymotic diseases increased, not only in houses with damp cellars, basements, and foundation-walls, but principally in those elaborately planned and richly furnished residences of the better class, where innumerable stationary washbowls, defective in arrangement and tightly enclosed by decorative cabinet-work, were scattered in bedrooms all over the house.[21]

The prevalence of damp cellars, foul odors, and leaky drains even in the best of homes convinced sanitarians that the "modern conveniences" of water closet and sink were as serious a public health hazard as was the more blatant uncleanliness of the laboring classes. They widened the association between contagion and defective household arrangements, which had formerly been limited to typhus and cholera, to include many common infectious ills that affected rich and poor alike. The expanded category of house diseases included typhoid, which was spread by fecal contamination of the water and air; diphtheria and other diseases involving sore throats, which were thought to be caused by foul air or sewer gas released by faulty plumbing; and highly contagious diseases such as measles and scarlet fever, which spread due to careless domestic nursing practices. (Although still regarded as a constitutional disease, consumption was also frequently linked with damp and poorly ventilated dwellings.)

Given the large role they accorded faulty household arrangements in the spread of such serious diseases, domestic sanitarians recognized voluntary reformation of the private sphere as one of the most direct and effective means of improving public health. This emphasis not only stemmed from the domestic sanitarians' conviction that pollution and contagion began in the home; it also reflected the realistic assessment that municipal authorities could not always be relied upon to uphold the highest sanitary standards. The householders needed to look after their own interests, particularly as their families became ever more dependent on common services such as sewers and water supplies for the maintenance of health. In an era when the state's public health powers were still rudimentary, domestic sanitary science restored a sense of control to the individual homeowner. As Joseph Edwards put it in 1882, "You cannot look into the sewer and see whether it is clean or not. But, into all the arrangements of your own individual house you can peer at all times, and can plainly see whether they are right or not."[22]

Popular health education advanced the sanitarians' broader political agenda as well, because a well-educated public would be more likely to support strong boards of health in their legislative battles to clean up the urban environment. Like reformers struggling against municipal corruption who needed "not only the sympathy, but also the active and hearty cooperation of the masses" to succeed, Edwards observed, it was "equally impossible for any board or boards of health to vanquish disease, unless they are thoroughly aided by the intelligent assistance of the public."[23]

METHODS OF POPULAR HEALTH EDUCATION

Fortunately for the sanitarians' purposes, their commitment to popular education coincided with the availability of increasingly inexpensive books, magazines, newspapers, and pamphlets to circulate sanitary information to the public. Beginning in the 1870s, scores of domestic hygiene manuals were written. Some, by prominent public health leaders, were published by major presses; others, by lay health enthusiasts, were privately printed. These manuals ranged from highly specialized technical tomes that throughly expounded the scientific rationale for the author's recommendations, to short, simple summaries of proper domestic behavior that supplied only the most rudimentary explanations for the advice offered.[24]

At one end of the spectrum were specialized treatises, aimed at both a professional and a lay audience, on topics such as sanitary plumbing or disinfection. For example, William Eassie's *Sanitary Arrangements for Dwellings* (1874), an English volume frequently cited on both sides of the Atlantic, was "intended for the use of officers of health, architects, builders, and householders." More general manuals, such as Henry Hartshorne's *Our Homes* (1880), which appeared in Blakiston's American Health Primers series, contained chapters on varied matters of concern to the householder, from building or choosing a home to creating a "home hospital" for the care of contagious illness. Domestic encyclopedias and family medical guides, such as *Wood's Household Practice of Medicine, Hygiene, and Surgery* (1880), which was intended for the use of "families, travelers, seamen, miners, and others," contained condensed versions of the same sanitarian advice.[25]

The concern with domestic hygiene carried over into the periodical literature as well, including ladies' magazines and popular science journals. The venerable *Godey's Lady's Book* contained short homilies on home health matters, and the more up-to-date *Ladies' Home Journal*, which began publication in 1883, had regular features on the prevention and management of infectious diseases. From its first issue in 1872, the *Popular Science Monthly* carried articles on sanitary plumbing, disinfection, and the germ theory of disease. In 1875, the *Atlantic Monthly* carried a widely read series of articles on domestic hygiene by the sanitary engineer George Waring. By the 1890s, even weekly religious newspapers such as the New York City *Independent* had columnists covering public health issues.[26]

The format and content of this published advice suggests that it was designed to provide guidance at certain common junctures of family life when health issues were particularly salient, such as when a family was deciding to build or rent a house, nursing a case of infectious disease in the home, or caring for a newborn baby. Clearly addressed to middle-class interests and pocketbooks, books and magazines furnished a constant flow of information about domestic hygiene to which individuals and families attended when the interest or need arose.

A second strand of popular education, far more episodic and intense in nature, was prompted by the fear of epidemics. The recurrent outbreaks of cholera spurred the most aggressive educational campaigns in American cities; periodic outbreaks of the much-dreaded childhood diseases scarlet fever and diphtheria also prompted special educational efforts. The outbreak of such fearful diseases presented unparalleled educational opportunities that public health authorities were quick to exploit; however fleetingly, epidemics brought the need for prevention to the whole community's attention.

These mass educational crusades arose on short notice and lasted only briefly, yet they reached more people and infused their message with a deeper sense of urgency than did the more voluminous advice literature. The chief form of "crisis education" was the circular, a brief fact sheet on disease prevention distributed *gratis* by municipal and state health departments. The circular's text was usually published in the daily newspapers as well. These two- or three-page tracts presented short, simple versions of the sanitarian gospel that assumed literacy, but little else, of their readers. Health department circulars and newspaper notices were probably the chief means by which detailed information about infectious diseases reached the working classes of the 19th-century American city.[27]

THE PRINCIPLES OF DOMESTIC HYGIENE

At the core of popular sanitarian writings about the home was a vision of life as an intricate process of respiration, consumption, excretion, and decay, in which the individual body figured prominently as a pollutant. Sanitarians believed the waste products of the body, particularly respired air and excrement, to be poisonous. As Mary F. Armstrong warned her readers in a tract written for the Hampton Institute, "everthing which is thrown out from the human body is unclean, and becomes at once dangerous to human life."[28] When human beings were packed too closely together, as in large cities, the sum total of their wastes was truly horrifying to contemplate. Henry Hartshorne, a Philadelphia medical professor and prominent hygiene authority, eloquently summed up this perspective in his domestic manual *Our Homes:*

> Apart from human interference, there is in nature a balance of formation and destruction, of life and death, food and waste, making a perfect natural economy everywhere. Man comes in with his artificial constructions, and sweeps away much of this economy of nature. . . . Hence comes foulness of the earth, water, and air; stench, miasma, pestilence. A guerilla warfare seems to be waged all around the invader of nature. . . . We must maintain or restore the original balance or primeval nature, by providing for the reappropriation of the products of life and the results of death and decay around us.[29]

In righting this balance, hygiene of the home played a critical role, for it was there that human beings spent the bulk of their time breathing, secreting, and excreting. Careful domestic hygiene was necessary to ensure the provision of pure "intake," that is, clean water, food, and air, and safe removal of the "outgo," or human wastes in the form of respired air and sewage. Otherwise, vitiated and corrupted air, poisonous exhalations, and dangerous discharges would be trapped in the home and given a deadly opportunity to work in its inhabitants, who ate and slept in blissful ignorance of the hazards surrounding them. The man who feels secure in his "castle," wrote the anonymous author of the *Bazar Book of Health* in 1873, "shuts himself up in it with his worst enemies." The prominent sanitarian F. S. B de Chaumont speculated in 1874 that "it might be a question whether or not in many cases it would be better to be without a house at all, than remain exposed to the numerous causes of disease arising within it."[30]

To guard the home against infection, sanitarians concentrated chiefly on four areas of domestic conduct: proper construction and maintenance of the house itself, especially ventilation and plumbing, to ensure pure air and an absence of dangerous sewer gases; careful home nursing of patients with contagious diseases, to prevent the spread of infectious material thrown off by their bodies; a specialized hygiene of the nursery, to protect children from the deadly diseases of childhood; and general housekeeping measures designed to ensure cleanliness.[31] Some sanitarian measures merely expanded and updated older hygienic conventions, such as the concern for pure air and the isolation of those ill with infectious diseases; others, such as the emphasis on sanitary plumbing and pure water, represented a response to new scientific information about the role of fecal contamination in causing cholera and typhoid.[32]

In addition, the sanitarians' prescriptions concerning domestic hygiene drew heavily upon their prior experience with institutions, particularly hospitals. The emphasis on careful building design to ensure healthful living conditions, the precise ratios of fresh air needed per person to disperse respired gases, the elaborate techniques for fumigating and disinfecting rooms, all can be found in the hospital reform literature of the period. Florence Nightingale's *Notes on Nursing*, which attempted to instruct women in "every day sanitary knowledge," was a deliberate attempt to take the lessons of the hospital into the home. In many less obvious ways, domestic sanitary science involved the transfer of technologies for ventilation and disinfection from large-scale to small-scale human habitations.[33]

The first and most important set of directives centered on building and maintaining the home: it must be sited properly, on dry soil; and so oriented as to ensure the maximum amount of what domestic sanitarians were fond of calling "natural disinfectants," namely fresh air and sunshine. In place of the advice simply to secure a "constant supply of fresh air," which sufficed in older hygiene manuals, painstaking attention was given to the proper proportioning of rooms, especially sleeping chambers, to ensure enough cubic feet of air per occupant to dilute the exhaled waste materials. In his book, Joseph Edwards urged that windows and doors be as large as possible, to replicate the healthy experience of living outdoors: "The larger you make your

openings, the nearer will your house approach a tent." Many manuals included instructions on how to rig windows with simple "ventilators" (e.g., by wedging the window open and placing a board in front of it to force the air upward, or by tilting an upper window in at the top) to ensure the circulation of fresh air without creating dangerous drafts.[34]

"Ours is the Age of Plumbing," emphasized Henry Hartshorne, and even the simplest hygiene manuals included lengthy and detailed discussions of the complexities of traps, water closets, and soil pipes. The essential goals of sanitary plumbing were straightforward—to get pure water into and human wastes out of the house without contamination of air or water—but the technological means to accomplish them were exceedingly complex. Domestic sanitarians considered the bare minimum of precautions to include complete separation of the drinking water and waste water systems; water closets that flushed thoroughly to prevent the putrefaction of wastes; watertight pipes to conduct wastes into the sewer; traps on all drains to prevent the discharge of sewer gas back into the room; and a special soil pipe running up the side of the house and venting above roof level, to allow the safe conduct of gases away from the home. Alfred Carroll, writing in a Staten Island missionary paper in 1878, confidently claimed that "with one of these simple appliances out-of-doors, a cellar tight and dry, and indoor drain pipes without material leakage, domestic life would be secure from the worst of its invaders."[35]

Sanitary authorities recommended hiring only the best plumbers and supervising their work carefully. To this end, J. Pridgin Teale designed his "pictorial guide to domestic sanitary defects" so that the homeowner might "test every sanitary point, one by one, and as he goes round book in hand, . . . catechise his plumber, his mason, or his joiner." Those renting or buying a home were advised to use the "peppermint test," which involved introducing oil of peppermint into a water closet and sniffing to see if the aromatic odor leaked out elsewhere in the house, a sure sign of faulty plumbing. An English physician recommended that his medical brethren write out directions for the test as they did a prescription, for though it was no substitute for a thorough plumbing inspection, it was "an admirably simple means of arousing a householder from the slumbers of a false security, from a fool's paradise, and of establishing the evident necessity of calling in skilled assistance."[36]

Similarly, the writers of manuals passed on detailed instructions for testing and purifying drinking water, warning that appearance, taste, and smell were not enough to determine the safety of water. In the summer, and during epidemics, they advised prudent householders to filter or boil all drinking water. Domestic manuals often included instructions for constructing simple home filters of sand, charcoal, and cloth. "But really suspicious water should, before using it for drinking or cooking, be boiled as well as filtered," advised Hartshorne in 1880.[37]

In an era when hospitals catered primarily to the poor and friendless, having a family member sicken with an infectious disease posed a special threat to the domestic environment. Hygiene manuals routinely included a chapter on home nursing, which spelled out the measures to be followed to create a home hospital that would simultaneously aid the patient's recovery and protect others in the household from contagion. A light and airy chamber was to be chosen and stripped of all carpeting and drapes, a sheet drenched in a strong disinfectant such as carbolic acid hung at the doorway, and liberal use of disinfectants made throughout the house. Last and most important, the patient's wastes were to be *immediately* disinfected and removed from the home.[38]

The proper use of disinfectants received lenghty explication in both domestic hygiene manuals and public health circulars. Public health authorities recommended employing disinfectants both as a daily precaution against disease and as a preventive in the sickroom. In place of the old, odoriferous techniques such as burning sulfur, which had sufficed for previous generations, domestic sanitarians extolled the virtues of the many new chemical disinfectants developed since the 1840s, including carbolic acid, sulfate of iron (also known as copperas), sulfate of zinc, chloride of iron, and permanganate of potash. Circulars and home hygiene manuals described at length the properties of the various disinfectants and gave recipies for inexpensive solutions to purify the air of a sickroom, rid the skin of contagious matter, disinfect the excreta of the ill, fumigate clothing and linens, and cleanse the plumbing system. Well into the early 1900s, dis-

infection was presented as one of the most important precautions against disease that householders could practice, "lest by neglect the health of the family may suffer," as a New Hampshire State Board of Health circular stated in 1885.[39]

Home nursing of childhood diseases required special precautions, since infectious diseases such as scarlet fever and diphtheria took such a high toll among the young. The writers of manuals advised parents to send their other children away immediately when a sibling fell ill, and to be very painstaking in washing and boiling the latter's bed linen and clothing. The contagious matter produced by eruptive diseases was believed to be especially hard to destroy and easily transmitted by ordinary objects. Joseph Perry told a standard cautionary tale: A cap was worn by a boy who developed scarlet fever; it hung in his sickroom, and after his funeral, it was "put away in a closely covered tin box without disinfecting." Two years later, the hat was removed and worn by the boy's younger brother, who within three days fell sick of the same disease, furnishing to Perry and his readers conclusive proof of the contagion's "tenacity." In a circular on scarlet fever published in 1888, the Massachusetts State Board of Health cautioned that the disease could be transmitted by "air, food, clothing, sheets, blankets, whiskers, hair, furniture, toys, library-books, wallpaper, curtains, cats, [and] dogs."[40]

Because of children's vulnerability to infectious diseases, the hygiene of the nursery necessitated special attention to ventilation, plumbing, disinfection, and pure water. Parents were told to place the nursery on an upper floor and keep its furnishings sparse. The importance of location was illustrated by a story in an 1888 manual about a woman whose children kept having "recurring diphtheric symptoms." The family doctor discovered that she kept them in a basement workroom during the day, and after he had her move the nursery upstairs, "the change was almost magical." The writers of manuals advised parents to keep their children out of doors as much as possible, and to place them in separate beds so that their "exhalations" would not mingle. Feeding utensils, especially for infants, were to be kept scrupulously clean and the water and milk for infants' use carefully boiled or filtered as a precaution against illness, particularly the dreaded "summer complaint," or infant diarrhea. Parents were also warned to check the health of the cows giving milk for their children, and to seek a source of clean, fresh milk that had not been allowed to stand for long.[41]

Finally, domestic hygiene manuals constantly emphasized the value of general cleanliness as a preventive against disease: the yard and cellar were to be kept dry and clear of "nuisances," dust and dirt to be swept up with a damp mop, the bedroom and bedclothes to be aired daily to rid them of "exhalations" (adults were also advised to sleep alone), and the house and plumbing periodically disinfected. Housewives were advised to replace heavy drapes and carpets with more easily cleaned curtains and rugs in order to prevent the accumulation of disease-producing dust. Warning her readers about the importance of such housekeeping details, Harriette Plunkett stated in her 1885 text *Women, Plumbers, and Doctors* that "eternal vigilance is the price of everything worth the having or keeping."[42]

These discussions of plumbing, disinfection, and general housekeeping accorded enormous importance to small details of behavior. An unsigned editorial entitled "Unconsidered Trifles," which appeared in *Godey's Lady's Book* in 1872, made this point quite dramatically:

> It cannot be too deeply impressed on people, especially the young, that very few real trifles exist in life; that is, there are very few actions, habits, or words which carry with them no consequences. . . . That ditch at the bottom of your garden—well, there is no denying that it smells badly enough in certain winds; but that is a mere trifle—a thing you cannot be expected to bother about when you have so much more important work on hand in the planning of your new conservatory. . . . The ditch is a trifle compared to the importance of such pursuits—a trifle, however, that brings diphtheria and typhus into your pretty house and takes off your children like sheep with the rot.[43]

Such heavy-handed attempts to inspire guilt and anxiety most likely did not affect the two sexes equally. A rough gender division of hygienic labor is evident in the texts: men were directed to look after the general construction and upkeep of the home, including the plumbing and the cellar, while women were expected to superintend the hygienic aspects of home nursing, child care, and house-

keeping. Of course, both sexes had ways in which they could delegate this responsibility: men could hire plumbers or architects to make the house safe, and women could have domestic servants do the heavy cleaning. However, both male and female authors tended to assume that their female readers had a greater responsibility for and interest in preserving health because their lives were more closely tied to home and family. As Benjamin Richardson, the patron saint of domestic sanitarians, explained, "The men of the house come and go," while "the women are conversant with every nook of the dwelling, from basement to roof, and on their knowledge, wisdom, and skill the physician rests his hopes."[44]

Given the continuing scientific controversy about how zymotic diseases actually originated and spread, domestic hygiene authors assigned to both individuals and famimes a surprisingly high level of responsibility for the prevention of infectious disease. The gospel of prevention emphasized that the vast majority of illnesses could be avoided by scrupulous adherence to a detailed hygienic code of behavior. "Many, very likely, will say that it is too much trouble to take the preventive measures advised," admitted Joseph Perry in his 1887 manual *Health in Our Homes*. "In the lives of all who study convenience so closely," he replied rhetorically, "there may come a time when, had those simple hints been observed, serious illnesses would have been averted, and possibly lives been spared."[45] In this fashion, the popularization of sanitary science intensified the public's belief in personal accountability for illness and reinforced their motivation to acquire ever more accurate information about preventing infectious diseases.

ASSIMILATION OF THE GERM THEORY

By the 1880s, when the germ theory of disease began to receive widespread explication in the popular press, domestic sanitarians had already instilled in the public a sense of responsibility for disease prevention. Not surprisingly, popular health writers were quick to report on the newest scientific theory, which John Shaw Billings neatly summarized in 1883 as the belief "that certain diseases are due to the presence and propagation in the system of minute organisms, which have no share or part in its normal economy."[46] Beginning in the late

1870s, popular writers began routinely to include germs in their list of household dangers that should be guarded against by proper domestic hygiene.[47] Some authors embraced the germ theory enthusiastically, as did Emma Hewitt, who declared in her 1888 text that "the study of the theory of germ proliferation has yielded amazing results in the way of furnishing the means of checking epidemics." Others took a more cautious position, emphasizing that the germ theory was still controversial, but at the same time insisting that its tenets supported the thrust of sanitarian recommendations. For example, after briefly summarizing the germ theory in his 1880 domestic hygiene manual, the English sanitarian George Wilson concluded, "Of far greater importance is to know that, whatever be the origin or mode of propagation of these diseases, they are to a very large extent controllable."[48]

Although older sanitarians tended to be distrustful of what they felt to be the oversimplifications of bacteriology and of experimental methods generally, their hygienic formulations were easily expanded to incorporate germs into the schema of household dangers. Physicians might debate among themselves about the ability of microorganisms to cause disease, but popular hygiene writers had little trouble, at the level they had to explain matters, in associating dirt, infection, and germs. Damp cellars, noisome garbage, sewers, corrupted air, all could easily be portrayed as breeding grounds for these minute living particles; likewise, the individual "emanations" and "effete matter" thought to lodge in bedclothes and carpets could be seen as laden with disease germs. The ability of microorganisms to produce dangerous toxins or poisons could be easily assimilated into older notions of decay and putrefaction as sources of infection.

Moreover, those popular hygiene authors who were not physicians, a category that included most of the women writers, had none of the professional investment in established scientific explanations of disease that made it so hard for some physicians to accept the germ theory. The actual cause of the disease mattered little to them, so long as the same actions were effective in preventing it. The germ theory was rapidly incorporated into popular advice literature precisely because it supported what

seemed to be common sense, that is, the already "proven" precautions of ventilation, disinfection, isolation of the ill, and general cleanliness. The public was given new information about germs in terms of what it already believed to be true about hygienic living conditions, and layers of old and new knowledge became tightly interlaced.

Thus the early popular understanding of germs closely followed the outlines of established sanitarian belief, suggesting that the dividing line between sanitarian and bacteriological conceptions of disease was never very sharp. The demand for usable knowledge about how to avoid infection outweighed the lack of scientific consensus about the etiology of disease. As a result, domestic sanitary science was not vanquished by bacteriology; rather it might more properly be said that the former appropriated the latter to its own uses. Contrary to the worst fears of sanitarians such as Benjamin Richardson and Florence Nightingale, the germ theory did not vitiate the connection between a clean, moral life and safety from disease. Instead, to quote the author Emma Hewitt, the germ theory "placed in the hands of every one, if not the power of destroying these germs, at least the power to prevent their proliferation" by the practice of "antiseptic cleanliness" in the home."[49]

THE PRACTICE OF DOMESTIC HYGIENE

While it can easily be shown that the principles of domestic hygiene were widely circulated through the print media, it is difficult to prove that the specific actions called for by the domestic sanitarians were ever taken. Even middle-class urban dwellers, who frequently confided their health concerns in diaries and letters, rarely commented upon plumbing improvements, disinfectant use, or sickroom procedures. Still, it is possible by more indirect evidence to arrive at some informed conjectures about how middle-class Americans changed their domestic behavior in order to avoid infectious diseases in the 1870s and the 1880s.

The sanitary reclamation of some famous homes provides one line of evidence that domestic hygiene was taken seriously. In 1861, Queen Victoria's husband Albert, the prince consort, died of typhoid, "like a common Terling peasant," in the words of an indignant sanitarian; barely a decade later, in 1872, the Prince of Wales, heir to the English throne, nearly died of the same disease. Balthazar Foster drew the obvious conclusion in a lecture, "The Prince's Illness: Its Lessons," delivered in 1872: "Truly, ignorance of sanitary science is not confined to the poor, but flourishes even in the highest places." The royal typhoid cases prompted a thorough investigation of the sanitary condition of Windsor Castle and other royal dwelling places.[50]

Americans were not spared their own national sanitary disgrace: when President Garfield was shot in 1881, some sanitarians attributed his failure to recover from his wounds to the unwholesome state of the White House. In November 1881, George Waring performed an inspection of the presidential mansion and found serious defects: disintegrated waste pipes soaked the basement with "foul matters," whitewash from the kitchen walls flaked off into the food, and a general state of damp and decay prevailed. Waring informed John Shaw Billings that "while they are free from some defects often found in the better class of houses in our cities, the plumbing appliances of the Executive Mansion do not conform to what are now accepted as the necessary sanitary requirements of a safe dwelling." President Chester Arthur wanted the White House torn down, but due to a combination of Congressional cheeseparing and sentimental attachment to the old mansion, he had to settle for a thorough plumbing overhaul.[51]

While the sanitary reclamation of such famous houses confirms that domestic hygiene was taken seriously, the question still remains: how did ordinary people change their everyday behavior to avoid infection, and what results might that behavioral change have produced, in psychological, social, and demographic terms? To answer that question, it is useful first to identify the most important measures advocated by the domestic sanitarian creed: ventilation, disinfection, plumbing, water purification, and general cleanlines. Regarding each of these points, increasing consumer demand for so-called sanitary goods suggests that middle-class Americans took the gospel of domestic sanitary science to heart in the late 19th century.[52]

Assuming that supply followed demand, the rush to develop and to patent sewer traps, toilet designs,

window ventilators, and water filtration systems in this period suggests that entrepreneurs found a lucrative market among householders anxious to safeguard their families against infection. Such products began to proliferate in the 1870s, and with the popularization of the germ theory their numbers showed an explosive increase in the 1880s and the 1890s.[53] In the 1880s, companies specializing in sanitary services such as disinfection and water purification appeared in city directories, suggesting that there was a living to be made by appealing to the fear of house diseases. Significantly, advertising brochures for these goods and services echoed, in both content and tone, the domestic manuals of the period: in promoting their goods, inventors and entrepreneurs assumed a familiarity with sanitarian precepts, indicating that such knowledge was indeed widespread, at least among the sort of people they expected to purchase their wares.

No doubt because of the sanitarian obsession with plumbing, devices aimed at protecting homes against sewer gas were among the earliest sanitary goods to appear on the market. Between 1870 and 1885, numerous sewer trap designs, as many as 22 a year, received patents, along with some remarkable designs for ornamental lamps that used sewer gas for illumination.[54] The marketing of sewer traps made effective use of the same kind of concepts and personal anecdotes employed by the domestic sanitarians. In an 1880 brochure advertising his "perfected sewer valve trap," William F. Downey recounted how he had been driven to invent the device—which he described proudly as "the most important invention of the 19th century"—after his wife almost died from diphtheria contracted in their poorly constructed new home.[55]

An even more active area of innovation was the redesign of toilets to ensure more prompt and complete removal of human wastes. Hundreds of designs for water closets and flushing devices were patented in the 1870s and 1880s; the 1884 catalogue of the Meyer-Sniffen Company presented no less than 33 different versions of the sanitary water closet, from the simplest to the most elaborate. For those families reluctant to invest in a whole new toilet, the enterprising patent solicitor R. D. O. Smith advertised "an odorless water closet" that a plumber could attach to an old-fashioned hopper toilet. Civil engineers such as William Paul Gerhard and George Waring took advantage of the new concern about plumbing to promote their careers as expert inspectors and improvers of house drainage.[56]

Not to be outdone by plumbers and sanitary engineers, manufacturers of domestic heating and ventilation devices emphasized the healthful aspects of their products. The designers of building and window ventilators were granted many patents. Henry Hartshorne noted in his 1880 text that "many different systems and apparatus for ventilation of rooms and houses have been invented, more or less ingenious and successful in attaining their end—Muir's, McKinnell's, Tobin's, Ruttan's, Hulin's, and a host of others." As did the designers of sewer traps, the promoters of ventilation devices used health claims to sell their wares. For example, in 1873 the Philadelphia firm of Charles Williams included in its promotion brochure for the "Golden Eagle" furnace a lengthy discussion of the dangers of overheated and poorly ventilated homes, stating, "That a pure atmosphere is necessary to preserve health we need not attempt to prove by reasoning—it is a truth universally known and acknowledged." *Boyd's Business Directory for Philadelphia, 1879* listed two businesses that specialized in ventilators, including one company that sold only Read's Patent Window Ventilator.[57]

Disinfectants represented an even more fertile field than ventilating devices for commercial development. By the late 1860s, patent preparations were already being marketed so aggressively that the American Medical Association's Committee on Disinfectants warned against them. "In general these patented compunds, which will no doubt be largely multiplied, are made to smell and to sell, and are not founded on any hidden scientific knowledge unknown to the profession, and are not as good as less bulky, more established disinfectants." By the early 1880s, when George Sternberg began testing the germicidal properties of commercial disinfectants for the American Public Health Association, there existed scores of proprietary solutions, with fanciful names, to be tested: Little's Soluble Phenyle, Bromo-Chloralum, Phenol Sodique, Withers Antizymotic Solution, and the ubiquitous Listerine. Druggists' catalogues in the 1870s and 1880s listed both generic and proprietary disinfectants in small quantities suitable for household use.[58]

Mechanisms for dispensing disinfectants were also a popular line for commercial development: during the 1870s and 1880s, inventors received patents for disinfectant pocket inhalers, disinfecting apparatus for toilets and slop jars, and disinfectant devices for sewers. Champions of chemical disinfection, such as the physician C. L. Cohn, who invented a device for toilets known as the Germicide, often spoke disparagingly of the "multiplying mechanical devices in plumbing, ventilation, etc." designed to protect the home. "*There is no safety or security in any other method,*" Cohn warned.[59]

As companies that specialized in sanitary engineering and ventilation became more numerous, business concerns that specialized in disinfection also began to appear in large eastern cities. In 1879, the Philadelphia business directory listed several individuals whose chief business was selling disinfectants. By the middle 1880s, companies such as Tayman's Disinfectant and Fumigating Company of Philadelphia offered home services using their own patented method. Like C. L. Cohn, the Tayman Company disparaged traps and other such devices, saying, "experience abundantly proves that mechanical devices are insufficient, that we must seek the aid of chemistry and obtain some agent that will antagonize and destroy the seeds of diseases."[60]

Moreover, in the 1880s and the 1890s companies began to market domestic filtration systems that saved the trouble of constantly having to boil drinking water. Private water companies served consumers in New York, Boston, and Philadelphia. The Hyatt Pure Water Company of Philadelphia marketed domestic water filters capable of cleaning from 1.5 to 8 gallons per minute. A list of patrons given in an 1890 brochure included such prominent Philadelphians as the publisher Alexander McClure and the banker Anthony J. Drexel. For less affluent customers, the Sub-Merged Filter Company of Philadelphia sold a simple filter made of charcoal and sand that could be fitted to a water cooler or a home reservoir. The company brochure assured the public that its filters could remove all the filth from Schuylkill river water, as well as "the innumerable minute worms" that throve there.[61]

The inventors and promoters of these various

sanitary devices frequently asserted that their prices were so low that they were "within the means of the most humble households," as Cohn said of his Germicide. In reality, the expense of devices such as the ornamental sewer gas flare, the siphon flush toilet, and the Hyatt water filtration system undoubtedly confined their use to wealthy households. However, so many domestic hygiene manuals included simple, do-it-yourself versions of sanitary devices that cost per se did not limit people's adherence to sanitary recommendations. For the conscientious homeowner determined to ventilate a room, disinfect a toilet, or install a water filter, a range of alternatives existed at different prices.[62]

While the marketing of sanitary goods such as sewer traps and ventilators began in the 1870s, well before the germ theory gained wide public circulation, the revelations about microscopic life greatly intensified the appeal of these products. As is evident in Cohn's and Tayman's brochures for disinfection, promoters were quick to present germs as a grave new danger that made their devices even more essential to the healthy home. The prudent homeowners who had thought their work (and expense) was finished when they installed mechanical devices such as traps and ventilators now had to upgrade their sanitary protection by chemical means.

Thus in one sense, the advertising copy for sanitary goods carried on the work of the advice manuals, popularizing new information about disease prevention while at the same time promoting new sanitary goods. Yet in putting that information to frankly commercial uses, businesses often distorted it. Entrepreneurs were prone to overstate and exaggerate both the dangers posed by infection and the benefits to be derived from their particular product. Public health authorities often spoke disapprovingly of the patent gadgetry being marketed as preventives against deadly gases and germs, and emphasized that strict household discipline and inexpensive homemade devices offered equal protection. Nevertheless, the very success of the sanitarians' popular campaign ensured that they lost control of the scientific content of their message: once popular anxieties had been raised, the public was easily manipulated by commercial interests whose overriding concern was making a profit. However ineffective many of these patent devices

were, though, the late 19th-century boom in sanitary supplies and services provides impressive testimony to the public's eagerness to purchase exemption from deadly infectious diseases.

FROM VOLUNTARY TO COMPULSORY INITIATIVES IN THE PUBLIC HEALTH MOVEMENT

The activities of municipal boards of health in the 1880s and the 1890s bear impressive testimony to the seriousness with which domestic hygiene came to be regarded. Invoking the authority of the domestic sanitarians, public health officers took an increasingly active and coercive role in enforcing sanitary regulation of the household. Instead of being private measures that conscientious homeowners might adopt for the safety of their families, sanitary plumbing, disinfection, and water filtration gradually became duties of the state. Some of the most important regulatory goals of the late 19th-century public health movement can be seen, then, as the natural extension of voluntary domestic reform.

The activities of the Philadelphia Board of Health illustrate this shift from private to public initiatives. In 1885, the city adopted a building code that required homeowners to provide minimum standards of sanitary plumbing. Individuals building new homes had to file plans and specifications for their plumbing. The law provided for the board of health to inspect house drainage and register master plumbers who might perform approved work. Thereafter the great majority of prosecutions undertaken by the board of health were not for "nuisances" in public spaces such as markets and streets, but for unsanitary home plumbing. Moreover, they harassed homeowners not just for blatant infractions, such as overflowing privies, but also for more subtle errors such as unventilated drainage, defective traps, and improperly connected soil pipes. Significantly, neighbors played an active role in reporting sanitary violations to the board, suggesting that they, too, perceived such conditions as serious threats to their health.[63]

Likewise, the board expanded the practice of compulsory quarantine and disinfection to regulate families who could not be trusted to follow the accepted procedures for the proper conduct of the "home hospital." An 1884 law provided for posting placards at homes where infectious diseases had been reported, but well into the 1890s the medical inspector was allowed "to set aside the practice in those cases where he was abundantly satisfied that the inmates of the house would faithfully carry out the printed instructions applicable to contagious diseases." This selective quarantine policy met with such resentment that after 1895 the board began to post placards at all homes, regardless of their inhabitants' cooperation.[64] In 1885, the Philadelphia Board of Health established the post of Disinfector, and began to require the fumigation of homes where contagious diseases had been reported. Public disinfection services expanded rapidly; by 1900, the city employed six men who disinfected almost seventy-five hundred Philadelphia homes in that year alone. While most domestic disinfections were involuntary, the city also provided the service for families who requested it, thus providing at no cost what commercial disinfection companies such as Tayman's offered the more affluent.[65]

Last but not least, the gradual expansion of municipal water purification capabilities made domestic filtration systems unnecessary. In response to repeated outbreaks of typhoid, Philadelphia slowly but surely began to improve its municipal sewer and water systems in the 1880s. Political factionalism and corruption dragged the process out for decades, but by 1912 the whole city had a filtered water supply. Thus, by governmental action, all citizens of early 20th-century Philadelphia acquired the filtered water supply that affluent families had previously purchased through private water companies.[66]

CONCLUSION

It is tempting to speculate, but impossible to prove, that the popularization of sanitary science, and the changes this campaign induced at the household level (e.g., improvements in home plumbing and purification of drinking water) may have contributed to the beginnings of the great mortality transition of the 1870s. What is more certain, and equally significant, is the way many late 19th-century public health authorities *interpreted* the falling rates of infectious disease to reinforce their long-standing belief in the importance of sanitary science and domestic hygiene.[67] The fact that the declining rates occurred more precipitously in affluent districts,

where the basics of household hygiene were more firmly entrenched, only confirmed the link between individual behavior and infectious disease forged by the pioneer generation of sanitarians. Now, with a generation of middle-class voters educated in the proper hygienic principles behind them, public health authorities were able more effectively to impose those same practices on the poor, under the aegis of state medicine.

The growing power of city and state boards of health to regulate the health conditions of individual homes did not diminish the public health movement's commitment to mass health education, however. As had their 19th-century predecessors, the advocates of the "new public health" realized that voluntary compliance with sanitarian precepts was essential to their success. Many public health workers continued to believe that if the poor could only be taught the same hygienic practices observed in affluent homes, they too could be saved from needless suffering. Groups that had high death rates from infectious diseases such as typhoid and tuberculosis—immigrant and working-class families, and poor white and black rural families— were targeted for intensive popular health education about domestic sanitation in the early 1900s.[68]

These early 20th-century health education campaigns preserved the older sanitarian beliefs about sunshine and fresh air even as they incorporated elements of a new, bacteriologically derived hygiene.[69] In her 1906 manual on housekeeping, Maria Parloa, a founder of the home economics movement, wrote about the combined threats of damp cellars, vitiated air, and bacteria-laden carpets with no sense of contradiction.[70] Well into the 1920s, domestic manuals continued to warn against the twin dangers of sewer gas and deadly germs.

Gradually, as the benefits of sanitary plumbing, municipal sewer systems, and large-scale water purification extended to more and more Americans, and mortality rates from infectious diseases continued to decline, house diseases lost their central role in public health campaigns, to be supplanted by measures such as school health programs, mass immunization, and industrial hygiene. The lessening emphasis on domestic sanitation probably occurred first among the middle and upper classes, who by the 1920s enjoyed an unprecedented freedom from the infectious diseases that had plagued their parents and grandparents. No doubt the relief from anxiety occurred much later among poorer families, who continued both to experience higher rates of disease and to be the focus of intense educational campaigns about personal and domestic hygiene.[71]

The psychological ramifications of popular beliefs about the link between domestic hygiene and illness may have lingered much longer. In an era of high infant and child mortality, the domestic sanitarians forged a powerful association between guilt and responsibility for infection. The heightened intensity with which Victorian parents mourned their children, an emotional trend that is usually attributed to smaller family size and the greater value placed on the individual child, may also reflect the sanitarian educational crusade and its efforts to cultivate greater parental, and chiefly maternal, vigilance against house diseases such as diphtheria and typhoid.[72]

Perhaps as a legacy of this painful period, the daughters of those Victorian mothers, who began child-rearing in the first and second decades of the 20th century, ingrained in their children an obsession with germs and cleanliness born of an earlier era of domestic sanitary hazards.[73] To those born after the "pax antibiotica," who never experienced the "bad old days" when debility and death from infectious diseases were commonplace, such fears may seem irrational; but the AIDS epidemic has suddenly diminished the complacency of the post–World War II generation and made the threat of infection very real once again. The persistence of health beliefs acquired in childhood helps to explain why many Americans responded to the AIDS crisis with fears concerning casual modes of transmission.[74]

Meanwhile, the rising concern about environmental toxins has reintroduced the notion of house diseases in a different conext. Although domestic hygiene no longer figures prominently in public health campaigns against *infectious* diseases, rising cancer rates have brought about a new set of concerns about domestic pollution. Today's "killer houses" are impregnated not with dangerous bacilli but with carcinogenic substances. The public is now demanding that their homes be secured from radon gas, asbestos, toxic waste dumps, and utility lines. The nature of the threat of disease has changed radically, reflecting the new realities of life and death in a polluted world; but the drive to find protective strategies—the radon test kit, the home water filter, and the like—has haunting similarities to the efforts of late 19th-century families to safeguard their domestic space from infection. However irrational and ineffective individual solutions to collective health threats may seem, the "private side" of public health has a persistent appeal born of the very human need to try to control a dangerous world.

NOTES

Earlier versions of the paper were presented at the 62nd annual meeting of the American Association for the History of Medicine, 28 April 1989; at the Francis C. Wood Institute for the History of Medicine and the University of Illinois at Champaign-- Urbana in April 1989; and at the history department colloquium, State University of New York at Stony Brook, in March 1990. I thank the audiences at those sessions for their helpful comments. I would also like to thank the following people for their helpful comments on this article: Joan Jacobs Brumberg, Gretchen Condran, Janet Golden, Christopher Sellers, Janet Tighe, and the anonymous reviewers for the *Bulletin of the History of Medicine*. The staffs of the Historical Collections of the College of Physicians of Philadelphia, Philadelphia, Pa., the National Library of Medicine, History of Medicine Division, Bethesda, Md., and the Hagley Museum and Library, Wilmington, Del., were exceptionally helpful in tracking down materials for me. Research for the article was conducted during my tenure as a Rockefeller Fellow at the Wood Institute of the College of Physicians of Philadelphia during 1988–89.

1   Leopold Brandeis, "Defective house drainage," *Sanitarian*, 1873–74, *1*: 447.

2   Charles V. Chapin, *Papers of Charles V. Chapin, M.D.*, ed. Clarence L. Scamman (New York: Commonwealth Fund; London: Oxford Univ. Press, 1934), p. 50; *Information Regarding the Germicide and Its Protective Influence* (Philadelphia: Pennsylvania Germicide Co., 1884); L. S. Beale, quoted in W. D. Foster, *A History of Medical Bacteriology and Immunology* (London: Heinemann, 1970), p. 16.

3   In this article, I use the term *infectious* to denote diseases that spread from person to person, including the category of contagious diseases. As the most cur-

sory reading of the literature shows, the distinction between the terms *infectious* and *contagious* was very confused in the 19th century. The English sanitarian George Wilson observed in 1880, "'infection' and 'contagion' are now used as synonymous terms, or, at all events, are used indiscriminately, and are intended to convey the same meaning." Geroge Wilson, *Health and Healthy Homes: A Guide to Domestic Hygiene,* ed. Joseph G. Richardson (Philadelphia: Presley Blakiston, 1880), p. 117. Most scientific discussions limited the term *contagious* to diseases that spread by direct, person-to-person contact, and used the term *infectious* to denote both contagious diseases, and diseases that could be spread by corrupted matter in the air or water. I have tried to stick to their usage.

4 I use the term *domestic sanitarians* to denote reformers who in the 1870s and the 1880s emphasized popular education and domestic hygiene as important goals of the public health movement. Concern about domestic hygiene was a relatively late development in the sanitarian movement. While they were concerned about the sanitary condition of urban slums, where squalid homes bred epidemics, the pioneer generation of English sanitarians, including Edwin Chadwick and Thomas Southwood Smith, were not much interested in popular education or domestic sanitary reform; their first concern was state medicine and broad-gauge environmental reform. I suspect that later, when the progress of state medicine seemed stymied by political and popular apathy, sanitarians began to invest more energy in finding private solutions to public health problems. This does *not* mean that they gave up the ideals of state medicine but rather that they believed in voluntary reform as a necessary intermediary step to acquire the political backing from voters to widen state power over health matters. By my definition, the writings of Benjamin Richardson and George Wilson in England and George Waring and Henry Hartshorne in the United States are good examples of domestic sanitarian argument.

5 The literature on the middle-class American family and its insatiable "manual mania," as Kathryn Kish Sklar terms it in her introduction to Catharine E. Beecher, *A Treatise on Domestic Economy* (1841; New York: Schocken Books, 1977), p. v, is extensive. Some useful works on the history of homelife and the family are Sylvia D. Hoffert, *Private Matters: American Attitudes toward Childbearing and Infant Nurture in the Urban North, 1800–1860* (Urbana: Univ. of Illinois Press, 1989); Mary P. Ryan, *Cradle of the Middle Class: The Family in Oneida County, New York, 1790–1865* (Cambridge and New York: Cambridge Univ. Press, 1981); and Kathryn K. Sklar, *Catharine Beecher: A*

Study in American Domesticity (New Haven: Yale Univ. Press, 1973). For the American obsession with staying healthy, see Susan E. Cayleff, *Wash and Be Healed: The Water-Cure Movement and Women's Health* (Philadelphia: Temple Univ. Press, 1987); Anita Clair Fellman and Michael Fellman, *Making Sense of Self: Medical Advice Literature in Late Nineteenth-Century America* (Philadelphia: Univ. of Pennsylvania Press, 1981); Harvey Green, *Fit for America: Health, Fitness, Sport, and American Society* (New York: Pantheon Books, 1986); Martha H. Verbrugge, *Able-Bodied Womanhood: Personal Health and Social Change in Nineteenth-Century Boston* (New York: Oxford Univ. Press, 1988); and James Whorton, *Crusaders for Fitness: The History of American Health Reformers* (Princeton: Princeton Univ. Press, 1982). The special appeal of health reform to women is explored in Regina Morantz, "Nineteenth century health reform and women: a program of self-help," in *Medicine without Doctors: Home Health Care in American History,* ed. Guenter B. Risse, Ronald L. Numbers, and Judith W. Leavitt (New York: Science History Publications, 1977), pp. 73–93; and *idem,* "Making women modern: middle-class women and health reform in nineteenth-century America," *J. Soc. Hist.,* 1977, *10*: 490–507.

6 On the decline in mortality rates from infectious diseases, see Gretchen A. Condran and Rose A. Cheney, "Mortality trends in Philadelphia: age- and cause-specific death rates 1870–1930," *Demography,* 1982, *19*: 97–123; Gretchen Condran, Henry Williams, and Rose Cheney, "The decline in mortality in Philadelphia from 1870 to 1930: the role of municipal services," *Penn. Mag. Hist. & Biography,* 1984, *108*: 153–177 (ch. 29, this book); and Frederick L. Hoffman, "American mortality progress during the last half century," in *A Half-Century of Public Health,* ed. Mazÿck P. Ravenal (1921; New York: Arno Press, 1970), pp. 94–117.

7 This gender-linked burden led middle-class women to play an active role in municipal sanitary reform, as Suellen M. Hoy shows in "'Municipal housekeeping': the role of women in improving urban sanitation practices, 1880–1917," in *Pollution and Reform in American Cities, 1870–1930,* ed. Martin V. Melosi (Austin: Univ. of Texas Press, 1980), pp. 173–198.

8 Norbert Elias, *The Civilizing Process: The History of Manners,* trans. of 1939 German ed. (Oxford: Basil Blackwell, 1978), p. 159. A more recent cultural analysis in the same spirit is Georges Vigarello, *Concepts of Cleanliness: Changing Attitudes in France since the Middle Ages,* trans. Jean Birrell (Cambridge and New York: Cambridge Univ. Press, 1988). On bathing and personal cleanliness, see Richard L. Bushman and Claudia L. Bushman, "The early history of cleanliness in America," *J. Am. Hist.,* 1988, *74*: 1213–1238;

and Jacqueline Wilkie, "Submerged sensuality: technology and the perception of bathing," *J. Soc. Hist.*, 1986, *19*: 649–664. On changing architectural and housekeeping standards in the 19th century, see Ruth Schwartz Cowan, *More Work for Mother: The Ironies of Household Technology from the Open Hearth to the Microwave* (New York: Basic Books, 1983); Faye E. Dudden, *Serving Women: Household Service in Nineteenth-Century America* (Middletown, Conn.: Wesleyan Univ. Press, 1983); and Gwendolyn Wright, *Moralism and the Model Home: Domestic Architecture and Cultural Conflict in Chicago, 1873–1913* (Chicago: Univ. of Chicago Press, 1980).

9　What I think of as the "narrowing" thesis is usually associated with Barbara G. Rosenkrantz's *Public Health and the State: Changing Views in Massachusetts, 1842–1936* (Cambridge, Mass.: Harvard Univ. Press, 1972). Rosenkrantz argues that the conception of *state* medicine became increasingly biomedical, as public health officials tried to put their profession on a solid scientific basis. However, she does not argue that the whole public health movement abandoned its interest in the broader social and economic determinants of disease (see esp. pp. 177–182). Elizabeth Fee makes a similar point in *Disease and Discovery: A History of the Johns Hopkins School of Hygiene and Public Health, 1916–1939* (Baltimore: Johns Hopkins Univ. Press, 1987). An excellent history of the public health movement in one city, which manages to be very sensitive to social and cultural context, is Judith Walzer Leavitt's *The Healthiest City: Milwaukee and the Politics of Health Reform* (Princeton: Princeton Univ. Press, 1982). Also John Duffy's valuable new survey, *The Sanitarians: A History of American Public Health* (Urbana: Univ. of Illinois Press, 1990), shows much more interest in popular education than older volumes on the subject.

10　Lloyd G. Stevenson, "Science down the drain: on the hostility of certain sanitarians to animal experimentation, bacteriology, and immunology," *Bull. Hist. Med.*, 1955, *29*:1–26. See also James H. Cassedy, "The flamboyant Colonel Waring: an anticontagionist holds the American stage in the age of Pasteur and Koch," *ibid.*, 1962, *36*: 163–176, which anticipates many of my arguments in this paper; and Charles E. Rosenberg, "Florence Nightingale on contagion: the hospital as moral universe," in *Healing and History: Essays for George Rosen*, ed. Charles E. Rosenberg (New York: Science History Publications, 1979), pp. 116–136.

11　Thomas McKeown, *The Modern Rise of Population* (New York: Academic Press, 1976).

12　I believe that the popular acceptance of sanitary science gave the physician a new confidence in dealing with patients and their families, which helped to im-

prove the profession's popular image in the 1880s and the 1890s, before the benefits of the newer, experimental medicine were much evident. This is an argument that I plan to make in the book I am now writing.

13　Historians have commented on the relatively rapid public assimilation of the germ theory, as well as the similarity between sanitarian and bacteriological prescriptions for preventive hygiene. However, no one has systematically explored how the popularization of sanitary science laid the groundwork for the acceptance of the germ theory. See Howard D. Kramer, "The germ theory and the early public health program in the United States," *Bull. Hist. Med.*, 1948, *22*: 233–247, esp. pp. 235, 241; and Andrew McClary, "Germs are everywhere: the germ threat as seen in magazine articles, 1890–1920," *J. Am. Culture*, 1980, *3*: 33–46. (For my observation here about the growing tension between elite and popular versions of scientific truth, I am indebted to John Burnham's very interesting study of popular science education, *How Superstition Won and Science Lost: Popularizing Science and Health in the United States* [New Brunswick, N.J.: Rutgers Univ. Press, 1987]).

14　I am grateful to Gretchen Condran for sharing her work in progress on this topic with me. She and Samuel Preston are preparing a review essay on the problem of personal hygiene and the mortality transition.

15　The choice of Philadelphia was dictated largely by the richness of its archival sources for the 19th century. I have no reason to believe that the evolution of popular attitudes and behavior in Philadelphia was significantly different from that in other large eastern cities of the period.

16　Owsei Temkin, "An historical analysis of the concept of infection," in his collection of essays *The Double Face of Janus* (Baltimore: Johns Hopkins Univ. Press, 1977), pp. 456–471, provides an excellent overview of traditional views of infection. Noting the consistent association between foul smells, decay, and disease, he notes, "It is remarkable how our modern terminology has remained within the orbit of ancient and medieval imagery" (p. 461). Carlo M. Cipolla, *Fighting the Plague in Seventeenth-Century Italy* (Madison: Univ. of Wisconsin Press, 1981), esp. pp. 76–78, discusses house quarantine of plague victims during the Tuscan epidemic of 1630–31. On the French hygienists, see Alain Corbin's fascinating account, *The Foul and the Fragrant: Odor and the French Social Imagination* (Cambridge, Mass.: Harvard Univ. Press, 1986). On hygienic traditions prior to the 1800s, see James C. Riley, *The Eighteenth-Century Campaign to Avoid Disease* (New York: St. Martin's Press, 1987); and Ginnie Smith, "Prescribing the rules of health," in *Patients and Practitioners: Lay Perceptions of Medicine in Pre-Industrial Society,* ed. Roy Porter (Cambridge and New York: Cambridge Univ. Press, 1985), pp. 249–282.

17 The Rush letter is quoted in Whitfield J. Bell, Jr., *The College of Physicians of Philadelphia: A Bicentennial History* (Canton, Mass.: Science History Publications, 1987), p. 28. My comparative remarks about pre-1850 hygiene manuals are based on a reading of volumes such as Bernhard C. Faust, *Catechism of Health for the Use of Schools, and for Domestic Instruction* (Dublin, 1794; New York: Arno Press, 1972); Robert W. Johnson, *Friendly Cautions to the Heads of Families and Others . . .* , 3d ed. (Philadelphia: James Humphreys, 1804); and Caleb B. Ticknor, *The Philosophy of Living; or, the Way to Enjoy Life and Its Comforts* (New York: Harper & Bros., 1836).

18 *The Home Book of Health and Medicine: A Popular Treatise on the Means of Avoiding and Curing Diseases . . .* (Philadelphia: Key & Biddle, 1835); quotations on pp. 382, 503. The advice given in general books on domestic affairs, such as Catharine E. Beecher's widely read *Treatise on Domestic Economy,* was even less detailed than that found in the family medical manuals of this type.

19 The literature on changing concepts of disease in this period is vast. For my discussion in the following paragraphs I am heavily indebted to the following works: John M. Eyler, *Victorian Social Medicine: The Ideas and Methods of William Farr* (Baltimore: Johns Hopkins Univ. Press, 1979); Margaret Pelling, *Cholera, Fever, and English Medicine, 1825–1865* (New York: Oxford Univ. Press, 1978); Charles E. Rosenberg, *The Cholera Years: The United States in 1832, 1849, and 1866* (Chicago: Univ. of Chicago Press, 1962); *idem,* "The therapeutic revolution" in Morris J. Vogel and Charles E. Rosenberg, eds., *The Therapeutic Revolution: Essays in the Social History of American Medicine* (Philadelphia: Univ. of Pennsylvania Press, 1979), pp. 3–25; Owsei Temkin, "The scientific approach to disease: specific entity and individual sickness," in *Double Face of Janus,* pp. 441–455; and John Harley Warner, *The Therapeutic Perspective: Medical Practice, Knowledge, and Identity in America, 1820–1885* (Cambridge, Mass.: Harvard Univ. Press, 1986).

20 F. S. B. de Chaumont, "Hygiene," *Sanitary Record,* 5 Dec. 1874, pp. 398–399, provides an excellent example of sanitarian eclecticism; in his table of the zymotic diseases, impure air, contaminated water, *and* direct contagion were implicated in the etiology of each disease. George Wilson's comment on the indiscriminate use of the terms *infection* and *contagion* (see n. 3 above) reinforces this observation. Margaret Pelling argues that by the mid-1860s, the majority of English physicians had adopted some version of "contingent contagionism." Rosenberg reached

much of the same conclusion in his study of cholera in the United States. See Pelling, *Cholera, Fever, and English Medicine,* esp. pp. 295–310; Rosenberg, *Cholera Years,* esp. pp. 192–199. The desire to straddle the middle ground explains the continued popularity of Muchison's "pythogenic theory" of fever and Pettenkofer's version of contingent contagionism, which allowed for both direct and indirect modes of contagion.

21 William Paul Gerhard, *The Drainage of a House* (Boston: Rand Avery Co., 1888), p. 4. For a good historical survey of the "wastewater crisis" of the late nineteenth century, see Joel Tarr, James McCurley, and Terry F. Yosie, "The development and impact of urban wastewater technology," in Melosi, ed., *Pollution and Reform,* pp. 59–82.

22 Joseph F. Edwards, *How We Ought to Live* (Philadelphia: H.C. Watts & Co., 1882), p. 151.

23 *Ibid.,* p. 407.

24 A few books and articles dating from before 1875 appeared under the heading "Habitations" in the first series of the *Index-Catalogue of the Library of the Surgeon-General's Office, United States Army* (Washington, D.C.; Government Printing Office, 1880–95), but the real flood began in the late 1870s and 1880s. Home hygiene was a popular topic among sectarians as well as "mainstream" physicians. See, for example, the hydropathic version of the sanitarian message included in *The Household Manual* (Battle Creek, Mich.: Health Reformer Office, 1875).

25 William Eassie, *Sanitary Arrangements for Dwellings, Intended for the Use of Officers of Health, Architects, Builders, and Householders* (London: Smith Elder & Co., 1874); Henry Hartshorne, *Our Homes* (Philadelphia: Presley Blakiston, 1880); Frederick A. Castle, ed., *Wood's Household Practice of Medicine, Hygiene, and Surgery,* 2 vols. (New York: William Wood & Co., 1880). The quotation from the latter volume is taken from the title page. Volume 1 has several chapters on house construction and domestic hygiene.

26 George Waring's articles were reprinted in book form as *The Sanitary Drainage of Houses and Towns* (New York: Hurd & Houghton, 1876). Charles Chapin was speaking particularly of Waring's series when he noted that it made New Englanders fear sewer gas "perhaps more than they did the Evil One." Quoted in Cassedy, "Flamboyant Colonel Waring," p. 166.

27 Numerous examples of these circulars, from boards of health in Boston, New York, Philadelphia, and Providence, can be found. For example, the Library of the College of Physicians of Philadelphia has *Sanitary and Preventive Measures: Disinfectants, How to Use Them, or What May Be Done by the Public to Guard against Yellow Fever and Diseases Common to Summer Months,* prepared by the Sanitary Committee of the Board of

Health of Philadelphia (Philadelphia: E. C. Markley & Son, 1878); the History of Medicine Division, National Library of Medicine, has one from the Massachusetts State Board of Health entitled *Suggestions for Preventing the Spread of Scarlet Fever* (n.p., n.d.). This brochure has no date, but is stamped as received at the Surgeon General's Office, U.S. Army, in 1888. Of course, broadsides and circulars were traditionally used by pre-modern health officials to educate the public in times of plague and other epidemic diseases.

28 Mary F. Armstrong, *Preventable Diseases,* Hampton Tracts for the People, Sanitary Series, no. 3 (Hampton, Va.: Hampton Institute Press, 1878), p. 5.

29 Hartshorne, *Our Homes,* p. 9. See Christopher Hamlin, "Providence and putrefaction: Victorian sanitarians and the natural theology of health and disease," *Victorian Studies,* 1985, *28:* 381–411, for an interesting discussion of the Victorian concept of putrefaction.

30 *The Bazar Book of Health* (New York: Harper & Bros., 1873), p. 17 (the author identified him- or herself as a physician); de Chaumont, "Hygiene," p. 397. Note the similarity between the laws seen to govern the home and the traditional economy of the individual body in health and disease, as described so well by Rosenberg in "Therapeutic revolution." Rosenberg uses the terms *intake* and *outgo* to describe the balance between individual and environment that was central to maintaining health.

31 The following summary of the principles of domestic hygiene is based on some 50 manuals published between 1870 and 1900 that I found in the Library of the College of Physicians of Philadelphia, the National Library of Medicine, the Drexel University Library, Philadelphia, Pennsylvania, and the Van Pelt Library of the University of Pennsylvania, Philadelphia. With few exceptions, I confined my survey to books written by Americans and American editions of English manuals.

32 See Temkin, "Concept of infection," and Rosenberg, "Therapeutic revolution." As Temkin points out, the concept of disease as pollution is a very ancient one. Mary T. Douglas, in *Purity and Danger: An Analysis of Concepts of Pollution and Taboo* (London: Routledge & Kegan Paul, 1966) argues that taboos about cleanliness and uncleanliness are a fundamental aspect of human societies.

33 Florence Nightingale, *Notes on Nursing: What It Is, and What It Is Not* (New York: D. Appleton & Co., 1865); quotation on p. 3. See Charles E. Rosenberg, *The Care of Strangers: The Rise of America's Hospital System* (New York: Basic Books, 1987), esp. ch. 5, for a discussion of the hospital reform movement.

34 Beecher, *A Treatise,* p. 273; Edwards, *How We Ought to Live,* p. 158. Edwards was probably invoking the Civil War experience that temporary tent hospitals had lower mortality rates than their permanent counterparts. For a typical description of a do-it-yourself window ventilator, see Roger S. Tracy, *Hand-book of Sanitary Information for Householders* (New York: D. Appleton & CO., 1884), p. 17.

35 Hartshorne, *Our Homes,* p. 101; Alfred Carroll, "The enemy in the air," *Sanitarian,* 1878, 6: 255. The latter article was reprinted from the *Messenger,* which the *Sanitarian*'s editor described as "an enterprising missionary paper of Staten Island." *Sanitarian,* 1878, 6: 253.

36 J. Pridgin Teale, *Dangers to Health: A Pictorial Guide to Domestic Sanitary Defects* (London: Churchill, 1879), p. 9; R. T. Hildyard, "Influence on sanitary progress which medical men might exercise in their private practice," *Transactions. Sanitary Institute of Great Britain,* 1883, *4:* 109.

37 Hartshorne, *Our Homes,* p. 100.

38 For a representative set of instructions on how to operate a home hospital, see Edwards, *How We Ought to Live,* pp. 395–401.

39 New Hampshire State Board of Health, *Disinfectants and Their Use* (Concord, N.H.: Parsons B. Cogswell, 1885), p. 5; see Hartshorne, *Our Homes,* pp. 130–136, for a standard discussion of disinfection. One of the distinctive tenets of the "new public health" was its rejection of the late 19th-century belief in disinfection. See Charles V. Chapin, "The fetich of disinfection," *J.A.M.A.,* 1906, *47:* 574–580; and *idem, The Sources and Modes of Infection* (New York: Wiley; London: Chapman and Hall, 1910).

40 Joseph F. Perry, *Health in Our Homes* (Boston: Thayer, 1887), p. 403; Massachusetts Board of Health, *Suggestions,* p. 1.

41 Emma C. Hewitt, *Queen of the Home* (Philadelphia: Miller Magee Co., 1888), p. 112.

42 Harriette Plunkett, *Women, Plumbers, and Doctors; or, Household Sanitation* (New York: D. Appleton & Co., 1885), p. 43.

43 "Unconsidered trifles," *Godey's Lady's Book,* 1872, *89:* 45. The author's invocation of typhus is rather odd, since it was most emphatically a disease associated with the homes of the very poor; families with conservatories would be more likely to fear typhoid, so perhaps she confused the two.

44 Quoted in Plunkett, *Women, Plumbers, and Doctors,* p. 11.

45 Perry, *Health in Our Homes,* p. 65.

46 Quoted in Plunkett, *Women, Plumbers, and Doctors,* p. 148.

47 The earliest mention of germs I found in my sample of manuals appeared in 1878 in Armstrong, *Preventable Diseases.* However, Charles Rosenberg has drawn

my attention to a very brief account, in Catharine E., Beecher and Harriet Beecher Stowe, *The American Woman's Home* (1869; New York: Arno Press, 1971), pp. 421–422, of how "microscopic plants" cause zymotic diseases. Beecher and Stowe do not use the term *germ*, however.

48   Hewitt, *Queen of the Home*, p. 225; Wilson, *Health and Healthy Homes*, p. 117. Wilson became a bitter critic of bacteriology in later life, but in this popular health manual he gives a short and respectful exposition of the germ theory.

49   Hewitt, *Queen of the Home*, p. 225.

50   S. Sneade Brown, *A Lay Lecture on Sanitary Matters* (Clifton, England: E. Austin, 1871), p. 15; Balthazar W. Foster, *The Prince's Illness: Its Lessons. A Lecture on the Prevention of Disease* (London: J. A. Churchill, 1872), p. 16. Windsor Castle was given a sanitary inspection in 1859, and its drainage was pronounced safe by Charles Murchison. See "The death of the prince consort," *Lancet*, 21 Dec. 1861, p. 599. The Prince of Wales's illness led to a meticulous inspection of the lodges where he had stayed before his attack. See "Report of the *Lancet* Sanitary Commission on the state of Londesborough Lodge & Sandringham, in relation to the illness of H.R.H. the Prince of Wales," *Lancet*, 9 Dec. 1871, pp. 828–831. The evidence found seemed to implicate faulty plumbing at Londesborough in the Prince's case.

51   George Waring to John Shaw Billings, "Report on the improvement of the sanitary condition of the executive mansion," 7 Dec. 1881, manuscript; History of Medicine Division, National Library of Medicine; William Seale, *The President's House: A History*, 2 vols. (Washington, D.C.: White House Historical Association, 1986) *1*: 536–538. Plunkett, *Women, Plumbers, and Doctors*, p. 230, refers to the sanitary scandal at the time of Garfield's wounding. Of course, it seems clear now that Garfield's condition was the result of his doctors' failure to observe antiseptic procedures while examining his wounds.

52   I believe (although I will not elaborate on my opinion here) that the late 19th-century revolution in home design and home furnishings should also be read, at least in part, as a response to sanitarian precepts. Reform movements championing the abandonment of ornamentation in favor of simple rectilinear lines invoked the need for cleanliness; less cluttered rooms offered fewer places hiding places for deadly dust. Likewise, a concern for dirt and dust prompted the shift in preference from room-sized carpets to area rugs that could easily be taken up and beaten outside; and to the use of tile, and later linoleum, in kitchens and baths. Note that Bushman and Bushman in "Early history of cleanliness," date the beginning of the "soap boom" that is, the use of soap

for bathing, to the post–Civil War period. They do not mention the rise of popular sanitary science in their explanations for this phenomenon, but it certainly seems plausible that the two were related.

53   My observations about the timing of these developments are based primarily on my research on patent applications done at the United States Patent Office in Alexandria, Virginia. In the patent search office, copies of all patents are categorized according to the type of process involved and filed together in boxes (e.g., Class 4, "Baths, Closets, Spittoons, and Sinks," Class 424, "Disinfection," and 410, "Water Filtration"). Within each class, the patents are filed by the year they were granted. A search of these and similar categories revealed that the products discussed here began to appear in the early 1870s, and their numbers increased dramatically in the 1880s and the 1890s. (I am deeply grateful to Don Garber for telling me about the U.S. Patent Office sources.)

54   I surveyed the *Annual Report of the Commissioner of Patents* (Washington, D.C.: Government Printing Office) at five-year intervals from 1870 to 1890; 1 sewer trap was patented in 1870, 10 in 1875, 9 in 1880, 22 in 1885, and 18 in 1890. Under Class 4-221, "Ventilation, Sewer, Burner," there are a number of remarkable devices for burning off sewer gas, including one lamp designed by J. Eckhardt in 1892. Eckhardt wrote in his patent application, "It would be superfluous at this day to call attention to the deadly nature of sewer-gas to prove the value of any device by which it may be rendered innoxious [*sic*] even if not turned to use, as in the present invention." The Hagley Museum and Library's collection of trade catalogues from the same period document the wide variety of traps and piping available. Note that plumbing supply companies began to call their wares "sanitary goods" in the 1870s. See, for example, J. L. Mott Iron Works, *Price List of the Plumbing and Sanitary Department* . . . (New York: E. D. Slater, 1881), Trade Catalog Collection, Hagley Museum and Library, Wilmington, Delaware (hereafter, HC). Joseph D. Galloway, *Gasfitters and Plumbers' Companion* (Philadelphia: By the author, 1875), p. 55, noted that lead pipe was rapidly becoming unpopular as a water pipe, and was being replaced by galvanized iron, tin, and lead pipe lined with tin. The best houses, Galloway observed, were being fitted with "seamless brass pipe." Although he does not specifically mention health concerns, these were changes that sanitarians were urging at the time.

55   [William F. Downey], *The Downey Perfected Automatic Sewer Valve Trap . . . The Deadly Enemy Conquered . . .* (Washington, D.C.: National Republican Printing House, 1880), title page, HC.

56   Meyer-Snifen Co., Ltd., *Illustrated Catalogue of Water-*

Closet and Bathing Arrangements for Public and Private Places (n.p., 1884), HC; R. D. O. Smith, *The Odorless Water Closet* (n.p., 2 Nov. 1875), HC; Sanitary Association of Philadelphia, *Guarding the Home: Skeletons of Our Homes*, Tract no. 1 (n.p., [1887]), HC. For an excellent account of how Waring turned sanitary science into an engineering career, see Cassedy, "Flamboyant Colonel Waring." For a general history of toilets, see Lawrence Wright, *Clean and Decent: The Fascinating History of the Bathroom and the Water Closet* . . . (New York: Viking Press, 1960), esp. pp. 200–216.

57 Hartshorne, *Our Homes*, pp. 72–73; Charles Williams Company, *Heating and Ventilating* (Philadelphia: Longacre & Co., 1873), p. 8, HC; *Boyd's Business Directory for Philadelphia, 1879* (Philadelphia, n.p., 1879), p. 932. From my reading of patent applications and reports, it appears that ventilators were first developed for industrial use and were then modified for the home, another example of a "technology transfer" from the public to the private sector.

58 "Report of committee on disinfectants," *Tr. A.M.A.*, 1866, *17*: 154. Preliminary reports of Sternberg's experiments were published in 1885, then reprinted in book form: The American Public Health Association, Committee on Disinfectants, *Disinfection and Disinfectants* (Concord, N.H.: Republic Press Association, 1888). In his study, Sternberg made a point of testing only substances commonly available from druggists in small packets, and scores of products were available in this manner. For a representative drug catalogue, see W. H. Schieffelin & Co., *General Prices Current of Foreign and Domestic Drugs, Medicines, and Chemicals* (New York: Holt Bros., 1885), HC. The disinfectant Labarraque's Solution was developed for use in Paris dissecting rooms, and carbolic acid gained in popularity after Lister's famous experiments in the operating room, providing additional examples of the "technology transfer" from hospital to home.

59 C. L. Cohn, *The Germicide Endorsed by Science and Experience* (n.p., [1882]), pp. 4, 5, HC. Italics are in the original. In 1875, patent no. 164,842 was awarded to George Jennings for an Apparatus for Disinfecting Water Closets; no. 166,135 went to Charles F. Parker for a Disinfecting Sick Room Slop Jar; and no. 168,972 went to Henry G. Dayton for a Pocket Disinfector and Inhaler. In 1877, patent no. 198,675 went to John H. Peterson for a Disinfecting Apparatus and Safety-Seat for Water-Closets. In 1879, patent no. 212,981 went to Abraham Rand for a Means for Ventilating Sewer-Pipes and Deodorizing the Foul Air Within. In examining the patent applications, I was struck by the number of simple disinfecting devices designed for privies, chamber pots, dry closets, and

the like, which suggest that people unable to afford a flush toilet were still trying to make their toilet facilities more hygienic.

60 Tayman's Disinfectant and Fumigating Co., *Tayman's Disinfectors and Fumigators* (Philadelphia: Privately printed, 1885), p. 4, HC. Under "Disinfectants," *Boyd's Business Directory for Philadelphia, 1879* lists two individuals in the business; in 1885, they were joined by the Reliable Disinfectant and Deodorizing Company. Tayman's company was incorporated that same year.

61 Sub-Merged Filter Company, Ltd., *A Perfect House Filter* (Philadelphia: n.p., [1885?]), p. 4, HC; Hyatt Pure Water Co., *The Hyatt System of Water Purification* (New York, n.d. [c. 1890]), HC. The Hyatt Company was based in New York City but had branches in other cities; about fifty private homes in Philadelphia are listed on pp. 50–52 of its promotional book. Brochures for the Gate City Stone Water Filter Co. of New York, and the Boston Water Purifier, can be found in the History of Medicine Division, National Library of Medicine. As in the case of ventilation, the technology of water filtration originated in the manufacturing sector and then was adapted for home use.

62 Cohn, *Germicide*, [p. 17]. The only device for which I did not find a do-it-yourself version was the sewer trap. Instructions for homemade ventilators and water filters were commonplace.

63 These observations are based on the Philadelphia Board of Health, Minutes of Meetings, 1888–1892, Board of Health Papers, Philadelphia City Archives, Philadelphia, Pennsylvania. The "Rules and Regulations Governing House Drainage, Ventilation and Cesspools in the City of Philadelphia . . . June 30, 1885," are tipped in the minutes for 20 Oct. 1891. The minutes are filled with listings of the "nuisances" reported in relation to private houses. A typical entry, dated 6 Oct. 1891, includes an unventilated drainage system, a defective connection between the house drain and the vertical soil pipe, a hopper water closet with insufficient water supply, and a defectively trapped water closet. *The Second Annual Message of Samuel H. Ashbridge, Mayor of the City of Philadelphia, with Annual Reports of . . . the Bureau of Health, for the Year ending December 31, 1900* (p. 119) gave statistical data about those who made complaints about nuisances; of approximately 19,000 complaints, 16,000 came from the public. In that report, the chief inspector of the Nuisance Division, Charles Kennedy, noted that despite improvements in the city's sanitary condition, the number of complaints continued to rise; he attributed this to "our people becoming better educated to the importance of sanitation" and looking to the board for "relief" (p. 119). For an

overview of the activities of the Philadelphia Board of Health in this period, see Edward Morman, "Scientific medicine comes to Philadelphia: public health transformed, 1854–1899" (Ph.D. diss., Univ. of Pennsylvania, 1986). Note that by 1885, Boston, Brooklyn, and New York City had plumbing regulations similar to Philadelphia's code. They are reprinted in *Plumbing Problems* (New York: Sanitary Engineer, 1885), pp. 225–236.

64   City of Philadelphia, *First Annual Message of Charles F. Warwick, Mayor of Philadelphia, with Annual Reports of . . . the Board of Health, for the year ending December 31, 1895*, p. 77.

65   "Annual Report of the Division of Disinfection," in *Second Annual Message . . . for 1900*, Appendix, p. 112. Morman, "Scientific medicine," pp. 196–198, discusses the activities of the disinfection division.

66   Michael P. McCarthy, *Typhoid and the Politics of Public Health in Nineteenth-Century Philadelphia* (Philadelphia: American Philosophical Society, 1987).

67   For data on declining rates of infectious diseases, see Condran and Cheney, "Mortality trends in Philadelphia."

68   Naomi Rogers, "Germs with legs: flies, disease, and the new public health," *Bull. Hist. Med.*, 1989, *63*: 599–617, and Richard A. Meckel, *Save the Babies: American Public Health Reform and the Prevention of Infant Mortality, 1850–1920* (Baltimore: Johns Hopkins Univ. Pres, 1990), discuss different aspects of the mass health education crusades of the early 1900s.

69   While there was a great deal of continuity in hygienic advice from 1870 to 1920, there were significant changes in the emphasis given certain practices, especially after tuberculosis became the chief "house disease." For example, the campaigns against spitting, common drinking cups, and flies reflect the influence of the threat of tuberculosis. I divide my study of popular hygiene into two periods, roughly 1870 to 1890, and 1890 to 1920, reflecting these changes in emphasis.

70   Maria Parloa, *Home Economics: A Guide to Household Management . . .*, new ed. (New York: Century Co., 1906).

71   Significantly, the new time-management ethos in homemaking, which became popular in the 1910s and 1920s, was devoid of overt concern about infectious disease. The one area in which sanitary concerns remained highly visible among urban middle-class women was in the shift from bulk to packaged goods. See, for example, Christine M. Frederick, *The New Housekeeping: Efficiency Studies in Home Management* (Garden City, N.Y.: Doubleday, Page, & Co., 1914).

72   Nancy Schrom Dye and Daniel Blake Smith, "Mother love and infant death, 1750–1920," *J. Am. Hist.*, 1986, *73*: 329–353. Schrom Dye and Smith suggest that medical writings that attributed infant mortality to poor mothering were one factor that may have increased maternal guilt. On Victorian mourning customs, see Martha V. Pike and Janice G. Armstrong, *A Time to Mourn: Expressions of Grief in Nineteenth-Century America* (Stony Brook, New York: Museums at Stony Brook, 1980).

73   This is an idea I heard my colleague Ruth Cowan express many years before I began this project. The personal testimonies of people in their fifties and sixties who have heard me talk on this subject strongly support this observation.

74   I develop this argument at more length in an unpublished paper, "Popular health education from tuberculosis to AIDS."

PUBLIC HEALTH AND MEDICAL THEORY

During the 1793 Philadelphia yellow fever epidemic physicians hotly debated the origins of the outbreak. On the one side were those who believed the epidemic spread from abroad and had been brought in with the immigrants. These "contagionists" advocated strict quarantines to keep disease at bay. The other side in the medical debate, the "anti-contagionists," believed that local causes, primarily contaminated water and rotting organic matter on the streets, caused the disease. They advocated sanitation projects, including water works, to improve the city's health. In both cases, a close connection between medical theory and desired public health action was evident.

Although physicians continued to believe that some diseases, such as smallpox, were transmitted directly from one person to another, the predominant 19th-century medical theory attributed rising mortality rates in American cities to the filthy urban environment. The popular "miasmatic" theory, linking dirt with disease, motivated much public health activity: street cleaning, garbage collection and disposal, water and sewer systems, and food regulation.

Polluted water sources presented a major public health problem. In 1801 Philadelphia opened its municipal water system, designed to bring fresh country water into the city. This action helped to save Philadelphia from devastation by cholera in 1832. New York and Boston, recognizing the benefits, followed suit with their own engineering projects in the 1840s that brought clean country water 40 or more miles into the cities. Other cities across the country, similarly basing their decisions on mainstream medical thinking about the relationship between· dirt and disease, copied the eastern models. Chicago, although located at the edge of Lake Michigan, an ample body of water, faced a peculiar drainage problem because of the city's flat terrain. Louis P. Cain describes the lengths to which that city went to supply its inhabitants with sanitary water and sewer systems. The Chicago hero Ellis Sylvester Chesbrough, like other engineers, achieved national prominence for his urban sanitation efforts.

As medical theories changed, and especially as the new science of bacteriology influenced thinking about the cause of infectious diseases, so, too, did public health responses. Sanitation lost its dominant position in the public health movement as germ theory changed the focus of health activity from cleaning the environment to tracking down specific bacteria. Polio, a disease of cleanliness and the product of earlier successes in sanitizing the cities, offers one example of the role medical thinking played in shaping responses to it. Typhoid fever and the new understanding of healthy carriers provides another. As Naomi Rogers explains for polio and Judith Walzer Leavitt demonstrates with regard to typhoid fever, however, medical theory itself was not sufficient explanation for determining public health approaches.

Today, medical and nonmedical factors continue to affect the success of public health activities, even in situations where medicine has proven efficacious. Penicillin, for example, can cure most forms of syphilis and gonorrhea, yet sexually transmitted diseases exist in epidemic proportions. Individual modesty or fear, sexual morality, the reluctance of physicians to report cases, and the question of individual rights all influence the incidence and spread of these diseases. Public health and medicine act and interact within a broad social context.

34

Raising and Watering a City:
Ellis Sylvester Chesbrough and Chicago's First Sanitation System

LOUIS P. CAIN

The engineers responsible for invention and mechanization in agriculture, manufacturing, and transportation are prominent historical figures, but few people are aware of the men who pioneered the sanitation systems so crucial to urbanization. As cities grew, their initial approaches to waste disposal and water supply proved unacceptable. As early as 1798 Benjamin Latrobe noted in his journal that the fresh groundwater which located the site of Philadelphia was befouled by the city's increasing population concentration. In Latrobe's opinion, Philadelphia's existing water-supply strategy was a major source of disease. Even before he assumed the responsibility for the city's new waterworks, Latrobe was convinced of the project's utility: "The great scheme of bringing the water of the Schuylkill to Philadelphia to supply the city is now become an object of immense importance, . . . though it is at present neglected from a failure of funds. The evil, however, which it is intended collaterally to correct is so serious and of such magnitude as to call loudly upon all who are inhabitants of Philadelphia for their utmost exertions to complete it."[1]

The emerging concentrations of population and manufacturing in the 19th century necessitated a reexamination of sanitation strategies. With urbanization, the haphazard approaches of the past could not guarantee pure water supplies and adequate waste disposal. Urban growth inevitably required the implementation of sanitation systems, and these systems, in turn, permitted further growth.

Students of Chicago's formative decades inevitably encounter the name of Ellis Sylvester Chesbrough; by studying Chesbrough, a student can focus on the truly unique character and contribution of Chicago's sanitation system. Chesbrough's works were the innovations most responsible for Chicago's unrestricted urban growth; they freed the city from the limitations imposed by an unfavorable natural topography. A flat, nonporous terrain, slightly elevated from Lake Michigan and the Chicago River, made drainage and absorption nearly impossible. In rainy weather, the topsoil became swamplike. Urban growth required a drainage system which could remove both surface water and household wastes. The natural depository for such a drainage system was Lake Michigan; however, the lake was simultaneously the city's natural water-supply source. Lake water had to be conserved if it was to be potable, and this meant it had to be protected from urban wastes. Fortunately, beginning in the 1850s, Chicago's city fathers recognized pollution as a serious threat to the city's health and took immediate action. This paper investigates how Chesbrough responded to Chicago's anomalous water-supply and waste-disposal needs in the 1850s and 1860s, and inquires into his engineering education to discover the antecedents of his innovative ideas.

I

Ellis Sylvester Chesbrough was born of Puritan ancestry in Baltimore County, Maryland, in July 1813. An unsuccessful business venture exhausted the family's means and suspended young Sylvester's education, and so, at nine years of age, he went to work. Between his ninth and fifteenth birthdays

LOUIS P. CAIN is Professor of Economics, Loyola University of Chicago, Chicago, Illinois.

Reprinted with permission from *Technology and Culture*, 1972, *13*: 353–372. Copyright © 1972 by the Society for the History of Technology, and published by the University of Chicago Press.

Chesbrough spent only a year in a classroom, but he did find time outside his counting-house duties to pursue his studies. Chesbrough acquired most of his basic education without the benefit of formal training or a regular teacher, and the same was true of his engineering education.

In 1828 Chesbrough's father took a job with a railroad engineering company employed by the Baltimore and Ohio Railroad Company. Through the father's influence, the son gained employment as a chainman with a similar company engaged in preliminary surveying work in and about Baltimore.[2] Chesbrough's company was under Lieutenant Joshua Barney, U.S. Army, and most of the engineers were army officers, many of them graduates of the U.S. Military Academy's practical, as opposed to theoretical, engineering course.[3] Chesbrough was fortunate in being affiliated with several of the army's most prominent engineers. In 1830–31 he worked as an assistant engineer to Colonel Stephen H. Long.[4] Near the end of 1831, Chesbrough joined the engineering corps of Captain William Gibbs McNeill, where he served immediately under Lieutenant George W. Whistler.[5]

The Panic of 1837 and the resulting depression dealt a hard blow to the country's internal improvement's bubble, and Chesbrough, like many other engineers, found himself out of work as the flow of funds dried up in the early 1840s. He went to his father's residence in Providence, Rhode Island, where, during the winter of 1842, he spent his leisure time in the workshop of a nearby railroad learning the practical use of tools. The following year he purchased a farm adjacent to one owned by his father in Niagara County, New York. His venture into farming was mercifully brief; after an unsuccessful year, Chesbrough gladly returned to engineering.

In 1846 Chesbrough was offered the position of chief engineer on the Boston Water Works' West Division. This position completed his engineering education. Up to this time, all his experience was related to railroad engineering, and he had mastered many civil engineering essentials, such as grading, tunneling, and surveying. Chesbrough was reluctant to accept the Boston position because he considered himself unacquainted with hydraulic engineering. His friends and Boston's water commissioners implored him to accept the position, and, after being assured John Jervis's counsel, Chesbrough assented.

There was good reason for Chesbrough to consider an association with Jervis valuable. Jervis had been active in every phase of engineering, particularly those dealing with hydraulics. Jervis was a product of the New York canal system and had learned hydraulic engineering on the job by working on the Erie Canal. In 1846 Jervis was appointed consulting engineer on the Boston Water Works, with Chesbrough the chief engineer. Jervis had the responsibility for designing both the Cochituate Aqueduct and the Brookline Reservoir; Chesbrough, the responsibility for supervising the execution of Jervis's plans.[6] In 1850 Chesbrough became sole commissioner of Boston's water works, and a year later, he became Boston's first city engineer.

The United States' early experience with internal improvements and the education of engineers coalesced in Chesbrough's career. He learned civil engineering from some of the army's most competent engineers. He learned hydraulic engineering from Jervis, perhaps the most competent engineer trained by the New York canal system. The education and experience which Chesbrough utilized in freeing Chicago from its topographical liabilities and in implementing an effective sanitation system grew out of his first-hand experience with many of the country's internal improvements.

II

In the early 1850s Chicago's random waste disposal methods led to a succession of cholera and dysentery epidemics. The Illinois legislature created the Chicago Board of Sewerage Commissioners on February 14, 1855, to combat what was generally conceded to be an intolerable situation.[7] The commissioners sought "the most competent engineer of the time who was available for the position of chief engineer."[8] Their selection, E. Sylvester Chesbrough, resigned his position as Boston's city engineer and came to Chicago.[9] Immediately after accepting the position, Chesbrough submitted a report in which he outlined his plan for a sewerage system designed to solve Chicago's drainage and waste-disposal problem. His plan represents the first comprehensive sewerage system undertaken by any major city in the United States. He had learned about sewer construction, grading, and "building-raising" from different sources. Now he merged them and "pulled Chicago out of the mud."

Prior to Chesbrough's arrival, Chicago's sewerage commissioners solicited the public for plans and suggestions. Thirty-nine proposals were received, and, although the board claimed Chesbrough utilized many of these suggestions, he did not use any of the proposals in its entirety.[10] Chesbrough's task was to construct a sewerage system whose main objective was to "improve and preserve" the city's health. In his opinion, the existing privy vaults and drainage sluices were "abominations that should be swept away as speedily as possible," and that "to construct the vaults as they should be, and maintain them even in a comparatively inoffensive condition, would be more expensive than to construct an entire system of sewerage for no other purpose, if the past experience of London and other large cities was any guide for the future of Chicago."[11]

Chesbrough's 1855 report to the Board of Sewerage Commissioners made several references to the sewers of New York, Boston, and Philadelphia. Additionally, the report showed that Chesbrough was familiar, through his reading, with the sewers of London, Paris, and other European cities. It is important to remember, however, that not one U.S. city at that time had a comprehensive sewerage system, even though most had sewers. Consequently, Chesbrough had to rely on his training and intuition in assessing sewerage system alternatives.

Chesbrough's 1855 report considered four possibilities: (1) drainage directly into the Chicago River and then into Lake Michigan; (2) drainage directly into Lake Michigan; (3) drainage into artificial reservoirs to be pumped and used as fertilizer (sewage farming); and (4) drainage directly into the Chicago River, and then by a proposed steamboat canal into the Des Plaines River. Although this fourth possibility was the method which Chicago eventually adopted (the Chicago Sanitary District's Sanitary and Ship Canal), the city's 80,000 inhabitants in 1855 did not warrant the expense which this alternative involved.

Chesbrough recommended the first plan.[12] This is not to say he failed to realize that his preferred method was a potential health hazard, particularly during the warmer months, and might obstruct river navigation by making the waterways shal-

lower.[13] Chesbrough discussed the objections to his recommended alternative:

> It is proposed to remove the first [health hazard] by pouring into the river from the lake a sufficient body of pure water into the North and South Branches to prevent offensive or injurious exhalations. . . . The latter objection [obstruct navigation] is believed to be groundless, because the substances to be conveyed through the sewers to the river could in no case be heavier than the soil of this vicinity, but would generally be much lighter. While these substances might, to some extent, be deposited there when there is little or no current, they would, during the seasons of rain and flood, be swept on by the same force that has hitherto preserved the depth of the river.[14]

Apparently, Chesbrough did not realize that spring freshets and floods might force the sewers' accumulations into the lake in such a way as to pollute the city's water supply. This is somewhat surprising, as the basic sanitation principle of the day was to locate the eventual sewage outlet as far from the water-supply source as possible.

Chesbrough had three objections to the second possibility, drainage directly into Lake Michigan. First, it would require a greater sewer length and, consequently, would incur greater cost. Second, he supposed that this plan would seriously affect the water supply, if any sewer outlets were located near the pumping station. At this time, Chicago's water-supply intake was located a short distance offshore at the Chicago Avenue lakefront, approximately one-half mile north of the Chicago River's mouth. Chesbrough did not elaborate on this objection. Third, he felt drainage into the lake would create difficulties in preventing sewer outlet injury during stormy weather, or snow and ice obstruction during winter.[15]

Sewer farming was rejected in part because of the uncertainty whether future fertilizer demand would be sufficient to cover distribution costs. Further, Chesbrough was uncertain as to both the needed reservoir capacity and the expense of building the necessary reservoirs. Finally, Chesbrough thought there would be a great health hazard created by foul odors emanating from sewage spread over a wide surface.

Chesbrough termed the use of a steamboat canal, not yet constructed to flush the sewage into the Des Plaines River, the fourth possibility, "too remote." Although he was aware of the "evils" which would result when raw sewage passed into Lake Michigan, Chesbrough felt it impossible to create an outlet to the southwest. Brown claims, however, that "he appears to have believed that this would be the ultimate solution of the sewerage problem," as, in fact, it was.[16]

> With regard to the fourth plan . . . which would divert a large and constantly flowing stream from Lake Michigan into the Illinois River, it is too remote a contingency to be relied upon for present purposes; besides the cost of it, or any other similar channel in that direction, sufficient to drain off the sewage of the city, would be not only far more than the present sewerage law provides for, but more than would be necessary to construct the sewers for five times the present population. Should the proposed steam-boat canal ever be made for commercial purposes the plan now recommended would be about as well adapted to such a state of things, as it is to the present.[17]

Certainly his plan was readily adaptable to such a scheme. The Sanitary District of Chicago was created in 1889 for the express purpose of implementing this fourth possibility. The Sanitary District then constructed the "proposed steamboat canal," which unquestionably was beyond the means of Chicagoans in 1855.

In December 1855 Chesbrough submitted his plan for Chicago's sewage disposal and drainage. Under this plan, all of the sewage of Chicago's west division, all the sewage of the north division except for the lakefront area, and about one-half the sewage of the south division was deposited in the Chicago River. This sewage passed from the river into Lake Michigan. The dividing line in the south division was State Street; the area east of State Street drained directly into the lake. As the area east of State Street was primarily residential, Chicago's business district was sewered into the river. This district, west of State Street, included the majority of Chicago's packinghouses, distilleries, and hotels. Thus, the river received large quantities of pollutants daily.[18]

The sewers themselves were outstanding phenomena. Brick sewers, three to six feet in diameter, were laid above the ground down the center of the street. Chicago's topography, being unusually flat, was unfavorable to sewer construction. The Chicago River banks were only two feet above the water level. Near the river's north and south branches, the ground level reached a maximum of 10 to 12 feet above the lake. In reality, the task of constructing underground sewers required raising the city.[19] From the beginning, Chesbrough insisted that a high grade was necessary for proper drainage and dry streets. Chicago lacked this high grade, and, thus, the decision to raise the city's level, concomitant with sewer installation, was one which solved the waste disposal and drainage problem in the context of Chicago's existing topography and future necessities.[20]

The Chesbrough plan called for an intercepting sewer system which emptied into the Chicago River. The sewers were to be constructed on the combined system; that is, they would collect sewage from both buildings and streets. This was consistent with the best contemporary thinking and practice. As sewer construction progressed away from the river, the streets had to be raised beneath the sewers. After the sewers were laid, earth was filled in around them, entirely covering them. The packed-down fill provided roadbeds for new, higher streets. These streets were rounded in the center, with gutter apertures leading to the sewer. Such streets would stay dry and could be paved, as contrasted to the mud which had plagued the city previously.

A second facet of Chesbrough's sewerage plan involved dredging the Chicago River. The river had been dredged previously, but it was still too small to handle the anticipated sewage load. Chesbrough planned to widen and deepen the river, as well as to straighten its meandering course. Contracts for this work had been let to the partnership of John P. Chapin and Harry Fox. It was Fox who suggested using the dredgings from the river as fill around the sewers.[21]

It is interesting to digress on the consequences of Chesbrough's plan to raise the city. Where vacant lots existed, they were filled to the new level. A few old frame buildings were torn down, and the lots filled. It proved relatively easy to raise frame buildings to the new level, if the owners could afford it. The city's newer buildings were brick and stone, however, and they were constructed on the old level. These newer buildings would not be torn down, and many of Chicago's homes and offices were to be left "in the hole."

When new buildings and sidewalks were constructed on the higher level, Chicago increasingly became a city built on two levels.[22] Legal attempts to maintain the lower level were uniformly settled in favor of the city and its new level.[23]

The raising of brick buildings proved to be a difficult proposition. George Pullman, who later became famous for his "Palace cars," devised and instituted a method to raise brick buildings.[24] Pullman first used his method in connection with the Erie Canal enlargement of the 1850s, so Chesbrough would have known that the problems concomitant with raising the city's grade were surmountable. One of Pullman's biographers described his activities during those years:

> He made contracts with the State of New York for raising buildings on the line of the enlargement of the Erie Canal, which occupied about four years in their completion. At the end of that time, in 1859, he removed to Chicago, and almost immediately entered upon the work, then just begun, of bringing our city up to grade by the raising of many of our most prominent brick and marble structures, including the Matteson and Tremont Houses, together with many of our heaviest South Water street blocks. He was one of the contractors for raising by one operation, the massive buildings of the entire Lake street front of the block between Clark and LaSalle streets, including the Marine Bank and several of our largest stores, the business of all these continuing almost unimpeded during the process — a feat, in its class, probably without a parallel in the world.[25]

The Tremont Hotel was the first brick building which Pullman raised in Chicago. Soon his method was utilized to raise all Chicago's brick buildings from their former muddy level. The work required years. No one knows the cost, but it has been estimated at $10,000,000.[26]

In December 1856 the sewerage commissioners sent Chesbrough to visit several European cities in order to discover if their sewage disposal techniques were relevant to Chicago's needs.[27] Chicago was taking an open-minded approach to this question, and, evidently, the city was prepared to adopt an unconventional approach if it proved to be the best solution. The report of this trip, which Chesbrough submitted in 1858, represents one of the first sanitary engineering treatises.[28] Chesbrough visited and reported on the sewerage of Liverpool, Manchester, Rugby, London, Amsterdam, Hamburg, Paris, Worthing, Croydon, Leicester, Edinburgh, Glasgow, and Carlisle. He concluded that none of these cities furnished an exact criterion to judge the effects of disposing sewage directly into the Chicago River, but he felt their collective experience suggested that it probably would be necessary to keep the river free of sewage accumulations.

Chesbrough ended his report by relating the European experience to Chicago's sewerage needs. Two points which Chesbrough made in this concluding section are worthy of special mention.[29] The first is the experience of Worthing, "a small watering town on the southern coast of England." At one time this town of 5,000 had drained directly into the sea, "but owing to offensive smells caused by this practice, and the consequent injury to the reputation of town as a watering place, upon which its prosperity very much depends," Worthing decided to find an alternative sewerage scheme.[30] Chesbrough concluded that Worthing's experience "shows that the mere discharge of filth into the sea gives no security against its being cast back in a more offensive state than ever, especially when the prevailing winds are toward shore," and that this suggests "the possibility of creating on the lake shore as great a nuisance as would be taken from the river."[31]

Second, Chesbrough included a prophetic paragraph which could serve as a summary to Chicago's sanitary history for a half-century thereafter:

> Under these circumstances it seems advisable to do nothing with regard to relieving the river at present, nor towards carrying out that portion of the plan which provides for forcing water from the lake into it, during the summer months. Should the Canal Company [the Illinois and Michigan Canal] not be obliged to pump enough during warm weather to keep the river from being offensive, it is understood that they would pump as much as they could for a reasonable compensation. This would furnish some criterion by which to judge of the probable effect of a still greater quantity driven in from the lake, according to the plan. The thorough [sic] cut for a steamboat canal, to the Illinois River, which the demands of commerce are calling more and more loudly for, if ever constructed, would give as perfect relief to Chicago as is proposed for London by the latest intercepting scheme.[32]

The Chicago River's south branch became quite polluted shortly after sewage was admitted into it. The Illinois and Michigan Canal's pumps, however, utilized south branch water to provide the canal's summit level, and, consequently, the pumps relieved a portion of the river's pollution load. The real significance of Chesbrough's statement lies in the fact that, as early as 1858, Chicagoans recognized the Illinois and Michigan Canal's sewage disposal potential.[33] In following years, the canal's pumps were used regularly to relieve the pollution load. Further, the canal itself was deepened and additional pumps were installed to increase the canal's capacity for handling sewage. Finally, the Chicago Sanitary District was formed in order to construct a new and enlarged canal to service Chicago's waste disposal needs, as Chesbrough had prophesied.

In 1861 the Board of Public Works was formed by incorporating the duties of the Board of Sewerage Commissioners, the Board of Water Commissioners, and other miscellaneous departments. Chesbrough was named chief engineer of this new board and, consequently, inherited the water-supply problem in addition to the waste-disposal problem. His inheritance was the "vicious circle" created by Lake Michigan's dual role as water supplier and eventual waste disposer.

III

Chicago's continued population growth through the decade of the 1850s, the new sewerage works, and the expansion of packinghouses and distilleries had increased the number of pollutants drained into the Chicago River. Lake Michigan soon became fouled by the river's influx, and Chicagoans began to complain of the public water supply's offensiveness and pollution. The existing water intake was a wooden pipe which extended a few hundred feet out into Lake Michigan, one-half mile north of the Chicago River's mouth. In 1859, one of Chicago's water commissioners "proposed to sink a wrought iron pipe . . . one mile out into the lake, to obtain the supply from a point which could not be affected by the river."[34] Chesbrough was asked to study and report on the commissioner's plan, and to do the same on "erecting additional pumping works, in such locality as shall secure a supply of pure water."

Chesbrough's report discussed several methods without making a specific recommendation. Even at this early date, however, he considered a tunnel under the lake to be the most desirable alternative. Chesbrough was not afraid to combine grading, tunneling, and hydraulic principles to create a new water-supply system. When he later offered plans for a lake tunnel, his innovative proposal drew considerable opposition at the start and unmitigated acclaim when it proved successful.

Shortly after its formation in 1861, the Board of Public Works adopted as its goal the acquisition of an unpolluted water supply. Consequently, the board requested Chesbrough to make a canvass of the various water-supply possibilities and to investigate several filtration methods. Chesbrough dismissed the existing filtration methods as inadequate; his studied opinion was that the tunnel method was the most desirable:

> The engineer of the Board [E. S. Chesbrough], after much doubt and careful examination of the whole subject, became more inclined to the tunnel plan than any other, as combining great directness to the nearest inexhaustible supply of pure water, with permanency of structure and ease of maintenance. The possibility, and, in the estimation of many, the probability of meeting insuperable difficulties in the nature of soil, or storms, or ice on the lake, were fully considered. One by one the objections appeared to be overcome, either by providing against them, or discovering that they had no real foundation.[35]

Chesbrough continued to explore the tunnel plan's potential. When he had worked out the details, a proposal was submitted to several engineers, all of whom considered the tunnel plan to be feasible. Nevertheless, the 1861 board was against adopting the project. After a new board was elected and additional soil examinations had been made, Chesbrough's water-supply tunnel plan was adopted. The new board reported:

> What is most to be desired by the city is, that the supply should be drawn from the deep water of the lake, two miles out from the present Water Works. . . . The careful investigation of the subject has satisfied us sufficiently to say, that with our present knowledge, we consider it practicable to extend a tunnel of five feet diameter the required distance under the bed of the lake, the mouth or inlet to such a conduit being the out-

most shaft, protected by a pier [crib], which will be used in the construction of the tunnel.[36]

In their 1863 report, the Board of Public Works noted that three projects had been considered, any one of which would have afforded Chicago a healthier and better protected water supply. These were (1) a two-mile lake tunnel, (2) a filtering or settling basin, and (3) a one-mile lake tunnel located five miles to the north.[37] The board had two principal objections to the second plan. First, they commented:

> For settling and filtering the water from sediment, we are of the opinion that the basin would be found effective, and would continue to be so, but that for filtration it is not safe to rely upon it. There have been filtering basins of the character in other places. Some of them appear to have continued to work well during long use, and others have failed and become useless.[38]

Second, the board objected to the basin scheme because the water supply intakes would still be in the shallow water close to shore, and would not be located in a deeper point where the water was considered to be better.

Chesbrough's 1863 report acknowledged that the board had considered the three most promising possibilities and had rejected one; he was to assess the remaining two. Almost immediately he dismissed, on the grounds of greater cost, any project which required moving the existing water works, such as the board's third proposal:

> Other projects, such as erecting a new pumping works at Winnetka, or going to Crystal Lake and bringing a supply thence by simple gravitation, as is done for cities of New York, Boston, Baltimore, and Albany, have been considered, but their great cost, as compared with that of obtaining an abundant supply of good and wholesome water at points much nearer the city, is deemed a sufficient apology for not discussing their details here.[39]

Chesbrough concerned himself only with those plans which would bring water from a point two miles east of the existing Chicago Avenue Water Works, and there were two of these:

> Of the plans proposed for obtaining water from the lake, where it will be free from not only the wash of the shore, but from the effects of the river,

two classes only have been considered; one, an *iron pipe with flexible joints*; and the other, *a tunnel under the bottom of the lake*.[40]

Although the cost of the iron pipe project was slightly less than the tunnel project, Chesbrough chose between them on other than an initial cost basis:[41]

> In consequence of the possibility of such a pipe being injured by anchors, by the sinking of a heavily loaded vessel over it, or by the effect of an unusual current in the lake moving it from its place, it has been thought preferable to attempt the construction of a tunnel under the bottom of the lake.[42]

His research had convinced him that the tunnel's construction would be less difficult than was generally supposed. Lill and Diversey's brewery, adjacent to the water works, was the site of artesian borings which showed that, between 25 and 100 feet deep, the ground at the lake shore was a clay which was also found on the lake bottom where the water was 25 feet deep. A tunnel could easily be constructed in this type of clay, if it were continuous. Chesbrough was confident that the clay was continuous, but he admitted he was uncertain whether beds of sand might not be interspersed with the clay.[43]

The lake shaft was to be formed by sinking iron cylinders to the desired depth. Chesbrough noted that this was not a difficult problem in that the pneumatic process had been successfully employed on "the Theiss bridge in Hungary, and the railroad bridge across the Savannah River . . . and recently the Harlem bridge in New York."[44]

In giving cost estimates for the tunnel project's component parts, Chesbrough clearly showed the sources of his research. The principal source was the Thames tunnel, and Chesbrough noted that the first thoughts of most people were the great construction difficulties and "enormous" costs which had been encountered on the Thames project. He was quick to refute these thoughts and countered that "as we have every reason to believe, the clay formation here would shield us from such inroads of water as were met within the Thames tunnel operation."[45] In estimating excavation costs, Chesbrough made the same point: "There is good reason to believe that nothing in the soil here would be more difficult than that through which the sewers of London are sometimes tunneled."[46]

Chesbrough also used the Thames experience, plus that of the Boston Water Works tunnel, to estimate masonry costs. Cribs had been used principally in pier and breakwater construction, and Chesbrough based his crib cost estimates on figures which had been made for a proposed breakwater in Michigan City, Indiana, at the bottom of Lake Michigan.

After reaching his cost estimate for masonry and excavation, Chesbrough compared it with figures which had been reached for other major tunnel projects.[47] In particular, Chesbrough referred to reports from (1) the commissioner of the Troy and Greenfield Railroad, and (2) the Hoosac Tunnel. Included in the commissioner's report was the report of Charles Storrow, who had been sent to investigate European tunnels. Because the tunnels which Storrow had studied were for railroads, they were all much larger than the one which Chesbrough was planning. Therefore, Chesbrough estimated the cost of each tunnel had it been constructed with a five-foot width. From these estimates, he concluded that his cost estimate for the proposed water tunnel was reasonable.

The engineering achievement involved in constructing the water-supply system was no less significant than that represented by Chesbrough's sewer system. As conceived, the task was to dig a shaft near the lake shore to a depth significantly below the lake bottom and then burrow two miles beneath the lake. A similar shaft was to be dug at the lake end and was to be protected by a crib. The engineering problem was to connect the shore and lake points by a straight line 69 feet below the surface of Lake Michigan. Contemporary compasses could not be used since, below ground level, local attraction rendered them inaccurate. To a worker in the tunnel, the only place where the direction of the line drawn between the two shafts could be observed was at the top of either shaft. Consequently, when the engineers attempted to run the tunnel's axis parallel to this imaginary line on the lake's surface, they ran into difficulties affecting the turn from shaft to tunnel.[48]

When the lake shaft was completed, workers were lowered to begin burrowing westward to meet with the other workers burrowing eastward. The tunnel was sloped two feet per mile from the lake end to the shore so that it could be emptied should repairs prove necessary; the water would be shut off at the lake end. Although the methods were primitive — the tunnel was dug entirely by manual labor — it was claimed that the workers caused the two tunnel sections to meet within one inch of achieving a perfectly smooth wall.[49]

Chesbrough's engineering competence was coupled with a sense of economic reality, and these traits combined to insure the reputation he earned in Chicago. His 1863 report contained a section on "plans for improving the Chicago river." Chesbrough knew that moving the water-supply intake farther into the lake would not improve the river's offensive condition. In the 1855 sewerage report, he had argued that flushing canals would be necessary in both the north and south branches to purify the river, and he restated this position in several reports thereafter. By raising the issue once again, Chesbrough not only demonstrated the completeness of his approach, but also what one memorialist called "the characteristic firmness of conviction and modest persistence of Mr. Chesbrough."[50]

As before, Chesbrough's methodology was to enumerate and evaluate the possibilities for improving the river: (1) north and south branch flushing canals, (2) Des Plaines River diversion into the south branch, and (3) drainage southwest into the Illinois River Valley. The first was preferred because Chesbrough felt it was "undoubtedly feasible, would be completely under the control of the city, and there is every reason to believe [it] would be effectual."[51] He considered the second plan "defective" in that the Des Plaines River's flow was least when the Chicago River's pollution was greatest. Although Chesbrough correctly assumed that the third project would be the ultimate solution, he rejected it as "requiring much larger means than the Board can at present control."[52] Chesbrough's attention to Chicagoans' ability to pay established him as a practical man and lent credence to his innovative ideas. His consideration of a sanitary canal connecting the Chicago and Illinois Rivers indicates Chesbrough had learned that water-supply and waste-disposal problems are interdependent and must be solved simultaneously.

IV

Chicago is an urban center which had, and still has, serious water pollution problems. Lake Michigan's present pollution problem is primarily the result of industrial discharge in the Calumet and Indiana Harbor areas and the discharge of inadequately treated sewage by the North Shore Sanitary District (Lake County, Illinois) and several Wisconsin cities. Under normal circumstances, the Metropolitan Sanitary District of Greater Chicago diverts the sewage and the treated effluent from Lake Michigan. Presently, Chicago is meeting its responsibility with respect to Lake Michigan pollution. On the other hand, both the Chicago River and the Illinois River valley are polluted because some industries in the Chicago area still discharge their wastes into the water and the Sanitary District falls short of 100 percent treatment. Approximately 10 percent of the sewage goes untreated at this time, but it is the district's stated objective to achieve 100 percent treatment in the 1970s.[53] While these few sentences oversimplify a very complex situation, the outline is apparent. Chicago must seek outside help to reduce Lake Michigan pollution and the consequent threat to the city's water supply. Chicago and its Cook County suburbs, by themselves, could significantly reduce pollution in the Chicago, Des Plaines, and Illinois rivers.

When faced with Lake Michigan and Chicago River pollution in the 1850s and 1860s, Chicagoans had sought the best solutions available. Cost considerations had entered the argument only in deciding among equally effective methods; Chicagoans were not reluctant to pay the price necessary to secure sanitary conditions. They indebted the city through bond issues and themselves through tax assessments in order to finance these public works. Muddy streets and impure water were manifest physical representations of the city's problems, and solutions to these benefitted the city's residents, individually and collectively. The public's acceptance of an increased tax burden to finance these works must be viewed as public recognition of the problems' dimensions. If the city's water supply had not been conserved, and if the

city's natural topography had not been improved, Chicago's urban growth would have been severely limited.

When the pollution problem is explored in a historical context, students will find that the objectives which Chesbrough sought — minimize pollution and obtain a pure water supply — are the same as today's objectives. Nineteenth-century engineers, however, were not faced with the imminent "death" of large bodies of water; they were faced only with protecting urban populations from polluted water supplies.

In studying Chesbrough's works in Chicago, one gets the impression that today's pollution problem is not the result of ignorance as to pollution's effects, but ignorance as to how deadly the pollution load has become. In many cases, techniques first utilized in the 1850s and 1860s are still used today. Although these techniques no longer solve the problems for which they were intended, their inadequacies did not become apparent until recently. Perhaps this is because the demands on these techniques were much less heavy during the earlier period than they now are. Perhaps it is because the engineers of Chesbrough's generation made such dramatic innovations that the declining effectiveness of these techniques and improvements just recently became evident to sanitary engineers and laymen. Or perhaps it is because the 20th-century sanitary engineers who recognize the problem are unable to communicate the necessity for action. While the technology and technicians have been available, an uninformed and apathetic public has not invested sufficient capital in pollution control. Whatever the case, through inaction, the cost of proper treatment has reached a price which may be greater than the public is willing to pay. Unfortunately, the 20th century has been unable to find a sanitary engineer with the same farsightedness in his method, and resoluteness in seeing his proposals adopted, as that characteristic of Ellis Sylvester Chesbrough.

NOTES

1   Benjamin Henry Latrobe, *The Journal of Latrobe* (New York, 1905), p. 98.
2   The engineering education of E. S. Chesbrough began in this company, and he quickly proved an apt student. See *Biographical Sketches of the Leading*

Men of Chicago, written by the Best Talent of the Northwest (Chicago, 1868), p. 192; see also *Proc. Am. Soc. Civil Engrs.,* November 1889, *15:* 161.
3   See Daniel H. Calhoun, *American Civil Engineer: Origins and Conflict* (Cambridge, Mass., 1960), p. 38;

and Forest Hill, *Roads, Rails and Waterways* (Norman, Okla., 1957), pp. 12 ff.

4   Of particular interest to Chesbrough's later career is the fact that Long had carried out extensive exploratory surveys in the West. In 1816 Long was asked to report to the federal government on the physiographic features in the region of a proposed canal between Lake Michigan and the Illinois River. Although it is only speculation, one wonders how much knowledge of Chicago's topographical peculiarities Long passed on to Chesbrough. It is known that Long prepared detailed reports of his visit to Chicago. See Richard George Wood, *Stephen Harriman Long* (Glendale, Calif., 1966).

5   The major supply of engineers developed from what Calhoun called "the persisting pattern of on-the-job training." The supply provided by the leading scholastic source, the U.S. Military Academy, and the leading civilian source, the New York State canal system, was insufficient. The engineers of that day were active builders; thus, some form of on-the-job training had to be inaugurated to increase the supply and meet the demand. What developed was a hierarchical engineering corps. Lacking any formal education, Chesbrough learned every phase of his job by working his way up the civil engineering hierarchy.

In addition to the books by Hill and Calhoun (see n. 3 above), other recent books which discuss the oral transmission of engineering knowledge are Stephen Salsbury, *The State, the Investor, and the Railroad: The Boston and Albany, 1825–1867* (Cambridge, Mass., 1967); Harry N. Scheiber, *Ohio Canal Era* (Athens, Ohio, 1969); and Ronald E. Shaw, *Erie Water West* (Lexington, Ky., 1966).

6   Chesbrough's role in the Cochituate works is mentioned in a study of the waterworks of Boston, New York, Philadelphia, and Baltimore by Nelson M. Blake (*Water for the Cities* [Syracuse, 1956]).

7   The board was empowered to (1) supervise the drainage and sewage disposal of Chicago's three natural divisions; (2) plan a coordinated system for the future; and (3) issue bonds, purchase lots, and erect buildings implementing their plan. The board's actions were made subject to the Chicago City Council's approval. The act is summarized in several works including G. P. Brown, *Drainage Channel and Waterway* (Chicago, 1894), p. 50.

8   A. T. Andreas, *History of Chicago from the Earliest Period to the Present Time* (Chicago, 1884), I, 191; Bessie Louis Pierce, *A History of Chicago* (New York, 1940), II, 330; Soper, Watson, and Martin, *A Report to the Chicago Real Estate Board on the Disposal of the Sewage and Protection of the Water Supply of Chicago,*

Illinois (Chicago, 1915), p. 69, hereafter referred to as the *C.R.E.B. Report.*

9   It is quite possible that Jervis played a significant role in Chicago's choice, for during the early 1850s Jervis was professionally engaged in the Chicago area. Chicago's city fathers would have been aware of Jervis's engineering reputation, and it is probable that he was consulted regarding chief engineer candidates. Because he had worked with Chesbrough just prior to this, it is likely that Jervis gave Chesbrough an excellent recommendation.

In 1881 Chesbrough, serving as consulting engineer of the New Croton Aqueduct, employed Jervis, who discussed the work with Chesbrough daily. This indicates the esteem in which Chesbrough held Jervis, for Jervis was then 86 years old. Chesbrough, at 68 years of age, belonged to another generation.

10 Although the commissioners' report mentions the public's proposals, it does not indicate what they were, or even which parts of Chesbrough's plan were adapted from these suggestions.

11 Brown, *Drainage Channel,* p. 53.

12 *Report and Plan of Sewerage for the City of Chicago, Illinois,* adopted by the Board of Sewerage Commissioners, Dec. 31, 1855, hereafter referred to as the *1855 Report.* Also quoted in *C.R.E.B. Report,* p. 71. Chesbrough had a systematic approach to costs, but a very general approach to benefits. This evidently was consistent with the approach adopted on other U.S. internal improvement projects. See Lawrence G. Hines, "The early nineteenth century internal improvement reports and the philosophy of public investment," *J. Econ. Issues,* December 1968, *2:* 384–392.

13 Chesbrough planned to pump sufficient lake water into the north and south branches of the Chicago River to flush offensive solid pollutants. He also proposed flushing the sewers as well. See reprinted article, Langdon Pearse, "Chicago's quest for potable water," *Water and Sewage Works,* May 1955, *3.*

14 *1855 Report.* Also quoted in Andreas, *History of Chicago,* I, 191.

15 *C.R.E.B. Report,* p. 72; Pearse, "Chicago's quest," p. 3.

16 Brown, *Drainage Channel,* p. 53. To be precise, the Sanitary District's Sanitary and Ship Canal was the last step in Chicago's adoption of the dilution method. Ultimately, Chicago's growth was sufficient to require sewage treatment in addition to dilution.

17 *1855 Report.* Also quoted in Brown, *Drainage Channel,* p. 55.

18 R. Isham Randolph, "A history of sanitation in Chicago," *J. Western Soc. Engrs.,* October 1939, *44:*

229; Richard S. Kirby and Philip G. Laurson, *The Early Years of Modern Civil Engineering* (New Haven, 1932), p. 234; George W. Rafter and M. N. Baker, *Sewage Disposal in the United States* (New York, 1894), pp. 169–170.

19 *C.R.E.B. Report,* p. 69.

20 The grade which the city council adopted was lower than Chesbrough advocated, but it was sufficiently high to permit the construction of 7–8 foot cellars. The council's decision was to raise the grade to 10 feet on streets adjacent to the river; Chesbrough's higher grade was rejected because the city fathers felt there would be difficulties in locating sufficient fill. See *C.R.E.B. Report,* p. 70.

21 *Biographical Sketches,* p. 482. Fox's company was responsible for almost every topographical improvement in the Chicago area. The company deepened the Chicago River, developed the Chicago Harbor, installed road and railroad bridges, dredged the Illinois and Michigan Canal, and then performed similar services throughout the Midwest.

22 Randolph, "Sanitation in Chicago," p. 229. For many years, some sewers lay wholly above the ground, at the same level or higher than adjoining buildings.

23 "Up from the mud: an account of how Chicago's streets and buildings were raised," compiled by Workers of the Writer's Program, W.P.A. in Illinois for Board of Education, 1941. The raising of cities was relatively common. It was pointed out to me that all of downtown Atlanta was "raised" by the construction of roadways.

24 *Ibid.*

25 *Biographical Sketches,* p. 472. See also Seymour Currey, *Chicago: Its History and Its Builders* (Chicago, 1962), Vol. III; and Stanley Buder, *Pullman: An Experiment in Industrial Order and Community Planning, 1880–1930* (New York, 1967).

26 Lloyd Wendt and Herman Kogan, *Give the Lady What She Wants* (Chicago, 1952), p. 57. Wendt and Kogan do not say how they arrived at this number, and give no reference. Pullman reportedly received $45,000 for raising the Tremont Hotel. At $45,000 per brick building, $10,000,000 will raise over 200 buildings. This is probably an overestimate of the number of buildings raised, but the large number of other expenditures, including Chicago River dredging and legal expenditures, suggest that the $10,000,000 figure is an underestimate.

27 *Report of the Results of Examinations Made in Relation to Sewerage in Several European Cities, in the Winter of 1856–57,* published in Chicago by the Board of Sewerage Commissioners (1858), p. 3, hereafter referred to as the *1858 Report.* See also Randolph,

"Sanitation in Chicago," p. 229; Brown, *Drainage Channel,* p. 57.

28 *1858 Report,* p. 92. Chesbrough's memorialist in the *Proc. Am. Soc. Civil Engrs.* (November 1889, *15:* 162), unhesitatingly assessed the significance of Chesbrough's European trip report: "The importance of this report and the influence it exerted . . . can hardly be estimated. At the time the report was written, there was not a town or city in the United States that had been sewered in any manner worthy of being called a system. This being, perhaps, the first really exhaustive study, which the subject had received on this side of the water, and Chicago being the first city to adopt a systematic sewerage system, the Chicago system soon became famous and Mr. Chesbrough, for twenty-five years, was the recognized head of sanitary engineering in this country." Modern usage would limit the term "sanitary engineer" to those men involved with water and sewage treatment. Apparently, the American Society of Civil Engineers at that time considered a sanitary engineer to be a man involved with sanitation works. Thus, while Chesbrough was not concerned with sanitary engineering as that discipline is currently defined, he must be considered a precursor of the modern sanitary engineer and, in fact, was called one by his peers and contemporaries.

29 On this trip Chesbrough visited Zaardam, near Amsterdam, to investigate the possibility of using windmills to pump flushing water for Chicago's sewers; he decided in favor of steam pumps (*1858 Report,* p. 29).

30 *1858 Report,* p. 39.

31 *Ibid.,* p. 93.

32 *Ibid.,* p. 94.

33 Nevertheless, in 1863, the Board of Public Works issued a report on purifying the Chicago River. This is discussed in Brown, *Drainage Channel,* ch. 6. The report recommended the construction of flushing canals along the lines of Fullerton Avenue and Sixteenth Street. Therefore, although the Illinois and Michigan Canal's potential was realized, city officials evidently were not ready to pursue it.

34 Brown, *Drainage Channel,* p. 32.

35 Reported in Brown, *Drainage Channel,* p. 33.

36 *Second Annual Report of the Board of Public Works to the Common Council of the City of Chicago* (April 1, 1863), p. 5, hereafter referred to as *1863 Report.*

37 Cost estimates for each of the projects were as follows: 2-mile lake tunnel exclusive of light house, $307,552; a filtering or settling basin, $300,575; a 1-mile lake tunnel 5 miles to the north, $380,000 (*ibid.,* p. 9).

38 *Ibid.,* p. 8.

39 *Ibid.,* p. 39.

40 *Ibid.*

41 Chesbrough roughly estimated the iron pipe scheme to cost $250,000. The choice seems to have been made on the basis of expected cost. *Ibid.,* pp. 40–41.

42 *Ibid.,* p. 41.

43 *Ibid.*

44 *Ibid.* Originally, Chesbrough planned on four shafts.

45 *Ibid.,* p. 43.

46 *Ibid.,* p. 45.

47 Chesbrough estimated the cost to be $13.54 per linear foot. *Ibid.,* p. 48.

48 J. M. Wing, *The Tunnels and Water System of Chicago* (Chicago, 1874), p. 33.

49 *Ibid.,* p. 76.

50 *Proc. Am. Soc. Civil Engrs.,* November 1889, *15*: 162.

51 *1863 Report,* p. 57. Chesbrough was concerned with a definite planning period which seems to reflect a longer time than the Marshallian short run, and a shorter time than the Marshallian long run.

52 *Ibid.*

53 It can now (1985) be noted that the district failed to achieve this objective.

35

Dirt, Flies, and Immigrants: Explaining the Epidemiology of Poliomyelitis, 1900–1916

NAOMI ROGERS

"The trees of New York City are infested with caterpillars," a Boston school teacher warned in a letter to Rockefeller scientist Simon Flexner. Poisonous caterpillars, she explained, were the cause of the poliomyelitis epidemic of 1916, paralyzing children, including her own. A San Francisco man blamed electric radio waves, which were causing "pressure on brain substance and madness"; a New Yorker urged his city's board of health to fumigate subway cars, which were "reeking with billions of germs caused by the filthy foreign element constantly using them." During the first two decades of the 20th century, as poliomyelitis epidemics began to appear with increasing severity, confused and frightened men and women also blamed the epidemic on, among other things, foul sewage odors, mouldy flour, infected milk bottles, Swedish gooseberries, and rubber diapers.[1]

Scientists were also confused, a sentiment disturbingly out of place in this era of scientific optimism. European and American researchers had established the viral etiology of polio by 1909, and many felt confident that etiologic precision would enable them to explain the spread of the disease. Their optimism reflected a broad faith in the successes that modern science had recently brought to medical understanding. Bacteriology had proven especially rewarding, making visible specific agents of disease in water, food, and blood, and, by the 1910s, providing physicians with precise tests to identify a variety of diseases, including the Widal

test for typhoid, the Schick for diphtheria, and the Wassermann for syphilis.

Bacteriology seemed to promise precision not only in the diagnosis but also in the prevention of disease. Proponents of the germ theory adopted a new approach to public health, rejecting the filth theory of disease as well as its public health practice, sanitary science. Public health officials such as Charles V. Chapin, health officer of Providence, Rhode Island, emphasized that germs were spread by personal contact, and urged his colleagues to instruct individuals in their responsibility for the prevention of disease. The work of Ronald Ross in malaria and Walter Reed in yellow fever showed that germs could also be spread by insects. Contemporaries could cite the successful campaign against the yellow fever mosquito in the building of the Panama Canal, and the Rockefeller Foundation's efforts against hookworm in the American South and Latin America.

Yet this optimism had met frustrations; the relationship between etiological knowledge and epidemiological explanation was not straightforward. Identifying germs and insect vectors did not always provide a clear guide to effective therapy or preventive action. As early as 1888 Chapin reminded his contemporaries that despite the great publicity over Robert Koch's identification of the tubercle bacillus the only effective tuberculosis therapy was still education.[2] Syphilis, similarly, with both a diagnostic blood test and the magic bullet Salvarsan, proved a continuing public health problem.[3]

As with syphilis and tuberculosis, the clarification of polio's etiologic agent had not helped to control the disease; in fact, its diagnosis and treatment remained essentially unchanged. Although some physicians had experimented with an immune

NAOMI ROGERS is Lecturer in the Women's Studies Program and Research Affiliate of the Section of the History of Medicine, Yale University School of Medicine, New Haven, Connecticut.

Reprinted with permission from *Journal of the History of Medicine and Allied Sciences*, 1989, *44*: 486–505.

serum during the 1916 epidemic, laboratory re-search had not resulted in a vaccine. Doctors' treat-ment of paralyzed children was restricted to pre-scribing drugs to reduce pain and fever. Most disappointing, the poliomyelitis virus was impos-sible to grow in culture. Thus, the diagnostic tech-niques available to physicians were limited to close clinical observation of symptoms, which often mim-icked other more common infant diseases, or spinal taps, which were rarely done until after paralysis had appeared.[4]

Scientists remained frustratingly unable to deter-mine how the virus was transmitted, although Flex-ner and his team at the Rockefeller Institute for Medical Research in New York had found it in the mucous membranes of the nose and throat, and believed that the disease was spread by sneezing, kissing, and other close contact. But even if polio spread like a respiratory disease, its epidemiologic pattern did not fit the familiar picture of an infec-tious disease. Even during widespread epidemics, for example, usually only one child in a family de-veloped paralysis. It was known as a children's dis-ease (infantile paralysis), yet adults were sometimes victims too. It appeared in immigrant tenements with poor hygiene, but also in the homes of some of the best families. Etiological precision had done little to clarify the epidemiology of poliomyelitis.

Although America epidemiologists remained committed to the ideas and techniques of scientific medicine, their epidemiological work suggests that they were uncomfortable with some of its theoreti-cal implications. In principle, the germ theory im-plied that contagion might spread disease more or less at random; but in practice researchers assumed a close relationship between disease, poverty, and filth. Traditional assumptions about the relation between disease and environment, I suggest, pro-foundly shaped the ways they interpreted their ex-perience of epidemic polio. In explaining what was predictable and what anomalous, epidemiologists linked poverty and filth with the spread of disease, even when their evidence suggested that polio did not fit this pattern. This is also, then, a case study of what happens when old theories do not fit. When researchers found that poverty and filth did not seem to predict the appearance and pattern of epi-demic poliomyelitis, they did not suggest that good sanitation and living conditions were better pre-dictors. Instead, they reinterpreted the appearance

of cases in clean suburban homes as random, and sought additional factors such as infected milk, in-sect vectors, and individual sanitary carelessness to reinforce their belief in the relationship between filth, poverty, and disease.

A NEW AND BAFFLING DISEASE

By the 1910s epidemic poliomyelitis had become a problem of national importance, debated by physi-cians, health officials, and ordinary citizens. Para-lytic polio outbreaks first appeared in Norway and Sweden during the 1870s and 1880s, and then in the United States.[5] Vermont, in 1894, had the largest epidemic then reported, with 132 cases. In the following decades health officials, particularly from the Northeast and Midwest, began to report outbreaks in growing numbers: in 1907 New York State had more than 2,000 cases; in 1910 Minne-sota had about 1,000 and Massachusetts over 800.[6] In 1916 America experienced its largest polio epi-demic, with 27,000 cases reported in 26 states, and over 6,000 deaths. No other epidemic disease—at least until the 1918 influenza pandemic—was more prominent in the public mind. After the 1916 epi-demic one federal official predicted accurately that polio's "menace for the future . . . is very real."[7]

Despite the popular perception, poliomyelitis was not a new disease; what was new was its epi-demic appearance. In fact, infection rarely pro-duces paralytic symptoms, usually only a mild fe-ver, and generally confers lifelong immunity. But if a child is protected from the virus early in life and infected at a later age, paralysis is more likely to develop. In the 1900s while most poor and immi-grant infants had probably become infected and immune, American children from clean middle-class homes remained at risk of contracting the par-alytic form of the disease.[8]

The intricacies of the poliomyelitis story were worked out slowly and with difficulty during the first half of this century.[9] By 1916 some elements were already known. The virus had not been con-sistently shown in the bloodstream, and Flexner's research emphasized instead the neurotropic na-ture of the virus, a finding that discouraged efforts to develop a vaccine. More promising were contem-porary epidemiological observations. In 1905, after careful study of outbreaks in rural and isolated communities, Swedish epidemiologist Ivar Wickman identified a large number of infected children with nonparalytic or abortive symptoms. He suggested that polio might not be solely a paralytic disease, and that abortive cases might be the most important way to understand its epidemiologic picture. Wickman argued convincingly that the disease was contagious and was generally spread by direct contact.[10]

Without a clear-cut diagnostic test most of Wickman's invisible cases remained, of course, unseen and uncounted; poliomyelitis cases were difficult enough to trace without worrying about healthy carriers or patients with the nonparalytic form of the disease. Even Wickman admitted that paralysis was still "the only sign which is characteristic of the disease; and . . . which conclusively establishes the diagnosis."[11] American epidemiologists were, however, influenced by Wickman's focus on small, rural outbreaks. One official, for example, urged epidemiologic attention to rural epidemics for he believed that "epidemiologic evidence would be clearer cut in smaller places than amongst the complications necessarily encountered in a big city."[12]

A new element was added to the epidemiological picture after a major epidemic in New York City in 1907, when the disease became associated with urban communities.[13] In large urban areas direct contagion was even more difficult to establish than for rural epidemics. Hidden nonparalytic cases continued to confuse investigators, and skewed case fatality rates so that a higher proportion of the cases identified were reported to be fatal. Abortive cases also made clinical diagnosis more difficult; now physicians and parents worried that every fever might be a danger sign, as they tried to distinguish the early symptoms of polio from other infant complaints.

MODELS IN POLIOMYELITIS EPIDEMIOLOGY, 1900–16

During this period there were few professional epidemiologists in the United States. Epidemiological work was usually done by medical men employed by local or state health departments, but those who studied polio also included pediatricians, entomologists, and some private practitioners. A few states became centers of poliomyelitis epidemiological research because of their extensive experience with the disease. Epidemics during the 1900s led Hib-

bert Winslow Hill, a Minnesota health official, to publish reports which earned him a place on the American Public Health Association's poliomyelitis committee in 1911.[14] Officials from the Massachusetts State Board of Health and physicians at Harvard Medical School wrote a series of studies between 1907 and 1912, which tried to link epidemiologic theory, clinical experience, and laboratory evidence.[15] In New York City epidemiological work was undertaken by the city's Department of Health, and viral research by scientists at the Rockefeller Institute.[16] But the most influential institution which employed professional epidemiologists during the period was the United States Public Health Service. Federal epidemiologists Wade Frost and Allen Freeman, who investigated New York's 1916 epidemic extensively, both later taught at the Johns Hopkins School of Public Health and Hygiene.[17]

The models used by investigators reflected their experience with other endemic and epidemic diseases. Enteric diseases such as typhoid fever and cholera had begun to be prevented by environmental sanitation measures such as the regulation of sewage, water, and milk. Similarly, environmental factors were emphasized in the study of epidemic polio. Reflecting the long-standing link between public health work and sanitary reform, researchers studied overcrowding, poor housing and living conditions, and other signs of poverty and filth. They believed that poor immigrant families were likely victims and carriers, particularly those newly arrived from eastern and southern Europe. This focus on urban slums countered Wickman's recommendation to study rural epidemics and reemphasized the danger of life in American cities. It reflected an implicit nativism, but also the hope that with proper regulation the American city could be made a pleasant and healthful place to live.

A second model was that of insect vectors. The successes of medical entomology excited researchers who admitted that bacteriology had somewhat undermined the premise of environmental sanitation, and they sought pragmatic methods to integrate the germ theory into public health work. Insect vectors offered a way to explain the specific mechanism of disease transmission, were visible to both physician and parent, and continued the link between public health work, sanitary reform, and the filth theory. Yet this entomological model was also part of the new public health; a dirty child surrounded by flies was clearly a sign of parental ignorance and carelessness.[18] Still, disease-carrying insects with wings could also explain the presumed random appearance of cases. Specially apt for explaining the spread of polio, this model allowed officials to account for the appearance of paralytic cases in middle-class homes otherwise assumed sanitary and safe.

The third model emphasized direct contagion. It drew on the scientific work of Flexner and other scientists who argued that polio was spread by respiratory secretions. Epidemiologists urged the importance of identifying and tracing individual cases, but, with no precise diagnostic test, finding evidence of direct contact proved difficult. Wickman had been able to undertake detailed clinical epidemiology in small Scandinavian towns, but most American investigators faced epidemics in large urban communities where such detailed work was almost impossible to carry out. The selection of target field groups thus became of great importance, and the choices investigators made reflected their assumptions that poverty and filth would be linked with the spread of this disease. Despite the integration of the new science of bacteriology, the long-standing fear lingered that the poor and immigrants were the sources of dirt and disease.

TENEMENTS AND IMMIGRANTS

In the 19th century health reformers and sanitarians had associated the appearance and spread of disease with poverty and city life. American public health reformers such as Stephen Smith and Lemuel Shattuck used their work to critique the housing, living conditions, and sanitation of the urban poor.[19] A focus on urban poverty seemed especially appropriate as epidemic polio began to appear in the major Northeastern cities in the early 1900s. But American physicians and health officials found many inconsistencies in their attempts to classify polio as an environmental disease.

Most epidemiologists expected to find the majority of families with polio living in an environment filled with dirt, animals, and vermin. They assumed that their housing would be old, poorly built, with inadequate sewerage, probably large tenements or multifamily dwellings. But epidemiological work showed instead that many affected families lived in recently built houses, single family dwellings, or small tenement buildings, and had comparatively good domestic hygiene.

Researchers who studied urban outbreaks in the 1910s sought to find consistency with the characteristics of earlier poliomyelitis outbreaks. They found a higher incidence of epidemic polio in low density neighborhoods than in the tenements of Manhattan's Lower East Side or the slums of South Philadelphia. Thus, in 1916, Manhattan and Brooklyn, highly congested areas of New York, experienced a relatively lower incidence of the disease than did more sparsely populated boroughs such as Queens and Richmond. Epidemics in cities like New York and Philadelphia offered a distinctly different picture from the rural farming communities Wickman had described, but researchers began to observe so-called rural characteristics of the disease even within the metropolis of New York.

The lower incidence of paralysis in children in dense urban districts also raised the issue of the probable immunity of adult and congested populations. In 1911 federal epidemiologist Wade Frost had discovered that children from rural communities who developed paralysis tended to be older than these from urban neighborhoods. This crucial insight led him to consider polio as not solely a children's disease. Frost compared polio to "certain other infectious diseases, notably measles, [which] are largely limited to children, not because they are essentially children's diseases, but because the adult population has been more or less immunized."[20]

Epidemiologists' commitment to an environmental model justified their choice of Staten Island as part of a detailed study of the 1916 New York epidemic by the Public Health Service. Staten Island (the borough of Richmond) had intrigued observers because it had seen the highest proportional incidence of epidemic polio of the five boroughs of New York. Yet Staten Island was relatively uncongested; its population density was as low as two inhabitants per acre, in contrast to 170 in Manhattan. A detailed study of 328 affected families found their houses and surroundings clean and usually with inside toilets, a sign of higher sanitary standards. Most of these families were native-born and lived "under suburban conditions." And, true to Frost's rural factor, they had the highest proportion of older children affected with paralytic polio. In fact, epidemiologist Allen Freeman believed, Richmond was "definitely separated, both geographically and in general character of life" from the rest of New York.[21]

Nor was the economic standing of these families unusually low. Federal officials could discover "nothing significant" about the occupations of the householders, the phrase suggesting that this measure of class standing was also confusing. Even more frustrating for the federal researchers was the evidence of limited direct contact, only one child in five. Epidemiologist Claude Lavinder was unable to determine polio's "precise mechanism of transmission and the avenues of infection," and refused to consider the implications of his environmental investigations. Despite strong evidence of a correlation between suburban living, good sanitation, and the epidemic, Lavinder argued that the incidence of the disease was "relatively independent of local conditions" such as "density, economic status, and housing of the population." Polio, his colleague Wade Frost agreed, seemed to be unrelated to hygienic or economic conditions, race, or nationality. "We are, in truth, quite ignorant as to the principles which underlie such phenomena," the Public Health Service researchers concluded.[22]

Thus, even in an anomalous suburban district, poliomyelitis researchers focused on environmental factors such as sanitation, class, housing, and living conditions. Although congestion and poverty did not seem to predict the appearance of epidemic polio, researchers nonetheless continued to target overcrowded slum neighborhoods. The expectation that outbreaks would occur in these areas also influenced their view of the ethnicity of likely victims and transmitters of the disease. The relationship between nativity and disease was a longstanding public health concern, heightened during these years of mass immigration. Although in practice epidemiologists found it almost impossible to identify carriers of the disease, their tendency to blame immigrants nonetheless suggests an underlying nativism. They expected the typical affected family to be poor southern or eastern European immigrants. But the families they found were often native-born or of established immigrant groups such as the Germans and the Irish. As well, these children had higher mortality rates than those from Italian, Russian, or Polish families. "Certainly," New York City's health commissioner reflected, "the social and economic conditions under which these people live are no more favorable than those under which the Americans, Germans, and the Irish live."[23]

City and federal investigators tried to explain these perceived discrepancies by assuming that the ignorance of recent immigrants skewed their results, and made their answers unreliable. Allen Freeman, for example, believed that it was impossible to get accurate clinical histories because the "large foreign born population" with their "character and habits of life" made such evidence difficult to obtain and trust.[24] A New York City official found that, of the 5,500 cases he studied, over 60% were from native-born families, and immigrant children with polio seemed to have been well cared for and rarely sick. But he discounted this information as unreliable because of the working-class origin of his informants: parents "of the poorer classes," he believed, "cannot give the same attention to detail as the well-to-do."[25]

Although the disease was known to be contagious, cases were difficult to link together, particularly as investigators studied mostly paralytic patients.[26] Despite their evidence that a higher proportion of cases of paralytic polio were native-born and middle-class children than immigrant children, investigators concluded that class and ethnicity played no part in determining the spread of the disease. That is, they defined class as the working-class, and ethnicity as only that of recent immigrants such as Poles and Italians. The evidence they did find suggesting a link between the disease and other nationalities did not fit their preconceptions, and so was largely ignored. An environmental model framed epidemiologists' questions, but proved inadequate to explain the patterns they found, for researchers were only willing to associate the disease with the presence, not the absence, of poverty and dirt.

THE APPEAL OF INSECTS

The new science of medical entomology offered a promising way to explain the spread of poliomyelitis. The idea of insects as disease carriers seemed consistent with the new public health emphasis on domestic sanitation and individual responsibility, and also had a practical appeal. In 1913 Harvard public health authority Milton Rosenau argued that if the fly were the major culprit, the suppression of polio would be relatively easy, compared to dealing with healthy carriers and missed cases where "the difficulties of the problem will be multiplied manifold." Meanwhile, he told public health workers, "the public must be given the benefit of the doubt, and the infection fought along all probable lines."[27]

Epidemics of poliomyelitis generally appeared in late summer and early fall, and diminished with the beginning of winter. These characteristics supported the theory that polio was spread by insects, perhaps flies, which breed and travel more rapidly in hot weather.[28] Inspired by the pathbreaking work of Ross and Reed, investigators hoped that medical entomology would be able to answer puzzling questions about the spread of disease, and at the same time involve the latest laboratory research methods. One researcher warned that the "notorious errors made in the epidemiology of malaria and yellow fever in the days before [the mosquito] . . . serve as warning to make no final or conclusive statement regarding the epidemiology of polio."[29]

In August 1916 New York City's health department hired an entomologist to study the city's poliomyelitis epidemic. Charles Brues, an assistant professor of entomology at Harvard, began his search for an "intermediate agent" which would help explain several "peculiar factors" about the epidemic. He drew analogies with other epidemic diseases which had been shown to involve insect or animal vectors, such as yellow fever and bubonic plague. The plague (which had appeared in California a decade before) particularly interested him, for it seemed to display epidemiological features strikingly similar to epidemic polio: little relation to the density of population, and a tendency for cases to appear near waterfronts and on the lower floors of apartment buildings.[30] Another researcher agreed that insects such as the rat flea might help to resolve polio's epidemiologic inconsistencies, adding that "whoever has seen a city slum street in summer-time cannot imagine more intimate personal contact than is enjoyed by the tenement children playing in the crowded, hot, dusty thoroughfare, and yet it appeared that cases of infantile paralysis might be numerous in the tenements on one side of a street, with no cases whatever in similar houses opposite."[31]

Brues considered the possibility of bedbugs, catfleas, sandflies, and houseflies, but was most convinced by the laboratory research on stable flies, citing his own experimental work at Harvard Medical School and its confirmation by federal workers at the Hygienic Laboratory in Washington. Brues warned that this research had not been fully confirmed, and so was "not free from possible error."[32] Still, he felt that an insect vector was the most promising way to explain an epidemic which had "involved a population living under entirely different conditions from those existing in places where previous epidemiological investigations have been made."[33]

An insect theory, then, helped to explain the appearance of polio among middle-class and nonimmigrant children. Insects, potentially, could spread disease randomly; it was not the fault of middle-class parents with a paralyzed child if a germ-carrying fly had traveled from the worst parts of the city. The disease, thus, could continue to be linked to dirt and the slums. Sometimes these assumptions about class were made explicit. In 1914 two physicians argued, in support of their theory of blood-sucking insects as carriers of polio, that "observation demonstrates that constant distribution of the bedbug among members of various social classes takes place." They found examples of cross-class infection in a doctor attending a slum case, a lawyer in court, and a maid who spent her half-day in a tenement. The daily paper, they reminded readers, was distributed by tenement dwellers, and hand laundry was washed there. In such ways, they feared, it was possible to observe the "invasion of the American home."[34]

Even after the popularity of the insect theory of poliomyelitis transmission began to diminish, physicians used the entomological model to reinforce the link between class, germs, and dirt. A Kansas physician, discussing polio orthopedic therapy during the 1916 epidemic, argued that doctors should trust the evidence of their own eyes, particularly when they found afflicted families surrounded by filth and insects. For, he urged, if "New York will screen its windows and doors you will possibly stop polio . . . and in spite of the fact that we have been told and I suppose it has been accepted as true that insects and flies have nothing to do with disease, still . . . if you will go on the Third Avenue [E]L and go down town to the Battery you will see I don't know how many million windows and doors without screens."[35]

The entomological model appealed to scientists,

health officials, and clinicians. Laboratory research on disease-carrying insects was considered at least as precise as virological study of the poliomyelitis virus. Scientists at Harvard Medical School, the federal Hygienic Laboratory, and at various city and state public health laboratories tried to identify vectors by testing flies and other insects. Although their results proved inconclusive, public health officials drew on this work to support fervent swat-the-fly campaigns during epidemics. And clinicians, denied a specific diagnostic technique such as they had for syphilis or typhoid, could search for insect bites on their patients and ask specific questions of parents and other household members about insects in their environment. The entomological model allowed epidemiologists to trace the direct contact of cases, but at the same time explain the so-called random appearance of cases in clean, middle-class families. Insects provided one way to integrate the environmental assumptions of researchers into the confusing picture of this disease.

DIRT AND DOMESTIC HYGIENE

In 1908 a Massachusetts health official argued that the discovery of polio's etiologic agent would clarify the way the disease spread. "Until the organism causing the disease is known," he commented, "it will be impossible to say whether the infection is carried directly to the patient or by means of food."[36] But his hope that etiologic identification would clarify epidemiology was, in the short term, ill-founded. Flexner's work had suggested that the virus was transmitted by personal contact such as sneezing or kissing. Some investigators welcomed this emphasis on direct contagion, for it associated the spread of the disease with personal sanitary behavior. This association was both practical as well as intuitive; clean and careful families could protect themselves from disease. If the poliomyelitis virus was spread through unsanitary habits, then epidemiologists and health officials could explain epidemics in a way that was both convincing to the lay public and scientifically precise. In any case, urging individuals to control their domestic hygiene and individual behavior was part of a familiar public health litany and not easily discarded. Yet investigators found it difficult to link the appearance of paralytic polio with a lack of cleanliness.

While an emphasis on germs may in theory have provided researchers with greater objectivity than a broad environmental model, investigators did not assume that the germs of polio spread randomly irrespective of ethnicity or class. At a 1916 Washington conference, for example, New York City's health commissioner Haven Emerson reported a case of a doctor's child who had died of the disease. He argued, however, that it was not an example of a physician bringing infection into his own family, for the child's Polish nurse was more likely at fault. "She, with her habits, coming from an infected case, appeared to us and to the doctor who lost the child as the more probable carrier," Emerson assured the conference delegates.[37] Similarly, Yale public health authority C.-E. A. Winslow explained the appearance of the disease in an isolated wealthy family in Connecticut by the revelation that the family's chauffeur had been secretly visiting his sister in Brooklyn, a center of polio cases.[38]

Even more striking than the appearance of cases in the best families were the limited numbers reported among children living in orphanages. During the 1916 epidemic in New York City, Haven Emerson found only 10 cases reported, from around 100 public institutions housing 21,000 children.[39] Only one investigator dared to suggest that this was due to improved hygienic conditions in the institutions, where, he claimed, "cleanliness and discipline are carefully observed."[40] Most other observers fell back on isolation as the explanation, suggesting that only institutionalized children were thoroughly protected from the outside world. Still, in New York these institutions were forced to undergo massive sanitary campaigns, including white muslin handkerchiefs provided for all children, and increased bathing.[41] And although a study of patients with poliomyelitis from 58 New York infant health stations found that almost all were healthy, well-nourished, and fed on pasteurized milk, this evidence of cleanliness did not dampen fears of dirt and contagion. No afflicted baby during the 1916 epidemic was permitted to attend an infant health station; health officials warned mothers about the importance of domestic hygiene; and no one from "infected premises" was allowed to "mingle unduly with the regular clientele" of any station.[42]

Unwilling to link the appearance of the disease with healthy children and good sanitation, investigators sought external factors which would explain the appearance of cases in families assumed to be sanitary and safe; the disease, they argued, must have entered homes from the outside. As poliomyelitis was consider a child's illness, researchers urged particular attention to milk as a probable source of infection. Infected milk and water were, after all, familiar epidemiological subjects, and had provided health officials with the means to control other dirt-related diseases such as cholera and typhoid. Parents were questioned closely about where children had obtained their food and drink; some feared immigrant peddlers were selling infected food to unwary middle-class children. A number of health officials adopted a theory of infected milk, and in New Jersey the Newark city health department placed its milk supply under close surveillance throughout the 1916 epidemic.[43]

These health officials tried to separate the germ theory from the older filth theory of disease, but found it difficult. Proponents of the new public health argued that the lay public needed to be educated to understand how diseases were spread, and the particular importance of personal hygiene. A singularly contagionist view was voiced by a New York health official whose study of the New York City 1916 epidemic was funded by the Rockefeller Foundation. Alvah Doty, a proponent of the new public health, believed in the overriding danger of personal secretions in the transmission of disease for, he argued, "the true media of infection are discharged from the body containing infectious organisms in their active state."[44]

Doty stridently attacked other theories that he found less scientific. "Nothing," he warned "has contributed more to the extension of infectious diseases than erroneous theories concerning media of infection for they have encouraged carelessness in detecting the real means by which these maladies are transmitted from one person to another." He considered the standard epidemiologic repertoire for poliomyelitis and rejected it completely: dust, horses, poultry, stable flies, the rat flea.[45] Unfortunately, Doty could find evidence that polio was spread by personal contact in only 10% of the 5,000 New York cases he studied. He justified this discrepancy, as Wickman had, by noting the difficulty of diagnosing the disease, particularly abortive and mild cases; after all, "until recently this disease has not been well understood or clearly defined."[46]

Doty believed that hygienic carelessness and ig-

norance allowed infection into the home, and specifically mentioned infected milk. But he found no evidence of infected milk in New York's general supply. To explain the fact that a large number of families with paralyzed children had used pasteurized milk he blamed irresponsible housewives. "The contamination of milk after it reaches the household is real and not imaginary," he warned, for it "easily become[s] a source of infection through want of proper attention to cleanliness." Immigrant and poor families, through "ignorance, overcrowding, bad sanitary conditions, and need of strick economy in the necessities of life" brought disease upon themselves.[47] For, he wrote, that "among the poorer classes the person who prepares the food is often the one who tends the sick baby. . . . [Her] hands are more or less constantly contaminated with discharges, particularly from the intestinal tract, and there is no reason to believe that any special effort is made to ensure proper cleanliness. Therefore, even milk, although pasteurized, may become contaminated after reaching the home."[48]

Doty warned that the lay community, particularly immigrants and the poor, did not understand the dangers of germs and the way they spread disease from person to person. He was disturbed to find that some members of the public still held a belief in the filth and miasmatic theories of disease. Like Chapin he agreed that one of the most effective weapons against disease was popular education. Perhaps poliomyelitis could be conquered this way. "Outbreaks of infectious diseases," he argued, "cannot be successfully dealt with without the cooperation of the public and this cannot be gained unless the people have definite and reliable knowledge regarding the means by which these affections are transmitted."[49]

Investigating sanitation in the home was both familiar and practical, for it provided health officials with specific ways to educate the public. Parents were warned to keep streets and alleyways clean and to ensure that food was free of dirt and germs. Officials were, not surprisingly, unwilling to believe that cleanliness, by protecting the very young from exposure to the virus, might help explain polio's epidemiologic picture. Doty's faith that a concerned housekeeper could guard her family from infection ignored previous epidemiologic evidence that poliomyelitis had attacked families with healthy, well-nourished children. One epidemiologist who studied outbreaks in New Jersey and Connecticut found that sanitary conditions "were certainly as good as the average and perhaps better, and this is the most striking thing shown." Yet he argued that there was nothing to suggest that "the spread of the disease bears any special relation to the sanitary conditions under which it occurred."[50] The discrepancy between dirt and disease was the most difficult break from the traditional epidemiological past. Researchers explained the appearance of polio in middle-class homes by invoking factors such as careless parents who had allowed dangerous germs to enter their homes. This etiologic model allowed investigators to transform dirt into infected filth, a sign that individuals, through carelessness and ignorance, had forsaken their responsibility for preventing the spread of disease.

CONCLUSION

This study of poliomyelitis epidemiology between 1916 and 1920 suggests that the integration of the new sciences into epidemiological practice was shaped by traditional assumptions about the relationship between disease, the environment, and the individual. Professional and popular acceptance of the germ theory existed uneasily with older ideas about the nature and cause of disease. In fact, the practice of American epidemiologists reflected their belief that for epidemic poliomyelitis the promise of the germ theory had not been fulfilled. The new tools of bacteriology had not helped to clarify or control the disease; there was no simple blood test or other precise way to distinguish infant summer complaints from the early or mild symptoms of this disease.

Despite Ivar Wickman's emphasis on the large numbers of abortive cases, paralysis remained polio's determining diagnostic sign. Epidemiologists who traced mainly paralytic cases found their maps confusing, for they did not always suggest a straightforward pattern of direct contagion. Drawing on their knowledge of other epidemic diseases,

researchers tried to explain this confusing epidemiologic picture by implicating housing and living conditions, class and ethnicity. But, unwilling to reject their environmentalist assumptions, epidemiologists chose to ignore correlations which did not seem to make sense. When they found cases among native-born, middle-class homes, they did not associate the disease with cleanliness or wealth. They expected poliomyelitis to appear in working-class families who were recent immigrants, living in overcrowded, filthy tenements. When cases were scattered among other members of the community, they were termed random.

To explain this random spread, researchers sought special factors that might have helped to carry infection to otherwise safe households. Their explanations drew on the new public health approach, and stressed the importance of individual responsibility in preventing the spread of disease. Health officials' emphasis on the importance of personal hygiene intensified their tendency to target the behavior of poor and immigrant families. Researchers such as Doty blamed the "random" appearance of cases in middle-class families on housekeepers who had clearly allowed into their homes tainted food, milk, or insects. Clearly, middle-class families were threatened by a dangerous outside environment as the daily interaction and services of immigrant families brought dirt and infection inside otherwise protected homes.

But in this case neither an appeal to germs nor dirt helped to explain the spread of epidemic poliomyelitis. Traditional theories of diseases spread by overcrowding and poor sanitation were consistently challenged by most of the major studies of this period. But, just as consistently, researchers concluded that hygiene, ethnicity, and class (as they defined these factors) seemed to have little or no relationship to the spread of this disease. In one way they were right: the pattern of poliomyelitis contradicted their given picture of epidemic disease. And no one chose to turn that picture upside down.

NOTES

I would like to acknowledge with thanks the editorial advice and insightful comments on drafts of this paper from Charles Rosenberg, June Factor, Michael Katz, Alan Morrison, Jim Patterson, Lisa Mae Robinson, and John Harley Warner.

1   Henriette G. to Simon Flexner, 15 Aug. 1916, "Polio" file C #2, Simon Flexner Papers, American Philosophical Society, Philadelphia; Michael C. to John Mitchel, 29 Nov. 1916, General Correspondence 1916, Mayor John Mitchel Papers, New York City

Municipal Archives; J. V. T. to John Mitchel, 18 July 1916, General Correspondence 1916, Mitchel Papers. For additional theories see other letters written to Flexner and Mitchel in these collections.

2 Barbara Gutmann Rosenkrantz, "Introductory essay: Dubos and tuberculosis, master teachers," in René & Jean DuBos, *The White Plague: Tuberculosis, Man, and Society* (New Brunswick, N.J. and London: Rutgers Univ. Press, 1987), pp. xxix–xxxi.

3 Allan M. Brandt, *No Magic Bullet: A Social History of Venereal Disease in the United States Since 1880* (New York: Oxford Univ. Press, 1987), pp. 40–47.

4 For a detailed study of polio see John R. Paul, *A History of Poliomyelitis* (New Haven and London: Yale Univ. Press, 1971); and for this period see also Naomi Rogers, "Screen the baby, swat the fly: polio in the northeastern United States, 1916," Ph.D. dissertation, Univ. of Pennsylvania, 1986.

5 These outbreaks suggest that the improved sanitary conditions of some families had begun to protect children from early exposure to infection. In 1916, for example, New York City officials found that while epidemic polio was paralyzing the largest number of victims ever, the city was also experiencing its lowest rate of infant mortality; Haven Emerson, *A Monograph on the Epidemic of Poliomyelitis (Infantile Paralysis) in New York City in 1916. Based on the Official Reports of the Bureaus of the Department of Health* (New York: Department of Health, 1917), p. 132. Hereafter *Monograph*.

6 James Warren Sever, "Anterior poliomyelitis: a review of the recent literature in regard to the epidemiology, etiology, modes of transmission, bacteriology and pathology," in *Infantile Paralysis in Massachusetts 1907–1912* (Boston: Wright & Potter, 1914), pp. 5–23. See also Robert W. Lovett, "The occurrence of infantile paralysis in the United States and Canada in 1910," *Transactions of the American Pediatric Society*, 1911, *23*: 181; and *Infantile Paralysis in Vermont, 1894–1922: A Memorial to Charles S. Caverly, M.D.* (Burlington, Vt.: State Department of Public Health, 1924). Although the incidence of poliomyelitis remained much smaller than the rates of other childhood illnesses such as measles, diphtheria, and scarlet fever, epidemic polio remained a major public health concern in the United States until widespread immunization through vaccination during the 1960s. Epidemic polio is more likely to cripple than kill, and the national morbidity rate throughout this century rarely exceeded 12 per 100,000 population. Paul states that the 1916 epidemic reached only 28.5 per 100,000; Paul, *A History*, p. 148; but see *Historical Statistics of the United States: Colonial Times to 1970*, 2 vols. (Washington, D.C.: Bureau of the Census, 1975), I, 77 (Table B291–304), stating the 1916

incidence was 41. The next highest incidence was in 1952 when polio reached 37 per 100,000. See also Saul Benison, "The history of polio research in the United States: appraisals and lesson," in Gerald Holton, ed., *The Twentieth-Century Sciences: Studies in the Biography of Ideas* (New York: W. W. Norton, 1972), pp. 308–343.

7 C. H. Lavinder, "The prevalence of poliomyelitis before 1916," in Lavinder, A. W. Freeman, and W. H. Frost, "Epidemiologic studies of poliomyelitis in New York City and the north-eastern United States during the year 1916," *Public Health Bulletin*, 1918, *91*: 35. Hereafter "Studies." Polio became reportable in Norway in 1904, Sweden in 1905, the United States in 1909, and England in 1911.

8 Dorothy M. Horstmann, "The poliomyelitis story: a scientific hegira," *Yale J. Biol. & Med.*, 1985, *58*: 79–80. Nonparalytic polio was probably endemic throughout the 19th century, and rarely recognized or recorded.

9 The mechanisms of polio's transmission were not finally resolved until American researchers in the 1940s and 1950s identified the virus in the bloodstream and showed that it is initially an intestinal disease, usually spread by infected feces as children play with one another, and only rarely by water, milk, or insects; Paul, *A History*, ch. 36.

10 Paul, *A History*, ch. 10; Ivar Wickman, *Acute Poliomyelitis (Heine-Medin's Disease)* [1907], trans. J. Wm. J. A. M. Maloney (New York: Journal of Nervous and Mental Diseases Publishing Co., monograph no. 16, 1913), p. 113. Wickman's study is still regarded as a classic for understanding polio's epidemiologic and clinical characteristics.

11 Wickman, *Acute Poliomyelitis*, p. 38. See also Paul, *A History*, ch. 13.

12 Wickman, *Acute Poliomyelitis*, p. 38; H. W. Hill, "The epidemiology of anterior poliomyelitis," *Journal of the Minnesota State Medical Association and Northwestern Lancet*, 1909, *29*: 371.

13 *Epidemic Polio: Report on the New York Epidemic of 1907 by the Collective Investigative Committee* (New York: Journal of Nervous and Mental Disease Publishing Co., monograph no. 6, 1910).

14 Hill was appointed head of the new division of epidemiology in the Minnesota State Board of Health in 1910; see Philip D. Jordan, *The People's Health: A History of Public Health in Minnesota to 1948* (Saint Paul, Minn.: Historical Society, 1953), pp. 91, 12, 88–89. See also Frost, Hill and Samuel Dixon, "Report of the committee of the American Medical Association on methods for the control of epidemic poliomyelitis," *Sixth Annual Report of the Commissioner of Health for the Commonwealth of Pennsylvania 1911* (Harrisburg: J. L. L. Kuhn, 1912), pp. 76–85.

15 See *Infantile Paralysis in Massachusetts in 1909* (Boston: Wright & Potter, 1910); and *Infantile Paralysis in Massachusetts 1907–1912*. See also Barbara Gutmann Rosenkrantz, *Public Health and the State: Changing Views in Massachusetts, 1842–1936* (Cambridge, Mass.: Harvard Univ. Press, 1972); and Henry K. Beecher and Mark D. Altschule, *Medicine at Harvard: The First Three Hundred Years* (Hanover, N.H.: Univ. Press of New England, 1977), pp. 357–358, 421, 140.

16 Paul, *A History*, pp. 118–120; and Francis W. Peabody, George Draper, and A. R. Dochez, *A Clinical Study of Acute Poliomyelitis* (New York: Rockefeller Institute of Medical Research, monograph no. 4, June 1, 1912). Paul argues that the study's epidemiological section was written by Flexner. See also G. W. Corner, *A History of the Rockefeller Institute, 1901–1953: Origins and Growth* (New York: Rockefeller Institute Press, 1962); A. McGehee Harvey, *Science at the Bedside: Clinical Research in American Medicine, 1905–1945* (Baltimore: Johns Hopkins Univ. Press, 1981), pp. 93–95; and Saul Benison, "Speculation and experimentation in early poliomyelitis research," *Clio Medica*, 1975, *10*: 1–22.

17 Elizabeth Fee, *Disease and Discovery: A History of the John Hopkins School of Public Health and Hygiene, 1916–1939* (Baltimore: Johns Hopkins Univ. Press, 1987), pp. 68, 70. See also Allen Wier Freeman's autobiography, *Five Million Patients: The Professional Life of a Health Officer* (New York: Charles Scribner's Sons, 1946); Thomas B. Turner, *Heritage of Excellence: The Johns Hopkins Medical Institutions, 1914–1947* (Baltimore: Johns Hopkins Univ. Press, 1974); and Victoria A. Harden, *Inventing the NIH: Federal Biomedical Research Policy, 1887–1937* (Baltimore: John Hopkins Univ. Press, 1986).

18 For a more detailed development of this point, see Naomi Rogers, "Germs with legs: flies, disease and the new public health," *Bull. Hist. Med.*, 1989, *63*: 599–617.

19 Rosenkrantz, *Public Health and the State*, ch. 1.

20 "Infantile paralysis: a round table discussion," *Am. J. Public Hlth.*, 1917, 7: 130. See also Matthias Nicoll, Jr., "Epidemiologic data in the poliomyelitis epidemic in New York State," *Tr. Assn. Am. Physicians*, 1917, *24*: 234; and Freeman, "The epidemic of poliomyelitis in New York City, 1916," in "Studies," pp. 108–113. The borough of Richmond had the highest percentage of children over five years (27%) compared to the city's overall figure of 22%.

21 Freeman, "Epidemic," in "Studies," pp. 81, 84, 97, 133. See also Lavinder, "Richmond," in "Studies," pp. 137–160. A study of 1,000 polio cases in New York State found over half the families economically in "moderate circumstances," and concluded that there was no relationship between economic status, sanitary conditions, and the appearance of the disease, Emerson, *Monograph*, p. 186.

22 Lavinder, *ibid.*, p. 151; Lavinder "Summary," in "Studies," pp. 211–212; Frost, "Problem of polio," in "Studies," p. 19; Lavinder, "Richmond," in "Studies," p. 159. Frost concluded that the epidemic was either "nonspecific" or the result of previous exposure of unrecognized infected cases, *ibid.*, p. 23.

23 Emerson, *Monograph*, p. 108.

24 Freeman, "The epidemic," in "Studies," p. 85.

25 Alvah H. Doty, "Special investigation of infantile paralysis" [unpublished ms.], "Doty" file, Flexner Papers, American Philosophical Society, Philadelphia, pp. 11, 15. This copy also has Flexner's editorial comments.

26 For a discussion of the increased difficulty of diagnosis with the concept of abortive cases see H. W. Hill to Simon Flexner, 16 Nov. 1909, "Polio" file, Flexner papers, American Philosophical Society, Philadelphia. Hill warned that since abortive cases "may or may not be poliomyelitis in fact . . . whether they are reported as such or not simply depends on the attitude of the physician concerned," p. 2.

27 Milton J. Rosenau, "The mode of transmission of poliomyelitis," *J.A.M.A.*, 1913, *60*: 1615.

28 See *Infantile Paralysis in Massachusetts 1907–1912;* see also Robert W. Lovett and Philip A. E. Sheppard, "The occurrence of infantile paralysis in Massachusetts in 1910," *Forty-Second Annual Report of the State Board of Health of Massachusetts* (Boston: Wright & Potter, 1911), pp. 423–435.

29 Hill, "Epidemiology," p. 373.

30 C. T. Brues, "Insects as carriers of infection: an entomological study of the 1916 epidemic," in Emerson, *Monograph*, pp. 136, 155. See also M. J. Rosenau and Charles T. Brues, "Some experimental observations upon monkeys concerning the transmission of poliomyelitis through the agency of *Stomoxys calcitrans*," *Monthly Bulletin* Massachusetts State Board of Health (Sept. 1912), reprinted in George Whipple, ed., *State Sanitation: A Review of the Work of the Massachusetts State Board of Health* (Cambridge, Mass.: Harvard Univ. Press, 1917), pp. 358–361.

31 Mark W. Richardson, "The rat and infantile paralysis—a theory," *Tr. Assn. Am. Physicians*, 1918, *25*: 166. Richardson's rat theory was pursued by Public Health Service researchers studying New Jersey and Connecticut outbreaks, and some "suggestive" evidence was found, but the federal researchers rejected the theory; see Freeman, "Outside New York City," in "Studies," pp. 190–191.

32 Brues, "Insects," pp. 151, 161, 134. See "Transactions of a special conference of state and territorial health officers with the United States Public Health Service, for the consideration of the prevention of

the spread of poliomyelitis: Held at Washington, D.C., August 17 and 18, 1916" *Public Health Bulletin*, 1917, *83*: 98. Hereafter "Transactions." See also John F. Anderson and W. H. Frost, "Transmission of poliomyelitis by means of the stable fly (*Stomoxys calcitrans*)," *Public Hlth. Rep.*, 1912, *27*: 1733–1735. For additional work on stable flies, particularly during 1913–1914, see Sever, "Anterior poliomyelitis," pp. 19–20.

33 Brues, "Insects," p. 137.

34 Henry Frauenthal and Jacolyn Van Vliet Manning, *A Manual of Infantile Paralysis with Modern Methods of Treatment Including Reports Based on the Treatment of Three Thousand Cases* (Philadelphia: F. A. Davis Co., 1914), p. 54.

35 "Conference on infantile paralysis held at the office of [the] Rockefeller Foundation, Saturday, August 5th, 1916, 10 o'clock A. M.," Simon Flexner Papers, #200/25/283, Rockefeller Archives Center, Tarrytown, New York, pp. 55–56.

36 Herbert C. Emerson, "An epidemic of infantile paralysis in western Massachusetts in 1908," in Lovett and Emerson, eds., *The Occurrence of Infantile Paralysis in Massachusetts in 1908* (Boston: Wright & Potter, 1909), pp. 25–26.

37 "Transactions," p. 81.

38 "Round Table," p. 142.

39 *Ibid.*, pp. 127–128. One Newark official noted that the greatest incidence of polio in 1916 had occurred in "one of our cleanest wards" which had been cleaned with special pains for three months previous to the epidemic, *ibid.* p. 125.

40 Doty, "Special investigation," p. 14.

41 Emerson, *Monograph*, p. 24.

42 Emerson, *Monograph*, pp. 23–24.

43 Stuart Galishoff, "Newark and the great polio epidemic of 1916," *New Jersey History*, 1976, *54*: 108. A Maine health officer told fellow officials at a Washington conference that no bottled milk was allowed in any polio sick room, only canned or breast milk, "Transactions," p. 141.

44 Doty, "Special investigation," p. 11.

45 Doty, *ibid.*, pp. 11, 61, 49ff, 54–56.

46 Doty, *ibid.*, pp. 13, 27. Freeman found only 5% of multiple cases in families, "The epidemic," in "Studies," p. 124; Lavinder found 20% of multiple cases in his study of Staten Island families, although excluding "suspicious" cases the figure was only 11%. He could also trace 21% direct contact cases, but from a sample of only 96 cases, "Richmond," in "Studies," pp. 152, 157.

47 Doty, *ibid.*, pp. 43, 46–47, 58. He concluded, nonetheless, that the most common means of spread were abortive cases, p. 57.

48 Doty, *ibid.*, pp. 38–39.

49 Doty, *ibid.*, pp. 58–59. He criticized the emphasis of the daily press on cleanliness for "while the value of this is obvious it does not enlighten the public concerning media of infection, *for uncleanliness and unhygienic surroundings do not generate pathogenic organisms and therefore cannot cause infectious disease*," p. 59.

50 Freeman, "Outside New York City," in "Studies," p. 191.

36

"Typhoid Mary" Strikes Back:
Bacteriological Theory and Practice in
Early 20th-Century Public Health

JUDITH WALZER LEAVITT

The science of bacteriology, with its emphasis on the role of microorganisms in causing disease, undoubtedly influenced the practice of public health in early 20th-century America. But how much did this new laboratory-based science narrow the scope of public health activities? Or, to put it another way, how reductionist was this new science in practice? These questions are important because many historians and some late 20th-century social critics have assumed, implicitly or explicitly, that germ theory, in contracting conceptions of the etiology of disease primarily to microorganisms, brought an end to the more holistic understandings of illness that preceded it. According to these scholars, the laboratory became the most important instrument for measuring degrees of sickness, substituting microbe hunting with test tubes and microscopes for the wide-ranging sanitation and social welfare programs that characterized 19th-century public health practices. Paul Starr, for example, has written that "the new outlook brought with it a radically reduced view of the requirements of public health. . . . The limitations on public health in the twentieth century were . . . profound. The early public health reformers of the nineteenth century, for all their moralism, were concerned with social welfare in a broad sense. Their twentieth-century successors adopted a more narrow and technical

JUDITH WALZER LEAVITT is Professor of the History of Medicine, History of Science, and Women's Studies, and Associate Dean for Faculty, University of Wisconsin–Madison.

Reprinted with permission from *Isis*, 1992, *83*: 608–629. Copyright © 1992 by the History of Science Society, Inc., and published by the University of Chicago Press.

view of their calling."[1] In examining early experiences with healthy typhoid fever carriers, I have tried to determine whether and to what extent such reductionism characterized bacteriologically oriented public health practices. My findings suggest that statements such as Starr's exaggerate the bacteriologically induced limitations on the scope of 20th-century public health activities and that the search for microorganisms took place in a context that continued to be characterized by broad social concerns.

In reaching this conclusion, I join a growing group of scholars whose recent and current work on this period of public health history has found important social and political components in the "new" public health. Allan Brandt's study of the political context of venereal disease control, *No Magic Bullet*, John Ettling's *The Germ of Laziness*, which emphasizes the cultural and religious aspects of the campaign against hookworm, and Nancy Tomes's current work on the popularization and domestication of germ theory epitomize this literature.[2] Looking at different aspects of the relationship between experimental science and public health and medical practice, these historians have moved us far along the road to understanding the complexity and diverse nature of early 20th-century public health. This essay provides another example of how the laboratory affected public health and, like the studies mentioned above, suggests that the coming of bacteriology neither simplified public health activities nor allowed those applying this science to detach themselves from the impinging social context.

Throughout the 19th century, epidemics of infectious diseases had threatened and scourged American cities, and the considerable death and destruction they caused had fostered systematic programs to protect the public. Cities and states organized health departments whose functions included planning programs to contend with the worst of the disasters that continued to occur.[3] Smallpox, cholera, and yellow fever ravaged populations, prompting health officers to develop sophisticated responses that included massive urban sanitation projects to bring clean water into the city and to institute efficient sewage disposal, garbage collection and disposal, vaccination programs, isolation hospitals, clinics, and dispensaries. Much of the work during the 19th century rested on the prevailing medical theory that dirt caused disease and emphasized keeping the city environment clean. By the last decades of the century, experimental work from the laboratories of Louis Pasteur and Robert Koch had revolutionized medical theory about the causes of epidemic disease, substituting microorganisms for dirt as the culprit to be excised. Narrowing notions of the etiology of disease from the whole urban environment to microscopic germs seemed to pinpoint the public health activity needed to eliminate the problems. Despite the obvious historical importance of the paradigm shift, historians have not yet determined the full extent to which the identification of microorganisms as the single cause of infectious diseases limited health practices to the dimension of finding and eliminating those germs.[4]

The effects that bacteriology had on public health theory and practices can be illustrated vividly through the activities of Charles V. Chapin, the Providence, Rhode Island, health officer during the crucial years around the turn of the 20th century. James H. Cassedy, Chapin's biographer, has portrayed the excitement of those years for people engaged in efforts to protect and promote the public's health.[5] Laboratory experiments led to frequent announcements of new microbes, which in turn provided links to understanding diseases that had baffled scientists and decimated populations. The hope and promise of these revelations carried Chapin and many other public health workers on a wave of optimism, as they scrambled to incorporate the findings into their daily tasks. Chapin became the nation's leading proponent of using the new science to move the focus of public health

work away from environmental sanitation. In 1902 he wrote an impassioned article claiming that bacteriology "drove the last nail in the coffin" of the old filth theory of disease: "Sanitary reform was engaged principally in protecting drinking water from organic contamination, in building sewers, in developing plumbing to a complicated and expensive art, in cleaning streets, in removing dead animals, in collecting garbage and removing household rubbish, in whitewashing and repairing tenements, in the regulation of offensive trades, and the general suppression of all nuisances affecting the sense of smell." None of this was any longer necessary, Chapin claimed, because it would "make no demonstrable difference in a city's mortality whether its streets are clean or not, whether the garbage is removed promptly or allowed to accumulate, or whether it has a plumbing law."[6]

Chapin attacked the old practices one after another: for example, in one of his most influential papers he decried as a "fetich" the common practice of fumigating with steam or formaldehyde the rooms in which the sick had suffered or died.[7] Chapin worked to replace citywide cleanup programs with efforts that focused on living human germ carriers (sick persons or those carrying infectious diseases) and on fostering personal habits that would protect individuals from the people around them. He particularly emphasized a new worry that had been uncovered by recent bacteriological studies, the risk posed by healthy carriers, people who were not sick themselves but could nonetheless infect others. Chapin spent a good portion of his time educating the public (and his peers) about these hidden dangers:

> Neither you, nor I, nor the Board of Health, know where these [carriers and missed cases] are. The occupant of the next seat may, for all one knows, be a diphtheria carrier, so may the saleslady who ties up the package, the conductor who gives the transfer, or the expressman who leaves a parcel at the door. The dirty man hanging on the car strap may be a typhoid carrier, or it may be that the fashionably dressed woman who used it just before was infected with some loathsome disease. If these people were sick in bed we would avoid them. As it is we cannot. Science has shown this new danger.

Chapin believed that as experimental and laboratory studies revealed new sources of exposure to infectious diseases, citizens had to assume an increasing responsibility to try to avoid them:

> Contact with the fresh secretions, or excretions, of human beings, is the most important source of infection for most of our common contagious diseases. By turning the face from the coughing and loud talking of our neighbors; by putting nothing in the mouth except clean food and drink; by never putting the fingers in the mouth, or nose; most contagious diseases can be avoided. Wash the hands well before eating and always after the use of the toilet. Teach this to the children by precept and especially by example.[8]

Chapin not only preached the new gospel of individualized public health; in his long tenure as superintendent of health for Providence, he modeled the new science in practice. To help change procedures, he created a "score card" delineating the relative value of various public health activities: on a point scale with the total of city activities at 100, among other items, he awarded communicable disease work 36 points and sanitation 9. He pleaded with his fellow health officers to adopt a "more rational perspective" by directing their efforts toward the isolation of people with infectious diseases, medical inspection, vaccination, and laboratory investigation of milk supplies, instead of to the abatement of nuisances and municipal housekeeping.[9]

To varying degrees and over time, other health officers and health departments around the country adopted many of these precepts. The health department in Milwaukee, Wisconsin, for example, had given up control over water and garbage to the department of public works by 1911. The focus of local health work changed from citywide sanitation and disease control to closer observation of individuals, their habits, and their contagiousness. Concomitant with this shift, bacteriological laboratories with their microbe-identifying capabilities became crucial in selecting the people and the problems to which health departments should attend. The "golden age of bacteriology," in the words of Charles-Edward Amory Winslow, one of the country's leading public health theoreticians, recognized "the laboratory . . . [as] the scientific foundation of the public health campaign in America."[10]

Did the new focus on laboratory-based bacteriology that Chapin and others brought into public health narrow the scope of the enterprise or merely change its emphasis? Did public health practitioners don blinders in the new century with an extreme focus on particular bacteria and their transmission, reducing the range of their operations to microbe-centered laboratory investigations? Through the lens of one of Chapin's favorite subjects, healthy carriers, specifically healthy carriers of typhoid fever, it is possible to observe the evolution from environmental sanitation to personal cleanliness, from the filth theory to the germ theory of disease, from populations to individuals—in other words, to evaluate the ways in which bacteriology changed public health activity.

In the mid-19th century, when scientists first distinguished typhoid fever from typhus, the former was already causing significant public health problems in American cities.[11] A water- and food-borne systemic bacterial infection, typhoid brought sustained fever, headache, malaise, and gastrointestinal problems (constipation more often than diarrhea) to its victims. Although many mild cases occurred, typhoid carried a case fatality rate of about 10 percent. It struck most harshly those cities that sent untreated lake or river water through the pipes; thus it responded well to the implementation of water-filtration systems and to sanitation efforts instituted in many cities during the last third of the 19th century or in the early years of the 20th century.[12] Often it took repeated epidemics to convince legislators or taxpayers to spend the large sums necessary to clean up the water supply. Milwaukee, for example, did not eliminate the "typhoid highball" drawn from the city water pipes until citizens, after suffering from a major bout of diarrhea in 1916, finally approved a bond for a new sewage-treatment plant.[13] Typhoid fever exemplified the effectiveness of sanitation practices based on the filth theory of disease, even as bacteriology supplanted that theory. When the *Salmonella typhi* bacillus was identified (1880) and traced to contaminated water supplies, it underscored the necessity of providing clean water in urban environments. As more and more cities responded in the early 20th century, typhoid fever diminished nationwide as a major cause of morbidity and mortality (see Fig. 1).

Yet typhoid fever did not disappear, no matter how sophisticated the filtration method nor how thorough the oversight of sewage disposal and urban cleanliness. In the first decade of the 20th century, with the help of bacteriological investigations,

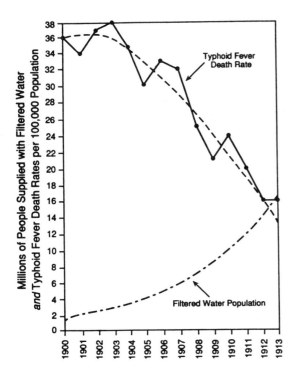

Fig. 1: Growth of water filtration and decrease in typhoid fever death rate in the registration cities of the United States. Graph from George A. Johnson, "The typhoid toll," *Journal of the American Water Works Association*, 1916, 3: 308.

medical scientists began to understand why there were still cases of typhoid after sanitation efforts had significantly lowered the incidence of the disease. "Residual" typhoid seemed to be caused by apparently healthy persons, either those recovered from the disease or those who could not remember being sick at all but who nevertheless harbored the bacillus and transmitted it to others. Labeled "germ distributors," "chronic carriers," or "healthy carriers," these individuals became important foci for bacteriologically oriented public health officials, potent illustrations of how the new science, by creating a new taxonomy, aided the fight against disease.

Friedrich Loeffler, the German bacteriologist, had posited the concept of a healthy carrier of disease as early as 1884 for diphtheria, and in 1893 Robert Koch did the same for cholera. In New York City, William H. Park and A. L. Beebe con-

vincingly established the carrier principle in cases of diphtheria in 1893, concluding that virulent diphtheria bacilli could be found in about 1 percent of the healthy throats in New York City.[14] By 1900 studies had revealed that typhoid fever, too, could be transmitted by healthy recovered persons. In 1902 Koch published a paper on the subject, and in subsequent years a few typhoid carriers were discovered and described in Europe.[15] By the time bacteriologists had identified the first healthy carrier to be charted and followed in North America, in 1907, scientists had firmly established the idea that healthy people could carry *Salmonella typhi* in their excreta.

As their experience with the carriers grew, researchers elaborated the details of how healthy people harbored the disease and transmitted it to others. By 1915 American scientists, familiar with the European work, shared data from various centers around the country, including San Francisco, St. Louis, Syracuse, Boston, Chicago, Wisconsin, and Minnesota. In these early decades of the 20th century American bacteriologists came to understand that the carrier state could follow either mild, even subclinical, cases or acute illnesses; that it seemed to develop in approximately 3 percent of recovered patients; and that more women than men seemed to be chronically affected. *Salmonella typhi* could be isolated from blood during the first week of sickness and in the urine or feces later on; the bacilli often lodged in the gallbladder. Carriers transmitted the disease through water or food contaminated by their feces or urine.[16]

Mary Mallon has the distinction of being the first typhoid fever carrier to be identified and charted in North America. An Irish-born cook, Mallon hired out her culinary services to wealthy New York–area families. In the summer of 1906 she found employment in the rented summer home of a New York banker, Charles Henry Warren, in Oyster Bay, Long Island. When typhoid fever struck six people in the household of eleven, the owner of the home, George Thompson of New York City, thinking he would be unable to rent the property again unless the mystery of these cases could be solved, hired George Soper, a civil engineer known for his detailed epidemiological analyses of typhoid fever epidemics, to investigate the outbreak.[17] Soper's inquiry first ruled out the common causes of such outbreaks, contaminated water and milk, and he systematically discarded other possible sources of infection, including clams from the bay and other foodstuffs consumed in the house and any contacts family members had with people outside. When none of these alternatives yielded viable explanations, Soper pressed family members to remember any other distinctive events that might have taken place during the period. Through close scrutiny Soper found his clue: he learned that the family had changed cooks during the weeks in question.[18]

Believing the cook who left the family three weeks after the outbreak to be a prime suspect, Soper worked next on trying to find her. Although the family claimed that the cook had been in perfect health, Soper, who undoubtedly knew about healthy carriers of other diseases and might have been reading (as he later claimed) literature from Germany and elsewhere in Europe about healthy typhoid fever carriers, expressed confidence that such a person was most likely to have caused this particular household outbreak.[19] By tracing the cook's job history before her arrival in Oyster Bay in August 1906, he found eight families who had previously employed her; in seven of them typhoid fever had followed her stay.[20] Soper was convinced that if Mary Mallon could be found, and her feces and urine tested, he could prove in the laboratory what his epidemiological study had already suggested: that she, although healthy, had transmitted typhoid fever to those who unsuspectingly ate the food she prepared.

Mary Mallon's apprehension in March 1907, in the Park Avenue home in which she was then employed, was dramatic and, for her, frightening. George Soper appeared without warning and tried to explain that Mallon was spreading disease and death through her cooking. The story seemed preposterous to the healthy Mallon, who promptly threw him out of the house.[21] But Soper persevered. He convinced the New York City health department, led by Hermann Biggs, that the epidemiological evidence implicating Mallon was strong enough that they should pursue her and gather specimens of her blood, feces, and urine. When approached by city health officer S. Josephine Baker, Mallon still did not understand the demands (how could she, being the first of her kind?) and refused to provide the evidence. Baker called in the police to help, and the officers took Mallon by force and

against her will to the Willard Parker Hospital, New York's receiving unit for those suffering from contagious diseases. There they subjected her excreta to careful laboratory analysis.[22] Finding high concentrations of typhoid bacilli in her feces, authorities kept her in health department custody, moving her to an isolation cottage on the grounds of the Riverside Hospital on North Brother Island. In 1909, more than two years after her apprehension, she unsuccessfully sued in court for release; in 1910 a new health commissioner unilaterally decided that the department had kept her long enough and freed her. She was arrested again in 1915 after officials traced a hospital outbreak of typhoid fever to her kitchen. This time she stayed in custody until her death in 1938. She lived in health department–imposed isolation for a total of 26½ years.

In some ways Mallon's case epitomized the health benefits that accrued from application of the new science of bacteriology, showing the importance of the laboratory as a supplement to or even substitute for physical examinations. By targeting the attack on typhoid to those individuals who could be identified as dangerous by laboratory methods, the health department streamlined its efforts. By getting Mary Mallon off the streets and out of the homes of citizens, health officials believed that they were protecting the public's health in the best possible way. Yet this protection came at the expense of Mallon, who was denied her liberty for over 26 years. This extreme control over the behavior of a single healthy carrier could not, for reasons both practical and political, be repeated for the hundreds of other carriers that would be identified in New York City. As the *Medical Record* put it: "It is evident that they cannot all be segregated and kept prisoners. . . . It would be difficult to obtain popular sanction to such interference with the liberty of apparently healthy individuals, and even if the measure were recognized as justifiable the number of bacilli carriers would render it difficult of execution."[23] Thus, using what they learned from their experience with Mallon, the health officials developed treatment protocols and regulatory guidelines that could be widely applied. The remainder of this essay examines the laboratory investigations, the treatment procedures and guidelines, and the concept of indefinite isolation to determine whether these activities, based heavily on the science of bacteriology, reduced the scope of health department work.

The laboratory to which Mallon's specimens were brought, part of the Division of Pathology, Bacteriology, and Disinfection within the health department, was the dream of Hermann Biggs; Charles Chapin described it as "perhaps the most important step in modernizing public health practice in the United States." Biggs hired William H. Park, a physician trained at Columbia University College of Physicians and Surgeons and in the scientific laboratories of Vienna, to organize the bacteriological laboratory in 1893, and he remained as its director until 1936. Park and his colleagues carried out studies that put New York in the forefront of bacteriological investigation in the United States. Among their more important works were their studies of healthy carriers.[24] Mary Mallon, because of her long stay under department auspices, provided much of their longitudinal data on typhoid fever carriers.[25]

Mallon's feces received close laboratory attention from 20 March 1907, when she was first brought to the hospital, until she sued for release, the last examination recorded in the court records occurring on 16 June 1909. More examinations were undoubtedly made after her second incarceration in 1915, but for these the data are not available. In the 28 months for which we have data, health officers collected 162 fecal specimens, an average of more than one a week. In Mallon's words; "When I first came here they took two Blood Cultures and feces went down three times per week say Monday Wednesday & Friday respectfully [sic] until the latter part of June after that they only got the feces once a week which was on Wednesday . . . when I first came here I was so nervous & almost prostrated with grief and trouble."[26]

The laboratory analyses revealed that Mallon was an intermittent carrier of typhoid fever. Repeatedly over this time her feces contained no typhoid bacilli at all; the laboratory reports were negative for 12 consecutive examinations from 16 September through 14 October 1907. In an on-again, off-again pattern over the 28 months, 119 of the 162 cultures tested positive, 43 negative. Her urine consistently tested negative.

While the city was conducting its laboratory tests, Mallon arranged for her urine and feces to be analyzed by a private company, the Ferguson Labora-

tories. The specimens were brought to them by Mr. A. Briehof, the man with whom Mallon lived when Soper first found her. George Ferguson conducted 10 tests on Mallon's urine and feces between 1 August 1908 and 30 April 1909, concluding, "I would state that *none* of the specimens submitted by you, of urine and feces, have shown typhoid colonies."[27] Comparison of the dates of Ferguson's analysis with those of the health department's analysis reveals that of the 10 negatives the private laboratory found, 8 were reported during weeks when Mallon's feces tested positive in the city laboratory. Mallon herself, even while denying the validity of the laboratory test to confirm the presence of a disease that she insisted she had never experienced, used the Ferguson Laboratories' negative findings to bolster her case in court.[28]

Similarly, the health department used the predominantly positive city laboratory reports to make its legal case, to insist to the court that it was necessary to isolate Mary Mallon. Dr. Fred S. Westmoreland, the resident physician at Riverside Hospital on North Brother Island, in whose care Mallon was placed, offered the following conclusion before the judge:

> A bacteriological examination revealed the fact that fully thirty percent of the bacteria voided with the feces were of typhoid bacilli; the urine was negative. . . . Weekly examinations of the stools have usually revealed large numbers of bacilli. . . . In view of the foregoing and owing to the large quantities of typhoid bacilli existing in the alimentary tract, or gall bladder of the patient and her occupation as a cook or the fact that she may at any time come in contact with people wherein they would be likely to be infected with the typhoid bacilli, the Department of Health concluded that the patient would be a dangerous person and a constant menace to the public health to be at large; and, consequently, . . . decided, after careful consideration and acting upon their examination of the patient, to place her in a contagious hospital and isolate her from the general public.

The laboratory reports formed the central argument in the health officials' case. Without them, the court probably would not have kept Mallon in isolation; with them, the case seemed clear to the Honorable Mitchell L. Erlanger, who ordered that the writ should be dismissed and "that the said petitioner, Mary Mallen [*sic*], be and she hereby is re-manded to the custody of the Board of Health of the City of New York."[29] Even though the labora-tories had issued contradictory reports, the seem-ingly incontrovertible evidence of repeated positive typhoid cultures in the city laboratory reports made Mary Mallon's danger palpable.

Given the laboratory evidence, and again illus-trating its narrowed focus on that evidence, the health department addressed the question of how to eliminate Mallon's infectiveness. According to Westmoreland's testimony, "Hexamethylenamin in doses gradually increasing from one hundred to one hundred and fifty grains a day has been given frequently with no apparent benefit. Attention to diet and mild laxative has caused the greatest re-duction but not [the bacilli's] disappearance." Mal-lon provided more details of her therapy, which she portrayed as punitive: "In spite of the medical staff Dr. Wilson ordered me Urotropin I got that on & off for a year sometimes the[y] had it & sometimes the[y] did not. I took the Urotropin for about 3 months all told during the whole year if I should have continued it would certainly have Killed me for it was very Severe[.]" Mallon indicated that the physicians had also tried brewer's yeast, but "at first I did not take it for Im a little afraid of the people & I have a good right for when I came to the Depart-ment the[y] said [the bacteria] were in my track later another said they were in the muscels [*sic*] of my bowels & laterly the[y] thought of the gall Bladder."[30] This lack of medical precision con-vinced Mallon that the doctors did not know what they were doing.

In addition to trying various drugs, health offi-cers urged Mallon to have her gallbladder surgi-cally removed:

> Dr. Studiford said to this man [Mallon's friend; perhaps Briehof] go and ask Mary Mallon & en-veigle [*sic*] her to have an Operation performed to have her Gall Bladder removed. She'll have the best Surgeon in town to do the Cutting. I said no[.] no Knife will be put on me I've nothing the matter with my gall bladder. Dr. Wilson asked me the very same question I also told him no then he replied it might not do you any good also the Supervising nurse asked me to have an operation performed. I also told her no & she made the remark would it not be better for you to have it done than remain here I told her no.

Although the physicians urging surgery on Mallon did not inform her of its poor record, in 1921 the department of health had followed five carriers who had agreed to the removal of their gallblad-ders, "all of them without success."[31] It seems that Mallon's skepticism was warranted.

The emphasis on laboratory findings, together with the drug therapy and proposed surgery, indi-cates the extent to which health department think-ing concentrated on the bacteria themselves rather than on a more comprehensive approach to elimi-nating the dangers Mary Mallon posed. Health of-ficers asked how they could kill the bacilli that Mal-lon was transmitting; they did not ask how to stop a carrier cook from cooking. If they had considered the second question, they might have focused on teaching Mallon ways to cut her infectivity by wash-ing her hands thoroughly and by not preparing raw food for anyone else and on helping her learn new skills. She was literate and spunky; her talents presumably could have been turned to other uses if health officials had put their minds to it. By the time of Mallon's second incarceration there was precedent for moving in this direction rather than insisting on strict isolation of healthy typhoid carri-ers. At the Pasteur Institute in Paris the bacteriolo-gist Ilya Metchnikoff had found employment in a library for a healthy carrier whose case interested him. In 1918 New York State began subsidizing the incomes of those carriers who were having diffi-culty finding adequate employment outside the food industry. Indeed, the New York City health department itself maintained the numerous car-riers it identified outside of isolation, with an increasingly elaborate set of guidelines governing their behavior.[32]

Health officials outside of New York City—in-cluding, most significantly, Charles Chapin of Prov-idence and Milton L. Rosenau of the U.S. Public Health Service's hygienic laboratory, both advo-cates for bacteriologically based changes in public health activity—objected to the stringent isolation of Mary Mallon. As Rosenau stated at the very first meeting at which George Soper had revealed Mal-lon's capture and isolation, in April 1907, "It is not necessary to imprison the bacillus carrier; it is suf-ficient to restrict the activities of such an individ-ual." Chapin concluded his 1910 analysis of how healthy carriers should be handled with an indict-ment of the incarceration of Mallon: "What result is secured by keeping her in confinement, other than the placing of discredit on public health work, it is difficult to see."[33] Both believed isolation too strong a penalty, as well as an impractical remedy, for healthy carriers. For these two prominent public health officials, bacteriology, while emphasizing the importance of germ hosts, did not reduce the problem to one of removing such carriers from society.[34]

Yet Mary Mallon herself remained in lifelong isolation. Walking the streets of New York City, she would have posed no health hazard; working in an an occupation other than cooking food for others, she would not have been a "constant menace to the public health." Nonetheless, there is no evidence that health officers tried to retrain Mallon until the final years of her isolation. When the new health commissioner released Mallon in 1910, he helped her find work in a laundry, but he did not succeed in tapping her interest or potential, and the change in occupation did not stick. Only in the last years of her life, immediately before her paralyzing stroke in 1932, did a physician at Riverside Hospital train Mallon to work in the hospital laboratory.[35] The energy and commitment of the particular bacteriologically guided health officials who supervised her isolation rested not with social rehabilitation but with the pathogenic bacilli—clearly indicating a narrowing of the focus of health-related work.

Elsewhere in New York's health department activity there is evidence of broader thinking about the problems that healthy carriers posed. Soon after Mallon's apprehension the city health department discovered many other healthy carriers of typhoid fever, most of whom were working in food-handling occupations. The numbers grew with the years; in 1908, the city watched only five carriers (including Mallon); by 1920, 85 were under observation; and by 1938, the year of Mallon's death, the health department listed 394 healthy typhoid fever carriers in the city.[36] In its efforts to limit the potential dangers such people posed, the health department implemented guidelines for their regulation, intended to curtail behaviors that might be dangerous to the public health. In 1916, the year following Mary Mallon's final incarceration, the rules were in place, although modifications continued over the years. The guidelines bear close examination. Through them we can understand how health

officials, after almost 10 years of experience with healthy carriers of typhoid fever, came to interpret the problems they posed. In these guidelines we add another dimension to our understanding of how germ theory affected public health work.

Health department rules forbade any person sick with an infectious disease to handle food. Although rarely enforced in the period before healthy carriers were known to exist, no doubt because of the enormity of the problem, this precept provided the precedent upon which regulation of healthy carriers rested. After Mallon's second incarceration, health inspectors took the potential danger of healthy carriers more seriously than they had previously. In 1916 the rules that had been written to curtail the activity of the sick began to be applied also to healthy carriers of infectious diseases. The health department Bureau of Preventable Diseases, through its occupational clinic, began requiring that the estimated ninety thousand food handlers in the city's hotels and restaurants produce "certificates of freedom from infectious disease" before they could be employed. This regulation included the mandate that fecal specimens of healthy people must be free from pathogenic bacteria.[37]

The health department organized its work of finding healthy typhoid carriers by following people sick with the disease as they recovered. Health officials declared well only those sufferers whose urine and feces tested negative for typhoid bacilli three consecutive times. Typhoid fever convalescents holding food-handling jobs could not return to work without health department permission, which was forthcoming only following laboratory analysis.[38] The names of those unfortunate few who continued to test positive (about 10 percent for three months and 2–5 percent chronically) were entered on a health department list of typhoid carriers. Health officers kept in close contact with these carriers, required them to submit specimens for laboratory analysis on a regular basis, and insisted that they not handle any food. The list grew rapidly, and by 1918 the health officers wrote: "The problem of supervising typhoid carriers requires constant watchfulness and study. At present we have a record of 70 chronic carriers, three of whom are detained forcibly in Department hospitals, the others being permitted to stay at home under constant supervision. The examination of food handlers for discovery of typhoid carriers is a most important activity." Despite the difficulties, the *Annual Report* optimistically claimed that "public health is purchasable."[39]

In 1919 health officers noted that only two typhoid fever carriers were restrained in city hospitals, that three had absconded from health department purview, and that the other 62 identified chronic carriers were living under conditions that "in all cases were excellent. They had been carefully instructed how to protect others, and they carefully observe these instructions." Although the list of chronic carriers continued to grow—in New York City, in the state, and in the country—it no longer seemed necessary forcibly to detain the food handlers among them. In 1922 an absconder from New York City, a carrier who had reportedly caused an outbreak of 87 cases and two deaths, was found by New Jersey health authorities, who blamed him for still another outbreak that had resulted in 35 cases and three deaths. Rather than incarcerating him for repeated violations and breaking parole, the health officers added him to the list of carriers and concluded the case with the remark, "This carrier is now employed in this City as a laborer in building construction work and is required to report to us weekly."[40] In 1924 Alphonse Cotils, a bakery and restaurant owner who had been identified as a healthy carrier and forbidden to prepare food in his own establishment, appeared in court after he had violated the terms of his agreement with the health department. Although finding him guilty of the charge that he was preparing food against health department instruction, the judge suspended sentence and let him remain free on the promise that he would no longer handle food. By 1928 the city had examined 270,000 food handlers and that year had traced only one outbreak of typhoid to a known carrier.[41] The rules governing healthy carriers, which allowed them to remain at home and at work as long as the latter did not involve food preparation, seemed to be working.

In 1929, a typical year of typhoid fever control work in New York City, the health department identified 29 new chronic carriers and added them to the official list. Of these, officers had discovered six at the time they had applied for food-handler's permits, for which they needed a routine stool examination. Health inspections had revealed eight cases among recently recovered persons whom the

health department had been tracking. The other 15 came to light during health department investigations of new infectious active cases.[42]

The guidelines for maintaining watch over chronic typhoid fever carriers stated that they "need not be retained in hospitals or institutions if not desired. They will be sent home if home conditions are satisfactory." The health department did not spell out the particulars of "satisfactory" home conditions, although it did elaborate enough to note, "The family must be intelligent and willing to carry out rules of the Department of Health."[43] The vagueness of these statements left room for significant maneuvering and permitted health officials to make decisions about healthy carriers that incorporated some social factors in addition to laboratory findings. Thus authorities could differentiate between a Mary Mallon, whom health officials did not trust to behave in the public's interest, and other healthy carriers whose home conditions or personal attitudes predicted better compliance with the health codes. Although two healthy carriers might have borne equally dangerous pathogenic bacteria, as identified in the laboratory, it was not required that they be treated equally in practice. According to the health department protocols, the carriers' social conditions and even their psychological responses could be considered alongside the laboratory reports in order to evaluate the dangers they presented and determine ways to protect the public from them.

These rules embodied reductionist notions of the role of the laboratory in determining the risk posed by typhoid carriers alongside more comprehensive ideas incorporating a broad appraisal of individual situations. Health officers, because of the magnitude of the job of regulating thousands of food handlers, and in light of growing numbers of healthy carriers among them, developed guidelines that encompassed both particular and general factors. Health officials needed to control disease and found ways to do this without isolating all who tested positive in the laboratory. Toward this end they judged housing conditions, sanitary facilities, and the individual's tractability as they determined the proper handling of typhoid bacilli-carrying healthy people.

The necessity of isolating refractory healthy carriers, such as Mary Mallon, remains to be considered in order to judge the extent of extralaboratory considerations in bacteriologically oriented public health work. The denial of Mallon's personal liberty in the interest of promoting public health was complete; a government agency determined that her total isolation, for the duration of her life, was necessary to protect the public. Why did the New York health department staff agree on this point and insist on keeping Mallon on an island in the East River for 26 years? Did bacteriology demand such extreme treatment?

Public health authorities distinguished carefully in their texts and articles between the handling of persons sick with typhoid fever and those recovered people who continued to be carriers of typhoid fever. Both could transmit the disease to others; yet health officials, from the time of their earliest guidelines, suggested different responses for the two groups. Some public health leaders believed that home isolation was sufficient for typhoid fever sufferers; others promoted the isolation hospital as the best place for treatment of the sick. As Milton J. Rosenau advised, "The proper place to care for typhoid fever is in a suitable hospital. A private home is a poor makeshift for a hospital, and it is unreasonable to turn a household into a hospital for four to eight weeks or longer." In the hospital patients could receive the latest treatments; stools, urine, and sputum could be properly disposed of; and medical and nursing attendance could be systematized. Convalescents could be kept until "the danger of bacillus carrying has passed. This may be determined only by bacteriologic examination of the stools and urine."[44]

While he recommended these strict standard procedures for persons sick with the disease, Rosenau offered different recommendations for healthy recovered people who continued to harbor the bacilli. His textbook *Preventive Medicine and Hygiene*, first appearing in 1913 and enjoying its seventh edition by 1951, became the standard for the field and influenced generations of public health workers. The 1935 edition, the last one Rosenau himself wrote, provided this advice:

> We cannot lightly imprison persons in good health, especially in the case of breadwinners, even though they be a menace to others. In some infections there are so many carriers that it would require military rule to carry out such a plan. Fortu-

nately in most cases absolute quarantine is not necessary. Sanitary isolation is sufficient. Thus the danger from a typhoid carrier may be neutralized if the person exercises scrupulous and intelligent cleanliness, and is not allowed to handle food intended for others. Such a person might well engage as carpenter, banker, seamstress, etc., without endangering his fellowmen. . . . The price of liberty is "good behavior."[45]

Rosenau's analysis identified the various factors he thought should protect healthy carriers from hospitalization or incarceration: their health; their economic value in the family; their frequency in the population; their behavior (habits and jobs), which could be effectively controlled outside the hospital. Despite obtaining identical laboratory reports for sick and healthy persons infected with typhoid bacilli, bacteriologically oriented public health officials did not recommend the same regulations for both groups. The sick should be isolated in homes or hospitals; the healthy carriers could be allowed to walk about on the city streets. The laboratory thus did not define the full scope of the public health problem or its solution.

While insisting that individuals sick with the disease remain in isolation, Rosenau did not believe that healthy carriers should be isolated. These people were not sick; they felt well. The experienced health officer knew that healthy citizens would not easily comply with orders to remove themselves from society. He realized that despite the power of laboratory definitions to portray the healthy as disease transmitters, the public would not support the reduction of public policy to a laboratory test determining the presence or absence of fecal bacilli.

The second part of Rosenau's formula reveals even more of the impact of social values on public health activity. A carrier's financial responsibilities made it inconceivable for a health officer to contemplate denying his (or her) livelihood. A person's social and economic position as head of a household provided some immunity against the state's authority to interfere. The suspended sentence Alphonse Cotils received when found guilty of violating the health codes probably reflected this viewpoint. The judge let him go because he was healthy and because he was an established businessman.[46]

Rosenau felt that the very numbers of people involved precluded the institutionalization of healthy carriers. By the 1930s New York City alone had identified almost 400 healthy carriers of typhoid fever, adding around 20 new cases each year, and many times that number of carriers of other diseases. New York State listed over 300 typhoid fever carriers in 1933, claiming to add approximately 60 new chronic carriers each year. Nationwide, estimates rose into the thousands for typhoid fever and into the tens and hundreds of thousands for other diseases such as diphtheria.[47] While it might have been physically possible to isolate these growing numbers of healthy carriers, such an approach was neither politically nor, in city health departments already operating on limited budgets, economically realistic.

Rosenau's precepts for healthy-carrier control reflected the common optimism of early 20th-century bacteriologists, with their faith in the potential for altering human behavior. He assumed that both scrupulous personal cleanliness and cooperation in changing jobs offered workable solutions to public health problems. His optimism was echoed in New York City health officers' statements about the level of cooperation they actually received from most carriers. Rosenau's interpretations of both the problem posed by healthy carriers and its solution led him to add social considerations to the laboratory findings.

Charles Chapin offered a similar analysis and reached a similar conclusion about the inadvisability of isolating healthy carriers of typhoid fever. In his 1910 text, *Sources and Modes of Infection,* he found the suggestion to isolate healthy carriers impossible, unjust, and ineffectual:

> There certainly would be most energetic opposition on the part of the public, which probably would ultimately be sustained by the courts. The health officer who attempted to isolate convalescents until bacilli were no longer to be found in their urine, would be in an awkward position if he allowed chronic carriers to go at large, and he would be in a still more awkward position if he attempted to isolate all chronic carriers indefinitely. . . . To attempt to isolate 6,000 carriers [his estimate nationwide of each year's new additions] would of course be futile. . . . To isolate the small fraction of carriers who can be discovered is practically useless, and therefore unjust. It may be, and

> probably is, wise to regulate the life [*sic*] of such carriers as may be discovered, and at times to forbid their engaging in certain occupations, such as those of cook, waitress and milk dealer.[48]

Chapin advocated retraining carriers to allow them to find jobs outside the food industry, rather than indefinite isolation; he applied this position to Mary Mallon and to all healthy typhoid fever carriers.

The limits of laboratory findings become evident in this example of the differential treatment of sick typhoid fever sufferers and healthy carriers of typhoid fever. Both had equally virulent and dangerous bacilli in their excreta; both could transmit the bacilli to others. Chapin argued that, although both healthy and sick were "equally dangerous potentially," in fact the "well person moving freely about may be more dangerous to the community than the sick person who is confined to the house."[49] Sick people felt too ill to prepare food for others or carry on their normal duties; the healthy continued their usual patterns. Chapin, Rosenau, and other public health officials advocated and practiced differential treatment. Public health workers sought control of carriers in ways that acknowledged their health, their place in the community, and the near impossibility of constraining them all, considerations well beyond the laboratory findings.

Notwithstanding such advice about the handling of healthy carriers, New York health officials continued to justify their long-term isolation of Mary Mallon, even while not insisting on similar treatment of other recalcitrant healthy carriers. Presumably aiming to make an example of her in order to discourage others from resisting public health authority, in their public utterances they emphasized her negative character traits above her chronic bacilli carrying. In the words of health officer S. Josephine Baker, the physician who first brought Mallon to the hospital in 1907, she was a "destroying angel" whose "own bad behavior . . . inevitably led to her doom. . . . [S]he never learned to listen to reason. . . . [S]he was constitutionally incapable of believing all this mystery about germs." "The only answer," Baker concluded, "was to keep her in the custody of the Department, out of contact with other people's food." Mary Mallon's "perversity," as Soper labeled it, indeed caused such frustration among those trying to work with her that it became an important factor in her continuing incarceration, even though by the time of her second apprehension her mood showed resignation and defeat more than feistiness and resistance. Mallon's social class also became a subject of discussion, affecting her isolation. Soper made disparaging remarks to the press about the cleanliness of cooks as a group in the context of discrediting Mallon, even though he clearly did not intend to lock up all of New York's cooks. Officials similarly used her seeming rejection of gender norms—the determined set of her jaw, her "masculine" mind and walk—in their justification of their treatment of her. These arguably extraneous traits—the degree of Mallon's femininity and the social status of cooks—became aspects of the abrogation of her liberty. As the *New York World-Telegram* put it, Mary Mallon "was not imbued with that sweet reasonableness which would have allowed her to listen to the explanations of learned men about her particular case."[50] Personal qualities as much as germ carrying became reasons for long-term isolation.

To Baker and to George Soper, a central reason for keeping Mary Mallon isolated was her refusal to accept the authority of bacteriological findings. She persisted in denying that she had ever been sick with typhoid and insisted that she was "in perfect physical condition" and was "not in any way or any degree a menace to the community or any part thereof." Mallon's denial of the validity of the bacteriological findings—perhaps because her own laboratory informed her that her stools showed no typhoid bacilli; perhaps because her rival world view credited experience above science—became part of the indictment against her. She did not cooperate because she did not believe what the doctors told her. She did not believe their claims, she said, at least in part because they kept changing them, sometimes telling her the bacilli lodged in her gallbladder and other times locating them in the muscles of her bowels. The confusion that she felt is understandable given that she was the first healthy carrier to be followed in America and that health officers were learning through their experience with her. Nonetheless, the health officials were quick to censor her for not accepting their scientific explanations, even though those explanations were contradictory or at best incomplete. They used her

lack of scientific belief to call upon social criteria to justify her incarceration. In the process they created a particularly negative public image of Mary Mallon, portraying her as a social deviant and an unfeminine woman, fabricating a caricature of "Typhoid Mary" in place of the real Mary Mallon.[51]

It was one thing to develop bacteriological explanations for public health activity; it was obviously another to apply that science in the public domain. Early 20th-century proponents of bacteriology could no more isolate disease from its environmental and social context than could their predecessors who were driven by the filth theory of disease. Bacteriologically oriented public health had the potential for restricting determinations of disease causation and prevention-oriented activities to the germ itself, and some people in the field indeed tried to focus primarily on the microbe. Certainly the laboratory became, in theory and in practice, a mainstay of public health work in the early 20th century. But the laboratory could not provide all the answers to health officers trying to protect the public from contagious diseases. The real world impinged on their deliberations, and of necessity the health officers in the field adopted a broad approach to disease control, one that in many respects resembled earlier practices. Bacteriology narrowed the theoretical underpinnings of contagious disease control, and it elevated the status of laboratory data in diagnosing diseases. But it did not to any significant extent contract the comprehensiveness of public health work in contagious disease control.

Mary Mallon, who would have remained anonymous without bacteriology, personally taught health officials more than did the impersonal laboratory findings. Through her rejection of scientific claims and explanations and her insistence on a broader, experientially based authority to define public health problems, she posed a dilemma for health authorities. Officials learned that, especially in the face of noncooperating citizens, they needed to define and try to solve public health problems through inclusive considerations that took into account laboratory findings and a range of personal and social characteristics. Science, when applied on the streets and in the tenements of urban America, became tainted by the values, limitations, and commitments of that world. Mary Mallon's historical significance as the classic healthy carrier of typhoid fever in America grows when we see her also as a reminder that a scientific revolution such as the one bacteriology began retains many roots in previous paradigms.[52]

The laboratory-based science of bacteriology profoundly affected clinical medicine and public health. A theoretical breakthrough of huge proportions and wide-ranging implications, germ theory constituted a genuinely new approach to understanding the etiology of many diseases and, as such, significantly altered subsequent prevention and therapeutic efforts. The caution introduced in this study is that germ seeking itself—despite its major importance—constituted only one of many determinants of public health activity. A labyrinth of cultural values, social and political contexts, and personal stories tempered any early impetus to reductionist approaches that microorganisms might have provided. Mary Mallon, in her epic battle with public health officials, ultimately won a pyrrhic victory; she paid a heavy personal cost, but she did make her point. The new public health theorists learned from her and from subsequent interactions with healthy carriers that laboratory findings needed to be moderated with the socially sensitive policies that were also characteristic of prebacteriological public health practice.

NOTES

I want to thank many people for their help and encouragement in my Mary Mallon project, of which this paper is a part. John Duffy and Daniel Fox, both of whom have written on public health in New York, were most generous in discussing sources and resources with me. David Rosner and Joel Howell offered helpful suggestions along the way. Joan Jacobs Brumberg, Susan Stanford Friedman, Linda Gordon, and John Harley Warner read drafts of earlier versions of the work and provided essential insights. R. Alta Charo, Hendrik Hartog, and Leslie Reagan provided needed advice concerning the legal sources. Research assistance by Dawn Corley and Sarah A. Leavitt was crucial to the completion of my first paper on the project in time for its earliest presentation at Yale University in February 1990. Lewis Leavitt offered consultation on the medical issues and critical judgments throughout. I am grateful for all the suggestions and questions I received from my colleagues at the University of Wisconsin, Yale University, Harvard University, and the Canadian Association for Medical History, when I presented various versions of my work on Mary Mallon. I am especially grateful for the informed readings and thoughtful criticisms I received from the four anonymous referees of this paper and the *Isis* editorial staff.

1 Paul Starr, *The Social Transformation of American Medicine* (New York: Basic Books, 1982), pp. 190, 196. See also, e.g., Barbara Gutmann Rosenkrantz, *Public Health and the State: Changing Views in Massachusetts, 1842–1936* (Cambridge, Mass.: Harvard Univ. Press, 1972); John Duffy, *The Sanitarians: A History of American Public Health* (Urbana: Univ. of Illinois Press, 1990); Charles-Edward Amory Winslow, *Conquest of Epidemic Disease: A Chapter in the History of Ideas* (Princeton, N.J.: Princeton Univ. Press, 1943; Madison: Univ. of Wisconsin Press, 1980); Frederic P. Gorham, "The history of bacteriology and its contribution to public health work," in *A Half Century of Public Health*, ed. Mazÿck Procher Ravenel (New York: Arno, 1970), pp. 66–93; and George Rosen, *A History of Public Health* (New York: MD Publications, 1958). For contemporary critics see Ivan Illich, *Medical Nemesis: The Expropriation of Health* (New York: Pantheon, 1976).

2 Allan M. Brandt, *No Magic Bullet: A Social History of Venereal Disease in the United States since 1880* (New York: Oxford Univ. Press, 1985); John Ettling, *The Germ of Laziness: Rockefeller Philanthropy and Public Health in the New South* (Cambridge, Mass.: Harvard Univ. Press, 1981); Nancy Tomes, "The private side of public health: sanitary science, domestic hygiene, and the germ theory, 1870–1900," *Bull. Hist. Med.*, 1990, *64*: 509–539 (ch. 33, this book); and Tomes, "The wages of dirt were death: women and domestic hygiene, 1870–1930," paper presented at the annual meeting of the Organization of American Historians, Louisville, Kentucky, 1991. See also Naomi Rogers, *Dirt and Disease: Polio in America before FDR* (New Brunswick, N.J.: Rutgers Univ. Press, 1992); and Barbara Bates, *Bargaining for Life: A Social History of Tuberculosis, 1876–1938* (Philadelphia: Univ. of Pennsylvania Press, 1992). On the connections between experimental science and clinical medicine see Gerald L. Geison, "'Divided we stand': physiologists and clinicians in the American context," in *The Therapeutic Revolution: Essays in the Social History of American Medicine*, ed. Morris J. Vogel and Charles E. Rosenberg (Philadelphia: Univ. of Pennsylvania Press, 1979), pp. 67–90 (ch. 7, this book); Russell C. Maulitz, "'Physician versus bacteriologist': the ideology of science in clinical medicine," *ibid.*, pp. 91–107; and John Harley Warner, "Ideals of science and their dis-

contents in late nineteenth-century American medicine," *Isis*, 1991, *82*: 454–478.

3   Nineteenth-century public health work can be followed best through local studies. See, e.g., John Duffy, *A History of Public Health in New York City*, 2 vols. (New York: Russell Sage, 1968, 1974); Judith Walzer Leavitt, *The Healthiest City: Milwaukee and the Politics of Health Reform* (Princeton, N.J.: Princeton Univ. Press, 1982); Stuart Galishoff, *Newark, the Nation's Unhealthiest City, 1832–1895* (New Brunswick, N.J.: Rutgers Univ. Press, 1988); and Galishoff, *Safeguarding the Public Health: Newark, 1895–1918* (Westport, Conn.: Greenwood, 1975).

4   Some historians have looked at parts of this issue. See, e.g., Lloyd Stevenson, "Science down the drain: on the hostility of certain sanitarians to animal experimentation, bacteriology, and immunology," *Bull. Hist. Med.*, 1955, *29*: 1–26; William Rothstein, "Bacteriology and the medical profession," in *American Physicians in the Nineteenth Century: From Sects to Science* (Baltimore: Johns Hopkins Univ. Press, 1972), pp. 261–281; Phyllis Allen Richmond, "American attitudes toward the germ theory of disease (1860–1880)," *J. Hist. Med.*, 1954, *9*: 428–454; Howard D. Kramer, "The germ theory and the early public health program in the United States," *Bull. Hist. Med.*, 1948, *22*: 233–247; and Barbara Gutmann Rosenkrantz, "Cart before horse: theory, practice, and professional image in American public health, 1810–1920," *J. Hist. Med.*, 1974, *29*: 55–73.

5   James H. Cassedy, *Charles V. Chapin and the Public Health Movement* (Cambridge, Mass.: Harvard Univ. Press, 1962). Chapin served as superintendent of health for the city of Providence from 1884 through 1931.

6   Charles V. Chapin, "Dirt, disease, and the health officer," in *Papers of Charles V. Chapin, M.D.: A Review of Public Health Realities*, selected by Frederic P. Gorham, ed. by Clarence L. Scamman, with a foreword by Haven Emerson (New York: Commonwealth Fund, 1934), pp. 20–26, quotations on pp. 21, 22.

7   Charles V. Chapin, "The fetich of disinfection," in *Papers*, pp. 65–75. Chapin first presented this paper as an address to the American Medical Association in Boston in June 1906. It was first published in *J. A.M.A.*, 1906, *47*: 574–577.

8   Charles V. Chapin, *How to Avoid Infection* (Harvard Health Talks) (Cambridge, Mass.: Harvard Univ. Press, 1918), pp. 21, 60–61.

9   Charles V. Chapin, "Effective health work," in *Papers*, pp. 37–45 (the scale appears on pp. 41–42); and Chapin, "Dirt, disease," p. 25.

10 C.- E. A. Winslow, *The Evolution and Significance of the Modern Public Health Campaign* (New Haven: Yale Univ. Press, 1923), p. 36 (quotation from 3rd printing, July 1984). On Milwaukee see Leavitt, *Healthiest City*.

11 See, e.g., Michael P. McCarthy, *Typhoid and the Politics of Public Health in Nineteenth-Century Philadelphia* (Philadelphia: American Philosophical Society, 1987). See also Ronald K. Huch, "'Typhoid' Truelsen, water, and politics in Duluth, 1896–1900," *Minnesota History*, 1981, *47*: 189–199; Reimert T. Ravenholt and Sanford P. Lehman, "History, epidemiology, and control of typhoid fever in Seattle," *Medical Times*, 1964, *92*: 342–352; and Terra Ziporyn, "Typhoid fever: a disease of the indifferent," in *Disease in the Popular American Press* (Westport, Conn.: Greenwood, 1988), pp. 71–111. On the medical understanding of the disease see Leonard G. Wilson, "Fevers and science in early nineteenth-century Medicine," *J. Hist. Med.*, 1978, *33*: 386–407; Lloyd G. Stevenson, "Exemplary disease: the typhoid pattern," *ibid.*, 1982, *37*: 159–181; and Dale C. Smith, "Gerhard's distinction between typhoid and typhus and its reception in America, 1833–1860," *Bull. Hist. Med.*, 1980, *54*: 368–385.

12 On typhoid characteristics see Abram S. Benenson, ed., *Control of Communicable Diseases in Man*, 14th ed. (Washington, D.C.: American Public Health Association, 1985), pp. 420–424. The available evidence suggests to me that typhoid fever was, in fact, one of the diseases most responsive to public health sanitation projects at the end of the 19th century. Of 21 cities analyzed by George A. Johnson in 1916, 20 showed significant reductions (between 28 and 85 percent) in mortality from typhoid after the introduction of water-filtration systems: see Johnson, "The typhoid toll," *Journal of the American Water Works Association*, 1916, *3*: 249–326, esp. pp. 304–310. See also Edward Meeker, "The improving health of the United States, 1850–1915," *Explorations Econ. Hist.*, 1972, *9*: 353–373; and Eric Ashby, "Reflections on the costs and benefits of environmental pollution," *Perspect. Biol. & Med.*, 1979, *23*: 7–24. For a slightly less optimistic reading of these data see Gerald N. Grob, "Disease and environment in American history," in *Handbook of Health, Health Care, and the Health Professions*, ed. David Mechanic (New York: Free Press, 1983), p. 18.

13 Leavitt, *Healthiest City*, p. 61. In coastal cities contaminated shellfish continued to constitute a risk even after city waterworks had been adequately protected.

14 Hermann M. Biggs, William H. Park, and Alfred L. Beebe, *Report on Bacteriological Investigations and Diagnosis of Diphtheria from May 4, 1893 to May 4, 1894*, Scientific Bulletin No. 1, Bacteriological Laboratory, Health Department, City of New York (New York: Martin B. Brown, 1895); rpt. 25 *The Carrier State* (New York: Arno, 1977).

15 The most succinct account of the work on healthy carriers is C.- E. A. Winslow, *Conquest of Epidemic Disease*, pp. 337–346. Although the original work identifying healthy carriers was carried out in Europe, Americans quickly joined the process. In the United States see, e.g., Charles Bolduan and W. Carey Noble, "A typhoid bacillus–carrier of forty-six years' standing, and a large outbreak of milk-borne typhoid fever traced to this source," *J.A.M.A.*, 1912, *58*: 7–9; C. W. Gould and G. L. Qualls, "A study of the convalescent carries of typhoid," *ibid.*, pp. 542–546; Frederick G. Novy, "Disease carriers," *Science*, n.s., 5 July 1912, *36*: 1–10; and C. L. Overlander, "The typhoid carrier problem," *Boston Med. & Surg. J.*, 1913, *169*: 37–40. See also Mazÿck Ravenel, "History of a typhoid carrier," *J.A.M.A.*, 1914, *62*: 2029–2030; O. McDaniel and E. M. Wade, "The significance of typhoid carriers in community life, with a practical method of detecting them," *Am. J. Public Hlth.*, 1915, *5*: 764–765; F. M. Meader, "The detection and control of typhoid carriers of disease," *Medical Times*, Sept. 1916, *44*: 278; and A. J. Chesley, H. A. Burns, W. P. Greene, and E. M. Wade, "Three years' experience in the search for typhoid carriers in Minnesota," *J.A.M.A.*, 1917, *68*: 1882–1885. For a lengthy exploration of this early work see John Andrew Mendelsohn, "Typhoid Mary: medical science, the state, and the 'germ carrier'" (undergraduate thesis, Harvard Univ., 1988). I thank Mr. Mendelsohn for his permission (granted through his advisor, Barbara Gutmann Rosenkrantz) for me to read this paper.

16 These understandings were gleaned from the sources in note 15. See also "Typhoid bacillus carriers: their importance and management," *J.A.M.A.*, 8 May 1909, *52*: 1501; and "Typhoid carriers," *ibid.*, 13 June 1908, *50*: 1986–1987. Other factors emerged by the 1940s. For example, while early investigators noted the predominance of women carriers, health officers usually attributed this to the fact that more women than men handled food: Chesley et al., "Three years' experience," p. 1884. In the 1940s studies began to document more women carriers in the population at large, not just among those found in food-handling jobs who were transmitting the disease. In a New York State study published in 1943 the investigators concluded, "The rate of development of the carrier state at all ages is almost twice as high for females as for males." The most striking sex difference found in that study occurred among those aged 40–49, in which 16 precent of cases in females and only 3.5 percent of cases in males resulted in the chronic carrier state: Wendell R. Ames and Morton Robins, "Age and sex as factors in the development of the typhoid carrier state, and a method for estimating carrier prevalence," *Am. J. Public Hlth.*, 1943,

33: 221–230, on p. 223. Medical science in the 1990s acknowledges similar sex and age differentials. I want to thank Dennis Maki, Head of the Section of Infectious Diseases, Department of Medicine, University of Wisconsin Medical School, and Herbert Dupont, Chief of Infectious Diseases at the University of Texas, Houston, for generously consulting with me on this issue.

17 See George Soper, ["The epidemic at Butler, Pa."], *Engineering News*, 1903, *50*: 542; Soper, "Filtration and typhoid," *Engineering Magazine*, 1904, *26*: 754–755; Soper, "The epidemic of typhoid fever at Ithaca, N.Y.," *Journal of the New England Water Works Association*, 1905, *18*: 431–461; and Soper, "The management of the typhoid fever epidemic at Watertown, N.Y., in 1904," *ibid.*, 1908, *21*: 87–163. The last was published after Soper's initial work on Mary Mallon, but the work reported in it was carried out and known earlier. A brief biography of Soper is available from his entry in *Who's Who in America*, 1946–1947, *24*: 2215. Many thanks to Judy Houck for locating this entry on short notice. Soper was a frequent consultant on typhoid fever matters around the country in the early decades of the century, later working on sewage problems, on air and ventilation in subways, and on street-cleaning methods. He received his engineering training at Columbia University (A.M. 1898, Ph.D. 1899). My conclusion that Mary Mallon is the first healthy carrier to be identified and followed in North America rests on Soper's work (see n. 18) and on subsequent renditions of the story. See, e.g., Milton J. Rosenau, *Preventive Medicine and Hygiene*, 6th ed. (New York: Appleton-Century, 1935), p. 141. I looked for the identification or charting of healthy carriers in the United States before Mallon and could not find any.

18 George A. Soper, "The work of a chronic typhoid germ distributor," *J.A.M.A.* 1907, *48*: 2019–2022, offers the first published account of tracing Mary Mallon. See also Soper, "The curious career of Typhoid Mary," *Bull. N.Y. Acad. Med.*, 1939, *15*: 698–712; and Soper, "Typhoid Mary," *Military Surgeon*, 1919, *45*: 1–15.

19 Soper had not sought healthy carriers in his earlier typhoid fever work in Ithaca, Butler, or Watertown, but in his first published article on Mary Mallon he did refer to the German word for chronic carriers, *Typhusbazillentragerin*, indicating his familiarity with the literature: Soper, "Work of a chronic typhoid germ distributor," p. 2022. When he first presented this work to the Biological Society of Washington, D.C., on 6 Apr. 1907, he called attention to Koch's work. See the discussion of the meeting in "Chronic Typhoid Fever Producer," *Science*, n.s., 1907, *25*: 863–865, esp.p. 863 (reports Soper's calling attention

to Koch). In his 1919 recounting of Mary Mallon's discovery Soper wrote: "Somewhat similar investigations had been made in Germany, and I make no claim of originality or for any other credit in her discovery. My interest and experience in the epidemiology of typhoid had been of long standing. I had read the address which Koch had delivered before the Kaiser Wilhelm's Akademie, November 28, 1902, and his investigation into the prevalence of typhoid at Trier, and thought it was one of the most illuminating of documents. In fact it had been the basis of much of the epidemic work with which I had been connected." See Soper, "Typhoid Mary," p. 14.

20 Seven of the places and dates of outbreaks that Soper attributed to Mary Mallon are listed at the end of the laboratory reports submitted by the health department to the New York Supreme Court at Mallon's habeas corpus hearing in July 1909. They were:

4 Sept. 1900, Marmaroneck, N.Y.	1 case
9 Dec. 1901, New York City	1 case
17 June 1902, Dark Harbor, Maine	9 cases
1 June 1904, Sands Point, N.Y.	4 cases
27 Aug. 1906, Oyster Bay, N.Y.	6 cases
21 Sept. 1906, Tuxedo, N.Y.	1 case
23 Jan. 1907, New York City	2 cases

The laboratory reports can be found in the New York County Court House files in New York City.

21 I will tell this story in greater detail and examine other issues in an article now in preparation and in my planned book on Mary Mallon. [See *Typhoid Mary: Captive to the Public's Health* (Boston: Beacon, Press, 1996).]

22 Accounts of the early years of Mallon's incarceration can be found in the Soper articles cited in note 18 and in William H. Park, "Typhoid bacilli carriers," *J.A.M.A.*, 1908, *51*: 981. The published health department reports for the year 1907 note merely that "special studies were made during the year on the so-called typhoid carriers.... A woman who had served as cook in various families during the past five years is known to have infected at least twenty-six people and has caused at least two deaths. This patient was examined from week to week": *Annual Report of the Board of Health of the Department of Health of the City of New York for the Year Ending December 31, 1907* (New York, 1908), p. 321 (hereafter Department of Health, City of New York, *Annual Report*). Any unpublished health department material for this period is either destroyed or unavailable at the present time.

23 "Typhoid bacilli carriers," *Med. Rec.*, 18 May 1907, *71*: 818–819.

24 Chapin was quoted in Winslow, *Conquest of Epidemic Disease*, p. 340. On William Park see W. W. Oliver, *The Man Who Lived for Tomorrow: A Biography of William Hallock Park, M.D.* (New York: Dutton, 1941); Winslow, *Conquest of Epidemic Disease;* Winslow, *The Life of Hermann M. Biggs* (Philadelphia: Lea & Febiger, 1929); and Hans Zinsser, "William Hallock Park, 1863–1939," *Journal of Bacteriology*, 1939, *38*: 1–3. For more on the significance of the laboratory see Jon M. Harkness, "The reception of Pasteur's rabies vaccine in America: an episode in the application of the germ theory of disease" (M.A. paper, History of Science Department, Univ. Wisconsin–Madison, 1987); and John Harley Warner, "The fall and rise of professional mystery: epistemology, authority, and the emergence of laboratory medicine in nineteenth-century America," in *The Laboratory Revolution in Medicine*, ed. Andrew Cunningham and Perry Williams (Cambridge: Cambridge Univ. Press, 1992), pp. 110–141.

25 Of course, New York City followed other carriers and used the data they provided as well. See, e.g., Bolduan and Noble, "Typhoid bacillus–carrier of forty-six years' standing."

26 Mary Mallon's letter, undated, in her own hand, is filed with the laboratory records and other legal papers in the New York County Court House.

27 George Ferguson to Mary Mallon, 30 Apr. 1909, New York County Court House file. My guess is that Briehof (once spelled Nriehof) was the man Soper visited in his efforts to learn more about Mary Mallon. See Soper, "Curious career," pp. 704–705.

28 There are a number of possible explanations for the discrepancy between the city and private laboratory reports. We do not know how Mallon's feces were transported to the private laboratory, and it is possible that the specimens were not fresh when they got there. There is also the possibility that the laboratory did not carry out the work expeditiously or carefully, or that it had incentives to carelessness. There is also a danger of false negatives. Since Mallon was an intermittent carrier, there were bound to be weeks during which her feces both carried and did not carry the bacteria. On laboratories and typhoid diagnoses see Thomas G. Hull, "The Widal test as carried out in public health laboratories," *Am. J. Public Hlth.*, 1926, *16*: 901–904; Fred Berry and R. E. Daniels, "Comparative studies in typhoid stool examinations," *ibid.*, 1928, *18*: 883–892; Marion B. Coleman, "Serological and bacteriological procedures in the diagnosis of enteric fevers," *ibid.*, 1935, *25*(suppl.): 147–151; and T. F. Sellers, "Practical procedures in the laboratory diagnosis of typhoid and clinically related fevers," *ibid.*, 1937, *27;* 659–666.

29 Fred S. Westmoreland, Return to Writ, in the matter of the application for a Writ of Habeas Corpus for the production of Mary Mallon, New York Supreme Court, 22 July 1909; and Proposed Order and No-

tice of Settlement, in the Matter of the Application for a writ of Habeas Corpus for the production of Mary Mallon, 19 July 1909, New York Supreme Court, New York County Court House file.

30 Westmoreland, Return to Writ; and Mallon, undated letter, New York County Court House file. Westmoreland and Mallon refer to the same drug; hexamethylenamin is methenamine, a condensation product of ammonia and formaldehyde, $(Ch_2)_6N_4$, a urinary antiseptic. Urotropin is a proprietary brand of methenamine.

31 Mallon, undated letter; and Department of Health, City of New York, *Annual Report*, 1921, p. 53. On the surgical cure for typhoid fever see Thomas J. Leary, "Surgical method of clearing up chronic typhoid carriers," *J.A.M.A.*, 1913, *60*: 1293–1294; H. J. Nichols et al., "The surgical treatment of typhoid carriers," *ibid.*, 1919, *73*: 680–684; Edwin Henes, "Surgical treatment of typhoid carriers," *ibid.*, 1920, *75*: 1771–1774; Walter H. Vosburgy and Anna E. Perkins, "The surgical treatment of typhoid carriers in the Gowanda State Hospital," *Surg., Gynec. & Obst.*, 1925, *40*: 404–406; George H. Bigelow and Gaylord W. Anderson, "Cure of typhoid carriers," *ibid.*, 1933, *101*: 348–352; and Herman F. Senftner and Frank E. Coughlin, "Typhoid carriers in New York State with special reference to gall bladder operations," *American Journal of Hygiene*, 1933, *17*: 711–723.

32 *New York Times*, 30 Mar. 1913 (on Metchnikoff); and Senftner and Coughlin, "Typhoid carriers," p. 718 (on subsidization of healthy carriers). These regulations are examined later in this paper; they are explained in the health department annual reports.

33 Rosenau, in "Chronic typhoid fever producer," p. 864 (Rosenau later became professor of preventive medicine and hygiene at Harvard Medical School); and Charles V. Chapin, *Sources and Modes of Infection* (New York: Wiley, 1910), p. 37. For more criticism of the New York health department's treatment of Mary Mallon see W. H. Hamer, "Typhoid carriers and contact infection: some difficulties suggested by study of recent investigations carried out on 'living lines,'" *Proceedings of the Royal Society of Medicine*, 1911, *4*: 105–146.

34 The work of epidemiology itself precluded narrow approaches to disease control. A science that emerged in the prebacteriological 1840s, it searched all aspects of communicable diseases in its efforts to determine the reasons for the occurrence of epidemics. Epidemiologists like George Soper added germ theory to their lists of factors to be considered at the turn of the 20th century. Although the historian William Coleman concluded that bacteriology "greatly reduced" the scope of epidemiological investigations, the Mallon case indicates that the reduction was not

so great. See William Coleman, *Yellow Fever in the North: The Methods of Early Epidemiology* (Madison: Univ. of Wisconsin Press, 1987), quotation on p. 173.

35 Mallon fondly remembered the physician, Dr. Alexandra Plavska, in her will; she left her $200. Mallon's will is filed with the Chief Clerk of the Surrogates Court of Bronx County, New York, no. 894 P, 1938.

36 Department of Health, City of New York, *Annual Reports*, 1908, 1920, 1939. The years for which I have been able to identify the number of healthy carriers of typhoid fever in New York City, and the numbers, are as follows (derived from *Annual Reports*):

1908	5	1921	107	1930	255
1915	23	1922	112	1931	275
1918	70	1926	172	1937	385
1919	67	1927	208	1938	394
1920	85	1929	235	1950	392

37 Department of Health, City of New York, *Annual Report*, 1915, p. 73.

38 Department of Health, City of New York, *Annual Report*, 1916, p. 56.

39 Department of Health, City of New York, *Annual Report*, 1918, pp. 56, 53. One of the three was Mary Mallon; the other two are at this point unknown, although it is clear from the records that they were not kept on an indefinite basis. In 1919 health officials voiced slightly less optimism: "Unfortunately with the facilities at our disposal, we have not been able to do more than merely scrape the surface in the examination of approximately three-quarter million of foodhandlers in this study." See Department of Health, City of New York, *Annual Report*, 1919, p. 82.

40 Department of Health, City of New York, *Annual Report*, 1919, p. 81; and *ibid.*, 1922, p. 92. The department list of chronic carriers included 112 persons. See also *New York Times*, 13 Oct. 1922, 21 Jan. 1923.

41 Department of Health, City of New York, *Annual Report*, 1928, p. 54.

42 Department of Health, City of New York, *Annual Report*, 1929, p. 93.

43 *Ibid.*, p. 56.

44 Rosenau, *Preventive Medicine and Hygiene*, pp. 156–158.

45 *Ibid.*, p. 638

46 See reports in the *New York Times*, 14 and 15 Mar. 1924.

47 *New York Times*, 17 Apr. 1933; and Department of Health, City of New York, *Annual Reports*. In 1933, 65,000 persons suffered from typhoid fever, and about 6,500 of them died. If approximately 3 percent of survivors became chronic carriers, 1,800 persons would be added to the nation's list of healthy typhoid fever carriers each year. The nationwide case number is provided by Rosenau, *Preventive Medicine and*

Hygiene, p. 138. Chapin's estimate was three times higher.

48 Chain, *Sources and Modes of Infection* (1910), p. 110; 2nd ed. (1912), p. 152.

49 *Ibid.* (1910), p. 93.

50 S. Josephine Baker, *Fighting for Life* (New York: Macmillan, 1939), quotations on pp. 76, 75, 76; and *New York World-Telegram,* 12 Nov. 1938, p. 26.

51 Writ of habeas corpus, 10 July 1909, New York County Court House file (quotation).

52 I use Kuhnian language to make this un-Kuhnian point, with thanks to Thomas S. Kuhn, *The Structure of Scientific Revolutions* , (Chicago: Univ. of Chicago Press, 1962).

Reference Material

A GUIDE TO FURTHER READING

The books in this Guide are arranged under the following headings: General, Allied Health Professions, Alternative Medicine, Diseases, Education, Health-Care Policy, Hospitals, Legal and Ethical Issues, Lifestyle, Mental Health, Military, Public Health, Race and Ethnicity, Science, Theory and Practice, and Women and Children.

GENERAL

Burbick, Joan. *Healing the Republic: The Language of Health and the Culture of Nationalism in Nineteenth-Century America.* Cambridge: Cambridge University Press, 1994.

Cassedy, James H. *American Medicine and Statistical Thinking, 1800–1860.* Cambridge, Mass.: Harvard University Press, 1984.

Cassedy, James H. *Demography in Early America: Beginnings of the Statistical Mind, 1600–1800.* Cambridge, Mass.: Harvard University Press, 1969.

Cassedy, James H. *Medicine and American Growth, 1800–1860.* Madison: University of Wisconsin Press, 1986.

Cassedy, James H. *Medicine in America: A Short History.* Baltimore: Johns Hopkins University Press, 1991.

Duffy, John. *The Healers: The Rise of the Medical Establishment.* New York: McGraw-Hill Book Co., 1976.

Haller, John S., Jr. *American Medicine in Transition, 1840–1910.* Urbana: University of Illinois Press, 1981.

Kaufman, Martin, Stuart Galishoff, and Todd L. Savitt, eds. *Dictionary of American Medical Biography.* 2 vols. Westport, Conn.: Greenwood Press, 1984.

Kett, Joseph F. *The Formation of the American Medical Profession: The Role of Institutions, 1780–1860.* New Haven: Yale University Press, 1968.

King, Lester S. *American Medicine Comes of Age,* 1840–1920. [Chicago]: American Medical Association, 1984.

Malmsheimer, Richard. *"Doctors Only": The Evolving Image of the American Physician.* Westport, Conn.: Greenwood Press, 1988.

Numbers, Ronald L., and Darrel W. Amundsen, eds. *Caring and Curing: Health and Medicine in the Western Religious Traditions.* New York: Macmillan, 1986.

Numbers, Ronald L., ed. *Medicine in the New World: New Spain, New France, and New England.* Knoxville: University of Tennessee Press, 1987.

Reverby, Susan, and David Rosner, eds. *Health Care in America: Essays in Social History.* Philadelphia: Temple University Press, 1979.

Rosen, George. *Preventive Medicine in the United States, 1900–1975: Trends and Interpretations.* New York: Prodist, 1976.

Rosenberg, Charles E. *Explaining Epidemics and Other Studies in the History of Medicine.* Cambridge: Cambridge University Press, 1992.

Rothstein, William G. *American Physicians in the 19th Century: From Sects to Science.* Baltimore: Johns Hopkins University Press, 1972.

Rothstein, William G., ed. *Readings in American Health Care: Current Issues in Socio-Historical Perspective.* Madison: University of Wisconsin Press, 1995.

Shryock, Richard Harrison. *Medicine and Society in America, 1660–1860.* New York: New York University Press, 1960.

Shryock, Richard Harrision. *Medicine in America: Historical Essays.* Baltimore: Johns Hopkins Press, 1966.

Starr, Paul. *The Social Transformation of American Medicine.* New York: Basic Books, 1982.

Stevens, Rosemary. *American Medicine and the Public Interest.* New Haven: Yale University Press, 1971.

Vogel, Moris, J., and Charles E. Rosenberg, eds.,

The Therapeutic Revolution: Essays in the Social History of American Medicine. Philadelphia: University of Pennsylvania Press, 1979.

ALLIED HEALTH PROFESSIONS

Borst, Charlotte G. *Catching Babies: The Professionalization of Childbirth, 1870–1820.* Cambridge, Mass.: Harvard University Press, 1995.

Donegan, Jane B. *Women and Men Midwives: Medicine, Morality, and Misogyny in Early America.* Westport, Conn.: Greenwood Press, 1978.

Higby, Gregory J. *In Service to American Pharmacy: The Professional Life of William Procter, Jr.* Tuscaloosa: University of Alabama Press, 1992.

Lagemann, Ellen Condliffe, ed. *Nursing History: New Perspectives, New Possibilities.* New York: Teachers College Press, 1983.

Litoff, Judy Barrett. *American Midwives: 1860 to the Present.* Westport, Conn.: Greenwood Press, 1978.

McCluggage, Robert W. *A History of the American Dental Association: A Century of Health Service.* Chicago: American Dental Association, 1959.

Marshall, Helen E. *Mary Adelaide Nutting: Pioneer of Modern Nursing.* Baltimore: Johns Hopkins University Press, 1972.

Melosh, Barbara. *"The Physician's Hand": Work Culture and Conflict in American Nursing.* Philadelphia: Temple University Press, 1982.

Mottus, Jane E. *New York Nightingales: The Emergence of the Nursing Profession at Bellevue and New York Hospital, 1850–1920.* Ann Arbor, Mich.: UMI Research Press, 1980.

Reverby, Susan M. *Ordered to Care: The Dilemma of American Nursing, 1850–1945.* Cambridge: Cambridge University Press, 1987.

Ulrich, Laurel Thatcher. *A Midwife's Tale: The Life of Martha Ballard, Based on Her Diary, 1785–1812.* New York: Alfred A. Knopf, 1990.

ALTERNATIVE MEDICINE

Cayleff, Susan E. *Wash and Be Healed: The Water-Cure Movement and Women's Health.* Philadelphia: Temple University Press, 1987.

Donegan, Jane B. *"Hydropathic Highway to Health": Women and Water-Cure in Antebellum America.* Westport, Conn.: Greenwood Press, 1986.

Fuller, Robert C. *Alternative Medicine and American*

Religious Life. New York: Oxford University Press, 1989.

Gevitz, Norman. *The D.O.'s: Osteopathic Medicine in America.* Baltimore: Johns Hopkins University Press, 1982.

Gevitz, Norman, ed. *Other Healers: Unorthodox Medicine in America.* Balitmore: Johns Hopkins University Press, 1988.

Haller, John S., Jr. *Medical Protestants: The Eclectics in American Medicine, 1825–1939.* Carbondale: Southern Illinois University Press, 1994.

Harrell, David Edwin, Jr. *All Things Are Possible: The Healing and Charismatic Revivals in Modern America.* Bloomington: Indiana University Press, 1975.

Kaufman, Martin. *Homeopathy in America: The Rise and Fall of a Medical Heresy.* Baltimore: Johns Hopkins Press, 1971.

Moore, J. Stuart. *Chiropractic in America: The History of a Medical Alternative.* Baltimore: Johns Hopkins University Press, 1993.

Richards, Evelleen. *Vitamin C and Cancer: Medicine or Politics?* New York: St. Martin's Press, 1991.

Wardell, Walter I. *Chiropractic: History and Evolution of a New Profession.* St. Louis: Mosby Year Book, 1992.

Young, James Harvey. *American Health Quackery: Collected Essays.* Princeton: Princeton University Press, 1992.

Young, James Harvey. *The Medical Messiahs: A Social History of Health Quackery in Twentieth-Century America.* Princeton: Princeton University Press, 1967.

Young, James Harvey. *The Toadstool Millionaires: A Social History of Patent Medicines in America before Federal Regulation.* Princeton: Princeton University Press, 1961.

DISEASES

Ackerknecht, Erwin H. *Malaria in the Upper Mississippi Valley.* Baltimore: Johns Hopkins Press, 1945.

Bates, Barbara. *Bargaining for Life: A Social History of Tuberculosis, 1876–1938.* Philadelphia: University of Pennsylvania Press, 1992.

Blake, John B. *Benjamin Waterhouse and the Introduction of Vaccination: A Reappraisal.* Philadelphia: University of Pennsylvania Press, 1957.

Brandt, Allan M. *No Magic Bullet: A Social History of*

Venereal Disease in the United States since 1880. New York: Oxford University Press, 1985.

Brumberg, Joan Jacobs. *Fasting Girls: The Emergence of Anorexia Nervosa as a Modern Disease.* Cambridge, Mass.: Harvard University Press, 1988.

Caldwell, Mark. *The Last Crusade: The War on Consumption, 1862–1954.* New York: Atheneum, 1988.

Carrigan, Jo Ann. *The Saffron Scourge: A History of Yellow Fever in Louisiana, 1796–1905.* Lafayette: Center for Louisiana Studies, University of Southwestern Louisiana, 1994.

Crosby, Alfred W., Jr. *Epidemic and Peace, 1918.* Westport, Conn.: Greenwood Press, 1976.

Dowling, Harry F. *Fighting Infection: Conquests of the Twentieth Century.* Cambridge, Mass.: Harvard University Press, 1977.

Duffy, John. *Sword of Pestilence: The New Orleans Yellow Fever Epidemic of 1853.* Baton Rouge: Louisiana State University Press, 1966.

Ellis, John H. *Yellow Fever & Public Health in the New South.* Lexington: University Press of Kentucky, 1992.

Etheridge, Elizabeth. *The Butterfly Caste: A Social History of Pellagra in the South.* Westport, Conn.: Greenwood Publishing Co., 1972.

Fee, Elizabeth, and Daniel M. Fox, eds. *AIDS: The Burdens of History.* Berkeley and Los Angeles: University of California Press, 1988.

Feldberg, Georgina D. *Disease and Class: Tuberculosis and the Shaping of Modern North American Society.* New Brunswick, N.J.: Rutgers University Press, 1995.

Gould, Tony. *A Summer Plague: Polio and Its Survivors.* New Haven: Yale University Press, 1995.

Hannaway, Caroline, Victoria A. Harden, and John Parascandola, eds. *AIDS and the Public Debate: Historical and Contemporary Perspectives.* Tokyo: IOS Press, 1995.

Harden, Victoria A. *Rocky Mountain Spotted Fever: History of a Twentieth-Century Disease.* Baltimore: Johns Hopkins University Press, 1990.

Humphreys, Margaret. *Yellow Fever and the South.* New Brunswick, N.J.: Rutgers University Press, 1992.

Maulitz, Russell C., ed. *Unnatural Causes: The Three Leading Killer Diseases in America.* New Brunswick, N.J.: Rutgers University Press, 1989.

McCarthy, Michael P. *Typhoid and the Politics of Public Health in Nineteenth-Century Philadelphia.* Philadelphia: American Philosphical Society, 1987.

Patterson, James T. *The Dread Disease: Cancer and Modern American Culture.* Cambridge, Mass.: Harvard University Press, 1987.

Poirier, Suzanne. *Chicago: War on Syphilis, 1937–1940: The Times, the Trib, and the Clap Doctor.* Urbana: University of Illinois Press, 1995.

Powell, J. H. *Bring Out Your Dead: The Great Plague of Yellow Fever in Philadelphia in 1793.* Philadelphia: University of Pennsylvania Press, 1949.

Proctor, Robert N. *Cancer Wars: How Politics Shapes What We Know and Don't Know about Cancer.* New York: Basic Books, 1995.

Rogers, Naomi. *Dirt and Disease: Polio before FDR.* New Brunswick, N.J.: Rutgers University Press, 1992.

Rosenberg, Charles E. *The Cholera Years: The United States in 1832, 1849, and 1866.* Chicago: University of Chicago Press, 1962.

Rosenberg, Charles, and Janet Golden, eds. *Framing Disease: Studies in Cultural History.* New Brunswick, N.J.: Rutgers University Press, 1992.

Rothman, Sheila M. *Living in the Shadow of Death: Tuberculosis and the Social Experience of Illness in American History.* New York: Basic Books, 1994.

Ryan, Frank. *The Forgotten Plague: How the Battle Against Tuberculosis was Won—and Lost.* Boston: Little, Brown & Co., 1993.

Teller, Michael E. *The Tuberculosis Movement: A Public Health Campaign in the Progressive Era.* Westport, Conn.: Greenwood Press, 1988.

Ziporyn, Terra. *Disease in the Popular American Press: The Case of Diphtheria, Typhoid Fever, and Syphilis, 1870–1920.* Westport, Conn.: Greenwood Press, 1988.

EDUCATION

Barzansky, Barbara, and Norman Gevitz, eds. *Beyond Flexner: Medical Education in the Twentieth Century.* Westport, Conn.: Greenwood Press, 1992.

Bell, Whitfield J., Jr. *John Morgan: Continental Doctor.* Philadelphia: University of Pennsylvania Press, 1965.

Blustein, Bonnie Ellen. *Educating for Health and Prevention: A History of the Department of Community and Preventive Medicine of the (Women's) Medical College of Pennsylvania.* Canton, Mass.: Science History Publications/USA, 1993.

Bonner, Thomas Neville. *American Doctors and German Universities: A Chapter in International Intellectual Relations, 1870–1914.* Lincoln: University of Nebraska Press, 1963.

Bonner, Thomas Neville. *Becoming a Physician: Medical Eduation in Britain, France, Germany, and the United States, 1750–1945.* New York: Oxford University Press, 1995.

Fee, Elizabeth, and Roy M. Acheson. *A History of Education in Public Health.* New York: Oxford University Press, 1991.

Fleming, Donald. *William H. Welch and the Rise of Modern Medicine.* Boston: Little, Brown and Company, 1954.

Kohler, Robert. *From Medical Chemistry to Biochemistry: The Making of a Biomedical Discipline.* Cambridge: Cambridge University Press, 1982.

Ludmerer, Kenneth M. *Learning to Heal: The Development of American Medical Education.* New York: Basic Books, 1985.

Norwood, William Frederick. *Medical Education in the United States before the Civil War.* Philadelphia: University of Pennsylvania Press, 1944.

Numbers, Ronald L., ed. *The Education of American Physicians.* Berkeley: University of California Press, 1979.

Rothstein, William G. *American Medical Schools and the Practice of Medicine: A History.* New York: Oxford University Press, 1987.

Wheatley, Steven C. *The Politics of Philanthropy: Abraham Flexner and Medical Education.* Madison: University of Wisconsin Press, 1988.

HEALTH-CARE POLICY

Berliner, Howard S. *A System of Scientific Medicine: Philanthropic Foundations in the Flexner Era.* New York: Tavistock Publications, 1985.

Brown, E. Richard. *Rockefeller Medicine Men: Medicine and Capitalism in America.* Berkeley and Los Angeles: University of California Press, 1979.

Burrow, James G. *AMA: Voice of American Medicine.* Baltimore: Johns Hopkins Press, 1963.

Burrow, James G. *Organized Medicine in the Progressive Era: The Move Toward Monopoly.* Baltimore: Johns Hopkins University Press, 1977.

Fox, Daniel M. *Health Policies, Health Politics: The British and American Experience, 1911–1965.* Princeton: Princeton University Press, 1986.

Hendricks, Rickey. *A Model for National Health Care:*

The History of Kaiser Permanente. New Brunswick, N.J.: Rutgers University Press, 1993.

Hirshfield, Daniel S. *The Lost Reform: The Campaign for Compulsory Health Insurance in the United States from 1932 to 1943.* Camridge, Mass.: Harvard University Press, 1970.

Hollingsworth, J. Rogers. *A Political Economy of Medicine: Great Britain and the United States.* Baltimore: Johns Hopkins University Press, 1986.

Numbers, Ronald L. *Almost Persuaded: American Physicians and Compulsory Health Insurance, 1912–1920.* Baltimore: Johns Hopkins University Press, 1978.

Numbers, Ronald L., ed. *Compulsory Health Insurance: The Continuing American Debate.* Westport, Conn.: Greenwood Press, 1982.

Okun, Mitchell. *Fair Play in the Marketplace: The First Battle for Pure Food and Drugs.* Dekalb: Northern Illinois University Press, 1986.

Poen, Monte M. *Harry S. Truman Versus the Medical Lobby: The Genesis of Medicare.* Columbia: University of Missouri Press, 1979.

Stevens, Robert, and Rosemary Stevens. *Welfare Medicine in America: A Case Study of Medicaid.* New York: Free Press, 1974.

Young, James Harvey. *Pure Food: Securing the Federal Food and Drugs Act of 1906.* Princeton: Princeton University Press, 1989.

HOSPITALS

Dowling, Harry F. *City Hospitals: The Undercare of the Underprivileged.* Cambridge, Mass.: Harvard University Press, 1982.

Drachman, Virginia G. *Hospital with a Heart: Women Doctors and the Paradox of Separation at the New England Hospital, 1862–1969.* Ithaca: Cornell University Press, 1984.

Fink, Leon, and Brian Greenberg. *Upheaval in the Quiet Zone: A History of Hospital Workers' Union, Local 1199.* Urbana: University of Illinois Press, 1989.

Howell, Joel D. *Technology in the Hospital: Transforming Patient Care in the Early Twentieth Century.* Baltimore: Johns Hopkins University Press, 1995.

Long, Diana Elizabeth, and Janet Golden, eds. *The American General Hospital: Communities and Social Contexts.* Ithaca: Cornell University Press, 1989.

Rosenberg, Charles E. *The Care of Strangers: The Rise of America's Hospital System.* New York: Basic Books, 1987.

Rosner, David. *A Once Charitable Enterprise: Hospitals and Health Care in Brooklyn and New York, 1885–1915.* Cambridge: Cambridge University Press, 1982.

Stevens, Rosemary. *In Sickness and in Wealth: American Hospitals in the Twentieth Century.* New York: Basic Books, 1989.

Vogel, Morris J. *The Invention of the Modern Hospital: Boston, 1870–1930.* Chicago: University of Chicago Press, 1980.

LEGAL AND ETHICAL ISSUES

De Ville, Kenneth Allen. *Medical Malpractice in Nineteenth-Century America; Origins and Legacy.* New York: New York University Press, 1990.

Haber, Samuel. *The Quest for Authority and Honor in the American Professions, 1750–1900.* Chicago: University of Chicago Press, 1991.

Lederer, Susan E. *Subjected to Science: Human Experimentation in America before the Second World War.* Baltimore: Johns Hopkins University Press, 1995.

Mohr, James C. *Doctors and the Law: Medical Jurisprudence in Nineteenth-Century America.* New York: Oxford University Press, 1993.

Pernick, Martin S. *A Calculus of Suffering: Pain, Professionalism, and Anesthesia in 19th-Century America.* New York: Columbia University Press, 1985.

Rosenberg, Charles E. *The Trial of the Assassin Guiteau: Psychiatry and Law in the Gilded Age.* Chicago: University of Chicago Press, 1968.

Rothman, David J. *Strangers at the Bedside: A History of How Law and Bioethics Transformed Medical Decision Making.* New York: Basic Books, 1991.

Shryock, Richard H. *Medical Licensing in America, 1650–1965.* Baltimore: Johns Hopkins Press, 1967.

LIFESTYLE

Belasco, Waren J. *Appetite for Change: How the Counterculture Took on the Food Industry.* Ithaca: Cornell University Press, 1993.

Burnham, John C. *Bad Habits: Drinking, Smoking, Taking Drugs, Gambling, Sexual Misbehavior, and Swearing in American History.* New York: New York University Press, 1993.

Cole, Thomas R. *The Journey of Life: A Cultural History of Aging in America.* Cambridge: Cambridge University Press, 1992.

Courtwright, David T. *Dark Paradise: Opiate Addiction in America before 1940.* Cambridge, Mass.: Harvard University Press, 1982.

D'Emilio, John, and Estelle B. Freedman. *Intimate Matters: A History of Sexuality in America.* New York: Harper & Row, 1988.

Green, Harvey. *Fit for America: Health, Fitness, Sport, and American Society.* New York: Pantheon Books, 1986.

Hoy, Suellen. *Chasing Dirt: The American Pursuit of Cleanliness.* New York: Oxford University Press, 1995.

Levenstein, Harvey. *Paradox of Plenty: A Social History of Eating in Modern America.* New York: Oxford University Press, 1993.

Levenstein, Harvey. *Revolution at the Table: The Transformation of the American Diet.* New York: Oxford University Press, 1988.

Marcus, Alan I. *Cancer from Beef: DES, Federal Food Regulation, and Consumer Confidence.* Baltimore: Johns Hopkins University Press, 1994.

Musto, David F. *The American Disease: Origins of Narcotic Control.* New Haven: Yale University Press, 1973.

Nissenbaum, Stephan. *Sex, Diet, and Debility in Jacksonian America: Sylvester Graham and Health Reform:* Westport, Conn.: Greenwood Press, 1980.

Numbers, Ronald L. *Prophetess of Health: A Study of Ellen G. White.* New York: Harper and Row, 1976.

Vinikas, Vincent. *Soft Soap, Hard Sell: American Hygiene in an Age of Advertisement.* Ames: Iowa State University Press, 1992.

Verbrugge, Martha H. *Able-Bodied Womanhood: Personal Health and Social Change in Nineteenth-Century Boston.* New York: Oxford University Press, 1988.

Whorton, James C. *Crusaders for Fitness: The History of American Health Reformers.* Princeton: Princeton University Press, 1982.

Williams, Marilyn Thornton. *Washington "The Great Unwashed": Public Baths in Urban American, 1840–1920.* Columbus: Ohio State University Press, 1991.

MENTAL HEALTH

Bayer, Ronald. *Homosexuality and American Psychiatry: The Politics of Diagnosis*. New York: Basic Books, 1981.

Bell, Leland V. *Treating the Mentally Ill: From Colonial Times to the Present*. New York: Praeger, 1980.

Burnham, John Chynoweth. *Psychoanalysis and American Medicine, 1894–1918: Medicine, Science, and Culture*. New York: International Universities Press, 1967.

Dain, Norman. *Clifford W. Beers: Advocate for the Insane*. Pittsburgh: University of Pittsburgh Press, 1980.

Dain, Norman. *Concepts of Insanity in the United States, 1789–1865*. New Brunswick, N.J.: Rutgers University Press, 1964.

Dwyer, Ellen. *Homes for the Mad: Life inside Two Nineteenth-Century Asylums*. New Brunswick, N.J.: Rutgers University Press, 1987.

Fox, Richard T. *So Far Disordered in Mind: Insanity in California, 1870–1930*. Berkeley and Los Angeles: University of California Press, 1978.

Friedman, Lawrence J. *Menninger: The Family and the Clinic*. New York: Alfred A. Knopf, 1990.

Geller, Jeffrey L., and Maxine Harris. *Women of the Asylum: Voices from behind the Walls, 1840–1945*. New York: Doubleday, 1994.

Gollaher, David. *Voice for the Mad: the Life of Dorothea Dix*. New York: Free Press, 1995.

Gosling, F. G. *Before Freud: Neurasthenia and the American Medical Community, 1870–1910*. Urbana: University of Illinois Press, 1987.

Grob, Gerald N. *Edward Jarvis and the Medical World of Nineteenth-Century America*. Knoxville: University of Tennessee Press, 1978.

Grob, Gerald N. *From Asylum to Community: Mental Health Policy in Modern America*. Princeton: Princeton University Press, 1991.

Grob, Gerald N. *The Mad Among Us: A History of the Care of America's Mentally Ill*. New York: Free Press, 1994.

Grob, Gerald N. *Mental Illness and American Society, 1875–1940*. Princeton: Princeton University Press, 1983.

Grob, Gerald N. *Mental Institutions in America: Social Policy to 1875*. New York: Free Press, 1973.

Grob, Gerald N. *The State and the Mentally Ill: A History of Worcester State Hospital in Massachusetts, 1830–1920*. Chapel Hill: University of North Carolina Press, 1966.

Hale, Nathan G., Jr. *Freud and the Americans: The Beginnings of Psychoanalysis in the United States, 1876–1917*. New York: Oxford University Press, 1971.

Hale, Nathan G., Jr. *The Rise and Crisis of Psychoanalysis in the United States: Freud and the Americans, 1917–1985*. New York: Oxford University Press, 1995.

Jimenez, Mary Ann. *Changing Faces of Madness: Early American Attitudes and Treatment of the Insane*. Hanover, N.H.: University Press of New England, 1987.

Kushner, Howard I. *Self-Destruction in the Promised Land: A Psychocultural Biology of American Suicide*. New Brunswick, N.J.: Rutgers University Press, 1989.

Lunbeck, Elizabeth. *The Psychiatric Persuasion: Knowledge, Gender, and Power in Modern America*. Princeton: Princeton University Press, 1994.

McCandless, Peter. *Moonlight, Magnolias, and Madness: Insanity in South Carolina from the Colonial Period to the Progressive Era*. Chapel Hill: University of North Carolina Press, 1996.

McGovern, Constance M. *Masters of Madness: Social Origins of the American Psychiatric Profession*. Hanover, N.H.: University Press of New England, 1985.

Noll, Steven. *Feeble-Minded in Our Midst: Institutions for the Mentally Retarded in the South, 1900–1940*. Chapel Hill: University of North Carolina Press, 1995.

Reilly, Philip R. *The Surgical Solution: A History of Involuntary Sterilization in the United States*. Baltimore: Johns Hopkins University Press, 1991.

Rothman, David J. *Conscience and Convenience: The Asylum and Its Alternatives in Progressive America*. Boston: Little, Brown and Company, 1980.

Rothman, David J. *The Discovery of the Asylum: Social Order and Disorder in the New Republic*. Boston: Little, Brown and Company, 1971.

Sicherman, Barbara. *The Quest for Mental Health in America, 1880–1917*. New York: Arno Press, 1980.

Tomes, Nancy. *A Generous Confidence: Thomas Story Kirkbride and the Art of Asylum-Keeping, 1840–1883*. Cambridge: Cambridge University Press, 1984.

Trent, James W., Jr. *Inventing the Feeble Mind: A History of Mental Retardation in the United States*. Berkeley and Los Angeles: University of California Press, 1994.

Tyor, Peter L., and Leland V. Bell. *Caring for the Retarded in America: A History.* Westport, Conn.: Greenwood Press, 1984.

MILITARY

Adams, George Washington, *Doctors in Blue: The Medical History of the Union Army in the Civil War.* New York: Henry Schuman, 1952.

Breeden, James O. *Joseph Jones, M.D.: Scientist of the Old South.* Lexington: University Press of Kentucky, 1975.

Cash, Philip. *Medical Men at the Siege of Boston, April, 1775–April, 1776: Problems of the Massachusetts and Continental Armies.* Philadelphia: American Philosophical Society, 1973.

Cowdrey, Albert E. *Fighting for Life: American Military Medicine in World War II.* New York: Free Press, 1994.

Cowdrey, Albert E. *War and Healing: Stanhope Bayne-Jones and the Maturing of American Medicine.* Baton Rouge: Louisiana State University Press, 1992.

Cunningham, H. H. *Doctors in Gray: The Confederate Medical Service.* Baton Rouge: Louisiana State University Press, 1958.

Gillett, Mary C. *The Army Medical Department, 1775–1818.* Washington: Center of Military History, 1981.

Gillett, Mary C. *The Army Medical Department, 1818–1865.* Washington: Center of Military History, 1987.

Langley, Harold D. *A History of Medicine in the Early U.S. Navy.* Baltimore: Johns Hopkins University Press, 1995.

Schroeder-Lein, Glenna R. *Confederate Hospitals on the Move: Samuel H. Stout and the Army of Tennessee.* Columbia: University of South Carolina Press, 1994.

PUBLIC HEALTH

Blake, John B. *Public Health in the Town of Boston, 1630–1882.* Cambridge, Mass.: Harvard University Press, 1959.

Blake, Nelson. *Water for the Cities: A History of the Urban Water Supply Problem in the United States.* Syracuse, N.Y.: Syracuse University Press, 1956.

Cain, Louis P. *Sanitation Strategy for a Lakefront Metropolis.* DeKalb: Northern Illinois University Press, 1978.

Cassedy, James H. *Charles V. Chapin and the Public Health Movement.* Cambridge, Mass.: Harvard University Press, 1962.

Derickson, Alan. *Workers' Health, Workers' Democracy: The Western Miners' Struggle, 1891–1925.* Ithaca: Cornell University Press, 1988.

Duffy, John. *Epidemics in Colonial America.* Baton Rouge: Louisiana State University Press, 1959.

Duffy, John. *A History of Public Health in New York City.* 2 vols. New York: Russell Sage Foundation, 1968, 1974.

Duffy, John. *The Sanitarians: A History of American Public Health.* Urbana: University of Illinois Press, 1990.

Dunlap, Thomas R. *DDT: Scientists, Citizens, and Public Policy.* Princeton: Princeton University Press, 1981.

Etheridge, Elizabeth W. *Sentinel for Health: A History of the Centers for Disease Control.* Berkeley and Los Angeles: University of California Press, 1992.

Ettling, John. *The Germ of Laziness: Rockefeller Philanthropy and Public Health in the New South.* Cambridge, Mass.: Harvard University Press, 1981.

Fee, Elizabeth. *Disease and Discovery: A History of the Johns Hopkins School of Hygiene and Public Health, 1916–1939.* Baltimore: Johns Hopkins University Press, 1987.

Galishoff, Stuart. *Newark: The Nation's Unhealthiest City, 1832–1895.* New Brunswick, N.J.: Rutgers University Press, 1988.

Galishoff, Stuart. *Safeguarding the Public Health: Newark, 1895–1918.* Westport, Conn.: Greenwood Press, 1975.

Jordan, Philip D. *The People's Health: A History of Public Health in Minnesota to 1848.* St. Paul: Minnesota Historical Society, 1953.

Leavitt, Judith Walzer. *The Healthiest City: Milwaukee and the Politics of Health Reform.* Princeton: Princeton University Press, 1982; Madison: University of Wisconsin Press, 1996.

Leavitt, Judith Walzer. *Typhoid Mary: Captive to the Public's Health.* Boston: Beacon Press, 1996.

Marcus, Alan I. *Plague of Strangers: Social Groups and the Origins of City Services in Cincinnati, 1819–1870.* Columbus: Ohio State University Press, 1991.

Melosi, Martin V. *Garbage in the Cities: Refuse, Reform, and the Environment, 1880–1980.* College Station: Texas A & M University Press, 1981.

Melosi, Martin V., ed. *Pollution and Reform in Ameri-*

can Cities, 1870–1930. Austin: University of Texas Press, 1980.

Mullen, Fitzhugh. Plagues and Politics: The Story of the United States Public Health Service. New York: Basic Books, 1989.

Rosenkrantz, Barbara Gutmann. Public Health and the State: Changing Views in Massachusetts, 1842–1936. Cambridge, Mass.: Harvard University Press, 1972.

Rosner, David, ed. Hives of Sickness: Public Health and Epidemics in New York City. New Brunswick, N.J.: Rutgers University Press, 1995.

Rosner, David, and Gerald Markowitz. Deadly Dust: Silicosis and the Politics of Occupational Disease in Twentieth-Century America. Princeton: Princeton University Press, 1991.

Rosner, David, and Gerald Markowitz, eds. Dying for Work: Workers' Safety and Health in Twentieth-Century America. Bloomington: Indiana University Press, 1987.

Tomes, Nancy. The Gospel of Germs: Men, Women, and the Microbe in American Life, 1870–1930. Cambridge, Mass.: Harvard University Press, 1997.

Whorton, James. Before Silent Spring: Pesticides and Public Health in Pre-DDT America. Princeton: Princeton University Press, 1974.

RACE AND ETHNICITY

Beardsley, Edward H. A History of Neglect: Health Care for Blacks and Mill Workers in the Twentieth-Century South. Knoxville: University of Tennessee Press, 1987.

Gamble, Vanessa Northington. Making a Place for Ourselves: The Black Hospital Movement, 1920–1945. New York: Oxford University Press, 1995.

Hine, Darlene Clark. Black Women in White: Racial Conflict and Cooperation in the Nursing Profession, 1890–1950. Bloomington: Indiana University Press, 1989.

Jones, James H. Bad Blood: The Tuskegee Syphilis Study Experiment. New York: Free Press, 1981.

Kiple, Kenneth F., and Virginia Himmelsteib King. Another Dimension to the Black Diaspora: Diet, Disease, and Racism. Cambridge: Cambridge University Press, 1981.

Kraut, Alan M. Silent Travelers: Germs, Genes and the "Immigrant Menace." New York: Basic Books, 1994.

Kunitz, Stephen J. Disease and Social Diversity: The European Impact on the Health of Non-Europeans. New York: Oxford University Press, 1994.

Kunitz, Stephen J. Disease Change and the Role of Medicine: The Navajo Experience. Berkeley and Los Angeles: University of California Press, 1983.

Lightfoot, Sara Lawrence. Balm in Gilead: Journey of a Healer. Reading, Mass.: Addison-Wesley, 1988.

McBride, David. From TB to AIDS: Epidemics among Urban Blacks since 1900. Albany: State University of New York Press, 1991.

McBride, David. Integrating the City of Medicine: Blacks in Philadelphia Health Care, 1910–1965. Philadelphia: Temple University Press, 1989.

Postell, William Dosite. The Health of Slaves on Southern Plantations. Baton Rouge: Louisiana State University Press, 1951.

Savitt, Todd L. Medicine and Slavery: The Diseases and Health Care of Blacks in Antebellum Virginia. Urbana: University of Illinois Press, 1978.

Smith, Susan L. Sick and Tired of Being Sick and Tired: Black Women and the National Negro Health Movement, 1915–1950. Philadelphia: University of Pennsylvania Press, 1995.

Summerville, James. Educating Black Doctors: A History of Meharry Medical College. University: University of Alabama Press, 1983.

Vogel, Virgil. American Indian Medicine. Norman: University of Oklahoma Press, 1970.

Young, T. Kue. The Health of Native Americans: Toward a Biocultural Epidemiology. New York: Oxford University Press, 1994.

SCIENCE

Achenbaum, W. Andrew. Crossing Frontiers: Gerontology Emerges as a Science. Cambridge: Cambridge University Press, 1995.

Benison, Saul. Tom Rivers: Reflections on a Life in Medicine and Science. Cambridge, Mass.: M.I.T. Press, 1967.

Benison, Saul, A. Clifford Barger, and Elin L. Wolfe. Walter B. Cannon: The Life and Times of a Young Scientist. Cambridge, Mass.: Harvard University Press, 1987.

Blustein, Bonnie Ellen. Preserve Your Love for Science: Life of William A. Hammond, American Neurologist. Cambridge: Cambridge University Press, 1991.

Bullough, Vern L. *Science in the Bedroom: A History of Sex Research*. New York: Basic Books, 1994.

Corner, George W. *A History of the Rockefeller Institute, 1901–1953: Origins and Growth*. New York: Rockefeller Institute Press, 1964.

Fye, W. Bruce, *The Development of American Physiology: Scientific Medicine in the Nineteenth Century*. Baltimore: Johns Hopkins University Press, 1987.

Harden, Victoria A. *Inventing the NIH: Federal Biomedical Research Policy, 1887–1937*. Baltimore: Johns Hopkins University Press, 1986.

Horsman, Reginald. *Frontier Doctor: William Beaumont, America's First Great Medical Scientist*. Columbia: University of Missouri Press, 1996.

Liebenau, Jonathan. *Medical Science and Medical Industry: The Formation of the American Pharmaceutical Industry*. Baltimore: Johns Hopkins University Press, 1987.

Parascandola, John. *The Development of American Pharmacology: John J. Abel and the Shaping of a Discipline*. Baltimore: Johns Hopkins University Press, 1992.

Rushton, Alan R. *Genetics and Medicine in the United States, 1800 to 1992*. Baltimore: Johns Hopkins University Press, 1994.

Shryock, Richard H. *American Medical Research: Past and Present*. New York: The Commonwealth Fund, 1947.

Strickland, Stephen P. *Politics, Science, and Dread Disease: A Short History of United States Medical Research Policy*. Cambridge, Mass.: Harvard University Press, 1972.

Swann, John P. *Academic Scientists and the Pharmaceutical Industry: Cooperative Research in Twentieth-Century America*. Baltimore: Johns Hopkins University Press, 1988.

THEORY AND PRACTICE

Bell, Whitfield J., Jr. *The Colonial Physician & Other Essays*. New York: Science History Publications, 1975.

English, Peter C. *Shock, Physiological Surgery, and George Washington Crile: Medical Innovation in the Progressive Era*. Westport, Conn.: Greenwood Press, 1980.

Estes, J. Worth. *Hall Jackson and the Purple Foxglove: Medical Practice and Research in Revolutionary America, 1760–1820*. Hanover, N.H.: University Press of New England, 1979.

Fye, Bruce W. *American Cardiology: The History of a Specialty and Its College*. Baltimore: Johns Hopkins University Press, 1996.

King, Lester S. *Transformations in American Medicine from Benjamin Rush to William Osler*. Baltimore: Johns Hopkins University Press, 1991.

Numbers, Ronald L., and Todd L. Savitt, eds. *Science and Medicine in the Old South*. Baton Rouge: Louisiana State University Press, 1989.

Reiser, Stanley Joel. *Medicine and the Reign of Technology*. Cambridge: Cambridge University Press, 1978.

Rosen, George. *The Structure of American Medical Practice, 1875–1941*. Philadelphia: University of Pennsylvania Press, 1983.

Savitt, Todd L., and James Harvey Young, eds. *Disease and Distinctiveness in the American South*. Knoxville: University of Tennessee Press, 1988.

Ward, Patricia Spain. *Simon Baruch: Rebel in the Ranks of Medicine, 1840–1921*. Tuscaloosa: University of Alabama Press, 1994.

Warner, John Harley. *Against the Spirit of System: The French Impulse in Nineteenth-Century American Medicine*. Princeton: Princeton University Press, 1996.

Warner, John Harley. *The Therapeutic Perspective: Medical Practice, Knowledge, and Professional Identity in America, 1820–1885*. Cambridge, Mass.: Harvard University Press, 1986.

Watson, Patricia A. *The Angelical Conjunction: The Preacher-Physicians of Colonial New England*. Knoxville: University of Tennessee Press, 1991.

WOMEN AND CHILDREN

Apple, Rima D. *Mothers and Medicine: A Social History of Infant Feeding, 1890–1950*. Madison: University of Wisconsin Press, 1987.

Apple, Rima D., ed. *Women, Health, and Medicine in America: A Historical Handbook*. New York: Garland, 1990.

Asbell, Bernard. *The Pill: A Biography of the Drug That Changed the World*. New York: Random House, 1995.

Brodie, Janet Farrell. *Contraception and Abortion in Nineteenth-Century America*. Ithaca: Cornell University Press, 1994.

Chesler, Ellen. *Woman of Valor: Margaret Sanger and the Birth Control Movement in America*. New York: Simon and Schuster, 1992.

Edwards, Margot, and Mary Waldorf. *Reclaiming Birth: History and Heroines of American Childbirth Reform*. Trumansburg, N.Y.: Crossing Press, 1984.

Gordon, Linda. *Woman's Body, Woman's Right: A Social History of Birth Control in America*. New York: Penguin, 1990.

Garrow, David J. *Liberty and Sexuality: The Right to Privacy and the Making of Roe v. Wade*. New York: MacMillan, 1994.

Haller, John S., and Robin M. Haller. *The Physician and Sexuality in Victorian America*. Urbana: University of Illinois Press, 1974.

Hoffert, Sylvia D. *Private Matters: American Attitudes toward Childbearing and Infant Nurture in the Urban North, 1800–1860*. Urbana: University of Illinois Press, 1989.

Joffe, Carole. *Doctors of Conscience: The Struggle to Provide Abortion before and after Roe v. Wade*. Boston: Beacon Press, 1995.

King, Charles R. *Children's Health in America: A History*. New York: Twayne Publishers, 1993.

Klaus, Alisa. *Every Child a Lion: The Origins of Maternal and Infant Health Policy; in the United States and France, 1890–1920*. Ithaca: Cornell University Press, 1993.

Leavitt, Judith Walzer. *Brought to Bed: Childbearing in America, 1750–1950*. New York: Oxford University Press, 1986.

Leavitt, Judith Walzer, ed. *Women and Health in America: Historical Readings*. Madison: University of Wisconsin Press, 1984.

Marsh, Margaret, and Wanda Ronner. *The Empty Cradle: Infertility in America from Colonial Times to the Present*. Baltimore: Johns Hopkins University Press, 1996.

May, Elaine Tyler. *Barren in the Promised Land: Childless Americans and the Pursuit of Happiness*. New York: Basic Books, 1995.

McCann, Carole R. *Birth Control Politics in the United States, 1916–1945*. Ithaca: Cornell University Press, 1994.

McLaughlin, Loretta. *The Pill, John Rock, and the Church: The Biography of a Revolution*. Boston: Little, Brown & Co., 1982.

McMillen, Sally G. *Motherhood in the Old South: Pregnancy, Childbirth, and Infant Rearing*. Baton Rouge: Louisiana State University Press, 1990.

Meckel, Richard A. *"Save the Babies": American Public Health Reform and the Prevention of Infant Mortality, 1850–1929*. Baltimore: Johns Hopkins University Press, 1990.

Mohr, James C. *Abortion in America: The Origins and Evolution of National Policy*. New York: Oxford University Press, 1977.

Moldow, Gloria. *Women Doctors in Gilded-Age Washington: Race, Gender, and Professionalization*. Urbana: University of Illinois Press, 1987.

Morantz-Sanchez, Regina Markell. *Sympathy and Science: Women Physicians in American Medicine*. New York: Oxford University Press, 1985.

Preston, Samuel H., and Michael R. Haines. *Fatal Years: Child Mortality in Late Nineteenth-Century America*. Princeton: Princeton University Press, 1991.

Reagan, Leslie J. *When Abortion Was A Crime: Women, Medicine, and Law, 1867–1973*. Berkeley: University of California Press, 1991.

Reed, James. *The Birth Control Movement and American Society: From Private Vice to Public Virtue*. Princeton: Princeton University Press, 1983.

Sandelowski, Margarete. *Pain, Pleasure, and American Childbirth: From the Twilight Sleep to the Read Method, 1914–1960*. Westport, Conn.: Greenwood Press, 1984.

Sicherman, Barbara. *Alice Hamilton: A Life in Letters*. Cambridge, Mass.: Harvard University Press, 1984.

Sklar, Kathryn Kish. *Catharine Beecher: A Study in American Domesticity*. New Haven: Yale University Press, 1973.

Solinger, Rickie. *The Abortionist: A Woman Against the Law*. New York: Free Press, 1994.

Solinger, Rickie. *Wake Up Little Susie: Single Pregnancy and Race before Roe v. Wade*. New York: Routledge, 1992.

Stage, Sara. *Female Complaints: Lydia Pinkham and the Business of Women's Medicine*. New York: W. W. Norton, 1979.

Walsh, Mary Roth. *"Doctors Wanted: No Women Need Apply": Sexual Barriers in the Medical Profession, 1835–1975*. New Haven: Yale University Press, 1977.

ABBREVIATIONS OF JOURNAL TITLES

Alabama Med. J. (Alabama Medical Journal)
A.M.A. Bull. (American Medical Association Bulletin)
Am. Hist. Rev. (American Historical Review)
Am. Econ. Rev. (American Economic Review)
Am. J. Dermatology & Genito-Urinary Dis. (American Journal of Dermatology and Genito-Urinary Diseases)
Am. J. Insanity (American Journal of Insanity)
Am. J. Med. Sci. (American Journal of the Medical Sciences)
Am. J. Psychiatry (American Journal of Psychiatry)
Am. J. Public Hlth. (American Journal of Public Health)
Am. J. Sociol. (American Journal of Sociology)
Am. Labor Legis. Rev. (American Labor Legislation Review)
Am. Med. Monthly (American Medical Monthly)
Am. Med. Times (American Medical Times)
Am. Medico-Surg. Bull. (American Medico-Surgical Bulletin)
Am. Mercury (American Mercury)
Am. Psychol. (American Psychologist)
Am. Quart. (American Quarterly)
Am. Rev. Resp. Dis. (American Review of Respiratory Disease)
Am. Rev. Tuberculosis & Pulmonary Dis. (American Review of Tuberculosis and Pulmonary Diseases)
Ann. Am. Acad. Polit. & Soc. Sci. (Annals of the American Academy of Political and Social Science)
Ann. Intern. Med. (Annals of Internal Medicine)
Ann. Med. Hist. (Annals of Medical History)
Ann. N.Y. Acad. Sci. (Annals of the New York Academy of Sciences)
Antitrust Bull. (Antitrust Bulletin)
Arch. Dermatology & Syphilis (Archives of Dermatology and Syphilis)
Arch. Med. (Archives of Medicine)
Arch. Neurol. & Psychiat. (Archives of Neurology and Psychiatry)
Atlanta J.-Rec. Med. (Atlanta Journal-Record of Medicine)
Boston J. Hlth. (Boston Journal of Health)
Boston Med. & Surg. J. (Boston Medical and Surgical Journal)
Brit. J. Exper. Path. (British Journal of Experimental Pathology)

Brit. Med. J. (British Medical Journal)
Bull. Hist. Med. (Bulletin of the History of Medicine)
Bull. Med. Lib. Assn. (Bulletin of the Medical Library Association)
Bull. N.Y. Acad. Med. (Bulletin of the New York Academy of Medicine)
Bull. Soc. Med. Hist. Chicago (Bulletin of the Society of Medical History [of Chicago])
Charleston Med. J. & Rev. (Charleston Medical Journal and Review)
Chicago Med. J. & Exam. (Chicago Medical Journal and Examiner)
Columbia Univ. Quart. Supp. (Columbia University Quarterly Supplement)
Edinburgh Rev. (Edinburgh Review)
Emory Univ. Quart. (Emory University Quarterly)
Explorations Econ. Hist. (Explorations in Economic History)
Films Rev. (Films Review)
Harper's Mag. (Harper's Magazine)
Hastings Cent. Rep. (Hastings Center Report)
Hastings Cent. Stud. (Hastings Center Studies)
Hosp. Admin. (Hospital Administration)
Johns Hopkins Hosp. Bull. (Johns Hopkins Hospital Bulletin)
J.A.M.A. (Journal of the American Medical Association)
J. Am. Culture (Journal of American Culture)
J. Am. Hist. (Journal of American History)
J. Assn. Am. Med. Colls. (Journal of the Association of American Medical Colleges)
J. Chron. Dis. (Journal of Chronic Diseases)
J. Econ. Issues (Journal of Economic Issues)
J. Exper. Med. (Journal of Experimental Medicine)
J. Florida Med. Assn. (Journal of the Florida Medical Association)
J. Gynec. Soc. Boston (Journal of the Gynecological Society of Boston)
J. Hist. Geography (Journal of Historical Geography)
J. Hist. Med. (Journal of the History of Medicine and Allied Sciences)
J. Hlth. Soc. Behav. (Journal of Health and Social Behavior)
J. Homosexuality (Journal of Homosexuality)
J. Hyg. & Herald Hlth. (Journal of Hygiene and Herald of Health)

J. Infect. Dis. (*Journal of Infectious Diseases*)
J. Interdisc. Hist. (*Journal of Interdisciplinary History*)
J. Maine Med. Assn. (*Journal of the Maine Medical Association*)
J. Med. Educ. (*Journal of Medical Education*)
J. Med. Soc. N.J. (*Journal of the Medical Society of New Jersey*)
J. Mich. St. Med. Soc. (*Journal of the Michigan State Medical Society*)
J. Nat. Med. Assn. (*Journal of the National Medical Association*)
J. Nerv. & Ment. Dis. (*Journal of Nervous and Mental Disease*)
J. Polit. Econ. (*Journal of Political Economy*)
J. Soc. Hist. (*Journal of Social History*)
J. Southern Hist. (*Journal of Southern History*)
J. Western Soc. Engrs. (*Journal of the Western Society of Engineers*)
Lit. & Med. (*Literature and Medicine*)
Long Island Med. J. (*Long Island Medical Journal*)
Los Angeles J. Eclectic Med. (*Los Angeles Journal of Eclectic Medicine*)
Med. & Surg. Reptr. (*Medical and Surgical Reporter*)
Med. Ann. District of Columbia (*Medical Annals of the District of Columbia*)
Med. Comm. Mass. Med. Soc. (*Medical Communications of the Massachusetts Medical Society*)
Med. Hist. (*Medical History*)
Med. News, N.Y. (*Medical News [New York]*)
Med. Rec. (*Medical Record*)
Med. Rev. Rev. (*Medical Review of Reviews*)
Mich. Med. (*Michigan Medicine*)
Milbank Mem. Fund Quart. Bull. (*Milbank Memorial Fund Quarterly Bulletin*)
Milbank Quart. (*Milbank Quarterly*)
Miss. Valley Hist. Rev. (*Mississippi Valley Historical Review*)
Modern Hosp. (*Modern Hospital*)
Monthly Steth. & Med. Reptr. (*The Monthly Stethoscope and Medical Reporter*)
Mt. Sinai J. Med. (*Mount Sinai Journal of Medicine*)
Nashville J. Med. & Surg. (*Nashville Journal of Medicine and Surgery*)
New Eng. J. Med. (*New England Journal of Medicine*)
New Orleans Med. & Surg. J. (*New Orleans Medical and Surgical Journal*)
New Orleans Med. J. (*New Orleans Medical Journal*)
New Orleans Med. News & Hosp. Gaz. (*New Orleans Medical News and Hospital Gazette*)
N.Y. Hist. Soc. Quart. (*New York Historical Society Quarterly*)
N.Y. J. Med. (*New York Journal of Medicine*)
N.Y. Med. J. (*New York Medical Journal*)
N.Y. Rev. Books (*New York Review of Books*)
N.Y. St. J. Med. (*New York State Journal of Medicine*)

North Am. Rev. (*North American Review*)
North Carolina Med. J. (*North Carolina Medical Journal*)
Ohio State Med. J. (*Ohio State Medical Journal*)
Penn. Mag. Hist. & Biography (*Pennsylvania Magazine of History and Biography*)
Penn. Med. J. (*Pennsylvania Medical Journal*)
Perspect. Am. Hist. (*Perspectives in American History*)
Perspect. Biol. & Med. (*Perspectives in Biology and Medicine*)
Philadelphia J. Med. & Phys. Sci. (*Philadelphia Journal of the Medical and Physical Sciences*)
Population Stud. (*Population Studies*)
Proc. Am. Medico-Psychological Assn. (*Proceedings of the American Medico-Psychological Association*)
Proc. Am. Soc. Civil Engrs. (*Proceedings of the American Society of Civil Engineers*)
Proc. Int. Cong. Hist. Sci. (*Proceedings of the International Congress of the History of Science*)
Public Hlth. Rep. (*Public Health Reports*)
Pubs. Col. Soc. Mass. (*Publications of the Colonial Society of Massachusetts*)
Quart. Bull. Northwestern Univ. Med. School (*Quarterly Bulletin of Northwestern University Medical School*)
Radical Hist. Rev. (*Radical History Review*)
Richmond & Louisville Med. J. (*Richmond and Louisville Medical Journal*)
St. Paul Med. J. (*St. Paul Medical Journal*)
Sci. Digest (*Science Digest*)
Scient. Am. (*Scientific American*)
Scient. Am. Suppl. (*Scientific American Supplement*)
Scribner's Mag. (*Scribner's Magazine*)
Soc. Forces (*Social Forces*)
Soc. Sci. Hist. (*Social Science History*)
Social Security Bull. (*Social Security Bulletin*)
South Atlantic Quart. (*South Atlantic Quarterly*)
Southern Med. & Surg. J. (*Southern Medical and Surgical Journal*)
Southern Med. J. (*Southern Medical Journal*)
Southern Med. Rep. (*Southern Medical Reports*)
Steth. & Va. Med. Gaz. (*The Stethoscope and Virginia Medical Gazette*)
Surg., Gynec. & Obst. (*Surgery, Gynecology & Obstetrics*)
Tenn. Hist. Quart. (*Tennessee Historical Quarterly*)
Texas Hlth. J. (*Texas Health Journal*)
Texas Med. News (*Texas Medical News*)
Tr. A.M.A. (*Transactions of the American Medical Association*)
Tr. Am. Hosp. Assn. (*Transactions of the American Hospital Association*)
Tr. Assn. Am. Physicians (*Transactions of the Association of American Physicians*)
Tr. Coll. Physicians Phila. (*Transactions and Studies of the College of Physicians of Philadelphia*)

Tr. Med. Assn. St. Ala. (Transactions of the Medical Association of the State of Alabama)

Tr. Med. Soc. St. N.C. (Transactions of the Medical Society of the State of North Carolina)

Tr. Med. Soc. St. N.Y. (Transactions of the Medical Society of the State of New York)

Tr. Med. Soc. St. Penn. (Transactions of the Medical Society of the State of Pennsylvania)

Tr. Med. Soc. St. Va. (Transactions of the Medical Society of the State of Virginia)

Tr. Southern Surg. & Gynec. Assn. (Transactions of the Southern Surgical and Gynecological Association)

Transylvania J. Med. (Transylvania Journal of Medicine and the Associated Sciences)

Va. Med. & Surg. J. (Virginia Medical and Surgical Journal)

Va. Med. J. (Virginia Medical Journal)

Va. Med. Mon. (Virginia Medical Monthly)

Water-Cure J. (Water-Cure Journal)

Western J. Med. & Surg. (Western Journal of Medicine and Surgery)

Western Penn. Hist. Mag. (Western Pennsylvania Historical Magazine)

Wis. Med. J. (Wisconsin Medical Journal)

Woman's Med. J. (Woman's Medical Journal)

Yale J. Biol. & Med. (Yale Journal of Biology and Medicine)

Yale Law J. (Yale Law Journal)

DATE DUE

OCT 18 '97		
APR 1 5 2002		
May 11 2002		
GAYLORD		PRINTED IN U.S.A.